Numerical methods are indispensable tools in the analysis of complex fluid flows, giving rise to the burgeoning field of computational fluid dynamics (CFD). This book focuses on computational techniques for high-speed gas flows, especially gas flows containing shocks and other steep gradients.

The book decomposes complicated numerical methods into simple modular parts, showing how each part fits and how each method relates to or differs from others. The text begins with a review of gasdynamics and computational techniques. Next come basic principles of computational gasdynamics. The last two parts cover basic techniques – including the Lax–Friedrichs method, the Lax–Wendroff method, MacCormack's method, and Godunov's method – and advanced techniques, such as TVD, ENO, flux-limited, and flux-corrected methods. Every method is tested on the same carefully constructed set of test problems, which helps to expose similarities and differences under actual performance conditions.

Senior- and graduate-level students, especially in aerospace engineering, as well as researchers and practicing engineers, will use this single source to find a wealth of invaluable information on high-speed gas flows.

Computational Gasdynamics

Computational Gasdynamics

CULBERT B. LANEY
University of Colorado

CAMBRIDGE UNIVERSITY PRESS
Cambridge, New York, Melbourne, Madrid, Cape Town, Singapore, São Paulo

Cambridge University Press
The Edinburgh Building, Cambridge CB2 8RU, UK

Published in the United States of America by Cambridge University Press, New York

www.cambridge.org
Information on this title: www.cambridge.org/9780521570695

First published 1998

A catalogue record for this publication is available from the British Library

Library of Congress Cataloguing in Publication data

Laney, Culbert B.

Computational gasdynamics / Culbert B. Laney.

p. cm.

Includes index.

ISBN 0-521-57069-7 (hb)

1. Gas dynamics – Mathematical models. I. Title.

QA930.L294 1998
533′.21′015194 – dc21 97-52717
CIP

ISBN 978-0-521-57069-5 hardback
ISBN 978-0-521-62558-6 paperback

Transferred to digital printing 2007

Contents

Preface

Welcome. This book concerns *computational gasdynamics*, a part of the broader field of *computational fluid dynamics* (*CFD*). More specifically, put in precise technical terms, this book concerns numerical methods for simulating high-speed flows of inviscid perfect gases, especially flows containing shocks. Computational gasdynamics falls across a number of traditional disciplines including aerospace engineering, mechanical engineering, chemical engineering, applied mathematics, numerical analysis, and physics. The book should appeal to practitioners of any of these disciplines, although it does not claim to have the theoretical rigor expected of mathematics texts nor the drive towards immediate full-scale applications found in many engineering books. This book presumes an understanding of calculus, differential equations, numerical analysis, fluid dynamics, and physics and a level of sophistication of the sort found in most seniors and first-year graduate students in science, mathematics, or engineering.

I have taught courses based on the material in this book numerous times in the past five years at the University of Colorado at Boulder. These courses have variously included juniors, seniors, graduate students, professors, post-docs, and practicing engineers in a variety of different disciplines. The feedback from my students has critically guided the development and evolution of the book, taking it from a handwritten collection of notes to the present form. Researchers in the field have also reviewed many parts of the book, although the needs of newcomers have always determined the final presentation. The book contains about a semester-and-a-half worth of material. To trim the material to a one-semester length, less advanced courses should use Part III selectively and omit Part V, while advanced courses should only sample Parts I and II.

I have tried to make the book both self-contained and self-teaching. The aim is to explain the simple philosophies underlying the sometimes complex material. If I have really done my job well, the reader should think "oh, of course, I could have discovered that myself." The book decomposes complicated numerical methods into simple modular parts, showing how each part fits, how each method relates to competing methods, how seemingly different methods are actually similar, and how seemingly similar methods are actually different. Some students would be happy with just examples and pictures. This book contains plenty of both. I have also taken great pains to make this book a useful reference, by including extensive citations to the research literature and to related material with the book itself and by taking extra care with the index.

Like all computational disciplines, CFD is relatively young. Most of the material seen in this book was developed between the mid-1950s and the late 1980s. As with any newer field, CFD has not yet fully established a standard repertoire and approach – researchers still disagree over the significance of various methods, principles, and theories, a situation exacerbated by the widely differing backgrounds of the engineers and mathematicians who populate the field. I have tried to navigate these shoals cautiously, but must beg the forbearance of those eminent researchers who prefer approaches other than those I have adopted.

It would take a great deal of space to mention everyone who has contributed to this book. However, I must at least mention some of the most critical figures who have offered encouragement, guidance, and support in my recent career, even if not directly related to the book. These include David Caughey, Brian Argrow, Richard Seebass, Robert Culp, Melvin Laney, and Carolyn Laney. I would also like to express gratitude to Phil Roe, Tony Jameson, Randy LeVeque, and so many others for their small but still much appreciated contributions.

Bert Laney

October 1996

CHAPTER 1

Introduction

Fluid dynamics concerns fluid motions. No single book, or even shelf of books, could hope to describe the full range of known fluid dynamics. By tradition, introductory books mainly concern three simplified categories of flow:

(1) *Gasdynamics* – compressible frictionless flows.
(2) *Viscous flows*, *boundary layers*, and *turbulence* – incompressible frictional flows.
(3) *Potential flows* – irrotational frictionless flows.

Books with generic titles like *fluid dynamics*, *fluid mechanics*, or *aerodynamics* generally survey all three basic categories of flow. Other books concern just gasdynamics or just viscous flows. These books sacrifice breadth for depth. Given the huge differences between compressible inviscid flows and viscous incompressible flows, single-topic treatments make sense for many readers, who need to know only one or the other.

Computational fluids dynamics (*CFD*) concerns computer simulations of fluid flows. Each type of fluid flow listed above has inspired a myriad of competing numerical methods, each with its own pros and cons, each demanding a good understanding of both traditional fluid dynamics and numerical analysis. As a result, the size of computational fluid dynamics far exceeds that of traditional fluid dynamics. At the present time, most CFD books survey numerical methods for all three of the basic flow categories. While such surveys work well for traditional fluid dynamics, the larger size of CFD requires more compromises and trade-offs. With so much to cover, even a two-volume survey can only describe the basic principles and techniques common to all three categories of flows, plus a limited and sometimes arbitrary sampling of the principles and techniques specific to each category of flow. Such condensation forces the reader to consult the original research literature for most of the details on most of the methods for each type of flow. But the expurgated surveys found in most CFD books do not provide enough background to face the typically complicated and highly mathematical CFD research papers.

This book takes a different approach – it focuses entirely on gasdynamics, and a limited type of gasdynamics at that. It omits computational methods for compressible potential flows. It omits computational methods for chemically reacting flows, real gas flows, and other important advanced topics. Instead, this book deals nearly exclusively with compressible, inviscid, unsteady flows of perfect gases. This approximately describes many flows of practical interest, especially aerodynamic flows. Furthermore, such flows serve as important *model problems* for, and constitute an essential first step towards, a broader range of practical compressible flows, including viscous and real gas flows. In keeping with the model problem concept, most of the book concerns one-dimensional flows; the last chapter explains how one-dimensional concepts carry over directly to multidimensions. The book also spends a lot of time discussing *scalar conservation laws*, such as Burgers' equation, which have precious few real-world applications, and whose popularity derives almost entirely from their success as model problems. Although a focus on model problems requires more patience early on, this author knows of no better or faster way to learn today's CFD.

This book does not claim to prepare readers to write state-of-the-art CFD codes. Writing CFD codes requires years of study, reading several books and dozens if not hundreds of research papers, preferably under the tutelage of a CFD expert. However, given the widespread availability of excellent commercial software and freeware, such as the RAMPANT code from the Fluent Corporation, most readers will never need to do large amounts of CFD coding. This book tries to prepare its readers to understand the CFD literature, or at least major chunks of it, should that become necessary. Also this book also tries to give its readers much of the background needed to intelligently evaluate and use the available CFD software.

Now let's look at the basic roadmap. The book has five parts. The first part reviews gasdynamics. The second part reviews numerical analysis. The third describes the basic principles of computational gasdynamics, including conservation, stability, upwinding, artificial viscosity, and so forth. The fourth part covers the *first-generation methods* of computational gasdynamics, also known as *nonadaptive* methods, including Lax–Wendroff methods, the Lax–Friedrichs method, and first-order upwind methods. The fifth part describes *second-* and *third-generation* methods of computational gasdynamics, also known as *adaptive* or *solution-sensitive* methods, including TVD, ENO, MUSCL, PPM, flux-corrected, and flux-limited methods. So fasten your seatbelts and prepare for a long lonesome ride across the heartland of CFD.

Part I

Gasdynamics Review

This part of the book concerns gasdynamics. One might think that the term "gasdynamics" could refer to any sort of flow of any sort of gas. However, by tradition, unless specifically stated otherwise, the terms *gasdynamics* or *compressible flow* refers to a relatively simple type of gas flow, affected only by pressure and flux, neither too dense nor too rare. A more precise definition appears in Chapter 2.

The treatment of gasdynamics found in Part I varies from the traditional gasdynamics treatment in several ways, due mainly to the demands of the numerical approximations studied later in this book, as opposed to the demands of the simple hand calculations studied in traditional gasdynamics texts. For example, traditional gasdynamics texts consider linearized potential flow approximations such as the the Prandtl–Glauert equation; while many people still use linear approximations, modern computing power has made them increasingly unnecessary. This book will not discuss linearized approximations. For another example, traditional gasdynamics texts focus mainly on steady two-dimensional flows, whereas this book focuses mainly on unsteady one-dimensional flows. These two model problems are equally difficult: Both model problems involve the same number of dependent and independent variables; and many solutions to one problem have an analogous solution in the other problem such as, for example, the steady two-dimensional expansion fan versus the unsteady one-dimensional expansion fan. The steady two-dimensional model problem has one major positive aspect: Many gas flows of practical interest are approximately steady and two dimensional. Unfortunately, the steady two-dimensional model problem has at least two critical cons. First, the nature of the steady problem changes dramatically when the flow shifts from subsonic to supersonic or vice versa, requiring similarly dramatic changes in the solution procedure. In particular, the equations governing steady two-dimensional supersonic flows exhibit a wavelike or *hyperbolic* behavior, whereas the equations governing steady subsonic two-dimensional flows exhibit a polar opposite nonwavelike or *elliptic* behavior. Because of this, solving steady two-dimensional gas flows requires a great deal of trouble and expense around the sonic lines separating subsonic and supersonic flows. As the second major con, techniques for solving steady two-dimensional flows do not readily extend to steady three-dimensional flows, nor to unsteady flows in any number of dimensions.

By contrast, consider the model problem of unsteady one-dimensional flow. This model problem has one major con: Few gas flows of practical interest are truly one dimensional. However, the model problem of unsteady one-dimensional flow has several major pros. First, unsteady one-dimensional flows have wavelike hyperbolic behaviors regardless of flow speed. Second, techniques for solving unsteady one-dimensional flows lead naturally to techniques for solving unsteady two- and three-dimensional flows and, in fact, most numerical techniques for simulating unsteady multidimensional flows are heavily based on numerical techniques for simulating unsteady one-dimensional flows, as discussed in Chapter 24. Third, if one can simulate unsteady flows in any given dimension, then one can also simulate steady flows in the same number of dimensions simply by letting the time run out to large values, until the unsteadiness settles out. Taken on balance, unless you need immediately useful solutions for a limited class of simple problems, the steady one-dimensional model problem proves far more fruitful than the unsteady two-dimensional model problem. Indeed, the unsteady one-dimensional model problem appears nearly universally in the modern computational gasdynamics literature as part of the basic construction and explanation of numerical methods, while the steady two-dimensional model problem usually appears only in passing, as part of the verification and testing phase.

Having decided on the model problem of unsteady one-dimensional flow, let us discuss some of the particulars. The governing equations of unsteady one-dimensional flows follow from the application of simple commonsense notions of *conservation*, as discussed in Chapter 2. Conservation of mass says that mass is neither created nor destroyed; conservation of momentum is Newton's second law; and conservation of energy is the first law of thermodynamics. *Characteristic* theory, coupled with *shock wave* theory, describes the wavelike nature of unsteady one-dimensional flows, as seen in Chapter 3. Chapter 4 describes simple models of unsteady one-dimensional flows called *scalar conservation equations*. Chapter 5 concerns the *Riemann problem*, which has uniform initial conditions punctuated by a single jump discontinuity. These simple initial conditions allow for an exact solution exemplifying most of the principles studied in Part I. The solution to the Riemann problem has assumed a primary role in modern computational gasdynamics, not only as a simple exact solution to compare numerical approximations against, but also as a fundamental element of the numerical methods themselves.

CHAPTER 2

Governing Equations of Gasdynamics

2.0 Introduction

The governing equations of gasdynamics are expressions of conservation and the second law of thermodynamics. *Conservation* requires that three fundamental quantities – mass, momentum, and energy – are neither created nor destroyed but are only redistributed or, excepting mass, converted from one form to another. For example, if momentum increases in one place, either momentum or an equivalent amount of energy must decrease someplace else. Conservation of momentum and energy is extremely complicated for a general physical system owing to the variety of mechanisms that can affect momentum and energy, such as viscosity, chemical reactions, gravity, electromagnetic forces, and so forth. To keep things simple, traditional gasdynamics as discussed in this book concerns inviscid flows of perfect gases, optionally free of forces except for pressure, which eliminates all influences on momentum and energy except for redistribution and pressure.

A companion principle to conservation, known as the *second law of thermodynamics*, requires that a fourth fundamental quantity called *entropy* should never decrease. The second law of thermodynamics restricts the redistributions and conversions of conserved quantities otherwise allowed by the conservation laws. As another supplement to conservation, the equations of state specify the nature and type of gas. Three conservation laws, two equations of state, and the second law of thermodynamics collectively constitute the *Euler equations* when expressed in a fixed coordinate system, or the *Lagrange equations* when expressed in a coordinate system that moves with the flow. This book mainly concerns the Euler equations, but see Problem 2.7 and Section 23.1 for an introduction to the Lagrange equations.

2.1 The Integral Form of the Euler Equations

2.1.1 *Conservation of Mass*

Conservation of mass states that the total mass of the universe is constant; in other words, mass is neither created nor destroyed but can only be moved from one place to another. Then if mass increases in one place it must decrease someplace else. For a one-dimensional flow, consider the fluid in a region $[a, b]$ during a time interval $[t_1, t_2]$. Then a precise statement of conservation of mass for one-dimensional flow is as follows:

♦ *change in total mass in $[a, b]$ in time interval $[t_1, t_2]$ = net mass passing through boundaries of $[a, b]$ in time interval $[t_1, t_2]$.*

This statement converts to mathematics immediately as follows:

♦ $$\int_a^b [\rho(x, t_2) - \rho(x, t_1)]\, dx = -\int_{t_1}^{t^2} [\rho(b, t)u(b, t) - \rho(a, t)u(a, t)]\, dt, \tag{2.1}$$

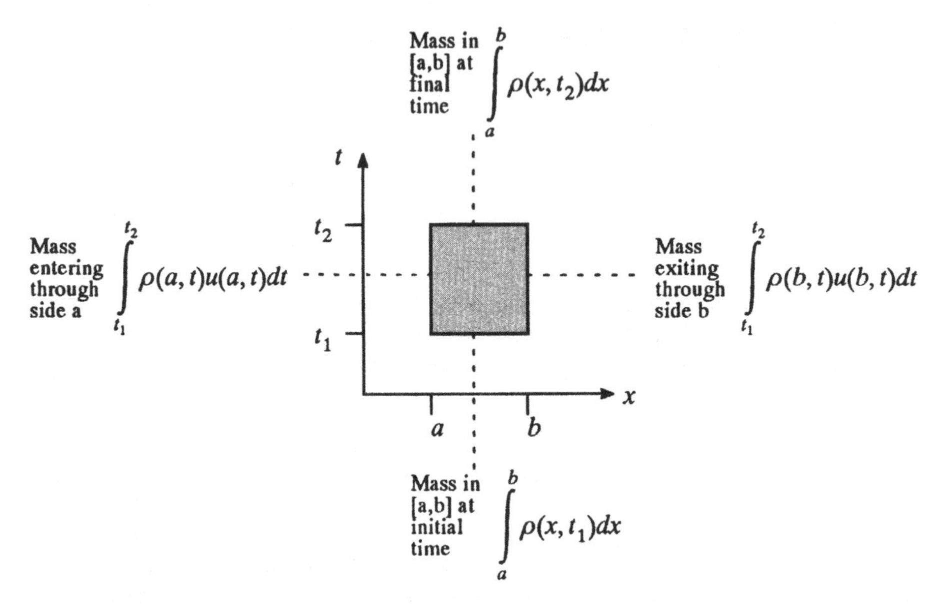

Figure 2.1 An illustration of conservation of mass.

where $\rho(x, t)$ is the mass per unit volume, $u(x, t)$ is the velocity, and $\rho(x, t)u(x, t)$ is the time rate of mass flow past point x. Then $\int_a^b \rho(x, t)\, dx$ is the total mass in $[a, b]$ at time t and $\int_{t_1}^{t_2} \rho(x, t)u(x, t)\, dt$ is the total mass passing x in the time interval $[t_1, t_2]$. The derivation of conservation of mass is illustrated in Figure 2.1.

Let us now introduce some standard fluids terminology. The mass per unit volume ρ is called *density*. A *flux* is a timed rate of flow of any property through any surface; then ρu is the instantaneous *mass flux* and $\int_{t_1}^{t_2} \rho(x, t)u(x, t)\, dt$ is the total mass flux through the surface $x = const.$ during the time interval $[t_1, t_2]$. The arbitrary spatial region $[a, b]$ is called a *control volume*. Similarly, the arbitrary region in the x–t plane defined by $[a, b]$ and $[t_1, t_2]$ is called a *space–time control volume*. Although the space–time control volume in Figure 2.1 is rectangular, it could just as well be triangular, quadrilateral, or some other shape. Nonrectangular space–time control volumes correspond to control volumes that contract, expand, or move in time.

2.1.2 *Conservation of Momentum*

Conservation of momentum says that momentum changes due to one of three factors – redistribution, conversion of momentum to or from energy, and force. In other words, if momentum increases in one place, either momentum or an equivalent amount of energy must decrease someplace else, or a force must act. Conservation of momentum can be extremely complicated owing to the variety of mechanisms that can affect momentum, such as viscosity, chemical reactions, gravity, electromagnetic forces, and so forth. However, classic gasdynamics concerns inviscid flows of perfect gases, optionally free of all forces except for pressure force, which eliminates all influences on momentum except for redistribution and pressure. In English, a precise statement of conservation of momentum for one-dimensional gasdynamics is as follows:

♦ *change in total momentum in $[a, b]$ in time interval $[t_1, t_2]$ = net momentum flow through boundaries of $[a, b]$ in time interval $[t_1, t_2]$ + net momentum change due to pressure on boundaries of $[a, b]$.*

Mathematically this statement converts to

$$\int_a^b [\rho(x, t_2)u(x, t_2) - \rho(x, t_1)u(x, t_1)]\, dx$$
$$= -\int_{t_1}^{t_2} [\rho(b, t)u^2(b, t) - \rho(a, t)u^2(a, t)]\, dt - \int_{t_1}^{t_2} [p(b, t) - p(a, t)]\, dt, \qquad (2.2)$$

where $\rho(x, t)u(x, t)$ is momentum per unit volume and $\rho(x, t)u^2(x, t)$ is the time rate of momentum flow or, in other words, the instantaneous *momentum flux*. Then $\int_a^b \rho(x, t)u(x, t)\, dx$ is the total momentum in $[a, b]$ at time t, $\int_{t_1}^{t_2} \rho(x, t)u^2(x, t)\, dt$ is the total momentum flowing past x in time interval $[t_1, t_2]$ (i.e., the total momentum flux), and $\int_{t_1}^{t_2} p(x, t)\, dt$ is the total momentum change at x due to pressure in time interval $[t_1, t_2]$.

2.1.3 *Conservation of Energy*

Conservation of energy says that energy change is due to one of three factors – redistribution, conversion of energy to or from momentum, or conversion to or from some other form of energy, heat, or work. In other words, if energy increases in one place, either energy or an equivalent amount of momentum must decrease someplace else, or heating or work must be done. Conservation of energy may be extremely complicated for a general physical system owing to the variety of mechanisms that can affect energy, such as viscosity, chemical reactions, gravity, electromagnetic forces, heating, and so forth. However, traditional gasdynamics concerns inviscid flows of perfect gases, optionally free of all sources of energy except for pressure, which eliminates all influences on energy except for redistribution and pressure.

Conservation of energy is most naturally stated in terms of *specific total energy*. Here and in general, *specific* means *per unit mass*. In the standard notation, specific quantities are denoted by lowercase letters, whereas nonspecific quantities – in other words, quantities that depend on the total amount of substance – are denoted by uppercase letters. For example, V is volume and $v = 1/\rho$ is *specific volume*. *Specific internal energy* $e(x, t)$ is the energy per unit mass contained in the microscopic motions of the individual fluid molecules, and the *specific kinetic energy* $u^2(x, t)/2$ is the energy per unit mass contained in the macroscopic motion of the overall fluid. Then the *specific total energy* $e_T(x, t) = e(x, t) + u^2(x, t)/2$ is the energy per unit mass stored in both microscopic and macroscopic motion.

In English, a precise statement of conservation of energy for one-dimensional gasdynamics is as follows:

♦ *change in total energy in $[a, b]$ in time interval $[t_1, t_2]$ = net energy flow through boundaries of $[a, b]$ in time interval $[t_1, t_2]$ + net energy change due to pressure on boundaries of $[a, b]$ in time interval $[t_1, t_2]$.*

This statement converts to mathematics immediately as follows:

$$\blacklozenge \quad \int_a^b [\rho(x,t_2)e_T(x,t_2) - \rho(x,t_1)e_T(x,t_1)]\,dx$$
$$= -\int_{t_1}^{t_2} [\rho(b,t)u(b,t)e_T(b,t) - \rho(a,t)u(a,t)e_T(a,t)]\,dt$$
$$-\int_{t_1}^{t_2} [p(b,t)u(b,t) - p(a,t)u(a,t)]\,dt, \tag{2.3a}$$

where $e_T(x,t)$ is the total energy per unit mass, $\rho(x,t)e_T(x,t)$ is the total energy per unit volume, and $\rho(x,t)e_T(x,t)u(x,t)$ is the instantaneous total *energy flux*. Then $E_T = \int_a^b \rho(x,t)e_T(x,t)\,dx$ is the total energy in $[a,b]$ at time t, $\int_{t_1}^{t_2} \rho(x,t)u(x,t)e_T(x,t)\,dt$ is the total energy flux (i.e., the total energy flowing past x in time interval $[t_1,t_2]$), and $\int_{t_1}^{t_2} p(x,t)u(x,t)\,dt$ is the total energy change at x in time interval $[t_1,t_2]$ due to pressure, which is also sometimes called *pressure work*. Equation (2.3a) omits body force terms, such as those due to gravity. It also omits heat transfer terms; in other words, the flow is assumed to be *adiabatic*. Indeed, as mentioned earlier, Equation (2.3a) neglects all energy effects except for pressure and redistribution.

Now let us consider an alternative form of Equation (2.3a). *Specific enthalpy* $h = e + p/\rho$ is the energy per unit mass found in the microscopic motions of the fluid molecules plus the potential energy per unit mass stored by compression. *Specific total enthalpy* $h_T = h + u^2/2 = e_T + p/\rho$ is the energy per unit mass stored in the microscopic *and macroscopic* motions of the fluid plus the potential energy per unit mass stored by compression. Then conservation of energy can be rewritten in terms of total enthalpy as follows:

$$\blacklozenge \quad \int_a^b [\rho(x,t_2)e_T(x,t_2) - \rho(x,t_1)e_T(x,t_1)]\,dx$$
$$= -\int_{t_1}^{t_2} [\rho(b,t)u(b,t)h_T(b,t) - \rho(a,t)u(a,t)h_T(a,t)]\,dt. \tag{2.3b}$$

Equation (2.3b) is certainly more compact than Equation (2.3a); this book will use both versions of Equation (2.3) at various times.

2.1.4 *Equations of State for a Perfect Gas*

We now have three equations in four unknowns. Specifically, the equations are conservation of mass, conservation of momentum, and conservation of energy and the unknowns are density, velocity, pressure, and total energy. A unique solution requires an equal number of equations and unknowns; thus this subsection introduces one more unknown and two more equations. The two additional equations are called *equations of state*. Equations of state specify the type of fluid; in other words, equations of state distinguish air from motor oil from any other sort of fluid.

A quick thermodynamics review is in order. *Mechanical properties* such as velocity and kinetic energy describe the macroscopic properties of a system, while *thermodynamic properties* such as internal energy and enthalpy describe the average microscopic properties of a system. Some properties, such as density and pressure, can be considered either mechanical or thermodynamic. The *thermodynamic state* is defined by three thermodynamic properties – for example, volume, mass, and internal energy – and all other thermodynamic properties can be expressed as functions of these three, that is, as *equations of state*. Three equations of state determine all other equations of state. All quantities in gasdynamics are traditionally expressed per unit mass – as *specific* quantities – which eliminates one variable, since mass per unit mass is always one. Then the *specific thermodynamic state* is defined by two specific thermodynamic properties – for example, density and specific internal energy – and all other specific thermodynamic properties can be expressed as functions of these two, that is, as *specific equations of state*. Two specific equations of state determine all other specific equations of state.

As it turns out, the equations of state are a result of conservation on the microscopic level. Assuming that large numbers of gas molecules interact only upon direct collision, conservation of momentum on the microscopic level yields the following equation of state known as the *ideal gas law* or the *thermal equation of state*:

♦ $$p = \rho R T, \tag{2.4}$$

where R is the *gas constant*. Although R is constant for a given gas, it is different for different gases. For sea-level air, a typical value is $R = 287\,\mathrm{N\cdot m/kg\cdot K}$. A fluid satisfying the thermal equation of state is called *thermally perfect*. Again assuming that large numbers of gas molecules interact only upon direct collision, conservation of energy on the microscopic level yields the following *caloric equation of state*:

♦ $$e = c_v T, \tag{2.5a}$$

where c_v is called the *constant volume specific heat*. An equivalent expression is as follows:

♦ $$h = c_p T, \tag{2.5b}$$

where c_p is the *constant pressure specific heat*. The specific heats are assumed constant, although the constant is different for different gases; typical values for sea-level air are $c_v = 717\,\mathrm{N\cdot m/kg\cdot K}$ and $c_p = 1{,}004\,\mathrm{N\cdot m/kg\cdot K}$. Either version of Equation (2.5) can optionally include an additive constant; the exact value of the constant is usually irrelevant and is taken here to be zero for convenience. Any fluid that satisfies the caloric equation of state, Equation (2.5), is called *calorically perfect*.

A fluid that satisfies both the thermal equation of state, Equation (2.4), and the caloric equation of state, Equation (2.5), is called a *perfect gas*; all other gases are known as *imperfect* or *real* gases. Real gases are either *dense* or *rarefied*. The microscopic particles in dense gases are close enough together that certain intermolecular forces, which are inversely proportional to powers of the separation distance, have significant strengths; in other words, the microscopic particles interact continuously instead of interacting only during direct collisions, as in a perfect gas. In contrast, the microscopic particles in rarefied gases are so far apart that the statistical averages used for perfect gases fail, since these statistical averages assume large numbers of particles per unit volume. This book concerns only perfect gases.

Equations of state are only required for compressible flows. For incompressible flows, there is no interaction between the microscopic and macroscopic levels, so that the nature of the substance as specified by the equations of state is irrelevant. You might think of this like banking. In incompressible flow, only the funds in the checking account are available; although there is a savings account, the funds in savings cannot be transferred to checking or back. By contrast, in compressible flow, the funds in checking and savings transfer back and forth freely. The equations of state describe how deposits, withdrawals, and transfers of funds affect various accounts.

Equations (2.1)–(2.5) are collectively known as the *integral form of the Euler equations*. In other words, the Euler equations equal conservation of mass, momentum, and energy for a perfect gas (and optionally the second law of thermodynamics as discussed in the next subsection).

This subsection ends with some useful perfect gas definitions and results.
The *ratio of specific heats* is defined as

$$\gamma = \frac{c_p}{c_v}. \tag{2.6}$$

The gas constant and specific heats are related by

$$c_p = R + c_v \tag{2.7}$$

or

$$c_p = \frac{\gamma R}{\gamma - 1}, \tag{2.8}$$

$$c_v = \frac{R}{\gamma - 1}. \tag{2.9}$$

Equations (2.4), (2.5), and (2.9) can be combined to yield a useful equation of state:

$$p = (\gamma - 1)\rho e = (\gamma - 1)\left(\rho e_{\mathrm{T}} - \frac{1}{2}\rho u^2\right). \tag{2.10}$$

The *speed of sound* is the speed at which small disturbances, especially acoustic disturbances, propagate through a substance measured relative to the movement of the substance. For a perfect gas, an equation of state for the speed of sound a is as follows:

$$a^2 = \gamma RT = \frac{\gamma p}{\rho}. \tag{2.11}$$

Equations (2.10) and (2.11) yield yet another useful equation of state:

$$\gamma e_{\mathrm{T}} = \frac{\gamma}{2}u^2 + \frac{1}{\gamma - 1}a^2. \tag{2.12}$$

This equation, the definition $h_{\mathrm{T}} = e_{\mathrm{T}} + p/\rho$, and Equation (2.11) imply

$$h_{\mathrm{T}} = \frac{1}{2}u^2 + \frac{1}{\gamma - 1}a^2. \tag{2.13}$$

Any number of other equations of state for perfect gases can be derived in a similar fashion, as needed. We shall see some more equations of state involving entropy in the next subsection.

2.1.5 *Entropy and the Second Law of Thermodynamics*

In addition to conservation laws, any physical system must also satisfy the second law of thermodynamics, which is stated in terms of entropy S or the specific entropy s. There are many ways to define entropy. For example, in traditional thermodynamics, entropy is defined as the ratio of reversible heat transfer to temperature. However, such a definition gives little insight into the true meaning of entropy. Intuitively, entropy indicates "disorder" or "loss of information." More specifically, for a given thermodynamic state, entropy measures how much is known about the positions, velocities, rotations, vibrations, etc. of the individual microscopic particles composing the fluid. Put another way, the thermodynamic state describes the average state of the microscopic particles, and entropy measures deviations from the average. Zero entropy corresponds to perfect knowledge of the microscopic states of the gas particles whereas increasing entropy corresponds to increasing uncertainty about the microscopic states of the gas particles.

Building on this intuitive definition of entropy, a formal definition of entropy is given by *Boltzmann's relation*:

$$S = k \ln \Omega, \tag{2.14}$$

where $k = 1.38 \times 10^{-23}$ J/K is a universal constant known as the *Boltzmann constant* and Ω is the number of ways in which the microscopic particles composing the gas can be arranged to give a specified thermodynamic state. For example, if there were just one way to arrange the fluid molecules to obtain a given thermodynamic state, then the entropy of that state would be $k \ln 1 = 0$. The quantity Ω grows exponentially with uncertainty; then $\ln \Omega$ grows linearly with uncertainty; then $S = k \ln \Omega$ also grows linearly with uncertainty, where the Boltzmann constant k is a traditional convenience factor.

For a perfect gas, the equation of state giving specific entropy s as a function of specific internal energy and density is

$$s = c_v \ln e - R \ln \rho + const. \tag{2.15a}$$

The exact value of the additive constant is usually unobtainable and usually unimportant. Equation (2.15a) can be rewritten in any number of ways using perfect gas relations. For example, an equivalent equation is as follows:

$$s = c_v \ln p - c_p \ln \rho + const. \tag{2.15b}$$

Suppose that the entropy is constant or, in other words, the flow is *homentropic*. This is often the case in inviscid flows, except at shocks. Then Equation (2.15) implies

$$p = (const.)\rho^{\gamma}, \tag{2.16a}$$

$$T = (const.)\rho^{\gamma-1}, \tag{2.16b}$$

$$a = (const.)\rho^{(\gamma-1)/2}, \tag{2.16c}$$

and many other similar relationships. For homentropic flow, any form of Equation (2.16) can replace Equations (2.3), (2.4), and (2.5), leaving only three equations in three unknowns in the Euler equations, rather than five equations in five unknowns.

Having defined entropy, the second law of thermodynamics can be expressed as follows: *The total entropy of the universe never decreases.* In other words, after any process, less is known about the microscopic state of the matter in the universe. If more is known about the microscopic particles in one place, then less must be known about the microscopic particles in another place, such that on balance less is known about the microscopic particles in the universe.

Without being specific about the source of entropy increase, the second law of thermodynamics for one-dimensional gasdynamics can be stated as

♦ *change in total entropy in $[a, b]$ in time interval $[t_1, t_2] \geq$ net entropy passing through boundaries of $[a, b]$ in time interval $[t_1, t_2]$.*

This statement converts to mathematics immediately as follows:

$$\begin{aligned} &\int_a^b [\rho(x, t_2)s(x, t_2) - \rho(x, t_1)s(x, t_1)]\, dx \\ &\qquad \geq - \int_{t_1}^{t_2} [\rho(b, t)u(b, t)s(b, t) - \rho(a, t)u(a, t)s(a, t)]\, dt, \end{aligned} \tag{2.17}$$

where $s(x, t)$ is the entropy per unit mass, $\rho(x, t)s(x, t)$ is the entropy per unit volume, and $\rho(x, t)s(x, t)u(x, t)$ is the instantaneous *entropy flux*. Then $S = \int_a^b \rho(x, t)s(x, t)\, dx$ is the total entropy in $[a, b]$ at time t and $\int_{t_1}^{t_2} \rho(x, t)u(x, t)s(x, t)\, dt$ is the total entropy flowing past x in time interval $[t_1, t_2]$, that is, the total entropy flux.

Although the second law of thermodynamics is sometimes lumped in with the Euler equations, it is more often considered separately. Even with the second law of thermodynamics, the Euler equations may have multiple solutions, at least in the special case of steady flows. For example, steady uniform supersonic flow over a wedge allows two solutions – a strong and a weak oblique shock solution. In fact, both solutions are observed in real life, although the weak shock is far more common. As another example, steady uniform supersonic flow over a cone also gives rise to strong and weak conical shock solutions. However, in this case, only the weak conical shock is observed in practice. As a third example, steady one-dimensional flow is always uniform except that it may allow a single normal shock anywhere in the flow, across which the flow switches from supersonic to subsonic; thus, there are infinitely many solutions, any of which can occur in real life. In short, the steady Euler equations allow multiple solutions in the presence of shocks and other entropy sources.

2.1.6 *Vector Notation*

Define the *vector of conserved quantities* as follows:

$$\mathbf{u} = \begin{bmatrix} \rho \\ \rho u \\ \rho e_T \end{bmatrix}. \tag{2.18}$$

The components of $\mathbf{u}$ will sometimes be referred to as u_1, u_2, and u_3, which represent mass per unit volume, momentum per unit volume, and total energy per unit volume, respectively. Mass, momentum, and energy are called the *conserved quantities*. The reader should take care not to confuse the conserved quantities u_1, u_2, and u_3 with the velocity u. Define the *flux vector* as follows:

$$\mathbf{f} = \begin{bmatrix} \rho u \\ \rho u^2 + p \\ (\rho e_T + p)u \end{bmatrix} = \begin{bmatrix} \rho u \\ \rho u^2 + p \\ \rho h_T u \end{bmatrix}. \tag{2.19}$$

The components of $\mathbf{f}$ will sometimes be referred to as f_1, f_2 and f_3, which represent mass flux, momentum flux plus pressure force, and total energy flux plus pressure work, respectively. Although $\mathbf{f}$ is called the flux vector, it includes pressure effects as well as fluxes. Strictly speaking the pressure effects on momentum and energy are not fluxes, but they can be and often are treated as such. However, if you prefer, the pressure and flux contributions can be separated as

$$\mathbf{f} = \begin{bmatrix} \rho u \\ \rho u^2 \\ \rho u e_T \end{bmatrix} + \begin{bmatrix} 0 \\ p \\ pu \end{bmatrix}. \tag{2.20}$$

Equations (2.1)–(2.3) can be rewritten compactly in terms of the flux vector and the vector of conserved quantities as follows:

♦ $$\int_a^b [\mathbf{u}(x,t_2) - \mathbf{u}(x,t_1)]\,dx = -\int_{t_1}^{t_2} [\mathbf{f}(b,t) - \mathbf{f}(a,t)]\,dt. \tag{2.21}$$

2.2 The Conservation Form of the Euler Equations

Historically, differential equations are preferred to integral equations, if for no other reason than that differential expressions are more compact. This section concerns a differential form of the Euler equations known as the *conservation form* of the Euler equations. If $\rho(x,t)$ is differentiable in time, then by the fundamental theorem of calculus

$$\rho(x,t_2) - \rho(x,t_1) = \int_{t_1}^{t_2} \frac{\partial \rho}{\partial t}\,dt.$$

Similarly, if $\rho(x,t)u(x,t)$ is differentiable in space then

$$\rho(b,t)u(b,t) - \rho(a,t)u(a,t) = \int_b^a \frac{\partial(\rho u)}{\partial x}\,dx.$$

Assuming integration in space is reversible with integration in time, Equation (2.1) becomes

$$\int_a^b \int_{t_1}^{t_2} \left[\frac{\partial \rho}{\partial t} + \frac{\partial(\rho u)}{\partial x}\right] dt\,dx = 0. \tag{2.22}$$

This integral is zero for any a, b, t_1, and t_2, which implies that the integrand is zero. Notice that this is true because the limits of integration are arbitrary. For example, $\int_0^{2\pi} \sin(x)\,dx = 0$ does not imply $\sin x = 0$; instead, the integral is zero because the positive and negative portions of $\sin x$ cancel. On the other hand, if $\int_a^b f(x)\,dx = 0$ for *all* a and b, then the integral cannot always be zero due to cancelation, and the integrand $f(x)$ must be identically zero, except for isolated points and other zero-measure regions.

Setting the integrand in Equation (2.22) equal to zero yields a differential form of conservation of mass. Differential forms of conservation of momentum, conservation of energy, and the second law of thermodynamics are obtained similarly. The results are summarized as follows:

♦ $$\frac{\partial \rho}{\partial t} + \frac{\partial(\rho u)}{\partial x} = 0, \tag{2.23}$$

♦ $$\frac{\partial(\rho u)}{\partial t} + \frac{\partial}{\partial x}(\rho u^2 + p) = 0, \tag{2.24}$$

♦ $$\frac{\partial(\rho e_T)}{\partial t} + \frac{\partial}{\partial x}(\rho u e_T + pu) = \frac{\partial(\rho e_T)}{\partial t} + \frac{\partial}{\partial x}(\rho u h_T) = 0, \tag{2.25}$$

♦ $$\frac{\partial(\rho s)}{\partial t} + \frac{\partial}{\partial x}(\rho u s) \geq 0. \tag{2.26}$$

Equations (2.23)–(2.25), (2.4), (2.5), and optionally (2.26) are known collectively as the *conservation form of the Euler equations*. This form is called the conservation form because the dependent variables are the conserved quantities. The term "conservation" is important to distinguish this differential form from other differential forms, such as the primitive variable form seen in the next section.

2.2.1 Vector and Vector–Matrix Notation

Using the vector notation of Section 2.16, the conservation form can be written as

♦ $$\frac{\partial \mathbf{u}}{\partial t} + \frac{\partial \mathbf{f}}{\partial x} = 0. \tag{2.27}$$

The flux vector $\mathbf{f}$ can be written as a function of the conserved quantities $\mathbf{u}$. For example, using Equation (2.10), f_3 can be written explicitly as a function of u_1, u_2, u_3:

♦ $$\begin{aligned} f_3 &= (\rho e_T + p)u = \left[\rho e_T + (\gamma - 1)\left(\rho e_T - \frac{1}{2}\rho u^2\right)\right]u \\ &= \left(\gamma \rho e_T - \frac{\gamma - 1}{2}\frac{\rho^2 u^2}{\rho}\right)\frac{\rho u}{\rho} \\ &= \left(\gamma u_3 - \frac{\gamma - 1}{2}\frac{u_2^2}{u_1}\right)\frac{u_2}{u_1} = \gamma\frac{u_2 u_3}{u_1} - \frac{\gamma - 1}{2}\frac{u_2^3}{u_1^2}. \end{aligned} \tag{2.28}$$

Then by the chain rule

$$\frac{\partial \mathbf{f}}{\partial x} = \frac{d\mathbf{f}}{d\mathbf{u}}\frac{\partial \mathbf{u}}{\partial \mathbf{x}},$$

where

$$\frac{d\mathbf{f}}{d\mathbf{u}} = \begin{bmatrix} \frac{\partial f_1}{\partial u_1} & \frac{\partial f_1}{\partial u_2} & \frac{\partial f_1}{\partial u_3} \\ \frac{\partial f_2}{\partial u_1} & \frac{\partial f_2}{\partial u_2} & \frac{\partial f_2}{\partial u_3} \\ \frac{\partial f_3}{\partial u_1} & \frac{\partial f_3}{\partial u_2} & \frac{\partial f_3}{\partial u_3} \end{bmatrix}, \tag{2.29}$$

which is called the *Jacobian matrix* of $\mathbf{f}$. For example, $\partial f_3/\partial u_1$ is computed from Equation (2.28) as follows:

$$\begin{aligned} \frac{\partial f_3}{\partial u_1} &= -\gamma\frac{u_2 u_3}{u_1^2} + (\gamma - 1)\frac{u_2^3}{u_1^3} = -\gamma\frac{\rho u \cdot \rho e_T}{\rho^2} + (\gamma - 1)\frac{(\rho u)^3}{\rho^3} \\ &= -\gamma u e_T + (\gamma - 1)u^3. \end{aligned}$$

If we call the Jacobian matrix A, Equation (2.27) becomes

♦ $$\frac{\partial \mathbf{u}}{\partial t} + A\frac{\partial \mathbf{u}}{\partial x} = 0, \tag{2.30}$$

where

♦ $$A = \begin{bmatrix} 0 & 1 & 0 \\ \frac{\gamma-3}{2}u^2 & (3-\gamma)u & \gamma - 1 \\ -\gamma u e_T + (\gamma - 1)u^3 & \gamma e_T - \frac{3}{2}(\gamma - 1)u^2 & \gamma u \end{bmatrix}, \tag{2.31a}$$

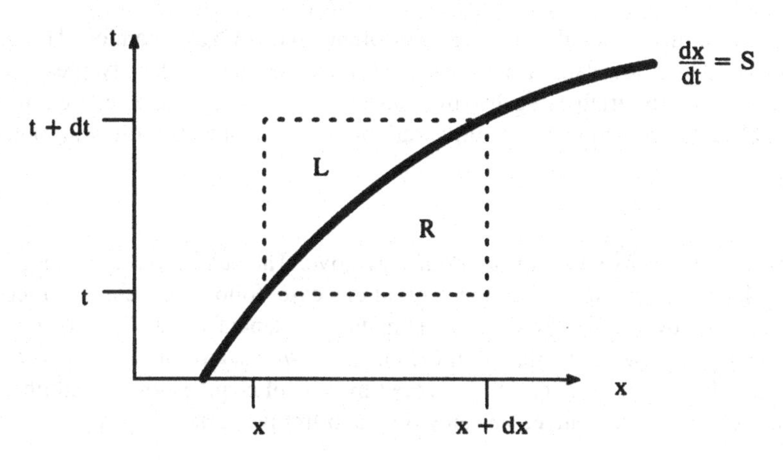

Figure 2.2 The derivation of the Rankine–Hugoniot relations.

or equivalently

$$A = \begin{bmatrix} 0 & 1 & 0 \\ \frac{\gamma-3}{2}u^2 & (3-\gamma)u & \gamma-1 \\ -uh_T + \frac{1}{2}(\gamma-1)u^3 & h_T - (\gamma-1)u^2 & \gamma u \end{bmatrix}. \tag{2.31b}$$

2.2.2 *Rankine–Hugoniot Relations*

The previous subsection glossed over a crucial issue – unlike the integral form of the Euler equations, the differential form requires differentiable solutions. But the true flow often contains large nondifferentiable jump discontinuities such as shocks and contact discontinuities which, apparently, the integral form of the Euler equations can model but the differential form cannot. Luckily, the mathematical theory of partial differential equations introduces the concept of *weak solutions* for just such occasions. To avoid complicated mathematics, let us just say that the differential form of the Euler equations can be considered shorthand for the integral form. Any nondifferentiable solutions are called *weak solutions of the differential form.*

Although the differential form gives a complete description of the flow in smooth regions, only the integral form can describe shocks and contact discontinuities. Thus, to provide a complete picture, the differential form is supplemented by a *jump condition* derived from the integral form. Consider a jump discontinuity separating states L and R traveling at speed S. Choose a rectangular space–time control volume surrounding the discontinuity, as shown in Figure 2.2. Apply the integral form of the Euler equations to the control volume to obtain the *Rankine–Hugoniot relations*:

$$\mathbf{f}_R - \mathbf{f}_L = S(\mathbf{u}_R - \mathbf{u}_L). \tag{2.32}$$

2.3 The Primitive Variable Form of the Euler Equations

This section concerns another differential form of the Euler equations, called the *primitive variable form.* While commonly used for incompressible, viscous, and turbulent fluid flows, the primitive variable form is less often used for gasdynamics. The *primitive variables* are those flow variables that can be directly measured – density, pressure, and velocity – as opposed to variables, such as the conserved variables, that are functions of measurable variables. Before stating the primitive variable form, consider the following notation:

$$\frac{D}{Dt} = \frac{\partial}{\partial t} + u\frac{\partial}{\partial x}. \tag{2.33}$$

This is known as the *substantial* or *material* derivative. The substantial derivative is the time rate of change following the substance or, in this case, following the gas. To see this, notice that the gas moves in the direction $(u, 1)$ in the x–t plane; then D/Dt is a directional derivative in the x–t plane in the direction $(u, 1)$. A *pathline* is any curve in the x–t plane everywhere parallel to $(u, 1)$. Then D/Dt is the time rate of change along a pathline.

Using the substantial derivative, we can write the primitive variable form as

♦ $$\frac{D\rho}{Dt} + \rho\frac{\partial u}{\partial x} = 0, \tag{2.34}$$

♦ $$\frac{Du}{Dt} + \frac{1}{\rho}\frac{\partial p}{\partial x} = 0, \tag{2.35}$$

♦ $$\frac{Dp}{Dt} + \rho a^2\frac{\partial u}{\partial x} = 0, \tag{2.36}$$

♦ $$\frac{Ds}{Dt} \geq 0. \tag{2.37}$$

Equation (2.34) is conservation of mass, Equation (2.35) is conservation of momentum, Equation (2.36) is conservation of energy, and Equation (2.37) is the second law of thermodynamics.

2.3.1 *Vector–Matrix Notation*

Define the vector of primitive variables as follows:

♦ $$\mathbf{w} = \begin{bmatrix} \rho \\ u \\ p \end{bmatrix}. \tag{2.38}$$

Then the primitive variable form of the Euler equations can be written in vector–matrix notation as

♦ $$\frac{\partial \mathbf{w}}{\partial t} + C\frac{\partial \mathbf{w}}{\partial x} = 0, \tag{2.39}$$

where

♦ $$C = \begin{bmatrix} u & \rho & 0 \\ 0 & u & \frac{1}{\rho} \\ 0 & \rho a^2 & u \end{bmatrix}. \tag{2.40}$$

Unlike the conservation form, the primitive variable form cannot be written in the form $\frac{\partial \mathbf{w}}{\partial t} + \frac{\partial \mathbf{f}(\mathbf{w})}{\partial x} = 0$, as the reader can show. Then, unlike the matrix A in the conservation form, the matrix C in primitive variable form is not the Jacobian of any flux function $\mathbf{f}(\mathbf{w})$.

There is a simple relationship between A and C. First notice that

$$d\mathbf{u} = Q\, d\mathbf{w}, \tag{2.41}$$

where

$$Q = \frac{d\mathbf{u}}{d\mathbf{w}} = \begin{bmatrix} 1 & 0 & 0 \\ u & \rho & 0 \\ \frac{1}{2}u^2 & \rho u & \frac{1}{\gamma-1} \end{bmatrix}. \tag{2.42}$$

Similarly

$$d\mathbf{w} = Q^{-1}\, d\mathbf{u}, \tag{2.43}$$

where

$$Q^{-1} = \frac{d\mathbf{w}}{d\mathbf{u}} = \begin{bmatrix} 1 & 0 & 0 \\ -\frac{1}{\rho}u & \frac{1}{\rho} & 0 \\ \frac{1}{2}(\gamma-1)u^2 & -(\gamma-1)u & \gamma-1 \end{bmatrix}. \tag{2.44}$$

Then Equation (2.30) can be written as

$$Q\frac{\partial \mathbf{w}}{\partial t} + AQ\frac{\partial \mathbf{w}}{\partial x} = 0$$

or

$$\frac{\partial \mathbf{w}}{\partial t} + Q^{-1}AQ\frac{\partial \mathbf{w}}{\partial x} = 0.$$

Comparing with Equation (2.39) we see that

$$C = Q^{-1}AQ. \tag{2.45}$$

In other words, A and C are *similar* matrices; the concept of similar matrices is discussed in any introductory text on linear algebra.

2.4 Other Forms of the Euler Equations

Of course, the conservative form and the primitive variable form are hardly the only ways to write the Euler equations. Any number of other ways follow from linear combinations of the Euler equations or, in other words, from multiplying the Euler equations by an invertible 3×3 matrix. For example, we can multiply both sides of Equation (2.30) by any invertible 3×3 matrix Q^{-1} to obtain the new form

$$Q^{-1}\frac{\partial \mathbf{u}}{\partial t} + Q^{-1}A\frac{\partial \mathbf{u}}{\partial x} = 0. \tag{2.46}$$

Notice that this procedure changes the equations but not the dependent variables, which are still the conservative variables. Similarly, Equation (2.39) can be multiplied by any invertible matrix Q^{-1} to obtain other new forms whose dependent variables are the primitive variables.

For convenience, new forms of the Euler equations often introduce new dependent variables. If we rewrite Equation (2.46) as

$$Q^{-1}\frac{\partial \mathbf{u}}{\partial t} + Q^{-1}AQ\, Q^{-1}\frac{\partial \mathbf{u}}{\partial x} = 0 \tag{2.47}$$

and then consider the linear change of dependent variables from $\mathbf{u}$ to $\mathbf{v}$, where

$$d\mathbf{v} = Q^{-1} d\mathbf{u}, \tag{2.48a}$$

or equivalently

$$\frac{\partial \mathbf{v}}{\partial x} = Q^{-1} \frac{\partial \mathbf{u}}{\partial x} \quad \text{and} \quad \frac{\partial \mathbf{v}}{\partial t} = Q^{-1} \frac{\partial \mathbf{u}}{\partial t}, \tag{2.48b}$$

then, using these new dependent variables, Equation (2.47) becomes

$$\frac{\partial \mathbf{v}}{\partial t} + Q^{-1} A Q \frac{\partial \mathbf{v}}{\partial x} = 0. \tag{2.49}$$

Notice that the matrix $Q^{-1}AQ$ is similar to A. In the next chapter, $\mathbf{v}$ is chosen such that $Q^{-1}AQ$ is diagonal.

References

Anderson, J. D. 1990. *Modern Compressible Flow with Historical Perspective*, 2nd ed., New York: McGraw-Hill.

Hirsch, C. 1990. *Numerical Computation of Internal and External Flows, Volume 2: Computational Methods for Inviscid and Viscous Flows*, Chichester: Wiley, Chapter 16.

Vincenti, W. G., and Kruger, C. H. 1965. *Physical Gas Dynamics*, New York: Wiley.

Problems

2.1 The following laws should be familiar to most readers from introductory courses. If not, please look up the laws in a relevant elementary text. In each case, state whether the laws express conservation of mass, conservation of momentum, conservation of energy, or no conservation law.
(a) Newton's laws of motion
(b) the first law of thermodynamics
(c) the continuity equation for fluids
(d) the continuity equation for electromagnetism ("conservation of charge")
(e) Chemical equations (e.g., $2HI \leftrightarrows H_2 + I_2$)
(f) $e = mc^2$

2.2 If $\mathbf{f}(\mathbf{u}) = \frac{d\mathbf{f}}{d\mathbf{u}}\mathbf{u}$ then $\mathbf{f}(\mathbf{u})$ is called a *homogeneous function of order one.*
(a) Find all *scalar* homogeneous functions of order one.
(b) Show that the flux function for the Euler equations, given by Equation (2.19), is a homogeneous function of order one.

2.3 In general, for any substance, the speed of sound a is defined as $a^2 = \frac{\partial p}{\partial \rho}|_{s=const.}$. Show that this general definition for the speed of sound implies Equation (2.11) in the case of perfect gases.

2.4 (a) For isothermal flow ($T = const.$), show that the Euler equations can be written as follows:

$$\frac{\partial}{\partial t}\rho + \frac{\partial}{\partial x}(\rho u) = 0,$$

$$\frac{\partial}{\partial t}(\rho u) + \frac{\partial}{\partial x}(\rho(u^2 + a^2)) = 0,$$

where $a^2 = (\frac{dp}{d\rho})_{T=const.} = RT$.

(b) Write the isothermal Euler equations in the forms $\frac{\partial \mathbf{u}}{\partial t} + \frac{\partial \mathbf{f}(\mathbf{u})}{\partial x} = 0$ and $\frac{\partial \mathbf{u}}{\partial t} + A\frac{\partial \mathbf{u}}{\partial x} = 0$.

(c) Homogeneous functions of order one were defined in Problem 2.2. Is the flux function $\mathbf{f}(\mathbf{u})$ for the isothermal Euler equations a homogeneous function of order one?

2.5 Assume that $u \neq 0$, $u \neq a$, and $u \neq -a$. Find an expression for the primitive variables (ρ, u, p) in terms of the components (f_1, f_2, f_3) of the conservative flux vector $\mathbf{f}$ given by Equation (2.19). Why is it impossible to find an expression for (ρ, u, p) in terms of (f_1, f_2, f_3) when $u = 0$, $u = a$, or $u = -a$?

2.6 (a) For any arbitrary control volume $[a, b]$ in steady one-dimensional gasdynamics, prove that

$$\rho_a u_a = \rho_b u_b,$$

$$p_a + \rho_a u_a^2 = p_b + \rho_b u_b^2,$$

$$\frac{1}{2}u_a^2 + \frac{1}{\gamma - 1}a_a^2 = \frac{1}{2}u_b^2 + \frac{1}{\gamma - 1}a_b^2.$$

(b) Use the results of part (a) to prove that

$$(u_b - u_a)a^{*2} = (u_b - u_a)u_a u_b,$$

where the *critical speed of sound* a^* is defined as follows:

$$a^{*2} = \frac{1}{2}u_a^2 + \frac{1}{\gamma - 1}a_a^2 = \frac{1}{2}u_b^2 + \frac{1}{\gamma - 1}a_b^2.$$

(c) Use the result of part (b) to argue that steady one-dimensional flow is completely uniform except that it can be punctuated by, at most, one shock. Show that the single steady shock, if it exists, satisfies *Prandtl's relation* $u_a u_b = a^{*2}$. Use the fact that a shock in steady flow always transitions between supersonic ($|u| > |a|$) and subsonic ($|u| < |a|$) flow. This is also true of moving shocks, provided that the shock moves at a constant speed and provided that the flow is expressed in a frame of reference fixed to the shock.

2.7 The Euler equations describe flows of compressible, inviscid, perfect gases in *stationary* coordinates. The *Lagrange equations* describe the same type of flow but in coordinates that *move with the fluid*. For the Lagrange equations, x is not an appropriate independent variable. Instead, the Lagrange equations require an independent variable that moves with the fluid. To define such a variable, first recall that any point that moves with the fluid is called a *fluid element*. You can visualize a fluid element as a tiny drop of dye that moves with the fluid. Choose any arbitrary reference fluid element 0. Then the *mass coordinate* m of any fluid element is defined as the mass between the fluid element and the reference $m = 0$. By conservation of mass, the mass between any two fluid elements is constant; thus the mass coordinate does not depend on time. The Lagrange coordinate m and the Eulerian coordinate x are related as follows:

$$m = \int_{x(0,t)}^{x(m,t)} \rho(x, t)\, dx$$

or

$$dx = \frac{\partial x}{\partial m} dm + \frac{\partial x}{\partial t} dt = v\, dm + u\, dt.$$

(a) Briefly explain why conservation of mass is trivial for the Lagrange equations. Instead, *conservation of volume* replaces conservation of mass for the Lagrange equations. Conservation of volume for the Lagrange equations serves the same purpose as conservation of mass for the Euler equations – both ensure that fluid neither mysteriously appears or disappears, and both are sometimes referred to as the *continuity equation*. In English, conservation of volume can be stated as follows:

> *change in total volume in* $[m_1, m_2]$ *during time interval* $[t_1, t_2]$ = *change in the x coordinate of* m_2 *minus the change in the x coordinate of* m_1.

Translate this expression to mathematics to obtain an integral form of conservation of volume for the Lagrange equations. Write the integral form in terms of specific volume $v(m, t)$ and velocity $u(m, t)$.

(b) Write conservation of momentum for the Lagrange equations in integral form. Write the integral in terms of velocity $u(m, t)$ and pressure $p(m, t)$.

(c) Write conservation of energy for the Lagrange equations in integral form. Write the integral in terms of velocity $u(m, t)$, pressure $p(m, t)$, and total specific energy $e_T(m, t)$.

(d) Write the integral form of the Lagrange equations in the following form:

$$\int_{m_1}^{m_2} [\mathbf{u}(m, t_2) - \mathbf{u}(m, t_1)]\, dm = -\int_{t_1}^{t_2} [\mathbf{f}(m_2, t) - \mathbf{f}(m_1, t)]\, dt.$$

(e) Derive a conservative differential form of the Lagrange equations. Write the result in the following form:

$$\frac{\partial \mathbf{u}}{\partial t} + \frac{\partial \mathbf{f}}{\partial x} = 0,$$

where

$$\mathbf{u} = \begin{bmatrix} v \\ u \\ e_T \end{bmatrix}, \qquad \mathbf{f} = \begin{bmatrix} -u \\ p \\ pu \end{bmatrix}.$$

(f) Show that the Rankine–Hugoniot relations for the Lagrange equations are as follows:

$$\Delta \mathbf{f} = W \Delta \mathbf{u},$$

where $W = dm_{\text{shock}}/dt$ is the *Lagrangean shock speed* (i.e., the shock speed in the m–t plane). For a right-running jump, the Lagrangean shock speed W is related to the Eulerian shock speed S by $W = \rho_R(S - u_R)$.

CHAPTER 3

Waves

3.0 Introduction

The last chapter explained changes in conserved quantities in terms of fluxes. This chapter explains changes in conserved quantities in terms of *waves*, a description every bit as complete and compelling as the flux description.

Everyone knows intuitively what waves are, but how are they actually defined? Tipler (1976) says "wave motion can be thought of as the transport of energy and momentum from one point in space to another without the transport of matter. In mechanical waves, e.g., water waves, waves on a string, or sound waves, the energy and momentum are transported by means of a disturbance in the medium which is propagated because the medium has elastic properties." Most other elementary physics books contain a similar definition. However, although this definition has some useful elements, it does not fully suffice in our applications. For example, this definition says that waves "travel without transport of matter." But mass transport constitutes a type of wave in fluid dynamics.

For a more precise understanding of waves, consider the following simple partial differential equation:

$$\frac{\partial u}{\partial t} + a\frac{\partial u}{\partial x} = 0,$$

where a is constant. This is known as the *linear advection equation*. Suppose that the spatial domain is $-\infty < x < \infty$ and suppose that the initial conditions are

$$u(x, 0) = u_0(x),$$

where $u_0(x)$ is any function. Then the solution to the linear advection equation is

$$u = u_0(x - at),$$

which is easily verified by substituting into the linear advection equation. According to this solution, the initial conditions propagate at a constant speed a to the right for $a > 0$ and to the left for $a < 0$. This is the simplest possible example of a wave solution. The lines $x - at = const.$ are called *wavefronts*, $u(x, t)$ is called the *signal* or *wave information*, a is called the *wave speed*, and $u_0(x)$ is called the *wave shape* or *wave form*.

As another example, discussed in most freshman physics books, consider the *one-dimensional wave equation*:

$$\frac{\partial^2 u}{\partial t^2} + a^2\frac{\partial^2 u}{\partial x^2} = 0,$$

where $a > 0$ is constant. The solution to this wave equation is

$$u = u_1(x - at) + u_2(x + at),$$

where the wave shapes u_1 and u_2 depend on the boundary and initial conditions. This solution is a superposition of two wave solutions – right-running waves with wavefronts

$x - at = const.$ and left-running waves with wavefronts $x + at = const.$ The superposition of two waves may obscure the waveforms since it is neither purely u_1 nor purely u_2 but some linear combination that may or may not resemble u_1 or u_2. Elementary physics texts often focus on sinusoidal waveforms where, for example, $u_1 = A_1 \cos(k_1 x - \omega_1 t)$ and $u_2 = A_2 \cos(k_2 x - \omega_2 t)$. Small disturbances in elastic substances tend to create sinusoidal waves; for example, a small lateral displacement of a taut string or a small compression of a solid bar will tend to create sinusoidal waves due to the elastic or "springy" properties of the string or bar; these waves may, in turn, communicate themselves to the surrounding air. However, whereas air and other gases allow sinusoidal waves, they also allow many other waveforms. Indeed, the waves that appear in gasdynamics generally are *not* sinusoidal. Small sinusoidal waves are the concerns of *acoustics* and *aeroacoustics*, which are specialized topics not discussed in this text. In general, any small wave or small disturbance in a gas that travels at the speed of sound a relative to the gas is called an *acoustic wave* or an *acoustic disturbance*, whether or not the waveform is sinusoidal.

In the linear advection equation and the one-dimensional wave equation, waves travel with a constant speed. However, in general, wave speeds may vary with position, time, and other factors. For example,

$$u = u_0(x - a(u)t) \tag{3.1}$$

and

$$u = u_0(x - a(u, x, t)t)$$

are perfectly valid wave solutions. Such waves *preserve* the initial waveform $u_0(x)$, although the initial waveform may be locally stretched or compressed due to variations in the wave speed a.

The linear advection equation and other equations preserve the initial waveform. However, more generally, the waveform may change in time due, for example, to viscosity or friction. In this case, the wave solution is generally no longer a function of the single variable $x - at$ but is instead a function of x and t separately, and thus the main advantage of the wave description is lost. At best, the solution may be the product of some function of t and some function of $x - at$ such as $e^{-\upsilon t} u_0(x - at)$, where υ is a constant associated with friction.

3.1 Waves for a Scalar Model Problem

As a simple scalar model problem, consider a first-order partial differential equation:

$$\frac{\partial u}{\partial t} + a \frac{\partial u}{\partial x} = 0, \tag{3.2}$$

where $u = u(x, t)$. This equation is more general than it first appears – if the coefficient of $\partial u / \partial t$ is anything other than one, simply divide both sides of the equation by the coefficient of $\partial u / \partial t$ to put the equation in form (3.2). Equation (3.2) is *linear* if $a = const.$ and *quasi-linear* if $a = a(u, x, t)$. Linear or quasi-linear, this equation has wave solutions. To see this, first notice that

$$\frac{\partial u}{\partial t} + a \frac{\partial u}{\partial x} = (1, a) \cdot \left(\frac{\partial u}{\partial t}, \frac{\partial u}{\partial x} \right) = (1, a) \cdot \nabla u.$$

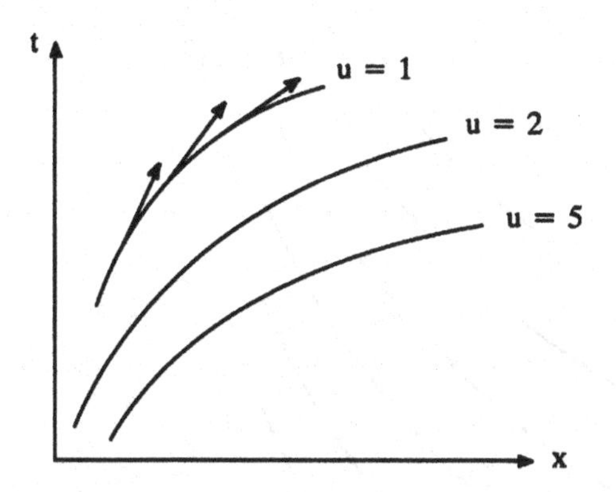

Figure 3.1 A typical wave diagram for a scalar model problem.

In other words, $\frac{\partial u}{\partial t} + a\frac{\partial u}{\partial x}$ is a directional derivative in the x–t plane in the direction $(1, a)$. Then, in English, Equation (3.2) says there is no change in the solution u in the direction of $(1, a)$ in the x–t plane. Now consider a curve $x = x(t)$ that is everywhere tangent to $(1, a)$ in the x–t plane. The slope of the vector $(1, a)$ is a and the slope of the curve $x = x(t)$ is dx/dt. Then

$$\frac{dx}{dt} = a.$$

Then Equation (3.2) is equivalent to the following:

♦ $$u = const. \quad \text{for} \quad \frac{dx}{dt} = a. \tag{3.3}$$

This is a wave solution – the curves $dx/dt = a$ are wavefronts, u is the signal or wave information, and a is the wave speed. The wavefronts $dx/dt = a$ are sometimes also called *characteristics curves* or simply *characteristics*. There are infinitely many wavefronts coating the entire x–t plane. A few select wavefronts are illustrated in Figure 3.1. The space–time vectors $(1, a)$ are also shown in a few instances to emphasize that the wavefronts are always parallel to $(1, a)$. Plots of wavefronts, such as Figure 3.1, are known as *wave diagrams*.

Example 3.1 Solve the following quasi-linear partial differential equation:

$$x\frac{\partial u}{\partial t} + \frac{\partial u}{\partial x} = 0.$$

Suppose the domain is $x \geq 0, t \geq 0$. Also, suppose the initial conditions are $u(x, 0) = c(x)$ and the boundary conditions are $u(0, t) = b(t)$.

Solution Divide the partial differential equation by x and apply Equation (3.3):

$$u = const. \quad \text{for} \quad \frac{dx}{dt} = \frac{1}{x}$$

or

$$u = const. \quad \text{for} \quad t = \frac{1}{2}x^2 + const.$$

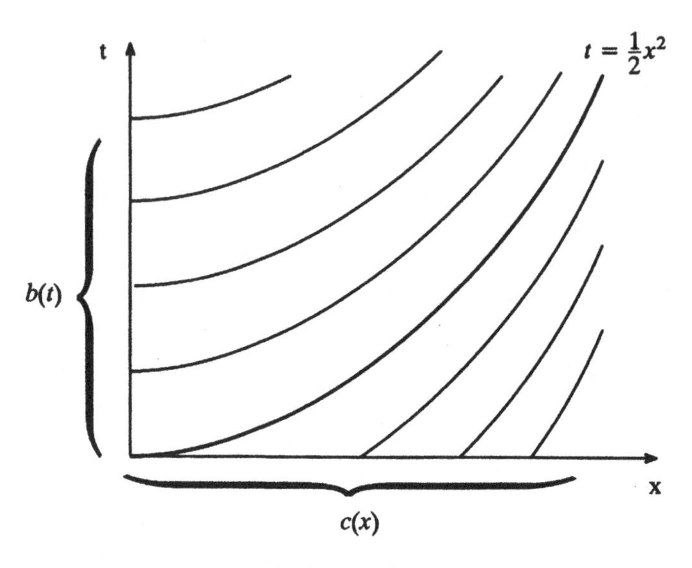

Figure 3.2 Wave diagram for Example 3.1.

The wavefronts $t = x^2/2 + const.$ are parabolas as illustrated in Figure 3.2. The easy part is done. To complete the solution, the value of u on each wavefront must be determined using the initial and boundary conditions. Notice that wavefronts with $t \leq x^2/2$ pass through the positive x axis and wavefronts with $t > x^2/2$ pass through the positive t axis. Now consider any point (x_1, t_1). The wavefront passing through (x_1, t_1) is

$$t - t_1 = \frac{1}{2}\left(x^2 - x_1^2\right).$$

If $t_1 \leq x_1^2/2$ this wavefront intersects the positive x axis at $x = x_0$, where

$$-t_1 = \frac{1}{2}\left(x_0^2 - x_1^2\right).$$

Solving this expression for x_0 gives

$$x_0 = \sqrt{x_1^2 - 2t_1}.$$

Then $u(x_1, t_1) = u(x_0, 0) = c(x_0)$, since any two points on the same wavefront have the same value of u, and this implies

$$u(x_1, t_1) = c\left(\sqrt{x_1^2 - 2t_1}\right).$$

Similarly, if $t_1 > x_1^2/2$ the wavefront intersects the positive t axis at $t = t_0$, where

$$t_0 - t_1 = -\frac{1}{2}x_1^2.$$

Solving this expression for t_0 gives

$$t_0 = t_1 - \frac{1}{2}x_1^2.$$

Then $u(x_1, t_1) = u(0, t_0) = b(t_0)$, since any two points on the same wavefront have the same value of u, and this implies

$$u(x_1, t_1) = b\left(t_1 - \frac{1}{2}x_1^2\right).$$

After dropping the subscripts, which are no longer needed, we obtain the final solution:

$$u(x,t) = \begin{cases} c\left(\sqrt{x^2 - 2t}\right) & t \leq \frac{1}{2}x^2, \\ b\left(t - \frac{1}{2}x^2\right) & t > \frac{1}{2}x^2. \end{cases}$$

In this example, and in general, it is extremely important to specify proper boundary and initial conditions. In particular, the boundary and initial conditions must be chosen so that each wavefront passes through one and only one boundary or initial condition. If a wavefront passes through more than one boundary or initial condition, the two conditions may yield conflicting values for u; in this case, the problem is *over specified*. However, if a wavefront does not pass through any boundary or initial condition, there is no way to determine the value of u on the characteristic; in this case, the problem is *under specified*. In either case, under specified or over specified, the problem is said to be *ill-posed*. Otherwise the problem is *well-posed*.

In the above example, it was relatively easy to solve $dx/dt = a$. In other cases, however, this ordinary differential equation may have no analytic solution such as, for example, $a = \cos(xt)$. In such cases, the wave description given by Equation (3.3) is not as useful.

The wavefronts illustrated in Figures 3.1 and 3.2 never intersect. However, as shown in Figure 3.3, this is not always the case. A conflict occurs when two wavefronts with different signals meet. For example, in Figure 3.3, $u = 1$ and $u = 2$ both cannot be true. Such conflicts can only be resolved by a jump discontinuity in the solution, known as a *shock wave*, shown by the bold curve in Figure 3.3. Shock waves are different from other waves. For example, shock waves are not described by Equation (3.3). Instead, shock waves are governed by jump relations and the theory of weak solutions. Shock waves are the "black hole" of waves – they

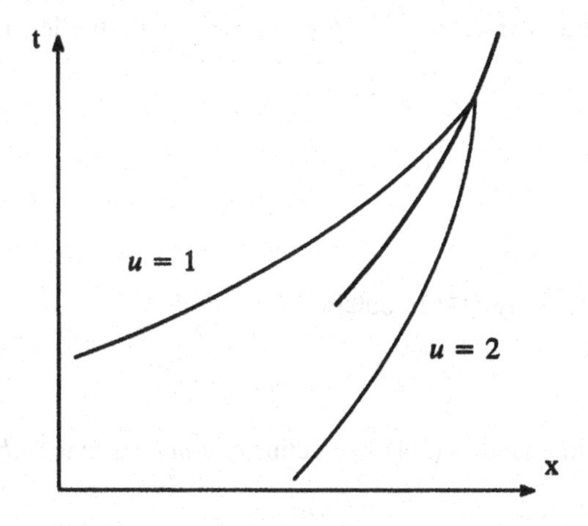

Figure 3.3 Wave diagram for a shock wave in a scalar model problem.

absorb any waves they meet, effectively destroying them and the signals they carry. Notice that shocks can occur any time wavefronts converge, despite the fact that the boundary and initial conditions may be completely smooth and continuous.

This section has introduced a number of important concepts. These concepts will be fleshed out in subsequent sections and will play a vital role in the remainder of the text.

3.2 Waves for a Vector Model Problem

Consider a system of first-order partial differential equations:

$$\frac{\partial \mathbf{u}}{\partial t} + A\frac{\partial \mathbf{u}}{\partial x} = 0, \tag{3.4}$$

where $\mathbf{u} = \mathbf{u}(x, t)$ and A is a square matrix. This equation is more general than it first appears – if the coefficient of $\partial \mathbf{u}/\partial t$ is an invertible matrix, simply multiply both sides of the equation by the inverse of the matrix to put the equation in form (3.4). This equation is *linear* if $A = const.$ and *quasi-linear* if $A = A(\mathbf{u}, x, t)$. An equation or system of equations with a complete wave description is sometimes called *hyperbolic*. For example, Equation (3.2) is always hyperbolic, as seen in the last section. On the other hand, the system of Equations (3.4) is hyperbolic if and only if A is diagonalizable. In other words, the system of Equations (3.4) is hyperbolic if and only if

$$Q^{-1}AQ = \Lambda \tag{3.5}$$

for some matrix Q, where Λ is a diagonal matrix. More specifically, Λ is a diagonal matrix whose diagonal elements λ_i are *characteristic values* or *eigenvalues* of A, Q is a matrix whose columns $\mathbf{r}_i$ are *right characteristic vectors* or *right eigenvectors* of A, and Q^{-1} is a matrix whose rows $\mathbf{l}_i$ are *left characteristic vectors* or *left eigenvectors* of A. As most readers will recall from elementary linear algebra, right characteristic vectors are defined as follows:

$$A\mathbf{r}_i = \lambda_i \mathbf{r}_i. \tag{3.6}$$

While less familiar, left characteristic vectors are defined in almost the same way as right characteristic vectors, except that left characteristic vectors multiply A on the left rather than on the right. In particular

$$\mathbf{l}_i^T A = \lambda_i \mathbf{l}_i^T \tag{3.7}$$

or, equivalently,

$$A^T \mathbf{l}_i = \lambda_i \mathbf{l}_i.$$

Multiply both sides of Equation (3.4) by Q^{-1} to obtain

♦ $$Q^{-1}\frac{\partial \mathbf{u}}{\partial t} + Q^{-1}A\frac{\partial \mathbf{u}}{\partial x} = 0. \tag{3.8}$$

This is called a *characteristic form* of equation (3.4). We define *characteristic variables* $\mathbf{v}$ as follows:

♦ $$d\mathbf{v} = Q^{-1}d\mathbf{u}. \tag{3.9}$$

Then the characteristic form becomes

$$\frac{\partial \mathbf{v}}{\partial t} + Q^{-1}AQ\frac{\partial \mathbf{v}}{\partial x} = 0$$

or

$$\blacklozenge \qquad \frac{\partial \mathbf{v}}{\partial t} + \Lambda\frac{\partial \mathbf{v}}{\partial x} = 0. \tag{3.10}$$

This is also called the *characteristic form*. This time the characteristic form is written in terms of the characteristic variables $\mathbf{v}$ rather than in terms of the conservative variables $\mathbf{u}$. (Compare the material in this section to the material in Section 2.4.)

The characteristic form is a wave form. To see this, consider the i-th equation in (3.10):

$$\frac{\partial v_i}{\partial t} + \lambda_i\frac{\partial v_i}{\partial x} = 0. \tag{3.11}$$

This is like Equation (3.2) except that, for quasi-linear systems of equations, λ_i depends on *all* of the characteristic variables and not just on the single characteristic variable v_i. Despite this difference, the same analysis applies so that by Equation (3.3)

$$\blacklozenge \qquad v_i = const. \quad \text{for} \quad \frac{dx}{dt} = \lambda_i. \tag{3.12}$$

The curves $dx = \lambda_i\, dt$ are called *wavefronts* or *characteristics*; the variables v_i are called *signals* or *information carried by the waves* or *characteristic variables*; and the characteristic values λ_i are called *wave speeds* or *characteristic speeds* or *signal speeds*. The term "characteristic" is used here because the analysis depends heavily on the characteristic values and characteristic vectors of matrix A. Using this terminology, Equation (3.12) says that the i-th characteristic variable is constant along the i-th characteristic curve.

For a system of N equations, there are N families of waves. Thus, there are N characteristics passing through each point in the x–t plane. This is illustrated in Figure 3.4 for $N = 3$. A point in the x–t plane is obviously only influenced by points at earlier times and can obviously only influence points at later times. However, because influence spreads in finite-speed waves, a point in the x–t plane is not influenced by every point at earlier times and does not influence every point at later times. Instead, a point in the x–t plane is influenced only by points in a finite *domain of dependence* and influences only points in a finite *range of influence*. The domain of dependence and range of influence are bounded on the right and left by the waves with the greatest and least speeds, respectively, as seen in Figure 3.4.

In a well-posed problem, the range of influence of the initial and boundary conditions should exactly encompass the entire flow in the x–t plane. Put in the reverse sense, in a well-posed problem, the domain of dependence of every point in the x–t plane should contain the boundary or initial conditions. Put yet another way, each of the three waves passing through a point in the x–t plane should carry a different signal from the initial or boundary conditions; a wave carries a signal from the initial or boundary conditions if it originates in the initial or boundary conditions or if it intersects another wave carrying a signal from the initial or boundary conditions.

Although intersections between characteristics of different families are routine, as seen in Figure 3.4, intersections between characteristics of the same family are not – any intersection between two characteristics from the same family creates a *shock wave*. Shock

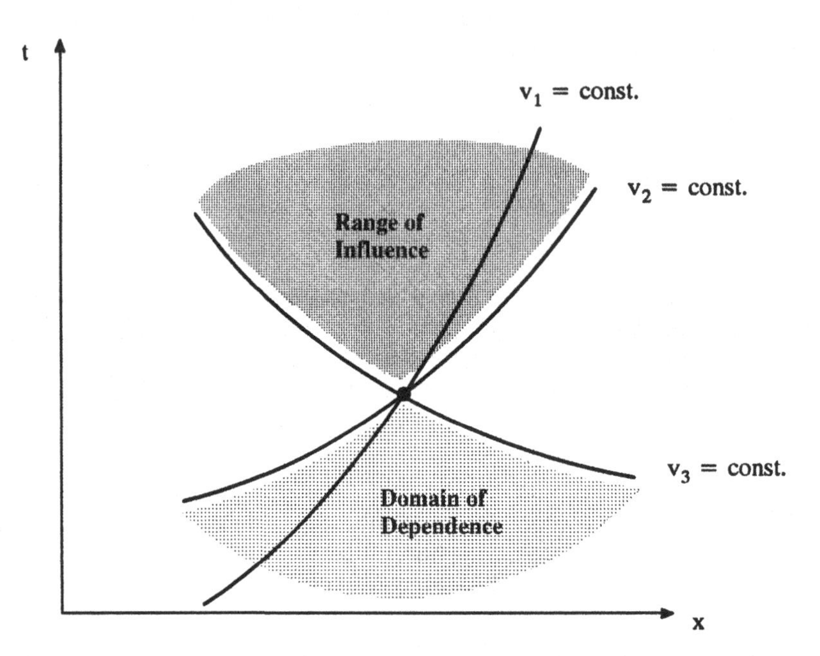

Figure 3.4 Typical wave diagram for a vector model problem.

waves are not governed by the ordinary characteristic equations, which originate in differential forms of the governing equations, but instead are governed by jump relations and the theory of weak solutions, which originate in integral forms of the governing equations.

The characteristic variables are defined by the equation $d\mathbf{v} = Q^{-1}d\mathbf{u}$. Unfortunately, this differential equation may not always have an analytic solution. Hence, whereas $d\mathbf{v}$ is always analytically defined, $\mathbf{v}$ may not be. For this reason, it is common to write Equation (3.12) as follows:

$$\blacklozenge \qquad dv_i = 0 \quad \text{for} \quad \frac{dx}{dt} = \lambda_i. \tag{3.13}$$

The relations $dv_i = 0$ are sometimes called *compatibility relations*. Similarly, in some cases, the ordinary differential equation $dx/dt = \lambda_i$ may not have an analytic solution. The wave description is of more limited use when either $dv_i = 0$ or $dx/dt = \lambda_i$ lack an analytical solution.

3.3 The Characteristic Form of the Euler Equations

The last chapter introduced the conservation form and the primitive variable form of the Euler equations. This section concerns a third form called the *characteristic form*. The unsteady Euler equations are hyperbolic, that is, they have a full wave description. To see this, first recall that the primitive variable form of the Euler equations can be written as

$$\frac{\partial \mathbf{w}}{\partial t} + C\frac{\partial \mathbf{w}}{\partial x} = 0, \tag{2.39}$$

where

$$\mathbf{w} = \begin{bmatrix} \rho \\ u \\ p \end{bmatrix} \tag{2.38}$$

and

$$C = \begin{bmatrix} u & \rho & 0 \\ 0 & u & \frac{1}{\rho} \\ 0 & \rho a^2 & u \end{bmatrix}. \tag{2.40}$$

As shown later, in Subsection 3.3.1, C is diagonalizable and

$$Q_C^{-1} C Q_C = \Lambda, \tag{3.14}$$

where

♦ $$Q_C = \begin{bmatrix} 1 & \frac{\rho}{2a} & -\frac{\rho}{2a} \\ 0 & \frac{1}{2} & \frac{1}{2} \\ 0 & \frac{\rho a}{2} & -\frac{\rho a}{2} \end{bmatrix}, \tag{3.15}$$

♦ $$Q_C^{-1} = \begin{bmatrix} 1 & 0 & -\frac{1}{a^2} \\ 0 & 1 & \frac{1}{\rho a} \\ 0 & 1 & -\frac{1}{\rho a} \end{bmatrix}, \tag{3.16}$$

and

♦ $$\Lambda = \begin{bmatrix} u & 0 & 0 \\ 0 & u+a & 0 \\ 0 & 0 & u-a \end{bmatrix}. \tag{3.17}$$

Then, by Equation (3.8), a characteristic form of the Euler equations is as follows:

$$Q_C^{-1} \frac{\partial \mathbf{w}}{\partial t} + Q_C^{-1} C \frac{\partial \mathbf{w}}{\partial x} = 0, \tag{3.18}$$

or

♦ $$\frac{\partial \rho}{\partial t} + u \frac{\partial \rho}{\partial x} - \frac{1}{a^2} \left(\frac{\partial p}{\partial t} + u \frac{\partial p}{\partial x} \right) = 0, \tag{3.18a}$$

♦ $$\frac{\partial u}{\partial t} + (u+a) \frac{\partial u}{\partial x} + \frac{1}{\rho a} \left(\frac{\partial p}{\partial t} + (u+a) \frac{\partial p}{\partial x} \right) = 0, \tag{3.18b}$$

♦ $$\frac{\partial u}{\partial t} + (u-a) \frac{\partial u}{\partial x} - \frac{1}{\rho a} \left(\frac{\partial p}{\partial t} + (u-a) \frac{\partial p}{\partial x} \right) = 0. \tag{3.18c}$$

By Equation (3.10), a characteristic form that involves characteristic rather than primitive variables is

$$\frac{\partial \mathbf{v}}{\partial t} + \Lambda \frac{\partial \mathbf{v}}{\partial x} = 0, \tag{3.19}$$

or

$$\frac{\partial v_0}{\partial t} + u\frac{\partial v_0}{\partial x} = 0, \tag{3.19a}$$

$$\frac{\partial v_+}{\partial t} + (u+a)\frac{\partial v_+}{\partial x} = 0, \tag{3.19b}$$

$$\frac{\partial v_-}{\partial t} + (u-a)\frac{\partial v_-}{\partial x} = 0, \tag{3.19c}$$

where

$$d\mathbf{v} = Q_C^{-1} d\mathbf{w}, \tag{3.20}$$

or

$$dv_0 = d\rho - \frac{dp}{a^2}, \tag{3.20a}$$

$$dv_+ = du + \frac{dp}{\rho a}, \tag{3.20b}$$

$$dv_- = du - \frac{dp}{\rho a}. \tag{3.20c}$$

In the last several equations, the characteristic variables are subscripted by $(0, +, -)$ rather than by $(1, 2, 3)$. Different sources order the characteristic variables differently. For example, many sources list the characteristic variables in order of ascending wave speeds, in which case $(1, 2, 3) = (-, 0, +)$, whereas other sources list the characteristic variables in order of descending waves speeds, in which case $(1, 2, 3) = (+, 0, -)$. Although $(1, 2, 3)$ may refer to different characteristics, $(0, +, -)$ always refers to the same characteristics, as defined above, by standard convention.

By Equation (3.13), the Euler equations can be written as

♦ $$dv_0 = d\rho - \frac{dp}{a^2} = 0 \quad \text{for} \quad dx = u\,dt, \tag{3.21a}$$

♦ $$dv_+ = du + \frac{dp}{\rho a} = 0 \quad \text{for} \quad dx = (u+a)\,dt, \tag{3.21b}$$

♦ $$dv_- = du - \frac{dp}{\rho a} = 0 \quad \text{for} \quad dx = (u-a)\,dt. \tag{3.21c}$$

Integrating the compatibility relations, these equations become

$$s = const. \quad \text{for} \quad dx = u\,dt, \tag{3.21a$'$}$$

$$v_+ = u + \int \frac{dp}{\rho a} = const. \quad \text{for} \quad dx = (u+a)\,dt, \tag{3.21b$'$}$$

$$v_- = u - \int \frac{dp}{\rho a} = const. \quad \text{for} \quad dx = (u-a)\,dt. \tag{3.21c$'$}$$

Therefore, in general, only the first compatibility relation is fully analytically integrable. To prove Equation (3.21a′), recall that

$$s = c_v \ln p - c_p \ln \rho + const. \tag{2.15b}$$

Then

$$ds = c_v \frac{dp}{p} - c_p \frac{d\rho}{\rho} = -\frac{c_p}{\rho}\left(d\rho - \frac{c_v}{c_p}\frac{\rho}{p}\,dp\right).$$

But Equation (2.6) says $\gamma = c_p/c_v$ and Equation (2.11) says $a^2 = \gamma p/\rho$. Thus

$$ds = -\frac{c_p}{\rho}\left(d\rho - \frac{dp}{a^2}\right) = -\frac{c_p}{\rho}\,dv_0.$$

Then $dv_0 = 0$ is equivalent to $ds = 0$, which is trivially integrated to obtain $s = const.$ This proves Equation (3.21a′). Also, Equation (3.19a) becomes

$$\blacklozenge \qquad \frac{Ds}{Dt} = \frac{\partial s}{\partial t} + u\frac{\partial s}{\partial x} = 0, \tag{3.19a′}$$

where D/Dt is the substantial derivative defined by Equation (2.33). Recall that the substantial derivative is the time rate of change following the fluid. Thus Equations (3.19a′) and (3.21a′) tell us that the entropy is constant following the fluid.

Compare Equation (3.19a′) with Equation (2.37). These equations are nearly identical: Equation (2.37) says that the substantial derivative of the entropy is *greater than or equal to* zero, while Equation (3.19a′) says that the substantial derivative of the entropy is *exactly equal to* zero. Which one is correct? Well, like all differential forms, the characteristic form applies everywhere except across jump discontinuities or, in particular, shocks. Then Equations (3.19a′) and (3.21a′) imply that *entropy is constant following the gas except across shocks*. This is a natural consequence of ignoring viscosity and other entropy-generating effects in the Euler equations (there is nothing in the Euler equations to generate entropy except for shocks). If entropy is constant following the gas, the flow is called *isentropic*. Then Equations (3.19a′) and (3.21a′) imply that the *flow is isentropic except at shocks*. Notice that the *Euler equations imply the second law of thermodynamics except at shocks* or, in other words, Equation (3.19a′) implies Equation (2.37) except at shocks, where Equation (3.19a′) does not apply. Only at shocks do the Euler equations need explicit supplementation by the second law of thermodynamics. At shocks, the second law of thermodynamics, as given by Equation (2.37), implies that the entropy increases following the gas, that is, shocks generate entropy. Shocks are discussed further in Section 3.6.

The characteristic form can also be derived from the conservation form of the Euler equations. Recall that the conservation form of the Euler equations can be written as

$$\frac{\partial \mathbf{u}}{\partial t} + A\frac{\partial \mathbf{u}}{\partial x} = 0, \tag{2.30}$$

where

$$\mathbf{u} = \begin{bmatrix} \rho \\ \rho u \\ \rho e_T \end{bmatrix} \tag{2.18}$$

and

$$A = \begin{bmatrix} 0 & 1 & 0 \\ \frac{\gamma-3}{2}u^2 & (3-\gamma)u & \gamma - 1 \\ -\gamma u e_T + (\gamma - 1)u^3 & \gamma e_T - \frac{3}{2}(\gamma - 1)u^2 & \gamma u \end{bmatrix}. \tag{2.31}$$

Then

$$Q_A^{-1} A Q_A = \Lambda, \tag{3.22}$$

where

♦ $$Q_A = \begin{bmatrix} 1 & \frac{\rho}{2a} & -\frac{\rho}{2a} \\ u & \frac{\rho}{2a}(u+a) & -\frac{\rho}{2a}(u-a) \\ \frac{u^2}{2} & \frac{\rho}{2a}\left(\frac{u^2}{2} + \frac{a^2}{\gamma-1} + au\right) & -\frac{\rho}{2a}\left(\frac{u^2}{2} + \frac{a^2}{\gamma-1} - au\right) \end{bmatrix}, \tag{3.23a}$$

♦ $$Q_A^{-1} = \frac{\gamma - 1}{\rho a} \begin{bmatrix} \frac{\rho}{a}\left(-\frac{u^2}{2} + \frac{a^2}{\gamma-1}\right) & \frac{\rho}{a}u & -\frac{\rho}{a} \\ \frac{u^2}{2} - \frac{au}{\gamma-1} & -u + \frac{a}{\gamma-1} & 1 \\ -\frac{u^2}{2} - \frac{au}{\gamma-1} & u + \frac{a}{\gamma-1} & -1 \end{bmatrix}, \tag{3.24}$$

and Λ is just as before. By Equation (2.13), Equation (3.23a) can also be written as

♦ $$Q_A = \begin{bmatrix} 1 & \frac{\rho}{2a} & -\frac{\rho}{2a} \\ u & \frac{\rho}{2a}(u+a) & -\frac{\rho}{2a}(u-a) \\ \frac{u^2}{2} & \frac{\rho}{2a}(h_T + au) & -\frac{\rho}{2a}(h_T - au) \end{bmatrix}. \tag{3.23b}$$

It can be shown that

$$d\mathbf{v} = Q_A^{-1}\, d\mathbf{u} = Q_C^{-1}\, d\mathbf{w}. \tag{3.25}$$

In other words, the characteristic variables are the same regardless of whether they are derived from the primitive variables or the conservative variables. Similarly, characteristic forms such as (3.19) and (3.21) are the same, regardless of whether they are derived from the primitive variable form or the conservation form of the Euler equations. Although the results are always the same, the conservation form requires more involved algebra than the primitive variable form, as we shall see later, in Subsection 3.3.1.

Consider the differential equations $dv_i = 0$ and $dx/dt = \lambda_i$ which appear in Equation (3.21). As seen above, only the first compatibility relation $dv_0 = dp - a^2 d\rho = 0$ can be analytically integrated. Also, λ_i is a function of u and a. Thus $dx/dt = \lambda_i$ cannot be integrated until u and a are known, and u and a are unknown until the problem is solved, which is a classic "Catch 22." The inability to analytically integrate $dv_i = 0$ and $dx/dt = \lambda_i$ limits the utility of the wave description of the Euler equations. The big exceptions to this situation are simple waves, as described in Section 3.4.

3.3.1 *Examples*

In this subsection, a few of the results seen above will be derived as examples. After studying these examples, the reader should be able to derive any of the results given above. For convenience, this subsection will use numbered indices $(1, 2, 3)$ rather than $(0, +, -)$.

Example 3.2 Derive Equation (3.17).

Solution The characteristic values of C are the solutions of $\det(\lambda I - C) = 0$. But

$$\det(\lambda I - C) = \begin{vmatrix} \lambda - u & -\rho & 0 \\ 0 & \lambda - u & -\frac{1}{\rho} \\ 0 & -\rho a^2 & \lambda - u \end{vmatrix}$$

$$= (\Lambda - u)\left[(\lambda - u)^2 - a^2\right]. \tag{3.26}$$

Then the characteristic values of C are $\lambda_1 = u$, $\lambda_2 = u + a$, and $\lambda_3 = u - a$, which lead immediately to Equation (3.17).

Example 3.3 Find the left and right characteristic vectors of matrix C associated with $\lambda_1 = u$, where C is given by Equation (2.40).

Solution To find the right characteristic vector associated with $\lambda_1 = u$, solve the following system of equations:

$$(uI - C)\mathbf{r} = \begin{bmatrix} 0 & -\rho & 0 \\ 0 & 0 & -\frac{1}{\rho} \\ 0 & -\rho a^2 & 0 \end{bmatrix} \begin{bmatrix} r_1 \\ r_2 \\ r_3 \end{bmatrix} = \begin{bmatrix} 0 \\ 0 \\ 0 \end{bmatrix}. \tag{3.27}$$

This is equivalent to

$$r_2 = r_3 = 0.$$

The value r_1 is arbitrary. Thus any right characteristic value associated with $\lambda_1 = u$ can be written as follows:

$$\mathbf{r}_{C1} = (const.) \begin{bmatrix} 1 \\ 0 \\ 0 \end{bmatrix}, \tag{3.28}$$

where "*const.*" is any functional factor. You might think that "*const.*" has to be a constant real number. However, when working in functional spaces, a "constant" may be any function, such as ρ or a^2.

To find the left characteristic vector associated with $\lambda_1 = u$, solve the following system of equations:

$$(uI - C)^T\mathbf{l} = \begin{bmatrix} 0 & 0 & 0 \\ -\rho & 0 & -\rho a^2 \\ 0 & -\frac{1}{\rho} & 0 \end{bmatrix} \begin{bmatrix} l_1 \\ l_2 \\ l_3 \end{bmatrix} = \begin{bmatrix} 0 \\ 0 \\ 0 \end{bmatrix}. \tag{3.29}$$

This is equivalent to

$$l_1 + a^2 l_3 = 0, \qquad l_2 = 0.$$

Thus any left characteristic value associated with $\lambda_1 = u$ can be written as follows:

$$\mathbf{l}_{C1} = (const.) \begin{bmatrix} 1 \\ 0 \\ -\frac{1}{a^2} \end{bmatrix}. \tag{3.30}$$

Example 3.4 Derive Equations (3.15) and (3.16).

Solution Using a process similar to that seen in the last example, the right and left characteristic vectors associated with $\lambda_2 = u + a$ are found to be

$$\mathbf{r}_{C2} = (const.) \begin{bmatrix} \frac{1}{\rho a} \\ 1 \\ \rho a \end{bmatrix} \tag{3.31}$$

and

$$\mathbf{l}_{C2} = (const.) \begin{bmatrix} 0 \\ 1 \\ \frac{1}{\rho a} \end{bmatrix} \tag{3.32}$$

and the right and left characteristic vectors associated with $\lambda_3 = u - a$ are found to be

$$\mathbf{r}_{C3} = (const.) \begin{bmatrix} -\frac{1}{\rho a} \\ -1 \\ \rho a \end{bmatrix} \tag{3.33}$$

and

$$\mathbf{l}_{C3} = (const.) \begin{bmatrix} 0 \\ 1 \\ -\frac{1}{\rho a} \end{bmatrix}. \tag{3.34}$$

To form the matrices Q_C and Q_C^{-1}, choose the functional factors "*const.*" in the characteristic vectors such that

$$\mathbf{r}_{Ci} \cdot \mathbf{l}_{Cj} = \delta_{ij} = \begin{cases} 1 & i = j, \\ 0 & i \neq j. \end{cases}$$

Then Q_C is a matrix whose columns are the right characteristic vectors. This is sometimes expressed as follows:

$$Q_C = [\mathbf{r}_{C1} \mid \mathbf{r}_{C2} \mid \mathbf{r}_{C3}].$$

Also Q_C^{-1} is a matrix whose rows are left characteristic vectors. This is sometimes expressed as follows:

$$Q_C^{-1} = [\mathbf{l}_{C1} \mid \mathbf{l}_{C2} \mid \mathbf{l}_{C3}]^T.$$

In general, Q_C and Q_C^{-1} can be written as

$$Q_C = \begin{bmatrix} \frac{1}{c_1} & \frac{\rho}{2ac_2} & -\frac{\rho}{2ac_3} \\ 0 & \frac{1}{2c_2} & \frac{1}{2c_3} \\ 0 & \frac{\rho a}{2c_2} & -\frac{\rho a}{2c_3} \end{bmatrix} \tag{3.35}$$

and

$$Q_C^{-1} = \begin{bmatrix} c_1 & 0 & -\frac{c_1}{a^2} \\ 0 & c_2 & \frac{c_2}{\rho a} \\ 0 & c_3 & -\frac{c_3}{\rho a} \end{bmatrix}, \tag{3.36}$$

where c_1, c_2, and c_3 are any functional factors. Equations (3.15) and (3.16) are obtained by choosing $c_1 = c_2 = c_3 = 1$.

Example 3.5 Derive Equation (3.23) for Q_A.

Solution Equation (3.23) can be derived directly just as in the last example. Alternatively, recall that

$$C = Q^{-1}AQ, \tag{2.45}$$

where

$$Q = \begin{bmatrix} 1 & 0 & 0 \\ u & \rho & 0 \\ \frac{1}{2}u^2 & \rho u & \frac{1}{\gamma-1} \end{bmatrix}. \tag{2.42}$$

Then

$$\Lambda = Q_C^{-1}CQ_C = Q_C^{-1}(Q^{-1}AQ)Q_C = (QQ_C)^{-1}A(QQ_C)$$

and thus

$$Q_A = QQ_C, \tag{3.37}$$

which yields Equation (3.23).

Example 3.6 Derive Equation (3.20a) from the conservation form of the Euler equations.

Solution The first row of $d\mathbf{v} = Q_A^{-1} d\mathbf{u}$ yields

$$dv_1 = \mathbf{l}_1 \cdot d\mathbf{u},$$

where $\mathbf{l}_1$ is the first row of Q_A^{-1}. Referring to Equations (2.18) and (3.24), we get

$$dv_1 = \mathbf{l}_1 \cdot d\mathbf{u} = -\frac{\gamma - 1}{a^2}\left[\frac{u^2}{2} - \frac{a^2}{\gamma - 1}, -u, 1\right] \cdot [d\rho, d(\rho u), d(\rho e_T)]$$

$$= -\frac{\gamma - 1}{a^2}\left[\left(\frac{u^2}{2} - \frac{a^2}{\gamma - 1}\right) d\rho - u d(\rho u) + d(\rho e_T)\right].$$

By Equation (2.10)

$$d(\rho e_{\mathrm{T}}) = d\left(\frac{1}{\gamma - 1} p + \frac{1}{2}\rho u^2\right) = \frac{1}{\gamma - 1} dp + \frac{1}{2} u^2 d\rho + \rho u \, du.$$

Then

$$dv_1 = -\frac{\gamma - 1}{a^2}\left[\left(\frac{u^2}{2} - \frac{a^2}{\gamma - 1}\right) d\rho - u(\rho \, du + u \, d\rho) \frac{1}{\gamma - 1} dp + \frac{1}{2} u^2 \, d\rho + \rho u \, du\right],$$

or

$$dv_1 = d\rho - \frac{dp}{a^2},$$

which agrees with Equation (3.20a) after accounting for the difference between the (1, 2, 3) and (0, +, −) indexing. Of course, the primitive variable form yields the same result for much less effort.

3.3.2 *Physical Interpretation*

There is a beautiful physical connection between the flow physics and the mathematics of characteristics. Consider the first characteristic family. The wave speed λ_1 equals the flow speed u, whereas the wavefronts $dx = \lambda_1 = u \, dt$ equal the pathlines. Then the first family of waves travels with the fluid. As seen in Equation (3.21a′), the signal is entropy; thus waves from the first family of characteristics are sometimes called *entropy waves*.

Now consider the other two families of characteristics. If $dx = u \, dt$ corresponds to travel with the local flow speed u, then $dx = (u + a) \, dt$ corresponds to travel at the local flow speed plus the local speed of sound, whereas $dx = (u - a)dt$ corresponds to travel at the local flow speed minus the local speed of sound. In either case, *the wave speed is the speed of sound relative to the flow*. As mentioned in Section 3.0, such waves are called *acoustic waves*. Unlike that of entropy waves, the signal carried by acoustic waves is not very easy to describe; in fact, the signal carried by acoustic waves does not correspond to any well-known physical quantity.

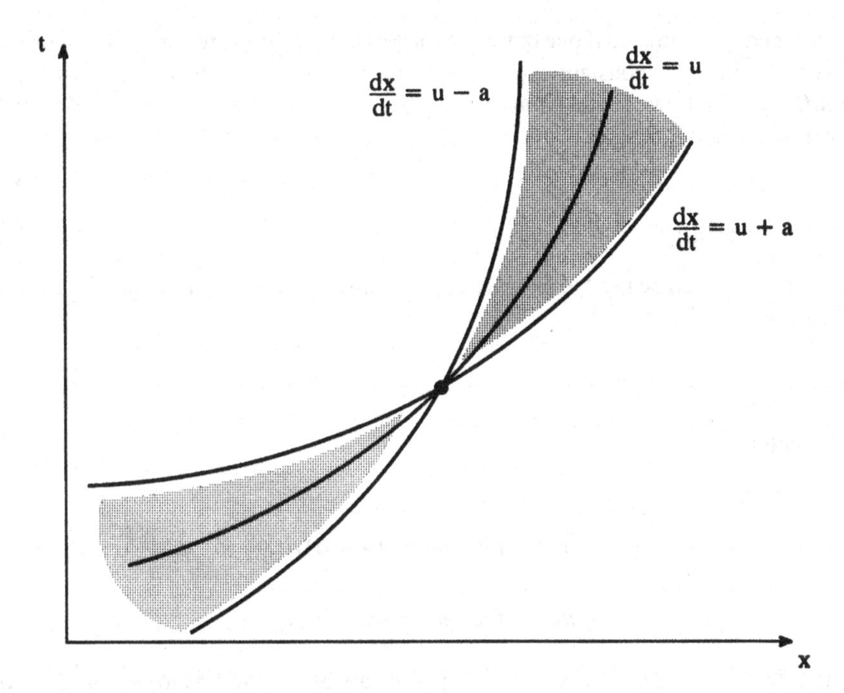

Figure 3.5 Typical wave diagram for the Euler equations.

The acoustic waves define a domain of dependence and range of influence for any point in the x–t plane. For example, Figure 3.5 illustrates the domain of dependence and range of influence for supersonic flow traveling in the positive x direction ($u < a < 0$). For supersonic flow traveling in the negative x direction ($u < -a < 0$), all wave speeds are negative, and thus all three characteristic curves slope to the left. For subsonic flow traveling in the positive x direction ($0 < u < a$), two wave speeds are positive and one wave speed is negative; and for subsonic flow traveling in the negative x direction ($-a < u < 0$), two wave speeds are negative and one wave speed is positive. Notice that supersonic flows force all waves to travel in one direction – all waves are swept downwind and none can travel upwind – whereas subsonic flows allow acoustic waves to travel in both directions. Thus, for example, it is impossible to have a conversation in supersonic flow; the upwind person can speak, and the downwind person can listen, but not vice versa.

In summary, the natural wave speeds in a compressible flow are the flow speed and the speed of sound relative to the flow. Except for shock waves (described in Section 3.6), all information travels through the fluid at one of these three natural wave speeds. Compare this situation to that of an incompressible flow. In an incompressible flow, the wave speeds are assumed to be infinite. Then all regions of fluid can communicate with all other regions of fluid instantly, and the domain of dependence and range of influence equal the entire fluid.

3.4 Simple Waves

The complicated nonlinear interactions between the three characteristics prevent analytical integration of $dx = \lambda_i dt$ and $dx/dt = \lambda_i$. However, these nonlinear interactions can be reduced or eliminated if one or two characteristic variables are constant. For example, suppose that entropy is constant everywhere, and not just along the characteristic curves $dx = udt$; then the flow is called *homentropic*. As seen in the last chapter, for homentropic flow of a perfect gas

$$p = (const.)\rho^{\gamma}, \tag{2.16a}$$

$$a = (const.)\rho^{(\gamma-1)/2}. \tag{2.16c}$$

These expressions can be used to express $dp/\rho a$ entirely in terms of a. In particular, a short calculation proves the following:

$$\int \frac{dp}{\rho a} = \frac{2a}{\gamma - 1} + const.$$

Then Equation (3.21) becomes

$$s = const., \tag{3.38a}$$

$$v_+ = u + \frac{2a}{\gamma - 1} = const. \quad \text{for} \quad dx = (u + a)\,dt, \tag{3.38b}$$

$$v_- = u - \frac{2a}{\gamma - 1} = const. \quad \text{for} \quad dx = (u - a)\,dt. \tag{3.38c}$$

The characteristic variables $v_\pm = u \pm 2a/(\gamma - 1)$ are also known as *Riemann invariants.*

Assuming constant entropy, $dv_i = 0$ can be analytically integrated but $dx = \lambda_i\,dt$ still cannot be analytically integrated. But suppose that, in addition to entropy, another characteristic variable is also constant. By assuming that two characteristic variables are constant, and that only one characteristic variable is not constant, the complicated nonlinear interactions among characteristic variables are completely eliminated, which allows a complete analytical integration of the characteristic equations. For example, suppose $s = const.$ and $v_- = u - 2a/(\gamma - 1) = const.$ As usual, $v_+ = u + 2a/(\gamma - 1) = const.$ along the characteristics $dx = (u + a)\,dt$. But then all three characteristic variables are constant along the characteristics $dx = (u + a)\,dt$. If all three characteristic variables are constant, then so are all three conservative variables, or all three primitive variables, or any other set of variables. In short, *all flow variables are constant along the characteristics* $dx = (u + a)\,dt$. In particular, u and a are constant along the characteristics $dx = (u + a)\,dt$. Then $dx = (u + a)\,dt$ can be trivially integrated, yielding

$$x = (u + a)t + const.$$

Thus *the characteristics are straight lines*. Of course, the other two families of characteristic curves are not straight lines; however, they are irrelevant, since the associated characteristic variables are constant everywhere, not just along those special characteristic curves.

To summarize:

♦ Assuming $s = const.$ and $v_- = u - 2a/(\gamma - 1) = const.$ then all flow properties are constant along the characteristic lines $x = (u + a)t + const.$ (3.39)

Similarly, one can show that:

♦ Assuming $s = const.$ and $v_+ = u + 2a/(\gamma - 1) = const.$ then all flow properties are constant along the characteristic lines $x = (u - a)t + const.$ (3.40)

Finally, one can show that:

♦ Assuming $v_- = const.$ and $v_+ = const.$ then all flow properties are constant along the characteristics lines $x = ut + const.$ (3.41)

Any region of flow governed by Equations (3.39), (3.40), or (3.41) is called a *simple wave.* More specifically, flow regions governed by Equation (3.39) or (3.40) are called *simple acoustic waves*, whereas flows governed by Equation (3.41) are called *simple entropy waves.* Two regions of steady uniform flow are always separated by simple waves, steadily moving shocks, or steadily moving contacts.

Consider simple acoustic waves. There are many alternatives to Equations (3.39) and (3.40). For example, one intriguing way to write Equation (3.39) is as follows:

$$\frac{\partial(u+a)}{\partial t} + (u+a)\frac{\partial(u+a)}{\partial x} = 0. \tag{3.42}$$

In this form, Equation (3.39) involves only the single variable $u + a$. Similarly, Equation (3.40) can be written as

$$\frac{\partial(u-a)}{\partial t} + (u-a)\frac{\partial(u-a)}{\partial x} = 0. \tag{3.43}$$

Now consider simple entropy waves. Simple entropy waves are defined by $v_+ = const.$ and $v_- = const.$ If these equations are added and subtracted, one obtains $u = const.$ and $a = const.$ Hence *the velocity and speed of sound are constant throughout a simple entropy wave.* Then Equation (3.41) can be written as

$$\frac{\partial s}{\partial t} + u\frac{\partial s}{\partial x} = 0, \tag{3.44}$$

where $u = const.$ There is nothing special about the choice of entropy in this equation. For example, this equation could just as well have been written in terms of density:

$$\frac{\partial \rho}{\partial t} + u\frac{\partial \rho}{\partial x} = 0. \tag{3.45}$$

Other variables can also be used, although probably u or a should be avoided since these are constant everywhere in the flow and not just along characteristics.

3.5 Expansion Waves

An *expansion wave* decreases pressure and density. For a one-dimensional flow of perfect gas, an expansion wave is any region in which the wave speed $\lambda_2 = u + a$ or $\lambda_3 = u - a$ increases monotonically from left to right. More specifically, an expansion occurs when

♦ $$u(x,t) + a(x,t) \leq u(y,t) + a(y,t), \qquad b_1(t) \leq x \leq y \leq b_2(t) \tag{3.46a}$$

or

♦ $$u(x,t) - a(x,t) \leq u(y,t) - a(y,t), \qquad b_1(t) \leq x \leq y \leq b_2(t). \tag{3.46b}$$

A typical expansion is illustrated in Figure 3.6. The figure shows only the characteristics in the family creating the expansion; the other two families of characteristics are not shown.

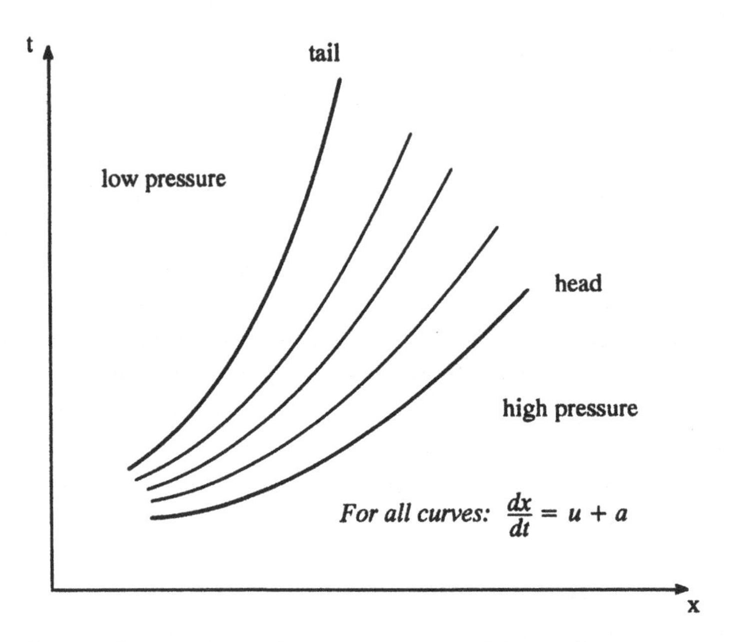

Figure 3.6 Wave diagram for an expansion in the Euler equations.

Now let us describe some of the properties of expansion waves. Expansion waves are always composed of acoustic waves – entropy waves cannot create expansions. An expansion wave is composed of characteristics; in particular, the boundaries $b_1(t)$ and $b_2(t)$ are characteristics. The boundary on the high-pressure side is called the *head* of the expansion. Similarly, the boundary on the low-pressure side is called the *tail* of the expansion. The first derivatives of pressure, density, and so forth may be discontinuous across the head and tail of an expansion; fortunately, these discontinuous derivatives can be handled naturally within the context of characteristics without resorting to weak solution theory.

A *simple expansion wave* is an expansion wave that is also a simple wave. Simple expansion waves separate regions of steady uniform flow. A *centered expansion fan* is an expansion wave in which all characteristics originate from a single point in the x–t plane, either a jump discontinuity in the initial conditions or an intersection between shocks or contact discontinuities. The term "fan" is used because, in a wave diagram, it looks like an old-fashioned hand fan. A *simple centered expansion fan* is, obviously, an expansion wave that is both simple and centered. A simple centered expansion fan is illustrated in Figure 3.7.

Example 3.7 Assume $s = const.$ and $u+2a/(\gamma-1) = const.$ Suppose a simple expansion fan centered on $(x, t) = (0, 0)$ connects the two steady uniform flows $\mathbf{u}_L$ and $\mathbf{u}_R$. Find u, a, and p in the expansion fan as functions of x and t.

Solution If $u + 2a/(\gamma - 1) = const.$ then

$$u + \frac{2a}{\gamma - 1} = u_L + \frac{2a_L}{\gamma - 1} = u_R + \frac{2a_R}{\gamma - 1}.$$

By Equation (3.40), all flow properties are constant along the characteristics $x = (u - a)t$ (notice that these characteristics all pass through the origin $(x, t) = (0, 0)$ as required). But

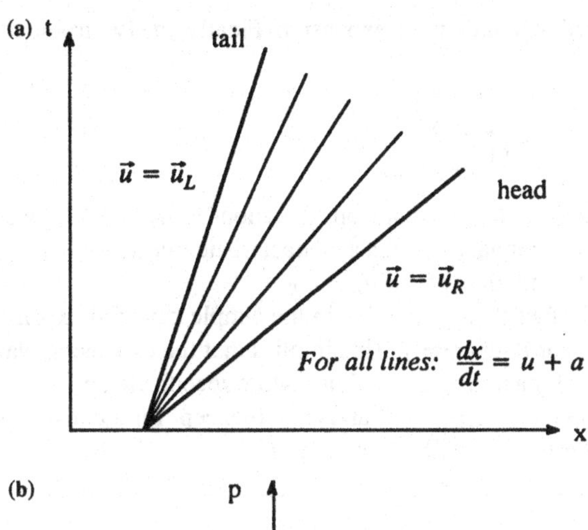

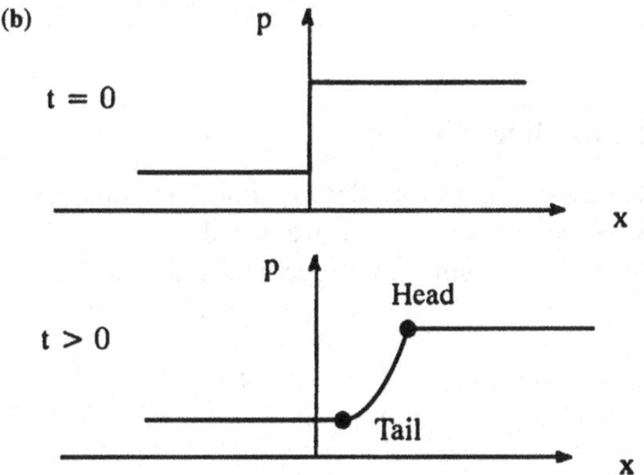

Figure 3.7 (a) Wave diagram for a simple centered expansion fan in the Euler equations. (b) Pressure as a function of x for a simple centered expansion fan in the Euler equations.

$x = (u - a)t$ implies $a = u - x/t$. Then

$$u + \frac{2}{\gamma - 1}\left(u - \frac{x}{t}\right) = u_L + \frac{2a_L}{\gamma - 1} = u_R + \frac{2a_R}{\gamma - 1}.$$

Solving for u yields

$$\begin{aligned} u(x,t) &= \frac{2}{\gamma + 1}\left(\frac{x}{t} + \frac{\gamma - 1}{2}u_L + a_L\right) \\ &= \frac{2}{\gamma + 1}\left(\frac{x}{t} + \frac{\gamma - 1}{2}u_R + a_R\right). \end{aligned} \tag{3.47}$$

This is the solution for the velocity in the expansion. Also, $a = u - x/t$ yields

$$\begin{aligned} a(x,t) &= \frac{2}{\gamma + 1}\left(\frac{x}{t} + \frac{\gamma - 1}{2}u_L + a_L\right) - \frac{x}{t} \\ &= \frac{2}{\gamma + 1}\left(\frac{x}{t} + \frac{\gamma - 1}{2}u_R + a_R\right) - \frac{x}{t}. \end{aligned} \tag{3.48}$$

This is the solution for the speed of sound in the expansion. Finally, the isentropic relations found in Equation (2.16) yield

$$p = p_L \left(\frac{a}{a_L} \right)^{2\gamma/(\gamma-1)} = p_R \left(\frac{a}{a_R} \right)^{2\gamma/(\gamma-1)}. \tag{3.49}$$

This is the solution for the pressure in the expansion. Equations (3.47), (3.48), and (3.49) hold for $b_1(t) \leq x \leq b_2(t)$. The boundary $b_1(t)$ is a characteristic line $x/t = u_L - a_L$ and the boundary $b_2(t)$ is a characteristic line $x/t = u_R - a_R$.

Notice that u, a, p, and all other flow properties in the simple centered expansion fan depend on x/t rather than on x or t separately. This implies that the expansion waveform remains the same in time except for a uniform stretching. More specifically, plots of $u(x, t_1)$ and $u(x, t_2)$ look exactly the same for all $t_1 > 0$ and $t_2 > 0$ except for a constant scaling factor in the x direction. Such waves are called *self-similar*.

3.6 Compression Waves and Shock Waves

A *compression wave* increases pressure and density. For a one-dimensional flow of perfect gas, a compression wave occurs when the wave speed $\lambda_2 = u + a$ or $\lambda_3 = u - a$ decreases monotonically from left to right. More specifically, a compression occurs when

♦ $$u(x,t) + a(x,t) \geq u(y,t) + a(y,t), \qquad b_1(t) \leq x \leq y \leq b_2(t) \tag{3.50a}$$

or

♦ $$u(x,t) - a(x,t) \geq u(y,t) - a(y,t), \qquad b_1(t) \leq x \leq y \leq b_2(t). \tag{3.50b}$$

Now let us describe some of the properties of compression waves. Compression waves are always composed of acoustic waves – entropy waves cannot create compressions. A compression wave is composed of characteristics; in particular, the boundaries $b_1(t)$ and $b_2(t)$ are characteristics. The first derivatives of pressure, density, and so forth may be discontinuous across the boundaries of a compression wave; fortunately, these discontinuous derivatives can be handled naturally within the context of characteristics without resorting to weak solution theory.

A *simple compression wave* is a compression wave that is also a simple wave. Simple compression waves separate regions of steady uniform flow. A *centered compression fan* is a compression wave in which all characteristics converge on a single point. A *simple centered compression fan* is, obviously, a compression wave that is both simple and centered.

As we have seen, the characteristics in an expansion diverge whereas the characteristics in a compression converge. Although divergence can continue forever, convergence cannot – instead, converging characteristics must eventually meet. An intersection between two or more characteristics from the same family creates a *shock wave*. A shock wave is a jump discontinuity governed by the Rankine–Hugoniot relations:

$$\mathbf{f}_R - \mathbf{f}_L = S(\mathbf{u}_R - \mathbf{u}_L), \tag{2.32}$$

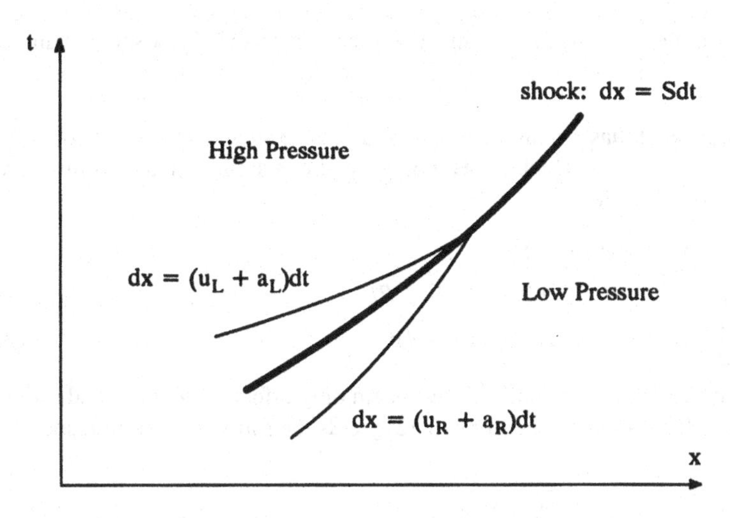

Figure 3.8 Wave diagram for a shock wave in the Euler equations.

where S is the shock speed, $\mathbf{f}_{L,R}$ are the flux vectors on the left- and right-hand sides of the shock, and $\mathbf{u}_{L,R}$ are the vectors of conserved quantities on the left- and right-hand sides of the shock. Furthermore, shocks must satisfy the following compression condition:

$$\blacklozenge \quad u_L + a_L \geq S \geq u_R + a_R \tag{3.51a}$$

or

$$\blacklozenge \quad u_L - a_L \geq S \geq u_R - a_R. \tag{3.51b}$$

Hence the wave speed just to the left of the shock is greater than the shock speed, which is, in turn, greater than the wave speed just to the right of the shock. A shock wave may originate in a jump discontinuity in the initial conditions or it may form spontaneously from a smooth compression wave. Figure 3.8 illustrates a typical shock wave.

Equation (3.51) can be seen as a natural consequence of the compression condition (3.50). Alternatively, Equation (3.51) can be seen as a natural consequence of the second law of thermodynamics. In smooth flow regions, the Euler equations imply the second law of thermodynamics, as seen in Equations (3.19a′) and (3.21a′). In other words, any smooth solution of the Euler equations automatically satisfies the second law of thermodynamics. In contrast, the Euler equations do not imply the second law of thermodynamics across shocks. Instead, a separate condition is required across shocks, such as Equation (3.51). In short, for the one-dimensional Euler equations, *the second law of thermodynamics amounts to inequality (3.51) at shocks.* The second law of thermodynamics and the perfect gas relations also imply that entropy, pressure, density, temperature, and the speed of sound all increase as fluid passes through a shock. Also, for a steadily moving shock, the second law of thermodynamics requires that the Mach number $M = u/a$ must decrease from greater than one to less than one in a coordinate system moving with the shock.

Real gases may reverse the shock relationships. For example, in a real gas, shocks may decrease rather than increase pressure and density – this is called an *expansion shock.* However, although the nature of the fluid as specified by the equations of state may affect other aspects of shocks, the second law of thermodynamics always requires that the wave speed just to the left of the shock must be greater than the shock speed, which must be, in turn, greater than the wave speed just to the right of the shock, just as in Equation (3.51).

Suppose that a shock has a constant speed S. Then the shock separates two regions of uniform flow $\mathbf{u}_L$ and $\mathbf{u}_R$. The Rankine–Hugoniot relations applied in a coordinate system moving with the shock yield

$$\rho_L(u_L - S) = \rho_R(u_R - S), \tag{3.52a}$$

$$\rho_L(u_L - S)^2 + p_L = \rho_R(u_R - S)^2 + p_R, \tag{3.52b}$$

$$h_L + \frac{1}{2}(u_L - S)^2 = h_R + \frac{1}{2}(u_R - S)^2. \tag{3.52c}$$

Notice that the movement of the coordinate system only affects velocity u and not scalars such as ρ and p. Manipulating Equations (3.52) yields the following useful result:

$$e_R - e_L = \frac{p_R + p_L}{2}\left(\frac{1}{\rho_L} - \frac{1}{\rho_R}\right), \tag{3.53}$$

which is known as the *Hugoniot relation*. This is proven, for example, in Anderson (1990).

It is often convenient to express shock properties in terms of the pressure ratio p_L/p_R. For example, for a steady right-running shock

$$\frac{T_L}{T_R} = \frac{a_L^2}{a_R^2} = \frac{p_L}{p_R}\frac{\frac{\gamma+1}{\gamma-1} + \frac{p_L}{p_R}}{1 + \frac{\gamma+1}{\gamma-1}\frac{p_L}{p_R}}, \tag{3.54}$$

$$u_L = u_R + \frac{a_R}{\gamma}\frac{\frac{p_L}{p_R} - 1}{\sqrt{\frac{\gamma+1}{2\gamma}\left(\frac{p_L}{p_R} - 1\right) + 1}}, \tag{3.55}$$

and

$$S = u_R + a_R\sqrt{\frac{\gamma+1}{2\gamma}\left(\frac{p_L}{p_R} - 1\right) + 1}. \tag{3.56}$$

These three equations (with a few minor differences) are derived in Anderson (1990) using the Hugoniot relation. There are a great number of other useful shock relations, as described in Anderson (1990) or in any basic text on compressible flow or gasdynamics. This section describes only the relations needed in the rest of the book.

3.7 Contact Discontinuities

Nearby characteristics must converge, diverge, or precisely parallel each other. As we have seen, convergence in acoustic characteristics creates compression waves or shock waves, whereas divergence in acoustic characteristics creates expansion waves. This section discusses a third possibility: parallel entropy waves creating neither compression nor expansion. In particular, a *contact discontinuity* occurs when the wave speed $\lambda_1 = u$ and pressure are continuous while other flow properties jump. In other words, for a contact discontinuity

$$u_L = u_R, \tag{3.57}$$

$$p_L = p_R. \tag{3.58}$$

In multidimensional flows, contact discontinuities are also called *slip lines* or *vortex sheets*.

Let us compare and contrast shocks with contact discontinuities. Like shocks, contact discontinuities are jump discontinuities. Unlike shocks, fluid does not pass through contact discontinuities; instead, since u is the same on both sides, contact discontinuities move with the fluid. Entropy may change across a contact discontinuity, as in shocks. However, contact discontinuities do not create entropy, unlike shocks, but simply separate regions with different entropies. The second law of thermodynamics says that entropy must increase following the fluid. However, no fluid passes through a contact; thus the second law does not apply across a contact; thus entropy, density, energy, and all other flow properties may either increase or decrease across a contact, unlike shocks. Like shocks, contact discontinuities obey the Rankine–Hugoniot relations. Unlike shocks, contacts cannot form spontaneously: They must originate either in the initial conditions or in the intersection of two shocks.

References

Anderson, J. D. 1990. *Modern Compressible Flow*, New York: McGraw-Hill, Chapter 7.

Courant, R., and Friedrichs, K. O. 1948. *Supersonic Flow and Shock Waves*, New York: Springer-Verlag, Chapter 3.

Hirsch, C. 1990. *Numerical Computation of Internal and External Flows, Volume 2: Computational Methods for Inviscid and Viscous Flows*, Chichester: Wiley, Chapter 16.

Paterson, A. R. 1983. *A First Course in Fluid Dynamics*, New York: Cambridge, Chapter 14.

Tipler, P. A. 1976. *Physics*, New York: Worth, Chapters 21–25.

Problems

3.1 The wave diagrams in Figure 3.9 each illustrate one family of characteristics. In each case, state whether the wave diagram is physically possible. Explain.

3.2 Consider the partial differential equation seen in Example 3.1. Suppose boundary and initial conditions are given along the bold curves shown in Figure 3.10. In each case, state whether the boundary and initial conditions are well-posed. Explain.

3.3 Consider the following partial differential equation:

$$e^x \frac{\partial u}{\partial t} + t\frac{\partial u}{\partial x} = 0.$$

Write this equation in the form "$u = const.$ for $t = t(x)$."

3.4 Consider the following system of partial differential equations:

$$\frac{\partial u_1}{\partial t} + \frac{\partial u_2}{\partial x} = 0,$$
$$\frac{\partial u_2}{\partial t} + \frac{\partial u_3}{\partial x} = 0,$$
$$\frac{\partial u_3}{\partial t} + 4\frac{\partial u_1}{\partial x} - 17\frac{\partial u_2}{\partial x} + 8\frac{\partial u_3}{\partial x} = 0.$$

Is this system of equations hyperbolic? In other words, does this system have a complete set of characteristics? If so, write the system of equations in the characteristic form $\frac{\partial \mathbf{v}}{\partial t} + \Lambda \frac{\partial \mathbf{v}}{\partial x} = 0$, where Λ is a diagonal matrix. Also, write the system of equations in the form "$v_i = const.$ for $x = \lambda_i t + const.$" Sketch the range of influence and domain of dependence for a typical point in the x–t plane.

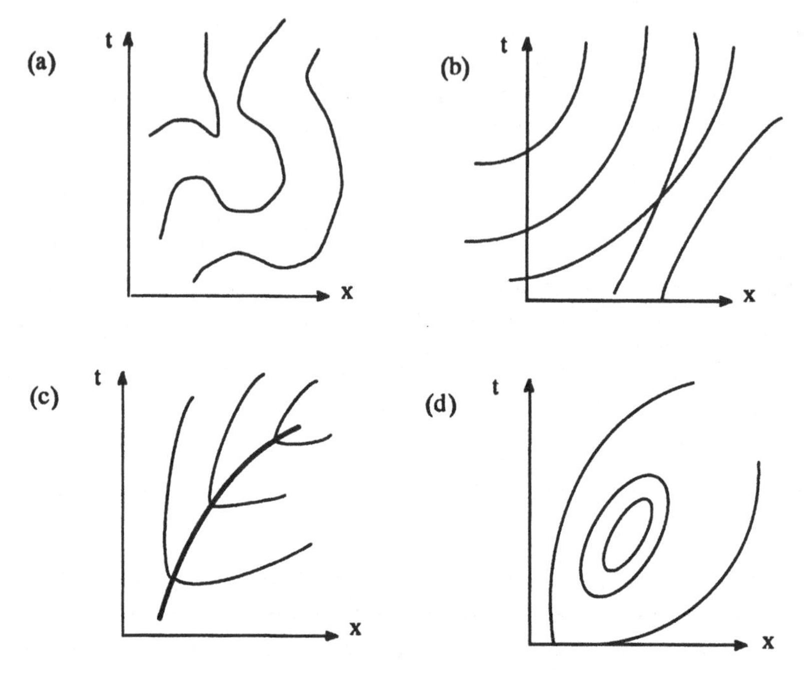

Figure 3.9 Wave diagrams for Problem 3.1.

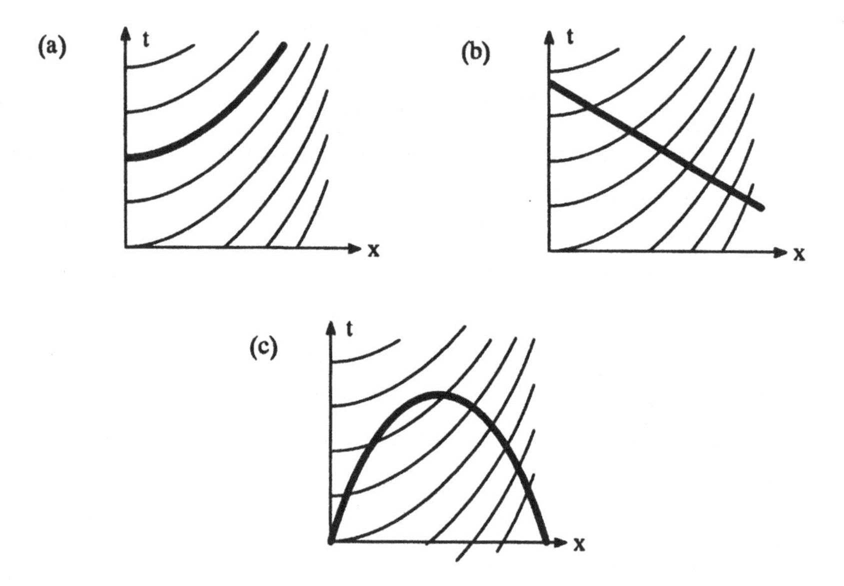

Figure 3.10 Wave diagrams for Problem 3.2.

3.5 Consider the Euler equations governing isothermal flow, given in Problem 2.4.

(a) For the Jacobian matrix A found in Problem 2.4b, find the characteristic values, left characteristic vectors, and right characteristic vectors.

(b) Using the results of part (a), find a matrix Q that diagonalizes A. Also find Q^{-1}.

(c) Using the results of parts (a) and (b), write the isothermal Euler equations in the form $\frac{\partial \mathbf{v}}{\partial t} + \Lambda \frac{\partial \mathbf{v}}{\partial x} = 0$, where Λ is a diagonal matrix.

(d) Using the results of parts (a)–(c), write the isothermal Euler equations in the form "$dv_i = 0$ when $dx = \lambda_i\, dt$." If possible, integrate the compatibility relations $dv_i = 0$.

3.6 Total enthalpy is constant along streamlines in steady adiabatic flow. In other words

$$\frac{1}{2}u^2 + \frac{1}{\gamma - 1}a^2 = const.$$

by Equation (2.13). In the Euler equations, replace conservation of energy by this algebraic equation. Then the conservation form of the Euler equations becomes

$$\frac{\partial}{\partial t}\rho + \frac{\partial}{\partial x}(\rho u) = 0,$$
$$\frac{\partial}{\partial t}(\rho u) + \frac{\partial}{\partial x}(\rho u^2 + p) = 0,$$
$$\frac{1}{2}u^2 + \frac{1}{\gamma - 1}a^2 = const.$$

Show that this system of equations is hyperbolic, that is, that it has a complete set of characteristics, just like the original Euler equations. The steady-state solutions of the original and modified systems are the same. Thus the steady-state solutions can be found by time evolving the modified system of equations, rather than the original system, to steady state. Replacing a differential equation by an algebraic equation substantially reduces the costs.

3.7 Consider a one-dimensional tube containing stagnant standard sea-level air. A piston is impulsively inserted into the tube, creating a normal shock wave moving with Mach number 1.75 relative to the stagnant standard sea-level air. What is the piston velocity? You may find it helpful to use the shock tables found in the appendices of most texts on compressible flow or gasdynamics, which list the ratios of various flow properties across a shock as functions of the preshock Mach number. Just make sure to calculate the preshock Mach number moving with the shock. In fact, the entire problem should be done in a coordinate system moving with the shock. In a standard sea-level atmosphere, $a = 340$ m/sec.

3.8 Consider the Lagrange equations, as seen in Problem 2.7. Show that the characteristic form of the Lagrange equations can be written as follows:

$$\frac{\partial v^+}{\partial t} = -c\frac{\partial v^+}{\partial m},$$
$$\frac{\partial v^-}{\partial t} = c\frac{\partial v^-}{\partial m},$$
$$\frac{\partial v^0}{\partial t} = 0,$$

where $c = \rho a$ is the *Lagrangean speed of sound* and where the characteristic variables are defined as follows:

$$dv^+ = du + \frac{1}{c}\,dp,$$
$$dv^- = du - \frac{1}{c}\,dp,$$
$$dv^0 = dp + c^2\,dv.$$

CHAPTER 4

Scalar Conservation Laws

4.0 Introduction

This chapter concerns simple scalar models of the Euler equations, called *scalar conservation laws*. Scalar conservation laws mimic the Euler equations, to the extent that any single equation can mimic a system of equations. In order to stress the parallels between scalar conservation laws and the Euler equations, the first part of this chapter essentially repeats the last two chapters, albeit in a highly abbreviated and simplified fashion. As a result, besides its inherent usefulness, the first part of this chapter also serves as a nice review and reinforcement of the last two chapters.

Like the Euler equations, scalar conservation laws can be written in integral or differential forms. In integral forms, scalar conservation laws look exactly like Equation (2.21) except that the vector of conserved quantities $\mathbf{u}$ is replaced by a single scalar conserved quantity u, and the flux vector $\mathbf{f}(\mathbf{u})$ is replaced by a single scalar flux function $f(u)$. Similarly, in differential forms, scalar conservation laws look exactly like Equations (2.27) and (2.30) except that the vector of conserved quantities $\mathbf{u}$ is replaced by a single scalar conserved quantity u, the flux vector $\mathbf{f}(\mathbf{u})$ is replaced by a single scalar flux function $f(u)$, and the Jacobian matrix $A(\mathbf{u})$ is replaced by a single scalar wave speed $a(u)$, not to be confused with the speed of sound.

Replacing several interlinked conserved quantities by a single conserved quantity dramatically simplifies the governing equations while retaining much of the essential physics. In particular, scalar conservation laws can model simple compression waves, simple expansion waves, shock waves, and contact discontinuities. Like simple waves in the Euler equations, scalar conservation laws have a complete analytical characteristic solution. Furthermore, like simple waves in the Euler equations, the characteristics of scalar conservation laws are always straight lines in the x–t plane. However, scalar conservation laws cannot model nonsimple waves, since nonsimple waves result from interactions between various families of characteristics, whereas scalar conservation laws only have one family of characteristics. The two most important scalar conservation laws are the linear advection equation, which models simple entropy waves and contacts, and Burgers' equation, which models simple acoustic waves and shocks.

The second law of thermodynamics selects physically relevant solutions of the Euler equations in the presence of shocks. To model physically relevant solutions, scalar conservation laws require an *entropy condition* analogous to the second law of thermodynamics. Entropy conditions can take the form of an integral or differential inequality analogous to Equations (2.17), (2.26), or (2.37). Alternatively, entropy conditions can take the form of an algebraic inequality at shocks analogous to Equation (3.51).

Although of little inherent interest, scalar conservation laws are prized as models of the Euler equations. In this capacity, scalar conservation laws serve a valuable pedagogical role throughout the rest of the book. Rather than explain methods and concepts for the Euler equations directly, the book will often first explain them for scalar conservation laws, and then explain the extension to the Euler equations. In other words, rather than taking one giant step, scalar conservation laws permit two baby steps.

4.1 Integral Form

Assume that there is a conserved quantity with density $u = u(x, t)$ and flux $f = f(u)$. In other words, assume that there is a conserved quantity that is only affected by flux; furthermore, assume that the flux is purely a function of the conserved quantity and does not depend independently on x, t, or on any other conserved quantities. The flux function f can be any continuously differentiable function of u. Specific choices for u and f will be discussed later in the chapter. (Be careful not to confuse the conserved quantity u seen in this chapter with the velocity u seen in the last two chapters.)

Consider an arbitrary one-dimensional region $[a, b]$ during time interval $[t_1, t_2]$. The scalar conservation law can be stated as follows:

♦ *change in total conserved quantity in* $[a, b]$ *in time interval* $[t_1, t_2]$ = *net flux through the boundaries of* $[a, b]$ *in time interval* $[t_1, t_2]$.

This translates to mathematics immediately as follows:

$$\blacklozenge \quad \int_a^b [u(x, t_2) - u(x, t_1)]\, dx = -\int_{t_1}^{t_2} [f(u(b, t)) - f(u(a, t))]\, dt, \tag{4.1}$$

which is the *integral form of the scalar conservation law*. This is analogous to the integral form of the Euler equations, as given by Equation (2.21).

4.2 Conservation Form

By the fundamental theorem of calculus

$$u(x, t_2) - u(x, t_1) = \int_{t_1}^{t_2} \frac{\partial u}{\partial t}\, dt$$

and

$$f(u(b, t)) - f(u(a, t)) = \int_a^b \frac{\partial f}{\partial x}\, dx.$$

Then Equation (4.1) becomes

$$\int_{t_1}^{t_2} \int_a^b \left(\frac{\partial u}{\partial t} + \frac{\partial f(u)}{\partial x} \right) dx\, dt = 0$$

or

$$\blacklozenge \quad \frac{\partial u}{\partial t} + \frac{\partial f(u)}{\partial x} = 0. \tag{4.2}$$

This is the *conservation form of the scalar conservation law* and is analogous to the conservation form of the Euler equations, as seen in Equation (2.27).

The solutions of the integral form (4.1) may contain jump discontinuities. Discontinuous solutions of the integral form (4.1) are called *weak solutions* of the differential form (4.2). Jump discontinuities in the differential form must satisfy a jump condition derived from the integral form. For a jump discontinuity traveling at speed S, the *jump condition* is

$$f(u_R) - f(u_L) = S(u_R - u_L), \tag{4.3}$$

which is analogous to the Rankine–Hugoniot relations, Equation (2.32).

Using the chain rule, the conservation form may be rewritten as

♦ $$\frac{\partial u}{\partial t} + a(u)\frac{\partial u}{\partial x} = 0, \tag{4.4}$$

where

♦ $$a(u) = \frac{df}{du}. \tag{4.5}$$

These expressions are analogous to Equations (2.30), (2.31), and (3.10).

4.3 Characteristic Form

Equation (4.4) is the *characteristic form of the scalar conservation law*. To see this, notice that Equation (4.4) is a quasi-linear partial differential equation in the form of Equation (3.2). By Equation (3.3)

$$u = const. \quad \text{for} \quad \frac{dx}{dt} = a(u).$$

But $a(u)$ is constant since u is constant. Thus $dx/dt = a(u)$ can be trivially integrated to obtain

♦ $$u = const. \quad \text{for} \quad x = a(u)t + const. \tag{4.6}$$

More specifically, if the characteristic passes through (x_0, t_0) then

$$u = u(x_0, t_0) \quad \text{for} \quad x - x_0 = a(u)(t - t_0). \tag{4.7}$$

This is a wave solution – the lines $x = a(u)t + const.$ are *wavefronts*, u is the *signal*, and a is the *wave speed* or *signal speed*. For any given time t, $u(x, t)$ can also be called the *waveform*. The wavefronts $x = a(u)t + const.$ are also called *characteristics*, u is also called the *characteristic variable*, and a is also called the *characteristic slope* or *characteristic speed*.

For scalar conservation laws, the characteristics comprise a single family of straight lines. In other words, the conserved variable u (which is also the characteristic variable) is constant along straight lines with slope $a(u)$ in the x–t plane. Because there is only one family of waves, the waves travel entirely to the right or entirely to the left at each point in the x–t plane, like supersonic flow in the Euler equations. The range of influence and domain of dependence are simple concepts for scalar conservation laws. The *range of influence* of (x, t) is the characteristic line passing through (x, t) for times greater than t. The *domain of dependence* of (x, t) is the characteristic line passing through (x, t) for times less than t. In other words, at each instant of time, each point influences exactly one point and is influenced by exactly one point.

Section 3.4 described simple waves for the Euler equations. In particular, simple waves involve two trivial families and one nontrivial family of characteristics. Scalar conservation laws only ever have one family of characteristics. Thus, leaving aside shock waves and contacts, scalar conservation laws support *only* simple waves. The advantages of simple waves are the same for scalar conservation laws as for the Euler equations: straight-line characteristics, analytically integrable compatibility relations, and so forth, leading to a simple analytical solution.

4.4 Expansion Waves

Expansion waves were discussed in Section 3.5. Scalar conservation laws support features analogous to simple expansion waves. For scalar conservation laws, an *expansion wave* is any region in which the wave speed $a(u)$ increases from left to right. More specifically, an expansion occurs where

♦ $$a(u(x,t)) \leq a(u(y,t)), \qquad b_1(t) \leq x \leq y \leq b_2(t). \tag{4.8}$$

Note the similarity to Equation (3.46). A *centered expansion fan* is an expansion wave where all characteristics originate at a single point in the x–t plane. Centered expansion fans must originate in the initial conditions or at intersections between shocks or contacts.

4.5 Compression Waves and Shock Waves

Compression and shock waves were discussed in Section 3.6. Scalar conservation laws support features analogous to simple compression and shock waves. For scalar conservation laws, a *compression wave* is any region in which the wave speed $a(u)$ decreases from left to right. More specifically, a compression wave occurs where

♦ $$a(u(x,t)) \geq a(u(y,t)), \qquad b_1(t) \leq x \leq y \leq b_2(t). \tag{4.9}$$

Note the similarity to Equation (3.50). A *centered compression fan* is a compression wave where all characteristics converge on a single point in the x–t plane.

The converging characteristics in a compression wave must eventually intersect, creating a *shock wave*. A shock wave is a jump discontinuity governed by Equation (4.3), which can be rewritten as follows:

$$S = \frac{f(u_R) - f(u_L)}{u_R - u_L}. \tag{4.10}$$

Thus the shock speed equals the slope of the secant line connecting u_L and u_R. By the mean value theorem, Equation (4.10) implies

$$S = f'(\xi) = a(\xi), \tag{4.11}$$

where ξ is between u_L and u_R. If you wish, you can think of $a(\xi)$ as an average of $a(u)$ for u between u_L and u_R. Then, in this sense, the shock speed equals an average wave speed. A shock wave may originate in a jump discontinuity in the initial conditions or it may form spontaneously from a smooth compression wave.

In addition to the jump condition (4.10), shock waves must satisfy the following condition:

♦ $$a(u_L) \geq S \geq a(u_R). \tag{4.12}$$

This says that the wave speed just to the left of the shock is greater than the shock speed, which is, in turn, greater than the wave speed just to the right of the shock. Note the similarity to Equation (3.51). If we interpret the wave speeds as slopes in the x–t plane, then Equation (4.12) implies that:

♦ *Waves (characteristics) terminate on shocks. Waves (characteristics) never originate in shocks.*

In other words, in keeping with the "black hole" notion of shocks, shocks only absorb waves; they never emit waves. Of course, the same holds true for shocks in the Euler equations. Equation (4.12) can be seen as a natural consequence of the compression condition (4.9). Alternatively, Equation (4.12) can be seen as a natural consequence of the second law of thermodynamics, or what passes for the second law of thermodynamics for scalar conservation laws, as discussed later in Section 4.10.

As defined in elementary calculus, a function $f(u)$ is *convex* or *concave up* if $f''(u) = a'(u) \geq 0$ for all u. Then a scalar conservation law is *convex* if its flux function is convex. For convex scalar conservation laws, (4.12) is equivalent to the following:

♦ $$u_L \geq u_R. \tag{4.13}$$

Convex scalar conservation laws model perfect gases, whereas nonconvex scalar conservation laws model real gases. For example, for a perfect gas, a jump discontinuity in the initial conditions of the Euler equations gives rise to at most one shock, one contact, and one simple centered expansion fan (i.e., one wave per conservation equation). Similarly, a jump discontinuity in the initial conditions of a convex scalar conservation law gives rise to at most one shock, or one contact, or one simple centered expansion fan (again, one wave per conservation equation). For a real gas, however, a jump discontinuity in the initial conditions may give rise to multiple shocks, multiple contacts, and multiple simple centered expansion fans. By the same token, a jump discontinuity in the initial conditions of a nonconvex scalar conservation law may give rise to multiple shocks, multiple contacts, or multiple expansion fans. Nonconvex scalar conservation laws will be discussed in Section 4.9.

4.6 Contact Discontinuities

Contact discontinuities were discussed in Section 3.7, Scalar conservation laws support features analogous to contact discontinuities. For scalar conservation laws, a contact discontinuity is a jump discontinuity from u_L to u_R such that

$$a(u_L) = a(u_R). \tag{4.14}$$

Note the similarity to Equation (3.57). Like contacts in the Euler equations, contacts in scalar conservation laws must originate in the initial conditions or at the intersections of shocks.

4.7 Linear Advection Equation

The next three sections concern possible choices for the flux function $f(u)$ in scalar conservation laws. The simplest possible choice for the flux function is

$$f(u) = au, \tag{4.15}$$

where $a = const.$ The resulting scalar conservation law is then

$$\blacklozenge \quad \frac{\partial u}{\partial t} + \frac{\partial (au)}{\partial x} = \frac{\partial u}{\partial t} + a\frac{\partial u}{\partial x} = 0. \tag{4.16}$$

This scalar conservation law is called the *linear advection equation* or the *linear convection equation*. (Recall that the linear advection equation was already discussed briefly in Section 3.0, to help introduce waves.) Notice that the linear advection equation is convex.

What is the relationship between the linear advection equation and the Euler equations? Recall that simple entropy waves are governed by Equations (3.41), (3.44), or (3.45). In particular, Equations (3.44) and (3.45) are

$$\frac{Ds}{Dt} = \frac{\partial s}{\partial t} + V\frac{\partial s}{\partial x} = 0,$$

$$\frac{D\rho}{Dt} = \frac{\partial \rho}{\partial t} + V\frac{\partial \rho}{\partial x} = 0,$$

where $V = const.$ is the velocity. Here V is used for velocity to avoid confusion with the variable u in the linear advection equation. Then *the linear advection equation governs simple entropy waves* where $a = V = const.$ and u equals any flow variable such as s or ρ.

Besides simple entropy waves, *the linear advection equations also allow contact discontinuities* – in fact, any jump discontinuity in the initial conditions obviously satisfies the contact condition (4.14). However, the linear advection equation does *not* allow shocks. This all makes sense since shocks are associated with acoustic waves, whereas contacts are associated with entropy waves. The linear advection equation models only entropy waves and thus models contacts but not shocks.

Suppose the initial conditions for the linear advection equation are $u(x, 0)$. By Equation (4.6), the linear advection equation has the following exact solution:

$$u(x, t) = u(x - at, 0). \tag{4.17}$$

Hence the initial conditions move as a unit to the right ($a > 0$) or to the left ($a < 0$) without stretching. The existence of a simple exact solution makes the linear advection equation ideal for testing numerical methods (without an exact solution to compare against, it is impossible to fully judge the quality of a numerical approximation). Of course, all scalar conservation laws have an exact solution given by Equations (4.3), (4.6), and (4.14). However, the exact solution of other scalar conservation laws is generally far more complicated; for example, see Theorem 3.1 of Lax (1973).

Despite its seeming simplicity, the linear advection equation poses some heady challenges for numerical approximations. Most numerical approximations introduce a modest amount of false smearing. For shocks, which are naturally compressive, physical compression and numerical smearing reach a rapid equilibrium. However, contacts are naturally neutral, being neither compressive nor expansive. With nothing to oppose numerical smearing, typical numerical approximations smear contacts progressively wider and wider as time increases, almost as if they were expansions. Furthermore, besides jump discontinuities, the linear advection equation also retains corners, spikes, and any other nonsmooth features of the initial conditions. Nonlinear equations do not tend to retain jumps, corners, and so forth, but instead tend to smear and smooth such features. Unfortunately, most numerical approximations tend to smear and smooth such features whether or not the governing equation is nonlinear.

As a word of caution, many modern numerical methods use the exact solution of the linear advection equation as an integral component, to circumvent the numerical difficulties mentioned in the last paragraph. Although there is nothing wrong with this practice per se, the linear advection equation no longer presents a completely fair test case for such methods – any method that knows the exact solution should have little trouble passing linear advection tests with flying colors.

4.8 Burgers' Equation

The simplest nonlinear choice of the flux function $f(u)$ is

$$f(u) = \frac{1}{2}u^2. \tag{4.18}$$

The resulting scalar conservation law is then

$$\blacklozenge \quad \frac{\partial u}{\partial t} + \frac{\partial}{\partial x}\left(\frac{1}{2}u^2\right) = \frac{\partial u}{\partial t} + u\frac{\partial u}{\partial x} = 0. \tag{4.19}$$

This scalar conservation law is called the *inviscid Burgers' equation*. As you might guess, given the appearance of the adjective "inviscid," there is also a viscous Burgers' equation. The viscous Burgers' equation will not be discussed in this book. Since there is no possibility for confusion, the inviscid Burgers' equation will usually be referred to simply as Burgers' equation. Notice that Burgers' equation is convex.

What is the relationship between the Burgers' equation and the Euler equations? Recall that simple acoustic waves are governed by Equations (3.39), (3.40), (3.42), and (3.43). In particular, Equations (3.42) and (3.43) are

$$\frac{\partial(V+a)}{\partial t} + (V+a)\frac{\partial(V+a)}{\partial x} = 0,$$
$$\frac{\partial(V-a)}{\partial t} + (V-a)\frac{\partial(V-a)}{\partial x} = 0,$$

where V is fluid velocity. As in the last section, V is used for velocity to avoid confusion with the variable u in Burgers' equation. Then *the inviscid Burgers' equation governs simple acoustic waves* where $u = V \pm a$.

Besides simple acoustic waves, *the inviscid Burgers' equation also allows shocks*. In fact, most solutions of Burgers' equation are quickly dominated by one or more shocks. However, Burgers' equation does not allow contacts; examining the contact condition (4.14), $a(u_L) = a(u_R)$ implies $u_L = u_R$, which is hardly a jump discontinuity. This all makes sense: Shocks are associated with acoustic waves, whereas contacts are associated with entropy waves. Burgers' equation models only acoustic waves and thus models shocks but not contacts. Burgers' equation and the linear advection equation, between the two of them, model both acoustic and entropy waves. Thus the linear advection equation and Burgers' equation model all three families of characteristics of the Euler equations.

Example 4.1 Solve Burgers' equation on an infinite domain with the following initial conditions:

$$u(x,0) = \begin{cases} 0 & \text{for } x < 0, \\ 1 & \text{for } x > 0. \end{cases}$$

Notice that the initial conditions are constant except for a single jump discontinuity; this is called a *Riemann problem.* All of the subsequent examples in this section and the next are also Riemann problems. The following chapter is entirely devoted to Riemann problems.

Solution Consider the following self-similar solution

$$u(x,t) = \begin{cases} 0 & \text{for } \frac{x}{t} < \frac{1}{2}, \\ 1 & \text{for } \frac{x}{t} > \frac{1}{2}. \end{cases} \tag{4.20}$$

The wave diagram for this solution is plotted in Figure 4.1. This solution clearly satisfies the initial conditions. This solution is constant away from the jump discontinuity; thus it clearly satisfies Burgers' equation away from the jump discontinuity. Furthermore, the jump discontinuity satisfies jump condition (4.10). This is proven as follows:

$$S = \frac{f(u_R) - f(u_L)}{u_R - u_L} = \frac{1/2 - 0}{1 - 0} = \frac{1}{2}.$$

Unfortunately, the jump in solution (4.20) does not satisfy conditions (4.12) or (4.13). Examining Figure 4.1, we see that the characteristics originate rather than terminate on the jump discontinuity, in violation of Equation (4.12). *Thus Equation (4.20) is not a valid solution.* Remember that every solution must satisfy four basic conditions: the initial and boundary conditions, the differential form of the governing equation away from jump discontinuities, the jump condition (4.10) at jump discontinuities, and compression conditions (4.12) or (4.13) at shocks.

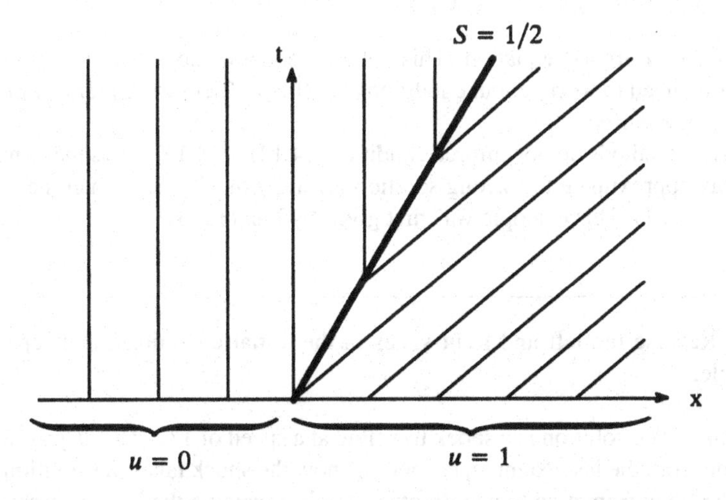

Figure 4.1 Wave diagram for the wrong solution of Example 4.1.

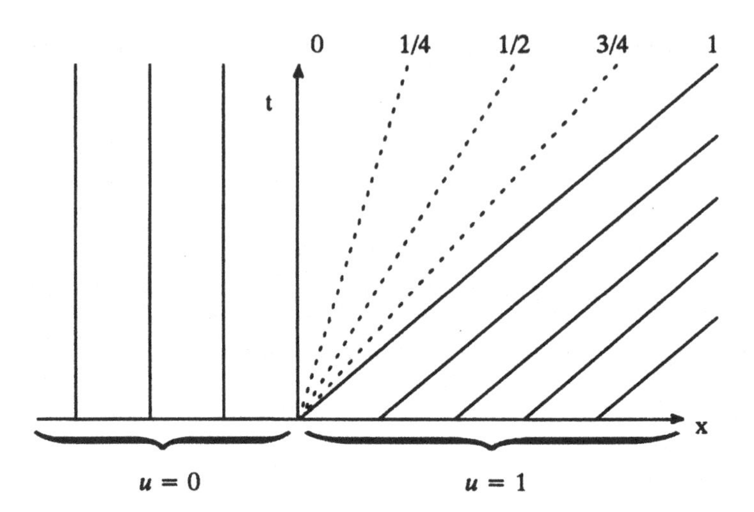

Figure 4.2 Wave diagram for the correct solution of Example 4.1.

Let us try again. Consider the following self-similar solution

$$u(x,t) = \begin{cases} 0 & \text{for } \frac{x}{t} < 0, \\ \frac{x}{t} & \text{for } 0 < \frac{x}{t} < 1, \\ 1 & \text{for } \frac{x}{t} > 1. \end{cases} \tag{4.21}$$

The wave diagram for this solution is plotted in Figure 4.2. This solution clearly satisfies the initial conditions. This solution is uniform outside the simple centered expansion fan and thus clearly satisfies Burgers' equation there. Inside the simple centered expansion fan

$$\frac{\partial u}{\partial t} + u\frac{\partial u}{\partial x} = \frac{\partial}{\partial t}\left(\frac{x}{t}\right) + \frac{x}{t}\frac{\partial}{\partial x}\left(\frac{x}{t}\right) = -\frac{x}{t^2} + \frac{x}{t}\frac{1}{t} = 0,$$

which again satisfies Burgers' equation. This solution contains no jump discontinuities, and thus there is no need to worry about conditions (4.10), (4.12), or (4.13). *Thus Equation (4.21) is the correct solution.*

Many numerical methods do not enforce conditions (4.12) or (4.13). Thus many numerical methods may approximate the wrong solution given by (4.20) rather than the correct solution given by (4.21). This example was first given by Lax (1973).

Example 4.2 Reverse the left and right states in the initial conditions, and repeat the previous example.

Solution The solution is a shock traveling at a speed of $1/2$. This is just like the "wrong" solution from the last example, except that now the shock takes the solution from one to zero, rather than from zero to one. In other words, reversing the left and right states makes the "wrong" solution "right."

Example 4.3 Solve Burgers' equation on an infinite domain for $t \leq 4/3$ with the following initial conditions:

$$u(x,0) = \begin{cases} 1 & \text{for } |x| < \frac{1}{3}, \\ 0 & \text{for } |x| > \frac{1}{3}. \end{cases}$$

Solution The jump at $x = -1/3$ creates a simple centered expansion fan; the jump at $x = 1/3$ creates a shock. Until the shock and expansion fan intersect, the exact piecewise-linear solution is as follows:

$$u(x,t) = \begin{cases} 0 & \text{for } -\infty < x < b_1, \\ \frac{x-b_1}{b_2-b_1} & \text{for } b_1 < x < b_2, \\ 1 & \text{for } b_1 < x < b_{\text{shock}}, \\ 0 & \text{for } b_{\text{shock}} < x < \infty, \end{cases} \tag{4.22}$$

where

$$b_1 = -\frac{1}{3},$$
$$b_2 = -\frac{1}{3} + t,$$
$$b_{\text{shock}} = \frac{1}{3} + \frac{1}{2}t.$$

Notice that $b_2 = b_{\text{shock}}$ for $t = 4/3$. Thus the shock and the expansion fan interact for $t > 4/3$, which complicates the solution. The wave diagram for the solution is plotted in Figure 4.3.

Example 4.4 Solve Burgers' equation on an infinite domain for $t \leq 2/3$ with the following initial conditions:

$$u(x,0) = \begin{cases} 1 & \text{for } |x| < \frac{1}{3}, \\ -1 & \text{for } |x| > \frac{1}{3}. \end{cases}$$

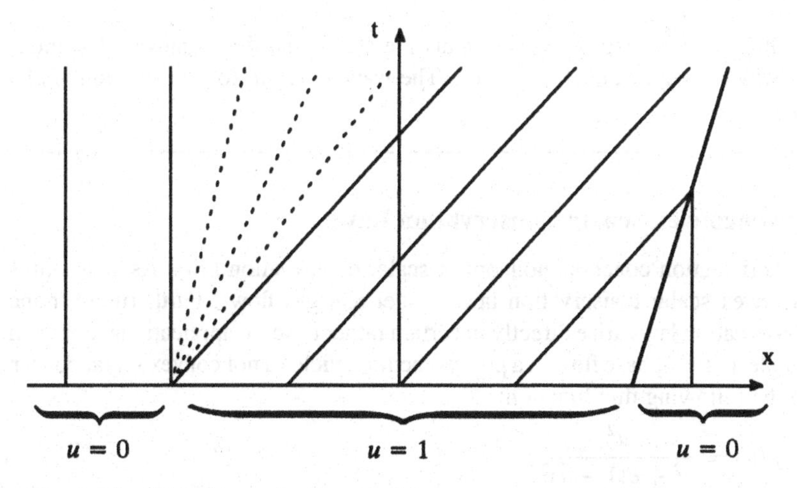

Figure 4.3 Wave diagram for Example 4.3.

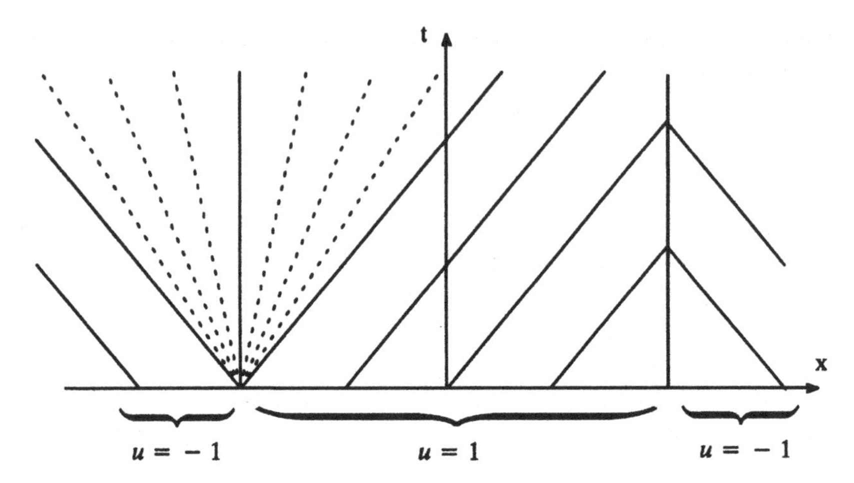

Figure 4.4 Wave diagram for Example 4.4.

Solution The jump at $x = -1/3$ creates a simple centered expansion fan; the jump at $x = 1/3$ creates a steady shock. Until the shock and expansion fan intersect, the exact piecewise-linear solution is as follows:

$$u(x,t) = \begin{cases} -1 & \text{for } -\infty < x < b_1, \\ -1 + 2\frac{x-b_1}{b_2-b_1} & \text{for } b_1 < x < b_2, \\ 1 & \text{for } b_2 < x < b_{\text{shock}}, \\ -1 & \text{for } b_{\text{shock}} < x < \infty, \end{cases} \tag{4.23}$$

where

$$b_1 = -\frac{1}{3} - t,$$

$$b_2 = -\frac{1}{3} + t,$$

$$b_{\text{shock}} = \frac{1}{3}.$$

Notice that $b_2 = b_{\text{shock}}$ for $t = 2/3$. Thus the shock and the expansion fan interact for $t > 2/3$, which complicates the solution. The wave diagram for the solution is plotted in Figure 4.4.

4.9 Nonconvex Scalar Conservation Laws

This section concerns nonconvex scalar conservation laws. As stated in Section 4.5, nonconvex scalar conservation laws model real gas flows. Furthermore, nonconvex scalar conservation laws arise directly in certain rather obscure applications. For example, a simple model of two-phase flow in a porous medium yields a nonconvex scalar conservation law with the following flux function:

$$f(u) = \frac{u^2}{u^2 + c(1-u)^2}, \tag{4.24}$$

where c is a constant. This scalar conservation law is called the *Bucky–Leverett equation.* Since this book mainly concerns perfect gases, we will accordingly mainly use convex conservation laws, which model perfect gas flows. However, nonconvex scalar conservation laws shed light on the nature of convex scalar conservation laws by showing what they are not. Also, the computational gasdynamics literature often mentions nonconvex scalar conservation laws, making this discussion essential for students of the literature.

The biggest difference between convex and nonconvex scalar conservation laws lies in the shock condition (4.12). For convex scalar conservation laws, Equation (4.12) is both necessary and sufficient. However, for nonconvex scalar conservation laws, Equation (4.12) is only necessary but not sufficient. In other words, nonconvex scalar conservation laws require something stronger than Equation (4.12). Specifically, for nonconvex scalar conservation laws, Equation (4.12) must be replaced by the following condition:

$$\blacklozenge \qquad \frac{f(u) - f(u_L)}{u - u_L} \geq S = \frac{f(u_R) - f(u_L)}{u_R - u_L} \geq \frac{f(u_R) - f(u)}{u_R - u} \tag{4.25}$$

for all u between u_L and u_R. This is called the *Oleinik entropy condition.* Unlike Equation (4.12), the Oleinik entropy condition applies both at shocks and at contacts. Notice that, by the mean value theorem, the Oleinik entropy condition implies Equation (4.12). The Oleinik entropy condition says that the flux function lies entirely above the secant line connecting u_L and u_R for $u_L < u_R$, and the flux function lies entirely below the secant line connecting u_L and u_R for $u_L > u_R$. Figure 4.5 illustrates this condition. The dashed lines are the secant lines connecting u_L and u_R to u, the solid line is the secant line connecting u_L and u_R, and the bold curve is the flux function. Notice that the flux function can do anything it likes between u_L and u_R – it can have maxima, minima, inflection points, and so forth – provided only that it does not cross the secant line connecting u_L and u_R.

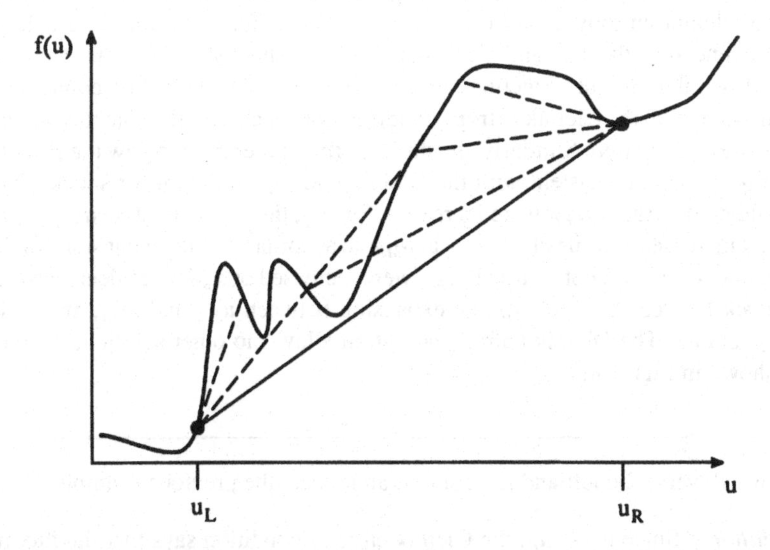

Figure 4.5 An illustration of the Oleinik entropy condition. The left and right states are connected by a single shock.

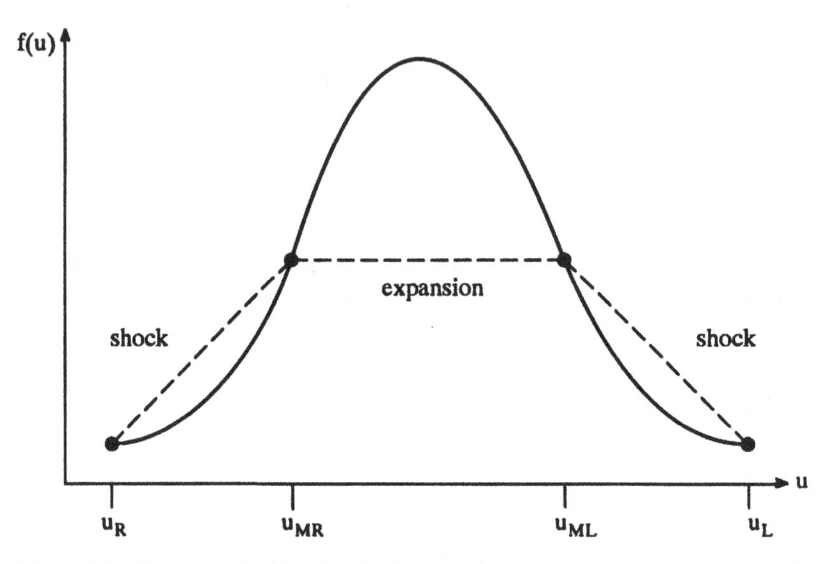

Figure 4.6 Nonconvex flux function for Example 4.5.

Example 4.5 Consider a scalar conservation law with the following initial conditions:

$$u(x,0) = \begin{cases} u_L & \text{for } x < 0, \\ u_R & \text{for } x > 0. \end{cases}$$

Assume the scalar conservation law has the nonconvex flux function illustrated in Figure 4.6. Sketch the wave diagram for the solution.

Solution In Figure 4.6, u_{ML} and u_{MR} are the inflection points of $f(u)$. Since $u_L > u_R$, the Oleinik entropy condition says that the flux function should lie below the secant lines connecting the left and right states of any shocks or contacts. Consulting Figure 4.6, we see that the flux function lies entirely below the secant line connecting u_L and u_{ML}, consistent with the Oleinik entropy condition for shocks. Notice that this would not be true if u_{ML} were decreased. Similarly, the flux function lies entirely below the secant line connecting u_{MR} and u_R, consistent with the Oleinik entropy condition for shocks. Notice that this would not be true if u_{MR} were increased. Finally, the flux derivative $a(u) = f'(u)$ monotonically increases as u runs from u_{ML} to u_{MR}, in accordance with expansion condition (4.8). Notice that this would not be true if u_{ML} were increased or u_{MR} were decreased. Thus there is a shock between u_L and u_{ML}, an expansion between u_{ML} and u_{MR}, and a shock between u_{MR} and u_R. The Oleinik entropy condition allows no other solutions. The wave diagram is shown in Figure 4.7.

Example 4.6 Reverse the left and right states, and repeat the previous example.

Solution Since $u_L < u_R$, the Oleinik entropy condition says that the flux function should lie above the secant lines connecting the left and right states of any shocks or

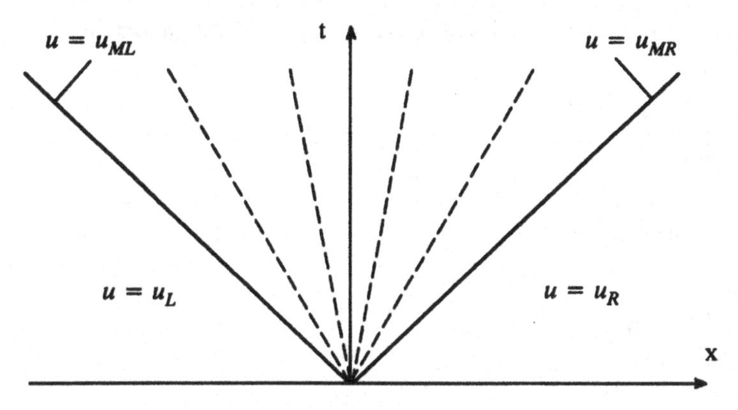

Figure 4.7 Wave diagram for Example 4.5.

contacts. In this case, the flux function lies entirely above the secant line connecting u_L and u_R. Furthermore, $a(u_L) = a(u_R) = 0$. Then the solution consists of a single contact connecting u_L and u_R. If u_L is moved slightly to the right, or u_R is moved slightly to the left, the solution would consist of a single shock between u_L and u_R.

4.10 Entropy Conditions

The Euler equations come paired with the second law of thermodynamics. Similarly, scalar conservation laws should come paired with entropy conditions analogous to the second law of thermodynamics. In the Euler equations, the second law of thermodynamics allows multiple solutions, at least for steady flows. By contrast, in scalar conservation laws, entropy conditions always select a unique solution.

In the applications of interest to us, the second law of thermodynamics and entropy conditions need not be written in terms of entropy. For example, for the Euler equations, the second law of thermodynamics amounts to an algebraic inequality (3.51) at shocks, which does not involve entropy. Similarly, for convex scalar conservation laws, the entropy condition amounts to an algebraic inequality (4.12) or (4.13) at shocks, which does not involve entropy. Finally, for nonconvex scalar conservation laws, the entropy condition amounts to an algebraic inequality (4.25) at shocks and contacts, which does not involve entropy.

Although entropy conditions need not explicitly involve entropy, it is interesting to write them in a form that does. Thus, whereas previous sections in this chapter put entropy conditions in the form of inequalities, this section puts entropy conditions in a differential form analogous to Equation (2.26), involving a scalar entropy function and a scalar entropy flux function. The scalar entropy and entropy flux are functions of u. Before proceeding further, as necessary background, the following example illustrates what happens to scalar conservation laws when u and f are replaced by functions of u.

Example 4.7 Section 2.4 described linear changes of dependent variables for the Euler equations. In a similar fashion, perform a change of dependent variables in Burgers' equations from u to $U = u^2$. Also, change the conservative flux so that the equation remains in conservation form. Show that this transformation alters the weak solutions.

Solution Multiply both sides of Burgers' equation by $2u$ to obtain

$$2u\frac{\partial u}{\partial t} + 2u^2\frac{\partial u}{\partial x} = 0,$$

which can be written as

$$\frac{\partial}{\partial t}(u^2) + \frac{\partial}{\partial x}\left(\frac{2}{3}u^3\right) = 0$$

or

$$\frac{\partial U}{\partial t} + \frac{\partial F}{\partial x} = 0,$$

where

$$U(x,t) = u^2(x,t)$$

and

$$F = \frac{2}{3}u^3 = \frac{2}{3}U^{3/2}.$$

Thus the dependent variable has been changed from u to $U = u^2$ and the flux has been changed from $f = u^2/2$ to $F = 2u^3/3$.

Unfortunately, the weak solutions of the transformed Burgers' equation differ substantially from the weak solutions of the original Burgers' equation. By Equation (4.10), the jump condition for Burgers' equation is

$$S = \frac{f(u_R) - f(u_L)}{u_R - u_L} = \frac{1}{2}\frac{u_R^2 - u_L^2}{u_R - u_L} = \frac{1}{2}(u_R + u_L),$$

whereas the jump condition for the transformed Burgers' equation is

$$S = \frac{F(U_R) - F(U_L)}{U_R - U_L} = \frac{2}{3}\frac{u_R^3 - u_L^3}{u_R^2 - u_L^2} = \frac{1}{2}(u_R + u_L) + \frac{1}{6}\frac{(u_R - u_L)^2}{u_R + u_L}.$$

Thus the transformed equation has a different shock speed. Although this example concerns Burgers' equation, in general, *changing u and f in scalar conservation laws always alters the weak solutions.*

In Section 2.4, we saw that changes of variables did not alter the weak solutions. However, in Section 2.4 we changed only the dependent variable $\mathbf{u}$ and not the flux $\mathbf{f}$. For scalar conservation laws, different flux functions f define different jump relations $f_R - f_L = S(u_R - u_L)$, causing the weak solutions to differ. Hence this example got us into trouble not because it changed u but because it changed f, and the altered flux function implied altered jump relations.

Consider the following properties of the second law of thermodynamics as seen in Subsections 2.1.5 and in Sections 2.2, 2.3, and 3.3:

- Entropy is a function of the conserved quantities.
- Entropy flux is a function of the conserved quantities.

- Except at shocks, entropy change is due entirely to entropy flux. In other words, the entropy following the fluid neither increases nor decreases except at shocks. Put yet another way, except at shocks, the Euler equations imply

  $$\frac{\partial(\rho s)}{\partial t} + \frac{\partial(\rho u s)}{\partial x} = 0.$$

  Thus the Euler equations automatically satisfy the second law of thermodynamics, except at shocks.
- Entropy increases across shocks. In other words, the entropy change across shocks is greater than the entropy change due to entropy flux. Put yet another way, across shocks, the second law of thermodynamics requires

  $$\frac{\partial(\rho s)}{\partial t} + \frac{\partial(\rho u s)}{\partial x} > 0.$$

Suppose $u(x, t)$ is any solution of a scalar conservation law. Transform the scalar conservation law as in Example 4.7. Then suppose any solution $u(x, t)$ satisfies

$$\frac{\partial U(u)}{\partial t} + \frac{\partial F(u)}{\partial x} = 0 \tag{4.26}$$

in smooth regions and

$$\frac{\partial U(u)}{\partial t} + \frac{\partial F(u)}{\partial x} > 0 \tag{4.27}$$

across shocks. In other words, combining the last two equations, suppose that

♦ $$\frac{\partial U(u)}{\partial t} + \frac{\partial F(u)}{\partial x} \geq 0 \tag{4.28}$$

for all weak solutions of a scalar conservation law. Then Equation (4.28) is analogous to the second law of thermodynamics. The function $U(u)$ is called an *entropy function* and the function $F(u)$ is called an *entropy flux function*.

As the reader can easily show, $\partial u/\partial t + \partial f/\partial x = 0$ implies Equation (4.26) in smooth regions if

$$\frac{dF}{du} = \frac{df}{du}\frac{dU}{du}. \tag{4.29}$$

With a good deal more effort, the reader can show that weak solutions of $\partial u/\partial t + \partial f/\partial x = 0$ satisfy (4.27) across shocks if

$$\frac{d^2U}{du^2} \leq 0. \tag{4.30}$$

Using standard calculus definitions, Equation (4.30) says that the entropy function U is *concave down* or simply *concave*. As shown by Merriam (1989), Equations (4.29) and (4.30) have the following simple general solution:

♦ $$U(u) = -u^2, \tag{4.31}$$

♦ $$\frac{dF}{du} = -2u\frac{df}{du}. \tag{4.32}$$

This solution is but one of many possible concave entropy functions and entropy fluxes. (Note that, in the literature, scalar entropy often decreases rather than increases, unlike physical entropy. In other words, Equation (4.28) is replaced by $\partial U/\partial t + \partial F/\partial x \leq 0$ and Equation (4.30) is replaced by $d^2U/du^2 \geq 0$, in which case the entropy function is *convex*. In this case, a general solution is $U(u) = u^2$ and $F(u) = 2u\, df/du$.)

Example 4.8 Consider Burgers' equation. Find a concave entropy function, entropy flux, and an entropy condition in conservation form.

Solution An entropy condition for Burgers' equation is as follows:

$$\frac{\partial U(u)}{\partial t} + \frac{\partial F(u)}{\partial x} \geq 0, \tag{4.33}$$

where

$$U(u) = -u^2 \tag{4.34}$$

and

$$F(u) = -\frac{2}{3}u^3. \tag{4.35}$$

For Burgers' equation, or for any convex scalar conservation law, entropy condition (4.13) says

$$u_L \geq u_R. \tag{4.36}$$

Believe it or not, conditions (4.33) and (4.36) mean exactly the same thing; this is proven in Example 3.4 of LeVeque (1992).

Unfortunately, there are major obstacles to enforcing entropy conditions in numerical approximations. First off, it is often impossible to enforce or verify entropy inequalities such as (4.12), (4.13), or (4.25). Thus, in practice, other forms are used, such as the differential form described in this section. Using these other forms, mathematicians have developed a body of complicated techniques for proving that numerical approximations to scalar conservation laws satisfy entropy conditions in the convergence limit $\Delta x \to 0$ and $\Delta t \to 0$. However, for ordinary finite values of Δx and Δt, Argrow and Cox (1993) have shown that most numerical approximations to Burgers' equation violate entropy conditions and that most numerical approximations to the Euler equations violate the second law of thermodynamics by producing locally negative entropy. Fortunately, local negative entropy production has surprisingly benign effects in practice. Rather than deal explicitly with entropy conditions, most code developers rely on a few tricks of the trade and extensive testing to ensure that their codes do not approximate unphysical solutions. In the hopefully rare cases where they occur, wildly wrong solutions can easily escape detection until compared with the results of experiments or of more accurate codes.

4.11 Waveform Preservation, Destruction, and Creation

Simple waves, shock waves, and contacts provide a complete wave description of the solutions to scalar conservation laws. As seen in this section, the wave description implies useful relationships between the solution $u(x, t)$ and the initial conditions $u(x, 0) = u_0(x)$. These relationships will be exploited heavily in Chapter 16.

To start with, suppose $u(x, t)$ is continuous for all x and t. If the solution is always continuous, the characteristics are always divergent or neutral, since convergent characteristics must eventually meet and form shocks. Hence for scalar conservation laws, any completely continuous solution is an expansion wave. By Equation (4.7)

$$u(x, t) = u_0(x - a(u)t).$$

Then, by definition, continuous solutions $u(x, t)$ *preserve* the initial waveform $u_0(x)$. Of course, the nonconstant wave speed $a(u)$ means that the initial waveform may stretch; however, this does not violate waveform preservation as defined here.

Now suppose the solution is continuous for all x and t, except for a single contact. Then the solution consists of two simple expansions separated by a contact discontinuity. To create the contact, the initial conditions must contain a jump from u_L to u_R such that $a(u_L) = a(u_R)$. Then the solution preserves the contact, although of course the contact moves to the left or right at speed $a(u_L) = a(u_R)$. The same argument applies to any number of contacts, assuming that the contacts never intersect. Thus continuous solutions, or solutions with any number of nonintersecting contacts, preserve the initial waveform. Waveform preservation is illustrated in Figure 4.8.

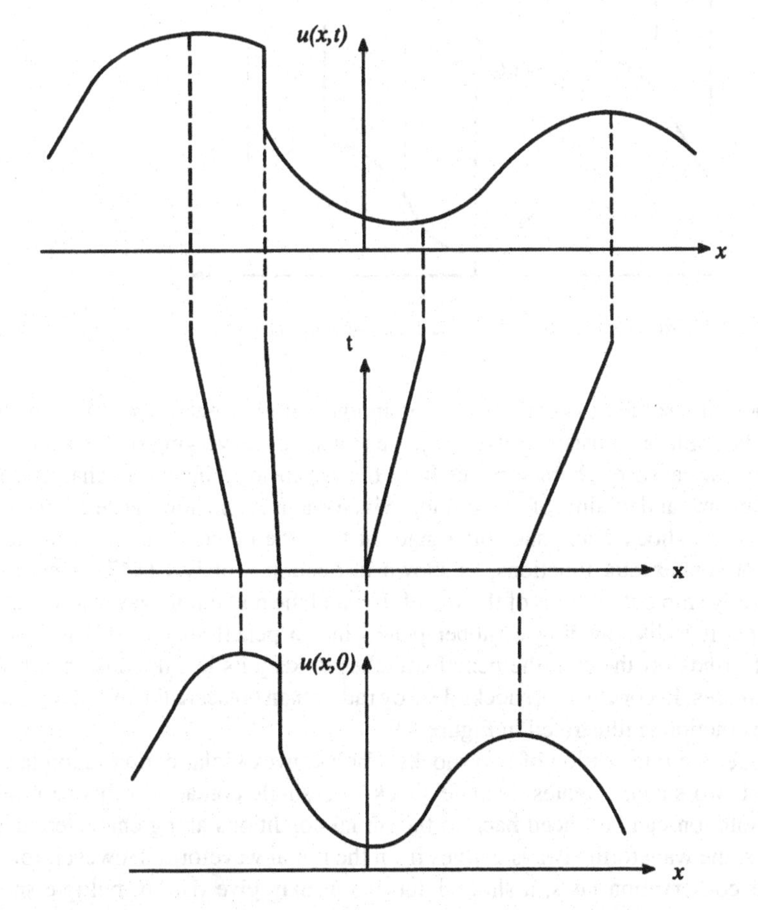

Figure 4.8 Waveform preservation for scalar conservation laws.

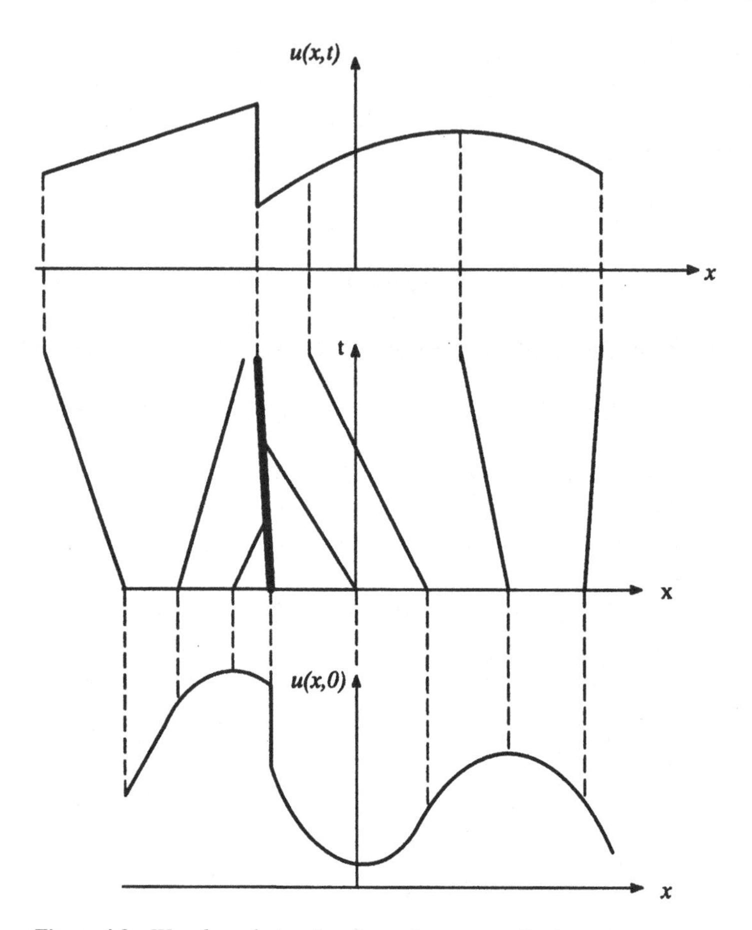

Figure 4.9 Waveform destruction for scalar conservation laws.

Now suppose the solution is completely continuous for all x and t, except for a single shock. Then the solution consists of two simple expansions converging on the shock. The shock continuously absorbs characteristics from the expansions, including characteristics carrying the maxima and minima of the solution. Since maxima and minima define the range of the solution, the shock diminishes the range. In fact, the range of the solution decays like $(t)^{-1/2}$ for convex flux functions, as shown in Section 4 of Lax (1973). Intuitively, shocks constantly snip out sections of the waveform, and the remaining waveform stretches to fill the gaps. It is like feeding a rubber pencil into a pencil sharpener – as much as the sharpener grinds off the end, the pencil stretches to keep its length constant while its thickness decreases. In conclusion, shocks destroy rather than preserve the initial waveform. Waveform destruction is illustrated in Figure 4.9.

Now consider the intersection of two shocks. For a convex scalar conservation law, the intersection of two shocks creates a single shock or a single contact. In this case, every value in the solution can be traced back to the initial conditions along characteristics or, in other words, the waveform always derives from the initial waveform. However, for non-convex scalar conservation laws, a shock intersection may give rise to multiple shocks,

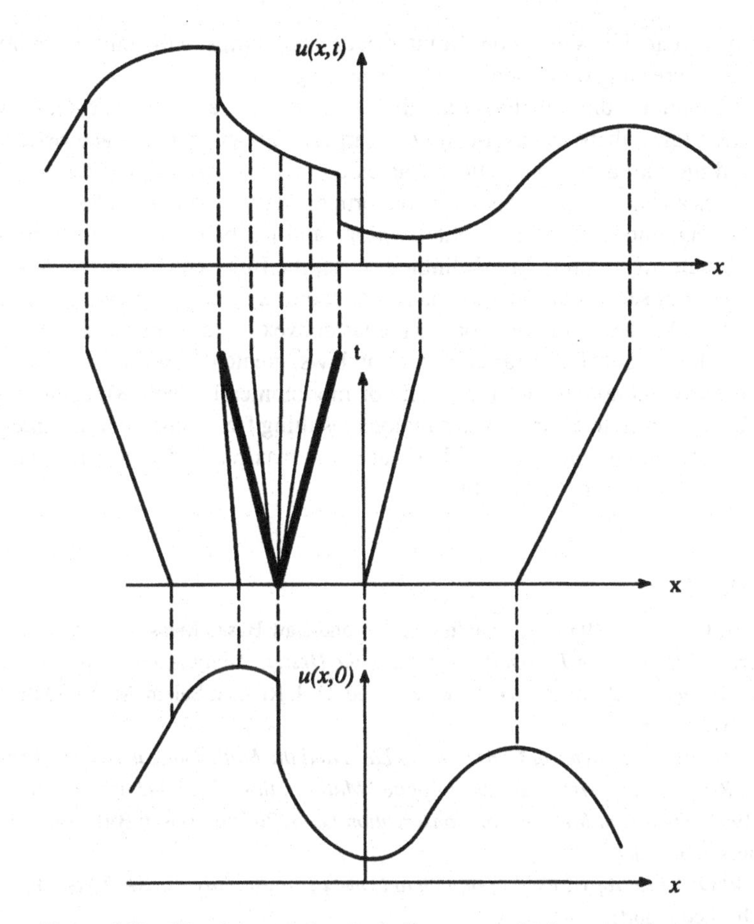

Figure 4.10 Waveform creation for scalar conservation laws.

multiple centered expansion fans, and multiple contacts. The values of the solution in the interior of such a multiwave region cannot be traced back to the initial conditions; hence the waveform in the interior of the multiwave region does *not* derive from the initial waveform. Instead, the solution must *create* a new waveform to fill the multiwave region. Similar considerations apply at the intersections of a shock and a contact, or a contact and a contact, or at jump discontinuities in the initial conditions. Any created waveforms are always monotonically increasing or monotonically decreasing. This important property will be used extensively in Chapter 16. Waveform creation is illustrated in Figure 4.10.

In summary, the solutions of scalar conservation laws consist entirely of smooth simple waves, shock waves, and contacts. For convex scalar conservation laws, smooth waves and contacts are always waveform preserving, whereas shock waves and shock/contact intersections are always waveform destroying; convex scalar conservation laws are never waveform creative. For nonconvex scalar conservation laws, smooth waves and contacts are always waveform preserving, shock waves are always waveform destroying, and shock/contact intersections and jumps in the initial conditions are sometimes waveform creating; any waveform sections created by shock/contact intersections or jumps in the initial conditions are monotonically increasing or monotonically decreasing.

For the Euler equations, the waveforms are the characteristic variables $v_i(x, t)$. As with scalar conservation laws, there are useful wave relationships between characteristic variables $v_i(x, t)$ and the initial characteristic variables $v_i(x, 0)$. As with scalar conservation laws, smooth waves and contacts are always waveform preserving, shock waves are always waveform destroying, and shock/contact intersections and jumps in the initial conditions are sometimes waveform creating. In this limited sense, the Euler equations are more like nonconvex than convex scalar conservation laws – nonconvex scalar conservation laws are waveform creating, like the Euler equations, whereas convex scalar conservation laws are not. However, unlike nonconvex scalar conservation laws, created waveforms in the Euler equations need not be monotonically increasing or monotonically decreasing. In fact, in general, waveforms created to fill multiwave regions resulting from shock/contact intersections or jumps in the initial conditions will contain new maxima and minima, caused by interactions between characteristic variables.

References

Argrow, B. M., and Cox, R. A. 1993. "A Quantitative, Second-Law Based Measure of Accuracy of Numerical Schemes." In *Thermodynamics and the Design, Analysis, and Improvement of Energy Systems (AES vol. 30 / HTD vol. 266)*, ed. H. J. Richter, Salem, MA: ASME, pp. 49–57.

Lax, P. D. 1973. *Hyperbolic Systems of Conservation Laws and the Mathematical Theory of Shock Waves. Regional Conference Series in Applied Mathematics*, 11, Philadelphia: SIAM.

LeVeque, R. J. 1992. *Numerical Methods for Conservation Laws*, 2nd ed., Basel: Birhäuser-Verlag, Chapters 2, 3, and 4.

Merriam, M. L. 1989. *An Entropy-Based Approach to Non-Linear Stability*, NASA-TM-101086 (unpublished report).

Problems

4.1 Consider Burgers' equation on an infinite domain with the following initial conditions:

$$u(x, 0) = \begin{cases} 2 & \text{for } |x| < \frac{1}{2}, \\ -1 & \text{for } |x| > \frac{1}{2}. \end{cases}$$

The two jump discontinuities create two waves. Find the time when the two waves first intersect. Solve Burgers' equation for all times less than the intersection time. Draw a wave diagram for the solution.

4.2 Write the Bucky–Leverett equation, defined by Equation (4.24), in characteristic form.

4.3 In each case, state whether the scalar conservation law is convex or nonconvex:

(a) $$\frac{\partial u}{\partial t} + \frac{\partial}{\partial x}\left(\frac{2}{3}u^{3/2}\right) = 0$$

(b) $e^{-u}\dfrac{\partial u}{\partial t} + \dfrac{\partial u}{\partial x} = 0$

(c) $\dfrac{\partial u}{\partial t} + \sin u \dfrac{\partial u}{\partial x} = 0$

4.4 Find conditions under which a scalar conservation law can have a stationary jump discontinuity. In particular, show that the flux function must have a maximum or a minimum. In other words, the wave speed must equal zero somewhere inside the shock. Put yet another way, the wind must change directions across the shock. Put yet another way, the flux function must have a *sonic point.*

4.5 Consider the following scalar conservation law:

$$\frac{\partial u}{\partial t} + \frac{\partial}{\partial x}\left(u^3 + \frac{u^2}{2}\right) = 0.$$

(a) Is this scalar conservation law convex?
(b) Write the entropy condition in the form of an algebraic inequality. Simplify as much as possible using relationships such as $a^3 - b^3 = (a-b)(a^2+ab+b^2)$.
(c) Find an entropy function and an entropy flux. Write the entropy condition as a partial differential inequality.

4.6 Consider a scalar conservation law with the following initial conditions:

$$u(x,0) = \begin{cases} u_L & \text{for } x < 0, \\ u_R & \text{for } x > 0. \end{cases}$$

Assume the scalar conservation law has the nonconvex flux function illustrated below.

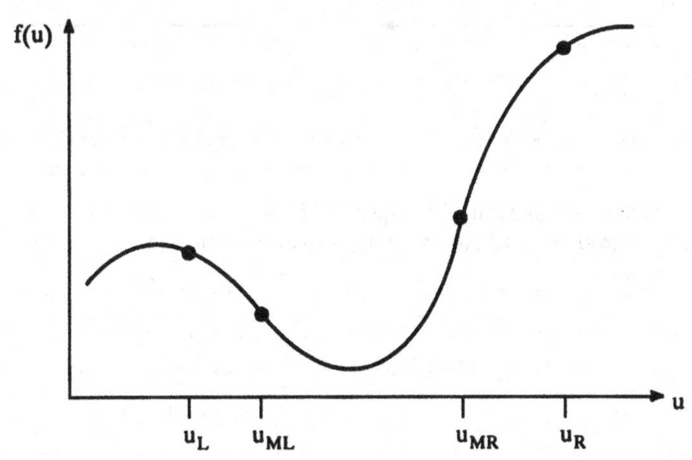

Problem 4.6

(a) Sketch the wave diagram for the solution. Explain.
(b) Reverse the left and right states. Sketch the wave diagram for the solution. Explain.

4.7 Consider the solutions to hypothetical scalar conservation laws shown in the figures. In each case, state whether the solution is waveform preserving, waveform destroying, waveform

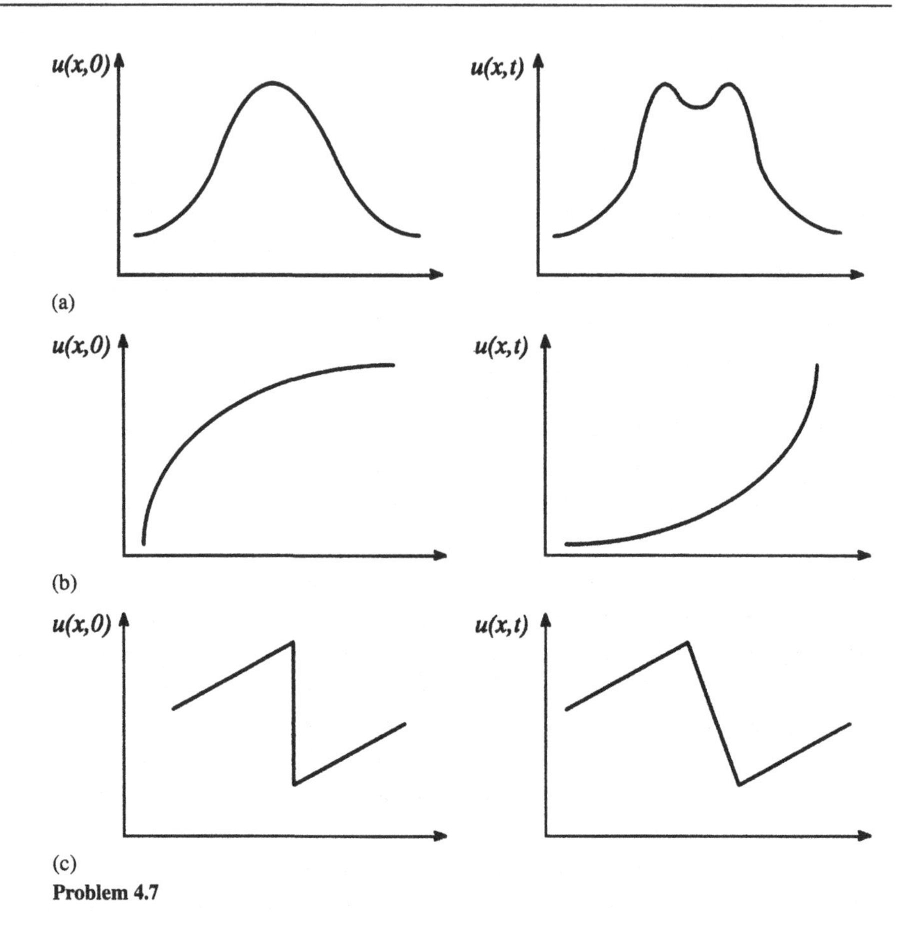

Problem 4.7

creating, or none of the above. Explain. You may find it helpful to draw some possible wave diagrams.

4.8 (a) Show that Equation (4.25) implies Equation (4.12).

(b) Show that Equations (4.12), (4.13), and (4.25) are equivalent for convex scalar conservation laws.

CHAPTER 5

The Riemann Problem

5.0 Introduction

For any equation, such as the one-dimensional Euler equations or scalar conservation equations, the *Riemann problem* has uniform initial conditions on an infinite spatial domain, except for a single jump discontinuity. For example, for the Euler equations, the Riemann problem centered on $x = x_0$ and $t = t_0$ has the following initial conditions:

$$\mathbf{u}(x, t_0) = \begin{cases} \mathbf{u}_L & x < x_0, \\ \mathbf{u}_R & x > x_0, \end{cases} \tag{5.1}$$

where $\mathbf{u}_L$ and $\mathbf{u}_R$ are constant vectors. For convenience, this chapter uses $x_0 = 0$ and $t_0 = 0$.

The Riemann problem has an exact analytical solution for the Euler equations, scalar conservation laws, and any linear system of equations. Furthermore, the solution is *self-similar* or *self-preserving*. In other words, the solution stretches uniformly in space as time increases but otherwise retains its shape, so that $\mathbf{u}(x, t_1)$ and $\mathbf{u}(x, t_2)$ are "similar" to each other for any two times t_1 and t_2. Put another way, the solution only depends on the single variable x/t rather than on x and t separately. Put yet another way, the solution is constant along any line $x = (const.)t$ passing through the origin in the x–t plane. More generally, a self-similar solution is any solution that depends on only one independent variable, such as x/t or $t/\sqrt{x}$, that is a function of the two original independent variables. By reducing the number of independent variables, self-similarity simplifies solution techniques and sometimes even leads to analytical solutions, as in the case of the Riemann problem. Other famous examples of self-similarity include steady laminar boundary layers, jets, and wakes, and Sedov's unsteady spherical, cylindrical, and planar shock waves.

There are three reasons for devoting an entire chapter to the Riemann problem. First, the last two chapters have concerned waves, and the Riemann problem provides an excellent example of waves. Indeed, for the Euler equations and scalar conservation laws, the solution of the Riemann problem involves the three basic types of waves described in the last two chapters: simple expansion waves, shock waves, and contacts. Second, since the Riemann problem has an exact solution, it makes a nice test case for numerical approximations. In fact, the solution to the Riemann problem is one of the very few exact solutions of the unsteady one-dimensional Euler equations. Third, and most importantly, many numerical methods incorporate an exact or approximate solution to the Riemann problem in order to provide superior wave capturing. Indeed, the Riemann problem tends to improve numerical modeling of both smooth waves (characteristics) as well as nonsmooth waves (shocks and contacts).

Numerical approximations based on the Riemann problem may solve the Riemann problem hundreds, thousands, or even millions of times to obtain a single solution. Thus the cost of solving the Riemann problem is a serious issue. When incorporated into numerical models, the solution to an approximate Riemann problem may prove almost as good as or even, in some ways, better than the solution to the true Riemann problem, often at a fraction of the cost. This chapter describes two approximate Riemann problem solvers that replace the true nonlinear flux function by a locally linearized approximation.

In most cases, numerical methods using Riemann solvers require only the flux along $x = 0$. Fortunately the flux is constant along the line $x = 0$, by self-similarity. For future reference, this chapter will, in many instances, derive expressions for the flux at $x = 0$. The chapter ends with a description of the Riemann problem for scalar conservation laws. Examples in Sections 4.8 and 4.9 illustrate some techniques for solving the Riemann problem for scalar conservation laws, both for convex and nonconvex flux functions. This chapter takes a different approach: The solution to the Riemann problem is written in terms of flux $f(u)$, which is especially convenient when the solution is used in numerical methods only requiring the flux at $x = 0$.

5.1 The Riemann Problem for the Euler Equations

Consider a one-dimensional tube containing two regions of stagnant fluid at different pressures. Suppose the two regions of fluid are initially separated by a rigid diaphragm. If the diaphragm is instantaneously removed, perhaps by a small explosion, the pressure imbalance causes a one-dimensional unsteady flow containing a steadily moving shock, a steadily moving simple centered expansion fan, and a steadily moving contact discontinuity separating the shock and expansion. The shock, expansion, and contact each separate regions of uniform flow. This setup is called a *shock tube*. The flow in a shock tube is illustrated in Figure 5.1.

The flow in a shock tube always has zero initial velocity. Removing this restriction, the shock tube problem becomes the Riemann problem and thus is a special case of the Riemann problem. Like the shock tube problem, the Riemann problem may give rise to a shock, a simple centered expansion fan, and a contact separating the shock and expansion; however, unlike the shock tube problem, one or two of these waves may be absent. Unfortunately, in general, nonzero velocities in the initial conditions of the Riemann problem are difficult

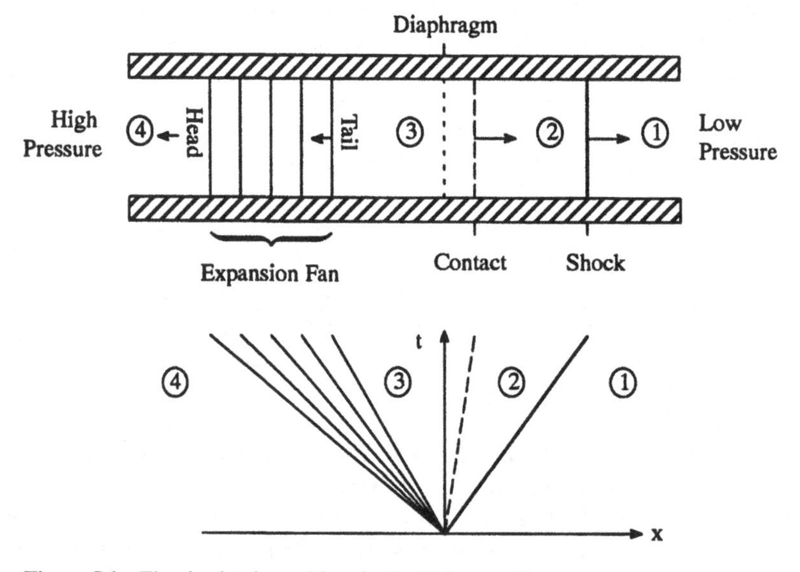

Figure 5.1 The shock tube problem for the Euler equations.

to produce experimentally. One approach is to burst two distant diaphragms, creating two shocks that move towards each other; the intersection of the two shocks instantaneously creates a Riemann problem. In other words, in one interpretation, *the Riemann problem equals the shock intersection problem*, or possibly the contact intersection problem or the shock–contact intersection problem.

To find the exact solution to the Riemann problem, first consider the shock. Let the states to the left and right of the shock be $\mathbf{u}_2$ and $\mathbf{u}_1 = \mathbf{u}_R$, respectively. From Section 3.6 recall that

$$\frac{a_2^2}{a_1^2} = \frac{p_2}{p_1}\frac{\frac{\gamma+1}{\gamma-1} + \frac{p_2}{p_1}}{1 + \frac{\gamma+1}{\gamma-1}\frac{p_2}{p_1}}, \tag{3.54}$$

$$u_2 = u_1 + \frac{a_1}{\gamma}\frac{\frac{p_2}{p_1} - 1}{\sqrt{\frac{\gamma+1}{2\gamma}\left(\frac{p_2}{p_1} - 1\right) + 1}}, \tag{3.55}$$

$$S = u_1 + a_1\sqrt{\frac{\gamma+1}{2\gamma}\left(\frac{p_2}{p_1} - 1\right) + 1}. \tag{3.56}$$

Next, consider the contact discontinuity. Let the states to the left and right of the contact be $\mathbf{u}_3$ and $\mathbf{u}_2$, respectively. From Section 3.7

$$u_2 = u_3, \tag{3.57}$$

$$p_2 = p_3. \tag{3.58}$$

Finally, consider the simple centered expansion fan. Let the states to the left and right of the expansion be $\mathbf{u}_4 = \mathbf{u}_L$ and $\mathbf{u}_3$, respectively. From Section 3.5

$$u(x,t) = \frac{2}{\gamma+1}\left(\frac{x}{t} + \frac{\gamma-1}{2}u_4 + a_4\right), \tag{3.47}$$

$$a(x,t) = u(x,t) - \frac{x}{t} = \frac{2}{\gamma+1}\left(\frac{x}{t} + \frac{\gamma-1}{2}u_4 + a_4\right) - \frac{x}{t}, \tag{3.48}$$

$$p = p_4\left(\frac{a}{a_4}\right)^{2\gamma/(\gamma-1)}. \tag{3.49}$$

Let us now combine the preceding shock, contact, and expansion results to find an equation for the unknown pressure ratio p_2/p_1 across the shock in terms of the known pressure ratio $p_4/p_1 = p_L/p_R$ across the Riemann problem. The simple wave condition $u + 2a/(\gamma - 1) = const.$ given in Equation (3.40) implies

$$u_3 + \frac{2a_3}{\gamma-1} = u_4 + \frac{2a_4}{\gamma-1}. \tag{5.2}$$

Combining Equations (3.49) and (5.2) yields

$$u_3 = u_4 + \frac{2a_4}{\gamma-1}\left[1 - \left(\frac{p_3}{p_4}\right)^{(\gamma-1)/2\gamma}\right]. \tag{5.3}$$

Combining Equations (3.57), (3.58), and (5.2) yields

$$u_2 = u_4 + \frac{2a_4}{\gamma - 1}\left[1 - \left(\frac{p_2}{p_4}\right)^{(\gamma-1)/2\gamma}\right]$$

or

$$u_2 = u_4 + \frac{2a_4}{\gamma - 1}\left[1 - \left(\frac{p_1}{p_4}\frac{p_2}{p_1}\right)^{(\gamma-1)/2\gamma}\right]. \tag{5.4}$$

Solving Equation (5.4) for p_4/p_1 yields

$$\frac{p_4}{p_1} = \frac{p_2}{p_1}\left[1 + \frac{\gamma - 1}{2a_4}(u_4 - u_2)\right]^{-2\gamma/(\gamma-1)}. \tag{5.5}$$

Finally, combining Equations (3.55) and (5.5) gives the desired result:

♦ $$\frac{p_4}{p_1} = \frac{p_2}{p_1}\left\{1 + \frac{\gamma - 1}{2a_4}\left[u_4 - u_1 - \frac{a_1}{\gamma}\frac{\frac{p_2}{p_1} - 1}{\sqrt{\frac{\gamma+1}{2\gamma}\left(\frac{p_2}{p_1} - 1\right) + 1}}\right]\right\}^{-2\gamma/(\gamma-1)}. \tag{5.6}$$

Notice that this equation is the wrong way around: Instead of expressing the unknown p_2/p_1 as a function of the known p_4/p_1, it expresses the known p_4/p_1 as a function of the unknown p_2/p_1. Unfortunately, there is no straightforward analytical way to invert (5.6). Any equation of the form $g(unknown) = f(known)$ is called *implicit* whereas any equation of the form $unknown = f(known)$ is called *explicit*. For another example of implicit equations, $A\mathbf{x} = \mathbf{b}$ is implicit whereas $\mathbf{x} = A^{-1}\mathbf{b}$ is explicit. Linear implicit systems of equations can be solved relatively easily using, for example, Gaussian elimination. However, nonlinear implicit equations, including Equation (5.6), must be solved using more expensive numerical techniques such as bisection, the method of false position, Newton's method, the secant method, or some other numerical procedure; such procedures are described in Mathews (1992) or any number of other numerical analysis texts under the heading of *root solvers*. For example, in the Riemann problem, if the first two guesses are $p_2/p_1 = 0.05p_4/p_1$ and $p_2/p_1 = 0.5p_4/p_1$, the secant method usually converges extremely rapidly; ten orders of magnitude in seven iterations is typical. However, attempts to solve Equation (5.6) may fail if the quantity in square brackets is negative and $2\gamma/(\gamma - 1)$ is not an integer. For example, $(-1)^7 = -1$ but $(-1)^{7.001}$ is undefined. This difficulty is easily overcome by simply taking both sides of Equation (5.6) to the power of $(\gamma - 1)/2\gamma$. This transfers the exponent to the pressure ratios, which are always positive and thus can always be raised to noninteger exponents.

Once p_2 is found using Equation (5.6), the rest of the solution follows easily. In particular, Equations (3.55) or (5.4) yield u_2, Equation (3.54) yields a_2, and Equation (3.56) yields the speed of the shock separating regions 1 and 2, thereby completely determining state 2. Similarly, Equation (3.57) yields u_3 (which is also the speed of the contact separating regions 2 and 3), Equation (3.58) yields p_3, and Equation (5.2) yields a_3, and thus state 3 is completely determined. Finally, Equation (3.47) yields u inside the expansion, Equation (3.48) yields a inside the expansion, and Equation (3.47) yields p inside the expansion. The expansion wave is bounded on the left by a characteristic of slope $u_4 - a_4$ and on the right by a characteristic of slope $u_3 - a_3$, both of which are known from the initial conditions and the contact and shock relations. Thus the expansion fan is completely determined. Notice that each flow region is specified in terms of u, a, and p; of course, any other flow property including any conservative flux can easily be determined from these three as required.

In some cases, the Riemann problem may yield only one or two waves, rather than three. To a large extent, the above solution procedure handles such cases automatically. For example, if the solution is a pure shock, then the exact Riemann solver will yield a shock, a very weak expansion, and a very weak contact. Unfortunately, for weak waves, in practice, the Riemann solver may yield incorrect wave speeds, which can lead to some rather alarming solutions, such as a very weak expansion wave whose tail travels faster than its head.

Example 5.1 Find the exact solution to the Riemann problem for the Euler equations at $t = 0.01$ s if $p_L = 100{,}000\,\text{N/m}^2$, $\rho_L = 1\,\text{kg/m}^3$, $u_L = 100\,\text{m/s}$ and $p_R = 10{,}000\,\text{N/m}^2$, $\rho_R = 0.125\,\text{kg/m}^3$, $u_R = -50\,\text{m/s}$. As usual, assume $\gamma = 1.4$ and $R = 287\,\text{N}\cdot\text{m/kg}\cdot\text{K}$.

Solution The Riemann problem yields an expansion, contact, and shock. The left-hand side of the expansion has velocity -274.2 m/s and the right-hand side of the expansion has velocity 1.247 m/s; the contact has velocity 329.5 m/s; and the shock has velocity 582.5 m/s. The solution at $t = 0.01$ s is plotted in Figure 5.2. By self-similarity, the solution at all other times is identical except for uniform stretching, which only affects the scale on the x axis.

5.2 The Riemann Problem for Linear Systems of Equations

The exact solution to the Riemann problem for the Euler equation is expensive, since the main equation (5.6) is nonlinear and implicit. This section concerns the exact solution to the Riemann problem for a *linear* system of equations. As with most problems, linearity simplifies the Riemann problem substantially. In fact, the main obstacle in this section is not finding the solution but finding the proper notation for expressing the solution; this section considers several common notations. Of course, this presentation is wasted unless there is a linear system of equations that approximates the nonlinear Euler equations, at least for the purposes of Riemann problems; this issue is discussed in following sections.

Consider the linear Riemann problem

$$\frac{\partial \mathbf{u}}{\partial t} + A\frac{\partial \mathbf{u}}{\partial x} = 0, \tag{5.7}$$

where

$$\mathbf{u}(x,0) = \begin{cases} \mathbf{u}_L & x < 0, \\ \mathbf{u}_R & x > 0, \end{cases}$$

and where A is a constant $N \times N$ matrix. Matrix characteristics and matrix diagonalization were described in detail in Chapter 3. Assume that A is diagonalizable:

$$A = Q\Lambda Q^{-1}, \tag{5.8}$$

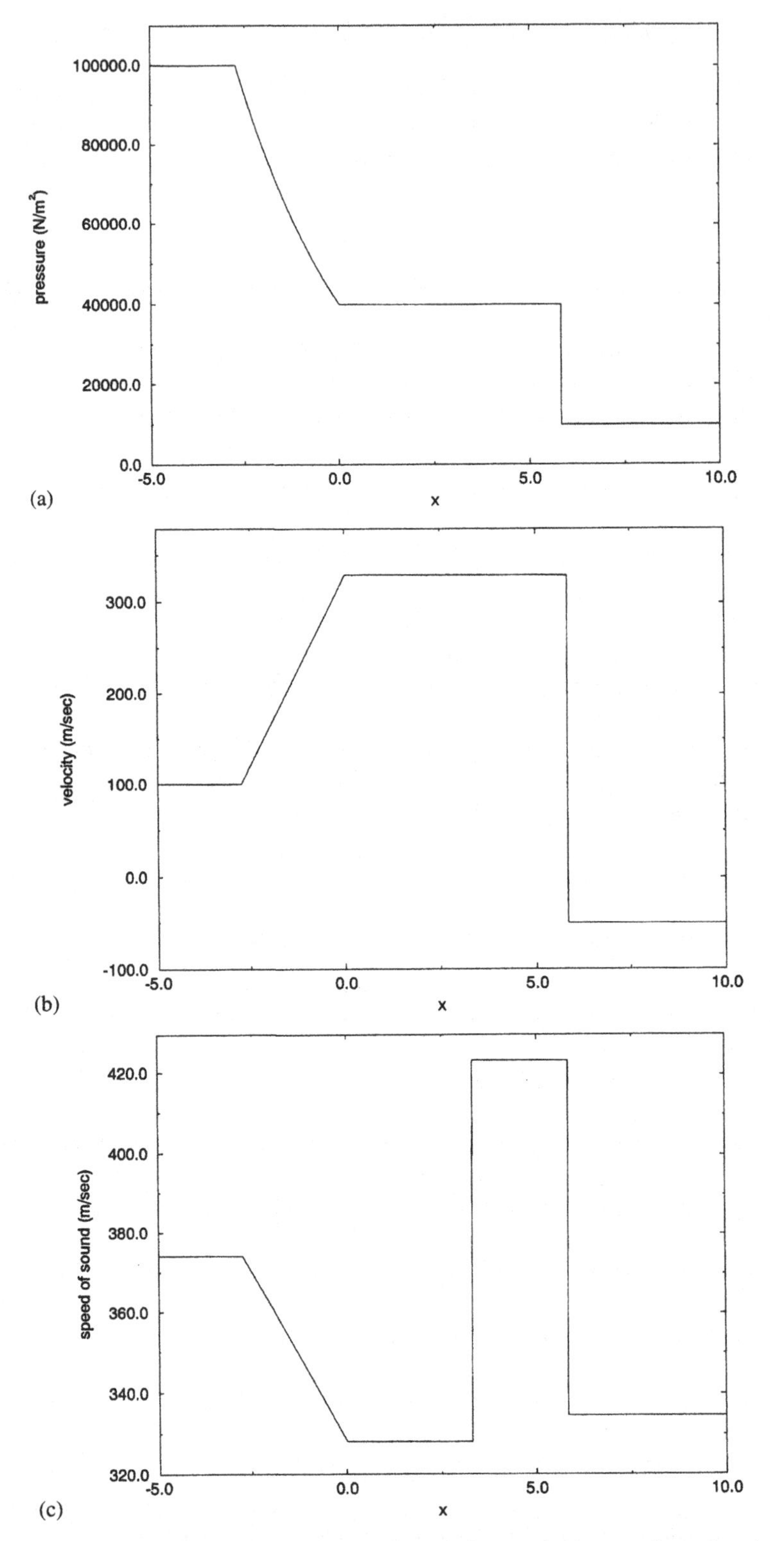

Figure 5.2 The solution to the Riemann problem for the Euler equations given in Example 5.1.

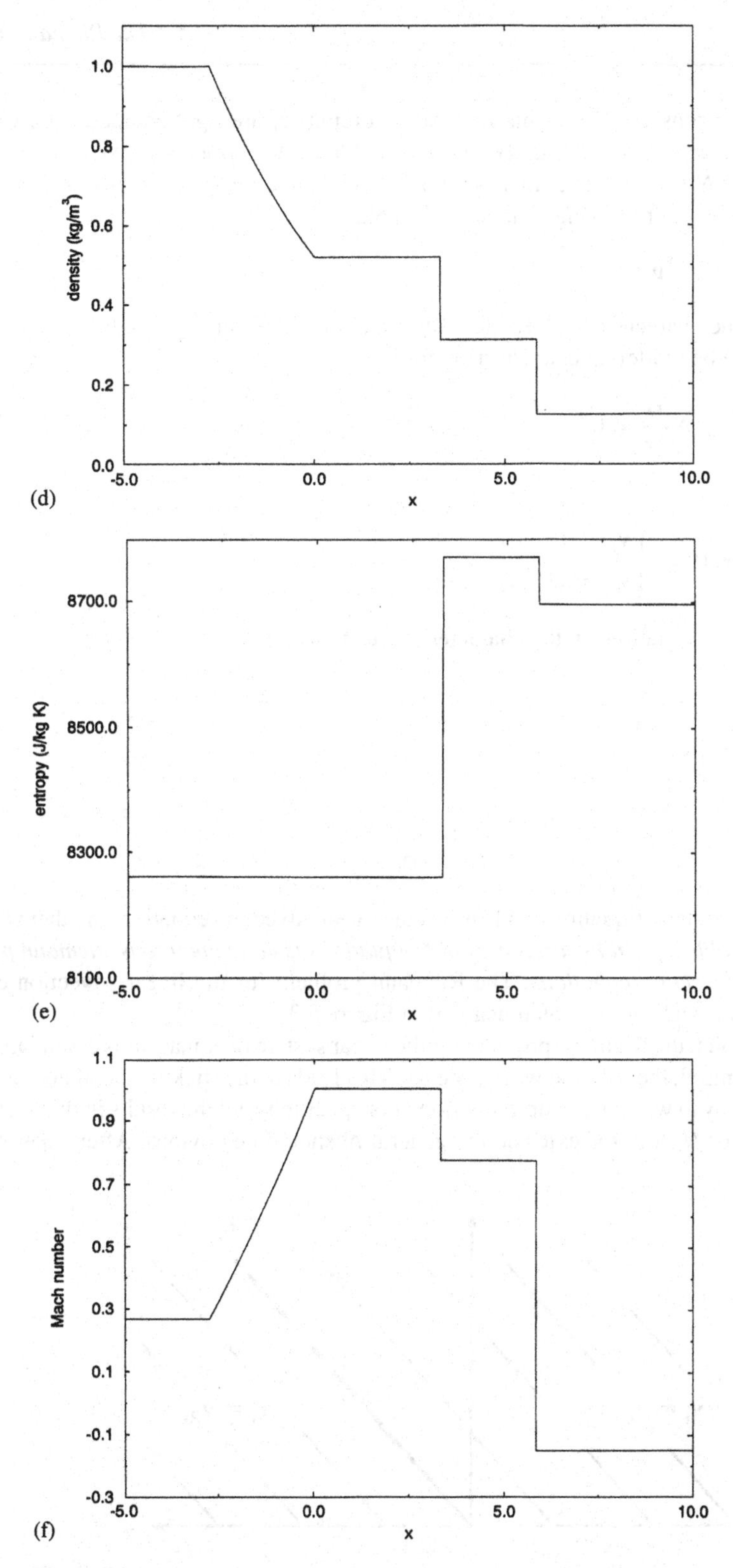

Figure 5.2 (*cont.*)

where Q is a constant $N \times N$ matrix whose columns $\mathbf{r}_i$ are right characteristic vectors of A, Q^{-1} is a constant $N \times N$ matrix whose rows $\mathbf{l}_i$ are left characteristic vectors of A, and Λ is a constant $N \times N$ diagonal matrix whose diagonal elements λ_i are characteristic values of A. Consider the following change of variables:

$$\mathbf{v} = Q^{-1}\mathbf{u}, \tag{5.9}$$

which are the characteristic variables. By Equations (3.9) and (3.10), the linear Riemann problem can be written in characteristic form as

$$\frac{\partial \mathbf{v}}{\partial t} + \Lambda \frac{\partial \mathbf{v}}{\partial x} = 0, \tag{5.10}$$

where

$$\mathbf{v}(x,0) = \begin{cases} \mathbf{v}_L = Q^{-1}\mathbf{u}_L & x < 0, \\ \mathbf{v}_R = Q^{-1}\mathbf{u}_R & x > 0. \end{cases}$$

The individual equations in the characteristic form are as follows:

$$\frac{\partial v_i}{\partial t} + \lambda_i \frac{\partial v_i}{\partial x} = 0, \tag{5.11}$$

where

$$v_i(x,0) = \begin{cases} v_{Li} = \mathbf{l}_i \cdot \mathbf{u}_L & x < 0, \\ v_{Ri} = \mathbf{l}_i \cdot \mathbf{u}_R & x > 0. \end{cases}$$

Since λ_i is constant, Equation (5.11) is just the linear advection equation. In other words, *the Riemann problem for a linear system of N equations is equivalent to N Riemann problems for linear advection equations.* The Riemann problem for the linear advection equation (5.11) has a trivial solution, as illustrated in Figure 5.3.

At this point, the Riemann problem for the linear system of equations is essentially solved and, as promised, the solution was very easy. This leads to the sticky issue of notation. What is the best way to write our simple solution? For specificity, all the results in this section will be written for $N = 3$; the extension to general N should be obvious. After superimposing

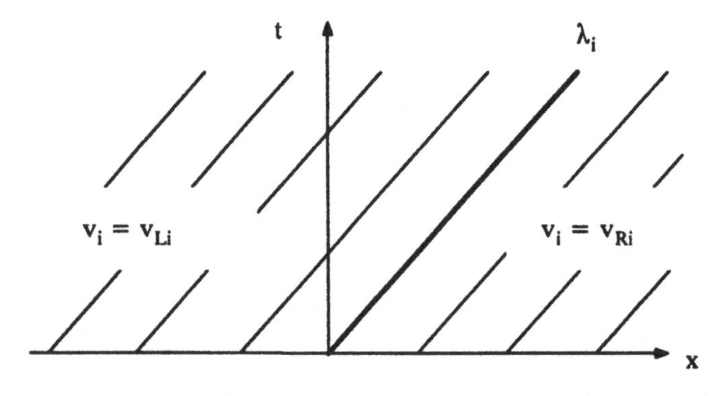

Figure 5.3 The solution to the Riemann problem for the linear advection equation.

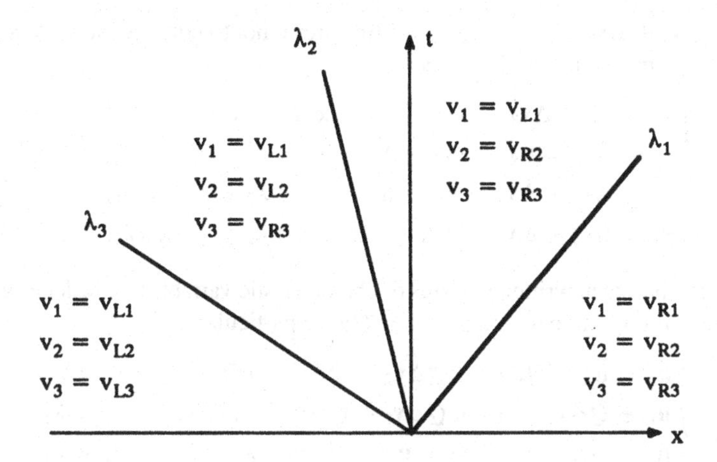

Figure 5.4 The solution to the Riemann problem for a linear system of three partial differential equations.

solutions to three linear advection equations, the solution to the characteristic form of the linear system of equations can be written as:

$$\mathbf{v}(x,t) = \mathbf{v}\left(\frac{x}{t}\right) = \begin{cases} (v_{L1}\ v_{L2}\ v_{L3})^T & x/t < \lambda_3, \\ (v_{L1}\ v_{L2}\ v_{R3})^T & \lambda_3 < x/t < \lambda_2, \\ (v_{L1}\ v_{R2}\ v_{R3})^T & \lambda_2 < x/t < \lambda_1, \\ (v_{R1}\ v_{R2}\ v_{R3})^T & x/t > \lambda_1. \end{cases} \tag{5.12}$$

The transposes T in Equation (5.12) serve simply to covert row vectors to column vectors since, in the standard notation of this book, all vectors are column vectors. The solution given by Equation (5.12) is illustrated in Figure 5.4. Figure 5.4 is obtained by overlaying three solutions of the type illustrated in Figure 5.3, just like overlaying transparencies.

Although Equation (5.12) is a perfectly fine way to write the solution to the Riemann problem for a linear system of equations, there are many other equally fine ways. In particular, let us rewrite Equation (5.12) to reflect the effects of each wave. Let $\Delta\mathbf{u} = \mathbf{u}_R - \mathbf{u}_L$ and $\Delta\mathbf{v} = \mathbf{v}_R - \mathbf{v}_L$. Notice that

$$\Delta\mathbf{v} = Q^{-1}\Delta\mathbf{u}. \tag{5.13}$$

Let $\Delta v_i = v_{Ri} - v_{Li}$ be the jump in the ith characteristic variable. In this context, Δv_i is sometimes called the *strength* or *amplitude* of the ith wave. In other words, the ith wave creates a jump Δv_i in the ith characteristic variable, leaving the other characteristic variables unchanged. The net effect of all three waves is a jump $\Delta\mathbf{v} = \mathbf{v}_R - \mathbf{v}_L$ in the characteristic variables or a jump $\Delta\mathbf{u} = \mathbf{u}_R - \mathbf{u}_L$ in the conservative variables. Let

♦ $$\Delta v_i = \mathbf{l}_i \cdot \Delta\mathbf{u}, \tag{5.14}$$

where $\mathbf{l}_i$ is the ith row of Q^{-1}. Also let

$$\Delta\mathbf{v}_1 = \begin{bmatrix} \Delta v_1 \\ 0 \\ 0 \end{bmatrix}, \quad \Delta\mathbf{v}_2 = \begin{bmatrix} 0 \\ \Delta v_2 \\ 0 \end{bmatrix}, \quad \Delta\mathbf{v}_3 = \begin{bmatrix} 0 \\ 0 \\ \Delta v_3 \end{bmatrix}.$$

Notice that $\Delta\mathbf{v}_1 + \Delta\mathbf{v}_2 + \Delta\mathbf{v}_3 = \Delta\mathbf{v} = \mathbf{v}_R - \mathbf{v}_L$. Be careful not to confuse the scalars Δv_i with the vectors $\Delta\mathbf{v}_i$. Then solution (5.12) becomes

$$\mathbf{v}\left(\frac{x}{t}\right) = \begin{cases} \mathbf{v}_L = \mathbf{v}_R - \Delta\mathbf{v}_3 - \Delta\mathbf{v}_2 - \Delta\mathbf{v}_1 & x/t < \lambda_3 < \lambda_2 < \lambda_1, \\ \mathbf{v}_L + \Delta\mathbf{v}_3 = \mathbf{v}_R - \Delta\mathbf{v}_2 - \Delta\mathbf{v}_1 & \lambda_3 < x/t < \lambda_2 < \lambda_1, \\ \mathbf{v}_L + \Delta\mathbf{v}_3 + \Delta\mathbf{v}_2 = \mathbf{v}_R - \Delta\mathbf{v}_1 & \lambda_3 < \lambda_2 < x/t < \lambda_1, \\ \mathbf{v}_L + \Delta\mathbf{v}_3 + \Delta\mathbf{v}_2 + \Delta\mathbf{v}_1 = \mathbf{v}_R & \lambda_3 < \lambda_2 < \lambda_1 < x/t. \end{cases} \tag{5.15}$$

So far, the solution has been written in terms of characteristic variables. Now let us write the solution in terms of the original variables $\mathbf{u} = Q\mathbf{v}$. In particular

$$\mathbf{u}\left(\frac{x}{t}\right) = \begin{cases} \mathbf{u}_L = \mathbf{u}_R - Q\Delta\mathbf{v}_3 - Q\Delta\mathbf{v}_2 - Q\Delta\mathbf{v}_1 & x/t < \lambda_3 < \lambda_2 < \lambda_1, \\ \mathbf{u}_L + Q\Delta\mathbf{v}_3 = \mathbf{u}_R - Q\Delta\mathbf{v}_2 - Q\Delta\mathbf{v}_1 & \lambda_3 < x/t < \lambda_2 < \lambda_1, \\ \mathbf{u}_L + Q\Delta\mathbf{v}_3 + Q\Delta\mathbf{v}_2 = \mathbf{u}_R - Q\Delta\mathbf{v}_1 & \lambda_3 < \lambda_2 < x/t < \lambda_1, \\ \mathbf{u}_L + Q\Delta\mathbf{v}_3 + Q\Delta\mathbf{v}_2 + Q\Delta\mathbf{v}_1 = \mathbf{u}_R & \lambda_3 < \lambda_2 < \lambda_1 < x/t. \end{cases} \tag{5.16}$$

But as the reader can easily show

$$Q\Delta\mathbf{v}_i = \mathbf{r}_i \Delta v_i, \tag{5.17}$$

where $\mathbf{r}_i$ is the ith column of Q. Then

♦ $$\mathbf{u}\left(\frac{x}{t}\right) = \begin{cases} \mathbf{u}_L = \mathbf{u}_R - \mathbf{r}_3\Delta v_3 - \mathbf{r}_2\Delta v_2 - \mathbf{r}_1\Delta v_1 & x/t < \lambda_3 < \lambda_2 < \lambda_1, \\ \mathbf{u}_L + \mathbf{r}_3\Delta v_3 = \mathbf{u}_R - \mathbf{r}_2\Delta v_2 - \mathbf{r}_1\Delta v_1 & \lambda_3 < x/t < \lambda_2 < \lambda_1, \\ \mathbf{u}_L + \mathbf{r}_3\Delta v_3 + \mathbf{r}_2\Delta v_2 = \mathbf{u}_R - \mathbf{r}_1\Delta v_1 & \lambda_3 < \lambda_2 < x/t < \lambda_1, \\ \mathbf{u}_L + \mathbf{r}_3\Delta v_3 + \mathbf{r}_2\Delta v_2 + \mathbf{r}_1\Delta v_1 = \mathbf{u}_R & \lambda_3 < \lambda_2 < \lambda_1 < x/t. \end{cases} \tag{5.18}$$

The solution has now been written in three different ways, once in terms of the conserved variables and twice in terms of the characteristic variables. Many numerical methods do not use the solution directly. Instead, many numerical methods use only the flux at $x = 0$. In this case, the flux function is $\mathbf{f}(\mathbf{u}) = A\mathbf{u}$. Thus, for future reference, let us compute the solution for $A\mathbf{u}(0)$. Again, there are many ways to write the solution for $A\mathbf{u}(0)$. First, multiply Equation (5.18) by A and recall that $\mathbf{r}_i$ is a right characteristic vector of A to obtain

$$A\mathbf{u}\left(\frac{x}{t}\right) = \begin{cases} A\mathbf{u}_L = A\mathbf{u}_R - \mathbf{r}_3\lambda_3\Delta v_3 - \mathbf{r}_2\lambda_2\Delta v_2 - \mathbf{r}_1\lambda_1\Delta v_1 & x/t < \lambda_3 < \lambda_2 < \lambda_1, \\ A\mathbf{u}_L + \mathbf{r}_3\lambda_3\Delta v_3 = A\mathbf{u}_R - \mathbf{r}_2\lambda_2\Delta v_2 - \mathbf{r}_1\lambda_1\Delta v_1 & \lambda_3 < x/t < \lambda_2 < \lambda_1, \\ A\mathbf{u}_L + \mathbf{r}_3\lambda_3\Delta v_3 + \mathbf{r}_2\lambda_2\Delta v_2 = A\mathbf{u}_R - \mathbf{r}_1\lambda_1\Delta v_1 & \lambda_3 < \lambda_2 < x/t < \lambda_1, \\ A\mathbf{u}_L + \mathbf{r}_3\lambda_3\Delta v_3 + \mathbf{r}_2\lambda_2\Delta v_2 + \mathbf{r}_1\lambda_1\Delta v_1 = A\mathbf{u}_R & \lambda_3 < \lambda_2 < \lambda_1 < x/t. \end{cases}$$

Then

$$A\mathbf{u}(0) = \begin{cases} A\mathbf{u}_L = A\mathbf{u}_R - \mathbf{r}_3\lambda_3\Delta v_3 - \mathbf{r}_2\lambda_2\Delta v_2 - \mathbf{r}_1\lambda_1\Delta v_1 & 0 < \lambda_3 < \lambda_2 < \lambda_1, \\ A\mathbf{u}_L + \mathbf{r}_3\lambda_3\Delta v_3 = A\mathbf{u}_R - \mathbf{r}_2\lambda_2\Delta v_2 - \mathbf{r}_1\lambda_1\Delta v_1 & \lambda_3 < 0 < \lambda_2 < \lambda_1, \\ A\mathbf{u}_L + \mathbf{r}_3\lambda_3\Delta v_3 + \mathbf{r}_2\lambda_2\Delta v_2 = A\mathbf{u}_R - \mathbf{r}_1\lambda_1\Delta v_1 & \lambda_3 < \lambda_2 < 0 < \lambda_1, \\ A\mathbf{u}_L + \mathbf{r}_3\lambda_3\Delta v_3 + \mathbf{r}_2\lambda_2\Delta v_2 + \mathbf{r}_1\lambda_1\Delta v_1 = A\mathbf{u}_R & \lambda_3 < \lambda_2 < \lambda_1 < 0. \end{cases}$$

Although this is a perfectly fine way to write $A\mathbf{u}(0)$, the expression is a bit long. It can be compacted by introducing special notation. In particular, let

$$\lambda_i^- = \min(0, \lambda_i), \tag{5.19}$$

$$\lambda_i^+ = \max(0, \lambda_i). \tag{5.20}$$

Then

$$A\mathbf{u}(0) = A\mathbf{u}_L + \sum_{i=1}^{3} \mathbf{r}_i \lambda_i^- \Delta v_i, \tag{5.21a}$$

$$A\mathbf{u}(0) = A\mathbf{u}_R - \sum_{i=1}^{3} \mathbf{r}_i \lambda_i^+ \Delta v_i. \tag{5.21b}$$

Averaging Equations (5.21a) and (5.21b) yields

$$A\mathbf{u}(0) = \frac{1}{2}A(\mathbf{u}_R + \mathbf{u}_L) - \frac{1}{2}\sum_{i=1}^{3} \mathbf{r}_i |\lambda_i| \Delta v_i. \tag{5.21c}$$

Keep in mind that Equations (5.21a), (5.21b), and (5.21c) are all the same result written in different ways.

Equation (5.21) is yet another perfectly fine way to write $A\mathbf{u}(0)$. However, sometimes it is more elegant to write $A\mathbf{u}(0)$ in terms of matrices, using a special matrix notation. In fact, the sums in Equation (5.21) can be viewed as matrix–vector products. Suppose Λ is a diagonal matrix with diagonal elements λ_i. Then Λ^+ is a diagonal matrix with diagonal elements $\lambda_i^+ = \max(0, \lambda_i)$, Λ^- is a diagonal matrix with diagonal elements $\lambda_i^- = \min(0, \lambda_i)$, and $|\Lambda|$ is a diagonal matrix with diagonal elements $|\lambda_i|$. Notice that

$$\Lambda^+ + \Lambda^- = \Lambda$$

and

$$\Lambda^+ - \Lambda^- = |\Lambda|.$$

Now suppose that A is any diagonalizable matrix such that

$$\Lambda = Q \Lambda Q^{-1}.$$

Then in standard notation we have

$$A^+ = Q\Lambda^+ Q^{-1}, \tag{5.22}$$

$$A^- = Q\Lambda^- Q^{-1}, \tag{5.23}$$

$$|A| = Q|\Lambda|Q^{-1}. \tag{5.24}$$

Notice that

$$A^+ + A^- = A, \tag{5.25}$$

and

$$A^+ - A^- = |A|. \tag{5.26}$$

In terms of this matrix notation, Equation (5.21) becomes

$$A\mathbf{u}(0) = A\mathbf{u}_L + A^-(\mathbf{u}_R - \mathbf{u}_L), \tag{5.27a}$$

$$A\mathbf{u}(0) = A\mathbf{u}_R - A^+(\mathbf{u}_R - \mathbf{u}_L), \tag{5.27b}$$

$$A\mathbf{u}(0) = \frac{1}{2}A(\mathbf{u}_R + \mathbf{u}_L) - \frac{1}{2}|A|(\mathbf{u}_R - \mathbf{u}_L). \tag{5.27c}$$

5.3 Three-Wave Linear Approximations – Roe's Approximate Riemann Solver for the Euler Equations

The exact solution to the Riemann problem for the Euler equations is too expensive for many applications. But suppose the Euler equations are replaced by a linear system of equations. More specifically, suppose the true nonlinear flux function is replaced by a locally linearized approximate flux function. Then the simple solution from the last section applies. Of course, the whole trick is to choose the linear approximation, which turns out to be a surprising amount of work, although the savings are often worth it. This section describes one possibility; the next section will discuss another.

5.3.1 *Secant Line and Secant Plane Approximations*

Consider any nonlinear scalar function $f(u)$. The two most popular linear approximations of $f(u)$ are tangent line and secant line approximations. The *tangent line* about u_L is defined as follows:

$$f(u) \approx a(u_L)(u - u_L) + f(u_L), \tag{5.28}$$

where $a(u) = f'(u)$. This approximation is best near u_L and gets progressively worse away from u_L. Similarly, the tangent line about u_R is defined as follows:

$$f(u) \approx a(u_R)(u - u_R) + f(u_R), \tag{5.29}$$

This approximation is best near u_R and gets progressively worse away from u_R. Instead of favoring u_L, u_R, or any other point, a linear approximation that clings to the function over the entire interval between u_L and u_R might be more desirable. In particular, the *secant line* is defined as follows:

$$f(u) \approx a_{RL}(u - u_L) + f(u_L), \tag{5.30a}$$

or equivalently,

$$f(u) \approx a_{RL}(u - u_R) + f(u_R), \tag{5.30b}$$

where

$$a_{RL} = \frac{f(u_R) - f(u_L)}{u_R - u_L}. \tag{5.31}$$

Whereas a tangent line is more accurate near a single point, the secant line is more accurate on average over the entire region between u_L and u_R. The mean value theorem implies the following relationship between tangent and secant lines:

$$a_{RL} = a(\xi) \tag{5.32}$$

for some ξ between u_L and u_R. In this sense, the secant line slope a_{RL} is an average tangent line slope $a(u)$, and thus the secant line is an average tangent line.

Now consider any nonlinear *vector* function $\mathbf{f}(\mathbf{u})$. The *tangent plane* approximation about $\mathbf{u}_L$ is defined as follows:

$$\mathbf{f}(\mathbf{u}) \approx A(\mathbf{u}_L)(\mathbf{u} - \mathbf{u}_L) + \mathbf{f}(\mathbf{u}_L), \tag{5.33}$$

where $A = d\mathbf{f}/d\mathbf{u}$ is a Jacobian matrix. (See Section 2.2.1 for a more detailed description of Jacobian matrices.) A *secant plane* is any plane containing the line connecting $\mathbf{u}_L$ and $\mathbf{u}_R$. There are infinitely many such planes. Secant plane approximations are defined as follows:

♦ $$\mathbf{f}(\mathbf{u}) \approx A_{RL}(\mathbf{u} - \mathbf{u}_L) + \mathbf{f}(\mathbf{u}_L), \tag{5.34}$$

where A_{RL} is any matrix such that

♦ $$\mathbf{f}(\mathbf{u}_R) - \mathbf{f}(\mathbf{u}_L) = A_{RL}(\mathbf{u}_R - \mathbf{u}_L). \tag{5.35}$$

If $\mathbf{f}(\mathbf{u})$ is a vector with N components, then A_{RL} is an $N \times N$ matrix with N^2 elements. Then Equation (5.35) consists of N equations in N^2 unknowns. Since there are more unknowns than equations, in general, Equation (5.35) yields infinitely many solutions for A_{RL} and thus, as promised, there are infinitely many secant planes. For example, two possible solutions are

$$A_{RL} = \begin{bmatrix} \frac{f_1(\mathbf{u}_R)-f_1(\mathbf{u}_L)}{u_{1,R}-u_{1,L}} & 0 & 0 \\ 0 & \frac{f_2(\mathbf{u}_R)-f_2(\mathbf{u}_L)}{u_{2,R}-u_{2,L}} & 0 \\ 0 & 0 & \frac{f_3(\mathbf{u}_R)-f_3(\mathbf{u}_L)}{u_{3,R}-u_{3,L}} \end{bmatrix}$$

and

$$A_{RL} = \frac{1}{3}\begin{bmatrix} \frac{f_1(\mathbf{u}_R)-f_1(\mathbf{u}_L)}{u_{1,R}-u_{1,L}} & \frac{f_1(\mathbf{u}_R)-f_1(\mathbf{u}_L)}{u_{2,R}-u_{2,L}} & \frac{f_1(\mathbf{u}_R)-f_1(\mathbf{u}_L)}{u_{3,R}-u_{3,L}} \\ \frac{f_2(\mathbf{u}_R)-f_2(\mathbf{u}_L)}{u_{1,R}-u_{1,L}} & \frac{f_2(\mathbf{u}_R)-f_2(\mathbf{u}_L)}{u_{2,R}-u_{2,L}} & \frac{f_2(\mathbf{u}_R)-f_2(\mathbf{u}_L)}{u_{3,R}-u_{3,L}} \\ \frac{f_3(\mathbf{u}_R)-f_3(\mathbf{u}_L)}{u_{1,R}-u_{1,L}} & \frac{f_3(\mathbf{u}_R)-f_3(\mathbf{u}_L)}{u_{2,R}-u_{2,L}} & \frac{f_3(\mathbf{u}_R)-f_3(\mathbf{u}_L)}{u_{3,R}-u_{3,L}} \end{bmatrix}.$$

In the scalar case, secant lines are average tangent lines. Hence, secant line slopes are average tangent line slopes. Then, in the vector case, suppose we require that secant planes be average tangent planes. Then, by analogy with Equation (5.32), A_{RL} should be an average of $A(u)$. This is certainly not true for either of the preceding matrices. Instead choose A_{RL} as follows:

♦ $$A_{RL} = A(\mathbf{u}_{RL}), \tag{5.36}$$

where $\mathbf{u}_{RL}$ is an average of $\mathbf{u}_L$ and $\mathbf{u}_R$. Now there are only N unknowns – the components of $\mathbf{u}_{RL}$. Then Equation (5.35) consists of N equations in N unknowns. With an equal number of equations and unknowns, Equation (5.35) now yields a unique solution. Besides yielding a unique solution, the choice (5.36) has another major advantage: Any expressions based on $A(\mathbf{u})$ also hold for $A_{RL} = A(\mathbf{u}_{RL})$ after substituting $\mathbf{u}_{RL}$ for $\mathbf{u}$. For example, the characteristic values of A are u, $u + a$, and $u - a$ while the characteristic values of A_{RL} are u_{RL}, $u_{RL} + a_{RL}$, and $u_{RL} - a_{RL}$.

Equation (5.36) is certainly not the only reasonable choice. Secant planes are used in numerous different contexts with numerous different slopes A_{RL}. For example, in the context of root solvers, the secant method uses secant lines or secant planes. Although there is a unique secant method for scalar functions based on the unique secant line, there are infinitely many secant methods for vector functions, depending on the choice of the secant plane (i.e. depending on the choice of A_{RL}). The most popular vector version of the secant method is called *Broyden's method.* As with all secant methods, Broyden's is an iterative method that requires an iterative sequence of secant planes. In Broyden's method, A_{RL} is first chosen equal to the Jacobian matrix A evaluated on an arbitrary initial guess. Thus the secant plane approximation begins as a tangent plane approximation. After the initial setting, every new A_{RL} is computed from an old A_{RL} – in particular, at each iteration, the new A_{RL} is chosen as close as possible to the old A_{RL} subject to condition (5.35); see Dennis and Schnabel (1983) for details. Although other applications use other secant planes, Equation (5.36) is the best choice for us. This section has concerned only linear approximations to functions. Part II will consider all sorts of polynomial approximations, including lines, quadratics, and cubics, although it will deal with mainly scalar functions.

5.3.2 *Roe Averages*

The last subsection described secant plane approximations to vector functions. This subsection derives a specific secant plane approximation to the flux vector found in the Euler equations or, in other words, this subsection derives an average Jacobian matrix A_{RL}. From Sections 2.1.6 and 2.2.1, recall that

$$\mathbf{u} = \begin{bmatrix} \rho \\ \rho u \\ \rho e_T \end{bmatrix}, \tag{2.18}$$

$$\mathbf{f} = \begin{bmatrix} \rho u \\ \rho u^2 + p \\ \rho h_T u \end{bmatrix}, \tag{2.19}$$

$$A = \begin{bmatrix} 0 & 1 & 0 \\ \frac{\gamma-3}{2}u^2 & (3-\gamma)u & \gamma-1 \\ -uh_T + \frac{1}{2}(\gamma-1)u^3 & h_T - (\gamma-1)u^2 & \gamma u \end{bmatrix}. \tag{2.31}$$

For this application, it is convenient to eliminate p and e_T in favor of h_T. Using the perfect gas relations found in Section 2.1.4, Equations (2.18) and (2.19) become

$$\mathbf{u} = \begin{bmatrix} \rho \\ \rho u \\ \frac{1}{\gamma}\rho h_T + \frac{1}{2\gamma}(\gamma-1)\rho u^2 \end{bmatrix}, \tag{5.37}$$

$$\mathbf{f} = \begin{bmatrix} \rho u \\ \frac{\gamma-1}{\gamma}\rho h_T + \frac{\gamma+1}{2\gamma}\rho u^2 \\ \rho u h_T \end{bmatrix}. \tag{5.38}$$

Assume that $A_{RL} = A(\mathbf{u}_{RL})$, as in Equation (5.36). Then Equation (2.31) implies

$$\blacklozenge \quad A_{RL} = \begin{bmatrix} 0 & 1 & 0 \\ \frac{\gamma-3}{2}u_{RL}^2 & (3-\gamma)u_{RL} & \gamma-1 \\ -u_{RL}h_{RL} + \frac{1}{2}(\gamma-1)u_{RL}^3 & h_{RL} - (\gamma-1)u_{RL}^2 & \gamma u_{RL} \end{bmatrix}. \tag{5.39}$$

This is called the *Roe-average Jacobian matrix.*

Let us solve for u_{RL} and h_{RL}. As a convenient notation, let $\Delta u = u_R - u_L$, $\Delta\rho = \rho_R - \rho_L$, $\Delta h_T = h_{T,R} - h_{T,L}$, and so forth. Then Equation (5.35) becomes

$$\begin{bmatrix} \Delta(\rho u) \\ \frac{\gamma-1}{\gamma}\Delta(\rho h_T) + \frac{\gamma+1}{2\gamma}\Delta(\rho u^2) \\ \Delta(\rho u h_T) \end{bmatrix}$$

$$= \begin{bmatrix} 0 & 1 & 0 \\ \frac{\gamma-3}{2}u_{RL}^2 & (3-\gamma)u_{RL} & (\gamma-1) \\ -u_{RL}h_{RL} + \frac{1}{2}(\gamma-1)u_{RL}^3 & h_{RL_T} - (\gamma-1)u_{RL}^2 & \gamma u_{RL} \end{bmatrix}$$

$$\times \begin{bmatrix} \Delta\rho \\ \Delta(\rho u) \\ \frac{1}{\gamma}\Delta(\rho h_T) + \frac{1}{2\gamma}(\gamma-1)\Delta(\rho u^2) \end{bmatrix}. \tag{5.40}$$

The first equation in (5.40) is the trivial identity $\Delta(\rho u) = \Delta(\rho u)$. Thus system (5.40) really consists of only two equations in two unknowns: u_{RL} and h_{RL}.

The second equation in system (5.40) is

$$\frac{\gamma-3}{2}u_{RL}^2 \cdot \Delta\rho + (3-\gamma)u_{RL} \cdot \Delta\rho u + (\gamma-1) \cdot \frac{1}{\gamma}\Delta\rho h_T$$

$$+ (\gamma-1) \cdot \frac{1}{2\gamma}(\gamma-1)\Delta\rho u^2 = \frac{\gamma-1}{\gamma}\Delta\rho h_T + \frac{\gamma+1}{2\gamma}\Delta\rho u^2.$$

After a bit of algebra, this reduces to

$$\Delta\rho u_{RL}^2 - 2\Delta(\rho u)u_{RL} + \Delta(\rho u^2) = 0,$$

which is a quadratic. Apply the quadratic formula to get

$$u_{RL} = \frac{\Delta(\rho u) \pm \sqrt{[\Delta(\rho u)]^2 - \Delta\rho \cdot \Delta(\rho u^2)}}{\Delta\rho}.$$

Now invoke the definitions of $\Delta\rho$, $\Delta(\rho u)$, and $\Delta(\rho u^2)$ to obtain

$$u_{RL} = \frac{\rho_R u_R - \rho_L u_L \pm \sqrt{(\rho_R u_R - \rho_L u_L)^2 - (\rho_R - \rho_L)\left(\rho_R u_R^2 - \rho_L u_L^2\right)}}{\rho_R - \rho_L}.$$

Simplification yields

$$u_{RL} = \frac{\rho_R u_R - \rho_L u_L \pm \sqrt{\rho_R \rho_L [u_R^2 - 2u_R u_L + u_L^2]}}{\rho_R - \rho_L}$$

$$= \frac{\rho_R u_R - \rho_L u_L \pm \sqrt{\rho_R \rho_L}(u_R - u_L)}{\rho_R - \rho_L}$$

$$= \frac{(\rho_R \pm \sqrt{\rho_R \rho_L})u_R - (\rho_L \pm \sqrt{\rho_R \rho_L})u_L}{\rho_R - \rho_L}$$

$$= \frac{\sqrt{\rho_R}(\sqrt{\rho_R} \pm \sqrt{\rho_L})u_R - \sqrt{\rho_L}(\sqrt{\rho_L} \pm \sqrt{\rho_R})u_L}{(\sqrt{\rho_R} + \sqrt{\rho_L})(\sqrt{\rho_R} - \sqrt{\rho_L})}.$$

The positive root yields an unphysical solution. Choose the negative root to find

♦ $$u_{RL} = \frac{\sqrt{\rho_R}\, u_R + \sqrt{\rho_L}\, u_L}{\sqrt{\rho_R} + \sqrt{\rho_L}}. \tag{5.41}$$

This is called the *Roe-average velocity.* Alternatively,

$$u_{RL} = \theta u_L + (1 - \theta) u_R, \tag{5.42}$$

where

$$\theta = \frac{\sqrt{\rho_R}}{\sqrt{\rho_R} + \sqrt{\rho_L}}.$$

Notice that $0 \le \theta \le 1$. Then u_{RL} is somewhere between u_L and u_R. If $0 \le \theta \le 1$ then averages such as (5.42) are called *linear averages, linear interpolations,* or *convex linear combinations.*

The third equation in system (5.40) is

$$\left(-u_{RL} h_{RL} + \frac{1}{2}(\gamma - 1) u_{RL}^3\right) \cdot \Delta\rho + \left(h_{RL} - (\gamma - 1) u_{RL}^2\right) \cdot \Delta\rho u$$
$$+ \gamma u_{RL} \cdot \frac{1}{\gamma} \Delta\rho h_T + \gamma u_{RL} \cdot \frac{1}{2\gamma}(\gamma - 1)\Delta\rho u^2 = \Delta(\rho u h_T).$$

Solving for h_{RL} yields

$$h_{RL} = \frac{\Delta(\rho u h_T) - u_{RL}\Delta(\rho h_T) - \frac{1}{2}(\gamma - 1) u_{RL}^3 \Delta\rho}{\Delta(\rho u) - u_{RL}\Delta\rho}$$
$$+ \frac{(\gamma - 1) u_{RL}^2 \Delta\rho u - \frac{1}{2}(\gamma - 1) u_{RL} \Delta(\rho u^2)}{\Delta(\rho u) - u_{RL}\Delta\rho}$$

or

$$h_{RL} = \frac{[\Delta(\rho u h_T) - u_{RL}\Delta(\rho h_T)] + \frac{1}{2}(\gamma - 1) u_{RL}^2 [\Delta(\rho u) - u_{RL}\Delta\rho]}{\Delta(\rho u) - u_{RL}\Delta\rho}$$
$$- \frac{\frac{1}{2}(\gamma - 1) u_{RL} [\Delta(\rho u^2) - u_{RL}\Delta(\rho u)]}{\Delta(\rho u) - u_{RL}\Delta\rho}. \tag{5.43}$$

For any flow variable v, Equation (5.41) implies

$$\Delta(\rho u v) - u_{RL}\Delta(\rho v) = \sqrt{\rho_L \rho_R} \cdot \Delta u \cdot \frac{\sqrt{\rho_R}\, v_R + \sqrt{\rho_R}\, \rho_L}{\sqrt{\rho_R} + \sqrt{\rho_L}}. \tag{5.44}$$

In particular, if $v = 1$,

$$\Delta(\rho v) - u_{RL}\Delta\rho = \sqrt{\rho_R \rho_L}\, \Delta u; \tag{5.45}$$

if $v = u$,

$$\Delta(\rho u^2) - u_{RL}\Delta(\rho u) = \sqrt{\rho_R \rho_L}\, \Delta u u_{RL}; \tag{5.46}$$

and if $v = h_T$,

$$\Delta(\rho u h_T) - u_{RL}\Delta(\rho h_T) = \sqrt{\rho_R \rho_L} \cdot \Delta u \cdot \frac{\sqrt{\rho_R}\, h_{T,R} + \sqrt{\rho_L}\, h_{T,L}}{\sqrt{\rho_R} + \sqrt{\rho_L}}. \tag{5.47}$$

Substituting the last three results into Equation (5.43) gives

♦ $$h_{RL} = \frac{\sqrt{\rho_R}\, h_{T,R} + \sqrt{\rho_L}\, h_{T,L}}{\sqrt{\rho_R} + \sqrt{\rho_L}}. \tag{5.48}$$

This is called the *Roe-average specific total enthalpy*. Then h_{RL} is somewhere between $h_{T,L}$ and $h_{T,R}$ just as u_{RL} is somewhere between u_L and u_R.

The usual perfect gas relationships hold between the Roe-averaged quantities. For example, Equation (2.13) becomes

$$h_{RL} = \frac{1}{2}u_{RL}^2 + \frac{a_{RL}^2}{\gamma - 1}, \tag{5.49}$$

which implies

$$a_{RL}^2 = \frac{\sqrt{\rho_R}\, a_R^2 + \sqrt{\rho_L}\, a_L^2}{\sqrt{\rho_R} + \sqrt{\rho_L}} + \frac{\gamma - 1}{2} \frac{\sqrt{\rho_R \rho_L}}{(\sqrt{\rho_R} + \sqrt{\rho_L})^2}(u_R - u_L)^2. \tag{5.50}$$

This is sometimes called the *Roe-average speed of sound*. This expression for a_{RL}^2 is certainly not as simple as expression (5.41) for u_{RL} or expression (5.48) for h_{RL}. In particular, a_{RL}^2 is not necessarily between a_L^2 and a_R^2. In fact, the expressions u_{RL} and h_{RL} are uniquely simple – the expressions for all other flow properties are more complicated. Thus it is usually most efficient to calculate u_{RL} and h_{RL} and to derive other averaged quantities as needed from these. Notice that although $p_{RL}/\rho_{RL} = a_{RL}^2/\gamma$ there are no expressions for p_{RL} or ρ_{RL} separately. This is not a serious problem since, no matter how it is written, A_{RL} never depends on p_{RL} and ρ_{RL} except in the ratio p_{RL}/ρ_{RL}. However, having said this, it is convenient and traditional to let

♦ $$\rho_{RL} = \sqrt{\rho_R \rho_L}. \tag{5.51}$$

This is sometimes called the *Roe-average density*. Then $p_{RL} = \rho_{RL} a_{RL}^2/\gamma$. This completes the derivation of the Roe-average matrix A_{RL}.

5.3.3 *Algorithm*

The secant plane approximation to the flux vector in the Euler equations is

$$\mathbf{f}(\mathbf{u}) \approx A_{RL}(\mathbf{u} - \mathbf{u}_L) + \mathbf{f}(\mathbf{u}_L),$$

where A_{RL} is the Roe-average flux Jacobian matrix found in the last subsection; see Equations (5.39), (5.41), and (5.48). Then *Roe's approximate Riemann problem* is

$$\frac{\partial \mathbf{u}}{\partial t} + \frac{\partial}{\partial x}(A_{RL}(\mathbf{u} - \mathbf{u}_L) + \mathbf{f}(\mathbf{u}_L)) = \frac{\partial \mathbf{u}}{\partial t} + A_{RL}\frac{\partial \mathbf{u}}{\partial x} = 0, \tag{5.52}$$

where

$$\mathbf{u}(x,0) = \begin{cases} \mathbf{u}_L & x < 0, \\ \mathbf{u}_R & x > 0. \end{cases}$$

Using the procedure from Section 5.2 to solve Equation (5.52) yields *Roe's approximate Riemann solver*. The following steps summarize Roe's approximate Riemann solver.

Step 1 Given the conserved variables $\mathbf{u} = (\rho, \rho u, \rho e_T)^T$ for the left and right states, compute $(u, p, h_T)^T$ for the left and right states. In particular, let $u = \rho u/\rho$, $p = (\gamma - 1)(\rho e_T - \rho u^2/2)$, and $h_T = (\rho e_T + p)/\rho$.

Step 2 Compute the Roe-averaged quantities as follows:

$$\rho_{RL} = \sqrt{\rho_R \rho_L},$$

$$u_{RL} = \frac{\sqrt{\rho_R}\, u_R + \sqrt{\rho_L}\, u_L}{\sqrt{\rho_R} + \sqrt{\rho_L}},$$

$$h_{RL} = \frac{\sqrt{\rho_R}\, h_{TR} + \sqrt{\rho_L}\, h_{TL}}{\sqrt{\rho_R} + \sqrt{\rho_L}},$$

$$a_{RL} = \sqrt{(\gamma - 1)\left(h_{RL} - \frac{1}{2}u_{RL}^2\right)}.$$

Step 3 Compute the Roe-average wave speeds. By Equation (3.17)

$$\lambda_1 = u_{RL}, \tag{5.53a}$$

$$\lambda_2 = u_{RL} + a_{RL}, \tag{5.53b}$$

$$\lambda_3 = u_{RL} - a_{RL}. \tag{5.53c}$$

Step 4 Compute the wave strengths. By Equation (3.20)

$$\Delta v_1 = \Delta\rho - \frac{\Delta p}{a_{RL}^2}, \tag{5.54a}$$

$$\Delta v_2 = \Delta u + \frac{\Delta p}{\rho_{RL} a_{RL}}, \tag{5.54b}$$

$$\Delta v_3 = \Delta u - \frac{\Delta p}{\rho_{RL} a_{RL}}, \tag{5.54c}$$

where $\Delta\rho = \rho_R - \rho_L$, $\Delta p = p_R - p_L$, and $\Delta u = u_R - u_L$.

Step 5 Compute the right characteristic vectors. By Equation (3.23)

$$\mathbf{r}_1 = \begin{bmatrix} 1 \\ u_{RL} \\ \frac{1}{2}u_{RL}^2 \end{bmatrix}, \tag{5.55a}$$

$$\mathbf{r}_2 = \frac{\rho_{RL}}{2a_{RL}} \begin{bmatrix} 1 \\ u_{RL} + a_{RL} \\ h_{RL} + a_{RL}u_{RL} \end{bmatrix}, \tag{5.55b}$$

$$\mathbf{r}_3 = -\frac{\rho_{RL}}{2a_{RL}} \begin{bmatrix} 1 \\ u_{RL} - a_{RL} \\ h_{RL} - a_{RL}u_{RL} \end{bmatrix}. \tag{5.55c}$$

Step 6 If necessary, compute the solution. Use Equation (5.18) to arrive at the solution from the left or the right, whichever direction has the least number of waves. If $x/t \leq \lambda_3$ then

$$\mathbf{u} = \mathbf{u}_L.$$

If $\lambda_3 \leq x/t \leq \lambda_1$ then

$$\mathbf{u} = \mathbf{u}_L + \mathbf{r}_3 \Delta v_3.$$

Therefore

$$\rho = \rho_L - \frac{\rho_{RL}}{2a_{RL}}\Delta v_3,$$

$$\rho u = \rho_L u_L - \frac{\rho_{RL}}{2a_{RL}}(u_{RL} - a_{RL})\Delta v_3,$$

$$\rho e_T = \rho_L e_{TL} - \frac{\rho_{RL}}{2a_{RL}}(h_{RL} - a_{RL}u_{RL})\Delta v_3.$$

If $\lambda_1 \leq x/t \leq \lambda_2$ then

$$\mathbf{u} = \mathbf{u}_R - \mathbf{r}_2 \Delta v_2.$$

Therefore

$$\rho = \rho_R - \frac{\rho_{RL}}{2a_{RL}}\Delta v_2,$$

$$\rho u = \rho_R u_R - \frac{\rho_{RL}}{2a_{RL}}(u_{RL} + a_{RL})\Delta v_2,$$

$$\rho e_T = \rho_R e_{TR} - \frac{\rho_{RL}}{2a_{RL}}(h_{RL} + a_{RL}u_{RL})\Delta v_2.$$

If $\lambda_2 \leq x/t$ then

$$\mathbf{u} = \mathbf{u}_R.$$

Many applications do not require the solution but instead require only the fluxes at $x = 0$, in which case this step can be safely skipped.

Step 7 If necessary, compute the fluxes at $x = 0$. By Equation (5.21), if $x/t = 0$, then the flux can be written as

$$\blacklozenge \qquad \mathbf{f}(\mathbf{u}(0)) \approx A_{RL}\mathbf{u}(0) = \mathbf{f}(\mathbf{u}_L) + \sum_{i=1}^{3} \mathbf{r}_i \min(0, \lambda_i)\Delta v_i, \tag{5.56a}$$

$$\blacklozenge \qquad \mathbf{f}(\mathbf{u}(0)) \approx A_{RL}\mathbf{u}(0) = \mathbf{f}(\mathbf{u}_R) - \sum_{i=1}^{3} \mathbf{r}_i \max(0, \lambda_i)\Delta v_i, \tag{5.56b}$$

$$\blacklozenge \qquad \mathbf{f}(\mathbf{u}(0)) \approx A_{RL}\mathbf{u}(0) = \frac{1}{2}(\mathbf{f}(\mathbf{u}_R) + \mathbf{f}(\mathbf{u}_L)) - \frac{1}{2}\sum_{i=1}^{3} \mathbf{r}_i |\lambda_i| \Delta v_i. \tag{5.56c}$$

For example, the right-hand side of Equation (5.56a) can be written in more detail as follows:

$$\begin{bmatrix} \rho_L u_L + \min(0, \lambda_1)\Delta v_1 + \frac{\rho_{RL}}{2a_{RL}}(\min(0, \lambda_2)\Delta v_2 - \min(0, \lambda_3)\Delta v_3) \\ \rho_L u_L^2 + p_L + u_{RL}\min(0, \lambda_1)\Delta v_1 + \frac{\rho_{RL}}{2a_{RL}}(\lambda_2 \min(0, \lambda_2)\Delta v_2 - \lambda_3 \min(0, \lambda_3)\Delta v_3) \\ \rho_L h_{TL} u_L + \frac{u_{RL}^2}{2}\min(0, \lambda_1)\Delta v_1 + \frac{\rho_{RL}}{2a_{RL}}((h_{RL} + a_{RL}u_{RL})\min(0, \lambda_2)\Delta v_2 \\ - (h_{RL} - a_{RL}u_{RL})\min(0, \lambda_3)\Delta v_3). \end{bmatrix}$$

The matrix expressions seen in Equation (5.27) can be used instead of these scalar expressions, but the matrix expressions are less computationally efficient.

5.3.4 *Performance*

Roe's approximate Riemann solver yields three equally spaced waves, just like the true Riemann solver. However, unlike the true Riemann solver, all three waves in Roe's approximate Riemann solver have zero spread – in other words, Roe's approximate Riemann solver cannot capture the finite spread of the expansion fan. Roe's approximate Riemann solver is illustrated in Figure 5.5.

Roe's approximate Riemann solver is roughly two and half times less expensive than the exact Riemann solver. While the price is right, how is the accuracy? First, suppose the

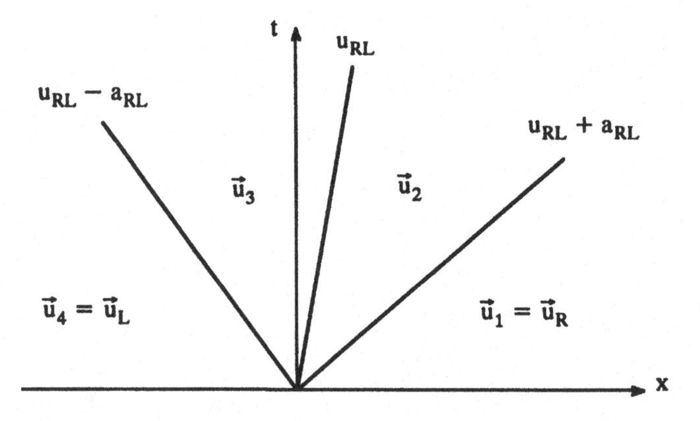

Figure 5.5 Roe's approximate Riemann solver for the Euler equations.

exact Riemann problem yields a single shock or a single contact with speed S. The shock or contact must satisfy the Rankine–Hugoniot relations

$$\mathbf{f}(\mathbf{u}_R) - \mathbf{f}(\mathbf{u}_L) = S(\mathbf{u}_R - \mathbf{u}_L). \tag{2.32}$$

Also, A_{RL} must satisfy the secant plane condition

$$\mathbf{f}(\mathbf{u}_R) - \mathbf{f}(\mathbf{u}_L) = A_{RL}(\mathbf{u}_R - \mathbf{u}_L). \tag{5.35}$$

Combining the last two equations gives

$$A_{RL}(\mathbf{u}_R - \mathbf{u}_L) = S(\mathbf{u}_R - \mathbf{u}_L), \tag{5.57}$$

which implies that S is a characteristic value of A_{RL} and that $\mathbf{u}_R - \mathbf{u}_L$ is a right characteristic vector of A_{RL}. More particularly, suppose $S = \lambda_i$ and $\mathbf{u}_R - \mathbf{u}_L = \mathbf{r}_j$ for any i and j. By Equation (5.14)

$$v_{Ri} - v_{Li} = \mathbf{l}_i \cdot (\mathbf{u}_R - \mathbf{u}_L) = \mathbf{l}_i \cdot \mathbf{r}_j = \delta_{ij} = \begin{cases} 1 & i = j, \\ 0 & i \neq j. \end{cases}$$

In other words, two of the three waves have zero strength. The single nontrivial wave makes the full transition between $\mathbf{u}_L$ and $\mathbf{u}_R$. Furthermore, the single nontrivial wave has speed $\lambda_i = S$. Then *for a single shock or a single contact, Roe's approximate Riemann solver yields the exact solution!* Notice that this result was obtained using only the general definition of a secant plane approximation, Equation (5.35), and did not require any special properties of Roe's secant plane approximation. Thus *for a single shock or a single contact, any secant plane approximation yields the exact solution.*

Except for a single shock or a single contact, Roe's approximate Riemann solver deviates substantially from the true solver. For example, suppose the exact Riemann solver yields a single centered expansion fan. Roe's approximate Riemann solver yields three waves: one strong wave located somewhere inside the true expansion fan and two weak but still significant waves located outside the true expansion fan. The strong wave acts like an expansion shock. An *expansion shock* is a jump discontinuity that satisfies the Rankine–Hugoniot relations; however, unlike a true shock for perfect gases, an expansion shock expands rather than compresses the flow. Although Roe's linear flux function allows expansion shocks, the true nonlinear flux function does not, since expansion shocks violate the second law of thermodynamics for perfect gases. The routine appearance of expansion shocks constitutes a major weakness of linearized approximations such as Roe's.

Example 5.2 Find the solution to Roe's approximate Riemann problem at $t = 0.01$ s if $p_L = 100{,}000\,\text{N/m}^2$, $\rho_L = 1\,\text{kg/m}^3$, $u_L = 100\,\text{m/s}$ and $p_R = 10{,}000\,\text{N/m}^2$, $\rho_R = 0.125\,\text{kg/m}^3$, $u_R = -50\,\text{m/s}$.

Solution Roe's approximate Riemann solver yields three waves. The first wave has speed -357.5 m/s and strength 584.3 m/s; the second wave has speed -10.82 m/s and strength $-0.1261\,\text{kg/m}^3$; the third wave has speed 335.9 m/s and strength -884.3 m/s. Notice that the strength of the "entropy" wave has units of density whereas the strengths of the "acoustic" waves have units of velocity. The solution at $t = 0.01$s is plotted in Figure 5.6. By self-similarity, the solution at all other times is identical except for uniform

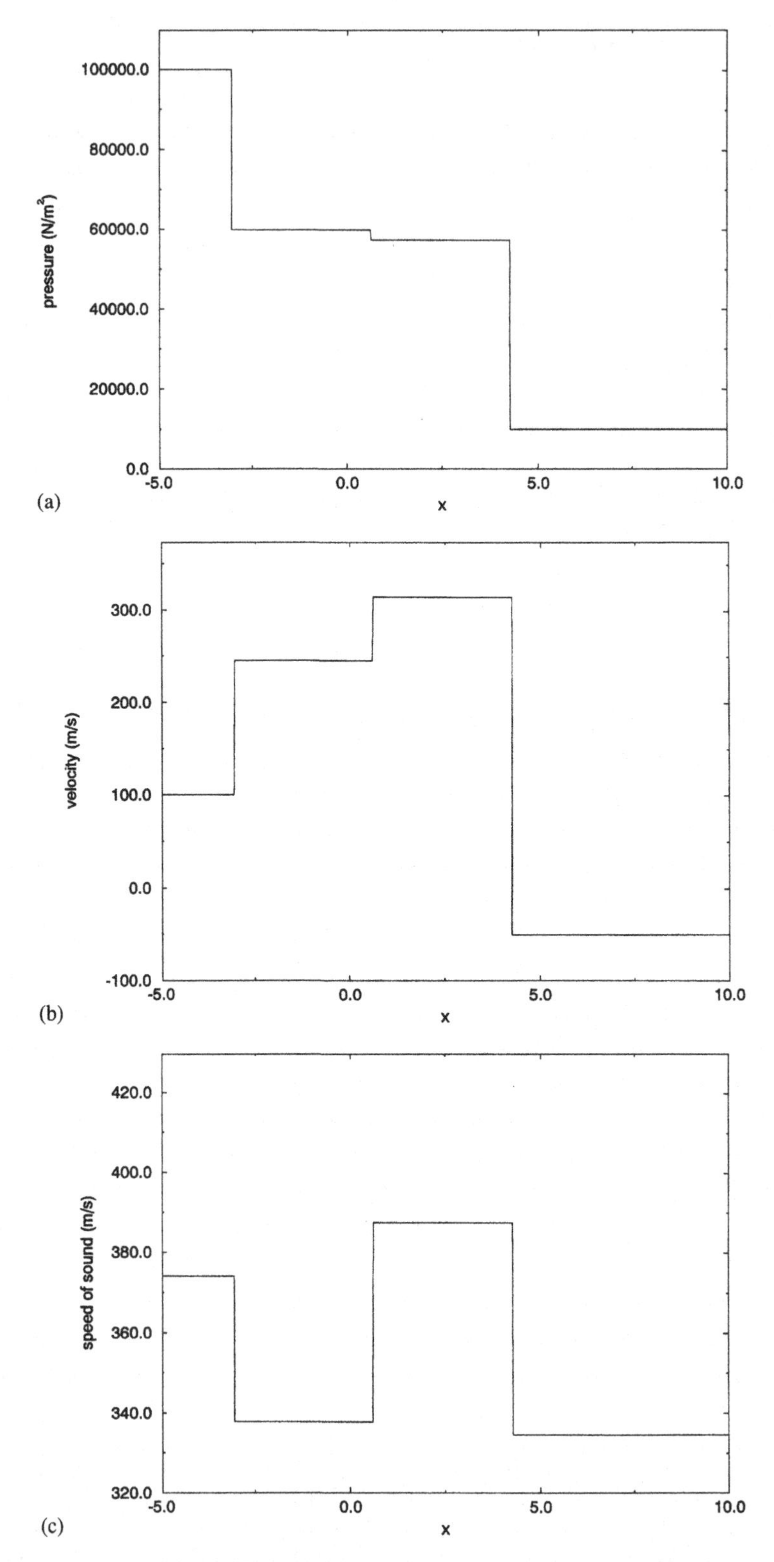

Figure 5.6 Roe's approximate solution to the Riemann problem for the Euler equations given in Example 5.2.

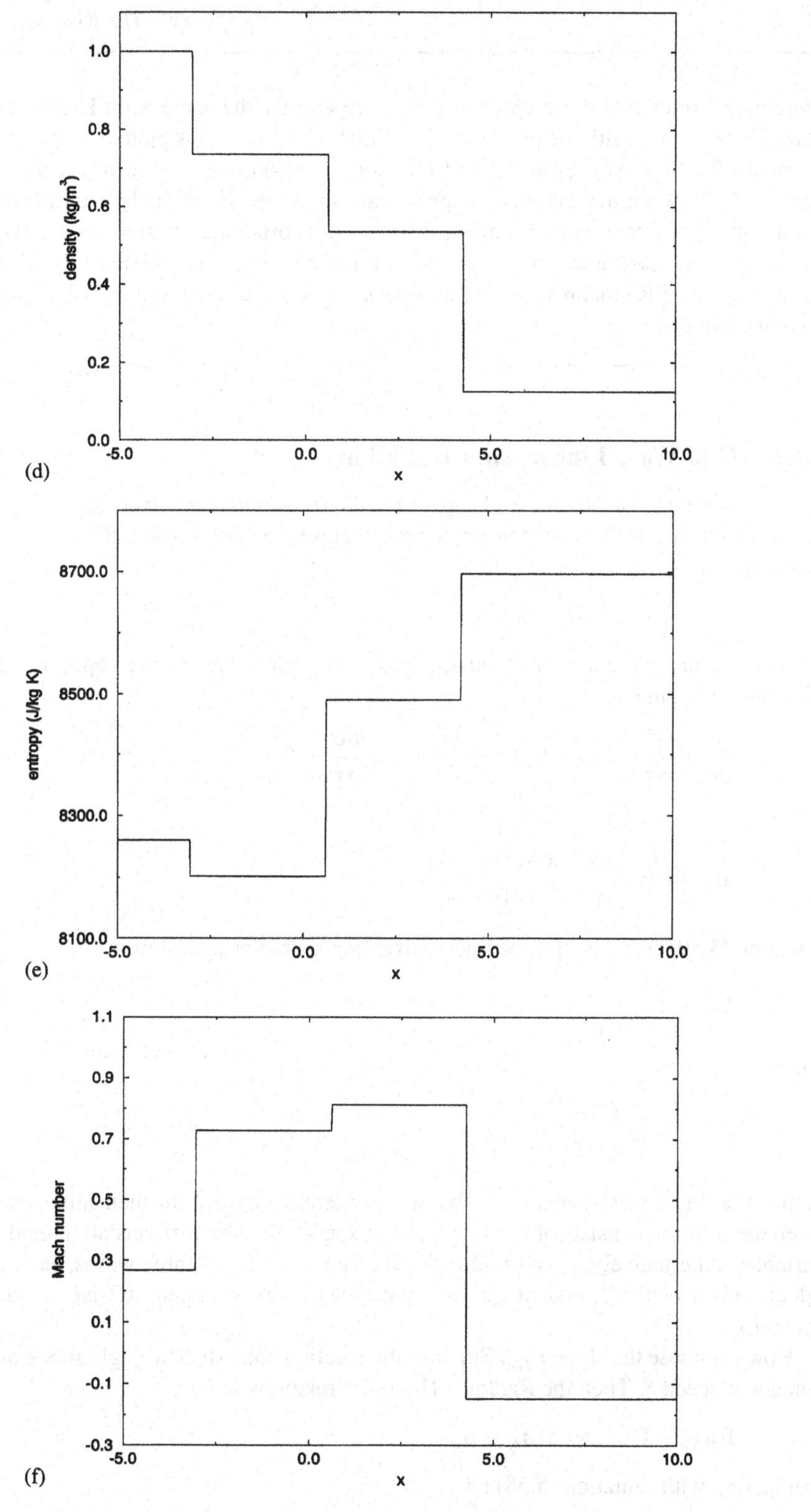

Figure 5.6 (*cont.*)

stretching. Notice that the initial conditions are exactly the same as in Example 5.1. For ease of comparison with the results of Example 5.1, Figure 5.6 is plotted on the same scale as Figure 5.2; however, even without this consideration, there is no mistaking the huge differences between the exact and approximate solutions. Fortunately, the differences are not usually significant when Riemann solvers are incorporated into numerical algorithms. In fact, it is misleading to consider Roe's approximate Riemann solver or, for that matter, any approximate Riemann solver separate from its application in numerical algorithms, as seen in Chapter 18.

5.4 One-Wave Linear Approximations

Although substantially cheaper than the true Riemann solver, Roe's approximate Riemann solver is still too expensive for many applications. Consider the following linear flux function:

$$\mathbf{f}(\mathbf{u}) \approx r_{RL}\mathbf{u} + \mathbf{b}_{RL}, \tag{5.58}$$

where r_{RL} is any constant scalar and $\mathbf{b}_{RL}$ is any constant vector. Then an approximate linear Riemann problem is

$$\frac{\partial \mathbf{u}}{\partial t} + \frac{\partial}{\partial x}(r_{RL}\mathbf{u} + \mathbf{b}_{RL}) = \frac{\partial \mathbf{u}}{\partial t} + r_{RL}\frac{\partial \mathbf{u}}{\partial x} = 0, \tag{5.59}$$

where

$$\mathbf{u}(x, 0) = \begin{cases} \mathbf{u}_L & x < 0, \\ \mathbf{u}_R & x > 0. \end{cases}$$

Equation (5.59) consists of three linear advection problems as follows:

$$\frac{\partial u_i}{\partial t} + r_{RL}\frac{\partial u_i}{\partial x} = 0,$$

where

$$u_i(x, 0) = \begin{cases} u_{iL} & x < 0, \\ u_{iR} & x > 0. \end{cases}$$

Notice that the linear advection problems are identical except for their initial conditions. Then the solution consists of a single wave of speed r_{RL} which affects all three dependent variables. Alternatively, if you prefer, the solution consists of three waves, each of which affects only a single dependent variable, and each of which happens to have the same wave speed r_{RL}.

How to choose the slope r_{RL}? Suppose the solution consists of a single shock or a single contact of speed S. Then the Rankine–Hugoniot relations say

$$\mathbf{f}(\mathbf{u}_R) - \mathbf{f}(\mathbf{u}_L) = S(\mathbf{u}_R - \mathbf{u}_L). \tag{2.32}$$

Comparing with Equation (5.58) let

$$r_{RL} = S. \tag{5.60}$$

Then, for a single shock or a single contact, the one-wave approximate Riemann problem (5.59) yields the exact solution. Hence with the right choice of r_{RL}, the simple one-wave model yields the exact solution in the same cases as Roe's three-wave model, and at a fraction of the cost. If the true solution contains more than a single shock or a single contact, the one-wave linear model is a poor approximation, but then so is Roe's three-wave linear model, albeit to a lesser extent.

Unfortunately, it is difficult to devise a general expression for r_{RL} consistent with Equation (5.60). For a single right-running shock, the shock speed S can be written as

$$S = u_{RL} + a_{RL},$$

where u_{RL} is the Roe-average velocity, a_{RL} is the Roe-average speed of sound, and $u_{RL}+a_{RL}$ is a Roe-average wave speed. Similarly, for a single left-running shock, the shock speed can be written as

$$S = u_{RL} - a_{RL}.$$

Finally, for a single contact, the contact speed can be written as

$$S = u_R = u_L = u_{RL}.$$

Notice that, in all cases, S is a characteristic value of the Roe-average Jacobian matrix A_{RL}. This suggests that, in all cases, r_{RL} should be a Roe-average wave speed or, in other words, a characteristic value of the Roe-average Jacobian matrix A_{RL}. Unfortunately, there is no simple way to choose between the three wave speeds. In practice, r_{RL} is usually set equal to the fastest wave speed. Thus

♦ $$r_{RL} = \rho(A_{RL}), \tag{5.61}$$

where $\rho(A_{RL})$ is the largest characteristic value of A_{RL} in absolute value. For example, if $u_{RL} = -200$ m/s and $a_{RL} = 300$ m/s then $\rho(A_{RL}) = u_{RL} - a_{RL} = -500$ m/s. The quantity $\rho(A_{RL})$ is sometimes called the *spectral radius* of A_{RL}. (Do not confuse the spectral radius $\rho(A)$ with density ρ.) Notice that Equation (5.61) satisfies Equation (5.60) for a single shock but not for a single contact – $\rho(A_{RL}) \neq u_{RL}$ since u_{RL} is always somewhere between $u_{RL} - a_{RL}$ and $u_{RL} + a_{RL}$.

In this context, using Roe averaging to find $\rho(A_{RL})$ is not necessarily worth the expense. Instead, the Roe-average Jacobian matrix can be replaced by any average Jacobian matrix. For example,

$$A_{RL} = A\left(\frac{\mathbf{u}_R + \mathbf{u}_L}{2}\right), \tag{5.62}$$

in which case

$$r_{RL} = \rho(A_{RL}) = u\left(\frac{\mathbf{u}_R + \mathbf{u}_L}{2}\right) \pm a\left(\frac{\mathbf{u}_R + \mathbf{u}_L}{2}\right). \tag{5.63}$$

For another example, if

$$A_{RL} = \frac{A(\mathbf{u}_R) + A(\mathbf{u}_L)}{2}, \tag{5.64}$$

then

$$r_{RL} = \rho(A_{RL}) = \frac{u(\mathbf{u}_R) + u(\mathbf{u}_L)}{2} \pm \frac{a(\mathbf{u}_R) + a(\mathbf{u}_L)}{2}. \tag{5.65}$$

Of course, Equation (5.60) is almost never satisfied when such simple averages replace the Roe average. Hence these simple averages cannot capture a single shock or a single contact exactly. On the positive side, these simple averages are less likely to admit expansion shocks.

5.5 Other Approximate Riemann Solvers

There are many other approximate Riemann solvers besides the two seen above. One of the most popular is *Osher's approximate Riemann solver*. This is a relatively complicated and costly approximate Riemann solver. Whereas the fluxes in Roe's approximate Riemann solver involve terms such as $\mathbf{r}_i \min(0, \lambda_i)$, as seen in Equation (5.56a), the fluxes in Osher's approximate Riemann solver involve terms such as $\int \mathbf{r}_i \min(0, \lambda_i) ds$, where the integration path parameterized by the variable s is specially chosen. In essence, Osher's approximate Riemann solver represents both the shock wave and the expansion wave by smooth waves with finite spread. As its main advantage, this approach captures the finite spread of the true expansion fan and thus prevents expansion shocks; on the other hand, this approach artificially induces shock spreading. In terms of complication and cost, Osher's approximate Riemann solver lies somewhere between Roe's approximate Riemann solver and the exact Riemann solver. See Osher and Solomon (1982) for details.

Harten, Lax, and Van Leer (1983) developed another well-known approximate Riemann solver. Sometimes known as the *HLL approximate Riemann solver*, this approximate Riemann solver is a two-wave linear model. In terms of complication, cost, and construction technique, it lies somewhere between a one-wave linear model and a three-wave linear model. Einfeldt (1988) later modified the HLL approximate Riemann solver in several ways, including modifications to prevent expansion shocks; this is sometimes known as the *HLLE approximate Riemann solver*. For more on HLL and HLLE approximate Riemann solvers see Einfeldt, Munz, Roe, and Sjögreen (1991) and Toro, Spruce, and Spears (1994). Harten derived yet a third approximate Riemann solver, based on a locally quadratic rather than a locally linear approximation to the flux function. Harten's approximate Riemann solver is described later in the book; see Subsection 17.3.3. There are certainly many other approximate Riemann solvers, although most are variants of the approximate Riemann solvers described here.

5.6 The Riemann Problem for Scalar Conservation Laws

Scalar conservation laws were described in Chapter 4. For convex flux functions, the Riemann problem always yields a single wave. In particular, if $a(u_L) > a(u_R)$, the Riemann problem yields a single shock with speed $S = (f(u_R) - f(u_L))/(u_R - u_L)$. If $a(u_L) = a(u_R)$, the Riemann problem yields a single contact with speed $S = a(u_L) = a(u_R)$. Finally, if $a(u_L) < a(u_R)$, the Riemann problem yields a single centered expansion fan whose left-hand side moves with speed $a(u_L)$ and whose right-hand side moves with speed $a(u_R)$.

Nonconvex flux functions introduce numerous complications; see Section 4.9 for several examples. Fortunately, a general implicit solution exists. First, let us find the implicit solution for $u(0, t)$. Consider the triangular region A in the x–t plane bounded by the characteristic

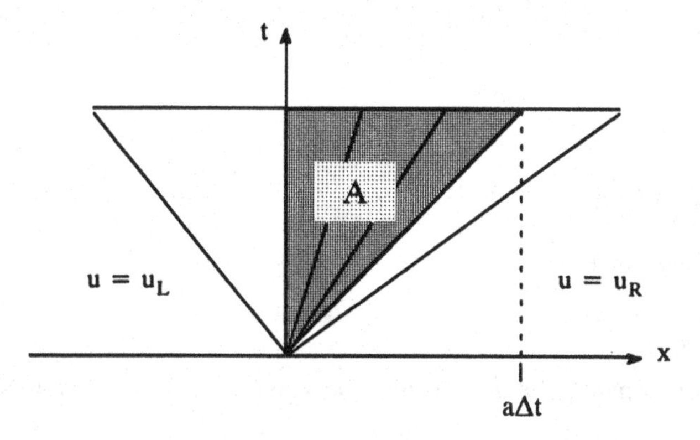

Figure 5.7 An illustration of the region required in the proof of Osher's exact solution of the Riemann problem for scalar conservation.

$x = 0$ on the one side, by the characteristic $x = at$ on the other side, and by $t = const.$ on the top, as illustrated in Figure 5.7. Notice that a and $t = const.$ can be anything you like. If $x = at$ happens to coincide exactly with a shock or a contact, it can be moved infinitesimally to the left or the right onto the immediately adjacent characteristic.

Integrate the scalar conservation law over region A as follows:

$$\int\int_A \left(\frac{\partial u}{\partial t} + \frac{\partial f}{\partial x}\right) dA = 0.$$

This yields

$$\int_{x=0}^{a\Delta t} \int_{t'=x/a}^{t} \frac{\partial u}{\partial t} dt'\, dx + \int_{t'=0}^{t} \int_{x=0}^{at} \frac{\partial f}{\partial x}\, dx\, dt' = 0$$

or

$$\int_{x=0}^{at} [u(x,t) - u(x, x/a)]\, dx + \int_{t'=0}^{t} [f(u(at', t')) - f(u(0, t'))]\, dt = 0.$$

Recall that the solution is always constant along characteristics. More specifically, suppose $u(x, 0) = u_0 = const.$ and $u(x, x/a) = u(at, t) = u_1 = const.$ Then

$$\int_{x=0}^{at} [u(x,t) - u_1]\, dx + \int_{t'=0}^{t} [f(u_1) - f(u_0)]\, dt = 0$$

or

$$f(u_0) - f(u_1) = \frac{1}{t} \int_{x=0}^{at} [u(x,t) - u_1]\, dx.$$

The discussion of waveform preservation, waveform destruction, and waveform creation seen in Section 4.11 implies that scalar conservation laws are always monotonicity preserving. In other words, if the initial conditions are monotone increasing ($u_R > u_L$) then the solution is monotone increasing for all time; and if the initial conditions are monotone decreasing ($u_R < u_L$) then the solution is monotone decreasing for all time. First suppose the solution is monotone increasing. Then $u_1 \geq u(x, t)$ for all $x \leq at$, which implies

$$f(u_0) - f(u_1) = \frac{1}{t} \int_{x=0}^{at} [u(x,t) - u_1]\, dx \leq 0$$

or

$$f(u_0) \le f(u_1).$$

Since u_1 is arbitrary,

$$f(u_0) \le \min_{u_L \le u_1 \le u_R} f(u_1).$$

In fact, a little thought shows that

$$f(u_0) = \min_{u_L \le u_1 \le u_R} f(u_1).$$

Now suppose the solution is monotone decreasing. The same reasoning as before shows that

$$f(u_0) = \max_{u_L \ge u_1 \ge u_R} f(u_1).$$

Dropping the subscripts, we can write the desired result as

♦ $$f(u(0,t)) = \begin{cases} \min\limits_{u_L \le u \le u_R} f(u) & \text{if } u_L < u_R, \\ \max\limits_{u_L \ge u \ge u_R} f(u) & \text{if } u_L > u_R. \end{cases} \tag{5.66}$$

Many numerical methods require only the flux at $x = 0$, as given in this equation.

Now suppose $x/t = s \ne 0$. The following change of variables rotates the line $x = st$ into the line $x = 0$, so that the previous result applies:

$$y = x - st \qquad \tau = t. \tag{5.67}$$

Then

$$\frac{\partial u}{\partial t} = \frac{\partial u}{\partial \tau}\frac{\partial \tau}{\partial t} + \frac{\partial u}{\partial y}\frac{\partial y}{\partial t} = \frac{\partial u}{\partial \tau} - s\frac{\partial u}{\partial y},$$

$$\frac{\partial f}{\partial x} = \frac{\partial f}{\partial \tau}\frac{\partial \tau}{\partial x} + \frac{\partial f}{\partial y}\frac{\partial y}{\partial x} = \frac{\partial f}{\partial y}.$$

Then the scalar conservation law $\partial u/\partial t + \partial f/\partial x = 0$ becomes

$$\frac{\partial u}{\partial \tau} - s\frac{\partial u}{\partial y} + \frac{\partial f}{\partial y} = 0$$

or

$$\frac{\partial u}{\partial \tau} + \frac{\partial}{\partial y}(f - su) = 0.$$

Notice that the change of independent variables does not affect the dependent variable u. Now apply Equation (5.66) to the flux function $f - su$:

$$f(u(y=0,\tau)) - su(y=0,\tau) = \begin{cases} \min\limits_{u_L \le u \le u_R} [f(u) - su] & \text{if } u_L < u_R, \\ \max\limits_{u_L \ge u \ge u_R} [f(u) - su] & \text{if } u_L > u_R. \end{cases}$$

But $y = x - st = 0$ implies $x = st$. Then

$$f(u(st,t)) - su(st,t) = \begin{cases} \min\limits_{u_L \le u \le u_R} [f(u) - su] & \text{if } u_L < u_R, \\ \max\limits_{u_L \ge u \ge u_R} [f(u) - su] & \text{if } u_L > u_R. \end{cases}$$

Then $s = x/t$ implies the final result:

$$\blacklozenge \qquad f(u(x,t)) - \frac{x}{t}u(x,t) = \begin{cases} \min\limits_{u_L \le u \le u_R} \left[f(u) - \frac{x}{t}u\right] & \text{if } u_L < u_R, \\ \max\limits_{u_L \ge u \ge u_R} \left[f(u) - \frac{x}{t}u\right] & \text{if } u_L > u_R. \end{cases} \tag{5.68}$$

The solution is in the implicit form $F(u(x/t), x/t) = 0$ rather than the explicit form $u(x/t) = F(x/t)$. The implicit form can be solved numerically using, for example, the secant method. Alternatively, an explicit solution can be found directly using the techniques seen in the examples of Sections 4.8 and 4.9. Equations (5.66) and (5.68) were first obtained by Osher (1984), although the proof given here is different.

Now consider approximate linearized Riemann solvers for scalar conservation laws. As shown in Section 5.3.1, there is always a unique secant line approximation for a scalar flux function. In particular,

$$f(u) \approx a_{RL}(u - u_L) + f(u_L), \tag{5.30a}$$

where

$$a_{RL} = \frac{f(u_R) - f(u_L)}{u_R - u_L}. \tag{5.31}$$

Then, for example, by Equation (5.66)

$$f(u(0,t)) \approx \begin{cases} \min\limits_{u_L \le u \le u_R} a_{RL}(u - u_L) + f(u_L) & \text{if } u_L < u_R, \\ \max\limits_{u_L \ge u \ge u_R} a_{RL}(u - u_L) + f(u_L) & \text{if } u_L > u_R \end{cases}$$

or

$$\blacklozenge \qquad f(u(0,t)) \approx \begin{cases} \min(f(u_L), f(u_R)) & \text{if } u_L < u_R, \\ \max(f(u_L), f(u_R)) & \text{if } u_L > u_R. \end{cases} \tag{5.69}$$

Many numerical methods require only the flux at $x = 0$, as given in this equation.

References

Anderson, J. D. 1990. *Modern Compressible Flow with Historical Perspective*, 2nd ed., New York: McGraw-Hill, Section 7.8.

Dennis, J. E., and Schnabel, R. B. 1983. *Numerical Methods for Unconstrained Optimization and Nonlinear Equations*, Englewood Cliffs, NJ: Prentice-Hall, Chapter 8.

Einfeldt, B. 1988. "On Godunov-Type Methods for Gas Dynamics," *SIAM Journal on Numerical Analysis*, 25: 294–318.

Einfeldt, B., Munz, C. D., Roe, P. L., and Sjögreen, B. 1991. "On Godunov-Type Methods near Low Densities," *Journal of Computational Physics*, 92: 273–295.

Harten, A., Lax, P., and Van Leer, B. 1983. "On Upstream Differencing and Godunov-Type Schemes for Hyperbolic Conservation Law," *SIAM Review*, 5: 1–20.

Mathews, J. H. 1992. *Numerical Methods for Mathematics, Science, and Engineering*, 2nd ed., Englewood Cliffs, NJ: Prentice-Hall, Chapter 2.

Osher, S. 1984. "Riemann Solvers, the Entropy Condition, and Difference Approximations," *SIAM Journal on Numerical Analysis*, 31: 217–235.

Osher, S., and Solomon, F. 1982. "Upwind Difference Schemes for Hyperbolic Systems of Equations," *Mathematics of Computation*, 38: 339–374.

Roe, P. L. 1981. "Approximate Riemann Solvers, Parameter Vectors, and Difference Schemes," *Journal of Computational Physics*, 43: 357–372.

Roe, P. L., and Pike, J. 1984. "Efficient Construction and Utilisation of Approximate Riemann Solutions." In *Computing Methods in Applied Science and Engineering, VI*, eds. R. Glowinski and J.-L. Lions, Amsterdam: North-Holland.

Toro, E. F., Spruce, M., and Speares, W. 1994. "Restoration of the Contact Surface in the HLL-Riemann Solver," *Shock Waves*, 4: 25–34.

Problems

5.1 Suppose that the initial conditions for the Riemann problem for the Euler equations contain only small jumps in all flow properties. Argue briefly and intuitively that the three waves in the solution act something like acoustic and entropy waves, i.e., like characteristics. Is the same thing true for Roe's approximate Riemann problem?

5.2 Argue that a solution containing a simple centered compression fan separating two regions of uniform flow is self-similar about the point where the characteristics converge. Are all simple waves self-similar?

5.3 Waveform creation was described in Section 4.11. Argue that the Riemann problem for the Euler equations is usually waveform creating. Also, argue that the created waveforms may contain new maxima and minima. Since entropy is a characteristic variable, or at least very closely related to a characteristic variable, you may wish to focus on the entropy plot in Figure 5.2.

5.4 (a) Argue that the exact solution to the Riemann problem for the Euler equations depends only on p_L/p_R rather than on p_L and p_R separately. Then, for example, the solution will be the same for $p_L = 100{,}000$ and $p_R = 10{,}000$ as for $p_L = 1$ and $p_R = 1/10$, regardless of the units used for pressure, assuming that density and velocity are kept the same in each case.

(b) Does the exact solution to the Riemann problem for the Euler equations depend only on ρ_L/ρ_R or does it depend on ρ_L and ρ_R separately?

(c) Does the exact solution to the Riemann problem for the Euler equations depend only on x/t or does it depend on x and t separately?

5.5 Consider the Riemann problem for the Euler equations.

(a) Suppose that the solution to the Riemann problem produces only a shock. Using the Hugoniot relation, Equation (3.53), find the relationship between p_R, p_L, ρ_R, and ρ_L. The plot of p_R versus $v_R = 1/\rho_R$ for a given p_L and $v_L = 1/\rho_L$ is called a *Hugoniot curve*. A Hugoniot curve illustrates all possible postshock states for a given preshock state. Prove that the Hugoniot curve is always a hyperbola.

(b) Suppose that the solution to the Riemann problem produces only a centered expansion fan. Find the relationship between p_R, p_L, ρ_R, and ρ_L.

(c) Suppose that the solution to the Riemann problem produces only a contact. Find the relationship between p_R and p_L and between u_R and u_L.

5.6 (a) Using your favorite programming language, write a code to solve Equation (5.6). You can use any root solver you like, including false position or the secant method. Suppose $p_L = 100{,}000\,\text{N/m}^2$, $\rho_L = 1\,\text{kg/m}^3$, $u_L = 50$ m/s and $p_R = 50{,}000\,\text{N/m}^2$, $\rho_R = 0.125\,\text{kg/m}^3$, $u_R = 25$ m/s. What is the pressure ratio p_2/p_1 across the shock?

(b) Use your root solver to write a code that finds the exact solution to the Riemann problem for the Euler equations. Test your code by verifying the results shown in Figure 5.2. Then solve the Riemann problem given in part (a). In particular, what are the pressure, density, and velocity to the left and right of the shock and the expansion?

(c) Use the states to the left and right of the shock found in part (b) as the left and right states for the Riemann problem. Describe the speeds and strengths of the waves in the solution. Do you obtain a pure shock solution?
(d) Use the states to the left and right of the expansion found in part (b) as the left and right states for the Riemann problem. Describe the speeds and strengths of the waves in the solution. Do you obtain a pure expansion?

5.7 Consider the Riemann problem for the following linear system of partial differential equations:

$$\frac{\partial u_1}{\partial t} + \frac{\partial u_2}{\partial x} = 0,$$

$$\frac{\partial u_2}{\partial t} + \frac{\partial u_3}{\partial x} = 0,$$

$$\frac{\partial u_3}{\partial t} + 4\frac{\partial u_1}{\partial x} - 17\frac{\partial u_2}{\partial x} + 8\frac{\partial u_3}{\partial x} = 0.$$

(a) Write the system of equations in the form $\partial \mathbf{u}/\partial t + A\partial \mathbf{u}/\partial x = 0$. Find A^-, A^+, and $|A|$.
(b) Write the general solution to the Riemann problem as in Equation (5.18).
(c) Write the general solution for $Au(0, t)$ in six different ways using Equations (5.21) and (5.27).

5.8 Consider the following Riemann problem:

$$\frac{\partial u}{\partial t} + \frac{\partial}{\partial x}\left(u^3 + \frac{u^2}{2}\right) = 0,$$

$$u(x, 0) = \begin{cases} -1 & \text{for } |x| < 0, \\ 1 & \text{for } |x| > 0. \end{cases}$$

(a) Plot the flux function for $-1 \leq u \leq 1$.
(b) Solve the Riemann problem using the techniques described in Chapter 4.
(c) Solve the Riemann problem using Equation (5.68).
(d) Approximately solve the Riemann problem using Equation (5.69).

5.9 Consider the isothermal Euler equations given in Problem 2.4. (You will need to complete Problem 2.4 before attempting this problem.)
(a) Find an exact solution to the Riemann problem for the isothermal Euler equations.
(b) Find a Roe-average matrix for the isothermal Euler equations. Use this result to find an approximate solution to the Riemann problem.
(c) Find an approximate one-wave solution to the Riemann problem for the isothermal Euler equations.

Part II

Computational Review

The terms *computational* or *numerical* apply to almost all computer calculations found in science, mathematics, and engineering. However, over the years, the terms *computational methods*, *numerical methods*, and *numerical analysis* have come to refer to a standard repertoire of basic numerical techniques including root solvers, linear systems of equations, least squares fitting, Fourier series, differentiation, numerical quadrature, optimization, Runge–Kutta methods, and so forth. There is nothing magical about the standard numerical repertoire; it is simply a grab bag selection of material that has proven generally useful in modern science, mathematics, and engineering.

This second part of the text reviews computational methods. Of course, it would be neither appropriate nor possible, given space constraints, to cover the entire standard computational repertoire in a text devoted specifically to computational gasdynamics. Instead, this part of the book concerns only the standard topics of greatest relevance to the rest of the book. Speaking in broadbrush terms, Part II deals with *polynomial* and *piecewise-polynomial approximations*. More specifically, Chapter 6 concerns computer numbers and numerical errors, which allow one to measure the quality of polynomial and piecewise-polynomial approximations. Chapter 7 describes orthogonal functions and Fourier series, which allow one to find the best possible polynomial approximation to a given function. Chapter 8 concerns polynomial interpolation. Putting aside Chapter 9 for a moment, Chapter 10 covers numerical differentiation and finite-difference formulae, derived from polynomial and piecewise-polynomial interpolations; numerical integration and numerical quadrature, also derived from polynomial and piecewise-polynomial interpolations; and numerical solutions of ordinary differential equations including Runge–Kutta methods, derived from numerical differentiation and integration techniques. Beyond these standard topics, Chapter 9 describes something unusual: piecewise-polynomial *essentially nonoscillatory* (*ENO*) *reconstruction* and *reconstruction via the primitive function*, both of which are heavily based on polynomial interpolation. Although developed specifically for CFD, these techniques have wide applicability. Most classic reconstruction and interpolation procedures exhibit large spurious oscillation nears jumps, such as shocks, but ENO reconstructions handle discontinuous and nonsmooth functions with aplomb. Ordinary reconstructions and interpolations transform samples into functions. Reconstruction via the primitive function extends these techniques in order to transform integral averages into functions.

CHAPTER 6

Numerical Error

6.0 Introduction

In the problems of interest to us, the solution is either a vector or a function. How can we measure the error of approximations to such solutions? One approach is to measure the error in each component of the vector or in each value of the function. Then the error is itself a vector or a function. However, such error vectors and functions provide a wealth of useless information, and may actually obscure critical evaluation. Thus, for most purposes, we prefer a single number measuring the overall size of the error. Such a measure is called a vector or functional *norm*.

After discussing error norms, this chapter describes the two basic categories of numerical errors: round-off errors and discretization errors. *Round-off errors* are any errors caused by the use of finite-precision real numbers, rather than the true infinite-precision real numbers. *Discretization errors* are any errors caused by using a finite-dimensional vector rather than the true infinite-dimensional vector, or a finite sequence rather than the true infinite sequence, or a finite series rather than the true infinite series, or a finite sequence or series such as a truncated Taylor series rather than the true function.

6.1 Norms and Inner Products

This section concerns vector and functional norms and inner products. Consider a scalar x. The absolute value $|x|$ measures the size of x. Recall that the absolute value has the following properties:

$$|x| \geq 0 \quad \text{and} \quad |x| = 0 \text{ if and only if } x = 0,$$
$$|\alpha x| = |\alpha||x| \quad \text{for any scalar } \alpha,$$
$$|x + y| \leq |x| + |y| \quad \text{for any } x \text{ and } y.$$

The last property is sometimes called the *triangle inequality*.

Consider a vector $\mathbf{x}$. The *norm* $\|\mathbf{x}\|$ measures the size of $\mathbf{x}$. More specifically, $\|\mathbf{x}\|$ is any scalar function of $\mathbf{x}$ such that

$$\|\mathbf{x}\| \geq 0 \quad \text{and} \quad \|\mathbf{x}\| = 0 \text{ if and only if } \mathbf{x} = \mathbf{0},$$
$$\|\alpha\mathbf{x}\| = |\alpha|\,\|\mathbf{x}\| \quad \text{for any scalar } \alpha,$$
$$\|\mathbf{x} + \mathbf{y}\| \leq \|\mathbf{x}\| + \|\mathbf{y}\| \quad \text{for any } \mathbf{x} \text{ and } \mathbf{y}.$$

Vector norms are the natural extensions of absolute value to vectors. However, whereas there is only one absolute value, there are infinitely many vector norms. The most popular vector norms are the *p-norms* or *l_p-norms* defined as follows:

$$\|\mathbf{x}\|_p = \left(|x_1|^p + |x_2|^p + \cdots + |x_N|^p\right)^{1/p}, \tag{6.1}$$

where $(x_1, \ldots, x_N)$ are the components of the vector $\mathbf{x}$ in any orthonormal basis, and p is any integer. For example, the 1-*norm* is defined as

$$\|\mathbf{x}\|_1 = |x_1| + |x_2| + \cdots + |x_N|. \tag{6.2}$$

For another example, the 2-*norm* or *Euclidean norm* is defined as

$$\|\mathbf{x}\|_2 = \sqrt{x_1^2 + x_2^2 + \cdots + x_N^2}. \tag{6.3}$$

As a final example, the ∞-*norm* or *maximum norm* is defined as

$$\|\mathbf{x}\|_\infty = \max(|x_1|, |x_2|, \ldots, |x_N|). \tag{6.4}$$

Consider any function f. The *norm* $\|f\|$ measures the size of f. More specifically, $\|f\|$ is any scalar function of f such that

$$\begin{aligned}
&\|f\| \geq 0 \quad \text{and} \quad \|f\| = 0 \text{ if and only if } f = 0,\\
&\|\alpha f\| = |\alpha| \|f\| \quad \text{for any scalar } \alpha,\\
&\|f + g\| \leq \|f\| + \|g\| \quad \text{for any } f \text{ and } g.
\end{aligned}$$

Functional norms are the natural extension of absolute value to functions. The most popular functional norms are the *p-norms* or L_p-*norms* defined as follows:

$$\|f\|_p = \left[\int_a^b |f(x)|^p \, dx\right]^{1/p}, \tag{6.5}$$

where $[a, b]$ is the domain of f and p is any integer. This definition obviously assumes that $|f(x)|^p$ is integrable over $[a, b]$. For example, the 1-*norm* is defined as

$$\|f\|_1 = \int_a^b |f(x)| \, dx, \tag{6.6}$$

assuming that $|f(x)|$ is integrable over $[a, b]$. For another example, the 2-*norm* is defined as

$$\|f\|_2 = \sqrt{\int_a^b f(x)^2 \, dx}, \tag{6.7}$$

assuming that $f(x)^2$ is integrable over $[a, b]$. As a final example, the ∞-*norm* or *maximum norm* is defined as

$$\|f\|_\infty = \operatorname*{ess\,sup}_{a \leq x \leq b} |f(x)|. \tag{6.8}$$

The term "ess sup" or "essential supremum" may not be familiar to many readers. The *supremum* or *sup* is the least upper bound. Although a function must attain its maximum it need not attain its supremum. Thus, for example, a horizontal asymptote is a supremum but not a maximum. For another example, if a function peaks at a jump discontinuity, such as a peak in the pressure function at a shock, then the peak is a supremum but not a maximum. An *essential supremum* or *ess sup* is like a supremum except that it ignores any strange or uncharacteristic points. For example, suppose a function suddenly leaps up at a single point and then leaps back down – this affects the maximum and supremum but not the essential supremum. Similarly, the *infemum* or *inf* is the greatest lower bound, and the *essential infemum* or *ess inf* is like an infemum except that it ignores any strange or uncharacteristic points. Having read this explanation, everyone except the mathematicians in the audience should now mentally replace "ess sup" by "max" in Equation (6.8).

Example 6.1 Suppose $f(x)$ is approximated by $g(x)$ with an error $e(x)$ as follows:

$$e(x) = f(x) - g(x) = 1 - x^2.$$

Measure the error in the 1-norm, 2-norm, and ∞-norm on the domain $[-2, 2]$.

Solution Assuming that the reader can perform the required integrations and maximizations, the answers are $\|e\|_1 = 4$, $\|e\|_2 = 2.477$, and

$$\|e\|_\infty = |e(2)| = |e(-2)| = 3.$$

Norms measure the size of vectors or functions. Norms also measure the distance between vectors or between functions. In particular, $\|\mathbf{x} - \mathbf{y}\|$ measures the *absolute difference* between vectors $\mathbf{x}$ and $\mathbf{y}$, while $\|\mathbf{x} - \mathbf{y}\|/\|\mathbf{x}\|$ or $\|\mathbf{x} - \mathbf{y}\|/\|\mathbf{y}\|$ measure the *relative* or *percentage difference* between vectors $\mathbf{x}$ and $\mathbf{y}$. Similar definitions apply to functional norms. Relative differences provide the most meaningful results. For example, if $\mathbf{x} = 100{,}000$ and $\mathbf{y} = 90{,}000$ the absolute difference is 10,000, which sounds large, but the relative difference is only 10%, which sounds smaller, and properly so.

Different norms are good for different things. For example, in one application, it might be easy to prove that there is a certain upper bound on the 1-norm but hard to prove that there is an upper bound on the 2-norm. For another example, it might be important to minimize the "worst case" ∞-norm error in one application, whereas it might be important to minimize the average error as measured in the 1-norm or 2-norm in another application. Fortunately, a number of inequalities exist that demonstrate that something large in one norm is also reasonably large in all other norms. For example,

$$\|\mathbf{x}\|_2 \le \|\mathbf{x}\|_1 \le \sqrt{N}\|\mathbf{x}\|_2,$$

where N is the vector dimension. In other words, relative judgements such as "large" and "small" are not usually terribly sensitive to the choice of norm.

This completes the discussion of vector and functional norms. The section ends with a brief discussion of vector and functional inner products. An ordinary product of real numbers has the following properties:

$$xx \ge 0 \quad \text{and} \quad xx = 0 \text{ if and only if } x = 0,$$
$$xy = yx \quad \text{for any } x \text{ and } y,$$
$$x(y + z) = xy + xz \quad \text{for any } x,\ y \text{ and } z.$$

A *vector inner product* or *scalar product* maps two vectors to a scalar such that

$$\mathbf{x} \cdot \mathbf{x} \ge 0 \quad \text{and} \quad \mathbf{x} \cdot \mathbf{x} = 0 \text{ if and only if } \mathbf{x} = 0,$$
$$\mathbf{x} \cdot \mathbf{y} = \mathbf{y} \cdot \mathbf{x} \quad \text{for any } \mathbf{x} \text{ and } \mathbf{y},$$
$$\mathbf{x} \cdot (\mathbf{y} + \mathbf{z}) = \mathbf{x} \cdot \mathbf{y} + \mathbf{x} \cdot \mathbf{z} \quad \text{for any } \mathbf{x},\ \mathbf{y},\ \text{and } \mathbf{z},$$
$$\alpha(\mathbf{x} + \mathbf{y}) = \alpha\mathbf{x} + \alpha\mathbf{y} \quad \text{for any } \mathbf{x},\ \mathbf{y},\ \text{and scalar } \alpha.$$

The most popular vector inner product is the Euclidean inner product

$$\mathbf{x} \cdot \mathbf{y} = x_1 y_1 + x_2 y_2 + \cdots + x_N y_N, \tag{6.9}$$

where $(x_1, \ldots, x_N)$ and $(y_1, \ldots, y_N)$ are the components of $\mathbf{x}$ and $\mathbf{y}$ in any orthonormal basis. In standard notation, the dot is often reserved for the Euclidean inner product; general inner products are denoted by $\langle \mathbf{x}, \mathbf{y} \rangle$ or $(\mathbf{x}, \mathbf{y})$. Every vector inner product is naturally associated with a norm as follows:

$$\mathbf{x} \cdot \mathbf{x} = \|\mathbf{x}\|^2. \tag{6.10}$$

For example, the Euclidean inner product is naturally associated with the 2-norm. There are many other possible vector products besides the inner product, including outer products and cross products. These will not be discussed here.

A *functional inner product* maps two functions to a scalar such that

$$\begin{aligned}
&f \cdot f \geq 0 \quad \text{and} \quad f \cdot f = 0 \text{ if and only if } f = 0,\\
&f \cdot g = g \cdot f \quad \text{for any } f \text{ and } g,\\
&f \cdot (g + h) = f \cdot g + f \cdot h \quad \text{for any } f, g, \text{ and } h,\\
&\alpha(f + g) = \alpha f + \alpha g \quad \text{for any } f, g, \text{ and scalar } \alpha.
\end{aligned}$$

The most popular functional inner products have the form

$$f \cdot g = \int_a^b w(x) f(x) g(x)\, dx, \tag{6.11}$$

where $w(x)$ is a nonnegative *weighting function*. Functional inner products of all varieties are more commonly denoted by $\langle f, g \rangle$ or (f, g). Every functional inner product is naturally associated with a norm as follows:

$$f \cdot f = \|f\|^2. \tag{6.12}$$

For example, if $w(x) = 1$, then the inner product defined by Equation (6.12) is naturally associated with the functional 2-norm.

6.2 Round-Off Error

This section very briefly discusses round-off error. A *binary digit* or *bit* is a digit in a base-two number system; it is either zero or one. A *byte* is eight bits. Computers do not deal in individual bits or bytes. Instead, computers deal in *words*, typically 32 bits or four-bytes long. A 32-bit word can represent roughly $2^{32} \approx 4.29 \times 10^9$ different objects. Those objects can be real numbers, integers, characters, and so forth. For example, consider integers. On a typical computer, the 2^{32} integers are more or less evenly divided between positive integers and negative integers. In other words, the integers run between $-2^{32}/2$ and $2^{32}/2$ give or take one. As another example, consider real numbers. Instead of the true infinite continuum of real numbers, there must be a maximum computer number, generally around 10^{39}, and any larger numbers cause *overflow*. There must also be a minimum computer number, generally around 10^{-39}, and any smaller numbers cause *underflow*. Finally, there must be a finite spacing between computer numbers. The spacing varies with the size of the numbers such that the *relative* spacing remains constant. Then the relative spacing between computer numbers, also called the *machine epsilon* ϵ, bounds the *relative round-off error* as follows:

$$\textit{relative round-off error} = \frac{|\textit{true number} - \textit{computer number}|}{|\textit{true number}|} \leq \epsilon.$$

Machine epsilon is typically about $\epsilon = 10^{-8}$; hence computer numbers have roughly eight decimal places of accuracy in base 10. As an alternative to single word representations, computers can also use two, four, or more words to represent objects such as real numbers; for example, double precision increases the maximum real number to something like 10^{309}, decreases the minimum real number to something like 10^{-309}, and decreases the machine epsilon, allowing roughly sixteen decimal places of accuracy in base 10.

Round-off error may seem insignificant, since the relative round-off error is always less than or equal to the machine epsilon ϵ. However, although round-off error starts small, there are at least three computational mechanisms that may magnify round-off errors. First, in a large sum, suppose that there is a wide range of numbers and that small numbers are repeatedly added to large numbers. For example, suppose that 1 and ϵ are added repeatedly. After each addition, the sum is still 1 since $1 + \epsilon = 1$ according to the computer, by the definition of ϵ. After $1/\epsilon$ such additions, the true sum should be 2 whereas the computer yields 1; thus the relative round-off error is 100%! The best solution to this problem is to always add numbers from smallest to largest, although this may require an expensive sorting procedure.

Another mechanism that magnifies round-off error is called *subtractive cancellation*. Suppose two nearly equal numbers are subtracted. For example, $44.55 - 44.54 = 0.01$. Although the original numbers have four decimal places of accuracy in base 10, their difference has only one decimal place of accuracy! Another example of subtractive cancellation follows.

Example 6.2 If $x = 1984$, find $\sqrt{x+1} - \sqrt{x}$ truncating all numbers to four decimal places in base 10. Suggest a way to avoid subtractive cancellation.

Solution If $x = 1984$ then

$$\sqrt{x+1} - \sqrt{x} = 44.55 - 44.54 = 0.01.$$

Alternatively,

$$\sqrt{x+1} - \sqrt{x} = \left(\sqrt{x+1} - \sqrt{x}\right)\frac{\sqrt{x+1} + \sqrt{x}}{\sqrt{x+1} + \sqrt{x}} = \frac{1}{\sqrt{x+1} + \sqrt{x}}.$$

Then

$$\sqrt{1985} - \sqrt{1984} = \frac{1}{\sqrt{1985} + \sqrt{1984}} = \frac{1}{44.55 + 44.54} = 0.01122.$$

This second result has four decimal places of accuracy rather than one.

A third mechanism that magnifies round-off error is solution sensitivity. In a sensitive problem, small changes in the problem lead to large changes in the solution. Round-off errors inevitably introduce small changes, which create larges changes in the solution. For example, a matrix A is called *ill-conditioned* if, for any vector $\mathbf{b}$, small changes in A or $\mathbf{b}$ cause large changes in $\mathbf{x}$, where $\mathbf{x}$ is the solution to $A\mathbf{x} = \mathbf{b}$. Matrices are often ill-conditioned when there is a large range of sizes in the elements of the matrix, or when one row of the matrix is nearly equal to some linear combination of other rows. See Golub and van Loan (1989) for a more detailed description of matrix ill-conditioning. Another example of sensitivity follows.

Example 6.3 Consider the following iterative equation:

$$u_{n+1} = 4\lambda u_n(1 - u_n),$$

where $0 \leq \lambda \leq 1$ is any constant and u_0 is any starting value. This simple quadratic equation is called the *logistic map*. Show that the solution to the logistic map is extremely sensitive to small errors.

Solution Notice that if $0 \leq u_0 \leq 1$ then $0 \leq u_n \leq 1$ for all n. However, if $u_0 > 1$ or $u_0 < 0$ then u_n quickly diverges to $\pm\infty$ as n increases. In this example, we shall choose $0 \leq u_0 \leq 1$ and $\lambda = 1$. In particular, let us compare the results using the three starting values $u_0 = 0.4 + 10^{-8}$, $u_0 = 0.4$, and $u_0 = 0.4 - 10^{-8}$. Using double-precision arithmetic, the three solutions are nearly equal for small n. However, the three solutions differ significantly for $n = 20$ and the three solutions are entirely unrelated for $n = 30$. This is a classic example of *chaos*. See Tabor (1989) for a more complete description of chaos. Chaos is but one interesting behavior studied under the heading of *dynamical systems*. Traditional computational gasdynamics does not allow particularly complicated dynamical behaviors or, in particular, chaos; however, computational gasdynamics with strong source terms, such as chemically reacting flows, allows all sorts of devious dynamical behaviors, which may pose extreme challenges for numerical models. For more information, see Yee, Sweby, and Griffiths (1991).

Sometimes extreme sensitivity is due to a mistake in the original problem statement. For example, the problem statement may not specify enough information or it may specify contradictory information; in other words, the problem may be *ill-posed*. In other cases, the problem statement is fine, and the extreme sensitivity is the result of a faulty numerical approximation. In yet other cases, the problem is simply inherently sensitive, and it is unreasonable to expect to approximate its details. For example, turbulent gas flows are constantly changing in a chaotic like fashion, and thus no approximation can be expected to capture the exact pattern of a given turbulent flow. This sort of sensitivity is sometimes called physical *instability*; see Chapter 14 for more discussion along these lines.

6.3 Discretization Error

The last section concerned round-off errors, that is, errors in the computer representation of numbers. This section concerns discretization error, that is, errors in the computer representation of functions and functional operators. As every reader will certainly recall, a *function* maps a domain to a range. A function is *continuously defined* if the domain is continuous; for example, the domain might equal the real numbers between a and b. A function is *discretely defined* if the domain is discrete; for example, a *sequence* is a discretely defined function whose domain is a set of consecutive integers, called the *index*. The term *vector* either refers to any sequence, especially when the index starts from one or zero, or to a sequence with special coordinate transformation properties, as discussed in Chapter 7.

The distinction between continuously defined and discretely defined functions is not as clear-cut as it first appears – in fact, continuously defined functions can often be naturally interchanged with discretely defined functions. For example, the discretely defined function $(a_0, a_1, a_2, a_3, \ldots)$ can be transformed to a continuously defined function $f(x)$ as follows:

$$f(x) = a_0 + a_1 x + a_2 x^2 + a_3 x^3 + \cdots.$$

Vice versa, a continuously defined function $f(x)$ can be transformed to a discretely defined function $(a_0, a_1, a_2, a_3, \ldots)$ using Taylor series, where

$$a_n = \frac{1}{n!}\frac{d^n f}{dx^n}(0).$$

How do computers represent functions? Computers can only perform the four basic arithmetic operations: addition, subtraction, multiplication, and division. Then, for example, a computer can easily represent $f(x) = ax^2 + bx + c$ since this function involves only basic arithmetic operations. In contrast $\ln x = \int_1^x \frac{dy}{y}$ involves integration in addition to basic arithmetic operations. However, consider the following series:

$$\ln x = \frac{2}{1}\frac{x-1}{x+1} + \frac{2}{3}\left(\frac{x-1}{x+1}\right)^2 + \frac{2}{5}\left(\frac{x-1}{x+1}\right)^5 + \cdots.$$

Then the continuously defined function $\ln x$ can be recovered from the discretely defined sequence $(2, 2/3, 2/5, \ldots)$ using only the four basic arithmetic operations.

Whereas many functions can be perfectly represented by *infinite* sequences and basic arithmetic operations, computers can only store *finite* sequences. Thus the true sequence representing the true function must be artificially truncated to a finite number of terms. To summarize: *Computers represent functions by finite sequences of numbers.* Many problems involve not only continuously defined functions but also operations on continuously defined functions such as integration and differentiation. If continuously defined functions are replaced by discretely defined functions, then all operators on continuously defined functions must also be replaced by operators on discretely defined functions. *Discretization* is the process of replacing all continuously defined functions and functional operators by discretely defined functions and functional operators. *Discretization error* is, naturally enough, any error caused by discretization. *Truncation* is the process of replacing infinite sequences or series by finite sequences or series, and *truncation error* is any error caused by truncation. Truncation is a type of discretization, since it is part of the process of putting functions and functional operators in a form suitable for computation; then truncation error is a type of discretization error.

Unfortunately, just as a finite number of bits can represent only a limited number of real numbers, a finite number of numbers $(a_0, a_1, \ldots, a_N)$ can represent only a limited number of functions. In other words, computers can rarely represent real numbers exactly, and real numbers are represented by the closest computer number; similarly, computers can rarely represent functions exactly, and functions are represented by the closest computer function.

To continue the analogy, round-off error is any error that results from replacing a real number by a computer number; similarly, discretization error is any error that results from replacing a function by a computer function (i.e., the error that results from replacing a function by a finite sequence of numbers). Unlike the computer representation of numbers, the computer representation of functions is completely under the user's control. The following examples illustrate some of the principles involved.

Example 6.4 If the required functions are always linear or nearly linear, a linear representation $f(x) \approx ax + b$ is the best choice. Linear representations require only two real numbers: a and b. However, unless the exact function is truly linear, a line can accurately represent the function only over a limited domain. Outside of the domain of accuracy, the approximating line strays progressively further and further from the true function; this occurs, for example, with tangent and secant lines. To approximate a nonlinear function over a wide region generally requires a representation with more than two numbers, such as a quadratic or cubic representation. Even within the domain of accuracy, for general functions, the accuracy of linear approximations is relatively low. Thus linear representations place severe limitations on both the accuracy and the domain of accuracy for general functions.

Example 6.5 If the required functions contain periodic oscillations, a trigonometric series may be the best choice. For example, $f(x) \approx a_0 + a_1 \cos x + b_1 \sin x$. This is a compact representation – it requires only three real numbers a_0, a_1, and b_1 – and captures the expected periodic oscillations. When using trigonometric series representations we assume that the computer has adequate representations for $\cos x$ and $\sin x$. Indeed, most programming languages have efficient built-in libraries of trigonometric functions.

Example 6.6 Suppose that, as part of some computer method, functions are frequently differentiated. This is extremely easy to do in a Taylor series form representation. In particular, suppose that

$$f(x) \approx a_0 + a_1(x-b) + a_2(x-b)^2 + \cdots + a_N(x-b)^N.$$

Then

$$\frac{df}{dx}(x) \approx a_1 + 2a_2(x-b) + \cdots + Na_N(x-b)^{N-1}.$$

Differentiation is more difficult with most other functional forms.

Suppose that a function $f(x)$ is replaced by an approximate function $g_N(x)$, where $g_N(x)$ is defined by a finite sequence of N numbers $(a_0, a_1, \ldots, a_N)$. The discretization error $e_N(x)$ is defined as follows:

$$e_N(x) = f(x) - g_N(x), \tag{6.13}$$

How does $e_N(x)$ depend on N? If $|e_N(x)| \to 0$ as $N \to \infty$ for all x, then the approximation function *pointwise converges* to the true function. If $\|e_N\|_2 \to 0$ as $N \to \infty$, then the approximation function *converges in the mean* to the true function; and if $\|e_N\|_\infty \to 0$ as

$N \to \infty$, then the approximation function *uniformly converges* to the true function. How does $e_N(x)$ depend on x? Suppose that $|e_N(x)| \le M|x^R|$ for all x, where M is some constant; then the approximation has *order of approximation* or *order of accuracy R*. In this case, a common notation is $e_N(x) = O(x^R)$.

Example 6.7 By Taylor series, a fourth-order accurate approximation to e^x is

$$e^x = 1 + x + \frac{x^2}{2!} + \frac{x^3}{3!} + O(x^4).$$

Also, a fifth-order accurate approximation to $\sin x$ is

$$\sin x = x - \frac{x^3}{3!} + O(x^5).$$

Then

$$\begin{aligned} e^x + \sin x &= 1 + x + \frac{x^2}{2!} + \frac{x^3}{3!} + O(x^4) + x - \frac{x^3}{3!} + O(x^5) \\ &= 1 + 2x + \frac{x^2}{2!} + O(x^4). \end{aligned}$$

Multiply the last two approximations to obtain the following approximation to $e^x \sin x$:

$$\begin{aligned} e^x \sin x &= \left(1 + x + \frac{x^2}{2!} + \frac{x^3}{3!} + O(x^4)\right)\left(x - \frac{x^3}{3!} + O(x^5)\right) \\ &= x - \frac{x^3}{3!} + O(x^5) + x^2 - \frac{x^4}{3!} + O(x^6) + \frac{x^3}{2!} - \frac{x^5}{2!3!} + O(x^7) \\ &\quad + \frac{x^4}{3!} - \frac{x^6}{3!3!} + O(x^8) + O(x^5) + O(x^7) + O(x^9) \\ &= x + x^2 + \left(\frac{1}{2!} - \frac{1}{3!}\right)x^3 + \left(\frac{1}{3!} - \frac{1}{3!}\right)x^4 + O(x^5) \\ &= x + x^2 + \frac{x^3}{3} + O(x^5). \end{aligned}$$

Rather than using a single function to represent all of $f(x)$, suppose that the domain of $f(x)$ is decomposed into cells and that a different function is used to represent $f(x)$ in each cell. This is illustrated by the following example:

Example 6.8 Consider a linear approximation

$$f(x) = ax + b + O(x^2).$$

This could be a secant line or a tangent line centered on $x = 0$. Notice that the error is large if $|x|$ is large. So, instead of a single line, suppose that the domain is decomposed into cells

$[x_0, x_1], [x_1, x_2], \ldots, [x_{N-1}, x_N]$ and that a different line is used to represent $f(x)$ in each cell. For example,

$$f(x) \approx \begin{cases} a_1(x - x_1) + b_1 + O[(x - x_1)^2] & x_0 \le x < x_1, \\ a_2(x - x_2) + b_2 + O[(x - x_2)^2] & x_1 \le x < x_2, \\ \vdots & \vdots \\ a_N(x - x_N) + b_N + O[(x - x_N)^2] & x_{N-1} \le x \le x_N, \end{cases}$$

where each line could be a local secant or tangent line approximation. Now suppose that $x_{i+1} = x_i + \Delta x$. Then

$$O(x - x_i) = O(\Delta x) \quad \text{for} \quad x_i \le x \le x_{i+1}.$$

Then

$$f(x) \approx \begin{cases} a_1(x - x_1) + b_1 + O(\Delta x^2) & x_0 \le x < x_1, \\ a_2(x - x_2) + b_2 + O(\Delta x^2) & x_1 \le x < x_2, \\ \vdots & \vdots \\ a_N(x - x_N) + b_N + O(\Delta x^2) & x_{N-1} \le x \le x_N. \end{cases}$$

Now the error depends mostly on the single constant parameter $\Delta x = (x_N - x_0)/N$ and much less on the variable x.

This last example illustrates *piecewise-polynomial* approximations. A piecewise-polynomial approximation *pointwise converges* if $e(x) \to 0$ as $\Delta x \to 0$ for all x. Similarly, a piecewise-polynomial approximation *converges in the mean* if $\|e\|_2 \to 0$ as $\Delta x \to 0$. Finally, a piecewise-polynomial approximation *converges uniformly* if $\|e\|_\infty \to 0$ as $\Delta x \to 0$. Rather than just requiring that the error go to zero as $\Delta x \to 0$, suppose we require that the error go to zero at some minimum rate; this leads us to new definitions for order of accuracy. In particular, a piecewise-polynomial approximation is *pointwise Rth-order accurate* if $|e(x)| \le M|\Delta x^R|$ for all x; *Rth-order accurate in the mean* if $\|e\|_2 \le M|\Delta x^R|$; and *uniformly Rth-order accurate* if $\|e\|_\infty \le M|\Delta x^R|$ for some constant M. Notice that all three definitions of order of accuracy imply convergence as $\Delta x \to 0$.

Unfortunately, the order of accuracy may vary from cell to cell. For example, referring to Example 6.8, if $f(x)$ varies linearly or nearly linearly in a certain cell, then the pointwise order of accuracy of the piecewise-linear approximation in that cell may be anything between three and infinity. In contrast, if $f(x)$ is discontinuous in a certain cell, then the pointwise order of accuracy in that cell may be very small, typically between one and zero. When a piecewise-polynomial approximation is called "Rth-order accurate," this usually means "the approximation has pointwise Rth-order accuracy except in certain cells." By tradition, some exceptions are explicitly noted while others are simply understood. Specifically, local losses in the order of accuracy caused by jump discontinuities in the function or its derivatives are usually understood and are not usually noted explicitly. However, any local losses in the order of accuracy near extrema are almost always explicitly noted. For example, many approximations in computational gasdynamics have second-order accuracy in most places but only first-order accuracy near smooth extrema, due to a phenomenon called clipping;

see Chapter 16 and Part V. In conclusion, as commonly used, "Rth-order accuracy," "order-of-accuracy R," and $O(\Delta x^R)$ are a bit vague – these measures depend on the choice of norm and may vary from place to place and from function to function.

Note that an approximation can have a high order of accuracy and yet have very poor absolute or relative accuracy. Thus the error for a given Δx may be quite large, regardless of the fact that the error would decrease rapidly if Δx were decreased. *Do not make the common mistake of confusing high accuracy with high order of accuracy.*

Remember that the order of accuracy refers to discretization error and does not include the effects of round-off error. Thus, in practice, the error of a numerical approximation based on piecewise-polynomials will decrease at the expected rate as Δx decreases, but only to a certain point, usually when Δx nears machine epsilon. Relative round-off error increases as Δx decreases so that eventually increases in relative round-off errors will completely offset decreases in the relative discretization error, and the overall relative error will stop decreasing unless the machine precision is increased.

It is important to be able to distinguish between discretization errors and round-off errors because their causes and cures are completely different. The classic technique is to increase the machine precision. Under ordinary circumstances, changing from single to double precision should only affect the results starting in the seventh or eighth decimal place. However, if the computation magnifies round-off error, the results may change even in the first or second decimal place. In this case, the machine precision should be doubled again and again, if possible, until the low-order decimal places stop changing. Then any error in the stable low-order decimal places must be discretization error.

References

Golub, G. H., and van Loan, C. F. 1989. *Matrix Computations*, 2nd ed., Baltimore, MD: Johns Hopkins University Press, Chapter 2.

Tabor, M. 1989. *Chaos and Integrability in Nonlinear Dynamics: An Introduction*, New York: Wiley.

Yee, H. C., Sweby, P. K., and Griffiths, D. F. 1991. "Dynamical Approach Study of Spurious Steady-State Numerical Solutions of Nonlinear Differential Equations. I. The Dynamics of Time Discretization and Its Implications for Algorithm Development in Computational Fluid Dynamics," *Journal of Computational Physics*, 97:249–310.

Problems

6.1 (a) Consider the following inner product:

$$f \cdot g = \int_{-1}^{1} f(x)g(x)\,dx.$$

Find $\|g\|$ in the natural norm and $f \cdot g$ where $f(x) = 1 - x + \frac{1}{3}x^2 + 5x^3$ and $g(x) = x - 3x^2$.

(b) Consider the following inner product:

$$f \cdot g = \int_{-1}^{1} \frac{f(x)g(x)}{\sqrt{1-x^2}}\,dx.$$

Find $\|g\|$ in the natural norm and $f \cdot g$ where $f(x) = \sin^{-1} x$ and $g(x) = x$. Please feel free to avail yourself of a table of integrals.

6.2 Suppose that $f(x)$ is approximated by $g(x)$ with a pointwise-error $e(x) = \sin x$.

(a) Measure the error in the 1-norm, 2-norm, and ∞-norm on the domain $[-\pi/3, \pi/4]$.
(b) Suppose $\|f\|_1 = 0.1$ and $\|g\|_1 = 0.523$. What are the relative errors in the 1-norm? Is the error large or small, as measured in the 1-norm?

6.3 In each of the following cases, underflow will occur on most machines using single-precision arithmetic. In each case, suppose that any quantities that underflow are replaced by zero. What are the resulting absolute and relative round-off errors? Based on the relative round-off error, in which cases is it reasonable to replace the quantities that underflow by zero?
(a) $x + y$ where $x = 10^{-50}$ and $y = 1$.
(b) $x + y$ where $x = y = 10^{-25}$.
(c) xy where $x = 10^{-100}$ and $y = 10^{-99}$.
(d) y/x where $x = 10^{-100}$ and $y = 10^{-99}$.

6.4 Consider the roots of a quadratic $ax^2 + bx + c = 0$. The quadratic equation yields the following two solutions:

$$x = \frac{-b \pm \sqrt{b^2 - 4ac}}{2a}.$$

For one of the roots, show that this expression may magnify round-off error when b^2 is much larger than $4ac$. What causes this error magnification? Which root is subject to round-off error magnification and which root is not? For the root subject to round-off error magnification, suggest an equivalent expression that avoids the round-off error.

6.5 Add 0.333, 2.31, 50.2, and 263 to find the exact value for the sum. Now add the numbers from largest to smallest, truncating all intermediate results to three decimal places in base 10. What is the relative round-off error of the sum? Now add the numbers from smallest to largest, again truncating all intermediate results to three decimal places in base 10. What is the relative round-off error of the sum?

6.6 In each case, suggest an equivalent expression that avoids round-off error magnification. Feel free to use trigonometric and logarithmic identities.
(a) $1 - \cos x$ when $x \approx 0$.
(b) $\ln(x + 1) - \ln x$ for large x.

6.7 Consider the following Taylor series approximations:

$$\cos x = 1 - \frac{1}{2}x^2 + O(x^4),$$

$$\sin x = x - \frac{1}{6}x^3 + O(x^5).$$

Using these expressions, find approximations for $\sec x = 1/\cos x$ and $\tan x = \sin x/\cos x$; what is the order of accuracy of these approximations? Use the fact that

$$\frac{1}{1+x} = 1 + x + x^2 + x^3 + x^4 + x^5 + \cdots.$$

6.8 Consider cells $[-\pi/16, \pi/16]$, $[\pi/16, 3\pi/16]$, $[3\pi/16, 5\pi/16]$, $[5\pi/16, 7\pi/16]$, $[7\pi/16, 9\pi/16]$, $[9\pi/16, 11\pi/16]$, $[11\pi/16, 13\pi/16]$, $[13\pi/16, 15\pi/16]$, $[15\pi/16, 17\pi/16]$.
(a) Consider a piecewise-constant approximation to $\sin x$ based on the value of $\sin x$ at the cell center. For each cell, what is the order of accuracy? Use Taylor series. Be especially careful in the last cell. Does the order of accuracy depend on the choice of norm?
(b) Consider a piecewise-linear approximation to $\sin x$ based on a Taylor series approximation to $\sin x$ about the cell center. For each cell, what is the order of accuracy? Be especially careful in the first cell. Does the order of accuracy depend on the choice of norm?

CHAPTER 7

Orthogonal Functions

7.0 Introduction

This chapter concerns orthogonal functions. A professor recently translated some of Elvis Presley's hit songs into Latin. The love songs were easy, but the professor had trouble with Elvis' rock hits – there are no Latin words for "blue suede shoes" or "hound dog." Computers have a similar difficulty when it comes to functions. As seen in the last chapter, computers represent functions by finite sequences. Unfortunately, most finite sequences cannot adequately express most discontinuous functions. For example, whereas an infinite-order polynomial can represent any piecewise-smooth function, even the best finite-order polynomial approximations exhibit substantial oscillations in the presence of jump discontinuities. Of course, the best polynomial depends on how you define "best." For example, the "best" polynomial could be the polynomial with the least error in the 1-norm, the 2-norm, the ∞-norm, or at some specified critical point. However, sometimes there are no good polynomials regardless of your criteria. This is quite often the case with discontinuous functions. Even completely smooth functions can suffer, as seen in the following example.

Example 7.1 Find the best quadratic approximation for a function with one maximum and one minimum.

Solution As seen in Figure 7.1, the error is enormous no matter which quadratic is chosen. The quadratic can model the maximum or the minimum but not both. There is simply no way to make a one-hump camel look like a two-hump camel. In this case, the choice of quadratic is largely arbitrary, and thus it is impossible to tell much about the original function by examining the chosen quadratic representation.

An Nth order polynomial has at most $N-1$ extrema. Then, as in the last example, a function represented by an Nth-order polynomial should have no more than $N-1$ extrema. Although this is a nice rule of thumb, the rest of this chapter describes a more rigorous approach based on choosing the best finite-length *vector* approximation to a function. From Section 6.3, recall that computers describe functions by sequences and, as defined in this chapter, vectors are sequences that retain their identities under coordinate transformations.

After a general introduction to orthogonal functions in Section 7.1, the rest of the chapter concerns specific orthogonal functions. In particular, Section 7.2 concerns Legendre polynomials, Section 7.3 concerns Chebyshev polynomials, and Section 7.4 concerns Fourier series. Fourier series prove invaluable later in the book (Chapter 15 uses discrete Fourier series heavily).

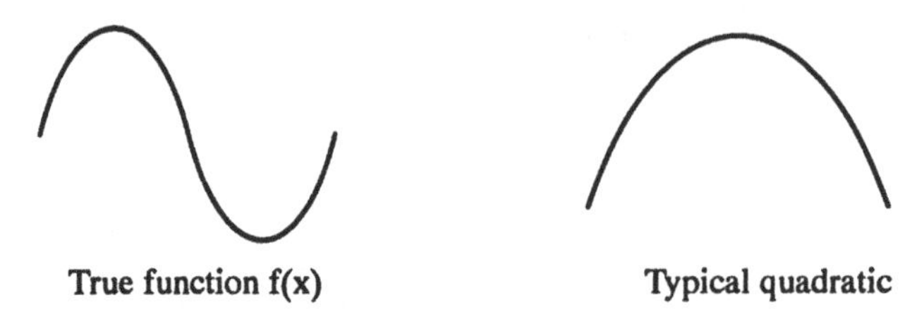

Figure 7.1 Quadratic approximation of a function with two extrema.

Spectral methods represent functions by coefficients in finite series of orthogonal functions; see, for example, Canuto et al. (1988). Although traditionally used for incompressible and turbulent flows, researchers have increasingly applied spectral methods to compressible flows. For example, Giannakouros and Karniadakis (1994) describe a spectral version of the flux-corrected transport method seen in Chapter 21. Although this book will not consider spectral methods, this chapter gives readers the fundamental background required to understand such methods. As two other applications of orthogonal functions relevant to this book, Van Leer (1977) uses Legendre polynomials and Sanders (1988) uses Hermite polynomials. Many books discuss orthogonal functions in terms of one of their most popular applications: the solution of ordinary differential equations. This book does not concern ordinary differential equations, and thus we will not discuss this application of orthogonal functions.

Putting aside the applications of orthogonal functions mentioned above, the main purpose of this chapter is to provide a reference standard for the next chapter. For example, this chapter shows how to find the "best" quadratic approximation to a given function over a given region. This chapter also illustrates the definitions of norm, inner product, and especially convergence seen in the last chapter.

7.1 Functions as Vectors

The last chapter showed that many functions could be interpreted as sequences or, more specifically, vectors. In fact, in a very concrete sense, many functions *are* vectors. This section gives a brief review of vector algebra, as seen in any basic linear algebra textbook, except that functions replace vectors.

A *linear combination* of functions $(f_0, \ldots, f_M)$ is defined as follows:

$$a_0 f_0 + \cdots + a_M f_M,$$

where $(a_0, \ldots, a_M)$ are real numbers. A set of functions $(f_0, \ldots, f_M)$ is *linearly independent* if

$$a_0 f_0 + \cdots + a_M f_M = 0 \quad \text{implies} \quad a_0 = \cdots = a_M = 0,$$

where the zero on the right-hand side of the linear combination refers to the zero function, rather than the real number zero. A *basis* for a functional space is a set of linearly independent functions such that every function in the space can be written as a linear combination of the basis functions. In other words, for any function f and any basis $(f_0, \ldots, f_M)$ there exist real numbers $(a_0, \ldots, a_M)$ such that

$$f = a_0 f_0 + \cdots + a_M f_M.$$

The numbers $(a_0, \ldots, a_M)$ are called the *components* of f in basis $(f_0, \ldots, f_M)$. The number $M+1$ is called the *dimension* of the space. Functional spaces are often infinite dimensional. Any basis for a space is said to *span* the space.

Recall functional inner products, as discussed in Section 6.1. Two functions f and g are *orthogonal* if $f \cdot g = 0$. An *orthogonal basis* is a basis of mutually orthogonal functions; thus

$$f_i \cdot f_j = 0 \qquad i \neq j.$$

As seen in Section 6.1, every inner product defines a natural norm as follows:

$$\|f\|^2 = \mathrm{f} \cdot \mathrm{f}. \tag{7.1}$$

An *orthonormal basis* is an orthogonal basis where $\|f_i\| = 1$ using the natural norm.

The components of f in an orthogonal basis $(f_0, \ldots, f_M)$ are computed as follows:

$$a_i = \frac{f \cdot f_i}{f_i \cdot f_i}. \tag{7.2}$$

Similarly, the components of f in an orthonormal basis $(f_0, \ldots, f_M)$ are computed as follows:

$$a_i = f \cdot f_i. \tag{7.3}$$

Suppose that the full space has dimension $M+1$. Consider a subspace of dimension $N+1 \leq M+1$ with orthogonal basis $(g_0, \ldots, g_N)$. Then, for every function f in the full space, the *orthogonal projection* of f onto the subspace is

$$g = a_0 g_0 + \cdots + a_N g_N, \tag{7.4}$$

where

$$a_i = \frac{f \cdot g_i}{g_i \cdot g_i}. \tag{7.5}$$

The orthogonal projection g is the function in the subspace closest to f as measured in the natural norm: If we let f be any function in the full space, g be the orthogonal projection of f onto the subspace, and h be any other function in the subspace, then

$$\|f - g\| \leq \|f - h\|, \tag{7.6}$$

where $\|f\| = \sqrt{f \cdot f}$. If the full space is three dimensional and the subspace is two dimensional, orthogonal projections can be visualized as in Figure 7.2.

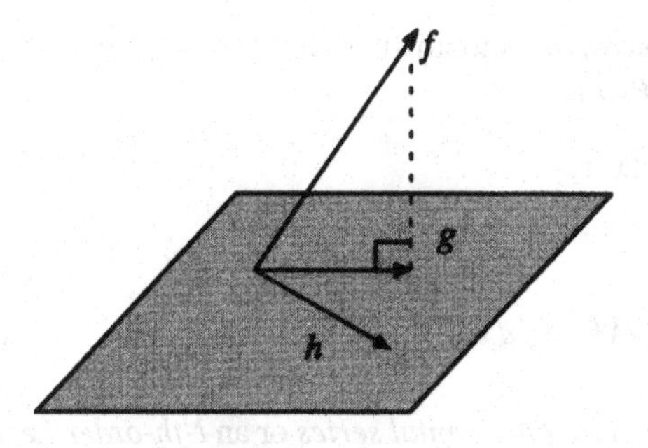

Figure 7.2 Orthogonal projection from three to two dimensions.

7.2 Legendre Polynomial Series

The Legendre polynomials $P_j(x)$ are defined as follows:

$$P_0(x) = 1, \qquad P_1(x) = x,$$
$$(j+1)P_{j+1}(x) = (2j+1)xP_j(x) - jP_{j-1}(x). \tag{7.7}$$

For example,

$$P_2(x) = \frac{1}{2}(3x^2 - 1), \qquad P_3(x) = \frac{1}{2}(5x^3 - 3x).$$

The Legendre polynomials P_j form a basis for the space of piecewise-smooth functions with domains $[-1, 1]$. This basis is orthogonal using the following inner product:

$$f \cdot g = \int_{-1}^{1} f(x)g(x)\,dx. \tag{7.8}$$

In particular,

$$P_i \cdot P_j = \int_{-1}^{1} P_i(x)P_j(x)\,dx = \begin{cases} 2/(2j+1) & i = j, \\ 0 & i \neq j. \end{cases} \tag{7.9}$$

Then, for any piecewise-smooth function $f(x)$ with domain $[-1, 1]$,

$$f(x) = \sum_{i=0}^{\infty} a_j P_j(x), \tag{7.10}$$

where

$$a_j = \frac{f \cdot P_j}{P_j \cdot P_j} = \frac{2j+1}{2} \int_{-1}^{1} f(x)P_j(x)\,dx. \tag{7.11}$$

This is called the *Legendre polynomial series* for $f(x)$. For functions whose domains do not equal $[-1, 1]$, use a linear mapping from the domain $[a, b]$ to $[-1, 1]$.

The first $N+1$ Legendre polynomials $(P_0, \ldots, P_N)$ span the space of polynomials of order N or less. A simpler but nonorthogonal basis for the same subspace is $(1, x, x^2, x^3, \ldots)$. Any nonorthogonal basis can be converted to an orthogonal basis using the Gram–Schmidt procedure, as described in any elementary linear algebra text such as Anton's (1981); in particular, the *Gram–Schmidt procedure* using the inner product of Equation (7.8) converts $(1, x, x^2, \ldots, x^N)$ to $(P_0, \ldots, P_N)$.

The orthogonal projection of any piecewise-smooth function f with domain $[-1, 1]$ onto the subspace spanned by $(P_0, \ldots, P_N)$ is

$$f(x) \approx a_0 P_0(x) + \cdots + a_N P_N(x), \tag{7.12}$$

where

$$a_j = \frac{f \cdot P_j}{P_j \cdot P_j} = \frac{2j+1}{2} \int_{-1}^{1} f(x)P_j(x)\,dx. \tag{7.13}$$

This is sometimes called a *truncated Legendre polynomial series* or an *Nth-order Legendre polynomial series* for $f(x)$.

The Nth-order Legendre polynomial series is the closest Nth-order polynomial to f as measured in the 2-norm. Therefore

$$\|f(x) - a_0 P_0(x) - \cdots - a_N P_N(x)\|_2 \leq \|f(x) - b_0 P_0(x) - \cdots - b_N P_N(x)\|_2$$

or

$$\int_{-1}^{1} (f(x) - a_0 P_0(x) - \cdots - a_N P_N(x))^2 dx$$
$$\leq \int_{-1}^{1} (f(x) - b_0 P_0(x) - \cdots - b_N P_N(x))^2 dx,$$

where the a_j are defined by Equation (7.13) and the b_j are any other real numbers. The Nth-order Legendre polynomial series is sometimes referred to as the Nth-order polynomial that best approximates f in the *least squares* sense. Compared with Legendre polynomials, other polynomials may reduce the error at a single point, such as a Taylor series taken about that point. Moreover, other polynomials may reduce the error over any subdomain of $[-1, 1]$; however, no other Nth-order polynomial can reduce the error over the entire domain $[-1, 1]$, as measured in the 2-norm, compared with the Nth-order Legendre polynomial series.

From Section 6.3, recall that there are at least three common definitions for functional convergence: pointwise convergence, convergence in the mean (2-norm), and uniform convergence (∞-norm). The Nth-order Legendre polynomial series pointwise converges to the true function as $N \to \infty$. In other words, if f is piecewise-smooth with domain $[-1, 1]$, then the Nth-order Legendre polynomial series pointwise converges to $f(x)$ for all $-1 \leq x \leq 1$ where f is continuous; furthermore, the Nth-order Legendre polynomial series pointwise converges to the average $(f_L + f_R)/2$, where f jumps from f_L to f_R. The Nth-order Legendre polynomial series also converges in the mean to the true function as $N \to \infty$. In other words:

$$\lim_{N\to\infty} \|f - a_0 P_0 - \cdots - a_N P_N\|_2 = 0$$

or

$$\lim_{N\to\infty} \int_{-1}^{1} (f(x) - a_0 P_0(x) - \cdots - a_N P_N(x))^2 dx = 0.$$

These convergence results are not quite so rosy as they may first appear. First off, the rate of convergence may be extremely low, especially in the presence of jump discontinuities. Furthermore, the Nth-order Legendre polynomial series may not converge uniformly. In other words, the maximum or ∞-norm error may remain large regardless of N.

Example 7.2 Find the second-order Legendre polynomial series for

$$f(x) = \begin{cases} 0 & -1 \leq x < 0, \\ x & 0 \leq x \leq 1. \end{cases}$$

Solution Assuming the reader can do the integrations seen in Equation (7.13), the required coefficients are $a_0 = 1/4$, $a_1 = 1/2$, and $a_2 = 5/16$. Then the second-order

Legendre polynomial series is

$$\frac{1}{4}P_0(x) + \frac{1}{2}P_1(x) + \frac{5}{16}P_2(x) = \frac{1}{4} + \frac{1}{2}x + \frac{5}{16}\cdot\frac{1}{2}(3x^2 - 1).$$

This result has been written in terms of the basis (P_0, P_1, P_2). However, $(1, x, x^2)$ is also a basis for the same space, albeit a nonorthogonal basis. In terms of basis $(1, x, x^2)$, the Legendre polynomial series becomes

$$\frac{15}{32}x^2 + \frac{1}{2}x + \frac{3}{32}.$$

Since Taylor series use the basis $(1, x, x^2)$, this is sometimes called the *Taylor series form* of the Legendre polynomial series. However, although this is in Taylor series form, the Legendre polynomial series does *not* equal the Taylor series! The Taylor series taken about b equals zero for b less than zero, equals x for b greater than zero, and is undefined for b equal to zero. Then the Taylor series captures the function perfectly on one side of $x = 0$, either the left or the right side, but gets the function completely wrong on the other side of $x = 0$, where $x = 0$ is the location of the jump discontinuity in the first derivative. Basically, Taylor series use very detailed information about the function at a single point; however, a single point cannot indicate the presence of a distant jump discontinuity in the function or its derivatives, so that the Taylor series becomes invalid for points on the other side of a jump discontinuity in the function or its derivatives. Thus Taylor series minimize the error at a single point while Legendre polynomial series and other series minimize the error over an entire region. Note that, unlike Legendre polynomial series, Taylor series are not vectors and thus do not properly belong in this discussion.

Example 7.3 Plot the 20th-order polynomial that has the minimum least-squares error approximating $f(x)$ where

$$f(x) = \begin{cases} 0 & 1/3 < |x| < 1, \\ 1 & |x| \le 1/3. \end{cases}$$

Solution The 20th-order Legendre polynomial series is plotted in Figure 7.3. The jump discontinuities at $x = \pm 1/3$ cause the Legendre polynomial series to oscillate about the correct function throughout the domain. The largest errors occur just to the left and right of the jump discontinuities and at the endpoints $x = \pm 1$. Because they are so narrow, the errors near the endpoints are a bit hard to see; however, they are just as large as the errors near the jump discontinuities.

As the order N increases, the number of oscillations increases, the period of the oscillations decreases, and the amplitude of the oscillations decreases except near the endpoints and jump discontinuities. There is always one large oscillation near each jump discontinuity and endpoint – the period of these oscillations decreases but the amplitudes do not decrease as $N \to \infty$. Thus there are always a few increasingly narrow bands where the error remains large. These spurious oscillations are sometimes called *Gibbs oscillations*. The Legendre polynomial in the limit of very large N appears roughly as shown in Figure 7.4.

Although the Legendre polynomial series may have the minimum error in a least squares sense, it does not really capture the essence of the square wave – in fact, just looking at the graph of the Legendre polynomial series, it could be hard to guess what function it was

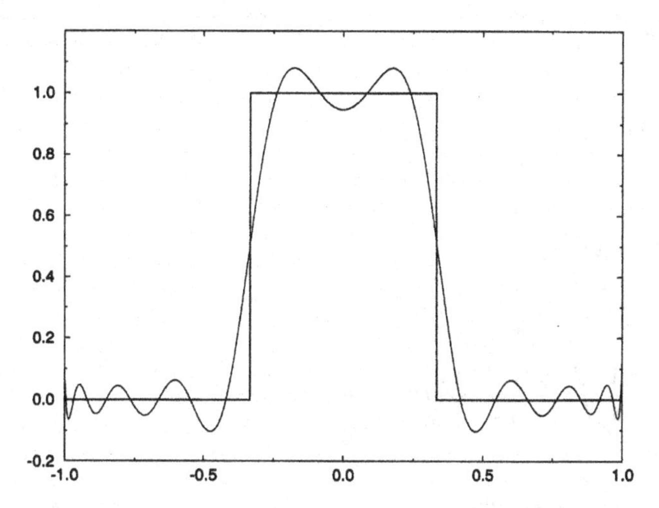

Figure 7.3 Twentieth-order Legendre polynomial series for a square wave.

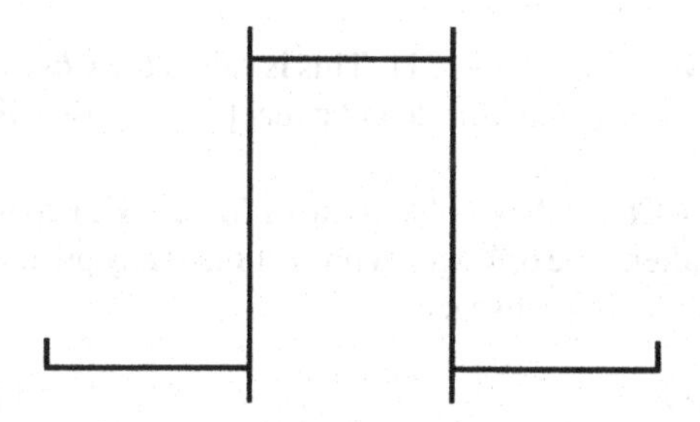

Figure 7.4 Very high-order Legendre polynomial series for a square wave.

trying to approximate, as least for small N. It might be well worth incurring a little extra 2-norm error to obtain an approximation that better captured the shape of the square wave and, in particular, was free of spurious overshoots and oscillations. This can be done using piecewise-polynomial approximations, as seen in Chapter 9.

7.3 Chebyshev Polynomial Series

The Chebyshev polynomials $T_j(x)$ are defined as follows:

$$T_0(x) = 1, \qquad T_1(x) = x,$$
$$T_{j+1}(x) = 2xT_j(x) - T_{j-1}(x). \tag{7.14}$$

For example,

$$T_2(x) = 2x^2 - 1, \qquad T_3(x) = 4x^3 - 3x.$$

Like the Legendre polynomials, the Chebyshev polynomials T_j form a basis for the space of piecewise-smooth functions with domains $[-1, 1]$. This basis is orthogonal using the following inner product:

$$f \cdot g = \int_{-1}^{1} \frac{f(x)g(x)}{\sqrt{1-x^2}} dx. \tag{7.15}$$

Then

$$f(x) = \sum_{i=0}^{\infty} a_j T_j(x), \tag{7.16}$$

where

$$a_0 = \frac{f \cdot T_0}{T_0 \cdot T_0} = \frac{1}{\pi} \int_{-1}^{1} \frac{f(x)T_j(x)}{\sqrt{1-x^2}} dx \tag{7.17}$$

and

$$a_j = \frac{f \cdot T_j}{T_j \cdot T_j} = \frac{2}{\pi} \int_{-1}^{1} \frac{f(x)T_j(x)}{\sqrt{1-x^2}} dx \tag{7.18}$$

for any piecewise-smooth function $f(x)$ with domain $[-1, 1]$. This is called the *Chebyshev polynomial series* for $f(x)$. For functions whose domains do not equal $[-1, 1]$, use a linear mapping from the domain $[a, b]$ to $[-1, 1]$.

Like Legendre polynomials, the first N Chebyshev polynomials $(T_0, \ldots, T_N)$ span the subspace of polynomials of order N or smaller. The orthogonal projection of any piecewise-smooth function f with domain $[-1, 1]$ onto this subspace is

$$f(x) \approx a_0 T_0(x) + \cdots + a_N T_N(x), \tag{7.19}$$

where the a_j are defined by Equations (7.17) and (7.18). This is sometimes called a *truncated Chebyshev polynomial series* or an *Nth-order Chebyshev polynomial series* for f. The Nth-order Chebyshev polynomial series is the closest Nth-order polynomial to f as measured in the natural norm $\|f\|^2 = f \cdot f$. In other words,

$$\int_{-1}^{1} \frac{(f(x) - a_0 T_0(x) - \cdots - a_N T_N(x))^2}{\sqrt{1-x^2}} dx \leq \int_{-1}^{1} \frac{(f(x) - b_0 T_0(x) - \cdots - b_N T_N(x))^2}{\sqrt{1-x^2}} dx,$$

where the coefficients a_j are defined by Equations (7.17) and (7.18) and where the coefficients b_j are any other real numbers.

The natural norm for Chebyshev polynomials includes a *weighting* function $w(x) = 1/\sqrt{1-x^2}$. By contrast, the natural norm for Legendre polynomial series is the 2-norm, which employs a uniform weighting function $w(x) = 1$. However, as we have seen, Legendre polynomial series tend to experience large errors near the endpoints. The weighting function $w(x) = 1/\sqrt{1-x^2}$ heavily penalizes errors near the endpoints ± 1 and thus results in a more uniform error distribution. For a given polynomial order, the *minimax* polynomial is the polynomial that minimizes the maximum error. In other words, the minimax polynomial minimizes the ∞-norm of the error. Unfortunately, there is no easy way to obtain the minimax polynomial, but in many cases the Chebyshev polynomial comes close.

The expressions (7.17) and (7.18) are a bit awkward unless $f(x)$ happens to include a factor of $\sqrt{1-x^2}$. Fortunately, there is a simple alternative. The nth-order Chebyshev polynomial $T_n(x)$ has exactly n roots, $z_{n1}, \ldots, z_{nn}$, where

$$z_{ni} = \cos\left(\frac{\pi(2i-1)}{2n}\right). \tag{7.20}$$

It can be shown that

$$a_0 = \frac{1}{N+1}\sum_{i=1}^{N+1} f(z_{N+1,i}) \tag{7.21}$$

and

$$a_n = \frac{2}{N+1}\sum_{i=1}^{N+1} f(z_{N+1,i})T_n(z_{N+1,i}). \tag{7.22}$$

Example 7.4 Find the second-order Chebyshev polynomial series for

$$f(x) = \begin{cases} 0 & -1 \le x < 0, \\ x & 0 \le x \le 1. \end{cases}$$

Solution By Equation (7.20), the roots of $T_3(x)$ are $\pm\sqrt{3}/2$ and 0. Assuming the reader can do the evaluations and sums seen in Equations (7.21) and (7.22), the required coefficients are $a_0 = a_2 = 1/(2\sqrt{3})$ and $a_1 = 1/2$. Then the second-order Chebyshev polynomial series is

$$\frac{1}{2\sqrt{3}}T_0(x) + \frac{1}{2}T_1(x) + \frac{1}{2\sqrt{3}}T_2(x) = \frac{1}{2}x + \frac{1}{\sqrt{3}}x^2,$$

where the expression on the right-hand side is the Taylor series form. The maximum error of the Chebyshev polynomial series occurs at $\pm\sqrt{3}/4 = \pm 0.433$ and is equal to $\sqrt{3}/16 = 0.108$. Compare this to the Legendre polynomial series found in Example 7.2, whose maximum error occurs at $\pm 8/15 = \pm 0.533$ and is equal to 0.110. Thus, in this case, the difference in the maximum error is only slight. In fact, overall, the Legendre polynomial series and the Chebyshev polynomial series are surprisingly similar in this case, illustrating that two polynomials may be similar even when their Taylor series coefficients differ radically; this will be an important point later in the text.

Example 7.5 Plot the 20th-order polynomial Chebyshev polynomial for

$$f(x) = \begin{cases} 0 & 1/3 < |x| < 1, \\ 1 & |x| \le 1/3. \end{cases}$$

Solution The 20th-order Chebyshev polynomial series is plotted in Figure 7.5. The discussion of Example 7.3 applies again here, except that the large spike in the Gibbs oscillations near the endpoints is eliminated. Then the Chebyshev polynomial in the limit of very large N appears roughly as shown in Figure 7.6.

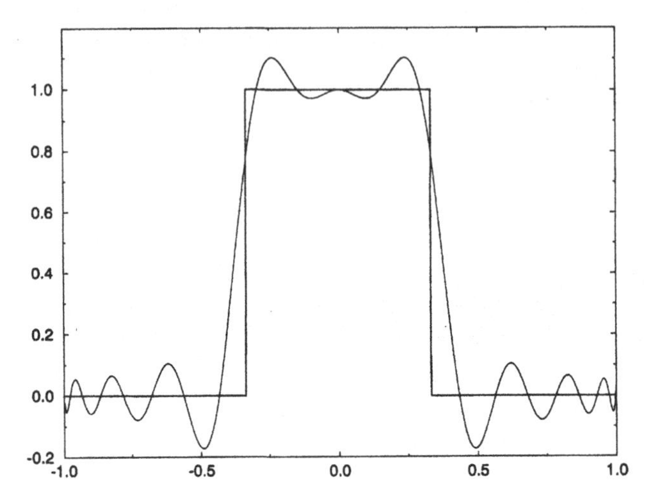

Figure 7.5 Twentieth-order Chebyshev polynomial series for a square wave.

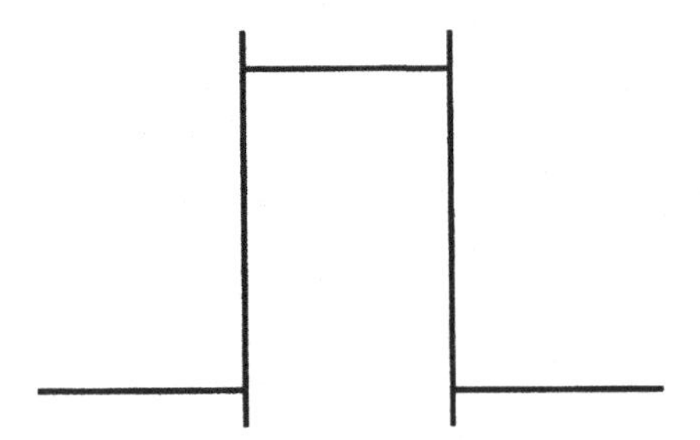

Figure 7.6 Very high-order Chebyshev polynomial series for a square wave.

7.4 Fourier Series

Consider the trigonometric functions 1, $\cos x$, $\cos 2x$, $\cos 3x, \ldots$ and $\sin x$, $\sin 2x$, $\sin 3x, \ldots$. These trigonometric functions form a basis for the space of piecewise-smooth functions with domains $[0, 2\pi]$. This basis is orthogonal using the following inner product:

$$f \cdot g = \int_0^{2\pi} f(x)g(x)\,dx. \tag{7.23}$$

Then

$$f(x) = a_0 + \sum_{n=1}^{\infty} a_n \cos nx + \sum_{n=1}^{\infty} b_n \sin nx, \tag{7.24}$$

where

$$a_0 = \frac{f \cdot 1}{1 \cdot 1} = \frac{1}{2\pi} \int_0^{2\pi} f(x)\,dx, \tag{7.25}$$

$$a_n = \frac{f \cdot \cos nx}{\cos nx \cdot \cos nx} = \frac{1}{\pi} \int_0^{2\pi} f(x) \cos nx\,dx, \tag{7.26}$$

and

$$b_n = \frac{f \cdot \sin nx}{\sin nx \cdot \sin nx} = \frac{1}{\pi} \int_0^{2\pi} f(x) \sin nx \, dx \tag{7.27}$$

for any piecewise-smooth function $f(x)$ with domain $[0, 2\pi]$. This is called a *Fourier series* for $f(x)$.

The first $2N + 1$ trigonometric functions ($1, \cos x, \ldots, \cos Nx$ and $\sin x, \ldots, \sin Nx$) form the basis for a $(2N+1)$-dimensional subspace of functions. The orthogonal projection of any piecewise-smooth function f with domain $[0, 2\pi]$ onto this subspace is

$$f(x) \approx a_0 + \sum_{n=1}^{N} a_n \cos nx + \sum_{n=1}^{N} b_n \sin nx, \tag{7.28}$$

where a_n and b_n are defined by Equations (7.25), (7.26), and (7.27). This is sometimes called a *truncated Fourier series* or a *2N-order Fourier series* for f. The truncated Fourier series is the closest function to f in the subspace spanned by ($1, \cos x, \ldots, \cos Nx$) and ($\sin x, \ldots, \sin Nx$) in a least squares sense. Whereas both Legendre polynomial series and Fourier series minimize 2-norm error, they minimize the 2-norm error over different subspaces.

The truncated Fourier series pointwise converges to the true function as $N \to \infty$. In other words, if f is piecewise-smooth with domain $[0, 2\pi]$ then the truncated Fourier series converges to $f(x)$ for all $0 \le x \le 2\pi$ where f is continuous; furthermore, the truncated Fourier series converges to the average $(f_L + f_R)/2$ where f jumps from f_L to f_R. The truncated Fourier series also converges in the mean to the true function as $N \to \infty$. However, the rate of convergence may be extremely low, especially in the presence of jump discontinuities. Furthermore, the truncated Fourier series may not converge uniformly: there may always be points where the error is large, regardless of N. This is yet another example of how convergence alone may not mean much – only a reasonably rapid rate of convergence in the ∞-norm guarantees a genuinely high quality approximation.

Outside of the domain $[0, 2\pi]$, Fourier series repeat themselves with period 2π. Therefore the domains $[2\pi, 4\pi]$, $[4\pi, 6\pi]$, $[-2\pi, 0]$, and $[-4\pi, -2\pi]$ all contain exact copies of the Fourier series on the domain $[0, 2\pi]$. In most cases, this repetition is an artifact of the Fourier series. However, some functions – called *periodic* functions – behave exactly this way. For example, all trigonometric functions are periodic. While particularly well suited for representing periodic functions, Fourier series can represent *any* function on the domain $[0, 2\pi]$. However, if the function is not periodic, the Fourier series will perceive any difference between $f(0)$ and $f(2\pi)$ as a jump discontinuity in the function. Also, the Fourier series will perceive any difference between $f'(0)$ and $f'(2\pi)$ as a jump discontinuity in the first derivative, and similarly for higher-order derivatives. Jump discontinuities in the function or its derivatives cause difficulties for Fourier series, just as for Legendre and Chebyshev polynomial series.

A function is called *bandlimited* if

$$f(x) = a_0 + \sum_{n=1}^{N} a_n \cos nx + \sum_{n=1}^{N} b_n \sin nx \tag{7.29}$$

for some N. In other words, for bandlimited functions, the truncated Fourier series is *exact* for some finite N. The analogous situation to bandlimiting for polynomial series, such as Legendre and Chebyshev polynomial series, is that a function is an Nth-order polynomial for some N. Although few physical functions are true polynomials, many physical functions are naturally bandlimited. Bandlimiting is often specified in terms of the shortest wavelength $2\pi/N$ or the highest frequency $N/2\pi$ rather than in terms of N. Discontinuous and nonsmooth functions cannot be bandlimited; then, according to the discussion in the previous paragraph, nonperiodic functions also cannot be bandlimited.

For functions whose domains do not equal $[0, 2\pi]$, we can use a linear mapping from the domain $[a, b]$ to $[0, 2\pi]$. Specifically, the trigonometric functions

$$\cos\left(2\pi n\frac{x-a}{b-a}\right), \qquad \sin\left(2\pi n\frac{x-a}{b-a}\right)$$

span the piecewise-smooth functions $f(x)$ with finite domains $[a, b]$. Then

$$f(x) \approx a_0 + \sum_{n=1}^{N} a_n \cos\left(2\pi n\frac{x-a}{b-a}\right) + \sum_{n=1}^{N} b_n \sin\left(2\pi n\frac{x-a}{b-a}\right), \tag{7.30}$$

where

$$a_0 = \frac{1}{b-a}\int_a^b f(x)\,dx, \tag{7.31}$$

$$a_n = \frac{2}{b-a}\int_a^b f(x)\cos\left(2\pi n\frac{x-a}{b-a}\right)dx, \tag{7.32}$$

and

$$b_n = \frac{2}{b-a}\int_a^b f(x)\sin\left(2\pi n\frac{x-a}{b-a}\right)dx. \tag{7.33}$$

Example 7.6 Plot the 20th-order Fourier series for

$$f(x) = \begin{cases} 0 & 1/3 < |x| < 1, \\ 1 & |x| \le 1/3. \end{cases}$$

Solution Notice that $N = 10$ for 20th-order Fourier series. The 20th-order Fourier series is plotted in Figure 7.7. Comparing with the Legendre polynomial series seen in Example 7.3 and the Chebyshev polynomial series seen in Example 7.5, we see that the Gibbs oscillations in the Fourier series are noticeably smaller. The Fourier series for $N \to \infty$ looks much like the Chebyshev polynomial series for $N \to \infty$, as seen in Figure 7.6, where the "hiccups" at each jump discontinuity are approximately 9% of the size of the jumps. The problem areas due to the Gibbs phenomenon in Fourier series and in the earlier polynomial series are like those of the ghetto – while you can make the ghetto as small as you like, things are still just as bad for the people left in the ghetto.

The trigonometric functions $(1, \cos x, \cos 2x, \cos 3x, \ldots)$ form a basis for the space of piecewise-smooth *even* functions with domains $[0, 2\pi]$; any even function expressed in this basis is called a *Fourier cosine series*. Similarly, the trigonometric functions $(\sin x, \sin 2x, \sin 3x, \ldots)$ form a basis for the space of piecewise-smooth *odd* functions with domains $[0, 2\pi]$; any odd function expressed in this basis is called a *Fourier sine series*. In

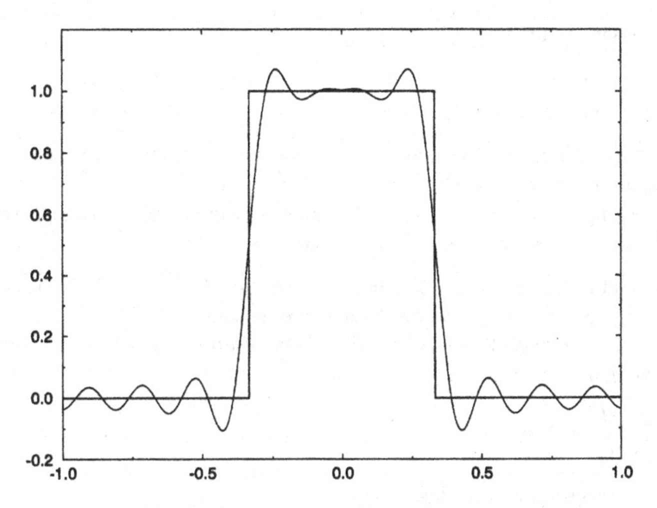

Figure 7.7 Twentieth-order Fourier series for a square wave.

general, any function expressed in terms of any basis of trigonometric functions is called a *trigonometric series*; Fourier series, Fourier cosine series, and Fourier sine series are all examples of trigonometric series.

This section and the previous two described several important sets of orthogonal basis functions. Of course, these are just a few examples among many. Some other well-known orthogonal basis functions include Laguerre polynomials, Hermite polynomials, and Jacobi polynomials; see the references below for further details. However, regardless of the choice of basis function, no single finite-length vector can adequately represent a discontinuous function. Instead, we require piecewise-polynomial approximations, as discused in Chapter 9.

References

Anton, H. 1981. *Elementary Linear Algebra*, 3rd ed., New York: Wiley.

Canuto, C., Hussaini, M. Y., Quarteroni, A., and Zang, T. A. 1988. *Spectral Methods in Fluid Dynamics*, Berlin: Springer-Verlag.

Davis, H. F. 1963. *Fourier Series and Orthogonal Functions*, New York: Dover.

Giannakouros, J., and Karniadakis, G. E. 1994. "A Spectral Element–FCT Method for the Compressible Euler Equations," *Journal of Computational Physics*, 115: 65–85.

Press, W. H., Teukolsky, S. A., Vetterling, W. T., and Flannery, B. P. 1992, *Numerical Recipes in C: The Art of Scientific Computing*, 2nd ed., New York: Cambridge University Press, Sections 4.5, 5.5, and 5.8.

Sanders, R. 1988. "A Third-Order Accurate Variation Nonexpansive Difference Scheme for Single Nonlinear Conservation Laws," *Mathematics of Computation*, 51: 535–558.

Van Leer, B. 1977. "Towards the Ultimate Conservative Difference Scheme. IV. A New Approach to Numerical Convection," *Journal of Computational Physics*, 23: 276–299.

Problems

7.1 Find the second-order Legendre polynomial series for $f(x) = \sin x$ on the domain $[-1, 1]$. Compare with the second-order Taylor series about $x = 0$.

7.2 Consider the following function:

$$f(x) = \begin{cases} 0 & 1 \leq |x| \leq 2, \\ 1 - x^2 & |x| < 1. \end{cases}$$

(a) Find the single quadratic that best approximates this function in the least squares sense over the function's entire domain.

(b) Roughly speaking, what happens to the Nth-order polynomial that best approximates this function in the least square sense as $N \to \infty$?

7.3 (a) Find the second-order Chebyshev polynomial series for $f(x) = e^{-x^2}$ on the domain $[-1, 1]$. Compare with the second-order Taylor series about $x = 0$.

(b) Show that the coefficients of the Legendre polynomial series can be written in terms of the *error function* defined as follows:

$$\operatorname{erf}(x) = \frac{2}{\sqrt{x}} \int_0^x e^{-y^2} dy.$$

7.4 Find the Fourier series for the following function:

$$f(x) = \begin{cases} 0 & -1 \leq x < 0, \\ x & 0 \leq x \leq 1. \end{cases}$$

Compare with the Chebyshev polynomial series found in Example 7.4. In particular, is the Chebyshev series more or less oscillatory than the Fourier series?

7.5 The *Hermite* polynomials $H_j(x)$ are defined as follows:

$$H_0(x) = 1,$$
$$H_1(x) = 2x,$$
$$H_{j+1}(x) = 2xH_j(x) - 2jH_{j-1}(x).$$

For example,

$$H_2(x) = 4x^2 - 2,$$
$$H_3(x) = 8x^3 - 12x.$$

(a) Prove that $H_0(x)$, $H_1(x)$, and $H_2(x)$ are orthogonal using the following inner product:

$$f \cdot g = \int_{-\infty}^{\infty} e^{-x^2} f(x)g(x)dx.$$

Also, find the norms of $H_0(x)$, $H_1(x)$, and $H_2(x)$ using the natural norm. You may wish to use a table of integrals.

(b) Assume that the full set of Hermite polynomials is orthogonal and that $\|H_j(x)\| = 2^j j! \sqrt{\pi}$. Find a general expression for any piecewise-smooth integrable function with domain $(-\infty, \infty)$ in terms of an infinite Hermite polynomial series. Also, for any such function, find the closest truncated Hermite polynomial series. In what sense is this finite-length polynomial series "closest" to the function? In other words, which errors does the weighting function e^{-x^2} tolerate and which errors does it penalize? Do you really expect Hermite polynomial series to accurately approximate a function over an entire infinite domain?

(c) Given their norms, what numerical problems would you expect when using Hermite polynomial series. Show that these problems can be eliminated by using *orthonormal* Hermite polynomials defined recursively as follows:

$$\tilde{H}_0(x) = \frac{1}{\pi^{1/4}}, \qquad \tilde{H}_1(x) = \frac{x}{\pi^{1/4}},$$

$$\tilde{H}_{j+1}(x) = x\sqrt{\frac{2}{j+1}}\tilde{H}_j(x) - \sqrt{\frac{j}{j+1}}H_{j-1}(x).$$

CHAPTER 8

Interpolation

8.0 Introduction

This chapter concerns polynomial and trigonometric interpolation. Polynomial interpolation leads to piecewise-polynomial reconstruction in Chapter 9 and to numerical differentiation and integration formulae in Chapter 10. As seen in Section 6.3, computers represent functions by sequences of real numbers. In the last chapter, the numbers were coefficients in functional forms such as Legendre polynomial series. In this chapter, the numbers are samples. In particular, consider any set of *sample points* $(x_0, x_1, \ldots, x_N)$ in the domain of a function $f(x)$. Then the *samples* $f_i = f(x_i)$ represent the function $f(x)$. Notice that the total number of samples is $N+1$ since the sample index starts from zero rather than one. The spacing between samples is $\Delta x_i = x_i - x_{i-1}$. If the sample spacing is constant then

$$\Delta x = \frac{x_N - x_0}{N}. \tag{8.1}$$

For example, six samples with $x_5 - x_0 = 5$ have the sample spacing $\Delta x = 1$. Samples are especially popular for solving differential equations, since differential equations are "pointwise" equations (i.e., they describe solutions in terms of rates of change at individual points). *Finite-difference* methods use samples as their primary representation.

Finite-difference methods often switch from samples to functional representations in order to differentiate or to perform other functional operations. Any function created from samples is called a *reconstruction*. Any reconstruction passing through the sample points is called an *interpolation* on the domain $[x_0, x_N]$ and an *extrapolation* on the domains $(-\infty, x_0)$ and (x_N, ∞). This chapter concerns interpolations that yield a single polynomial. The next chapter describes interpolations and other sorts of reconstructions that yield piecewise-polynomials.

The quality of a sampling is gauged by how well the original function can be reconstructed from the samples. Of course, like all finite-length sequences, samples cannot adequately represent just any arbitrary function.

Example 8.1 Suppose that a function has three extrema but only two samples. Then the two samples must inevitably miss important information about the function, as illustrated in Figure 8.1. Specifically, in every case, the two samples lose critical information about the size or location of an extremum. Thus it is impossible to accurately reconstruct the original function from only two samples. This example shows that, at the very least, there should be one sample near every extremum. Not counting the endpoints of the function, the function in this example requires at least three samples.

Example 8.2 Given only samples, and no other information about the function, it may be impossible to accurately reconstruct the original function. For example, any number

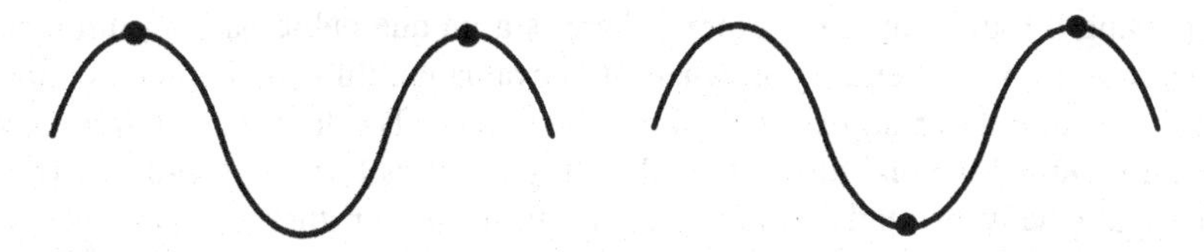

Figure 8.1 Two samples cannot represent three extrema.

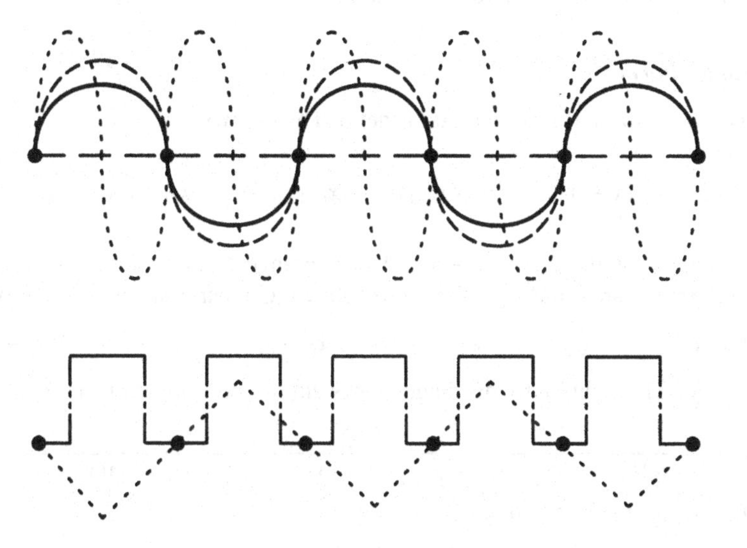

Figure 8.2 Different functions have the same samples.

of functions have uniformly zero samples $f(x_i) = 0$ as illustrated in Figure 8.2. Of course, referring to the previous example, there is absolutely no chance of reconstructing the true function if it is anything other than $f(x) = 0$ – for all of the other possible functions shown, the samples clearly miss critical information about the function.

As seen in these examples, one must know something more about a function other than its samples. In keeping with the spirit of the last chapter, one can restrict the possible functions to a limited finite-dimensional subspace, such that the samples uniquely specify a function from the subspace. In this case, the known subspace constitutes the "additional information" about the function required to make sampling work. As one possibility, suppose we know that the true function is a finite-length polynomial. Then the Nth-order polynomial interpolation equals the true function for large enough N. As another possibility, suppose we know that the true function is bandlimited or, in other words, that the true function equals a finite-length Fourier series. Then an Nth-order trigonometric interpolation equals the true function for large enough N.

8.1 Polynomial Interpolation

There is a unique Nth-order polynomial passing through any set of $N + 1$ samples. For example, there is a unique line passing through any two points; there is a unique quadratic passing through any three points; there is a unique cubic passing through any four points; and so on. Depending on where it is evaluated, this polynomial is either an interpolation polynomial or an extrapolation polynomial. The last chapter introduced a number of different polynomial forms, including Taylor series form, Legendre polynomial series form, and Chebyshev polynomial series form. However, these are not necessarily the best forms for expressing interpolation polynomials; this section introduces two new polynomial forms that are well suited for interpolation.

8.1.1 *Lagrange Form*

The *Lagrange form* of a polynomial is defined as follows:

$$p_N(x) = \sum_{i=0}^{N} a_i(x - x_0)(x - x_1)\cdots(x - x_{i-1})(x - x_{i+1})\cdots(x - x_{N-1})(x - x_N).$$

Notice that the sum skips the factor $x - x_i$. The Lagrange form coefficients a_i of the interpolation polynomial are found by solving the following linear systems of equations:

$$f(x_i) = a_i(x_i - x_0)(x_i - x_1)\cdots(x_i - x_{i-1})(x_i - x_{i+1})\cdots(x_i - x_{N-1})(x_i - x_N),$$

where $i = 0, \ldots, N$. The solution to this diagonal system of linear equations is trivial:

$$a_i = \frac{f(x_i)}{(x_i - x_0)(x_i - x_1)\cdots(x_i - x_{i-1})(x_i - x_{i+1})\cdots(x_i - x_{N-1})(x_i - x_N)}.$$

Consider the following standard notation:

$$\prod_{i=1}^{N} x_i = x_1 \cdots x_N, \tag{8.2}$$

where the capital pi stands for product. Then the Lagrange form of the interpolation polynomial can be written as

$$p_N(x) = \sum_{i=0}^{N} f(x_i) \prod_{\substack{j=0 \\ j\neq i}}^{N} \frac{x - x_j}{x_i - x_j}. \tag{8.3}$$

The Lagrange form is easy to derive and easy to remember but may be difficult to work with (for example, integration and differentiation are difficult in Lagrange form). The Lagrange form is discussed here mainly for completeness.

Example 8.3 Find the Lagrange form of the interpolation polynomial passing through the following points: $(-1, 1)$, $(0, 2)$, $(3, 101)$, $(4, 246)$.

Solution The Lagrange form of the interpolation cubic is

$$p_3(x) = -\frac{1}{20}x(x-3)(x-4) + \frac{1}{6}(x+1)(x-3)(x-4)$$
$$- \frac{101}{12}x(x+1)(x-4) + \frac{123}{10}x(x+1)(x-3).$$

8.1.2 Newton Form

The *Newton form* of a polynomial is defined as follows:

$$p_N(x) = a_0 + a_1(x - x_0) + a_2(x - x_0)(x - x_1) + a_3(x - x_0)(x - x_1)(x - x_2) + \cdots + a_N(x - x_0)\cdots(x - x_{N-1})$$

or

$$p_N(x) = a_0 + \sum_{i=1}^{N} a_i \prod_{j=0}^{i-1}(x - x_j).$$

The Newton form coefficients a_i of the interpolation polynomial are found by solving the following triangular linear systems of equations:

$$\begin{aligned}
f(x_0) &= a_0,\\
f(x_1) &= a_0 + a_1(x_1 - x_0),\\
f(x_2) &= a_0 + a_1(x_2 - x_0) + a_2(x_2 - x_0)(x_2 - x_1),\\
&\vdots\\
f(x_N) &= a_0 + a_1(x_N - x_0) + a_2(x_N - x_0)(x_N - x_1)\\
&\quad + \cdots + a_N(x_N - x_0)(x_N - x_1)\cdots(x_N - x_{N-1}).
\end{aligned}$$

Finding the solutions for a_0 and a_1 is relatively easy. In particular $a_0 = f(x_0)$ and

$$a_1 = \frac{f(x_1) - a_0}{x_1 - x_0} = \frac{f(x_1) - f(x_0)}{x_1 - x_0}.$$

Determining the solutions for a_2, a_3, and so on becomes progressively more difficult. Luckily, as it turns out, there is a simple recursive solution. To state this solution, we must first define the *Newton divided differences*. In particular, the *zeroth Newton divided differences* are defined as follows:

$$f[x_i] = f(x_i),$$

where $i = 0, \ldots, N$. The *first Newton divided differences* are then defined as

$$f[x_i, x_{i+1}] = \frac{f[x_{i+1}] - f[x_i]}{x_{i+1} - x_i} = \frac{f(x_{i+1}) - f(x_i)}{x_{i+1} - x_i},$$

where $i = 0, \ldots, N - 1$. The *second Newton divided differences* are defined as

$$\begin{aligned}
f[x_i, x_{i+1}, x_{i+2}] &= \frac{f[x_{i+1}, x_{i+2}] - f[x_i, x_{i+1}]}{x_{i+2} - x_i}\\
&= \frac{1}{x_{i+2} - x_i}\left(\frac{f(x_{i+2}) - f(x_{i+1})}{x_{i+2} - x_{i+1}} - \frac{f(x_{i+1}) - f(x_i)}{x_{i+1} - x_i}\right),
\end{aligned}$$

where $i = 0, \ldots, N - 2$. The *third Newton divided differences* are defined as

$$f[x_i, x_{i+1}, x_{i+2}, x_{i+3}] = \frac{f[x_{i+1}, x_{i+2}, x_{i+3}] - f[x_i, x_{i+1}, x_{i+2}]}{x_{i+3} - x_i},$$

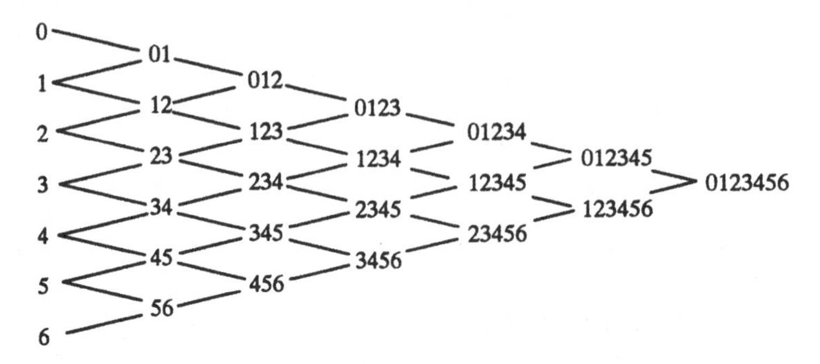

Figure 8.3 Formation of Newton divided differences.

where $i = 0, \ldots, N - 3$. In general, the *nth Newton divided differences* are defined as follows:

$$♦ \quad f[x_i, \ldots, x_{i+n}] = \frac{f[x_{i+1}, \ldots, x_{i+n}] - f[x_i, \ldots, x_{i+n-1}]}{x_{i+n} - x_i}, \tag{8.4}$$

where $i = 0, \ldots, N - n$. As a way to help remember this definition, notice that the denominator uses the two sample points omitted from one or the other of the divided differences in the numerator. Using this recursive definition, the Newton divided differences form a triangle such as the one for $f[x_0, x_1, x_2, x_3, x_4, x_5, x_6]$ seen in Figure 8.3.

For future reference, it is interesting to note the connection between the Newton divided differences and functional derivatives. If $f(x)$ has n continuous derivatives in $[x_i, x_{i+n}]$ then

$$f[x_i, \ldots, x_{i+n}] = \frac{1}{n!}\frac{d^n f}{dx^n}(\xi) \tag{8.5}$$

for some $x_i < \xi < x_{i+n}$. If the pth derivative of $f(x)$ has a jump discontinuity at z in $[x_i, x_{i+n}]$, but otherwise $f(x)$ is continuously differentiable, then

$$f[x_i, \ldots, x_{i+n}] = O\left(\frac{1}{\Delta x^{n-p}}\left(\frac{d^p f(z_R)}{dx^p} - \frac{d^p f(z_L)}{dx^p}\right)\right) \tag{8.6}$$

for all $n \geq p$, where z_L and z_R refer to the left- and right-hand limits of the pth derivative, respectively. Thus the nth Newton divided difference is proportional to the jump in the pth derivative divided by Δx^{n-p}.

The coefficients of the Newton form of the interpolation polynomial are easily defined in terms of Newton divided differences as follows:

$$a_i = f[x_0, x_1, \ldots, x_i]. \tag{8.7}$$

In other words, the Newton form of the interpolation polynomial is as follows:

$$♦ \quad p_N(x) = f[x_0] + \sum_{i=1}^{N} f[x_0, \ldots, x_i] \prod_{j=0}^{i-1} (x - x_j). \tag{8.8}$$

Sensibly enough, the ordering of the sample points does not affect the Newton divided differences nor the interpolation polynomial.

The Newton form of the interpolation polynomial becomes particularly simple if the spacing between the samples is a constant Δx. Suppose $x_{i+1} - x_i = \Delta x = const.$ Then

$$f[x_i, x_{i+1}] = \frac{f(x_{i+1}) - f(x_i)}{\Delta x},$$

$$f[x_i, x_{i+1}, x_{i+2}] = \frac{f(x_{i+2}) - 2f(x_{i+1}) - f(x_i)}{2\Delta x^2},$$

$$f[x_i, x_{i+1}, x_{i+2}, x_{i+3}] = \frac{f(x_{i+3}) - 3f(x_{i+2}) + 3f(x_{i+1}) - f(x_i)}{6\Delta x^3},$$

and so on. Define *zeroth differences* by

$$\Delta_i^0 f = f(x_i),$$

where $i = 0, \ldots, N$, *first differences* by

$$\Delta_i^1 f = \Delta_{i+1}^0 f - \Delta_i^0 f = f(x_{i+1}) - f(x_i),$$

where $i = 0, \ldots, N-1$, *second differences* by

$$\Delta_i^2 f = \Delta_{i+1}^1 f - \Delta_i^1 f = f(x_{i+2}) - 2f(x_{i+1}) + f(x_i),$$

where $i = 0, \ldots, N-2$, *third differences* by

$$\Delta_i^3 f = \Delta_{i+1}^2 f - \Delta_i^2 f = f(x_{i+3}) - 3f(x_{i+2}) + 3f(x_{i+1}) - f(x_i),$$

where $i = 0, \ldots, N-3$, and *fourth differences* by

$$\Delta_i^4 f = \Delta_{i+1}^3 f - \Delta_i^3 f = f(x_{i+4}) - 4f(x_{i+3}) + 6f(x_{i+2}) - 4f(x_{i+1}) + f(x_i),$$

where $i = 0, \ldots, N-4$. In general, we can define *nth differences* as follows:

$$\Delta_i^n f = \Delta_{i+1}^{n-1} f - \Delta_i^{n-1} f, \tag{8.9}$$

where $i = 0, \ldots, N-n$. Then

$$f[x_i, \ldots, x_{i+n}] = \frac{\Delta_i^n f}{n!\Delta x^n} \tag{8.10}$$

and the Newton form of the interpolation polynomial is

$$\begin{aligned} p_N(x) = f(x_0) &+ \frac{\Delta_0^1 f}{\Delta x}(x - x_0) + \frac{1}{2}\frac{\Delta_0^2 f}{\Delta x^2}(x - x_0)(x - x_1) \\ &+ \frac{1}{6}\frac{\Delta_0^3 f}{\Delta x^3}(x - x_0)(x - x_1)(x - x_2) \\ &+ \cdots + \frac{1}{N!}\frac{\Delta_0^N f}{\Delta x^N}(x - x_0)(x - x_1)\cdots(x - x_{N-1}), \end{aligned}$$

which, not coincidentally, looks quite a bit like a Taylor series.

Example 8.4 Find the Newton form of the interpolation polynomial passing through the following points: $(-1, 1)$, $(0, 2)$, $(3, 101)$, $(4, 246)$.

Solution The first divided differences are

$$f[x_0, x_1] = \frac{f(x_1) - f(x_0)}{x_1 - x_0} = \frac{2-1}{0-(-1)} = 1,$$

$$f[x_1, x_2] = \frac{f(x_2) - f(x_1)}{x_2 - x_1} = \frac{101-2}{3-0} = 33,$$

$$f[x_2, x_3] = \frac{f(x_3) - f(x_2)}{x_3 - x_2} = \frac{246-101}{4-3} = 145.$$

The second divided differences are

$$f[x_0, x_1, x_2] = \frac{f[x_1, x_2] - f[x_0, x_1]}{x_2 - x_0} = \frac{33-1}{3-(-1)} = 8,$$

$$f[x_1, x_2, x_3] = \frac{f[x_2, x_3] - f[x_1, x_2]}{x_3 - x_1} = \frac{145-33}{4-0} = 28.$$

The third divided difference is

$$f[x_0, x_1, x_2, x_3] = \frac{f[x_1, x_2, x_3] - f[x_0, x_1, x_2]}{x_3 - x_0} = \frac{28-8}{4-(-1)} = 4.$$

Then the Newton form of the interpolation polynomial is

$$p_3(x) = 1 + (x+1) + 8x(x+1) + 4x(x+1)(x-3).$$

Although it may look different, this polynomial is the same as the one found in Example 8.3.

8.1.3 *Taylor Series Form*

Although better than the Lagrange form, the Newton form is still awkward for many functional operations including integration and differentiation. This section introduces the Taylor series form of the interpolation polynomial. The Taylor series form is the hardest to create, costing three times as much to create as the Newton form, but it is the easiest to differentiate or integrate. The Taylor series form of a polynomial is

$$p_N(x) = a_0 + a_1(x-b) + \cdots + a_N(x-b)^N.$$

Then the Taylor coefficients a_i of the interpolation polynomial are found by solving the following linear systems of equations:

$$f(x_i) = a_0 + a_1(x_i - b) + a_2(x_i - b)^2 + \cdots + a_N(x_i - b)^N,$$

where $i = 0, \ldots, N$. The solution involves coefficients d_{ij} defined as follows:

$$d_{0j}(b) = 1$$

for $j = 0, \ldots, N$. Also,

$$d_{i0}(b) = (b - x_{i-1})d_{i-1,0}(b)$$

for $i = 0, \ldots, N$. In general,

♦ $$d_{ij}(b) = d_{i,j-1}(b) + (b - x_{i+j-1})d_{i-1,j}(b) \tag{8.11}$$

for $i = 1, \ldots, N$ and $j = 1, \ldots, N$. The coefficients d_{ij} may be considered as elements of an $(N+1) \times (N+1)$ matrix D. The only caution is that the elements of D are indexed starting from 0 rather than 1. The elements in the zeroth row of D are all 1 whereas the elements in the zeroth column of D are $1, b-x_0, (b-x_0)(b-x_1), (b-x_0)(b-x_1)(b-x_2)$, and so on. In general, d_{ij} is formed by adding the element $d_{i,j-1}$ on the left and $(b-x_{i+j-1})$ times the element $d_{i-1,j}$ above.

To get a further sense of d_{ij}, consider the following: *$d_{ij}(b)$ is the sum of all possible products of i distinct factors chosen from the set of $i+j$ elements* $\{(b-x_0), (b-x_1), \ldots, (b-x_{i+j-1})\}$. Although elegant, this way of thinking about d_{ij} should never be used computationally for large N, since it would be extremely expensive compared to the recursive definition of Equation (8.11).

The coefficients of the Taylor series form of the interpolation polynomial can be written as follows:

$$\blacklozenge \qquad a_j = \sum_{i=0}^{N-j} d_{ij}(b) f[x_0, \ldots, x_{i+j}]. \tag{8.12}$$

Recall that the *diagonal* of a matrix runs from the upper left to the lower right. Similarly, the *antidiagonal* of a matrix runs from the lower right to the upper left. Thinking in terms of matrices, only the elements of D above the antidiagonal are required to form a_j – the elements below the antidiagonal can either be ignored or considered to be zero.

Example 8.5 Find a general expression for the Taylor series form of the interpolation polynomial passing through three samples $(x_0, f(x_0)), (x_1, f(x_1)), (x_2, f(x_2))$.

Solution In this case, $N = 2$ and D is the following 3×3 matrix:

$$D = \begin{bmatrix} d_{00} = 1 & d_{01} = 1 & d_{02} = 1 \\ d_{10} = b - x_0 & d_{11} = b - x_0 + b - x_1 & 0 \\ d_{20} = (b-x_0)(b-x_1) & 0 & 0 \end{bmatrix}.$$

For example, d_{11} is formed as follows:

$$d_{11} = d_{10} + (b - x_1)d_{01} = b - x_0 + (b - x_1) \times 1.$$

Then a general Taylor series form of the quadratic interpolation polynomial is

$$p_2(x) = a_2(x-b)^2 + a_1(x-b) + a_0,$$

where $a_2 = f[x_0, x_1, x_2]$,

$$a_1 = f[x_0, x_1] + (b - x_0 + b - x_1) f[x_0, x_1, x_2],$$

and

$$a_0 = f[x_0] + (b - x_0) f[x_0, x_1] + (b - x_0)(b - x_1) f[x_0, x_1, x_2].$$

Example 8.6 For $N = 3$, show that a general expression for D is

$$\begin{bmatrix} 1 & 1 & 1 & 1 \\ b - x_0 & b - x_0 + b - x_1 & \begin{matrix} b - x_0 + b \\ - x_1 + b - x_2 \end{matrix} & 0 \\ (b - x_0)(b - x_1) & \begin{matrix} (b - x_0)(b - x_1) \\ + (b - x_0)(b - x_2) \\ + (b - x_1)(b - x_2) \end{matrix} & 0 & 0 \\ (b - x_0)(b - x_1)(b - x_2) & 0 & 0 & 0 \end{bmatrix}.$$

Solution The element d_{21} is formed as follows:

$$\begin{aligned} d_{21} &= d_{20} + (b - x_2)\, d_{11} \\ &= (b - x_0)(b - x_1) + (b - x_2)(b - x_0 + b - x_1). \end{aligned}$$

Notice that d_{21} is the sum of all possible products of two distinct factors chosen from the set $\{(b - x_0), (b - x_1), (b - x_2)\}$. The other elements are formed similarly.

Example 8.7 Find a general expression for a third-order interpolation polynomial in Taylor series form.

Solution Any third-order interpolation polynomial can be written as follows:

$$p_3(x) = a_3(x - b)^3 + a_2(x - b)^2 + a_1(x - b) + a_0.$$

Using the matrix D from the previous example, the zeroth column of D yields

$$\begin{aligned} a_0 = {} & f[x_0] + (b - x_0) f[x_0, x_1] + (b - x_0)(b - x_1) f[x_0, x_1, x_2] \\ & + (b - x_0)(b - x_1)(b - x_2) f[x_0, x_1, x_2, x_3]. \end{aligned}$$

The first column of D yields

$$\begin{aligned} a_1 = {} & f[x_0, x_1] + (b - x_0 + b - x_1) f[x_0, x_1, x_2] \\ & + [(b - x_0)(b - x_1) + (b - x_0)(b - x_2) \\ & + (b - x_1)(b - x_2)] f[x_0, x_1, x_2, x_3]. \end{aligned}$$

The second column of D yields

$$a_2 = f[x_0, x_1, x_2] + (b - x_0 + b - x_1 + b - x_2) f[x_0, x_1, x_2, x_3].$$

Finally, the third column of D yields $a_3 = f[x_0, x_1, x_2, x_3]$.

Example 8.8 Find the Taylor series form of the interpolation polynomial with $b = 0$ passing through the following points:

$$(-1, 1), \quad (0, 2), \quad (3, 101), \quad (4, 246).$$

Solution From Example 8.4 recall that $f[x_0, x_1] = 1$, $f[x_0, x_1, x_2] = 8$, and $f[x_0, x_1, x_2, x_3] = 4$. Consider the expression for D given in Example 8.6. If $b = 0$,

$x_0 = -1$, $x_1 = 0$, $x_2 = 3$, and $x_3 = 4$ then

$$D = \begin{bmatrix} 1 & 1 & 1 & 1 \\ 1 & 1 & -2 & 0 \\ 0 & -3 & 0 & 0 \\ 0 & 0 & 0 & 0 \end{bmatrix}.$$

The coefficients of the Taylor series form polynomial are

$$\begin{aligned} a_0 &= d_{00} f(x_0) + d_{10} f[x_0, x_1] + d_{20} f[x_0, x_1, x_2] + d_{30} f[x_0, x_1, x_2, x_3] \\ &= (1 \times 1) + (1 \times 1) + (0 \times 8) + (0 \times 4) = 2, \\ a_1 &= d_{01} f[x_0, x_1] + d_{11} f[x_0, x_1, x_2] + d_{21} f[x_0, x_2, x_3] \\ &= (1 \times 1) + (1 \times 8) - (3 \times 4) = -3, \\ a_2 &= d_{02} f[x_0, x_1, x_2] + d_{12} f[x_0, x_1, x_2, x_3] \\ &= (1 \times 8) - (2 \times 4) = 0, \end{aligned}$$

and

$$a_3 = d_{03} f[x_0, x_1, x_2, x_3] = 1 \times 4 = 4.$$

The Taylor series form of the interpolation polynomial is then

$$p_3(x) = a_0 + a_1 x + a_2 x^2 + a_3 x^3 = 2 - 3x + 4x^3.$$

Having worked this example three times now, in three different forms, we see that this is certainly the simplest form of the three. However, in general, the Taylor series form is *not* always the simplest form.

Example 8.9 Find a general expression for D when $N = 4$.

Solution The zeroth column of D is

$$\begin{bmatrix} 1 \\ b - x_0 \\ (b - x_0)(b - x_1) \\ (b - x_0)(b - x_1)(b - x_2) \\ (b - x_0)(b - x_1)(b - x_2)(b - x_3) \end{bmatrix}.$$

This column leads to the expression for a_0. The first column of D is

$$\begin{bmatrix} 1 \\ b - x_0 + b - x_1 \\ (b - x_0)(b - x_1) + (b - x_0)(b - x_2) + (b - x_1)(b - x_2) \\ (b - x_0)(b - x_1)(b - x_2) + (b - x_0)(b - x_1)(b - x_3) \\ + (b - x_0)(b - x_2)(b - x_3) + (b - x_1)(b - x_2)(b - x_3) \\ 0 \end{bmatrix}.$$

This column leads to the expression for a_1. The second column of D is

$$\begin{bmatrix} 1 \\ b - x_0 + b - x_1 + b - x_2 \\ (b - x_0)(b - x_1) + (b - x_0)(b - x_2) \\ + (b - x_1)(b - x_2) + (b - x_0)(b - x_3) \\ + (b - x_1)(b - x_3) + (b - x_2)(b - x_3) \\ 0 \\ 0 \end{bmatrix}.$$

This column leads to the expression for a_2. The third column of D is

$$\begin{bmatrix} 1 \\ b - x_0 + b - x_1 + b - x_2 + b - x_3 \\ 0 \\ 0 \\ 0 \end{bmatrix}.$$

This column leads to the expression for a_3. The fourth and last column of D has 1 as its first element and 0 everywhere else. This column leads to the expression for a_4, namely $a_4 = f[x_0, x_1, x_2, x_3, x_4]$.

8.1.4 *Accuracy of Polynomial Interpolation*

Most readers will be familiar with Taylor's remainder theorem for Taylor series. There is a similar result for polynomial interpolation. In particular, let the $N+1$ interpolation points be $(x_0, f(x_0)), \ldots, (x_N, f(x_N))$. Suppose $f(x)$ has $N+1$ continuous derivatives. Then the absolute error $e_N(x) = f(x) - p_N(x)$ of the interpolation polynomial $p_N(x)$ is

$$e_N(x) = (x - x_0)(x - x_1)\cdots(x - x_N)\frac{1}{(N+1)!}\frac{d^{N+1}f(\xi)}{dx^{N+1}} \tag{8.13}$$

for all $\min x_i \leq x \leq \max x_i$ and where $\xi = \xi(x)$ is some number $\min x_i \leq \xi \leq \max x_i$. Notice that Equation (8.13) looks like the next term in the Newton form of the interpolation polynomial, except that the Newton divided difference is replaced by a derivative, as in Equation (8.5).

Suppose that $f(x)$ has infinitely many continuous derivatives. According to Equation (8.13), the error will decrease as N increases if the $(N+1)$th derivative of $f(x)$ grows slower than $1/(N+1)!$ times $(x - x_0)\cdots(x - x_N)$. Then the interpolation polynomial $p_N(x)$ converges to the exact function $f(x)$ as $N \to \infty$. However, according to Equation (8.13), the error may increase as N increases if the $(N+1)$th derivative of $f(x)$ grows faster than $1/(N+1)!$ times $(x - x_0)\cdots(x - x_N)$. Then the interpolation polynomial $p_N(x)$ may diverge from $f(x)$ as $N \to \infty$. If $f(x)$ does not have infinitely many continuous derivatives, then Equation (8.13) does not apply. In this case, the error increases rapidly as N increases, at least for large enough N, and the interpolation polynomial $p_N(x)$ diverges from $f(x)$ as $N \to \infty$.

Large polynomial interpolation errors, including those that prevent convergence, tend to take the form of spurious oscillations called the *Runge phenomenon.* The Runge phenomenon is similar to the Gibbs oscillations found in Lagrange, Chebyshev, and Fourier series, as seen in the last chapter. However, unlike Gibbs oscillations, the Runge phenomenon can occur even when the function is completely smooth. The period and amplitude of Gibbs oscillations decrease everywhere as N increases except for the one oscillation next to each jump discontinuity, whose amplitude remains steady for large N. By contrast, the amplitude of Runge oscillations can increase throughout the domain as N increases, even for perfectly smooth functions.

Example 8.10 Consider the function

$$f(x) = -\sin \pi x.$$

Suppose the function is sampled using $N+1$ evenly spaced samples on the domain $[-1, 1]$. Find the interpolation polynomials $p_N(x)$ for $N=1, \ldots, 6$.

Solution The samples are $x_i = -1 + 2i/N$. Then $p_1(x) = p_2(x) = 0$,

$$p_3(x) = 2.923x(x^2 - 1),$$
$$p_4(x) = 2.667x(x^2 - 1),$$
$$p_5(x) = -3.134x + 4.962x^3 - 1.827x^5,$$

and

$$p_6(x) = -3.118x + 4.871x^3 - 1.754x^5.$$

In this case, the function is extremely smooth and thus the interpolation is extremely successful.

Example 8.11 Consider the following square-wave function:

$$f(x) = \begin{cases} 0 & -1 \le x < -1/3, \\ 1 & -1/3 \le x \le 1/3, \\ 0 & 1/3 < x \le 1. \end{cases}$$

Suppose the function is sampled using $N+1$ evenly spaced samples on the domain $[-1, 1]$. Find the interpolation polynomials $p_N(x)$ for $N=1, \ldots, 5$ and $N=20$.

Solution The first five interpolation polynomials are

$$p_1(x) = 0,$$
$$p_2(x) = 1 - x^2,$$
$$p_3(x) = 1.125(1 - x^2),$$
$$p_4(x) = 4x^4 - 5x^2 + 1,$$

and

$$p_5(x) = 3.255x^4 - 4.427x^2 + 1.172.$$

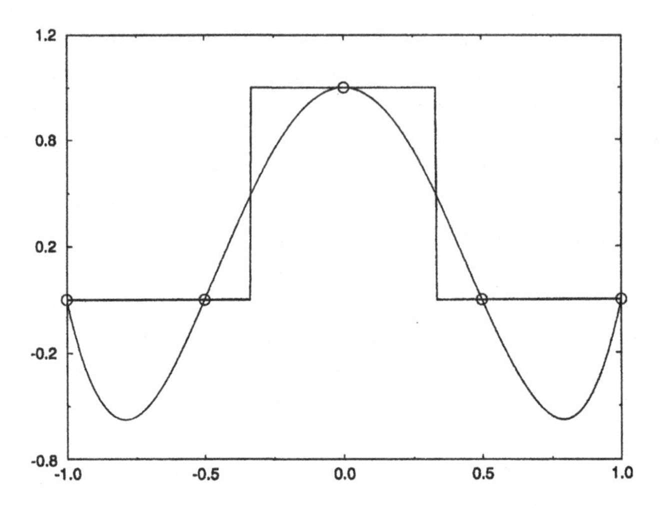

Figure 8.4 Fourth-order polynomial interpolation for a square wave.

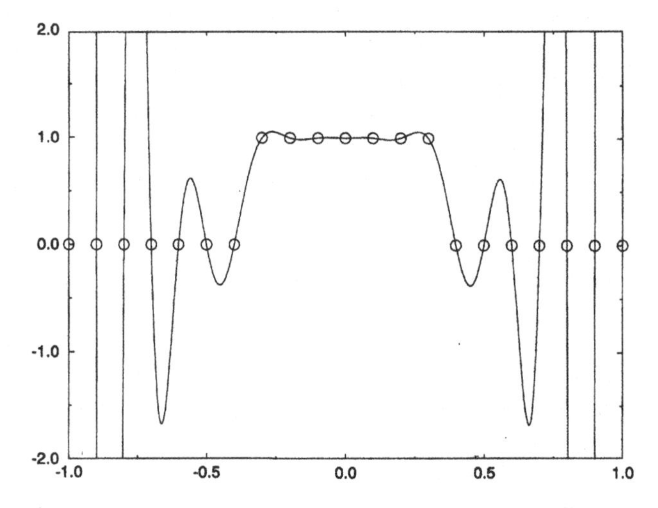

Figure 8.5 Twentieth-order polynomial interpolation for a square wave.

The interpolation polynomials $p_4(x)$ and $p_{20}(x)$ are plotted in Figures 8.4 and 8.5, respectively. The Runge oscillations in this example are incredibly severe. While the very center of the square wave improves as N increases, the edges blow up. This example shows that higher-order interpolation should be used cautiously, if at all, in the presence of jump discontinuities. Comparing Examples 7.3 and 7.5, we see that interpolation polynomials can be far worse than Legendre or Chebyshev polynomials. In other words, the Nth-order interpolation polynomial is much worse than the best Nth-order polynomial as measured in the 2-norm, the ∞-norm, or any other norm. In fact, in this case, even the best Nth-order polynomial is not especially good, and the interpolation polynomial does not even come close to the best. The interpolation polynomial gets worse as N increases whereas Legendre and Chebyshev polynomial series get better.

8.1.5 *Summary of Polynomial Interpolation*

The properties of polynomial interpolation are summarized as follows:

- Error tends to decrease as N increases unless $|d^{N+1}f/dx^{N+1}|$ increases rapidly with N or unless $d^{N+1}f/dx^{N+1}$ is discontinuous. If $|d^{N+1}f/dx^{N+1}|$ is reasonably small for all N, the error decreases as N^{-N}.
- Error tends to take the form of spurious oscillations, called the Runge phenomenon. The error tends to be greatest near the edges of the interpolation domain and least near the center. The Runge phenomenon often grows with N. As a result, most experts advise against using interpolations with orders greater than four. This rule of thumb has proven reliable both in general and in the specific context of computational gasdynamics.
- Large and discontinuous derivatives tend to cause the most error when located near the center of the interpolation domain and the least error when located near the edges of the domain.
- On a domain $[-1, 1]$, the 2-norm interpolation error is nearly minimized by choosing the samples x_i to be the roots of the Legendre polynomial $P_{N+1}(x)$ defined in Section 7.2. Although this sampling nearly minimizes the 2-norm error among interpolation polynomials, the error may still be far worse than in a Legendre polynomial series.
- On a domain $[-1, 1]$, the maximum interpolation error is nearly minimized by choosing the samples x_i to be the roots of a Chebyshev polynomial $T_{N+1}(x)$, as given by Equation (7.20).
- Error generally increases extremely rapidly outside of the interpolation domain $[a, b]$. Hence even when interpolation error is small, extrapolation error may be extremely large.

8.2 Trigonometric Interpolation and the Nyquist Sampling Theorem

Most numerical methods studied in this book use polynomial interpolation rather than trigonometric interpolation. Even still, trigonometric interpolation sheds an interesting light on the limitations of sampling, which justifies a brief consideration here. Assume that a reconstruction has the same form as a Fourier series. Given $2N + 1$ samples, the $2N + 1$ coefficients are determined by the following system of equations:

$$f(x_i) = a_0 + \sum_{n=1}^{N} a_n \cos\left(2\pi n \frac{x_i - a}{b - a}\right) + \sum_{n=1}^{N} b_n \sin\left(2\pi n \frac{x_i - a}{b - a}\right)$$

for $i = 0, \ldots, 2N$. This system appears to have $2N + 1$ equations in the $2N + 1$ unknowns a_n and b_n. Unfortunately, relations such as

$$\sin\theta = \cos(\theta - \pi/2) = -\sin(\theta - \pi) = -\cos(\theta - 3\pi/2)$$

imply that as many as half of the equations may be duplicates. Hence, despite appearances, there are generally more unknowns than equations, and thus the system yields *infinitely* many solutions for the Fourier series form coefficients. There are two ways to deal with this situation. One approach is to roughly double the number of samples, from $2N + 1$ to $4N + 1$. Indeed, $4N + 1$ or more samples always yield a unique solution for the coefficients

a_n and b_n. Alternatively, given any number of samples $M+1$ less than $4N+1$, find the coefficients a_n and b_n that minimize the 2-norm error of the $2N$-order trigonometric interpolation. As it turns out, both approaches yield the same expressions for the coefficients a_n and b_n. In particular, if $f(x)$ has domain $[0, 2\pi]$ then both approaches yield

$$f(x) = a_0 + \sum_{n=1}^{N} a_n \cos(nx) + \sum_{n=1}^{N} b_n \sin(nx), \tag{8.14}$$

where

$$a_0 = \frac{1}{M+1} \sum_{i=0}^{M} f(x_i), \tag{8.15}$$

$$a_n = \frac{2}{M+1} \sum_{i=0}^{M} f(x_i) \cos(nx_i), \tag{8.16}$$

and

$$b_n = \frac{2}{M+1} \sum_{i=0}^{M} f(x_i) \sin(nx_i), \tag{8.17}$$

and where $M+1$ is the number of samples and $2N$ is the order of the trigonometric series. Even though the trigonometric polynomial is written in Fourier series form, the trigonometric interpolation equals the true Fourier series only if $M \geq 4N$.

Even when the trigonometric interpolation equals the "best case" Fourier series, it may still exhibit large Gibbs oscillations, as discussed in Section 7.4. But suppose we know that the true function $f(x)$ is bandlimited, as defined in Section 7.4. In other words, suppose the true function $f(x)$ exactly equals its Fourier series:

$$f(x) = a_0 + \sum_{n=1}^{N} a_n \cos\left(2\pi n \frac{x-a}{b-a}\right) + \sum_{n=1}^{N} b_n \sin\left(2\pi n \frac{x-a}{b-a}\right) \tag{8.18}$$

for some N. Then the unique $2N$th-order trigonometric interpolation passing through any $4N+1$ samples equals the true function. Let us rewrite this result in terms of the sample spacing Δx instead of N. Assume that $x_0 = a$ and $x_N = b$. Then, substituting $4N$ for N in Equation (8.1), we get $\Delta x = (b-a)/4N$. Notice that the shortest wavelength in Equation (8.18) is $(b-a)/N$, that is, the shortest wavelength in Equation (8.18) equals $4\Delta x$. Then *samples spaced apart by Δx perfectly represent functions whose shortest wavelengths are* $4\Delta x$. This is called the *Nyquist sampling theorem.*

Example 8.12 If $f(x_i) = 0$ and the function is bandlimited according to the Nyquist sampling theorem, then the function must be $f(x) = 0$. This is the natural choice, as seen in Example 8.2.

Example 8.13 The shortest wavelength that can be captured by sampling with spacing Δx is $2\Delta x$. In particular, for $2\Delta x$-oscillations, also known as *odd–even oscillations*, every sample is alternately a maximum and a minimum. However, the Nyquist sampling theorem says that $4\Delta x$ is the shortest wavelength that can be *accurately* captured. In particular, consider the samplings shown in Figure 8.6; by the Nyquist sampling theorem, only the

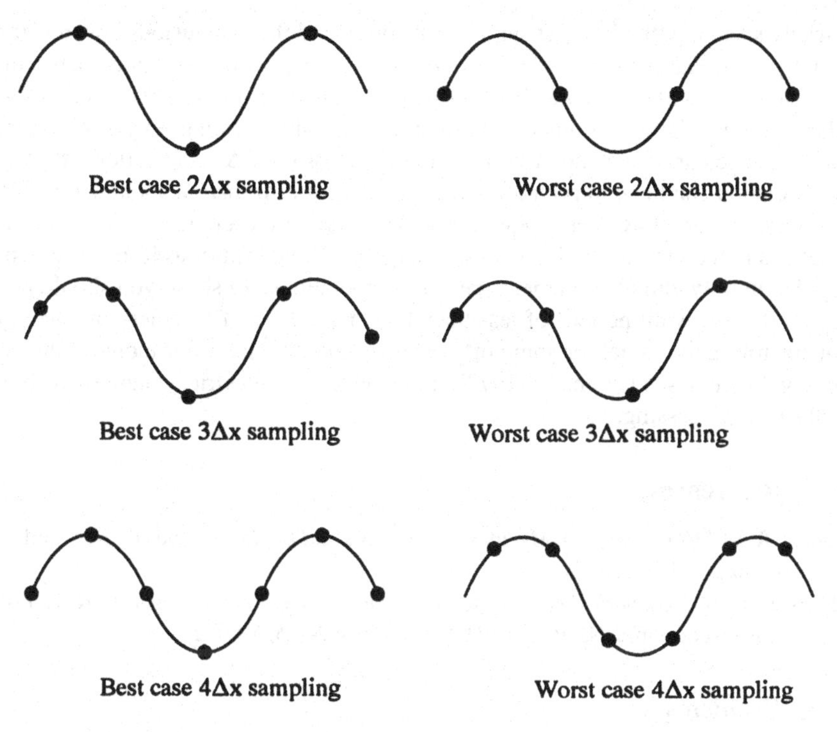

Figure 8.6 An illustration of the Nyquist sampling theorem.

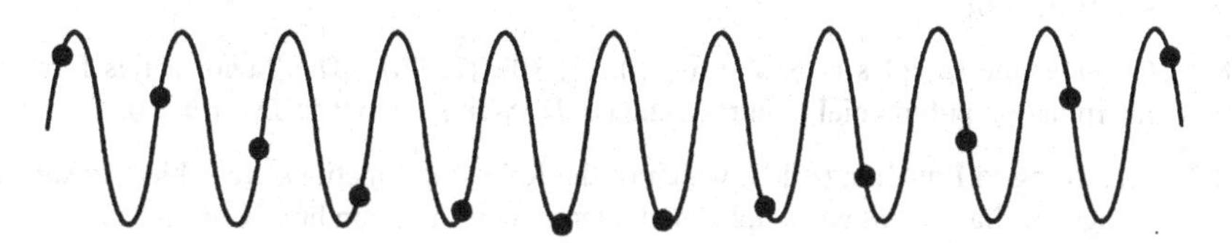

Figure 8.7 An illustration of aliasing.

$4\Delta x$-samplings always adequately represent the sinusoid. Intuitively, $4\Delta x$-samplings always place a sample at or near each extremum, whereas $2\Delta x$- and $3\Delta x$-samplings may or may not. Thus, among other things, the Nyquist sampling theorem requires that there be at least one sample near every extremum, so that the sampled local maximum is only slightly less than the true local maximum and the sampled local minimum is only slightly more than the true local minimum.

Example 8.14 When the Nyquist sampling theorem is violated, short-wavelength components in the original function may appear to be long-wavelength components in the sampling – this is known as *aliasing*. For example, the worst-case $2\Delta x$ sampling seen in Figure 8.6 represents a short-wavelength component by a very long wavelength; in fact, it turns a sinusoid into a constant, and a constant has an *infinite* wavelength! As another example, consider the sampling shown in Figure 8.7. The samples still vary sinusoidally, like the original function, but the samples have a much longer wavelength.

Strictly speaking, the Nyquist sampling theorem and the conclusions drawn therefrom, especially regarding the dangers of $2\Delta x$-oscillations, apply only to trigonometric interpolation. However, without too much effort, you can convince yourself that $2\Delta x$ "odd–even" oscillations pose threats to other sorts of interpolation, including polynomial interpolations. In particular, if a sampling misses the peaks of the $2\Delta x$-oscillations, then no sort of interpolation can hope to accurately recover the lost information about the oscillations. However, if the sampling does capture the peaks of the $2\Delta x$-oscillations, the resulting rapid oscillations in the samples will tend to spur large-amplitude spurious Runge oscillations in interpolation polynomial and other types of interpolations. In short, you should probably avoid oscillations with periods of less than $4\Delta x$, regardless of whether you intend to use trigonometric, polynomial, or some other sort of interpolation. For more on the Nyquist sampling theorem, see Hamming (1973) or any number of electrical engineering books on digital signal processing.

References

Hamming, R. W. 1973. *Numerical Methods for Scientists and Engineers*, 2nd ed., New York: Dover, Chapters 14, 31, and 32.

Mathews, J. H. 1992. *Numerical Methods for Mathematics, Science, and Engineering*, 2nd ed., Englewood Cliffs, NJ: Prentice-Hall, Sections 4.2–4.5, 5.4, 6.2.

Problems

8.1 Suppose $f[-5, -1, 0, 3, 5] = 24$, $f[0, 3, 5, 6] = 12$, and $f[-1, 0, 3, 5] = 15$. Find $f[-5, -1, 0, 3, 5, 6]$.

8.2 Consider the samples $(-5, 25)$, $(0, 5)$, $(1, 10)$, $(2, 20)$. The Taylor series form of the interpolation polynomial requires a matrix D. Write the matrix D for $b = 0$.

8.3 According to Equation (8.13), which of the following functions are subject to substantial Runge oscillations when sampled and interpolated? Explain briefly in each case.
(a) $\sin^2(\pi x)$ (b) $1/(1 + 10x^2)$
(c) e^{-8x^2} (d) $\ln 5x^2$

8.4 Consider $f(x) = 3\sin^2(\pi x/6)$. Sample the function at $x = 0, 1, 2$, and 4. Show all work (this is not a programming exercise).
(a) Find the interpolation polynomial in Lagrange form.
(b) Find the Newton divided differences of the samples.
(c) Find the interpolation polynomial in Newton form.
(d) For $b = 3$, find the coefficient matrix D.
(e) Find the interpolation polynomial in the form of a Taylor series about $b = 3$.

8.5 Consider the following samples: $f(-1) = 0$, $f(-0.2) = 0.1$, $f(0) = 1$, $f(0.2) = 0.1$, and $f(1) = 0$. Show all work (this is not a programming exercise).
(a) Find the interpolation polynomial in Lagrange form.
(b) Find the Newton divided differences of the samples.
(c) Find the interpolation polynomial in Newton form.
(d) For $b = 0$, find the coefficient matrix D.
(e) Find the interpolation polynomial in the form of a Taylor series about $b = 0$.

8.6 Draw at least five different functions through the samples shown below. The functions should represent a variety of behaviors. In particular, at least one of the functions should be

Figure 8.8 Samples for Problem 8.6.

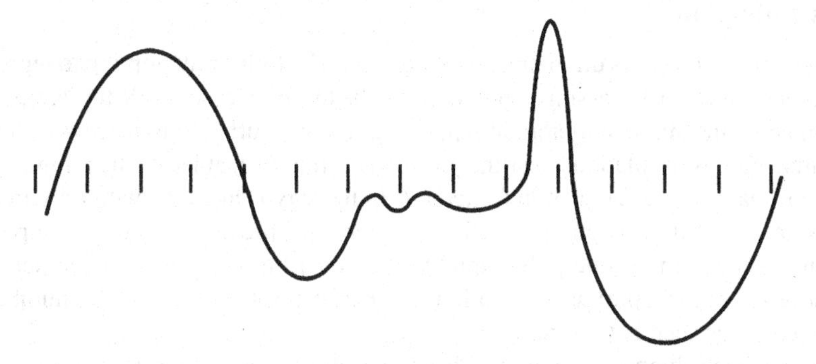

Figure 8.9 Function for Problem 8.7.

nonoscillatory. Will any of the functions passing through the samples satisfy the Nyquist sampling theorem?

8.7 Consider the function shown in the figure.

(a) Using the reasoning of Example 8.1, what is the minimum order of a polynomial that could adequately represent this function?

(b) Would you expect this function to be bandlimited? If so, roughly, what is the minimum wavelength in the trigonometric series that could adequately represent this function? If not, what features of the function create the unlimited short wavelengths?

(c) If the samples must be evenly spaced, what is the minimum number of samples required, according to the Nyquist sampling theorem?

(d) Suppose you were allowed uneven sample spacings, and you wanted to make a rough sketch of the function. Argue that the minimum number of samples required for an adequate rough sketch of the function equals one sample for each maximum and minimum, assuming that the samples are placed exactly on the maxima and minima.

8.8 Consider the following function:

$$f(x) = \sin x(\sin x + \cos x + \sin 3x + \cos 7x).$$

(a) This function is periodic. What is its period?

(b) Using trigonometric identities, write this in the form of a Fourier series over one period.

(c) What is the minimum number of samples required to represent one period of this function, according to the Nyquist sampling theorem?

CHAPTER 9

Piecewise-Polynomial Reconstruction

9.0 Introduction

The last two chapters dramatically demonstrate the folly in attempting to represent a discontinuous function by a single polynomial. In the best case, with the entire true function available, the single polynomial representation will suffer from narrow width but large-amplitude Gibbs oscillations near the jump discontinuities, at least when minimizing the error in ordinary norms, as seen in Chapter 7. In more typical cases, with only limited information about the true function available or, more specifically, with only samples of the function available, the single polynomial will suffer from the Runge phenomenon, a relatively severe form of spurious oscillation that can increase rapidly as the number of samples increases, as seen in Chapter 8.

To overcome the problems associated with single-polynomial reconstructions, this chapter will consider *piecewise*-polynomial reconstructions, which were introduced earlier in Section 6.3, especially in Example 6.8. In piecewise-polynomial reconstructions, instead of representing the entire function by a single polynomial, we represent different local regions or *cells* by different polynomials. Figure 9.1 illustrates a typical piecewise-polynomial representation. By using separate and independent polynomials for each cell, only the cells containing jump discontinuities need suffer from large spurious oscillations, rather than the entire representation. Furthermore, piecewise-polynomial representations naturally allow jump discontinuities: the simplest reconstructions allow jump discontinuities only at cell edges, whereas the *subcell resolution* techniques discussed in Section 9.4 allow jump discontinuities to occur anywhere, including the insides of cells. Of course, piecewise-polynomial reconstructions cost more to build and evaluate and require more storage space than a single polynomial reconstruction; however, for discontinuous functions, the accuracy improvements easily justify the additional costs.

The best known piecewise-polynomial reconstructions are called *splines*. Although splines are fine for reconstructing continuous smooth functions, they are not appropriate for the discontinuous functions found in computational gasdynamics. Splines expend a great deal of effort towards achieving smooth and continuous reconstructions, which is completely counterproductive if the true function is not smooth or continuous. To explain further, splines link the reconstructions in adjacent cells together to achieve continuity in the function and its derivatives across cell edges; then local problems caused by discontinuities in the function or its derivatives can pollute the entire spline, and not just the cells containing the discontinuities. Nevertheless, splines are still generally more successful than single polynomials when it comes to reconstructing discontinuous functions. Since splines are not appropriate for our applications, we shall not discuss them further. The details can be found in any number of numerical analysis texts, including Mathews (1992).

Instead of splines, this chapter concerns a class of piecewise-polynomial reconstructions sometimes referred to as *essentially nonoscillatory* or *ENO* reconstructions. The ENO reconstructions are heavily based on polynomial interpolations as studied in Chapter 8.

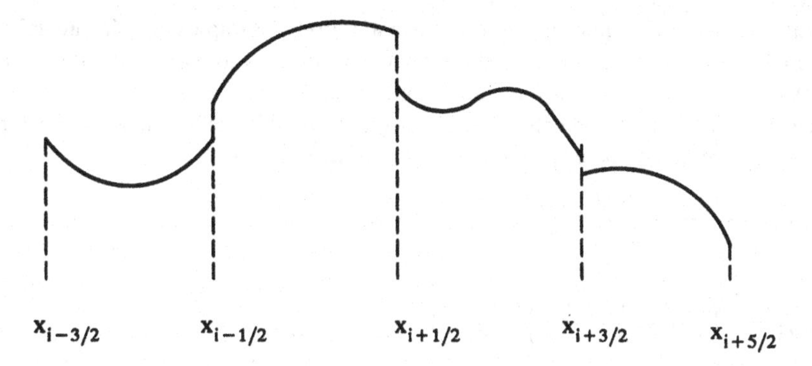

Figure 9.1 A typical piecewise-polynomial reconstruction.

In Section 9.1, each cell chooses an interpolation polynomial passing through neighboring samples; in Section 9.2, each cell averages interpolation polynomials passing through neighboring samples. The trick is to choose the right interpolation polynomial from among the many possible candidates, or to choose the right average of interpolation polynomials.

Many methods in computational gasdynamics approximate cell integral averages rather than samples. Thus, besides reconstructions from samples, this chapter also concerns reconstructions from cell integral averages. Section 9.3 introduces a clever technique called *reconstruction via the primitive function*, which turns any sample reconstruction technique into a cell-integral-average reconstruction technique.

Now let's introduce some nomenclature and notation. Suppose that the functional domain $[a, b]$ is divided into *cells* $[x_{-1/2}, x_{1/2}]$, $[x_{1/2}, x_{3/2}]$, and so on up to $[x_{N-1/2}, x_{N+1/2}]$. The points $x_{i+1/2}$ are called the *cell edges* and the points $x_i = (x_{i+1/2} + x_{i-1/2})/2$ are called the *cell centers*. The difference between the cell edges $\Delta x_i = x_{i+1/2} - x_{i-1/2}$ is called the *cell width*. Depending on the application, either the cell edges or the cell centers are aligned with the boundaries of the domain $[a, b]$. Therefore either $x_{-1/2} = a$ and $x_{N+1/2} = b$ or $x_0 = a$ and $x_N = b$; this is discussed further in Section 19.0. Incidentally, it is sometimes convenient to reverse the notation, so that cell edges are indexed by integers x_i and cell centers are indexed by half-integers $x_{i+1/2}$, as in Example 6.8.

Given cells, a function $f(x)$ can be represented by cell-centered samples such as $f(x_i)$. Alternatively, a function $f(x)$ can be represented by *cell-integral averages* defined by

$$\blacklozenge \qquad \bar{f}_i = \frac{1}{\Delta x_i} \int_{x_{i-1/2}}^{x_{i+1/2}} f(x)\, dx, \tag{9.1}$$

where $i = 0, \ldots, N$. As you might expect, cell-integral averages are especially popular for approximating integral equations whereas samples are especially popular for approximating differential equations. *Finite-difference* methods use samples as their primary representation whereas *finite-volume* methods use cell-integral averages as their primary representation. Of course, although finite-difference methods use samples as their primary representation, they sometimes switch to piecewise-polynomial representations. Similarly, although finite-volume methods use cell-integral averages as their primary representation, they sometimes switch to piecewise-polynomial representations. By contrast with finite-difference and finite-volume methods, *finite-element* methods use piecewise-polynomials as their primary representation. In other words, finite-element methods carry around the polynomial coefficients for each cell, either in place of or in addition to samples or cell-integral averages, and thus finite-element methods do not necessarily need to reconstruct the solution. This book concerns only finite-volume and finite-difference methods.

Besides cell-integral averages there are many other possible cell averages. For example, the cell average could be some type of algebraic average such as

$$\bar{f}_i = \frac{f(x_{i+1/2}) + f(x_{i-1/2})}{2}$$

or

$$\bar{f}_i = f\left(\frac{x_{i+1/2} + x_{i-1/2}}{2}\right) = f(x_i).$$

However, the differences between cell-integral averages and these other cell averages are small provided that Δx_i is small and f is smooth and continuous. In particular, one can show that

$$\frac{1}{\Delta x_i} \int_{x_{i-1/2}}^{x_{i+1/2}} f(x)\, dx = \frac{f(x_{i+1/2}) + f(x_{i-1/2})}{2} + O\left(\Delta x_i^2\right) \tag{9.2}$$

and

$$\frac{1}{\Delta x_i} \int_{x_{i-1/2}}^{x_{i+1/2}} f(x)\, dx = f(x_i) + O\left(\Delta x_i^2\right). \tag{9.3}$$

This last result says that *cell-centered samples equal cell-integral averages to within second-order accuracy*. Equation (9.3) is proven in Example 10.1, and Equation (9.3) implies Equation (9.2). Many numerical methods are only second-order accurate, in which case many distinctions between samples and cell-integral averages disappear, at least in one dimension.

9.1 Piecewise Interpolation-Polynomial Reconstructions

Most traditional reconstruction techniques, including splines and polynomial interpolations, are subject to large oscillatory errors in the presence of jump discontinuities. In the past, researchers have thoroughly investigated the nature of these oscillations while largely neglecting to suggest reconstruction techniques that could avoid them. Thus, when the connection between reconstruction and computational gasdynamics became widely appreciated in the 1970s, the computational gasdynamics community had to devise new reconstruction techniques that could handle the shocks and contacts found in their applications. Although many of the earliest such reconstruction techniques were applicable only in the context of computational gasdynamics, because they depended on parameters such as the CFL number specific to computational gasdynamics, some of the more recent reconstruction techniques are completely generic. Therefore, despite the fact that they were developed in the context of computational gasdynamics, the reconstruction techniques seen in the rest of this chapter are completely general and can be used in any context. These general reconstructions are sometimes called *essentially nonoscillatory (ENO) reconstructions*. ENO reconstruction was first developed by Harten, Engquist, Osher, and Chakravarthy (1987).

This section concerns piecewise-continuous piecewise-polynomial reconstructions composed strictly of interpolation polynomials. Consider a set of sample points $(x_0, x_1, \ldots, x_N)$. Define a set of cells whose centers are samples, that is, the cells are $[x_{-1/2}, x_{1/2}], \ldots,$

$[x_{N-1/2}, x_{N+1/2}]$ and the cell centers are $x_i = (x_{i+1/2} + x_{i-1/2})/2$. Then the simplest possible reconstruction is piecewise-constant, such that the reconstruction on cell $[x_{i-1/2}, x_{i+1/2}]$ equals $f(x_i)$. Away from jump discontinuities, a piecewise-constant reconstruction is first-order accurate. Although the piecewise-constant reconstruction has jumps at each cell edge, those jumps are only first order, at most. The only exception is when the true function has a jump discontinuity. Then the reconstruction may have zeroth-order jumps at cell edges near the true jump.

The next simplest reconstruction is piecewise-linear. Assuming the reconstruction is composed strictly of linear interpolations, the reconstruction on cell $[x_{i-1/2}, x_{i+1/2}]$ can be the line through $(x_{i-1}, f(x_{i-1}))$ and $(x_i, f(x_i))$ or the line through $(x_i, f(x_i))$ and $(x_{i+1}, f(x_{i+1}))$. Thus the reconstruction on cell $[x_{i-1/2}, x_{i+1/2}]$ is either

$$f(x_i) + f[x_{i-1}, x_i](x - x_i) = f(x_i) + \frac{f(x_i) - f(x_{i-1})}{x_i - x_{i-1}}(x - x_i)$$

or

$$f(x_i) + f[x_i, x_{i+1}](x - x_i) = f(x_i) + \frac{f(x_{i+1}) - f(x_i)}{x_{i+1} - x_i}(x - x_i).$$

As seen in Equations (8.5) and (8.6), large divided differences imply large or discontinuous derivatives, which in turn imply poor interpolation accuracy according to Equation (8.13). By this reasoning, the best linear interpolation is the one with the least divided difference in absolute value. Then the reconstruction on cell $[x_{i-1/2}, x_{i+1/2}]$ is

$$f(x_i) + f[x_j, x_{j+1}](x - x_i), \tag{9.4}$$

where

$$j = \begin{cases} i & |f[x_i, x_{i+1}]| \leq |f[x_{i-1}, x_i]|, \\ i - 1 & |f[x_i, x_{i+1}]| > |f[x_{i-1}, x_i]|. \end{cases} \tag{9.5}$$

For an equivalent and more compact expression, define the following function:

$$m(x, y) = \begin{cases} x & |x| \leq |y|, \\ y & |x| > |y|. \end{cases} \tag{9.6}$$

In other words, $m(x, y)$ is the minimum of x and y in absolute value. Then the reconstruction on cell $[x_{i-1/2}, x_{i+1/2}]$ is

$$f(x_i) + S_i(x - x_i), \tag{9.7}$$

where

$$S_i = m(f[x_{i-1}, x_i], f[x_i, x_{i+1}]) \tag{9.8}$$

and where the letter S stands for "slope." The piecewise-linear reconstruction has second-order accuracy and $O(\Delta x^2)$ jump discontinuities across each cell edge except, of course, when the true function has a jump discontinuity.

Now consider piecewise-quadratic reconstruction. Assuming the reconstruction is composed strictly of quadratic interpolations, the reconstruction on cell $[x_{i-1/2}, x_{i+1/2}]$ can be as follows: the quadratic passing through $(x_{i-2}, f(x_{i-2}))$, $(x_{i-1}, f(x_{i-1}))$, and $(x_i, f(x_i))$; the quadratic passing through $(x_{i-1}, f(x_{i-1}))$, $(x_i, f(x_i))$, and $(x_{i+1}, f(x_{i+1}))$; or the quadratic passing through $(x_i, f(x_i))$, $(x_{i+1}, f(x_{i+1}))$, and $(x_{i+2}, f(x_{i+2}))$. Thus, in Newton form, the reconstruction on cell $[x_{i-1/2}, x_{i+1/2}]$ is

$$f(x_i) + f[x_{i-1}, x_i](x - x_i) + f[x_{i-2}, x_{i-1}, x_i](x - x_i)(x - x_{i-1}), \tag{9.9}$$

$$\begin{aligned} &f(x_i) + f[x_{i-1}, x_i](x - x_i) + f[x_{i-1}, x_i, x_{i+1}](x - x_i)(x - x_{i-1}) \\ &\quad = f(x_i) + f[x_i, x_{i+1}](x - x_i) + f[x_{i-1}, x_i, x_{i+1}](x - x_i)(x - x_{i+1}), \end{aligned} \tag{9.10}$$

or

$$f(x_i) + f[x_i, x_{i+1}](x - x_i) + f[x_i, x_{i+1}, x_{i+2}](x - x_i)(x - x_{i+1}). \tag{9.11}$$

As before, large Newton divided differences imply large or discontinuous derivatives, which in turn imply poor interpolation accuracy. The above expressions involve two first-divided differences and three second-divided differences. In general, the polynomial with the least first-divided difference will not have the least second-divided difference and, vice versa, the polynomial with the least second-divided difference will not have the least first-divided difference. Thus there must be some sort of compromise between minimizing the first- and second-divided differences. As it turns out, in many cases (but certainly not in all cases), large first-divided differences are worse than large second-divided differences. Thus, the first-divided difference will be minimized first and the second-divided difference will be minimized second. Notice that Equation (9.10) involves both of the first-divided differences, depending on how it is written. Thus Equation (9.10) always makes the first cut, and the first-divided difference is used to eliminate either Equation (9.9) or (9.11) (in particular, to eliminate whichever of these polynomials has the greater first-divided difference in absolute value). After this, eliminate whichever of the two remaining equations has the greater second-divided difference in absolute value. Put another way, the piecewise-quadratic selection procedure chooses between the two lines $f(x_i) + f[x_{i-1}, x_i](x - x_i)$ and $f(x_i) + f[x_i, x_{i+1}](x - x_i)$, whichever has the least slope, just as in the piecewise-linear reconstruction. Then the piecewise-quadratic selection procedure chooses between two quadratics, whichever one is closest to the previously chosen line.

In summary, the reconstruction on cell $[x_{i-1/2}, x_{i+1/2}]$ is as follows:

$$f(x_i) + f[x_j, x_{j+1}](x - x_i) + f[x_k, x_{k+1}, x_{k+2}](x - x_j)(x - x_{j+1}), \tag{9.12}$$

where j is chosen as in Equation (9.5) and

$$k = \begin{cases} j & |f[x_j, x_{j+1}, x_{j+2}]| \leq |f[x_{j-1}, x_j, x_{j+1}]|, \\ j - 1 & |f[x_j, x_{j+1}, x_{j+2}]| > |f[x_{j-1}, x_j, x_{j+1}]|. \end{cases} \tag{9.13}$$

This piecewise-quadratic selection procedure is illustrated by the flow chart seen in Figure 9.2. The piecewise-quadratic reconstruction has third-order accuracy and $O(\Delta x^3)$ jump discontinuities at each cell edge except, of course, where the true function has a jump discontinuity.

Similarly, for a piecewise-cubic reconstruction, the reconstruction on cell $[x_{i-1/2}, x_{i+1/2}]$ is as follows:

$$\begin{aligned} &f(x_i) + f[x_j, x_{j+1}](x - x_i) + f[x_k, x_{k+1}, x_{k+2}](x - x_j)(x - x_{j+1}) \\ &\quad + f[x_l, x_{l+1}, x_{l+2}, x_{l+3}](x - x_k)(x - x_{k+1})(x - x_{k+2}), \end{aligned} \tag{9.14}$$

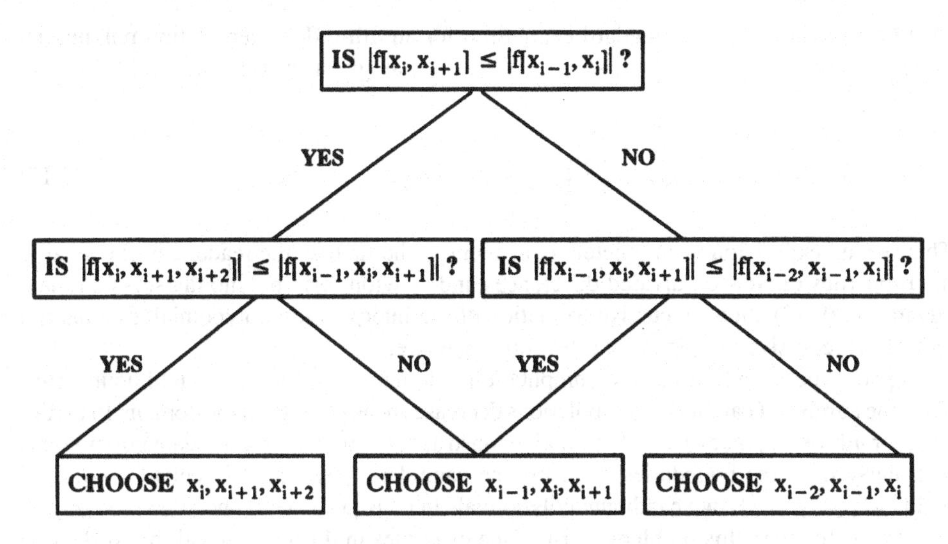

Figure 9.2 The selection procedure for a piecewise-quadratic ENO reconstruction.

where j and k are chosen as in Equations (9.5) and (9.13), respectively, and

$$l = \begin{cases} k & |f[x_k, x_{k+1}, x_{k+2}, x_{k+3}]| \leq |f[x_{k-1}, x_k, x_{k+1}, x_{k+2}]|, \\ k-1 & |f[x_k, x_{k+1}, x_{k+2}, x_{k+3}]| > |f[x_{k-1}, x_k, x_{k+1}, x_{k+2}]|. \end{cases} \tag{9.15}$$

As one way to view this selection procedure, notice that it chooses the line with the least slope, then the quadratic closest to the line, then the cubic closest to the quadratic. The piecewise-cubic reconstruction has fourth-order accuracy and $O(\Delta x^4)$ jump discontinuities across each cell edge except, of course, where the true function has a jump discontinuity.

The above reconstruction procedure can be generalized to any order of accuracy. Thinking recursively, suppose that the first $m+1$ interpolation points for cell $[x_{i-1/2}, x_{i+1/2}]$ have already been selected, where the left-most interpolation point is $l_m(i)$. In other words, suppose the interpolation points selected so far are $(x_{l_m(i)}, x_{l_m(i)+1}, \ldots, x_{l_m(i)+m})$. Then add a point to the left or to the right depending on which selection yields the least $(m+1)$th divided difference in absolute value. This selection procedure can be written in algorithmic form as

$l_0(i) = i$

for $m = 0, \ldots, n-1$

$$l_{m+1}(i) = \begin{cases} l_m(i) & \left|f\left[x_{l_m(i)}, \ldots, x_{l_m(i)+m+1}\right]\right| \leq \left|f\left[x_{l_m(i)-1}, \ldots, x_{l_m(i)+m}\right]\right|, \\ l_m(i)-1 & \left|f\left[x_{l_m(i)}, \ldots, x_{l_m(i)+m+1}\right]\right| > \left|f\left[x_{l_m(i)-1}, \ldots, x_{l_m(i)+m}\right]\right|. \end{cases}$$

Then, in Newton form, the nth-order polynomial reconstruction on cell $[x_{i-1/2}, x_{i+1/2}]$ is

$$\sum_{j=0}^{n} f\left[x_{l_n(i)}, \ldots, x_{l_n(i)+j}\right] \prod_{k=0}^{j-1} \left(x - x_{l_n(i)+k}\right). \tag{9.16}$$

This expression is just the standard expression for an nth-order interpolation polynomial passing through points $(x_{l_n(i)}, x_{l_n(i)+1}, \ldots, x_{l_n(i)+n})$, as given by Equation (8.8). Equivalently,

$$\sum_{j=0}^{n} f\left[x_{l_j(i)}, \ldots, x_{l_j(i)+j}\right] \prod_{k=0}^{j-1} \left(x - x_{l_{j-1}(i)+k}\right). \tag{9.17}$$

This second expression is completely equivalent to the first, as the reader can show using the properties of Newton divided differences and Newton form polynomials. As its main advantage, (9.17) allows recursive formation of the interpolation polynomial, as a natural part of the recursive sample-point selection procedure.

Suppose the boundaries of the computational domain are solid or far-field boundaries. Then the number of candidate interpolations decreases near the edge of the domain. In particular, for nth-order piecewise-polynomial reconstruction, there are n candidate polynomials well away from the boundaries but only one candidate polynomial in cells $[x_{-1/2}, x_{1/2}]$ and $[x_{N-1/2}, x_{N+1/2}]$, two candidate polynomials in cells $[x_{1/2}, x_{3/2}]$ and $[x_{N-3/2}, x_{N-1/2}]$, and so on. To avoid this problem, most of the examples in this chapter will be worked on periodic domains, where $x_{N+1} = x_1, x_{N+1/2} = x_{1/2}$, and so forth. On a periodic domain, any missing values can be found by wrapping around to the other side of the domain.

The above reconstruction technique is usually surprisingly effective, especially near jump discontinuities. Given the candidates, this procedure often chooses the best and rarely chooses the worst candidate. At the very least, it usually chooses candidates that do not cross jump discontinuities and thus do not suffer from large spurious oscillations. However, there are two exceptions. First, in the boundary cells $[x_{-1/2}, x_{1/2}]$ and $[x_{N-1/2}, x_{N+1/2}]$, there may only be one candidate, as mentioned in the last paragraph, and thus there is no way to avoid interpolating across jumps if jumps occur in boundary cells. Second, suppose two jump discontinuities occur right next to each other. For cells in between the two jumps, all candidate nth-order interpolation polynomials may cross one or the other of the jumps. This is more likely to occur as n increases. For example, if two jumps are separated by two cells, then piecewise-linear interpolations based on two points can avoid both jumps, but all candidate piecewise-quadratic interpolations based on three points pass through one jump or the other. If a jump discontinuity cannot be avoided using nth-order interpolations, one can reduce the order of the interpolation. Lower-order interpolations can more easily avoid jumps; furthermore, even if they cannot avoid jumps, lower-order interpolations suffer less severe oscillations when they do cross jumps, as seen in Chapter 8. Techniques for adjusting the interpolation order will not be discussed here.

Example 9.1 Consider the function

$$f(x) = -\sin \pi x.$$

Suppose the function is sampled using $N + 1$ evenly spaced samples on the periodic domain $[-1, 1]$. Figure 9.3 shows the piecewise-quadratic reconstruction, as described in this section, where the samples are taken at the cell centers and $N = 5$. The piecewise-quadratic reconstruction is far less accurate than a single fifth-order interpolation polynomial, as seen in Example 8.10. The large jumps at the cell edges are particularly annoying.

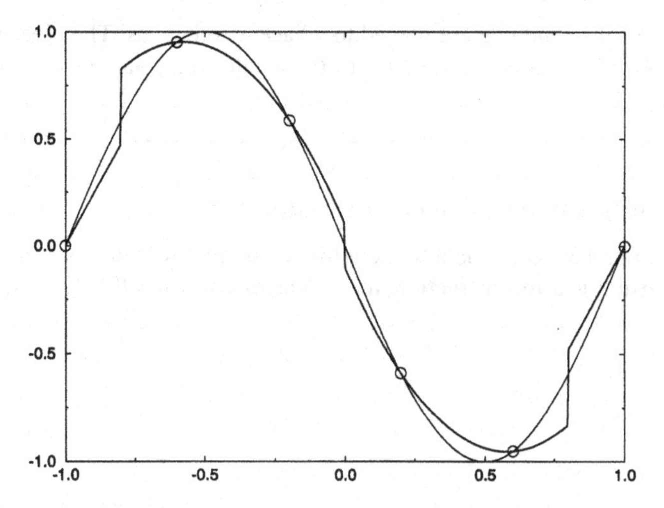

Figure 9.3 Piecewise-quadratic interpolation ENO reconstruction of a sine wave.

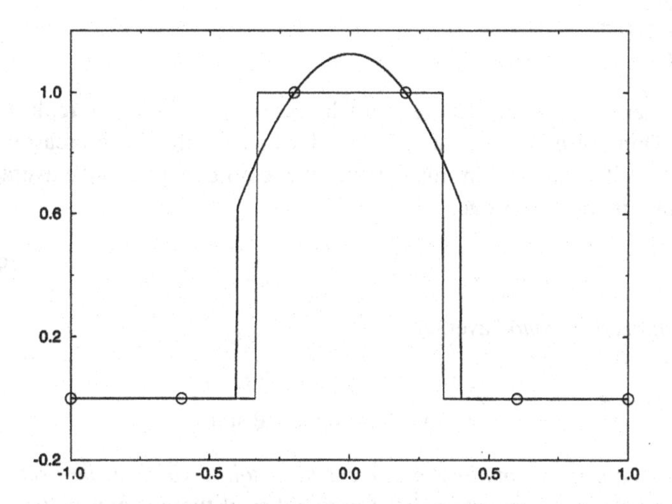

Figure 9.4 Piecewise-quadratic interpolation ENO reconstruction of a square wave.

Example 9.2 Consider the following square-wave function:

$$f(x) = \begin{cases} 0 & -1 \leq x < -1/3, \\ 1 & -1/3 \leq x \leq 1/3, \\ 0 & 1/3 < x \leq 1, \end{cases}$$

on a periodic domain. Suppose the function is sampled using $N + 1$ evenly spaced samples on the periodic domain $[-1, 1]$. Figure 9.4 shows the piecewise-quadratic reconstruction for $N = 5$. For $N = 10$, the reconstruction is essentially perfect, save for slight mislocations in the jump discontinuities. Thus, for $N = 10$ or above, the reconstruction is perfectly zero where it should be zero and perfectly one where it should be one; the only discrepancy is that the jumps between zero and one are moved to adjacent cell edges. The piecewise-quadratic reconstruction is obviously far superior to the single interpolation polynomial seen in Example 8.11.

9.2 Averaged Interpolation-Polynomial Reconstructions

In the last section, the reconstructions were formed purely from interpolations. In particular, for a piecewise-linear reconstruction, the reconstruction on cell $[x_{i-1/2}, x_{i+1/2}]$ was as follows:

$$f(x_i) + S_i(x - x_i), \tag{9.7}$$

where

$$S_i = m(f[x_{i-1}, x_i], f[x_i, x_{i+1}]) \tag{9.8}$$

and where

$$m(x, y) = \begin{cases} x & |x| \le |y|, \\ y & |x| > |y|. \end{cases} \tag{9.6}$$

If you think about it, there is absolutely no reason why $m(x, y)$ could not be replaced by some other function. Think of $m(x, y)$ as an average of x and y – there is no reason why an average must choose either x or y. In this context, two common alternative averaging functions are the usual arithmetic average,

$$\frac{x + y}{2}, \tag{9.18}$$

and the *minimum modulus* or *minmod* average,

$$minmod\,(x, y) = \begin{cases} m(x, y) & x \text{ and } y \text{ have same sign}, \\ 0 & x \text{ and } y \text{ have opposite sign}. \end{cases} \tag{9.19}$$

The minmod function chooses the smallest argument in absolute value if the arguments have the same sign and chooses zero otherwise. Using either of these averages to replace $m(x, y)$, the reconstruction on cell $[x_{i-1/2}, x_{i+1/2}]$ is no longer necessarily a pure linear interpolation; instead, the reconstruction is a line that falls somewhere between the two candidate interpolation lines. In this context, averaging functions such as $m(x, y)$ and $minmod(x, y)$ are often called *slope limiters.*

For piecewise-quadratic reconstructions, there are three candidate quadratic interpolations, as seen in the last section. The three candidates can be averaged; alternatively, we can eliminate the worst candidate and average the remaining two candidates. Let us pursue the latter idea. First, choose between candidates (9.9) and (9.10); in particular, choose whichever one has the least second-divided difference in absolute value. Then, in Newton form, one candidate is

$$\begin{aligned} &f(x_i) + f[x_{i-1}, x_i](x - x_i) \\ &\quad + m(f[x_{i-2}, x_{i-1}, x_i], f[x_{i-1}, x_i, x_{i+1}])(x - x_i)(x - x_{i-1}). \end{aligned} \tag{9.20}$$

Alternatively, this candidate can be written in Taylor series form as follows:

$$f(x_i) + S_i^-(x - x_i) + C_i^-(x - x_i)^2, \tag{9.21}$$

where

$$S_i^- = f[x_{i-1}, x_i] + C_i^-(x_i - x_{i-1}), \tag{9.22}$$

$$C_i^- = m(f[x_{i-2}, x_{i-1}, x_i], f[x_{i-1}, x_i, x_{i+1}]) \tag{9.23}$$

and where S stands for "slope," as before, and C stands for "curvature." Similarly, choose between candidates (9.10) and (9.11); in particular, choose whichever one has the least second-divided difference in absolute value. Then, in Newton form, another candidate is

$$f(x_i) + f[x_i, x_{i+1}](x - x_i) + m(f[x_{i-1}, x_i, x_{i+1}], f[x_i, x_{i+1}, x_{i+2}])(x - x_i)(x - x_{i+1}). \tag{9.24}$$

Alternatively, this candidate can be written in Taylor series form as follows:

$$f(x_i) + S_i^+(x - x_i) + C_i^+(x - x_i)^2, \tag{9.25}$$

where

$$S_i^+ = f[x_i, x_{i+1}] - C_i^+(x_{i+1} - x_i), \tag{9.26}$$

$$C_i^+ = m(f[x_{i-1}, x_i, x_{i+1}], f[x_i, x_{i+1}, x_{i+2}]). \tag{9.27}$$

In some cases, both candidates may be the same as Equation (9.10), in which case Equation (9.10) is the final winner; otherwise, average the two candidates. For example, one can minmod average the two candidates in Taylor series form. Then the reconstruction on cell $[x_{i-1/2}, x_{i+1/2}]$ is

$$f(x_i) + S_i(x - x_i) + C_i(x - x_i)^2, \tag{9.28}$$

where

$$S_i = minmod(S_i^+, S_i^-), \tag{9.29}$$

$$C_i = minmod(C_i^+, C_i^-). \tag{9.30}$$

There is nothing sacred about this averaging procedure. For example, $m(x, y)$ and *minmod* (x, y) could be replaced by other averaging functions. Also, the averaging could be done on the Newton form coefficients rather than on the Taylor series form coefficients. Large functional derivatives indicate inaccurate interpolations, and both the Newton and Taylor series coefficients approximate functional derivatives, albeit in somewhat different ways.

The preceding averaging procedure can be generalized to any arbitrary order of accuracy. The "−" candidate is chosen from among all nth-order interpolation polynomials passing through $(x_i, f(x_i))$ and $(x_{i-1}, f(x_{i-1}))$ using the procedure described in the last section; similarly, the "+" candidate is chosen from among all nth-order interpolation polynomials passing through $(x_i, f(x_i))$ and $(x_{i+1}, f(x_{i+1}))$, again using the procedure described in the last section. Then the two candidates are averaged in Taylor series form using minmod or something else.

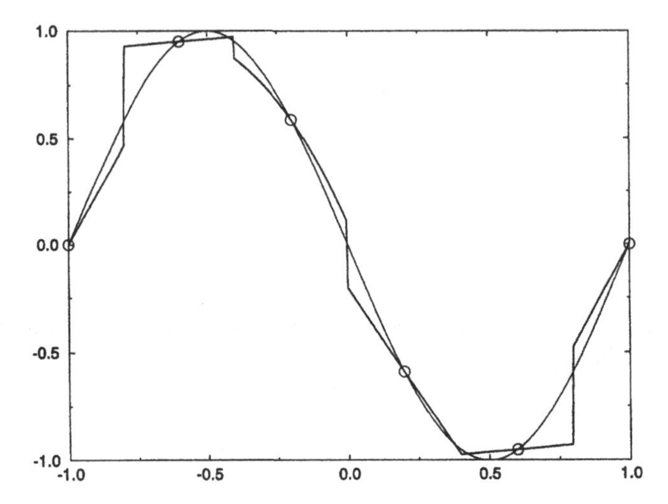

Figure 9.5 Average-quadratic interpolation ENO reconstruction of a sine wave.

Example 9.3 Consider the function

$$f(x) = -\sin \pi x.$$

Suppose the function is sampled using $N + 1$ evenly spaced samples on the periodic domain $[-1, 1]$. Figure 9.5 shows the piecewise-quadratic reconstruction, as described in this section, where the samples are cell centered and $N = 5$. The piecewise-quadratic reconstruction is far less accurate than the single fifth-order interpolation polynomial seen in Example 8.10. The minmod averaging zeros out the quadratic term in most of the cells, so that the reconstruction is linear in most of the cells.

Example 9.4 Consider the following square-wave function:

$$f(x) = \begin{cases} 0 & -1 \le x < -1/3, \\ 1 & -1/3 \le x \le 1/3, \\ 0 & 1/3 < x \le 1. \end{cases}$$

Suppose the function is sampled using $N+1$ evenly spaced samples on the periodic domain $[-1, 1]$. Figure 9.6 shows the piecewise-quadratic reconstruction for $N = 5$ found using the procedure described in this section. The minmod averaging zeros out the quadratic term in all of the cells, making the reconstruction piecewise-linear. For $N = 10$ or greater, the reconstruction is essentially perfect, save for slight mislocations in the jump discontinuities. The piecewise-quadratic reconstruction is far more accurate than the single interpolation polynomial seen in Example 8.11.

The previous section defined a set of candidate polynomial interpolations for each cell and eliminated all but one. Similarly, this section defined a set of candidate polynomial interpolations for each cell, eliminated all but two, and averaged the two finalists. Continuing

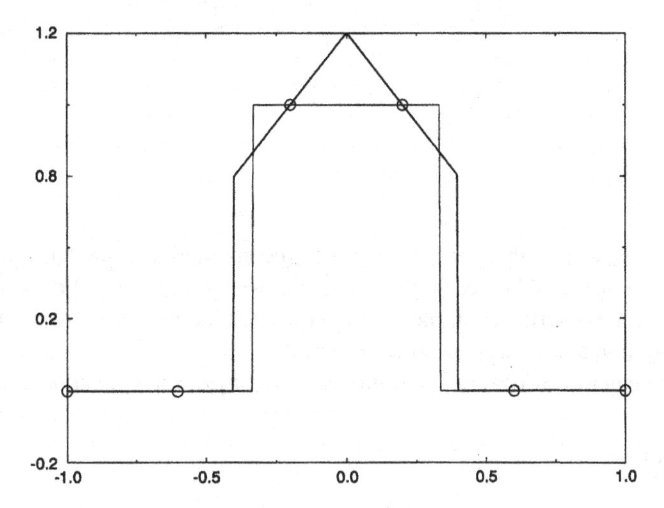

Figure 9.6 Average-quadratic interpolation ENO reconstruction of a square wave.

the progression, one can eliminate fewer and fewer candidates or, in other words, average more and more candidates to obtain the final reconstruction. In the most extreme case, one can eliminate only the candidates that appear to overlap shocks and average all of the remaining candidates. By keeping more candidates, one can improve the formal order of accuracy in smooth regions. Remember that N samples yield a formal order of accuracy as high as $N-1$; the ENO reconstructions explored above generally achieve a much lower order of accuracy, since they discard most sample points. Averaging more candidates also reduces the sensitivity of the reconstruction to small changes in the samples; when two candidates involve divided differences of nearly equal size, the ENO reconstructions described above may change their final choice abruptly with only small changes in the divided differences. Averaging more candidates reduces sensitivity and, in particular, reduces the effects of round-off errors on the reconstruction.

Along these lines, Liu, Osher, and Chan (1994) described *weighted ENO* (*WENO*) *reconstructions*. WENO reconstructions form convex linear combinations of the candidate polynomial interpolations, where the size of the coefficients varies inversely with the size of the local divided differences. The linear interpolation involves more arithmetic and complication, but it increases the order of accuracy to as high as $N/2 - 1$ in smooth regions, decreases the sensitivity of the reconstruction to small changes in the samples, and eliminates all of the logical "if" statements that appear in ordinary ENO methods as part of the candidate selection process, which allows WENO reconstructions to run much faster on parallel machines. For a variation on WENO reconstruction, and a detailed description of the benefits of WENO reconstruction, see Jiang and Shu (1996). This section has described reconstructions that average polynomial interpolations; however, to keep things simple, the rest of the book will focus on reconstructions that use one carefully chosen polynomial interpolation in each cell, as described in the previous section.

9.3 Reconstruction via the Primitive Function

Chapter 8 showed how to find the unique Nth-order polynomial satisfying $p_N(x_i) = f(x_i)$ for $i = 0, \ldots, N$. This section shows how to find the unique Nth-order polynomial satisfying

$$\frac{1}{\Delta x_i}\int_{x_{i-1/2}}^{x_{i+1/2}} p_N(x)\,dx = \frac{1}{\Delta x_i}\int_{x_{i-1/2}}^{x_{i+1/2}} f(x)\,dx,$$

or equivalently,

$$\bar{p}_{N,i} = \bar{f}_i$$

for $i = 0, \ldots, N$. In other words, this section describes reconstruction from cell-integral averages. You might think it would be necessary to start all over from scratch. Fortunately, this is not the case. Here we describe a simple technique for recycling everything we have learned about reconstruction and interpolation from samples.

As seen in any basic calculus textbook, the *antiderivative* or *primitive function* $F(x)$ of $f(x)$ is defined as

$$f(x) = \frac{dF(x)}{dx},$$

or equivalently,

$$F(x) = \int_{x_{-1/2}}^{x} f(y)\,dy,$$

where the lower limit of integration has been chosen such that

$$\sum_{j=0}^{i} \Delta x_j \bar{f}_j = \sum_{j=0}^{i}\int_{x_{j-1/2}}^{x_{j+1/2}} f(y)\,dy = \int_{x_{-1/2}}^{x_{i+1/2}} f(y)\,dy = F(x_{i+1/2}).$$

This expression says that *the partial sums of the cell integrals are samples of the primitive function*. The following three steps convert $N+1$ cell-integral averages into the desired reconstruction:

Step 1 Find $F(x_{i+1/2}) = \sum_{j=0}^{i} \Delta x_j \bar{f}_j$ for $i = -1, \ldots, N$. Note that $F(x_{-1/2}) = F(a) = 0$.

Step 2 Find the interpolation polynomial $P(x)$ passing through $(x_{i+1/2}, F(x_{i+1/2}))$ for $i = -1, \ldots, N$. The interpolation polynomial should be found in Taylor series form in order to ease the differentiation in the next step. Of course, trigonometric or any other sort of interpolation can be used in place of polynomial interpolation. In this sense, this algorithm is completely general and extends any sort of interpolation or reconstruction procedure from samples to cell-integral averages.

Step 3 Since $f(x) = dF/dx$ and $F(x) \approx P(x)$, a function approximating $f(x)$ is $p(x) = dP(x)/dx$.

This process is known as *reconstruction via the primitive function* or simply *RP*. RP was discovered by Colella and Woodward (1984). There is an alternative reconstruction technique known as *reconstruction via deconvolution* or *RD*. Unfortunately, RD only works for constant Δx. For constant Δx, reconstruction via deconvolution yields exactly the same results as reconstruction via the primitive function, albeit for less effort. See Harten, Engquist, Osher, and Chakravarthy (1987) for a description of reconstruction via deconvolution.

Example 9.5 Suppose that the region [0, 5] is divided into four cells, [0, 1], [1, 2], [2, 3], and [3, 5]. Suppose that the cell-integral averages are $\bar{f}_0 = 1.14166$, $\bar{f}_1 = 1.07166$, $\bar{f}_2 = 1.22166$, and $\bar{f}_3 = 1.58666$, respectively. Perform reconstruction via the primitive function based on polynomial interpolation.

Solution The samples of the primitive function at the cell edges are

$$F(0) = 0,$$
$$F(1) = \Delta x_0 \bar{f}_0 = 1.14166,$$
$$F(2) = \Delta x_1 \bar{f}_1 + F(1) = 1 \cdot 1.07166 + 1.14166 = 2.21333,$$
$$F(3) = \Delta x_2 \bar{f}_2 + F(2) = 1 \cdot 1.22166 + 2.21333 = 3.435,$$
$$F(5) = \Delta x_3 \bar{f}_3 + F(3) = 2 \cdot 1.58666 + 3.435 = 6.60833.$$

As the reader can easily verify, the interpolation polynomial passing through (0, 0), (1, 1.14166), (2, 2.21333), (3, 3.435), and (5, 6.60833) is

$$P_4(x) = -0.005x^4 + 0.06666666x^3 - 0.2x^2 + 1.28x.$$

Then a cubic reconstruction is as follows:

$$p_3(x) = \frac{dP_4}{dx} = -0.02x^3 + 0.2x^2 - 0.4x + 1.28.$$

Reconstruction via the primitive function can be combined with the piecewise-polynomial ENO reconstruction techniques seen in Sections 9.1 and 9.2; in fact, the original descriptions of ENO reconstructions all concern just such combinations. The rest of this section describes the combination of reconstruction via the primitive function with the material of Section 9.1; the combination with the material of Section 9.2 is similar, and thus the details are omitted.

While the last algorithm was phrased to work for any type of interpolation, the following algorithm is phrased specifically for piecewise-polynomial interpolation, which allows for some extra efficiencies.

Step 1 Find the Newton divided differences of $F(x_{i+1/2})$. For example,

$$F[x_{i-1/2}, x_{i+1/2}] = \frac{F(x_{i+1/2}) - F(x_{i-1/2})}{x_{i+1/2} - x_{i-1/2}}$$

$$= \frac{\sum_{j=0}^{i} \Delta x_j \bar{f}_j - \sum_{j=0}^{i-1} \Delta x_j \bar{f}_j}{\Delta x_i} = \frac{\Delta x_i \bar{f}_i}{\Delta x_i} = \bar{f}_i. \tag{9.31}$$

For another example,

$$F[x_{i-1/2}, x_{i+1/2}, x_{i+3/2}] = \frac{F[x_{i+1/2}, x_{i+3/2}] - F[x_{i-1/2}, x_{i+1/2}]}{x_{i+3/2} - x_{i-1/2}}$$

$$= \frac{\bar{f}_{i+1} - \bar{f}_i}{x_{i+3/2} - x_{i-1/2}} = \frac{\bar{f}_{i+1} - \bar{f}_i}{\Delta x_{i+1} + \Delta x_i}. \tag{9.32}$$

Notice that it is not necessary to actually form the sums $F(x_{i+1/2}) = \sum_{j=0}^{i} \Delta x_{ij} \bar{f}_j$, since they are never needed in formulas such as the preceding two.

Step 2 Choose the interpolation points for each cell. Notice that the interpolation should not be constant (the interpolation is differentiated in step 3, and the derivative of a constant is zero, which is not a very helpful approximation). Then assume that the interpolation in each cell is linear or some higher-order polynomial. For cell $[x_{i-1/2}, x_{i+1/2}]$, start with the line $F_{i-1/2} + F[x_{i-1/2}, x_{i+1/2}](x - x_{i-1/2})$ passing through the cell-edge samples, that is, start with the two interpolation points $l_0(i) = i + 1/2$ and $l_1(i) = i - 1/2$. Then the remaining interpolation points are chosen recursively as follows:
for $m = 1, \ldots, n$

$$l_{m+1}(i) = \begin{cases} l_m(i) & \left|F\left[x_{l_m(i)}, \ldots, x_{l_m(i)+m+1}\right]\right| \leq \left|F\left[x_{l_m(i)-1}, \ldots, x_{l_m(i)+m}\right]\right|, \\ l_m(i) - 1 & \left|F\left[x_{l_m(i)}, \ldots, x_{l_m(i)+m+1}\right]\right| > \left|F\left[x_{l_m(i)-1}, \ldots, x_{l_m(i)+m}\right]\right|. \end{cases}$$

Step 3 Find the Taylor series form of the interpolation polynomial in each cell, where the Taylor series is taken about the cell center x_i. Then the reconstruction $P_{n+1,i}$ on cell $[x_{i-1/2}, x_{i+1/2}]$ is as follows:

$$P_{n+1,i}(x) = \sum_{j=0}^{n+1} a_j (x - x_i)^j.$$

By Equation (8.12)

$$a_j = \sum_{k=0}^{n-j+1} d_{kj} F\left[x_{l_{n+1}(i)}, \ldots, x_{l_{n+1}(i)+j+k}\right],$$

and by Equation (8.11)

$$\begin{aligned} d_{0j} &= 1, \\ d_{k0} &= \left(x_i - x_{l_{n+1}(j)+k-1}\right) d_{k-1,0}, \\ d_{kj} &= d_{k,j-1} + \left(x_i - x_{l_{n+1}(i)+k+j-1}\right) d_{k-1,j}. \end{aligned}$$

Step 4 Since $f(x) = dF/dx$ and $F(x) \approx P_{n+1}(x)$, an nth-order polynomial approximating $f(x)$ is $p_n(x) = dP_{n+1}(x)/dx$. In particular, the reconstruction on cell $[x_{i-1/2}, x_{i+1/2}]$ is as follows:

$$p_{n+1,i}(x) = \sum_{j=1}^{n+1} j a_j (x - x_i)^{j-1}.$$

In general, the reconstruction is discontinuous across the cell edges $x_{i+1/2}$. To approximate $f(x)$ at cell edges, you can average the left- and right-hand limits of the reconstruction at the cell edges, using any sort of averaging function you like.

Example 9.6 Find a general expression for a piecewise-linear ENO reconstruction using reconstruction via the primitive function.

Solution The reconstruction on cell $[x_{i-1/2}, x_{i+1/2}]$ is

$$p_{1,i}(x) = a_1 + 2a_2(x - x_i).$$

Let

$$j = l_2(i) + \frac{1}{2}$$
$$= \begin{cases} i & |F[x_{i-1/2}, x_{i+1/2}, x_{i+3/2}]| \le |F[x_{i-3/2}, x_{i-1/2}, x_{i+1/2}]|, \\ i-1 & |F[x_{i-1/2}, x_{i+1/2}, x_{i+3/2}]| > |F[x_{i-3/2}, x_{i-1/2}, x_{i+1/2}]|. \end{cases}$$

Then

$$a_2 = \frac{\bar{f}_{j+1} - \bar{f}_j}{\Delta x_{j+1} + \Delta x_j} = \frac{1}{2}\frac{\bar{f}_{j+1} - \bar{f}_j}{x_{j+1} - x_j}$$

and

$$a_1 = \bar{f}_j + 2(x_i - x_j)a_2.$$

But if $j = i$ or $j = i - 1$ then

$$a_1 = \bar{f}_i.$$

Then the reconstruction on cell $[x_{i-1/2}, x_{i+1/2}]$ becomes

$$p_{1,i}(x) = \bar{f}_i + \frac{\bar{f}_{j+1} - \bar{f}_j}{x_{j+1} - x_j}(x - x_i).$$

For constant Δx

$$p_{1,i}(x) = \bar{f}_i + \frac{\bar{f}_{i+1} - \bar{f}_i}{\Delta x}(x - x_i) \tag{9.33}$$

or

$$p_{1,i}(x) = \bar{f}_i + \frac{\bar{f}_i - \bar{f}_{i-1}}{\Delta x}(x - x_i). \tag{9.34}$$

Notice that the same results are obtained if the cell-integral averages are treated like samples. In fact, since this is a second-order accurate approximation, there is no need to distinguish between samples and cell-integral averages.

Example 9.7 Find a general expression for a piecewise-quadratic reconstruction using reconstruction via the primitive function.

Solution The reconstruction on cell $[x_{i-1/2}, x_{i+1/2}]$ is

$$p_{2,i}(x) = a_1 + 2a_2(x - x_i) + 3a_3(x - x_i)^2.$$

Let

$$j = l_2(i) + \frac{1}{2},$$
$$k = l_3(i) + \frac{1}{2}.$$

Notice that $j = i, i-1$ and $k = j, j-1 = i, i-1, i-2$. Then

$$a_3 = \frac{1}{\Delta x_k + \Delta x_{k+1} + \Delta x_{k+2}} \left[\frac{\bar{f}_{k+2} - \bar{f}_{k+1}}{\Delta x_{k+2} + \Delta x_{k+1}} - \frac{\bar{f}_{k+1} - \bar{f}_k}{\Delta x_{k+1} + \Delta x_k} \right],$$

$$a_2 = \frac{\bar{f}_{k+1} - \bar{f}_k}{\Delta x_{k+1} + \Delta x_k} + (3x_i - x_{k-1/2} - x_{k+1/2} - x_{k+3/2})a_3,$$

$$a_1 = \bar{f}_i - \frac{\Delta x_i^2}{4} a_3.$$

The derivation of the preceding expression for a_1 requires a good deal of algebra, which is omitted.

For constant Δx, the above expressions simplify quite a bit and we get

$$a_3 = \frac{\bar{f}_{k+2} - 2\bar{f}_{k+1} + \bar{f}_k}{6\Delta x^2},$$

$$a_2 = \frac{\bar{f}_{k+1} - \bar{f}_k}{2\Delta x} + \frac{\bar{f}_{k+2} - 2\bar{f}_{k+1} + \bar{f}_k}{2\Delta x}\left(i - k - \frac{1}{2}\right),$$

$$a_1 = \bar{f}_i - \frac{1}{24}(\bar{f}_{k+2} - 2\bar{f}_{k+1} + \bar{f}_k).$$

The resulting expressions are

$$f(x) \approx \bar{f}_i - \frac{1}{24}(\bar{f}_i - 2\bar{f}_{i-1} + \bar{f}_{i-2}) + \frac{1}{2\Delta x}(3\bar{f}_i - 4\bar{f}_{i-1} + \bar{f}_{i-2})(x - x_i)$$
$$+ \frac{1}{2\Delta x^2}(\bar{f}_i - 2\bar{f}_{i-1} + \bar{f}_{i-2})(x - x_i)^2 \qquad (9.35)$$

or

$$f(x) \approx \bar{f}_i - \frac{1}{24}(\bar{f}_{i+1} - 2\bar{f}_i + \bar{f}_{i-1}) + \frac{1}{2\Delta x}(\bar{f}_{i+1} - \bar{f}_{i-1})(x - x_i)$$
$$+ \frac{1}{2\Delta x^2}(\bar{f}_{i+1} - 2\bar{f}_i + \bar{f}_{i-1})(x - x_i)^2 \qquad (9.36)$$

or

$$f(x) \approx \bar{f}_i - \frac{1}{24}(\bar{f}_{i+2} - 2\bar{f}_{i+1} + \bar{f}_i) - \frac{1}{2\Delta x}(3\bar{f}_i - 4\bar{f}_{i+1} + \bar{f}_{i+2})(x - x_i)$$
$$+ \frac{1}{2\Delta x^2}(\bar{f}_{i+2} - 2\bar{f}_{i+1} + \bar{f}_i)(x - x_i)^2. \qquad (9.37)$$

Example 9.8 Rework Examples 9.2 and 9.4 using cell-integral averages rather than samples. Make sure that the cell edges are the same as in Examples 9.2 and 9.4, that is, make sure the cell centers are the same as the sample points.

Solution The result of piecewise interpolation-polynomial reconstruction combined with reconstruction via the primitive function is shown in Figure 9.7. Compared with

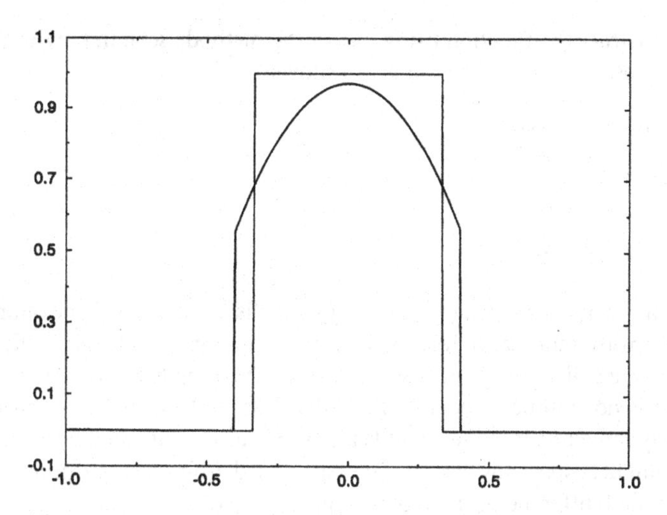

Figure 9.7 Piecewise-quadratic interpolation ENO reconstruction via the primitive function for a square wave.

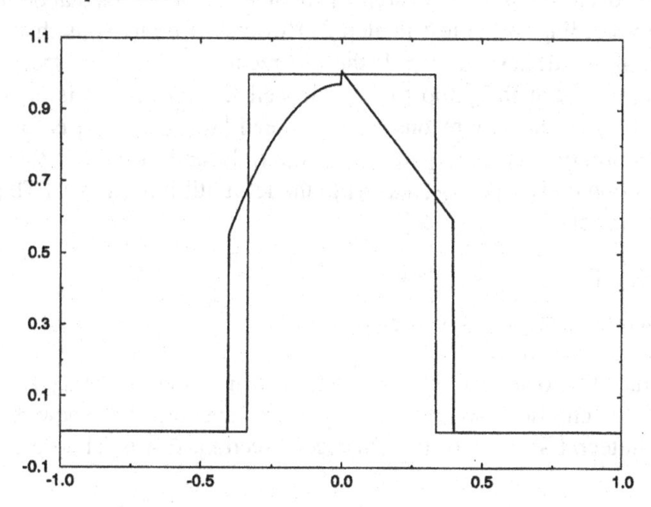

Figure 9.8 Average-quadratic interpolation ENO reconstruction via the primitive function for a square wave.

Figure 9.4, the results are somewhat inferior. Similarly, the result of average interpolation-polynomial reconstruction combined with reconstruction via the primitive function is shown in Figure 9.8. Compared with Figure 9.6, the results again are somewhat inferior.

9.4 Reconstructions with Subcell Resolution

The piecewise-continuous reconstructions seen so far can capture jump discontinuities, but only at cell edges. To overcome this weakness, consider the following set of steps:

Step 1 Reconstruct the function using one of the methods seen in previous sections. In particular, suppose the reconstruction is as follows:

$$\begin{cases} p_0(x) & x_{-1/2} \le x < x_{1/2}, \\ p_1(x) & x_{1/2} \le x < x_{3/2}, \\ \vdots & \vdots \\ p_N(x) & x_{N-1/2} \le x \le x_{N+1/2}. \end{cases}$$

Step 2 Identify cells suspected of harboring jump discontinuities. For example, a cell might contain a jump discontinuity if the Newton form or Taylor series form coefficients of the reconstruction in the cell are large in absolute value. The jump test should err on the side of safety for there is no significant accuracy penalty if the test yields a false positive, but there is a potentially large one if the test yields a false negative. For example, as a very conservative test, you might suspect any cell whose first-divided differences are greater than or equal to the first-divided differences in the two neighboring cells.

Step 3 The reconstruction found in step 1 is retained for any cells that definitely do not contain a jump according to the test in step 2. However, for any cells that might contain a jump according to the test in step 2, the old reconstruction is replaced by a new reconstruction. Suppose that the jump location in cell $[x_{i-1/2}, x_{i+1/2}]$ is θ_i, where θ_i is estimated in step 4. Then the new reconstruction in cell $[x_{i-1/2}, x_{i+1/2}]$ is found by extending the reconstruction on cell $[x_{i-1/2}, x_{i-3/2}]$ to the right until it reaches θ_i and by extending the reconstruction on cell $[x_{i+1/2}, x_{i+3/2}]$ to the left until it reaches θ_i. Thus the new reconstruction p_i^{SR} on cell $[x_{i-1/2}, x_{i+1/2}]$ is

$$p_i^{SR}(x) = \begin{cases} p_{i-1}(x) & x_{i-1/2} \le x < \theta_i, \\ p_{i+1}(x) & \theta_i \le x < x_{i+1/2}. \end{cases}$$

Step 4 Estimate the location θ_i of the shock in each cell $[x_{i-1/2}, x_{i+1/2}]$ identified in step 2. In particular, choose θ_i so that the cell-integral average of the new reconstruction equals the cell-integral average of the old reconstruction, that is, choose θ_i such that

$$\bar{p}_i^{SR} = \bar{p}_i,$$

or equivalently,

$$\frac{1}{\Delta x_i}\int_{x_{i-1/2}}^{\theta_i} p_{i-1}(x)\,dx + \frac{1}{\Delta x_i}\int_{\theta_i}^{x_{i+1/2}} p_{i+1}(x)\,dx = \frac{1}{\Delta x_i}\int_{x_{i-1/2}}^{x_{i+1/2}} p_i(x)\,dx. \tag{9.38}$$

Note that if p_i were formed from cell-integral averages using reconstruction via the primitive function, then $\bar{p}_i = \bar{f}_i$, and the right-hand side of Equation (9.38) would be $\bar{f}_i$. If Equation (9.38) has no real solutions, or if all solutions θ_i lie outside cell $[x_{i-1/2}, x_{i+1/2}]$, then the cell is assumed not to contain any jump discontinuities after all, and the reconstruction reverts to the original choice $p_i(x)$.

The above procedure is called *reconstruction with subcell resolution*. Subcell resolution was first proposed by Harten (1989). Equation (9.38) is at the heart of subcell resolution.

For example, given cell-integral averages $\bar{f}_i$ and a piecewise-constant reconstruction, Equation (9.38) implies

$$\frac{1}{\Delta x_i}((\theta_i - x_{i-1/2})\bar{f}_{i-1} + (x_{i+1/2} - \theta_i)\bar{f}_{i+1}) = \bar{f}_i$$

or

$$\theta_i = x_{i-1/2} + \frac{\bar{f}_{i+1} - \bar{f}_i}{\bar{f}_{i+1} - \bar{f}_{i-1}}\Delta x_i. \tag{9.39}$$

For another example, given cell-integral averages $\bar{f}_i$ and a piecewise-linear reconstruction, Equation (9.38) implies

$$\theta_i = x_{i-1/2} + d_i \Delta x_i,$$

where

$$\frac{1}{2}(S_{i+1} - S_{i-1})\Delta x_i d_i^2 + \left(\bar{u}_{i+1} - \bar{u}_i - \frac{3S_{i+1} + S_{i-1}}{2}\Delta x_i\right)d_i + S_{i+1}\Delta x_i - (\bar{u}_{i+1} - \bar{u}_i) = 0, \tag{9.40}$$

which can be solved using the quadratic equation. Unfortunately, for piecewise-quadratic and higher-order piecewise-polynomial reconstructions, there is no way to find an explicit expression for θ_i. One can either make do with less accurate expressions such as Equations (9.39) or (9.40) or use a numerical root solver such as Newton's method or the secant method as described in almost any basic text on numerical analysis.

References

Colella, P., and Woodward, P. R. 1984. "The Piecewise Parabolic Method (PPM) for Gas-Dynamical Simulations," *Journal of Computational Physics*, 54: 174–201.

Harten, A. 1989. "ENO Schemes with Subcell Resolution," *Journal of Computational Physics*, 83: 148–184.

Harten, A., Engquist, B., Osher S., and Chakravarthy, S. R. 1987. "Uniformly High Order Accurate Essentially Non-Oscillatory Schemes, III," *Journal of Computational Physics*, 71: 231–303.

Jiang, G.-S., and Shu, C.-W. 1996. "Efficient Implementation of Weighted ENO Schemes," *Journal of Computational Physics*, 126: 202–228.

Liu, X.-D., Osher, S., and Chan, T. 1994. "Weighted Essentially Non-Oscillatory Schemes," *Journal of Computational Physics*, 115: 200–212.

Mathews, J. H. 1992. *Numerical Methods for Mathematics, Science, and Engineering*, 2nd ed., Englewood Cliffs, NJ: Prentice-Hall.

Problems

9.1 Minmod can average any number of numbers. In particular, if all of the numbers are positive then minmod chooses the least number, if all of the numbers are negative then minmod chooses the greatest number, and otherwise minmod chooses zero. Using this definition, find $minmod(-1, -5.5, -\pi)$, $minmod(14.2, -0.1, 7, -14.2, 91)$, and $minmod(1, 0.1, 7)$.

9.2 Consider the following continuous piecewise-quadratic function:

$$f(x) = \begin{cases} (x+1)^2 & -1 \le x \le 0, \\ (x-1)^2 & 0 \le x \le 1, \end{cases}$$

Suppose that this function is sampled at $x = 0, \pm 1/2$, and ± 1.

(a) Consider the cell $[-1/4, 1/4]$. Find all possible interpolation quadratics passing through the cell center 0. Which of these quadratics best represents $f(x)$ over the entire cell and why?

(b) Use the piecewise-quadratic reconstruction technique described in Section 9.1 to find a quadratic reconstruction on cell $[-1/4, 1/4]$. Does this quadratic equal the best choice found in part (a)?

(c) Use the piecewise-quadratic reconstruction technique described in Section 9.2 to find a quadratic reconstruction on cell $[-1/4, 1/4]$. Is this reconstruction one of the interpolation polynomials found in part (a) or something else? Is this reconstruction better or worse than the best interpolation polynomial found in part (a)? Is this reconstruction better or worse than the interpolation polynomial found in part (b)?

(d) Suppose the reconstructions in parts (b) and (c) are modified to allow subcell resolution, as described in Section 9.4. That is, suppose the reconstruction can change from one polynomial to any other at some point θ *inside* cell $[-1/4, 1/4]$. What should θ be? Argue that subcell resolution, used properly, will yield an exact reconstruction of the original function.

9.3 Prove the following using trigonometric identities:

$$\int_{x_i-\Delta x/2}^{x_i+\Delta x/2} \sin x \, dx = 2 \sin \frac{1}{2} \Delta x \sin x_i .$$

Is this result consistent with Equation (9.3)? Explain.

9.4 Suppose that the cell-integral average of $f(x)$ on cell $[-3, -1]$ is 25, the cell-integral average of $f(x)$ on cell $[-1, 0]$ is 3.25, the cell-integral average of $f(x)$ on cell $[0, 3]$ is 40.25, and the cell-integral average of $f(x)$ on cell $[3, 4]$ is 167.25. Find a single polynomial approximation to $f(x)$ using reconstruction via the primitive function. (Hint: the interpolation polynomial passing through $(-3, 0)$, $(-1, 50)$, $(0, 53.25)$, $(3, 174)$, and $(4, 341.25)$ is $\frac{1}{4}x^4 + 3x^3 + \frac{3}{2}x^2 + 2x + 53.25$.)

9.5 (a) Prove Equation (9.39).
(b) Prove Equation (9.40).

9.6 The confusing thing about minmod is that it can be written in many different ways. This problem will help the reader become accustomed to some of these expressions. To do the proofs in each part, simply show that the expressions are equal if both x and y are positive, if both x and y are negative, and if x and y have opposite signs.

(a) Show that

$$\begin{aligned} minmod\,(x, y) &= sign\,(x)\, max\,[0, min\,(x\, sign\,(x), y\, sign\,(x))] \\ &= sign\,(x)\, max\,[0, min\,(|x|, y\, sign\,(x))], \end{aligned}$$

where

$$sign\,(x) = \begin{cases} -1 & x < 0, \\ 0 & x = 0, \\ 1 & x > 0. \end{cases}$$

(b) Show that

$$minmod\,(x, y) = y\, minmod\left(1, \frac{x}{y}\right) = y\, max\left[0, min\left(1, \frac{x}{y}\right)\right].$$

9.7 Consider the following periodic function:

$$f(x) = \begin{cases} \sqrt{1 - 4x^2} & |x| \leq 1/2, \\ 0 & 1/2 < |x| \leq 1. \end{cases}$$

For $|x| \leq 1/2$, this function is shaped like an ellipse. Suppose the function is sampled using 21 evenly spaced samples on the periodic domain $[-1, 1]$, including the two endpoints of the domain.

(a) Find the unique interpolation polynomial passing through all of the samples. The polynomial can be in Lagrange, Newton, or Taylor series form.

(b) Find a piecewise quadratic-interpolation reconstruction using the techniques of Section 9.1.

(c) Find an average quadratic-interpolation reconstruction using the techniques of Section 9.2.

(d) Would subcell resolution improve the results found in parts (b) and (c)? If so, where should θ be chosen in each of the cells where subcell resolution is helpful? If not, why not? Would your answer to this question change if the number of samples changed?

CHAPTER 10

Numerical Calculus

10.0 Introduction

This chapter concerns numerical differentiation, numerical integration, and the numerical solution of ordinary differential equations using Runge–Kutta methods. More specifically, the first two sections consider numerical approximations to derivatives and integrals based on samples $f(x_i)$ where $x_0 \le x_1 \le \cdots \le x_{N-1} \le x_N$. The differentiation and integration formulae found in this chapter are the foundation of the finite-difference and finite-volume methods used in computational gasdynamics, seen starting in Chapter 11. The last section of this chapter concerns Runge–Kutta methods for the numerical solution of ordinary differential equations. Of course, the equations seen in computational gasdynamics are partial differential equations rather than ordinary differential equations. In other words, the gasdynamic equations such as the differential forms of the Euler equations involve two or more independent variables rather than a single independent variable. However, after discretizing all but one independent variable, usually time, partial differential equations may be approximated by ordinary differential equations as seen in Subsection 11.2.1.

10.1 Numerical Differentiation

If a function is represented by samples, the samples must first be converted to a functional representation using interpolation or some other reconstruction procedure. The functional representation can then be differentiated using the usual rules of calculus. This section concerns differentiation formulae based on polynomial interpolation.

10.1.1 *Linear Approximations*

The line passing through any two samples $(x_i, f(x_i))$ and $(x_{i+1}, f(x_{i+1}))$ is as follows:

$$f(x) \approx p_1(x) = f(x_i) + \frac{f(x_{i+1}) - f(x_i)}{x_{i+1} - x_i}(x - x_i). \tag{10.1}$$

Then

$$\frac{df}{dx}(x) \approx \frac{dp_1}{dx}(x) = \frac{f(x_{i+1}) - f(x_i)}{x_{i+1} - x_i}. \tag{10.2}$$

Notice that this approximation is constant. It is called a *backward difference* if $x = x_{i+1}$, a *forward difference* if $x = x_i$, and a *central difference* if $x = (x_{i+1} + x_i)/2$. However, only the name of the approximation changes with x – the approximation itself is constant and does not depend on x.

Assuming that $f(x)$ has a continuous second derivative, Equation (8.13) gives the following expression for the error of Equation (10.1):

$$e_1(x) = f(x) - p_1(x) = (x - x_i)(x - x_{i+1})\frac{1}{2}\frac{d^2 f(\xi)}{dx^2}$$

for all $x_i \leq x \leq x_{i+1}$, where $x_i \leq \xi(x) \leq x_{i+1}$. Then the error of Equation (10.2) is

$$\frac{de_1}{dx}(x) = \left(x - \frac{x_{i+1} + x_i}{2}\right)\frac{d^2f(\xi)}{dx^2} + \frac{1}{2}(x - x_i)(x - x_{i+1})\frac{d^3f(\xi)}{dx_3}\frac{d\xi}{dx}. \tag{10.3}$$

In particular, for backward differences

$$\frac{de_1}{dx}(x_{i+1}) = \frac{x_{i+1} - x_i}{2}\frac{d^2f(\xi)}{dx^2} = O(x_{i+1} - x_i), \tag{10.4}$$

for forward differences

$$\frac{de_1}{dx}(x_i) = -\frac{x_{i+1} - x_i}{2}\frac{d^2f(\xi)}{dx^2} = O(x_{i+1} - x_i), \tag{10.5}$$

and for central differences

$$\begin{aligned}\frac{de_1}{dx}\left(\frac{x_{i+1} + x_i}{2}\right) &= \frac{1}{2}\left(\frac{x_{i+1} + x_i}{2} - x_i\right)\left(\frac{x_{i+1} + x_i}{2} - x_{i+1}\right)\frac{d^3f(\xi)}{dx_3}\frac{d\xi}{dx} \\ &= O((x_{i+1} - x_i)^2).\end{aligned} \tag{10.6}$$

Then for constant Δx we have

♦ $$\frac{df}{dx}(x_{i+1}) = \frac{f(x_{i+1}) - f(x_i)}{\Delta x} + O(\Delta x), \tag{10.7}$$

♦ $$\frac{df}{dx}(x_i) = \frac{f(x_{i+1}) - f(x_i)}{\Delta x} + O(\Delta x), \tag{10.8}$$

♦ $$\frac{df}{dx}\left(\frac{x_{i+1} + x_i}{2}\right) = \frac{f(x_{i+1}) - f(x_i)}{\Delta x} + O(\Delta x^2). \tag{10.9}$$

Notice that the central-difference approximation has a higher order of accuracy than the forward- and backward-difference approximations. This will be discussed further in the next subsection.

10.1.2 *Quadratic Approximations*

Using the results of Example 8.5, the quadratic passing through any three samples $(x_{i-1}, f(x_{i-1}))$, $(x_i, f(x_i))$, $(x_{i+1}, f(x_{i+1}))$ can be written as

$$p_2(x) = a_2(x - x_{i-1})^2 + a_1(x - x_{i-1}) + f(x_{i-1}), \tag{10.10}$$

where

$$a_2 = \frac{1}{x_{i+1} - x_{i-1}}\left(\frac{f(x_{i+1}) - f(x_i)}{x_{i+1} - x_i} - \frac{f(x_i) - f(x_{i-1})}{x_i - x_{i-1}}\right), \tag{10.11}$$

$$a_1 = \frac{f(x_i) - f(x_{i-1})}{x_i - x_{i-1}} - (x_i - x_{i-1})a_2. \tag{10.12}$$

Then

$$\frac{df}{dx}(x) \approx \frac{dp_2}{dx}(x) = 2a_2(x - x_{i-1}) + a_1. \tag{10.13}$$

For example, if $x = x_i$ then

$$\frac{df}{dx}(x_i) \approx \frac{dp_2}{dx}(x_i) = 2a_2(x_i - x_{i-1}) + a_1$$
$$= 2a_2(x_i - x_{i-1}) + \frac{f(x_i) - f(x_{i-1})}{x_i - x_{i-1}} - (x_i - x_{i-1})a_2$$
$$= a_2(x_i - x_{i-1}) + \frac{f(x_i) - f(x_{i-1})}{x_i - x_{i-1}}$$

or

$$\frac{df}{dx}(x_i) \approx \frac{x_i - x_{i-1}}{x_{i+1} - x_{i-1}}\left(\frac{f(x_{i+1}) - f(x_i)}{x_{i+1} - x_i} - \frac{f(x_i) - f(x_{i-1})}{x_i - x_{i-1}}\right) + \frac{f(x_i) - f(x_{i-1})}{x_i - x_{i-1}}$$

or

$$\frac{df}{dx}(x_i) \approx \theta_i \frac{f(x_{i+1}) - f(x_i)}{x_{i+1} - x_i} + (1 - \theta_i)\frac{f(x_i) - f(x_{i-1})}{x_i - x_{i-1}}, \tag{10.14}$$

where

$$\theta_i = \frac{x_i - x_{i-1}}{x_{i+1} - x_{i-1}}$$

and where $0 \leq \theta_i \leq 1$. The attentive reader may recall that a similar expression appeared in Equation (5.42). As with Equation (5.42), the expression in Equation (10.14) is called a *linear average, linear interpolation*, or a *convex linear combination* of first divided differences.

For constant Δx, Equation (10.14) becomes

$$\frac{df}{dx}(x_i) \approx \frac{f(x_{i+1}) - f(x_{i-1})}{2\Delta x}, \tag{10.15}$$

which is almost exactly the same as Equation (10.9). The main difference is the factor of two in the denominator, which reflects the fact that the samples x_{i-1} and x_{i+1} are spaced apart by $2\Delta x$ rather than by Δx. Unfortunately, approximation (10.15) wastefully ignores the middle sample $f(x_i)$. In particular, if the index i is even, then Equation (10.15) uses only the neighboring odd-indexed samples; similarly, if the index i is odd, then Equation (10.15) uses only the neighboring even-indexed samples. Then the even- and odd-indexed approximations draw from two completely different sets of samples, which can lead to a separation between the even- and odd-indexed approximations. Typical symptoms include large odd–even $2\Delta x$-wave oscillations in the approximate derivative. This phenomenon is sometimes called *odd–even decoupling*. Although the more general expression given by Equation (10.14) does not ignore the middle sample, and thus technically does not suffer from odd–even decoupling, it may behave similarly in many cases.

Much as in the last section, one can perform an order-of-error analysis using Equation (8.13); the details are omitted. For constant Δx,

♦ $$\frac{df}{dx}(x_i) = \frac{f(x_{i+1}) - f(x_{i-1})}{2\Delta x} + O(\Delta x^2). \tag{10.16}$$

Similarly,

$$\blacklozenge \qquad \frac{df}{dx}(x_{i-1}) = \frac{-f(x_{i+1}) + 4f(x_i) - 3f(x_{i-1})}{2\Delta x} + O(\Delta x^2) \tag{10.17}$$

and

$$\blacklozenge \qquad \frac{df}{dx}(x_{i+1}) = \frac{3f(x_{i+1}) - 4f(x_i) + f(x_{i-1})}{2\Delta x} + O(\Delta x^2). \tag{10.18}$$

These approximations are sometimes called *second-order forward-difference* and *second-order backward-difference*, respectively.

Quadratic interpolation can also be used to approximate second derivatives. In particular,

$$\frac{d^2 f}{dx^2}(x) \approx \frac{d^2 p_2}{dx^2}(x) = 2a_2. \tag{10.19}$$

For constant Δx,

$$a_2 = \frac{f(x_{i+1}) - 2f(x_i) + f(x_{i-1})}{2\Delta x^2},$$

and then one can show that

$$\blacklozenge \qquad \frac{d^2 f}{dx^2}(x_i) = \frac{f(x_{i+1}) - 2f(x_i) + f(x_{i-1})}{\Delta x^2} + O(\Delta x^2). \tag{10.20}$$

By differentiating cubics, quartics, and higher-order interpolation polynomials, one can obtain higher-order differentiation formulae; the details are omitted. For a complete listing of differentiation formulae for the first through the sixth derivative using three to six samples, see Table 25.2 of Abramowitz and Stegun (1964).

10.2 Numerical Integration

If a function is represented by samples, the samples must first be converted to a functional representation using interpolation or some other reconstruction procedure. The functional representation can then be integrated using the usual rules of calculus. Before proceeding, some standard terminology is introduced. Suppose an integration formula can be written in the following form:

$$\int_a^b f(x)\,dx = \sum_{i=0}^{N} w_i f(x_i) + E_N, \tag{10.21}$$

where E_N is the error, w_i are constant *weightings*, and x_i are the sample points, also known as *nodes* in this context. Any such integration formula is called a *quadrature*. *Quadratures* based on polynomial interpolation are called *Newton–Cotes quadratures*. A Newton–Cotes quadrature is *closed* if the first and last samples equal the limits of integration, that is, if $a = x_0$ and $b = x_N$.

The error analysis in this section requires a result sometimes referred to as the second integral mean value theorem. Most readers will be familiar with the ordinary or *first integral integral mean value theorem*:

$$\int_a^b f(x)\,dx = f(c)\int_a^b dx = f(c)(b-a) \quad \text{for some} \quad a \le c \le b. \tag{10.22}$$

Now suppose $g(x)$ is a function that does not change sign on $[a, b]$. Then the *second integral mean value theorem* is

$$\int_a^b f(x)g(x)\,dx = f(c)\int_a^b g(x)\,dx \quad \text{for some} \quad a \le c \le b. \tag{10.23}$$

Notice that the second integral mean value theorem implies the first integral mean value theorem.

10.2.1 *Constant Approximations*

The constant passing through a single sample $(x_0, f(x_0))$ is, of course, $p_0(x) = f(x_0)$. Then

$$\int_a^b f(x)\,dx \approx \int_a^b p_0(x)\,dx = \int_a^b f(x_0)\,dx = (b-a)f(x_0). \tag{10.24}$$

Assuming that $f(x)$ has a continuous first derivative, we have

$$e_0(x) = f(x) - p_0(x) = f(x) - f(x_0) = (x - x_0)\frac{df(\xi)}{dx},$$

where $\xi(x)$ is somewhere between x and x_0. Then the error of Equation (10.24) is

$$E_0 = \int_a^b e_0(x)\,dx = \int_a^b (f(x) - p_0(x))\,dx = \int_a^b (x - x_0)\frac{df(\xi)}{dx}dx.$$

Divide the integration up into two regions such that $x - x_0$ does not change sign in either region to find

$$E_0 = \int_a^{x_0} (x - x_0)\frac{df(\xi)}{dx}dx + \int_{x_0}^b (x - x_0)\frac{df(\xi)}{dx}dx.$$

Apply the second integral mean value theorem, Equation (10.23), to find

$$E_0 = \frac{df(\xi(c_1'))}{dx}\int_a^{x_0} (x - x_0)\,dx + \frac{df(\xi(c_2'))}{dx}\int_{x_0}^b (x - x_0)\,dx.$$

Let $c_1 = \xi(c_1')$ and $c_2 = \xi(c_2')$. Then the final result is

$$E_0 = -\frac{1}{2}(x_0 - a)^2\frac{df(c_1)}{dx} + \frac{1}{2}(x_0 - b)^2\frac{df(c_2)}{dx}, \tag{10.25}$$

where c_1 and c_2 are both between a and b.

Suppose $b - a = \Delta x$. Notice that $E_0 = O(\Delta x^2)$ for any $a \le x_0 \le b$. Then

$$\blacklozenge \qquad \int_a^b f(x)\,dx = \Delta x f(x_0) + O(\Delta x^2) \tag{10.26}$$

for any $a \le x_0 \le b$. As seen in the following example, there is one exception where the order of accuracy is three rather than two.

Example 10.1 As with numerical differentiation, centering can improve the order of accuracy of numerical integration. In particular, suppose the limits of integration are $a = x_0 - \Delta x/2$ and $b = x_0 + \Delta x/2$. Show that Equation (10.26) is third-order accurate.

Solution By Equation (10.25), the error in Equation (10.26) is

$$E_0 = -\frac{1}{2}\left(\frac{\Delta x}{2}\right)^2 \frac{df(c_1)}{dx} + \frac{1}{2}\left(\frac{\Delta x}{2}\right)^2 \frac{df(c_2)}{dx} = \frac{1}{8}\Delta x^2\left(\frac{df(c_2)}{dx} - \frac{df(c_1)}{dx}\right).$$

By the mean value theorem,

$$E_0 = \frac{1}{8}\Delta x^2 (c_2 - c_1)\frac{d^2 f(c_3)}{dx^2} \leq \frac{1}{8}\Delta x^3 \frac{d^2 f(c_3)}{dx^2} = O(\Delta x^3),$$

where c_1, c_2, and c_3 are between a and b and $b - a = \Delta x$. Then Equation (10.26) becomes

$$\int_{x_0-\Delta x/2}^{x_0+\Delta x/2} f(x)\,dx = \Delta x f(x_0) + O(\Delta x^3). \tag{10.27}$$

Note that after division by Δx, this result proves Equation (9.3).

As seen in any elementary calculus book, a large integration can be decomposed into a number of smaller integrations as follows:

$$\int_a^b f(x)\,dx = \int_a^{y_1} f(x)\,dx + \int_{y_1}^{y_2} f(x)\,dx + \cdots + \int_{y_{N-2}}^{y_{N-1}} f(x)\,dx + \int_{y_{N-1}}^{b} f(x)\,dx, \tag{10.28}$$

where the limits of integration y_i may or may not equal the sample points x_i. Then each of the small integrations can be approximated by Equation (10.26). This is known as a *Riemann sum.*

10.2.2 *Linear Approximations*

The line passing through any two samples $(x_0, f(x_0))$ and $(x_1, f(x_1))$ is as follows:

$$f(x) \approx p_1(x) = f(x_0) + \frac{f(x_1) - f(x_0)}{x_1 - x_0}(x - x_0). \tag{10.29}$$

Then

$$\int_a^b f(x)\,dx \approx \int_a^b p_1(x)\,dx = \int_a^b \left(f(x_0) + \frac{f(x_1) - f(x_0)}{x_1 - x_0}(x - x_0)\right) dx$$

or

$$\int_a^b f(x)\,dx \approx f(x_0)(b - a) + \frac{1}{2}\frac{f(x_1) - f(x_0)}{x_1 - x_0}[(b - x_0)^2 - (a - x_0)^2]. \tag{10.30}$$

One can analyze the order of accuracy of this expression much as in the previous subsection; the details are omitted.

Suppose $a = x_0$, $b = x_1$, and $b - a = x_1 - x_0 = \Delta x$. Then

$$\blacklozenge \quad \int_{x_0}^{x_1} f(x)\,dx = \frac{\Delta x}{2}(f(x_1) + f(x_0)) + O(\Delta x^3), \tag{10.31}$$

which is called the *trapezoid rule*, since the area under the curve $f(x)$ is approximated by a trapezoid. A large integration can be decomposed into small integrations as in Equation (10.28). Then each of the small integrations can be approximated using the trapezoid rule, Equation (10.31), which is known as the *composite trapezoid rule*. By integrating quadratics, cubics, quartics, and so on, one obtains higher-order integration formulae such as Simpson's rule, Simpson's $3/8$ rule, and Boole's rule; the details are omitted.

The properties of the integration formulae seen in this section are similar to the properties of the polynomial interpolations that birthed them. For example, as seen in Section 8.1.5, interpolation error is minimized by choosing the interpolation points to be the roots of Legendre or Chebyshev polynomials. Then integration error can also be minimized by choosing the samples to be the roots of Legendre or Chebyshev polynomials. In particular, if the samples are the roots of a Legendre polynomial, the integration is called a *Gauss–Legendre quadrature*. Similarly, if the samples are the roots of a Chebyshev polynomial, the integration is called a *Gauss–Chebyshev quadrature*. Tables 25.4–25.8 in Abramowitz and Stegun (1964) list weightings and nodes for Gauss–Legendre and Gauss–Chebyshev quadratures.

10.3 Runge–Kutta Methods for Solving Ordinary Differential Equations

Consider the following system of ordinary differential equations:

$$\frac{d\mathbf{u}}{dt} = \mathbf{R}(\mathbf{u}, t), \tag{10.32}$$

where, for future reference, the independent variable is now t instead of x, and the dependent variables are now $\mathbf{u}$ instead of $\mathbf{f}$. Using forward differences, as defined by Equation (10.8), we can write a very simple first-order accurate numerical approximation as

$$\frac{\mathbf{u}^{n+1} - \mathbf{u}^{n}}{\Delta t} = \mathbf{R}(\mathbf{u}^{n}, t^{n})$$

or

$$\mathbf{u}^{n+1} = \mathbf{u}^{n} + \Delta t\mathbf{R}(\mathbf{u}^{n}, t^{n}), \tag{10.33}$$

where $t^{n+1} = t^{n} + \Delta t$, $\mathbf{u}^{n+1} \approx \mathbf{u}(t^{n+1})$, and $\mathbf{u}^{n} \approx \mathbf{u}(t^{n})$. This is called the *forward-Euler approximation*.

The forward-Euler approximation can also be derived using numerical integration rather than numerical differentiation. To see this, integrate both sides of the ordinary differential equation to obtain the following integral equation:

$$u(t^{n+1}) - u(t^{n}) = \int_{t^{n}}^{t^{n+1}} \mathbf{R}(\mathbf{u}, t)\, dt. \tag{10.34}$$

Then apply Equation (10.26) to the right-hand side to obtain Equation (10.33). To save space, the derivations of the rest of the formulae in this section are omitted. However, rest assured that they can be obtained using numerical integration, numerical differentiation, or Taylor series.

Now consider the following second-order two-stage method:

$$\mathbf{u}^{(1)} = \mathbf{u}^n + \Delta t \mathbf{R}(\mathbf{u}^n, t^n), \tag{10.35a}$$

$$\mathbf{u}^{n+1} = \mathbf{u}^n + \frac{1}{2}\Delta t \mathbf{R}(\mathbf{u}^n, t^n) + \frac{1}{2}\Delta t \mathbf{R}\left(\mathbf{u}^{(1)}, t^{n+1}\right), \tag{10.35b}$$

which is called the *improved Euler approximation.* Notice that the initial guess found in the first stage is the forward-Euler approximation; the first stage is sometimes called a *predictor.* The second stage refines the initial guess, improving the order of accuracy by one; the second stage is sometimes called a *corrector.* As an alternative to the improved Euler method, consider the following second-order accurate two-step procedure:

$$\mathbf{u}^{(1)} = \mathbf{u}^n + \frac{1}{2}\Delta t \mathbf{R}(\mathbf{u}^n, t^n), \tag{10.36a}$$

$$\mathbf{u}^{n+1} = \mathbf{u}^n + \Delta t \mathbf{R}\left(\mathbf{u}^{(1)}, t^n + \frac{1}{2}\Delta t\right). \tag{10.36b}$$

This is known as the *modified Euler approximation.*

In general, consider any two-stage method of the following form:

$$\mathbf{u}^{(1)} = \mathbf{u}^n + \Delta t a_{10} \mathbf{R}(\mathbf{u}^n, t^n + c_1 \Delta t), \tag{10.37a}$$

$$\mathbf{u}^{n+1} = \mathbf{u}^n + \Delta t (a_{20} \mathbf{R}(\mathbf{u}^n, t^n + c_1 \Delta t) + a_{21} \Delta t \mathbf{R}(\mathbf{u}^{(1)}, t^n + c_2 \Delta t)). \tag{10.37b}$$

It can be shown that this two-stage method is second-order accurate if

$$c_1 = 0, \tag{10.38}$$

$$c_2 = a_{10}, \tag{10.39}$$

$$a_{20} + a_{21} = 1, \tag{10.40}$$

$$a_{21} a_{10} = \frac{1}{2}. \tag{10.41}$$

There are four equations in five unknowns, which leaves one degree of freedom. For example, the improved Euler method has $a_{10} = 1$, $a_{20} = 1/2$, $a_{21} = 1/2$, $c_1 = 0$, and $c_2 = 1$. For another example, the modified Euler method has $a_{10} = 1/2$, $a_{20} = 0$, $a_{21} = 1$, $c_1 = 0$, and $c_2 = 1/2$.

If Equation (10.40) holds, then Equation (10.37) can be written as

$$\mathbf{u}^{n+1} = a_{20}\mathbf{y}_1 + (1 - a_{20})\mathbf{y}_2,$$

where

$$\mathbf{y}_1 = \mathbf{u}^n + \Delta t \mathbf{R}(\mathbf{u}^n, t^n + c_1 \Delta t), \tag{10.42}$$

$$\mathbf{y}_2 = \mathbf{u}^n + \Delta t \mathbf{R}\left(\mathbf{u}^{(1)}, t^n + c_2 \Delta t\right). \tag{10.43}$$

Thus, assuming $0 \leq a_{20} \leq 1$, $\mathbf{u}^{n+1}$ is a *convex linear combination* of two guesses $\mathbf{y}_1$ and $\mathbf{y}_2$, where both guesses look like the forward-Euler method, the only difference being the starting times and the starting values for $\mathbf{u}$. This is now the third time we have seen convex linear combinations (the last time was in Section 10.1.2).

Let us generalize the preceding approach to any number of stages. An *explicit m-stage Runge–Kutta method* is defined as follows:

$$\begin{aligned}
\mathbf{u}^{(1)} &= \mathbf{u}^{(0)} + \Delta t a_{10}\mathbf{R}^{(0)},\\
\mathbf{u}^{(2)} &= \mathbf{u}^{(0)} + \Delta t\left(a_{20}\mathbf{R}^{(0)} + a_{21}\mathbf{R}^{(1)}\right),\\
\mathbf{u}^{(3)} &= \mathbf{u}^{(0)} + \Delta t\left(a_{30}\mathbf{R}^{(0)} + a_{31}\mathbf{R}^{(1)} + a_{32}\mathbf{R}^{(2)}\right),\\
&\vdots\\
\mathbf{u}^{(m)} &= \mathbf{u}^{(0)} + \Delta t\left(a_{m0}\mathbf{R}^{(0)} + a_{m1}\mathbf{R}^{(1)} + \cdots + a_{m,m-1}\mathbf{R}^{(m-1)}\right),
\end{aligned} \tag{10.44}$$

where

$$\mathbf{u}^{(0)} = \mathbf{u}^{n},$$
$$\mathbf{u}^{(m)} = \mathbf{u}^{n+1},$$

and where

$$\mathbf{R}^{(i)} = R\left(\mathbf{u}^{(i)}, t^n + c_{i+1}\Delta t\right).$$

The coefficients a_{ij} and c_i are called the *Runge–Kutta coefficients.* The Runge–Kutta coefficients typically satisfy

$$a_{i0} + a_{i1} + \cdots + a_{i,i-1} = c_{i+1} \tag{10.45}$$

for $i = 1, \ldots, m-1$ and

$$a_{m0} + a_{m1} + \cdots + a_{m,m-1} = 1. \tag{10.46}$$

The Runge–Kutta coefficients also typically satisfy $0 \le a_{ij} \le 1$ and $0 \le c_i \le 1$. In the rare cases where the Runge–Kutta coefficients are negative, they usually at least satisfy $|a_{ij}| \le 1$ and $|c_i| \le 1$. Finally, the Runge–Kutta coefficient c_1 typically satisfies $c_1 = 0$.

If Equation (10.46) holds, the last stage in the Runge–Kutta method can be written as

$$\mathbf{u}^{n+1} = a_{m0}\mathbf{y}_1 + a_{m1}\mathbf{y}_2 + \cdots + a_{m,m-1}\mathbf{y}_m, \tag{10.47}$$

where

$$\mathbf{y}_i = \mathbf{u}^n + \Delta t\mathbf{R}^{(i)}. \tag{10.48}$$

If $0 \le a_{mj} \le 1$ and Equation (10.46) holds, then $\mathbf{u}^{n+1}$ is a *convex linear combination* of m guesses $\mathbf{y}_1, \ldots, \mathbf{y}_m$, where all the guesses look like the forward-Euler method, the only difference being the starting times and the starting values for $\mathbf{u}$.

As the best interpretation, the Runge–Kutta method is nothing more than a simple general form. In other words, a Runge–Kutta method is any method that can be written in the form of Equation (10.44), where each guess is a linear combination of previous guesses, and the final guess is a convex linear combination of all m guesses. Many useful multistep methods can be written in Runge–Kutta form. Unfortunately, many useful methods cannot be written in Runge–Kutta form, whereas many useless methods can be.

Example 10.2 A three-stage Runge–Kutta method is third-order accurate if

$$\begin{aligned}
c_1 &= 0, \\
c_2 &= a_{10}, \\
c_3 &= a_{20} + a_{21}, \\
a_{30} + a_{31} + a_{32} &= 1, \\
a_{31}c_2 + a_{32}c_3 &= \frac{1}{2}, \\
a_{31}c_2^2 + a_{32}c_3^2 &= \frac{1}{3}, \\
a_{32}a_{21}c_2 &= \frac{1}{6}.
\end{aligned}$$

This is proven in Lambert (1991). Lambert also proves similar results for four- and five-stage Runge–Kutta methods.

To compactly specify a Runge–Kutta method, the Runge–Kutta coefficients are often arranged into an array as follows:

$$\begin{array}{c|ccccc}
c_1 & & & & & \\
c_2 & a_{10} & & & & \\
c_3 & a_{20} & a_{21} & & & \\
\vdots & \vdots & \vdots & & & \\
c_m & a_{m-1,0} & a_{m-1,1} & \cdots & a_{m-1,m-2} & \\
\hline
 & a_{m0} & a_{m1} & & a_{m,m-2} & a_{m,m-1}
\end{array} \tag{10.49}$$

which is called a *Butcher array*.

Example 10.3 The Butcher array for the improved Euler method is as follows:

$$\begin{array}{c|cc}
0 & & \\
1 & 1 & \\
\hline
 & 1/2 & 1/2
\end{array}$$

Example 10.4 Consider the following fourth-order-accurate four-stage Runge–Kutta method:

$$\begin{aligned}
\mathbf{u}^{(1)} &= \mathbf{u}^{(0)} + \frac{1}{2}\Delta t\mathbf{R}(\mathbf{u}^{(0)}, t^n), \\
\mathbf{u}^{(2)} &= \mathbf{u}^{(0)} + \frac{1}{2}\Delta t\mathbf{R}\left(\mathbf{u}^{(1)}, t^n + \frac{1}{2}\Delta t\right), \\
\mathbf{u}^{(3)} &= \mathbf{u}^{(0)} + \Delta t\mathbf{R}\left(\mathbf{u}^{(2)}, t^n + \frac{1}{2}\Delta t\right), \\
\mathbf{u}^{(4)} &= \mathbf{u}^{(0)} + \frac{1}{6}\Delta t\mathbf{R}(\mathbf{u}^{(0)}, t^n) + \frac{1}{3}\Delta t\mathbf{R}\left(\mathbf{u}^{(1)}, t^n + \frac{1}{2}\Delta t\right) \\
&\quad + \frac{1}{3}\Delta t\mathbf{R}\left(\mathbf{u}^{(2)}, t^n + \frac{1}{2}\Delta t\right) + \frac{1}{6}\Delta t\mathbf{R}(\mathbf{u}^{(2)}, t^{n+1}).
\end{aligned}$$

The Butcher array for this method is as follows:

$$\begin{array}{c|cccc} 0 & & & & \\ 1/2 & 1/2 & & & \\ 1/2 & 0 & 1/2 & & \\ 1 & 0 & 0 & 1 & \\ \hline & 1/6 & 1/3 & 1/3 & 1/6 \end{array}$$

This Runge–Kutta method is so popular that it is sometimes called *the* Runge–Kutta method. Thus if someone refers to the Runge–Kutta method without any specifics, they often mean this particular Runge–Kutta method. Despite its popularity, this Runge–Kutta method does not have any unique advantages over other four-stage Runge–Kutta methods.

Since the Runge–Kutta method is just a general form, its success depends entirely on the choice of the Runge–Kutta coefficients. As shown in Chapter 5 of Lambert (1991), the order of accuracy of an m-stage Runge–Kutta method is governed by the following results:

- Suppose $1 \leq m \leq 4$. Then an m-stage Runge–Kutta method can have mth-order accuracy, at best. Vector Runge–Kutta methods always have the same order of accuracy as scalar Runge–Kutta methods. In other words, Runge–Kutta methods yield the same order of accuracy for single ordinary differential equations as for systems of ordinary differential equations.
- Suppose $m \geq 5$. Then an m-stage Runge–Kutta method can have no more than mth-order accuracy but, in fact, the maximum order of accuracy may be less than m. For example, 5-stage Runge–Kutta methods are fourth-order accurate, at best. Unfortunately, the maximum attainable order of accuracy for arbitrary m is unknown. Surprisingly, for $m \geq 5$, vector Runge–Kutta methods may have a lower order of accuracy than scalar Runge–Kutta methods. These results imply that the Runge–Kutta form is not entirely natural for $m \geq 5$. However, in computational gasdynamics, Runge–Kutta methods with $1 \leq m \leq 4$ are usually quite sufficient.

The preceding results speak to the order of accuracy of Runge–Kutta methods. However, methods with the same order of accuracy may have wildly different accuracies. For example, as seen later in Chapters 11, 15, and 16, instability can result in unbounded oscillations regardless of the order of accuracy. Thus, in addition to order of accuracy constraints, Runge–Kutta coefficients are often chosen to optimize stability. For example, see Section 5.12 of Lambert (1991) for a linear stability analysis.

This section has concerned only *explicit* Runge–Kutta methods. It is also possible to construct *implicit* Runge–Kutta methods as follows:

$$\mathbf{u}^{(i)} = \mathbf{u}^{(0)} + \Delta t\left(a_{i0}\mathbf{R}^{(0)} + a_{i1}\mathbf{R}^{(1)} + \cdots + a_{i,m-1}\mathbf{R}^{(m-1)}\right). \tag{10.50}$$

Then each $\mathbf{u}^{(i)}$ depends on itself as well as every other value $\mathbf{u}^{(j)}$. Whereas the Butcher arrays for explicit Runge–Kutta methods are triangular, the Butcher arrays for implicit Runge–Kutta methods are square. Implicit Runge–Kutta methods are discussed in Lambert (1991).

There are, of course, many other numerical methods for solving ordinary differential equations beside Runge–Kutta methods; once again, see Lambert (1991). However, Runge–Kutta methods are far and away the most popular choice in computational gasdynamics.

In fact, some of the most famous and efficient numerical methods in computational gasdynamics rely heavily on the Runge–Kutta method. For two prominent examples of the use of Runge–Kutta methods in computational gasdynamics, see Sections 21.4 and 22.3.

References

Abramowitz, M., and Stegun, A. 1964. *Handbook of Mathematical Functions with Formulas, Graphs, and Mathematical Tables*, New York: Dover, Chapter 25.

Lambert, J. D. 1991. *Numerical Methods for Ordinary Differential Systems: The Initial Value Problem*, Chichester: Wiley, Chapter 5.

Mathews, J. H. 1992. *Numerical Methods for Mathematics, Science, and Engineering*, 2nd ed., Englewood Cliffs, NJ: Prentice-Hall, Chapters 6, 7, and 9.

Problems

10.1 Suppose you are given the samples $(x_0, f(x_0))$, $(x_0+h, f(x_0+h))$, and $(x_0+3h, f(x_0+3h))$.
(a) Approximate $f'(x_0 + h)$. What is the order of accuracy of the approximation?
(b) Approximate $f''(x_0 + 2h)$.
(c) Approximate $\int_{x_0}^{x_0+2h} f(x)dx$. Is this integration formula a Newton–Cotes quadrature? If so, is this integration formula a *closed* Newton–Cotes quadrature?

10.2 (a) Approximate $f'(x)$, $f''(x)$, and $f'''(x)$ given the samples $(-3, 0)$, $(-1, 50)$, $(0, 53.25)$, $(3, 174)$, and $(4, 341.25)$. The interpolation polynomial passing through these samples is $\frac{1}{4}x^4 + 3x^3 + \frac{3}{2}x^2 + 2x + 53.25$. What sort of accuracy would you expect from your approximation of $f''(7)$?
(b) Find an approximation to the primitive function $f(x) = \int_{-1}^{x} f(y)dy$.

10.3 Consider Example 8.10. Using the expressions for $p_N(x)$ for $N = 2, \ldots, 6$, approximate $f'(1/4)$ and $f''(1/4)$. Give a table listing the error of $f'(1/4)$ and $f''(1/4)$ as a function of N for $N = 2, \ldots, 6$.

10.4 (a) Prove Equation (10.17).
(b) Generalize Equation (10.17) to find an expression that works for nonconstant sample spacings.

10.5 Consider interpolation polynomials with Runge oscillations such as, for example, the interpolation polynomial for a square wave seen in Figure 8.5. Notice that, in all cases, the interpolation polynomial has the least error (zero error) at the sample points.
(a) Where does the derivative of the interpolation polynomial have the least error? The greatest error? Argue that the error of the derivative is out of phase with the error of the interpolation polynomial, that is, the error of the derivative is roughly the greatest where the error of the interpolation polynomial is the least, and vice versa.
(b) If the lower limit of integration is any sample point, where does the primitive function of the interpolation polynomial have the least error? The greatest error? Argue that the error of the primitive function has maxima and minima roughly at sample points, in an odd–even fashion. Argue that the primitive function has less error than the interpolation polynomial, due to cancellation of positive and negative error.

10.6 Consider Problem 9.7. Use your solutions for parts (a) though (c) to estimate the derivative $f'(x)$ and the primitive function $f(x) = \int_{-1}^{x} f(y)\,dy$. In each case, plot the true derivative versus the approximation based on interpolation, and the true primitive function versus the approximation based on interpolation, on the domain $[-1, 1]$. Also, in each case, find the error of the derivative and integral approximation on the domain $[-1, 1]$ in the 1-norm, the 2-norm, and the ∞-norm.

10.7 Consider the following ordinary differential equation:

$$\frac{du}{dt} = t - u,$$

where $u(0) = 0.9$.

(a) Approximate $u(5)$ using the forward-Euler method for $\Delta t = 0.5$ and $\Delta t = 0.125$.
(b) Approximate $u(5)$ using the improved Euler method for $\Delta t = 0.5$ and $\Delta t = 0.125$.
(c) Approximate $u(5)$ using the modified Euler method for $\Delta t = 0.5$ and $\Delta t = 0.125$.
(d) Approximate $u(5)$ using the Runge–Kutta method given in Example 10.4 for $\Delta t = 0.5$ and $\Delta t = 0.125$.

Part III

Basic Principles of Computational Gasdynamics

Computational gasdynamics combines the gasdynamics described in Part I with the computational techniques described in Part II. Part III concerns the basic principles of computational gasdynamics including conservation, the CFL condition, artificial viscosity, stability, and upwinding. The rest of the book relies heavily on the theory seen in Part III. Some readers may find a few passages in this part of the book a bit dry; they have permission to skip ahead and come back as necessary.

CHAPTER 11

Conservation and Other Basic Principles

11.0 Introduction

This chapter describes a founding principle of shock-capturing numerical methods called *numerical conservation*. It also introduces other fundamental concepts including implicit methods, explicit methods, stencil width, the numerical domain of dependence, consistency, convergence, stability, formal order of accuracy, semidiscrete approximations, and the method of lines. The material in this chapter applies equally to the Euler equations and to scalar conservation laws.

As seen in Sections 2.2 and 4.2, the physical principle of conservation leads to a description of fluid flow in terms of fluxes. This chapter concerns the analogous principle of numerical conservation, and its intimate relationship to conservative numerical fluxes. By mimicking the flux behavior of the integral and conservation form of the Euler equations, conservative numerical methods obtain the following advantage: correct shock placement. By contrast, nonconservative numerical methods consistently under- or overestimate shock speeds, so that numerical shocks increasingly lag or lead the true shocks as time progresses.

11.1 Conservative Finite-Volume Methods

Suppose that space is divided into *cells* $[x_{i-1/2}, x_{i+1/2}]$, where $x = x_{i+1/2}$ is called a *cell edge*. Also, suppose that time is divided into *time intervals* $[t^n, t^{n+1}]$, where $t = t^n$ is called a *time level*. Apply the integral form of the conservation law, Equation (2.21) or (4.1), to each cell during each time interval to obtain

$$\int_{x_{i-1/2}}^{x_{i+1/2}} [u(x, t^{n+1}) - u(x, t^n)]\, dx = -\int_{t^n}^{t^{n+1}} [f(u(x_{i+1/2}, t)) - f(u(x_{i-1/2}, t))]\, dt.$$

This leads immediately to the following numerical *conservation form*:

♦ $$\bar{u}_i^{n+1} = \bar{u}_i^n - \lambda\left(\hat{f}_{i+1/2}^n - \hat{f}_{i-1/2}^n\right), \tag{11.1}$$

where

♦ $$\bar{u}_i^n \approx \frac{1}{\Delta x}\int_{x_{i-1/2}}^{x_{i+1/2}} u(x, t^n)\, dx, \tag{11.2}$$

♦ $$\hat{f}_{i+1/2}^n \approx \frac{1}{\Delta t}\int_{t^n}^{t^{n+1}} f(u(x_{i+1/2}, t))\, dt, \tag{11.3}$$

and

♦ $$\lambda = \frac{\Delta t}{\Delta x}, \tag{11.4}$$

and where $\Delta x = x_{i+1/2} - x_{i-1/2}$ and $\Delta t = t^{n+1} - t^n$. In the above expressions, an overbar indicates spatial cell-integral averages, as in earlier chapters, while a hat indicates

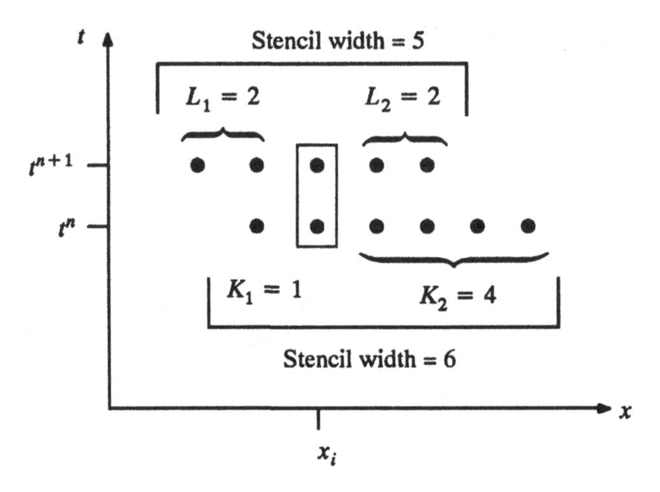

Figure 11.1 A typical stencil diagram.

time-integral averages. If numerical methods can be written in conservation form, as in Equation (11.1), then they are called *conservative* and the quantities $\hat{f}^n_{i+1/2}$ are called *conservative numerical fluxes.*

Equation (11.1) describes a *time step* from time level n to time level $n+1$. After n time steps, a numerical method knows the solution at all time levels less than n but does not know that solution at any time levels greater than n. In an *implicit* method, the unknown solution at time level $n+1$ depends on itself or on the unknown solution at later time levels. In particular, in a typical implicit method, $\bar{u}^{n+1}_i$ depends on $(\bar{u}^n_{i-K_1}, \ldots, \bar{u}^n_{i+K_2})$ and $(\bar{u}^{n+1}_{i-L_1}, \ldots, \bar{u}^{n+1}_{i+L_2})$, where $K_1 \geq 0$, $K_2 \geq 0$, $L_1 \geq 0$, and $L_2 \geq 0$ are any integers. In other words,

$$\bar{u}^{n+1}_i = \bar{u}\left(\bar{u}^n_{i-K_1}, \ldots, \bar{u}^n_{i+K_2}; \bar{u}^{n+1}_{i-L_1}, \ldots, \bar{u}^{n+1}_{i+L_2}\right). \tag{11.5}$$

Recall the earlier definition of implicit systems of equations, seen following Equation (5.6). An implicit numerical method must solve an implicit system of equations, using Gaussian elimination for linear systems or roots solvers such as such as Newton's method or Broyden's method for nonlinear systems.

The terms in equation (11.5), $(\bar{u}^n_{i-K_1}, \ldots, \bar{u}^n_{i+K_2})$ and $(\bar{u}^{n+1}_{i-L_1}, \ldots, \bar{u}^{n+1}_{i+L_2})$, are called the *stencil* or the *direct numerical domain of dependence* of $\bar{u}^{n+1}_i$. The quantities $K_1 + K_2 + 1$ and $L_1 + L_2 + 1$ are called *stencil widths.* Stencils are often illustrated by *stencil diagrams* such as the one shown in Figure 11.1. In order to avoid clutter, future stencil diagrams will include only the dots.

Example 11.1 Suppose that

$$\bar{u}^{n+1}_i = \bar{u}^n_i - \frac{\lambda}{2}\left(f\left(u^{n+1}_{i+1}\right) - f\left(u^{n+1}_{i-1}\right)\right) + \frac{\lambda}{2}\left(f\left(\bar{u}^n_{i+2}\right) - 4f\left(\bar{u}^n_{i+1}\right) + 3f\left(\bar{u}^n_i\right)\right).$$

Then $K_1 = 0$, $K_2 = 2$, $L_1 = 1$, and $L_2 = 1$. The stencil width at time level n is three and the stencil width at time level $n+1$ is also three.

Phrased in terms of conservative numerical fluxes, Equation (11.5) becomes

$$\blacklozenge \qquad \hat{f}^n_{i+1/2} = \hat{f}\left(\bar{u}^n_{i-K_1+1}, \ldots, \bar{u}^n_{i+K_2}; \bar{u}^{n+1}_{i-L_1+1}, \ldots, \bar{u}^{n+1}_{i+L_2}\right). \tag{11.6}$$

Compared with Equation (11.5), Equation (11.6) adds one to the lower indices $i - K_1$ and $i - L_1$. To understand why, suppose that Equation (11.6) did not add one to the lower indices, that is, suppose that

$$\hat{f}^n_{i+1/2} = \hat{f}\left(\bar{u}^n_{i-K_1}, \ldots, \bar{u}^n_{i+K_2}; \bar{u}^{n+1}_{i-L_1}, \ldots, \bar{u}^{n+1}_{i+L_2}\right).$$

Then

$$\hat{f}^n_{i-1/2} = \hat{f}\left(\bar{u}^n_{i-K_1-1}, \ldots, \bar{u}^n_{i+K_2-1}; \bar{u}^{n+1}_{i-L_1-1}, \ldots, \bar{u}^{n+1}_{i+L_2-1}\right)$$

and then

$$\bar{u}^{n+1}_i = u^n_i - \lambda\left(\hat{f}^n_{i+1/2} - \hat{f}^n_{i-1/2}\right).$$

The domain of dependence of u^{n+1}_i is the union of u^n_i, the domain of dependence of $\hat{f}^n_{i+1/2}$, and the domain of dependence of $\hat{f}^n_{i-1/2}$. However, this would imply

$$\bar{u}^{n+1}_i = \bar{u}\left(\bar{u}^n_{i-K_1-1}, \ldots, \bar{u}^n_{i+K_2}; \bar{u}^{n+1}_{i-L_1-1}, \ldots, \bar{u}^{n+1}_{i+L_2}\right),$$

which does not agree with Equation (11.5).

Implicit methods require costly solutions to implicit systems of equations. By contrast, in explicit methods, the unknown solution at time level $n + 1$ depends only on the known solution at time level n or earlier and does not depend on the unknown solution at time level $n + 1$ or later time levels. Typical explicit methods satisfy

$$\bar{u}^{n+1}_i = \bar{u}\left(\bar{u}^n_{i-K_1}, \ldots, \bar{u}^n_{i+K_2}\right), \tag{11.7}$$

or equivalently,

$$\blacklozenge \qquad \hat{f}^n_{i+1/2} = \hat{f}\left(\bar{u}^n_{i-K_1+1}, \ldots, \bar{u}^n_{i+K_2}\right). \tag{11.8}$$

Referring to Equation (11.7), $(\bar{u}^n_{i-K_1}, \ldots, \bar{u}^n_{i+K_2})$ is called the *stencil* or the *direct numerical domain of dependence* of $\bar{u}^{n+1}_i$, and $K_1 + K_2 + 1$ is called the *stencil width*. Each time step in an explicit method requires the solution of an explicit system of equations, which costs far less than the solution of an implicit system of equations. Although each time step in an explicit method costs less than in an implicit method, explicit methods usually have to take more time steps. The next chapter details the trade-offs between implicit and explicit methods.

Proper numerical approximations become perfect in the limit $\Delta x \to 0$ and $\Delta t \to 0$. An approximate equation is *consistent* if it equals the true equation in the limit $\Delta x \to 0$ and $\Delta t \to 0$. Also, a solution to an approximation equation is *convergent* if it equals the true solution to the true equation in the limit $\Delta x \to 0$ and $\Delta t \to 0$. Consistency is not the same thing as convergence; consistency refers to the discrete approximation whereas convergence refers to its solutions. Just because the discrete and exact equations are equal in the limit $\Delta x \to 0$ and $\Delta t \to 0$ does not imply that their solutions are also equal – even infinitesimal differences in an equation may create large differences in the solution, especially if the differences create instability.

When is a conservative approximation consistent? As a general rule, the smaller Δx and Δt, the smaller the differences between the solution values in the stencil. In fact, away from jumps, every solution value in the stencil becomes equal in the limit $\Delta x \to 0$ and $\Delta t \to 0$. Thus

$$\bar{u}^n_{i-K_1} = \cdots = \bar{u}^n_{i+K_2} = \bar{u}^{n+1}_{i-L_1} = \cdots = \bar{u}^{n+1}_{i+L_2} = u$$

in the limit $\Delta x \to 0$ and $\Delta t \to 0$. Then Equations (11.3), (11.6), and (11.8) imply the following *consistency* condition:

♦ $$\hat{f}(u, \ldots, u) = f(u), \tag{11.9}$$

assuming only that $\hat{f}$ is reasonably smooth and continuous. Then the conservative numerical flux $\hat{f}$ is said to be *consistent* with the physical flux f. Consistency between the numerical flux and the physical flux is necessary but not sufficient for consistency between the numerical approximation and the true governing equation.

Although the limiting case $\Delta x \to 0$ and $\Delta t \to 0$ motivates the definition of consistent numerical flux, a consistent numerical flux satisfies $\hat{f}^n_{i+1/2} = f(u)$ anytime the numerical solution equals a constant u throughout the stencil, whether the constancy is achieved for finite values of Δx and Δt or only in the limit $\Delta x \to 0$ and $\Delta t \to 0$. By one interpretation, $\hat{f}^n_{i+1/2}$ is a time integral-average of f. If all of the values in an average are equal, any "reasonable" average should preserve the common value. In this sense, the consistent flux condition is just a commonsense averaging condition.

Example 11.2 Suppose that the conservative numerical flux is

$$\hat{f}^n_{i+1/2} = \frac{f(\bar{u}^n_{i+1}) + f(\bar{u}^n_i)}{2}.$$

If $\bar{u}^n_{i+1} = \bar{u}^n_i = u$ then

$$\hat{f}^n_{i+1/2} = \frac{f(u) + f(u)}{2} = f(u).$$

Thus the conservative numerical flux is consistent with the physical flux.

Conservative numerical methods have the following fundamental property:

♦ *Conservative numerical methods automatically locate shocks correctly.*

Notice that this result only speaks to the location of the shock and not its shape. In fact, like all numerical methods, conservative methods may experience large spurious oscillations and smearing near shocks. However, averaging out oscillations and smearing, conservative methods always place shocks correctly. Conversely, even small deviations from strict conservative flux differencing typically result in markedly incorrect shock speeds unless, of course, the Rankine–Hugoniot jump relations given by Equation (2.32) are explicitly enforced, which requires costly shock-tracking logic. Methods that explicitly enforce the Rankine–Hugoniot relations are called *shock-fitting methods* or, sometimes, *shock-tracking methods*; methods that do not are called *shock-capturing methods*. Shock-capturing methods have dominated shock-fitting methods since the 1960s. This text concerns shock-capturing methods exclusively. Shock-capturing methods must be conservative. Modern shock-fitting methods are also often conservative.

What explains the unique shock-capturing abilities of conservative methods? Many sources cite the following result:

♦ *Consider a conservative numerical method. Assume that the conservative numerical flux is consistent with the physical flux. If the numerical solution converges as $\Delta x \to 0$ and $\Delta t \to 0$, then it converges to an exact solution of the integral form of the conservation law or, in other words, to an exact weak solution of the differential form of the conservation law.*

Put another way, the Lax–Wendroff theorem says that conservative numerical methods capture shock speeds (not to mention every other aspect of the solution) perfectly in the limiting case $\Delta x \to 0$ and $\Delta t \to 0$ assuming only that such a limiting solution exists. This result was first proven by Lax and Wendroff (1960) and is known as the *Lax–Wendroff theorem.* Also see LeVeque (1992) for a modern mathematically rigorous statement.

The Lax–Wendroff theorem has several limitations. First, it only says what happens if the solution converges as $\Delta x \to 0$ and $\Delta t \to 0$ – the Lax–Wendroff theorem never says that the solution actually *does* converge. While conservation is one element in convergence, convergence requires stability and consistency in addition to conservation, as discussed in Sections 15.4 and 16.11. Second, assuming that the solution converges, the Lax–Wendroff theorem guarantees perfection only in the limiting case $\Delta x \to 0$ and $\Delta t \to 0$, which may or may not imply good behavior for finite values of Δx and Δt. For example, the pointwise and 2-norm convergence result for Legendre polynomial series, Chebyshev polynomial series, and Fourier series do not prevent these series from experiencing large spurious oscillations and overshoots in the presence of jump discontinuities, as seen in Chapter 7. In short, lacking a decreasing upper bound on the maximum error, convergence results alone may not assure quality in ordinary calculations. Third, assuming that the solution converges, the Lax–Wendroff theorem allows *any* weak solution – it does not ensure that the weak solution satisfies the second law of thermodynamics or other entropy conditions. Fourth and finally, Shu and Osher (p. 452, 1988) indicate that the conclusions of the Lax–Wendroff theorem apply to a class of *nonconservative* methods. This class of methods allows $O(\Delta x^r)$ deviations from strict conservative flux differencing, where r is any positive exponent. Since these deviations disappear in the converged solution (i.e., in the limit $\Delta x \to 0$), the results of the Lax–Wendroff theorem still apply. However, these nonconservative methods tend to seriously mislocate shocks for any realistic values of Δx and Δt. Thus, although interesting and important, the Lax–Wendroff theorem alone does not explain the special shock-capturing abilities of conservative methods.

So what *does* explain the special shock-capturing abilities of conservative methods? First, consider a conservative numerical approximation to the total amount of conserved quantity in cells M through N as follows:

$$\begin{aligned}
\sum_{i=M}^{N} \bar{u}_i^{n+1} &= \sum_{i=M}^{N} \bar{u}_i^n - \sum_{i=M}^{N} \lambda\left(\hat{f}^n_{i+1/2} - \hat{f}^n_{i-1/2}\right) \\
&= \sum_{i=M}^{N} \bar{u}_i^n - \lambda\left(\hat{f}^n_{N+1/2} - \hat{f}^n_{N-1/2} + \hat{f}^n_{N-1/2} - \hat{f}^n_{N-3/2} + \cdots \right. \\
&\qquad \left. + \hat{f}^n_{M+3/2} - \hat{f}^n_{M+1/2} + \hat{f}^n_{M+1/2} - \hat{f}^n_{M-1/2}\right) \\
&= \sum_{i=M}^{N} \bar{u}_i^n - \lambda\left(\hat{f}^n_{N+1/2} - \hat{f}^n_{M-1/2}\right).
\end{aligned}$$

Hence, the conservative numerical fluxes cancel in the sum except for the flux through the right edge of cell N and the flux through the left edge of cell M. This is called the *telescoping flux* property. The term "telescoping" is evocative of a pocket telescope, which collapses in on itself. Most freshman calculus books discuss *telescoping series*. The term telescoping means that same thing here: Every term in the series cancels, except for the first and the last. The telescoping flux property is a direct consequence of strict flux differencing; strict flux differencing ensures that the numerical flux from cell i to cell $i+1$ is equal and opposite to the numerical flux from cell $i+1$ to cell i for all i, which implies the required cancellations in the telescoping flux property.

Suppose that a conservative method knows the exact values of $\Sigma_{i=M}^{N}\bar{u}_i^n$, $\hat{f}^n_{M-1/2}$, and $\hat{f}^n_{N+1/2}$. Then this method yields the exact value for $\Sigma_{i=M}^{N}\bar{u}_i^{n+1}$ even when the cells between M and N contain shocks and contacts. Thus, although the numerical solution may oscillate about the correct shock profile or smear the shock over several grid cells, the numerical solution still obtains the correct integral-average across cells M through N. Of course, this argument assumes that the method has exact values for $\Sigma_{i=M}^{N}\bar{u}_i^n$, $\hat{f}^n_{M-1/2}$, and $\hat{f}^n_{N+1/2}$, which is rarely true. However, these quantities tend to be nearly correct provided that the initial conditions $\bar{u}_i^0$ are correct, and provided that cells M and N are in smooth regions away from shocks.

In summary, *strict conservative flux differencing implies the telescoping flux property, which in turn implies correct shock locations, on average.* Strict conservative flux differencing is the key. Methods with strict flux differencing tend to locate shocks correctly, on average, whereas methods with even minor deviations from strict flux differencing do not.

The behavior of conservative numerical methods versus nonconservative numerical methods is illustrated in Figure 11.2. The conservative method obtains the correct integral-average $\Sigma_{i=M}^{N}\bar{u}_i^{n+1}$ (some values are too high, some values are too low, but the positive and negative errors cancel in the sum). The nonconservative method underestimates $\Sigma_{i=M}^{N}\bar{u}_i^{n+1}$ because it underestimates the shock speed. In this example, the conservative method has a

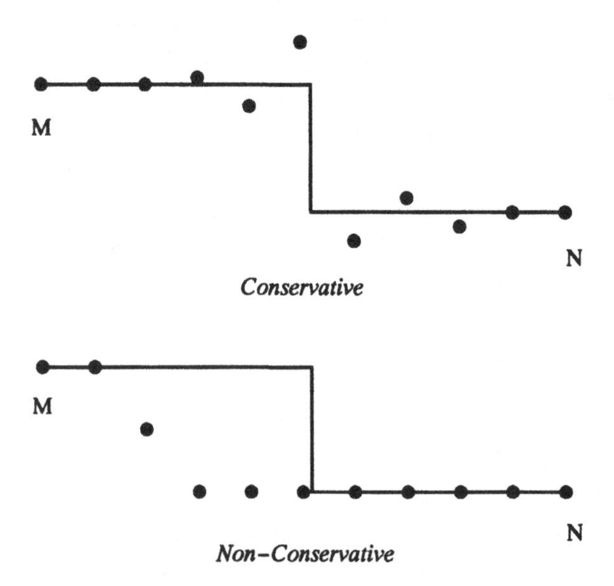

Figure 11.2 Conservative vs. nonconservative methods.

better shock placement whereas the nonconservative method has a better shock shape. However, in other cases, the conservative method will have both a better position and a better shape. The only consistent difference between conservative and nonconservative shock-capturing methods is that conservative methods correctly capture the speeds and locations of shocks, contacts, and other large-gradient flow features.

As we have seen, shock-capturing methods may exhibit large spurious oscillations near shocks. At the very least, shock-capturing methods spread shocks across several cells. For sinusoidal waves, the x direction is called phase while the y direction is called amplitude. Employing this terminology for shock waves, computational gasdynamics has long sought a simple condition that ensures proper shock amplitudes in the same way that conservation ensures proper shock phase, as discussed under the heading of nonlinear stability in Chapter 16. In their favor, shock-fitting methods do not generally suffer large amplitude errors at shocks. However, the cost required to treat amplitude errors in shock-capturing methods is usually less than the cost of shock tracking.

Lest the reader take conservation too much for granted, consider the case of chemically reacting gas flows with strong source terms. Yee and Shinn (1989) examined solutions consisting of a uniform flow punctuated by a single unsteady shock. In the presence of strong source terms, they found that supposedly conservative approximations propagated the shock at the wrong speed. Fortunately, this is impossible in our applications.

As examples of conservation and the other principles above, the remainder of the section will describe nine simple finite-volume approximations based directly on the conservation form. The integration formulae from Section 10.2 will approximate the flux integral seen in Equation (11.3).

11.1.1 *Forward-Time Methods*

To start with, use the simplest possible numerical integration formula. In particular, use the following constant *forward-time* extrapolation:

$$f(u(x_{i+1/2}, t)) = f(u(x_{i+1/2}, t^n)) + O(\Delta t)$$

for $t^n \leq t \leq t^{n+1}$. In other words, use Riemann integration as given by Equation (10.26) to find

$$\frac{1}{\Delta t}\int_{t^n}^{t^{n+1}} f(u(x_{i+1/2}, t))\,dt = f(u(x_{i+1/2}, t^n)) + O(\Delta t). \tag{11.10}$$

The right-hand side involves the unknown $f(u(x_{i+1/2}, t^n))$ rather than knowns such as $f(\bar{u}_i^n)$ or $f(\bar{u}_{i+1}^n)$. However, $f(u(x_{i+1/2}, t^n))$ is easily approximated in terms of $f(\bar{u}_i^n)$ and $f(\bar{u}_{i+1}^n)$ using interpolation, extrapolation, and reconstruction via the primitive function. For example, using constant *forward-space* extrapolation, we get

$$f(u(x_{i+1/2}, t^n)) = f(u(x_{i+1}, t^n)) + O(\Delta x).$$

Similarly, using constant *backward-space* extrapolation gives

$$f(u(x_{i+1/2}, t^n)) = f(u(x_i, t^n)) + O(\Delta x).$$

Finally, using linear *central-space* interpolation gives

$$f(u(x_{i+1/2}, t^n)) = \frac{f(u(x_{i+1}, t^n)) + f(u(x_i, t^n))}{2} + O(\Delta x^2).$$

These expressions still involve $u(x_i, t^n)$ rather than $\bar{u}_i^n$. In general, reconstruction via the primitive function converts $\bar{u}_i^n$ to $u(x_i, t^n)$, as seen in Chapter 9. Luckily, there is no need for anything so sophisticated in this case, since the order of accuracy is so low. In particular, by Equation (9.3), cell-centered samples equal cell-integral averages to second-order accuracy. Then

$$f(u(x_{i+1/2}, t^n)) = f(\bar{u}_{i+1}^n) + O(\Delta x),$$
$$f(u(x_{i+1/2}, t^n)) = f(\bar{u}_i^n) + O(\Delta x),$$
$$f(u(x_{i+1/2}, t^n)) = \frac{f(\bar{u}_{i+1}^n) + f(\bar{u}_i^n)}{2} + O(\Delta x^2).$$

In summary, the *forward-time forward-space* (*FTFS*) method is as follows:

♦ $$\bar{u}_i^{n+1} = \bar{u}_i^n - \lambda(\hat{f}_{i+1/2}^n - \hat{f}_{i-1/2}^n), \tag{11.11a}$$

♦ $$\hat{f}_{i+1/2}^n = f(\bar{u}_{i+1}^n), \tag{11.11b}$$

or equivalently,

♦ $$\bar{u}_i^{n+1} = \bar{u}_i^n - \lambda(f(\bar{u}_{i+1}^n) - f(\bar{u}_i^n)). \tag{11.12}$$

FTFS is formally first-order accurate in time and space. Notice that $K_1 = 0$ and $K_2 = 1$.

Similarly, the *forward-time backward-space* (*FTBS*) method is as follows:

♦ $$\bar{u}_i^{n+1} = \bar{u}_i^n - \lambda(\hat{f}_{i+1/2}^n - \hat{f}_{i-1/2}^n), \tag{11.13a}$$

♦ $$\hat{f}_{i+1/2}^n = f(\bar{u}_i^n), \tag{11.13b}$$

or equivalently,

♦ $$\bar{u}_i^{n+1} = \bar{u}_i^n - \lambda(f(\bar{u}_i^n) - f(\bar{u}_{i-1}^n)). \tag{11.14}$$

FTBS is formally first-order accurate in time and space. Notice that $K_1 = 1$ and $K_2 = 0$.

Finally, the *forward-time central-space* (*FTCS*) method is as follows:

♦ $$\bar{u}_i^{n+1} = \bar{u}_i^n - \lambda(\hat{f}_{i+1/2}^n - \hat{f}_{i-1/2}^n), \tag{11.15a}$$

♦ $$\hat{f}_{i+1/2}^n = \frac{f(\bar{u}_{i+1}^n) + f(\bar{u}_i^n)}{2}, \tag{11.15b}$$

or equivalently,

♦ $$\bar{u}_i^{n+1} = \bar{u}_i^n - \frac{\lambda}{2}(f(\bar{u}_{i+1}^n) - f(\bar{u}_{i-1}^n)). \tag{11.16}$$

FTCS is formally first-order accurate in time and second-order accurate in space. Notice that $K_1 = 1$ and $K_2 = 1$. The stencil diagrams for FTFS, FTBS, and FTCS are shown in Figure 11.3.

Some errors arise from the component approximations; for example, jump discontinuities cause spurious oscillations in Legendre polynomials, Fourier series, interpolation polynomials, numerical differentiation, numerical integration, and so on. Other errors arise from combinations of component approximations. In particular, by one definition, *instability* refers to errors arising from interactions between various space and time approximations.

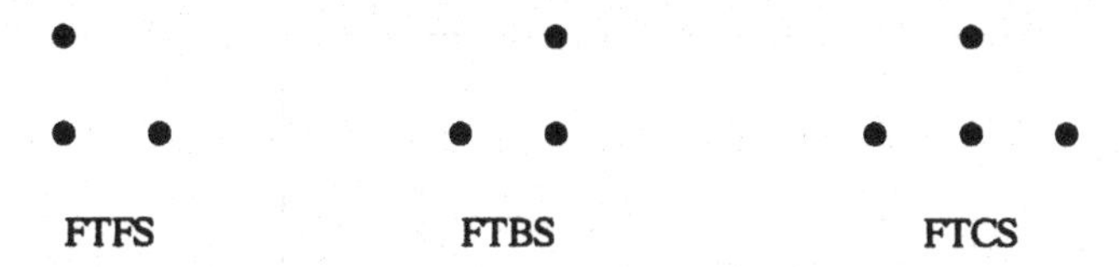

Figure 11.3 Stencil diagrams for forward-time methods.

Like the errors in the individual components, instability tends to take the form of large spurious oscillations. In the worst case, unstable methods "blow up" as time increases regardless of initial conditions and parameters such as Δx and Δt. In other words, the amplitude of the spurious oscillations increases without bound as time increases. In particular, FTFS tends to blow up when applied to flows with right-running waves, FTBS tends to blow up when applied to flows with left-running waves, and FTCS tends to blow up when applied to *any* flow. Chapters 15 and 16 discuss stability exhaustively.

The term "order of accuracy" is somewhat vague for single approximations, as discussed in Section 6.3, because it depends on whether the error is measured pointwise or in some norm, and because it depends on local behaviors of the solution such as jump discontinuities. The term "order of accuracy" is even more cloudy for combinations of approximations such as those found in FTBS, FTFS, and FTCS, because of instability. For example, consider FTBS applied to an equation with positive wave speeds. Suppose Δx is decreased while everything else is fixed, including Δt, the number of time steps, the initial conditions, and the boundary conditions. In general, the error decreases with Δx for large enough Δx (as you would expect from first-order spatial accuracy) but then the error dramatically and catastrophically *increases* after Δx drops below a certain threshold, due to instability. In fact, as we shall see later, error is sensitive not only to Δx and Δt separately, but also to the ratio $\lambda = \Delta t/\Delta x$. If the ratio $\lambda = \Delta t/\Delta x$ becomes too large relative to the local wave speeds, the error becomes large and grows rapidly with each time step, at least in explicit methods. In the individual integration and interpolation formulae, *R*th-order accuracy implies convergence as $\Delta x \to 0$ or $\Delta t \to 0$. However, convergence and other order-of-accuracy properties often fail when used in combinations such as FTBS because of instability. The orders of accuracy of the component formulae are sometimes called *formal* orders of accuracy. Other possible definitions of order of accuracy will be discussed in Subsection 11.2.2.

Example 11.3 Consider the following linear advection problem on a periodic domain $[-1, 1]$:

$$\frac{\partial u}{\partial t} + \frac{\partial u}{\partial x} = 0,$$

$$u(x,0) = \begin{cases} 1 & |x| \leq 1/3, \\ 0 & |x| > 1/3. \end{cases}$$

Approximate $u(x, 2)$ using FTFS, FTBS, and FTCS with 20 cells and $\lambda = \Delta t/\Delta x = 0.8$.

Solution A little background is in order before performing the required calculations. By Equation (4.17), the exact solution is

$$u(x,t) = u(x-t, 0).$$

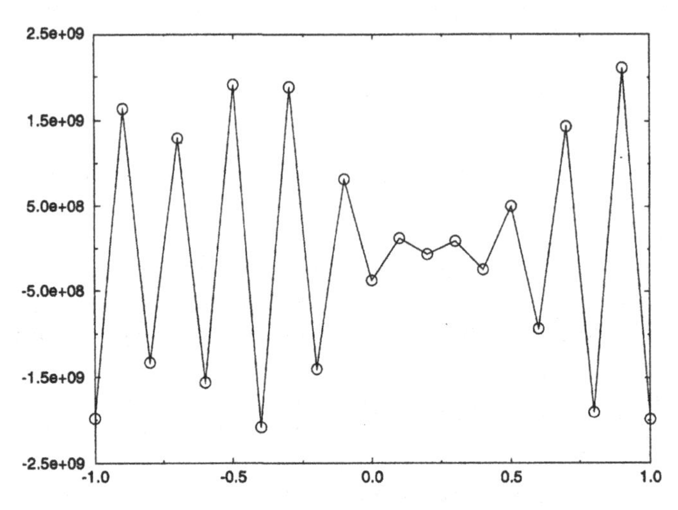

Figure 11.4 Linear advection of a square wave approximated by FTFS.

By definition of a periodic domain of width 2,

$$u(x, t) = u(x - 2, t).$$

Then $u(x, 2) = u(x - 2, 0) = u(x, 0)$. Thus the true solution travels about the periodic domain exactly once. The cell widths are $\Delta x = 2/20 = 1/10$ and the cell centers are $x_i = -1 + i\Delta x$ for $i = 0, \ldots, 20$. The solution in the 0th cell and the *N*th cell is equal by periodicity; in other words, $\bar{u}_0 = \bar{u}_N$. Similarly, $\bar{u}_1 = \bar{u}_{N+1}$. The time step is $\Delta t = 0.8\Delta x = 2/25$, and thus exactly $2/\Delta t = 25$ time steps are required to reach the final time $t = 2$. From Chapter 4, remember that jump discontinuities in the solution of the linear advection equation are contact discontinuities. Thus this example illustrates contact-capturing abilities rather than shock-capturing abilities. However, in general, contacts are more difficult to capture than shocks, and a method that captures contacts is liable to capture shocks even better.

The solutions found using FTFS, FTBS, and FTCS are shown in Figures 11.4, 11.5, and 11.6, respectively. Both FTFS and FTCS are unstable, although FTFS is many orders of magnitude more unstable than FTCS. By contrast FTBS yields a reasonably good solution, although the corners of the square wave are severely rounded off and the magnitude of the square wave is reduced by about 10%.

11.1.2 *Backward-Time Methods*

In the previous subsection, Equation (11.6) was approximated using a simple constant forward-time extrapolation. In this subsection, Equation (11.6) is approximated by a simple constant *backward-time* extrapolation as follows:

$$f(u(x_{i+1/2}, t)) = f(u(x_{i+1}, t^{n+1})) + O(\Delta t)$$

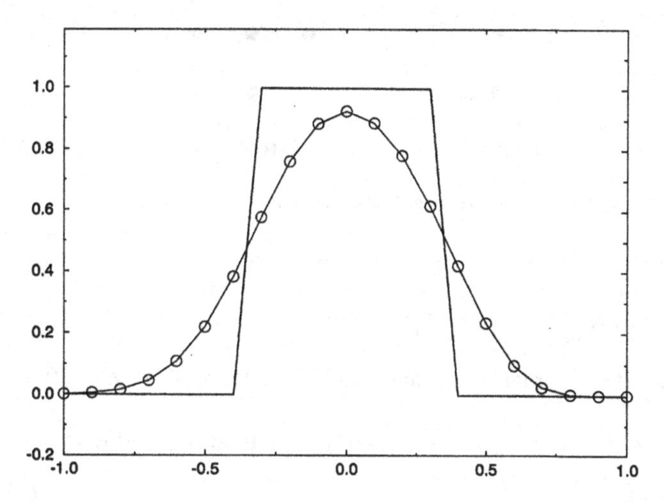

Figure 11.5 Linear advection of a square wave approximated by FTBS.

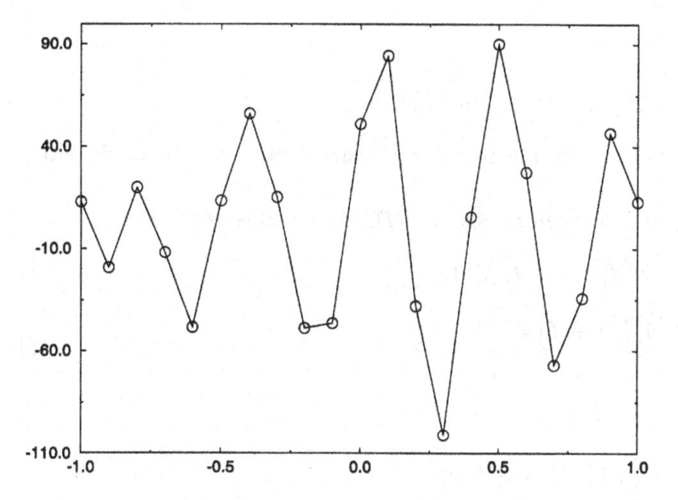

Figure 11.6 Linear advection of a square wave approximated by FTCS.

for $t^n \leq t \leq t^{n+1}$. In other words, use Riemann integration as given by Equation (10.26) to find

$$\frac{1}{\Delta t}\int_{t^n}^{t^{n+1}} f\big(u(x_{i+1/2}, t)\big)\, dt = f\big(u(x_{i+1/2}, t^{n+1})\big) + O(\Delta t). \tag{11.17}$$

Then $f(u(x_{i+1/2}, t^{n+1}))$ is approximated from $f(\bar{u}_i^{n+1})$ and $f(\bar{u}_{i+1}^{n+1})$ just as in the last subsection. The details are essentially the same as before and are therefore omitted.

The *backward-time forward-space* (*BTFS*) method is as follows:

$$\bar{u}_i^{n+1} = \bar{u}_i^n - \lambda\big(\hat{f}_{i+1/2}^n - \hat{f}_{i-1/2}^n\big), \tag{11.18a}$$

$$\hat{f}_{i+1/2}^n = f\big(\bar{u}_{i+1}^{n+1}\big), \tag{11.18b}$$

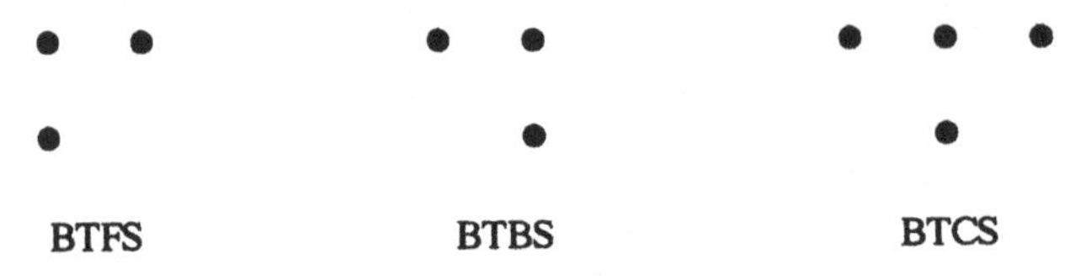

Figure 11.7 Stencil diagrams for backward-time methods.

or equivalently,

$$\bar{u}_i^{n+1} = \bar{u}_i^n - \lambda\left(f\left(\bar{u}_{i+1}^{n+1}\right) - f\left(\bar{u}_i^{n+1}\right)\right). \tag{11.19}$$

BTFS is formally first-order accurate in time and space. Notice that $K_1 = K_2 = 0$, $L_1 = 0$, and $L_2 = 1$.

Similarly, the *backward-time backward-space* (*BTBS*) method is as follows:

$$\bar{u}_i^{n+1} = \bar{u}_i^n - \lambda\left(\hat{f}_{i+1/2}^n - \hat{f}_{i-1/2}^n\right), \tag{11.20a}$$

$$\hat{f}_{i+1/2}^n = f\left(\bar{u}_i^{n+1}\right), \tag{11.20b}$$

or equivalently,

$$\bar{u}_i^{n+1} = \bar{u}_i^n - \lambda\left(f\left(\bar{u}_i^{n+1}\right) - f\left(\bar{u}_{i-1}^{n+1}\right)\right). \tag{11.21}$$

BTBS is formally first-order accurate in time and space. Notice that $K_1 = K_2 = 0$, $L_1 = 1$, and $L_2 = 0$.

Finally, the *backward-time central-space* (*BTCS*) method is as follows:

♦ $$\bar{u}_i^{n+1} = \bar{u}_i^n - \lambda\left(\hat{f}_{i+1/2}^n - \hat{f}_{i-1/2}^n\right), \tag{11.22a}$$

♦ $$\hat{f}_{i+1/2}^n = \frac{f\left(\bar{u}_{i+1}^{n+1}\right) + f\left(\bar{u}_i^{n+1}\right)}{2}, \tag{11.22b}$$

or equivalently,

♦ $$\bar{u}_i^{n+1} = \bar{u}_i^n - \frac{\lambda}{2}\left(f\left(\bar{u}_{i+1}^{n+1}\right) - f\left(\bar{u}_{i-1}^{n+1}\right)\right). \tag{11.23}$$

BTCS is formally first-order accurate in time and second-order accurate in space. Notice that $K_1 = K_2 = 0$, $L_1 = 1$, and $L_2 = 1$. BTCS is also known as the *implicit Euler* or the *backward Euler* method. The stencil diagrams for BTFS, BTBS, and BTCS are shown in Figure 11.7. Notice that all of the methods in this subsection are implicit whereas all of the methods in the previous subsection were explicit.

Example 11.4 Write the implicit Euler (BTCS) method for the linear advection equation on a periodic domain as a linear system of equations. Describe a simple numerical method for solving this linear system of equations.

Solution BTCS for the linear advection equation is as follows:

$$\bar{u}_i^{n+1} = \bar{u}_i^n - \frac{\lambda a}{2}\left(\bar{u}_{i+1}^{n+1} - \bar{u}_{i-1}^{n+1}\right)$$

or

$$\bar{u}_i^{n+1} + \frac{\lambda a}{2}\left(\bar{u}_{i+1}^{n+1} - \bar{u}_{i-1}^{n+1}\right) = \bar{u}_i^n.$$

Assume the cells are indexed from 1 to N. If $i = 1$ we have

$$\bar{u}_1^{n+1} + \frac{\lambda a}{2}\left(\bar{u}_2^{n+1} - \bar{u}_0^{n+1}\right) = \bar{u}_1^n.$$

But $u_0^n = u_N^n$ by the definition of a periodic domain. Then

$$\bar{u}_1^{n+1} + \frac{\lambda a}{2}\left(\bar{u}_2^{n+1} - \bar{u}_N^{n+1}\right) = \bar{u}_i^n.$$

If $i = N$ then

$$\bar{u}_N^{n+1} + \frac{\lambda a}{2}\left(\bar{u}_{N+1}^{n+1} - \bar{u}_{N-1}^{n+1}\right) = \bar{u}_N^n.$$

But $u_{N+1}^n = u_1^n$ by the definition of a periodic domain. Then

$$\bar{u}_N^{n+1} + \frac{\lambda a}{2}\left(\bar{u}_1^{n+1} - \bar{u}_{N-1}^{n+1}\right) = \bar{u}_N^n.$$

The desired result is

$$\begin{bmatrix} 1 & \lambda a/2 & & & & -\lambda a/2 \\ -\lambda a/2 & 1 & \lambda a/2 & & & \\ & -\lambda a/2 & 1 & \lambda a/2 & & \\ & & & & & \\ & & & -\lambda a/2 & 1 & \lambda a/2 \\ \lambda a/2 & & & & -\lambda a/2 & 1 \end{bmatrix} \begin{bmatrix} \bar{u}_1^{n+1} \\ \bar{u}_2^{n+1} \\ \bar{u}_3^{n+1} \\ \vdots \\ \bar{u}_{N-1}^{n+1} \\ \bar{u}_N^{n+1} \end{bmatrix} = \begin{bmatrix} \bar{u}_1^n \\ \bar{u}_2^n \\ \bar{u}_3^n \\ \vdots \\ \bar{u}_{N-1}^n \\ \bar{u}_N^n \end{bmatrix}.$$

The matrix is tridiagonal except for the elements in the upper right and lower left corners. The elements in the corners are a direct result of periodic boundary conditions; different boundary conditions would yield different results, although the matrix interior is always tridiagonal.

The above "periodic tridiagonal" system of equations may be solved using Gaussian elimination. Specifically, the subdiagonal may be eliminated by adding a multiple of the first row to the second, and then a multiple of the second row to the third, and so on. The resulting system of equations is upper bidiagonal, except that the far-right column is nonzero and the lower left-hand element is nonzero. Furthermore, the resulting system of equations is upper triangular except for the lower left-hand element. Adding a multiple of the first row to the last row creates a zero in the first element in the last row; however, it also creates a nonzero entry in the second element in the last row. Then adding a multiple of the second row to the last row creates a zero in the second element in the last row; however, it also creates a nonzero entry in the third element in the last row. The process continues, chasing the nonzero element across the last row from left to right until it finally lands in the last column of the last row, when the system of equations becomes entirely upper triangular. Once the system is upper triangular, it is easily solved using back-substitution, as described in any elementary text on linear algebra.

Example 11.5 Consider the following linear advection problem on a periodic domain $[-1, 1]$:

$$\frac{\partial u}{\partial t} + \frac{\partial u}{\partial x} = 0,$$

$$u(x, 0) = \begin{cases} 1 & |x| \leq 1/3, \\ 0 & |x| > 1/3. \end{cases}$$

Approximate $u(x, 2)$ using BTFS, BTBS, and BTCS with 20 cells and $\lambda = \Delta t/\Delta x = 0.8$.

Solution See Example 11.3 for a background discussion on this test problem. BTFS is highly unstable, and the results are omitted. The solution for BTBS is shown in Figure 11.8. Like FTBS, BTBS exhibits smearing and smoothing, although to a much greater degree. The solution for BTCS is shown in Figure 11.9. Unlike FTCS, seen in

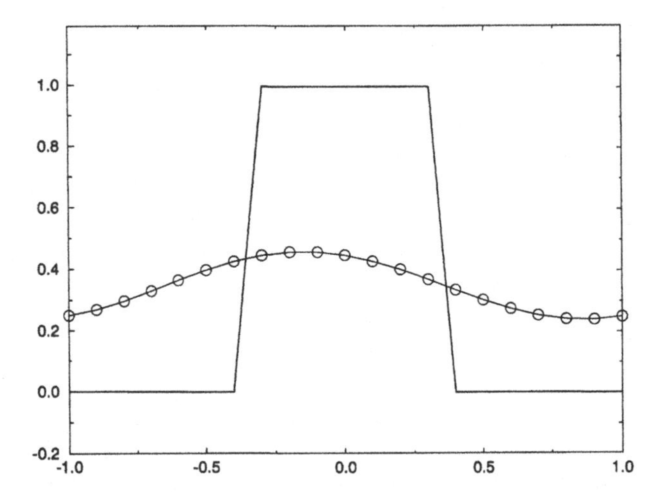

Figure 11.8 Linear advection of a square wave approximated by BTBS.

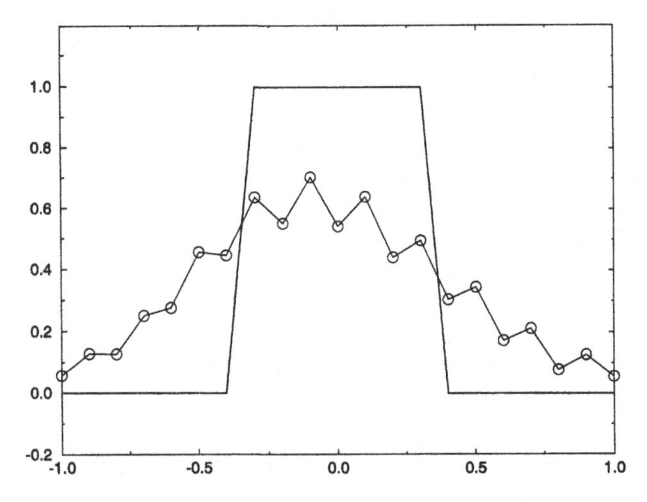

Figure 11.9 Linear advection of a square wave approximated by BTCS.

Example 11.3, BTCS does not blow up. However, BTCS still exhibits substantial error in the form of odd–even $2\Delta x$-wave oscillations. Furthermore, the size of the square wave has been eroded by perhaps 40%.

11.1.3 *Central-Time Methods*

This subsection concerns a common variant on the standard conservation form. In this variant, the time interval $[t^n, t^{n+1}]$ is replaced by the time interval $[t^{n-1}, t^{n+1}]$. Thus we apply the integral form of the conservation law, Equation (2.21) or (4.1), to each cell $[x_{i-1/2}, x_{i+1/2}]$ during the time interval $[t^{n-1}, t^{n+1}]$ to obtain

$$\int_{x_{i-1/2}}^{x_{i+1/2}} [u(x, t^{n+1}) - u(x, t^{n-1})]\, dx = -\int_{t^{n-1}}^{t^{n+1}} [f(u(x_{i+1/2}, t)) - f(u(x_{i-1/2}, t))]\, dt.$$

This leads immediately to the following alternative conservation form:

$$\bar{u}_i^{n+1} = \bar{u}_i^{n-1} - 2\lambda\left(\hat{f}_{i+1/2}^n - \hat{f}_{i-1/2}^n\right), \tag{11.24}$$

where

$$\bar{u}_i^{n\pm1} \approx \frac{1}{\Delta x}\int_{x_{i-1/2}}^{x_{i+1/2}} u(x, t^{n\pm1})\, dx, \tag{11.25}$$

$$\hat{f}_{i+1/2}^n \approx \frac{1}{2\Delta t}\int_{t^{n-1}}^{t^{n+1}} f(u(x_{i+1/2}, t))\, dt, \tag{11.26}$$

and

$$\lambda = \frac{\Delta t}{\Delta x}. \tag{11.27}$$

Like ordinary conservative methods, these *central-time conservative methods* automatically locate shocks correctly, since they involve strict flux differencing and thus obtain the telescoping flux property.

Approximate Equation (11.26) using Equation (10.27):

$$\frac{1}{2\Delta t}\int_{t^{n-1}}^{t^{n+1}} f(u(x_{i+1/2}, t))\, dt = f(u(x_{i+1/2}, t^n)) + O(\Delta t^2). \tag{11.28}$$

Then $f(u(x_{i+1/2}, t^n))$ is approximated by $f(\bar{u}_i^n)$ and $f(\bar{u}_{i+1}^n)$ just as in the last two subsections.

The *central-time forward-space* (*CTFS*) method is as follows:

$$\bar{u}_i^{n+1} = \bar{u}_i^{n-1} - 2\lambda\left(\hat{f}_{i+1/2}^n - \hat{f}_{i-1/2}^n\right), \tag{11.29a}$$

$$\hat{f}_{i+1/2}^n = f(\bar{u}_{i+1}^n), \tag{11.29b}$$

or equivalently,

$$\bar{u}_i^{n+1} = \bar{u}_i^{n-1} - 2\lambda\left(f(\bar{u}_{i+1}^n) - f(\bar{u}_i^n)\right). \tag{11.30}$$

CTFS is formally second-order accurate in time and first-order accurate in space.

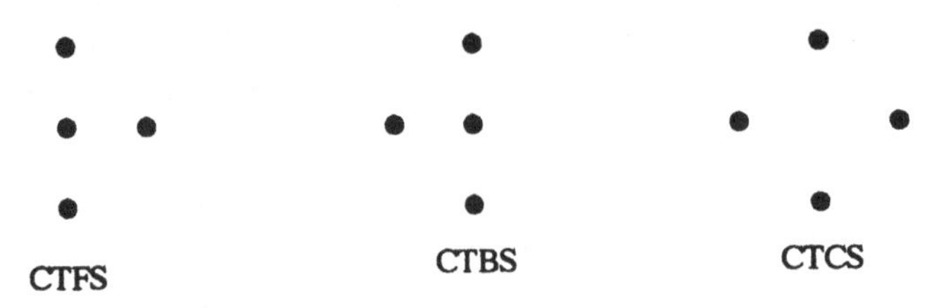

Figure 11.10 Stencil diagrams for central-time methods.

Similarly, the *central-time backward-space* (*CTBS*) method is as follows:

$$\bar{u}_i^{n+1} = \bar{u}_i^{n-1} - 2\lambda\left(\hat{f}_{i+1/2}^n - \hat{f}_{i-1/2}^n\right), \tag{11.31a}$$

$$\hat{f}_{i+1/2}^n = f\left(\bar{u}_i^n\right), \tag{11.31b}$$

or equivalently,

$$\bar{u}_i^{n+1} = \bar{u}_i^{n-1} - 2\lambda\left(f\left(\bar{u}_i^n\right) - f\left(\bar{u}_{i-1}^n\right)\right). \tag{11.32}$$

CTBS is formally second-order accurate in time and first-order accurate in space.

Finally, the *central-time central-space* (*CTCS*) method is as follows:

♦ $$\bar{u}_i^{n+1} = \bar{u}_i^{n-1} - 2\lambda\left(\hat{f}_{i+1/2}^n - \hat{f}_{i-1/2}^n\right), \tag{11.33a}$$

♦ $$\hat{f}_{i+1/2}^n = \frac{f\left(\bar{u}_{i+1}^n\right) + f\left(\bar{u}_i^n\right)}{2}, \tag{11.33b}$$

or equivalently,

♦ $$\bar{u}_i^{n+1} = \bar{u}_i^{n-1} - \lambda\left(f\left(\bar{u}_{i+1}^n\right) - f\left(\bar{u}_{i-1}^n\right)\right). \tag{11.34}$$

CTCS is formally second-order accurate in both time and space. CTCS is better known as the *leapfrog method*. The stencil diagrams for CTFS, CTBS, and CTCS are shown in Figure 11.10. The stencil diagram for CTCS helps explain the moniker "leapfrog": CTCS "leapfrogs" over the point (x_i, t^n).

All of the methods in this subsection are explicit and second-order accurate in time. Also, all of the methods in this subsection require two sets of initial conditions u_i^0 and u_i^1 rather than just one set u_i^0. The initial conditions u_i^1 can be generated from u_i^0 by another method such as, for example, BTCS.

For central-time methods, the odd-indexed time levels depend mainly on even-indexed time levels, while the even-indexed time levels depend mainly on odd-indexed time levels. This allows separation between odd- and even-indexed time levels called *temporal odd–even decoupling*. Also, in the leapfrog method, odd-indexed spatial points depend mainly on even-indexed spatial points, whereas even-indexed spatial points depend mainly on odd-indexed spatial points. This allows separation between odd- and even-indexed spatial points called *spatial odd–even decoupling*. Odd–even decoupling in time or space typically creates spurious odd–even $2\Delta x$-waves in time or space. For example, although the leapfrog method correctly locates shocks, it exhibits severe oscillations about shocks. Odd–even decoupling in the leapfrog method stems directly from odd–even decoupling in the underlying central difference approximation, as discussed in Subsection 10.1.2. Other methods based on the central difference approximation, such as FTCS and BTCS, may also exhibit signs of odd–even decoupling.

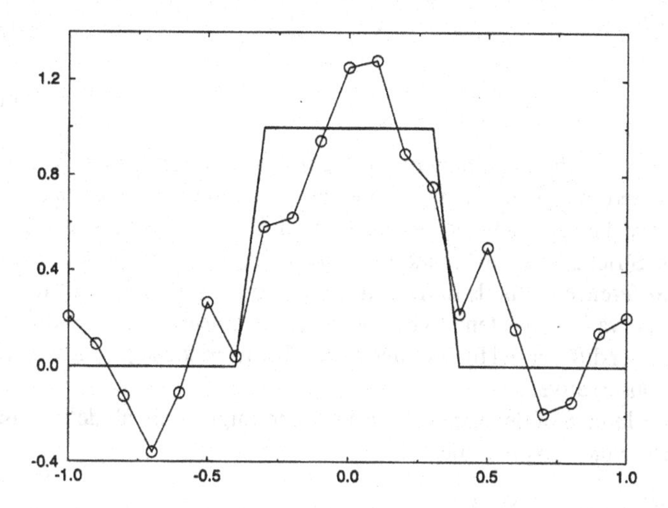

Figure 11.11 Linear advection of a square wave approximated by CTCS.

Example 11.6 Consider the following linear advection problem on a periodic domain $[-1, 1]$:

$$\frac{\partial u}{\partial t} + \frac{\partial u}{\partial x} = 0,$$

$$u(x, 0) = \begin{cases} 1 & |x| \leq 1/3, \\ 0 & |x| > 1/3. \end{cases}$$

Approximate $u(x, 2)$ using CTFS, CTBS, and CTCS with 20 cells and $\lambda = \Delta t/\Delta x = 0.8$.

Solution See Example 11.3 for a background discussion on this problem. CTFS and CTBS are unstable and the results are omitted. The solution for CTCS is shown in Figure 11.11. Compared with Examples 11.3 and 11.5, CTCS falls somewhere between BTCS and FTCS. CTCS does not blow up like FTCS; however, it does experience large spurious oscillations and overshoots, certainly larger than those found in BTCS.

11.2 Conservative Finite-Difference Methods

Like conservative finite-volume methods, conservative finite-difference methods imitate the integral and conservation forms of the Euler equations. In particular, just like the conservation form for finite-volume methods, the conservation form for finite-difference methods is defined as follows:

$$\blacklozenge \qquad u_i^{n+1} = u_i^n - \lambda\left(\hat{f}^n_{i+1/2} - \hat{f}^n_{i-1/2}\right), \tag{11.35}$$

where

$$u_i^n \approx u(x_i, t^n) \tag{11.36}$$

and

$$\lambda = \frac{\Delta t}{\Delta x}. \tag{11.37}$$

Not every finite-difference method can be written in conservation form; those which can are called *conservative* and the quantities $\hat{f}^n_{i+1/2}$ are called *conservative numerical fluxes.* Conservation has exactly the same advantages for finite-difference methods as it had for finite-volume methods: Strict conservative flux differencing implies correct shock and contact locations. Finite-difference methods derived from the conservation form of the Euler equations or scalar conservation laws tend to be conservative; conversely, finite-difference methods derived from other differential forms, such as the characteristic or primitive variable forms, tend not to be conservative.

Terms such as direct domain of dependence, explicit, and implicit are all defined just as before. In particular, for typical explicit methods

$$\hat{f}^n_{i+1/2} = \hat{f}\left(u^n_{i-K_1+1}, \ldots, u^n_{i+K_2}\right), \tag{11.38}$$

and for typical implicit methods

$$\hat{f}^n_{i+1/2} = \hat{f}\left(u^n_{i-K_1+1}, \ldots, u^n_{i+K_1}; u^{n+1}_{i-L_1+1}, \ldots, u^{n+1}_{i+L_2}\right). \tag{11.39}$$

The conservative finite-volume methods seen in the last section can all be rederived in a finite-difference context. The conservation form of the Euler equations or scalar conservation laws is as follows:

$$\frac{\partial u}{\partial t} + \frac{\partial f}{\partial x} = 0. \tag{11.40}$$

By Equation (10.8) we have

$$\frac{\partial u}{\partial t}(x, t^n) = \frac{u(x, t^{n+1}) - u(x, t^n)}{\Delta t} + O(\Delta t) \tag{11.41}$$

and

$$\frac{\partial f}{\partial x}(x_i, t) = \frac{f(u(x_{i+1}, t)) - f(u(x_i, t))}{\Delta x} + O(\Delta x), \tag{11.42}$$

which are *forward-time* and *forward-space* approximations, respectively. Also, by Equation (10.7),

$$\frac{\partial u}{\partial t}(x, t^{n+1}) = \frac{u(x, t^{n+1}) - u(x, t^n)}{\Delta t} + O(\Delta t) \tag{11.43}$$

and

$$\frac{\partial f}{\partial x}(x_i, t) = \frac{f(u(x_i, t)) - f(u(x_{i-1}, t))}{\Delta x} + O(\Delta x), \tag{11.44}$$

which are *backward-time* and *backward-space* approximations, respectively. Finally, by Equation (10.15),

$$\frac{\partial u}{\partial t}(x, t^n) = \frac{u(x, t^{n+1}) - u(x, t^{n-1})}{2\Delta t} + O(\Delta t^2) \tag{11.45}$$

and

$$\frac{\partial f}{\partial x}(x_i, t) = \frac{f(u(x_{i+1}, t)) - f(u(x_{i-1}, t))}{2\Delta x} + O(\Delta x^2), \tag{11.46}$$

which are *central-time* and *central-space* approximations, respectively. Then the three time discretizations and the three space discretizations can be paired in any combination for a total of nine methods. In particular, this approach yields forward-time forward-space (FTFS), forward-time backward-space (FTBS), forward-time central-space (FTCS), backward-time forward-space (BTFS), backward-time backward-space (BTBS), backward-time central-space (BTCS or implicit Euler), central-time forward-space (CTFS), central-time backward-space (CTBS), and central-time central-space (CTCS or leapfrog). All these methods are exactly the same as before. The only difference is that these methods now approximate cell-centered samples rather than cell-integral averages; however, since cell-centered samples equal cell-integral averages to within second-order accuracy, and since all of the methods in question are second-order accurate or less, there is no need to distinguish between cell-centered samples and cell-integral averages. The finite-difference derivations are far more common than the finite-volume derivations, but this only reflects a matter of taste and tradition.

11.2.1 *The Method of Lines*

The derivational approach described in the previous subsection involves arbitrary pairings of space and time discretizations. This approach can be generalized and formalized into the following two-step design procedure:

(1) Spatial Discretization Discretize the spatial derivative as follows:

$$\frac{\partial f}{\partial x}(x_i, t) \approx \frac{\hat{f}_{s,i+1/2}(t) - \hat{f}_{s,i-1/2}(t)}{\Delta x}.$$

In other words, freeze time and discretize space. Then Equation (11.40) becomes

$$♦ \quad \frac{du}{dt}(x_i, t) \approx -\frac{\hat{f}_{s,i+1/2}(t) - \hat{f}_{s,i-1/2}(t)}{\Delta x}. \tag{11.47}$$

This is called a *semidiscrete* finite-difference approximation and $\hat{f}_s$ is called the *semidiscrete conservative numerical flux*. The semidiscrete approximation comprises a system of ordinary differential equations. In many cases, the semidiscrete approximation is only needed at discrete time levels, in which case the semidiscrete approximation can be written as

$$\frac{du_i^n}{dt} \approx -\frac{\hat{f}^n_{s,i+1/2} - \hat{f}^n_{s,i-1/2}}{\Delta x}, \tag{11.48}$$

where $u_i^n = u(x_i, t^n)$ and $\hat{f}^n_{s,i+1/2} = \hat{f}_{s,i+1/2}(t^n)$.

(2) Temporal Discretization Starting with the semidiscrete flux $\hat{f}_s$, use a Runge–Kutta method or other ordinary differential equation solver to find an $\hat{f}$ such that

$$♦ \quad \frac{u_i^{n+1} - u_i^n}{\Delta t} = -\frac{\hat{f}^n_{i+1/2} - \hat{f}^n_{i-1/2}}{\Delta x}. \tag{11.49}$$

In other words, freeze space and discretize time. This is called a *fully discrete* finite-difference approximation and $\hat{f}$ is called the *fully discrete conservative numerical flux.* Notice that the fully discrete approximation (11.49) is the same as Equation (11.35); thus, by definition, the fully discrete approximation is conservative.

This two-stage design procedure is sometimes called the *method of lines*, where the lines are the coordinate lines $x = const.$ and $t = const.$ All of the forward-time and backward-time methods seen in this section were derived using the method of lines, although the derivations did not employ any of the above notation or formalism. The following examples explain how the finite-difference derivations of FTFS and BTCS fit into the formalism of the method of lines.

Example 11.7 For FTFS, the space discretization is the forward-space approximation with $\hat{f}_{s,i+1/2}(t^n) = f(u^n_{i+1})$ and the time discretization is the forward-time approximation with $\hat{f}^n_{i+1/2} = \hat{f}_{s,i+1/2}(t^n) = f(u^n_{i+1})$. The forward-time discretization is a Runge–Kutta method – specifically, the forward-time discretization equals the explicit forward-Euler approximation described in Section 10.3.

Example 11.8 For BTCS, the space discretization is the central-space approximation with $\hat{f}_{s,i+1/2}(t^n) = (f(u^n_{i+1}) + f(u^n_i))/2$ and the time discretization is the backward-time approximation with $\hat{f}^n_{i+1/2} = \hat{f}_{s,i+1/2}(t^{n+1})$. The backward-time discretization is a Runge–Kutta method – specifically, the backward-time discretization equals the implicit backward-Euler approximation.

Unfortunately, random "mix and match" pairings of temporal and spatial discretizations in the method of lines are often unstable. For example, FTCS, CTBS, and CTFS are unconditionally unstable. After choosing the spatial discretization, one can arbitrarily choose the temporal discretization, hoping against the odds for compatibility, as we have done in this section. Alternatively, after choosing the spatial discretization, one can choose a *class* of temporal discretizations – such as a four-stage explicit Runge–Kutta time discretization with ten free parameters – and then select from among the discretizations in the class using stability analysis, order of accuracy, numerical tests, and so on.

The right-hand side $R_i(t)$ of the semidiscrete approximation is sometimes called the *residual*,

$$R_i(t) = -\frac{\hat{f}_{s,i+1/2}(t) - \hat{f}_{s,i-1/2}(t)}{\Delta x}. \tag{11.50}$$

The term "residual" originates in the context of steady-state solutions; the residual is zero for steady-state solutions, and thus any norm of the residual measures the remaining or "residual" unsteadiness. This subsection has considered only finite-difference methods. However, the method of lines also applies to finite-volume methods; see Problem 11.11.

11.2.2 *Formal, Local, and Global Order of Accuracy*

Before ending this section, let us revisit the concept of "order of accuracy." Formal order-of-accuracy measures the order of accuracy of the individual space and time approximations. However, due to instability, formal order of accuracy may have little to do with the actual performance of the overall method. What are the alternatives? Besides formal order of accuracy, one way to measure the order of accuracy is to reduce Δx and Δt simultaneously, while fixing $\lambda = \Delta t/\Delta x$, the final time, and the initial and boundary conditions. Then a method has *global Rth-order accuracy in time and space* if

$$\| e \|_\infty \leq K \Delta x^R = K' \Delta t^R \tag{11.51}$$

for some constant K, where $e_i = u(x_i, t^n) - u_i^n$ is the absolute error and where $K' = \lambda^R K$. Of course, other error measures result if the ∞-norm is replaced by the 1-norm, the 2-norm, or any other vector norm, or if the error is measured pointwise. See Section 6.1 for a discussion on vector norms. Unfortunately, the global order of accuracy is difficult to predict theoretically. Instead, it is usually determined using numerical tests. Assuming equality rather than inequality in Equation (11.51), the order of accuracy R can be estimated by comparing the numerical solutions for two different values of Δx as follows:

$$R = \frac{\ln(\| e_2 \|_\infty / \| e_1 \|_\infty)}{\ln(\Delta x_2/\Delta x_1)} = \frac{\ln(\| e_2 \|_\infty / \| e_1 \|_\infty)}{\ln(\Delta t_2/\Delta t_1)}, \tag{11.52}$$

where $(e_1)_i = |u(x_i, t^n) - u_i^n|$ is the absolute error for Δx_1 and $\Delta t_1 = \lambda \Delta x_1$ and where $(e_2)_i = |u(x_i, t^n) - u_i^n|$ is the absolute error for Δx_2 an $\Delta t_2 = \lambda \Delta x_2$. Of course, other error measures result if the ∞-norm is replaced by the 1-norm, the 2-norm, or any other vector norm.

Order-of-accuracy measures such as Equation (11.52) may be complex functions of Δx_1, Δx_2, λ, the final time, the initial conditions, and the boundary conditions. For example, even when a numerical method has a high order of accuracy on smooth solutions, the order of accuracy on shocked solutions is typically first order or less, as measured in the ∞-norm or as measured pointwise near the shock. However, it is often possible to find a fairly representative value for the order of accuracy for a useful range of Δx and Δt assuming smooth solutions.

Another way to measure the order of accuracy is to measure the error caused by a single time step. Imagine that the numerical solution is *perfect* at time level n. That is, suppose that $u_i^n = u(x_i, t^n)$ for all i, where $u(x, t)$ is the exact solution. This is usually true for $n = 0$, since u_i^0 is usually found by sampling the exact known initial conditions $u(x, 0)$. Now take one more time step to time level $n + 1$. The *local truncation error* is the error introduced by the single time step from time level n to $n + 1$. More specifically, the local truncation error is defined as

$$t.e._i = \frac{u(x_i, t^{n+1}) - u_i^{n+1}}{\Delta t}. \tag{11.53}$$

In most cases, the term "local" means "local in space." However, here the term "local" means "local in time." Now reduce Δx and Δt simultaneously, while fixing $\lambda = \Delta t/\Delta x$, the final time, and the initial and boundary conditions. Then a method has *local Rth-order accuracy* if

$$\| t.e. \|_\infty \leq K \Delta x^R = K' \Delta t^R \tag{11.54}$$

for some constant K'. Of course, other error measures result if the ∞-norm is replaced by the 1-norm, the 2-norm, or any other vector norm, or if the truncation error is measured pointwise. Unlike the global order of accuracy, the local order of accuracy is relatively easy to predict theoretically; see Section 10.3 of LeVeque (1992) for details.

To this point, the book has defined consistency as follows: A consistent discrete equation equals the true governing equation in the limit $\Delta x \to 0$ and $\Delta t \to 0$. We now state a more restrictive definition: A consistent discrete equation has zero local truncation error in the limit $\Delta x \to 0$ and $\Delta t \to 0$. Unfortunately, this more restrictive definition depends not only on the discrete equation but also on the solution. In particular, by this definition, most numerical methods are consistent for smooth solutions but not for discontinuous solutions. For discontinuous solutions, the local truncation error tends to be $O(1)$ and thus the local truncation error does not go to zero as $\Delta x \to 0$ and $\Delta t \to 0$.

As a general rule, the formal order of accuracy is greater than or equal to the local order of accuracy, which is in turn greater than or equal to the global order of accuracy assuming, of course, that all three measures use the same norm. For stable smooth solutions, the three measures should be nearly the same, but otherwise the three measures may be quite different. Unfortunately, in the literature, the term "order of accuracy" can refer to the formal, local, or global order of accuracy in any norm. In this book, "order of accuracy" will usually refer to the *pointwise formal order of accuracy* unless explicitly stated otherwise.

11.3 Transformation to Conservation Form

Conservative numerical methods need not be written in conservation form. For example, Equation (11.16) is a nonconservation form of FTCS, Equation (11.23) is a nonconservation form of BTCS, and Equation (11.34) is a nonconservation form of CTCS. This section describes techniques for transforming numerical methods to conservation form. These transformation techniques distinguish conservative from nonconservative methods. Remember that, by definition, a method is conservative if and only if it can be written in conservation form. Conversely, a method is not conservative if and only if it cannot be written in conservation form or, in other words, if the transformation techniques described in this section fail.

There are many legitimate reasons for using nonconservation forms. First, different nonconservation forms expose different physical aspects of a method; whereas conservation form exposes conservation and flux aspects, other forms expose wave aspects, viscous aspects, and so on. By exposing certain critical physical aspects, nonconservation forms aid in the design and analysis of numerical methods. Second, different nonconservation forms expose different numerical aspects. For example, certain forms may instantly expose numerical stability properties, or at least simplify stability analysis. Third, nonconservation forms may characterize stencils better than conservation form. For example, Equation (11.22) appears to involve $f(\bar{u}_i^n)$ but Equation (11.23) clearly does not. Finally, nonconservation forms are sometimes more esthetically pleasing, more compact, or more elegant than conservation forms. Chapters 13 and 14 will introduce important nonconservation forms.

Although there are any number of good reasons for using various nonconservation forms, the variety of forms sometimes leads to confusion. Unfortunately, the same method often appears completely different when written in different forms. Without some experience with the variety of forms commonly seen in the literature, similar or even identical methods may seem completely unrelated based solely on superficial differences in form. Transformation techniques, such as the ones described below, are essential for putting methods in the same form in order to determine genuine similarities and differences. Even when using nonconservation forms, the conservation form often makes a nice intermediary for transforming between nonconservation forms. Thus rather than transform a method between forms A and

B directly, it is sometimes easier to transform a method from form A to conservation form, and then from conservation form to form B. The transformation techniques in this section will be described entirely by example.

Example 11.9 Consider FTCS in the following nonconservation form:

$$\bar{u}_i^{n+1} = \bar{u}_i^n - \frac{\lambda}{2}\left(f(\bar{u}_{i+1}^n) - f(\bar{u}_{i-1}^n)\right). \tag{11.16}$$

Rewrite FTCS in conservation form.

Solution You already know the answer from Equation (11.15). However, for purposes of this example, imagine that you have never seen Equation (11.15). Judging strictly by Equation (11.16), you might innocently try something like this:

$$\bar{u}_i^{n+1} = \bar{u}_i^n - \lambda\left(\hat{f}_{i+1/2}^n - \hat{f}_{i-1/2}^n\right),$$

where

$$\hat{f}_{i+1/2}^n = \frac{1}{2} f(\bar{u}_{i+1}^n),$$
$$\hat{f}_{i-1/2}^n = \frac{1}{2} f(\bar{u}_{i-1}^n).$$

However, this is *not* conservation form! In particular, the indexing makes no sense: Adding one to the index of $\hat{f}_{i-1/2}^n$ does not give $\hat{f}_{i+1/2}^n$, and subtracting one from the index of $\hat{f}_{i+1/2}^n$ does not give $\hat{f}_{i-1/2}^n$. Physically, this means that cells $[x_{i-1/2}, x_{i+1/2}]$ and $[x_{i+1/2}, x_{i+3/2}]$ see different fluxes through their common cell edge $x_{i+1/2}$, contradicting the basic notion of conservation.

Thus, although conservation form involves a flux difference, not just any flux difference will do. To put FTCS in conservation form, add and subtract $\lambda g_i^n/2$ on the right-hand side as follows:

$$u_i^{n+1} = u_i^n - \frac{\lambda}{2}\left(f(u_{i+1}^n) + g_i^n - g_i^n - f(u_{i-1}^n)\right).$$

Then FTCS is conservative if and only if there exists g_i^n such that

$$\hat{f}_{i+1/2}^n = \frac{1}{2}\left(f(u_{i+1}^n) + g_i^n\right)$$

and

$$\hat{f}_{i-1/2}^n = \frac{1}{2}\left(f(u_{i-1}^n) + g_i^n\right).$$

Requiring $\hat{f}_{i+1/2}^n = \hat{f}_{(i-1/2)+1}^n$ leads immediately to the following result:

$$\hat{f}_{i+1/2}^n = \frac{1}{2}\left(f(u_{i+1}^n) + g_i^n\right) = \frac{1}{2}\left(f(u_i^n) + g_{i+1}^n\right)$$

or

$$g_{i+1}^n - g_i^n = f(u_{i+1}^n) - f(u_i^n).$$

The obvious solution is

$$g_{i+1}^n = f(u_{i+1}^n),$$

or equivalently,

$$g_i^n = f(u_i^n).$$

In fact, this solution is unique to within an additive constant. Then the conservative numerical flux of FTCS is

$$\hat{f}_{i+1/2}^n = \frac{1}{2}(f(u_{i+1}^n) + g_i^n) = \frac{1}{2}(f(u_{i+1}^n) + f(u_i^n)).$$

Of course, this result agrees perfectly with Equation (11.15). In going from Equation (11.15) to (11.16), the term $g_i^n = f(u_i^n)$ cancelled out. This example has simply reconstructed g_i^n.

Example 11.10 Derive the following finite-difference method:

$$u_i^{n+1} = u_i^n - \frac{\lambda}{2}(-f(u_{i+2}^n) + 4f(u_{i+1}^n) - 3f(u_i^n)).$$

Write the method in conservation form.

Solution By Equation (10.17), the spatial discretization is

$$\frac{\partial f}{\partial x}(x_i, t^n) = \frac{-f(u_{i+2}^n) + 4f(u_{i+1}^n) - 3f(u_i^n)}{2\Delta x} + O(\Delta x^2).$$

By Equation (10.8), the time discretization is

$$\frac{\partial u}{\partial t}(x_i, t^n) = \frac{u_i^{n+1} - u_i^n}{\Delta t} + O(\Delta t).$$

The combination of the time and space discretization immediately yields the desired method. This method is formally second-order accurate in space and first-order accurate in time.

Now we move to the main task: writing the method in conservation form. A slightly different and more intuitive approach will be used in this example, as compared with the last example. First, notice that the term $-f(u_{i+2}^n)$ belongs in $\hat{f}_{i+1/2}^n$. To see why, imagine that $-f(u_{i+2}^n)$ were instead part of $\hat{f}_{i-1/2}^n$. Then $\hat{f}_{i+1/2}^n = \hat{f}_{(i-1/2)+1}^n$ implies that $\hat{f}_{i+1/2}^n$ contains $-f(u_{i+3}^n)$. But u_{i+3}^n lies outside the numerical domain of dependence and thus $\hat{f}_{i+1/2}^n$ should *not* depend on u_{i+3}^n. So, as claimed, $-f(u_{i+2}^n)$ belongs in $\hat{f}_{i+1/2}^n$. Then $\hat{f}_{i-1/2}^n = \hat{f}_{(i+1/2)-1}^n$ implies that $\hat{f}_{i-1/2}^n$ contains $-f(u_{i+1}^n)$. By similar reasoning, $3f(u_i^n)$ belongs in $\hat{f}_{i-1/2}^n$, and then $\hat{f}_{i+1/2}^n$ contains $3f(u_{i+1}^n)$. In conclusion,

$$\hat{f}_{i+1/2}^n = \frac{1}{2}(-f(u_{i+2}^n) + 3f(u_{i+1}^n)),$$

or equivalently,

$$\hat{f}_{i-1/2}^n = \frac{1}{2}(-f(u_{i+1}^n) + 3f(u_i^n)),$$

This example can also be done in the same fashion as the previous example by adding and subtracting $\lambda g_i^n/2 = \lambda f(u_{i+1}^n)/2$.

References

Boris, J. P., and Book, D. L. 1973. "Flux-Corrected Transport I. SHASTA, a Fluid Transport Algorithm that Works," *Journal of Computational Physics*, 11: 28–69.

Lax, P. D., and Wendroff, B. 1960. "Systems of Conservation Laws," *Communications on Pure and Applied Mathematics*, 13: 217–237.

LeVeque, R. J. 1992. *Numerical Methods for Conservation Laws*, 2nd ed., Basel: Birkhäuser-Verlag, Chapters 10, 11, and 12.

Shu, C.-W., and Osher, S. 1988. "Efficient Implementation of Essentially Non-Oscillatory Shock-Capturing Schemes," *Journal of Computational Physics*, 77: 439–471.

Yee, H. C., and Shinn, J. L. 1989. "Semi-Implicit and Fully-Implicit Shock-Capturing Methods for Hyperbolic Conservation Laws with Stiff Source Terms," *AIAA Journal*, 27: 299–307.

Problems

11.1 Find the missing quantities in the figure below.

$\hat{f}^n_{i-3/2} = 2$ $\quad \hat{f}^n_{i-1/2} = ?$ $\quad \hat{f}^n_{i+1/2} = 12$ $\quad \hat{f}^n_{i+3/2} = 4$

$\bar{u}^{n+1}_{i-1} = 9$	$\bar{u}^{n+1}_{i} = 2$	$\bar{u}^{n+1}_{i+1} = 9$
•	•	•
$\bar{u}^{n}_{i-1} = 7$	$\bar{u}^{n}_{i} = ?$	$\bar{u}^{n}_{i+1} = 1$

$x_{i-3/2}$ $\quad x_{i-1/2}$ $\quad x_{i+1/2}$ $\quad x_{i+3/2}$

Problem 11.1

11.2 (a) Derive the following finite-difference approximation:

$$\frac{u_i^{n+1} - u_i^n}{\Delta t} + \frac{-f\left(u_{i+2}^{n+1}\right) + 8f\left(u_{i+1}^{n+1}\right) - 8f\left(u_{i-1}^{n+1}\right) + f\left(u_{i-2}^{n+1}\right)}{12\Delta x} = 0.$$

(b) Draw the stencil diagram for this method. What are L_1 and L_2?

(c) Is this method conservative? If so, write the method in conservation form.

(d) Is this method implicit or explicit?

(e) What is the method's formal order of accuracy in time and space?

11.3 Consider the following numerical method:

$$u_i^{n+1} - u_i^{n-1} = \frac{\lambda}{2}\left(u_{i+1}^n - u_{i-1}^n\right)\left(u_{i+1}^n + u_{i-1}^n\right).$$

Is this method central-time conservative as defined in Subsection 11.1.3? If so, write the method in central-time conservation form. What type of approximation is this? What scalar conservation law does it approximate?

11.4 The simplest averages are linear averages. Thinking of $\hat{f}^n_{i+1/2}$ as an average, suppose that $\hat{f}^n_{i+1/2}$ is a linear combination of $(f(\bar{u}^n_{i-K_1}), \ldots, f(\bar{u}^n_{i+K_2}))$. Show that $\hat{f}^n_{i+1/2}$ is consistent with $f(u)$ if the linear combination is convex. Remember that, by definition, the coefficients in a convex linear combination are between zero and one, and the sum of the coefficients is equal to one.

11.5 Consider the following initial value problem for Burgers' equation:

$$\frac{\partial u}{\partial t} + u\frac{\partial u}{\partial x} = 0,$$

$$u(x,0) = \begin{cases} u_L & x < 0, \\ u_R & x > 0. \end{cases}$$

Notice that Burgers' equation is in characteristic rather than conservation form. Now consider the following finite-difference approximation:

$$u_i^{n+1} = u_i^n - \lambda u_i^n \left(u_i^n - u_{i-1}^n\right),$$

$$u_i^0 = \begin{cases} u_L & i < 0, \\ u_R & x \geq 0. \end{cases}$$

(a) Is this a conservative numerical approximation? If so, what is the conservative numerical flux $\hat{f}^n_{i+1/2}$?

(b) Assume that $u_L > u_R$, in which case the exact solution consists of a single steady shock. Does the numerical approximation yield correct shock speeds? Justify your answer using numerical results. Plot the exact versus the approximate solutions for at least two different choices of u_L and u_R.

11.6 Consider the following two-step method:

$$u_i^{(1)} = u_i^n - \frac{\lambda}{2}\left(f\left(u_{i+1}^n\right) - f\left(u_{i-1}^n\right)\right),$$

$$u_i^{n+1} = u_i^n - \frac{\lambda}{4}\left(f\left(u_{i+1}^n\right) - f\left(u_{i-1}^n\right)\right) - \frac{\lambda}{4}\left(f\left(u_{i+1}^{(1)}\right) - f\left(u_{i-1}^{(1)}\right)\right).$$

(a) Derive this method using the method of lines. What is $\hat{f}^n_{s,i+1/2}$? Use an appropriate Runge–Kutta method for the time discretization.

(b) What is the formal order of accuracy of the method in time and space?

(c) Draw the stencil diagram. What are K_1 and K_2?

(d) Write the method in conservation form.

(e) Prove that the conservative numerical flux $\hat{f}^n_{i+1/2}$ is consistent with the physical flux $f(u)$.

11.7 In 1973, Boris and Book suggested a first-order upwind method for scalar conservation laws. In the original paper, the first-order upwind method was written in the following nonconservation form:

$$u_i^{n+1} = \left(q^+_{i+1/2} + q^-_{i-1/2}\right)u_i^n + \frac{1}{2}\left(q^+_{i+1/2}\right)^2\left(u_{i+1}^n - u_i^n\right) - \frac{1}{2}\left(q^-_{i-1/2}\right)^2\left(u_i^n - u_{i-1}^n\right),$$

where

$$q^+_{i+1/2} = \frac{\frac{1}{2} - \lambda a(u_i)}{1 + \lambda(a(u_{i+1}) - a(u_i))},$$

$$q^-_{i+1/2} = \frac{\frac{1}{2} + \lambda a(u_{i+1})}{1 + \lambda(a(u_{i+1}) - a(u_i))}.$$

Notice that restrictions such as $|\lambda a(u)| < \frac{1}{2}$, $0 \leq \lambda a(u) < 1$, or $-1 < \lambda a(u) \leq 0$ will prevent zero denominators in the expressions for the coefficients q^+ and q^-.

(a) Show that this method can be written in conservation form as follows:

$$u_i^{n+1} = u_i^n - \left(\hat{f}^n_{i+1/2} - \hat{f}^n_{i-1/2}\right),$$

$$\hat{f}^n_{i+1/2} = -\frac{1}{2}\left(q^+_{i+1/2}\right)^2 u_{i+1}^n + \frac{1}{2}\left(q^-_{i+1/2}\right)^2 u_i^n.$$

As a helpful hint, notice that $q^+_{i+1/2} + q^-_{i+1/2} = 1$.

(b) Show that this method is linear when applied to the linear advection equation.
(c) Show that the conservative numerical flux found in part (a) is consistent with the true flux if and only if the method is applied to the linear advection equation.

11.8 In Example 11.4, BTCS for the linear advection equation was written as a "periodic tridiagonal" linear system of equations. Similarly, write BTBS and BTFS in terms of "periodic bidiagonal" linear system of equations.

11.9 Suppose Δx is not constant. Find expressions for FTFS, FTBS, FTCS, BTFS, BTBS, BTCS, CTFS, CTBS, and CTCS using the expressions given in Chapter 10. Is there any difference between the finite-volume and finite-difference expressions?

11.10 Consider the following linear advection problem on a periodic domain $[-1, 1]$:

$$\frac{\partial u}{\partial t} + \frac{\partial u}{\partial x} = 0,$$

$$u(x, 0) = \begin{cases} 1 & |x| \le 1/3, \\ 0 & |x| > 1/3. \end{cases}$$

Approximate $u(x, 2)$ using the following methods:
(a) FTFS, FTBS, and FTCS;
(b) BTFS, BTBS, BTCS;
(c) CTFS, CTBS, CTCS.
In each case, let $\lambda = \Delta t/\Delta x = 0.8$. Also, use 150, 300, and 600 evenly spaced grid points. Estimate the global order of accuracy using Equation (11.52). Repeat, substituting the 1-norm for the ∞-norm in Equation (11.52). In each case, how does the global order of accuracy compare with the formal order of accuracy? Is there a relationship between stability and the global order of accuracy?

11.11 Subsection 11.2.1 described the method of lines for finite-difference methods. This problem concerns the method of lines for finite-volume methods.
(a) Let $t_2 \to t_1$ in the integral form of any scalar conservation law to find the following integro-differential form:

$$\frac{d}{dt}\int_a^b u(x, t)\, dt = -[f(u(b, t)) - f(u(a, t))].$$

(b) Apply the integro-differential form of the conservation law found in part (a) to each cell $[x_{i-1/2}, x_{i+1/2}]$ at time $t = t^n$ to obtain the following natural form for a conservative semidiscrete finite-volume method:

$$\frac{d\bar{u}_i^n}{dt} = -\frac{\hat{f}^n_{s,i+1/2} - \hat{f}^n_{s,i-1/2}}{\Delta x},$$

where

$$\bar{u}_i^n \approx \frac{1}{\Delta x}\int_{x_{i-1/2}}^{x_{i+1/2}} u(x, t^n)\, dx,$$

and show that the semidiscrete conservative numerical flux is

$$\hat{f}^n_{s,i+1/2} \approx f(u(x_{i+1/2}, t^n)).$$

CHAPTER 12

The CFL Condition

12.0 Introduction

Waves imply finite *physical* domains of dependence, as discussed in Chapters 3 and 4. The last chapter introduced *numerical* domains of dependence. This chapter and the next flesh out the notion of numerical domains of dependence, and discuss the relationships between numerical and physical domains of dependence. This chapter and the next concern only finite-difference methods; however, the same principles apply to finite-volume methods, albeit with subtle modifications.

As seen in the Chapter 11, u_i^{n+1} depends on the solution at discrete points collectively called the *stencil* or the *direct numerical domain of dependence*. For example, for a typical implicit method, u_i^{n+1} depends on $(u_{i-K_1}^n, \ldots, u_{i+K_2}^n)$ and $(u_{i-L_1}^{n+1}, \ldots, u_{i+L_2}^{n+1})$. Of course, each of these values depends on other values. For instance, $u_{i-K_1}^n$ depends on $(u_{i-2K_1}^{n-1}, \ldots, u_{i-K_1+K_2}^{n-1})$ and $(u_{i-K_1-L_1}^n, \ldots, u_{i-K_1+L_2}^n)$. The *full* or *complete numerical domain of dependence* of u_i^{n+1} consists of $(u_{i-K_1}^n, \ldots, u_{i+K_2}^n)$ and $(u_{i-L_1}^{n+1}, \ldots, u_{i+L_2}^{n+1})$ plus the points that these values depend upon.

Example 12.1 Sketch the direct and full numerical domains of dependence for an implicit finite-difference method with $K_1 = K_2 = 2$ and $L_1 = L_2 = 1$. For the Euler equations, compare the numerical domain of dependence with the physical domain of dependence for a typical subsonic right-running flow.

Solution Figure 12.1 contrasts typical physical and numerical domains of dependence. In particular, in Figure 12.1, the lightly shaded area is the physical domain of dependence of u_i^{n+1} while the closed circles are the direct numerical domain of dependence of u_i^{n+1}. In Figure 12.1, the closed circles depend on the open circles. Similarly, the open circles depend on other circles, not shown, which depend on other circles, and so forth, until the circles dot the entire region beneath the line $t = t^{n+1}$. As a convenient definition, the numerical domain of dependence includes not only the discrete points but also the regions between the discrete points. Think of putting a rubber band around the outermost ring of points and considering everything inside the rubber band as part of the full numerical domain of dependence. Figure 12.1 represents the full numerical domain of dependence as the union of the lightly and darkly shaded regions. Then the full numerical domain of dependence contains the lightly shaded physical domain of dependence.

Consider the following condition:

- ♦ *The full numerical domain of dependence must contain the physical domain of dependence.*

This is called the *Courant–Friedrichs–Lewyor* (*CFL*) condition. Any numerical method that violates the CFL condition misses information affecting the true solution. If the missing

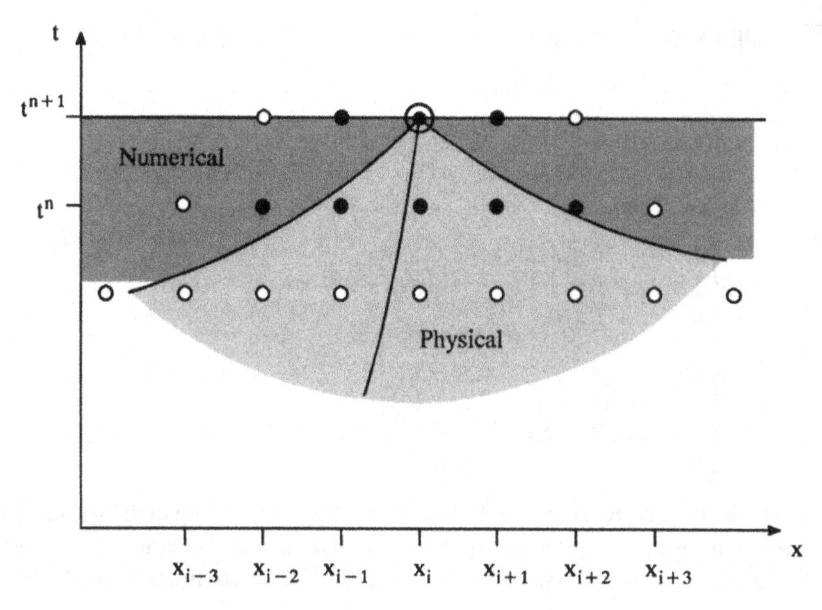

Figure 12.1 Numerical vs. physical domain of dependence.

information is critical, the numerical solution will deviate substantially from the true solution or, in the worst case, blow up to infinity. A method that violates the CFL condition is like a car with blind spots in its rearview mirrors, a horse with blinders on, or a person with tunnel vision, each of which can lead to serious accidents.

12.1 Scalar Conservation Laws

This section describes the CFL condition for scalar conservation laws, beginning with a series of examples.

Example 12.2 This example shows what can happen when a numerical method violates the CFL condition, thus missing vital information contained in the true domain of dependence but not in the numerical domain of dependence. Consider the following linear advection problem:

$$\frac{\partial u}{\partial t} + a\frac{\partial u}{\partial x} = 0,$$

$$u(x,0) = \begin{cases} 1 & x < 0, \\ 0 & x \geq 0. \end{cases}$$

Assume that $a > 0$. The exact solution is

$$u(x,t) = u(x-at,0) = \begin{cases} 1 & x - at < 0, \\ 0 & x - at \geq 0. \end{cases}$$

The FTFS approximation is

$$u_i^{n+1} = (1 + \lambda a)u_i^n - \lambda a u_{i+1}^n,$$

$$u_i^0 = u(i\Delta x, 0) = \begin{cases} 1 & i < 0, \\ 0 & i \geq 0, \end{cases}$$

where $\lambda = \Delta t/\Delta x$. Then

$$u_i^1 = \begin{cases} 1 & i \leq -2, \\ 1 + \lambda a & i = -1, \\ 0 & i \geq 0, \end{cases}$$

$$u_i^2 = \begin{cases} 1 & i \leq -3, \\ (1 + \lambda a)(1 - \lambda a) & i = -2, \\ (1 + \lambda a)(1 + \lambda a) & i = -1, \\ 0 & i \geq 0, \end{cases}$$

and so forth. As the first two time steps show, FTFS moves the jump in the wrong direction (left rather than right) and produces spurious oscillations and overshoots. Furthermore, the exact solution yields $u(0, \Delta t) = 1$ whereas the FTFS approximation yields $u_0^1 = 0$ for any λ.

What explains the poor performance of FTFS in this example? As illustrated in Figure 12.2, FTFS violates the CFL condition. In particular, at $(0, \Delta t)$, FTFS cannot see the jump discontinuity in the initial conditions. In other words, the true domain of dependence of $(0, \Delta t)$ contains only the value 1, whereas the numerical domain of dependence of $(0, \Delta t)$ contains only the value 0. After making the initial mistake, FTFS digs itself in deeper and deeper with every subsequent time step. In short, the radical differences between the true and numerical domains of dependence result in radical differences between the true and numerical solutions. In fact, FTFS generally blows up for positive wave speeds regardless

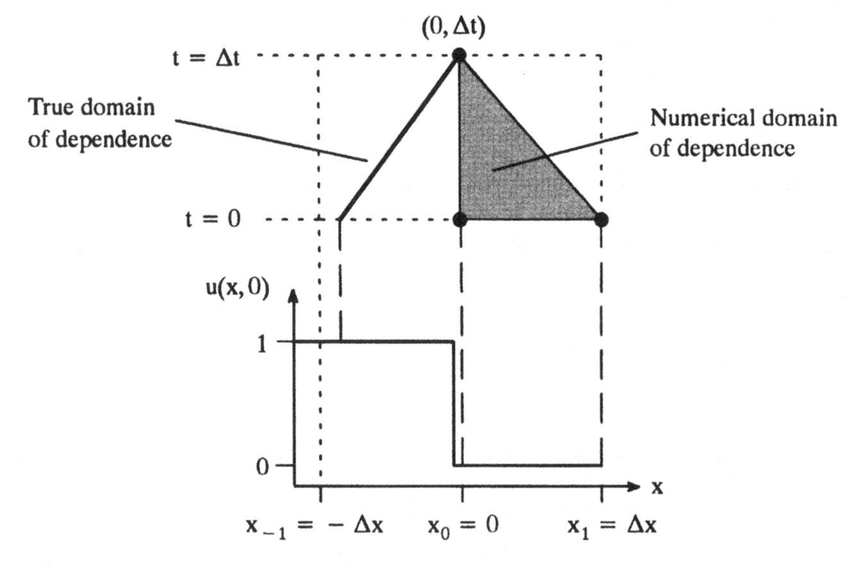

Figure 12.2 FTFS violates the CFL condition for positive wave speeds.

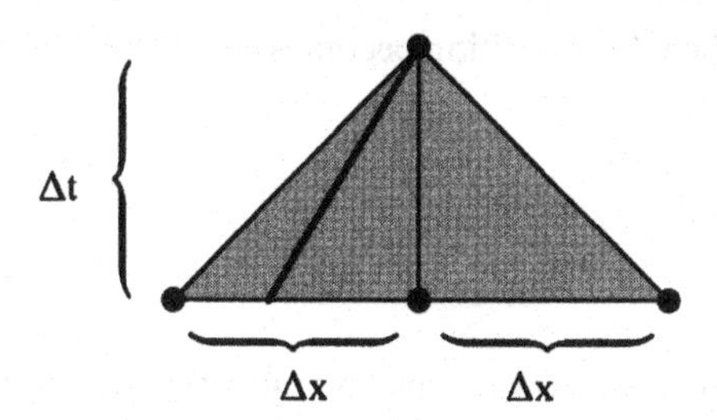

Figure 12.3 FTCS satisfies the CFL condition.

of initial conditions. Any scheme violating the CFL condition is *unstable* in the sense that it permits large errors, although CFL-violating schemes do not always blow up like FTFS.

Example 12.3 The CFL condition is necessary *but not sufficient* for stability. For example, consider FTCS. In the t–x plane, the true domain of dependence is a line of slope a whereas the numerical domain of dependence is the triangular region bounded by lines of slope $\pm\Delta x/\Delta t = \pm 1/\lambda$. Cocking your head, in the x–t plane, the true domain of dependence is a line of slope $1/a$ while the numerical domain of dependence is the triangular region bounded by lines of slope $\pm\lambda$, as seen in Figure 12.3. Then FTCS satisfies the CFL condition for $-1 \leq \lambda a \leq 1$. Despite the fact that FTCS satisfies the CFL condition, FTCS almost always blows up, as seen in Example 11.3.

Example 12.4 Consider the implicit Euler (BTCS) method:

$$u_i^{n+1} = u_i^n - \frac{\lambda}{2}\left(f\left(u_{i+1}^{n+1}\right) - f\left(u_{i-1}^{n+1}\right)\right).$$

The direct numerical domain of dependence of u_i^{n+1} is u_i^n, u_{i+1}^{n+1}, and u_{i-1}^{n+1}. But u_{i+1}^{n+1} depends on u_{i+1}^n, u_{i+2}^{n+1}, and u_i^{n+1}. In turn, u_{i+2}^{n+1} depends on u_{i+2}^n, u_{i+3}^{n+1}, u_{i+1}^{n+1}, and so on. This iterative reasoning shows that u_i^{n+1} depends on all of the grid points at time levels n and $n+1$, which then depend on all the grid points at all earlier time levels. In other words, the full numerical domain of dependence of u_i^{n+1} equals every grid point at every earlier time. Since the numerical domain of dependence for BTCS includes everything earlier, it certainly includes the physical domain of dependence. Thus BTCS always satisfies the CFL condition regardless of Δx, Δt, or λ.

For scalar conservation laws, the CFL condition translates into a simple inequality restricting the wave speed. For an explicit forward-time method, suppose that u_i^{n+1} depends on $(u_{i-K_1}^n, \ldots, u_{i+K_2}^n)$. In the x–t plane, the true domain of dependence is a line of slope $1/a(u_i^n)$, while the numerical domain of dependence is a triangle bounded on the left by a line of slope $\Delta t/(K_1\Delta x) = \lambda/K_1$ and on the right by a line of slope $-\Delta t/(K_2\Delta x) = -\lambda/K_2$. Alternatively, in the t–x plane, the true domain of dependence is a line of slope $a(u_i^n)$, while the numerical domain of dependence is a triangle bounded below by a line of slope $-K_2/\lambda$

and above by a line of slope K_1/λ. Then the CFL condition becomes

$$-\frac{K_2}{\lambda} \leq a \leq \frac{K_1}{\lambda}$$

or

$$\blacklozenge \qquad -K_2 \leq \lambda a \leq K_1. \tag{12.1}$$

Physically speaking, Equation (12.1) requires that waves travel no more than K_1 points to the right or K_2 points to the left during a single time step. If $K_1 = K_2 = K$, this becomes

$$\blacklozenge \qquad \lambda|a| \leq K. \tag{12.2}$$

Because of its intimate association with the CFL condition, λa is called the *CFL number* or the *Courant number.*

In Equations (12.1) and (12.2), a usually depends on u. Then inequalities (12.1) and (12.2) depend on the solution's range. For example, consider Burgers' equation with $a(u) = u$. Suppose that $|u_i^n| \leq M$ for all i and n. Then Equation (12.2) becomes $\lambda M \leq K$. Notice that the upper bound K/M on λ decreases as the upper bound M on $|u_i^n|$ increases. Suppose that a numerical approximation of Burgers' equation overshoots, creating overly large values of $|u_i^n|$. Such an approximation suffers a "double whammy" of overshoots plus reduced CFL numbers and time steps. By contrast, the linear advection equation has a constant wave speed a. Then the CFL limit depends only on the stencil via K and not on the numerical solution's range.

The true domain of dependence lies to the left if $a > 0$ and to the right if $a < 0$. Then the CFL condition implies $K_1 \geq 1$ if $a > 0$ and $K_2 \geq 1$ if $a < 0$. In other words, the CFL condition requires the minimum stencil $K_1 = 1$, $K_2 = 0$ for $a > 0$ or $K_1 = 0$, $K_2 = 1$ for $a < 0$. Figure 12.4 illustrates this minimum stencil, sometimes called an *upwind* stencil. The next chapter discusses minimum-width and other upwind stencils in detail.

Now consider the CFL condition for implicit methods. In particular, suppose that u_i^{n+1} depends on $u_{i-K_1}^n, \ldots, u_{i+K_2}^n$ and $u_{i-L_1}^{n+1}, \ldots, u_{i+L_2}^{n+1}$. If $L_1 > 0$ and $L_2 = 0$ then the full numerical domain of dependence of u_i^{n+1} includes everything to the left of $x = x_i$ and beneath $t = t^{n+1}$ in the x–t plane. Similarly, if $L_1 = 0$ and $L_2 > 0$ then the full numerical domain of dependence of u_i^{n+1} includes everything to the right of $x = x_i$ and beneath $t = t^{n+1}$ in the x–t plane. Finally, if $L_1 > 0$ and $L_2 > 0$ then the full numerical domain of dependence of u_i^{n+1} includes everything in the entire x–t plane beneath $t = t^{n+1}$. In short, *implicit methods avoid CFL restrictions by using the entire computational domain,* provided that the stencil includes at least one point to the left and one to the right; this includes BTCS but not BTFS or BTBS. Of course, explicit methods can also reduce or avoid CFL restrictions by using large or infinite values of K_1 and K_2, so that u_i^{n+1} depends on most or all of the points at time level n.

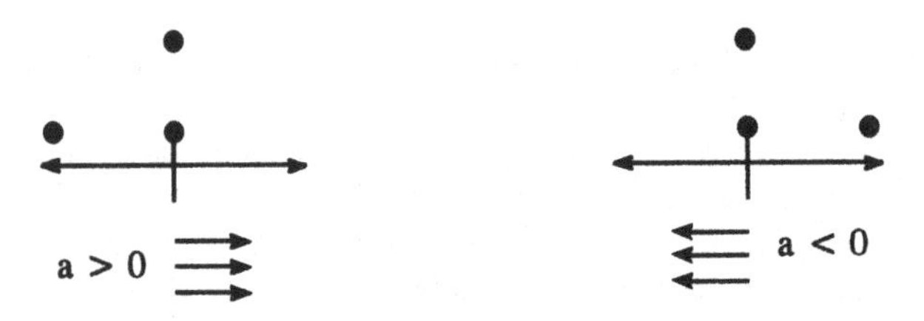

Figure 12.4 The smallest stencil allowed by the CFL condition.

This is a convenient time to compare implicit and explicit methods. As discussed in the last chapter, each time step in an implicit method costs much more than in an explicit method. On the other hand, lacking a CFL limit on the time step, implicit methods can take much larger time steps, bounded only by the desired time accuracy, and thus they can reach any given time in fewer time steps.

Consider a grid containing many large cells and a few small cells. For example, grids often cluster small cells near solid boundaries, shocks, and other high-gradient regions, which constitute only a small fraction of the total flow domain. For unsteady flows, the large cells in explicit approximations have to take a time step dictated by the few small cells. For example, for a symmetric explicit method, the CFL condition requires $\Delta t \leq \min(K|a|\Delta x_i)$. By contrast, implicit methods can take time steps as large as they like, constrained only by the fact that time accuracy decreases as the time step increases.

Consider again the case of a grid with a few small cells, but now think about steady-state calculations. In explicit methods, which no longer require time accuracy or synchronization, different cells can take different time steps, allowing the larger cells to use times steps proportional to their size. This is called *local time stepping*. While steadiness reduces the time-step restrictions on explicit methods, it has an even greater effect on implicit methods. Steadiness removes the only limitation on the time step – time accuracy – allowing implicit methods to take extremely large time steps relative to most explicit methods. Implicit methods commonly use CFL numbers ranging from the tens to the thousands on steady flows. In summary, explicit methods tend to take smaller cheaper time steps; implicit methods tend to take larger costlier time steps.

Chapters 15 and 16 discuss stability. Depending on your point of view, the CFL condition counts as a stability condition. However, when viewed in this way, the CFL condition is necessary *but not sufficient* for stability; numerical methods that satisfy the CFL condition may still blow up or exhibit other less drastic instabilities, as seen in Example 12.3. In general, other stability conditions are more restrictive than the CFL condition. In fact, like the CFL condition, many other stability conditions amount to bounds on the CFL number. For example, the CFL condition for an explicit centered method with a nine-point stencil is $|\lambda a| \leq 4$. However, another stability condition might require $|\lambda a| \leq 1$. The reader should resist the temptation to refer to all stability bounds on the CFL number as CFL conditions.

There is a trade-off between stability and the formal order of accuracy. Increased stencil width can increase the formal order of accuracy or increase the stability limits on the CFL number, but not both. Put another way, for a fixed stencil size, the stability limits tend to decrease as formal order of accuracy increases, and vice versa. This explains why a method with a nine-point stencil might operate under a stability restriction such as $|\lambda a| \leq 1$: The wide stencil has increased the formal order of accuracy rather than increased the stability limits.

12.2 The Euler Equations

This section concerns vector conservation laws such as the Euler equations. The Euler equations have three families of waves that define the physical domain of dependence. For example, consider the method shown in Figure 12.5. This method violates the CFL condition, since the true domain of dependence is partly outside of the numerical domain of dependence.

Just like scalar conservation laws, the CFL condition for vector conservation laws implies a simple restriction on wave speeds. For an explicit method, suppose that $\mathbf{u}_i^{n+1}$ depends

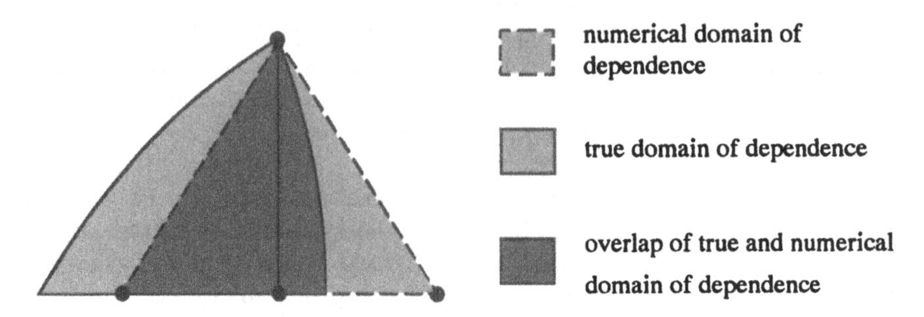

Figure 12.5 A method for the Euler equations that violates the CFL condition.

on $\mathbf{u}^n_{i-K_1}, \ldots, \mathbf{u}^n_{i+K_2}$. In the x–t plane, the true domain of dependence is the area bounded on the left by a characteristic curve of slope $1/(u+a)$ and on the right by a characteristic curve of slope $1/(u-a)$, while the numerical domain of dependence is a triangle bounded on the left by a line of slope $\Delta t/(K_1 \Delta x) = \lambda/K_1$ and on the right by a line of slope $-\Delta t/(K_2 \Delta x) = -\lambda/K_2$. Alternatively, in the t–x plane, the true domain of dependence is the area bounded below by a characteristic curve of slope $u-a$ and above by a characteristic curve of slope $u+a$, while the numerical domain of dependence is a triangle bounded below by a line of slope $-K_2/\lambda$ and above by a line of slope K_1/λ. Then the CFL condition becomes

$$-\frac{K_2}{\lambda} \leq u - a,$$

$$u + a \leq \frac{K_1}{\lambda}$$

or

$$\begin{aligned} -K_2 &\leq \lambda(u-a), \\ \lambda(u+a) &\leq K_1. \end{aligned} \tag{12.3}$$

Physically speaking, Equation (12.3) requires that waves travel no more than K_1 points to the right or K_2 points to the left during a single time step.

Suppose that $K_1 = K_2 = K$. Also suppose $\rho(A)$ is the larger of $u-a$ and $u+a$ in absolute value, where $u-a$ and $u+a$ are the least and greatest characteristic values of A, and A is the Jacobian matrix of the conservative flux vector. In general, for any matrix A, the quantity $\rho(A)$ is the largest characteristic value of A in absolute value; $\rho(A)$ is sometimes called the *spectral radius* of A. Then the CFL condition becomes

$$\lambda \rho(A) \leq K. \tag{12.4}$$

Because of its intimate association with the CFL condition, $\lambda\rho(A)$ is called the *CFL number* or the *Courant number*.

For supersonic flows, all waves travel in the same direction, either left or right. Then the minimum stencil allowed by the CFL condition contains either $\mathbf{u}^n_{i-1}$ and $\mathbf{u}^n_i$ for right-running supersonic flow or $\mathbf{u}^n_i$ and $\mathbf{u}^n_{i+1}$ for left-running supersonic flow. For subsonic flows, waves travel in both directions, and the minimum stencil always contains $\mathbf{u}^n_{i-1}$, $\mathbf{u}^n_i$, and $\mathbf{u}^n_{i+1}$. The other observations from the last section, including the contrast between explicit and implicit methods, carry over to vector conservation laws essentially unchanged.

Example 12.5 For subsonic flow, both FTFS and FTBS violate the CFL condition. Although FTFS satisfies the CFL condition for at least one family of waves, it also violates the CFL condition for at least one other family of waves. The same is true of FTBS. However, FTBS satisfies the CFL condition for supersonic right-running flow, whereas FTFS satisfies the CFL condition for supersonic left-running flow.

References

Courant, R., Friedrichs, K. O., and Lewy, H. 1928. "Uber die Partiellen Differenzengleichungen der Mathematisches Physik," *Math. Ann.*, 100: 32–74. English translation, 1967. "On the Partial Difference Equations of Mathematical Physics," *IBM Journal*, 11: 215–234.

Problems

12.1 Find the direct numerical domain of dependence, the full numerical domain of dependence, and the CFL condition for the following numerical approximations of scalar conservation laws:

(a) $$\frac{u_i^{n+1}-u_i^n}{\Delta t}+\frac{-f\left(u_{i+2}^{n+1}\right)+8f\left(u_{i+1}^{n+1}\right)-8f\left(u_{i-1}^{n+1}\right)+f\left(u_{i-2}^{n+1}\right)}{12\Delta x}=0$$

(b) $$u_i^{n+1}=u_i^n-\frac{\lambda}{2}\left(-f\left(u_{i+2}^n\right)+4f\left(u_{i+1}^n\right)-3f\left(u_i^n\right)\right)$$

(c) $$u_i^{n+1}=u_i^{n+1}-\frac{\lambda}{2}\left(-f\left(u_{i+2}^{n+1}\right)+4f\left(u_{i+1}^{n+1}\right)-3f\left(u_i^{n+1}\right)\right)$$

12.2 Find the CFL condition for the following numerical approximations of the Euler equations:

(a) $$\frac{\mathbf{u}_i^{n+1}-\mathbf{u}_i^n}{\Delta t}+\frac{-\mathbf{f}\left(\mathbf{u}_{i+2}^{n+1}\right)+8\mathbf{f}\left(\mathbf{u}_{i+1}^{n+1}\right)-8\mathbf{f}\left(\mathbf{u}_{i-1}^{n+1}\right)+\mathbf{f}\left(\mathbf{u}_{i-2}^{n+1}\right)}{12\Delta x}=0$$

(b) $$\mathbf{u}_i^{n+1}=\mathbf{u}_i^n-\frac{\lambda}{2}\left(-\mathbf{f}\left(\mathbf{u}_{i+2}^n\right)+4\mathbf{f}\left(\mathbf{u}_{i+1}^n\right)-3\mathbf{f}\left(\mathbf{u}_i^n\right)\right)$$

(c) $$\mathbf{u}_i^{n+1}=\mathbf{u}_i^n-\frac{\lambda}{2}\left(-\mathbf{f}\left(\mathbf{u}_{i+2}^{n+1}\right)+4\mathbf{f}\left(\mathbf{u}_{i+1}^{n+1}\right)-3\mathbf{f}\left(\mathbf{u}_i^{n+1}\right)\right)$$

12.3 Consider the following vector conservation law:

$$\frac{\partial \mathbf{u}}{\partial t}+\frac{\partial \mathbf{f}(\mathbf{u})}{\partial x}=0,$$

$$\mathbf{u}=\begin{bmatrix}u_1\\u_2\\u_3\end{bmatrix},\qquad \mathbf{f}(\mathbf{u})=\begin{bmatrix}u_2\\u_3\\4u_1-17u_2+8u_3\end{bmatrix}.$$

(a) Suppose that an explicit numerical approximation to this system of partial differential equations has a centered stencil containing $2K+1$ points. Find the CFL condition.

(b) Suppose that an uncentered explicit approximation has a stencil containing K_1 points to the left and K_2 points to the right. Find the CFL condition.

CHAPTER 13

Upwind and Adaptive Stencils

13.0 Introduction

This chapter describes the basic principles of stencil selection. To keep the discussion to a reasonable length, the focus will be on explicit finite-difference methods, except for the last section, which concerns explicit finite-volume methods. Some methods blindly choose the same stencil regardless of circumstances, including the nine simple methods seen in Chapter 11. However, this naive simplicity exacts a toll on accuracy, stability, or both. This chapter introduces methods with *adjustable* stencils, also called *adaptive* or *solution-sensitive* stencils. This chapter provides only an introduction; a full account will have to wait until later, especially Chapter 18 and Part V.

The governing principles of adjustable stencil selection are as follows:

- The numerical domain of dependence should model the physical domain of dependence. For example, the physical domain of dependence always lies entirely *upwind*; in other words, the physical domain of dependence lies to the right for left-running waves and to the left for right-running waves. Upwind stencils model this physical behavior by using only upwind points or, at least, more upwind than downwind points. However, in some cases, centered methods using equal numbers of upwind and downwind points actually model the physical domain of dependence better than upwind methods, as discussed below. Even downwind-biased stencils may provide good enough modeling, providing that they contain at least one upwind point; as seen in the last chapter, the CFL condition requires that every stencil contain at least one upwind point.
- The numerical domain of dependence should avoid shocks and contacts whenever possible. As seen in Part II, jump discontinuities can severely disrupt the component approximations, and the combinations used in computational gasdynamics sometimes further magnify the disruption.

Sections 13.1 and 13.2 describe general principles for upwind and adaptive stencil selection. The other four sections in this chapter introduce standard design techniques used to implement the principles seen in the first two sections. Section 13.3 introduces flux-averaging techniques including flux limiting, flux correction, and self-adjusting hybrids. Flux averaging addresses the second principle of stencil selection, repelling stencils from shocks and onto smooth solution regions. Sections 13.4 and 13.5 introduce flux splitting and wave speed splitting, respectively. Flux splitting and wave speed splitting address the first principle of stencil selection – they introduce upwinding, at least enough to satisfy the CFL condition. Many methods combine flux averaging and flux splitting, with flux averaging used to avoid shocks and to enforce nonlinear stability conditions, and flux splitting used for upwinding and/or to enforce the CFL condition. Finally, Section 13.6, introduces reconstruction–evolution, an alternative way to address the two principles of stencil selection. The reconstruction stage allows numerical methods to avoid shocks, while the evolution stage introduces upwinding, at least enough to satisfy the CFL condition. Finite-difference methods traditionally use flux averaging and flux splitting, whereas finite-volume methods traditionally use reconstruction–evolution. Although flux splitting, wave speed splitting, and reconstruction–evolution start out trying to model physical waves, they often end up as artificial numerical constructs. For example, reconstruction–evolution methods often use partially unphysical approximate Riemann solvers, such as those seen in Chapter 5, rather than the true Riemann solver, to improve efficiency and numerical accuracy.

13.1 Scalar Conservation Laws

There are only two directions in one-dimension – left and right. Rather than using left or right, we shall often specify directions relative to the wind. In particular, for right-running waves, right is the *downwind* direction and left is the *upwind* direction; similarly, for left-running waves, left is the *downwind* direction and right is the *upwind* direction. A numerical stencil contains more points to the right, more points to the left, or an equal number of points to the right and left. Equivalently, a stencil contains more points in the upwind direction, more points in the downwind direction, or an equal number of points in the upwind and downwind directions. Then every numerical approximation to a scalar conservation law belongs in one of three categories:

Centered: The stencil contains equal numbers of points in both directions.
Upwind: The stencil contains more points in the upwind direction.
Downwind: The stencil contains more points in the downwind direction.

Notice that, as defined here, upwind methods need not use upwind points exclusively; they can use downwind points provided that they do not outnumber the upwind points. Some people use the terms *upwind-biased* or *upstream* in place of upwind, the terms *downwind-biased* or *downstream* in place of downwind, and the term *symmetric* in place of centered.

Example 13.1 Categorize the methods seen in Chapter 11 as upwind, downwind, or centered. Comment on any correlations between upwinding and stability.

Solution The central-space methods FTCS, BTCS, and CTCS are centered. The backward-space methods FTBS, BTBS, and CTBS are upwind if $a > 0$ and downwind if $a < 0$. The forward-space methods FTFS, BTFS, and CTFS are upwind if $a < 0$ and downwind if $a > 0$.

Now consider the correlations between upwinding and stability. FTBS, FTFS, BTBS, and BTFS are unstable when they are downwind and stable when they are upwind, at least for small enough CFL numbers, as seen in Chapter 11. In contrast, CTBS and CTFS are highly unstable regardless of whether they are upwind or downwind and regardless of the CFL number. Turning to the centered methods, FTCS is unconditionally unstable, BTCS is unconditionally stable, and CTCS is stable for small enough CFL numbers. It seems that upwinding tends to enhance stability, at least for very simple methods with forward- or backward-time stepping.

Upwind and downwind stencils belong in the general category of *adjustable* or *adaptive* or *solution-sensitive* stencils. Upwind and downwind methods test for wind direction and then, based on the results of that test, select either a right- or left-biased stencil. Other adjustable stencils use other selection criteria in addition to wind direction, such as nonlinear stability conditions and first or second differences of the solution, which may indicate the presence of shocks or other large derivatives. Some methods always use the same fixed stencil regardless of shocks, wind direction, or other solution features, making them inherently simpler and less costly. However, intelligently designed adaptive stencils yield benefits far outweighing their increased complexity and cost.

As a class, upwind methods have an excellent reputation for shock capturing. This reputation stems mainly from the observations made in the last example. In other words, among very simple forward- or backward-time methods, upwind methods dramatically outdo centered methods. However, looking beyond such elementary examples, higher-order upwind methods often have no special advantages over higher-order centered methods. Specifically, upwind stencils may or may not model the true domain of dependence better than symmetric stencils. Furthermore, upwind stencils may or may not enhance accuracy or stability in the presence of shocks as compared with symmetric stencils. As one ingredient in their success, upwind methods absolutely must model wave physics, at least enough to determine wave directions; however, although not forced to, centered methods can also model waves in many of the same ways as traditional upwind methods using, for example, Riemann solvers, to obtain similar benefits. Therefore, when reading this chapter, make sure to distinguish between upwinding and other ways of modeling wave physics.

The rest of this section compares upwind, downwind, and centered methods for scalar conservation laws. As the first basis of comparison, consider the relationship between the physical and numerical domains of dependence. A numerical method is called *physical* if the numerical domain of dependence closely models the physical domain of dependence.

Example 13.2 Consider an explicit approximation to (x_i, t^{n+1}). Find the most physical two-point stencil for a smooth solution if $\lambda|a| \leq 1$.

Solution We wish to choose the two points at time level n closest to the domain of dependence of (x_i, t^{n+1}). For $0 \leq \lambda a \leq 1$, the answer is (x_i, t^n) and (x_{i-1}, t^n), as illustrated in Figure 13.1. Similarly, for $-1 \leq \lambda a \leq 0$, the answer is (x_i, t^n) and (x_{i+1}, t^n). In either case, the best stencil is the upwind stencil. Although still relatively close to the physical domain of dependence, the two-point downwind stencil seen in Figure 13.1 violates the CFL condition, making it unacceptably unphysical.

The two-point upwind stencil models the physical domain of dependence better than any other stencil – additional points beyond the two must lie relatively distant from the physical domain of dependence. On the other hand, as seen in Part II, two point approximations only have first-order accuracy at best; second-order accuracy requires at least three points as in the next example.

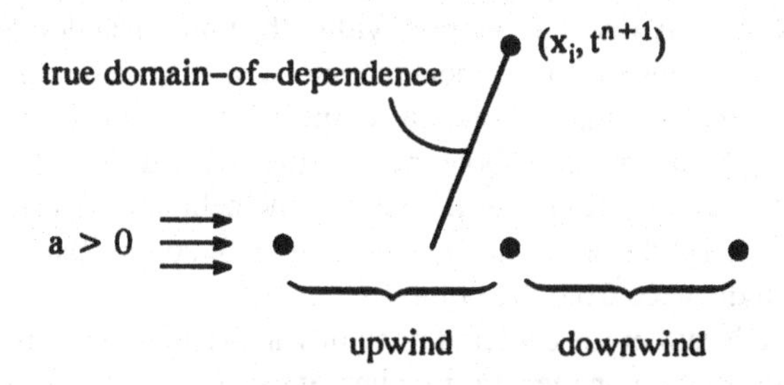

Figure 13.1 The true domain of dependence is always upwind.

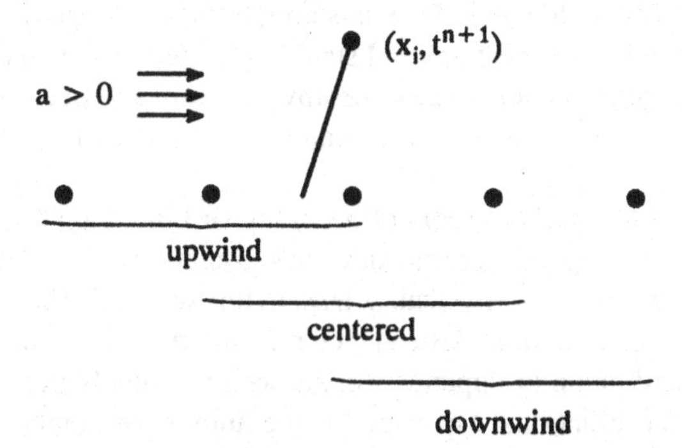

Figure 13.2 Upwind, downwind, and centered three-point stencils.

Example 13.3 Consider an explicit approximation to (x_i, t^{n+1}). Find the most physical three-point stencil for a smooth solution if $\lambda|a| \leq 3/2$.

Solution We must choose the three points at time level n closest to the domain of dependence of (x_i, t^{n+1}). For $-1/2 \leq \lambda a \leq 1/2$, the answer is the centered stencil (x_{i-1}, t^n), (x_i, t^n), (x_{i+1}, t^n), as illustrated in Figure 13.2. For $1/2 \leq \lambda a \leq 3/2$, the answer is the upwind stencil (x_{i-2}, t^n), (x_{i-1}, t^n), (x_i, t^n). Finally, for $-3/2 \leq \lambda a \leq -1/2$, the answer is the upwind stencil (x_i, t^n), (x_{i+1}, t^n), (x_{i+2}, t^n). Thus, for a three-point stencil, the upwind stencil always lies closest to the true domain of dependence for $1/2 \leq \lambda|a| \leq 3/2$, and the centered stencil always lies closest to the true domain of dependence for $\lambda|a| \leq 1/2$; the downwind stencil seen in Figure 13.2 always lies furthest from the true domain of dependence. More importantly, the centered stencil satisfies the CFL condition for $\lambda|a| \leq 1$, the upwind stencil satisfies the CFL condition for $\lambda|a| \leq 2$, whereas the downwind stencil never satisfies the CFL condition. This comparison favors centered methods for smaller CFL numbers, upwind methods for larger CFL numbers, and completely rejects the downwind stencil.

How do you justify using points relatively distant from the true domain of dependence, as in the last example? Beyond practical considerations such as the order of accuracy, consider this: whereas the true solution sees everything within the true domain of dependence, finite-difference approximations see only discrete samples within the numerical domain of dependence. Thus it only makes good sense to supplement this spotty view with points from outside the true domain of dependence. In smooth regions, points outside the true domain of dependence often contain valuable information about the solution inside the true domain of dependence, much in the same way that Taylor series can use the behavior of a smooth function at a single point to characterize the entire function, although, of course, the quality of the information decreases with distance from the point.

In general, for odd-width stencils, either the centered or an upwind stencil lies closest to the true domain of dependence, depending on the CFL number. Stencils with two, four, six, and other even widths do not allow centered stencils; thus an upwind stencil is closest to the true domain of dependence. Although downwind stencils are never as close to the true domain of dependence as the centered or the best upwind stencil, downwind stencils may still prove acceptable provided that they contain at least one upwind point and thus satisfy the CFL condition for small enough CFL numbers, which can occur for stencil widths of four or larger.

Wider stencils allow large time steps, higher orders of accuracy, or both. However, for wider stencils, the numerical domain of dependence may deviate substantially from the true domain of dependence, making the method less physical, at least in this sense. Furthermore, as a practical matter, a large numerical domain of dependence runs the risk of overlapping shocks that lie well outside the true domain of dependence. As seen repeatedly in Part II, shocks tend to cause large spurious oscillations, especially as the number of points in the approximation increases. Points on either side of a shock contain almost no information about each other, at least not without invoking the Rankine–Hugoniot relations, and any attempts to use points on the other side of a shock typically result in large oscillatory errors. Thus wide stencils potentially face a shock-capturing "double whammy": an increased risk of overlapping a shock and an increased error when they do.

Example 13.4 How does the answer to Example 13.3 change in the presence of shocks?

Solution To start with, suppose that the wind changes direction across a shock, which typically occurs at steady and slowly moving shocks. As seen in Figure 13.3, the upwind stencils always lie entirely to one side of the shock, whereas the centered stencils contain points on both sides of the shock. In this special case, the upwind stencil is clearly superior. But suppose the shock does *not* reverse the wind direction. As seen in Figure 13.4, an upwind stencil to the left of the shock avoids the shock, while a downwind stencil to the right of the shock avoids the shock. To summarize, unless shocks reverse the wind direction, upwinding does not ensure shock avoidance. In general, to avoid shocks, a method should have a leftward bias to the left of shocks and a rightward bias to the right of shocks, regardless of whether the upwind direction is left or right. Of course, the downwind stencil seen in Figure 13.4 violates the CFL condition, which generally creates larger errors than crossing the shock would. In general, the stencil must contain at least one upwind point, even when that upwind point causes the stencil to cross shocks.

This section indicates that centered methods sometimes outperform upwind methods both in terms of physicality and accuracy. The ENO methods discussed in Section 23.5 incorporate the principles described above, often selecting centered or downwind stencils

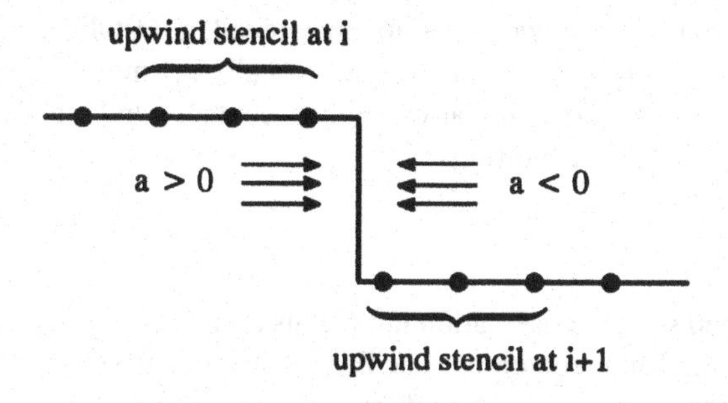

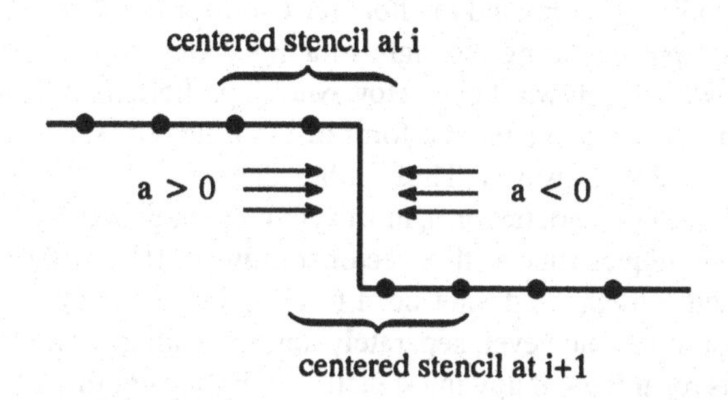

Figure 13.3 Upwind versus centered stencils at steady or slowly moving shocks containing compressive sonic points.

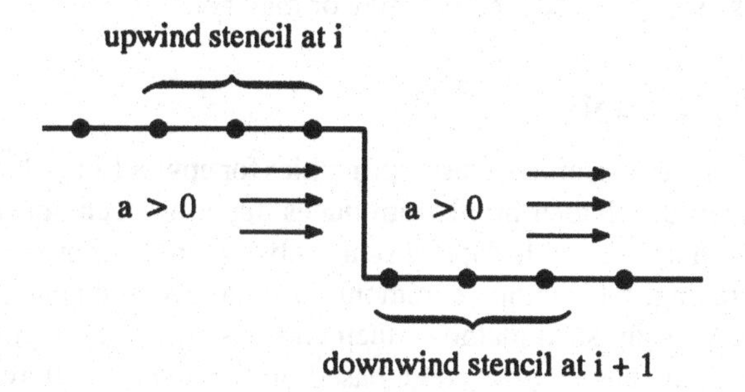

Figure 13.4 For steady shocks, upwind stencils avoid the shock on one side whereas downwind stencils avoid the shock on the other side.

over upwind stencils. For those still reluctant to give up their belief in the inherent superiority of upwind methods, consider the classic centered Lax–Wendroff method versus the Beam–Warming second-order upwind method, which basically differ only in their choice of stencil. As seen in Chapter 17, the centered Lax–Wendroff method outperforms the Beam–Warming method, at least for scalar conservation laws. The opposite holds true for the Euler equations, but remember that the Beam–Warming second-order upwind method for the Euler equations uses flux splitting or some other trick to separate the different families of waves, whereas the Lax–Wendroff method does not, and that gives the Beam–Warming method the edge, rather than its stencil's bias direction.

As a final point, an approximation cannot use points that do not exist (i.e., points lying outside of the computational domain). For example, a stencil must always have a leftward bias near a right-hand boundary, regardless of whether left is the upwind or downwind direction. This is discussed further in Chapter 19.

In conclusion, starting with the minimum two-point upwind stencil, adjustable stencils should select additional points symmetrically or nearly symmetrically in smooth regions, and otherwise select points to stay away from boundaries, shocks, contacts, and other steep gradients to the extent allowed by the CFL condition.

13.2 The Euler Equations

The last section concerned scalar conservation laws. This section briefly concerns vector conservation laws such as the Euler equations. The Euler equations have three families of waves. Subsonic flows have both right- and left-running waves of waves, which complicates the concept of upwinding. An upwind method must somehow decompose the solution into component waves, approximating the right-running waves with a leftward bias and the left-running waves with a rightward bias. However, the differential equations $dv_i = 0$ and $dx = \lambda_i dt$ appearing in the characteristic form of the Euler equations do not have analytic solutions, except for simple waves. Thus, rather than attempting to access characteristics directly through the characteristic form, most vector upwind methods access characteristics indirectly using techniques such as flux vector splitting or Riemann solvers. Unlike upwind methods, symmetric methods do not need to access characteristics using Riemann solvers or flux vector splitting; however, separately approximating characteristic families may yield improvements regardless of upwind stencil bias. Do not confuse upwinding with devices for separating different families of characteristics. As used in this book, upwinding refers only to the bias direction of the stencil. Upwind and downwind methods must separate characteristic families; centered methods may or may not.

13.3 Introduction to Flux Averaging

Up until now, this chapter has concerned general principles for upwind and adaptive stencil selection. The remainder of the chapter briefly introduces important techniques for designing numerical methods with upwind and adaptive stencils based on these principles. This section concerns flux averaging, a technique commonly used to avoid shocks. Flux averaging starts with two or more established methods, then chooses one of the methods, or averages the methods, varying the choice or average based on the divided differences of the solution. The following example shows why flux averaging works better for shock avoidance than for upwinding.

Example 13.5 Design a first-order upwind method for the linear advection equation that chooses between FTBS and FTFS.

Solution For $a > 0$, FTFS is highly unstable, so use FTBS. Conversely, for $a < 0$, FTBS is highly unstable, so use FTFS. A first-order upwind method for the linear advection equation is

$$u_i^{n+1} = u_i^n - \lambda a \begin{cases} u_i^n - u_{i-1}^n & \text{for } a > 0, \\ u_{i+1}^n - u_i^n & \text{for } a < 0. \end{cases}$$

This is sometimes called the *Courant–Isaacson–Rees* (*CIR*) method, after the authors of a 1952 paper. However, this book will usually refer to it using the longer but more descriptive phrase *the first-order upwind method for the linear advection equation.*

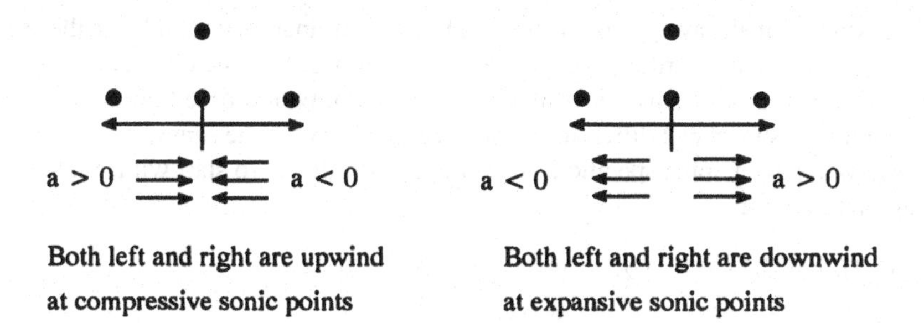

Figure 13.5 Problems with upwind methods at sonic points.

The above approach is problematic for nonlinear scalar conservation laws and vector conservation laws. For nonlinear scalar conservation laws, suppose that the wind direction changes from right to left; this occurs at *compressive sonic points*. Then both FTBS and FTFS are upwind. However, suppose that the wind direction changes from left to right, which occurs at *expansive sonic points*. Then both FTBS and FTFS are downwind. This is illustrated in Figure 13.5. As seen later in the book, expansive sonic points often cause numerical troubles, such as expansion shocks. Fortunately, most methods handle compressive sonic points with aplomb. For vector conservation laws, such as the Euler equations, the winds run in both directions in subsonic flow, even away from sonic points. Then both FTBS and FTFS violate the CFL condition, and either choice is unstable. In general, flux averaging fails whenever the component methods all share the same flaw. In this specific example, the flux averaging fails when both FTFS and FTBS are unstable, which occurs when waves travel in both directions, either because there are multiple families of waves traveling in both directions or because a single family of waves switches directions at a sonic point.

Rather than choosing one method or the other, as in the last example, why not compromise on something somewhere between the two? A flux average can be written in the following form:

$$\hat{f}^n_{i+1/2} = avg^n_{i+1/2}\left(\hat{f}^{(1)}_{i+1/2}, \hat{f}^{(2)}_{i+1/2}\right), \tag{13.1}$$

where $avg^n_{i+1/2}$ is any *averaging function*, $\hat{f}^{(1)}_{i+1/2}$ is the conservative numerical flux of a numerical method such as a first-order upwind method with a two-point stencil, and $\hat{f}^{(2)}_{i+1/2}$ is the conservative numerical flux of another method such as a second-order centered method. The subscripts and superscripts on the averaging function $avg^n_{i+1/2}$ indicate that it might depend on time and space, usually via first- or second-divided differences of the solution u^n_i. Flux-averaged methods typically do two things: (1) They choose an appropriate stencil by choosing between first- and second-order accurate methods and (2) they enforce nonlinear stability conditions such as those discussed in Chapter 16. For example, flux averaging leads directly to the popular *TVD methods*, which use the freedom of flux averaging to enforce the nonlinear stability conditions seen in Sections 16.2, 16.3, 16.4, or 16.5.

There are many different ways to write the same average. As a simple example, consider

$$\frac{1}{2}(x+y) = x + \frac{1}{2}(y-x) = y - \frac{1}{2}(y-x).$$

In the first expression, the average is written as a linear combination of x and y. In the second expression, the average is written as x plus a correction based on the difference $y - x$. In the third expression, the average is written as y plus a correction based on the difference $y - x$. Despite the superficial differences, the average is always the same.

Let us now consider some specific flux-averaging notations. To start with, consider the following flux average:

$$avg^n_{i+1/2}(x, y) = \theta^n_{i+1/2}x + \left(1 - \theta^n_{i+1/2}\right)y. \tag{13.2a}$$

In other words,

$$\hat{f}^n_{i+1/2} = \theta^n_{i+1/2}\hat{f}^{(1)}_{i+1/2} + \left(1 - \theta^n_{i+1/2}\right)\hat{f}^{(2)}_{i+1/2}. \tag{13.2b}$$

If $0 \le \theta^n_{i+1/2} \le 1$ then $\hat{f}_{i+1/2}$ is called a *linear average*, a *linear interpolation*, or a *convex linear combination* of $\hat{f}^{(1)}_{i+1/2}$ and $\hat{f}^{(2)}_{i+1/2}$. The convexity condition $0 \le \theta^n_{i+1/2} \le 1$ ensures the following reasonable condition:

$$\min\left(\hat{f}^{(1)}_{i+1/2}, \hat{f}^{(2)}_{i+1/2}\right) \le \hat{f}^n_{i+1/2} \le \max\left(\hat{f}^{(1)}_{i+1/2}, \hat{f}^{(2)}_{i+1/2}\right).$$

Thus $\hat{f}_{i+1/2}$ is always somewhere between $\hat{f}^{(1)}_{i+1/2}$ and $\hat{f}^{(2)}_{i+1/2}$. In this context, the averaging parameter $\theta^n_{i+1/2}$ is sometimes called a *shock switch* and the averaged method is sometimes called a *self-adjusting hybrid* method. This form of flux averaging was first suggested by Harten and Zwas (1972). For more details on self-adjusting hybrids and shock switches, see Chapter 22.

An equivalent way to write Equation (13.2a) is as follows:

$$avg^n_{i+1/2}(x, y) = x + \phi^n_{i+1/2}(y - x), \tag{13.3a}$$

where $\phi^n_{i+1/2} = 1 - \theta^n_{i+1/2}$. In other words;

$$\hat{f}^n_{i+1/2} = \hat{f}^{(1)}_{i+1/2} + \phi^n_{i+1/2}\left(\hat{f}^{(2)}_{i+1/2} - \hat{f}^{(1)}_{i+1/2}\right). \tag{13.3b}$$

In this context, the averaging parameter $\phi^n_{i+1/2}$ is sometimes called a *flux limiter* and the averaged method is sometimes called a *flux-limited method*. This form of flux averaging was first suggested by Van Leer (1974), although his notation differs somewhat. For more details on flux-limited methods, see Chapter 20.

In many cases, the averaging function is exchanged for a *differencing function*. Specifically,

$$avg^n_{i+1/2}(x, y) = x + diff^n_{i+1/2}(x, y). \tag{13.4}$$

For example, suppose that

$$diff^n_{i+1/2}(x, y) = \phi^n_{i+1/2}(y - x),$$

where $\phi^n_{i+1/2}$ is a flux limiter. Then let

$$\hat{f}^{(C)}_{i+1/2} = diff^n_{i+1/2}\left(\hat{f}^{(1)}_{i+1/2}, \hat{f}^{(2)}_{i+1/2}\right) = \phi^n_{i+1/2}\left(\hat{f}^{(2)}_{i+1/2} - \hat{f}^{(1)}_{i+1/2}\right).$$

Then we have

$$\hat{f}^n_{i+1/2} = \hat{f}^{(1)}_{i+1/2} + \hat{f}^{(C)}_{i+1/2} = \hat{f}^{(1)}_{i+1/2} + diff^n_{i+1/2}\left(\hat{f}^{(1)}_{i+1/2}, \hat{f}^{(2)}_{i+1/2}\right). \tag{13.5}$$

In this context, $\hat{f}^{(C)}_{i+1/2}$ is called a *flux-correction* and the averaged method is called a *flux-corrected method.* This form of flux averaging was first proposed by Boris and Book (1973). For more details on flux-corrected methods, see Chapter 21. Equation (13.5) is exactly the same as Equations (13.2) and (13.3), although Equations (13.2) and (13.3) are phrased in terms of averages, whereas Equation (13.5) is phrased in terms of differences. In fact, despite the differences in terminology and form, there is no fundamental distinction between any of the flux averages or flux differences seen above. Despite similar starting points, traditional distinctions have arisen between self-adjusting hybrid, flux-limited, and flux-corrected methods in terms of their component methods, their averaging or differencing functions, their extension from scalar conservation laws to the Euler equations, sonic point treatments, stability treatments, and so forth, as discussed in Chapters 20, 21, and 22.

13.4 Introduction to Flux Splitting

As seen in the last section, flux averaging can generate sophisticated new methods from established simple methods. But where do these "simple methods" come from? Specifically, most flux-averaging methods use upwind methods but, as seen in Example 13.5, flux averaging itself does not naturally generate upwind methods. This section and the next concern classic *splitting* techniques for constructing finite-difference upwind methods. Many methods combine flux averaging and flux splitting, using flux splitting to ensure that a stencil has the minimum amount of upwinding required by the CFL condition, and using flux averaging to further adapt the stencil to avoid shocks.

The flux and wave descriptions of gas flows pull numerical methods in two different directions. On the one hand, numerical methods must use fluxes to satisfy conservation as described in Chapters 2 and 11. On the other hand, numerical methods must use waves to satisfy the CFL condition as described in Chapters 3 and 12. Since numerical methods must use both descriptions, they need a simple way to transition from fluxes to waves. Imagine that right-running waves cause flux f^+ and left-running waves cause flux f^- such that

$$f(u) = f^+(u) + f^-(u) \tag{13.6}$$

and

$$\frac{df^+}{du} \geq 0, \qquad \frac{df^-}{du} \leq 0. \tag{13.7}$$

This is called *flux splitting*. For the Euler equations, the flux vector derivatives yield Jacobian matrices, and the inequalities refer to the signs of the characteristic values of the Jacobian matrices. Written in terms of flux splitting, the governing conservation law becomes

$$\frac{\partial u}{\partial t} + \frac{\partial f^+}{\partial x} + \frac{\partial f^-}{\partial x} = 0, \tag{13.8}$$

which is called the *flux split form*. Then $\partial f^+/\partial x$ can be discretized conservatively using at least one point on the left, and $\partial f^-/\partial x$ can be discretized conservatively using at least one point on the right, thus obtaining conservation and the CFL condition.

Unfortunately, flux splitting cannot describe the true relationship between fluxes and waves unless all waves run in the same direction. Specifically, when all the waves are right-running, the unique physical flux splitting is $f^+ = f$ and $f^- = 0$. Similarly, when all the waves are left-running, the unique physical flux splitting is $f^- = f$ and $f^+ = 0$. However, waves only run in the same direction away from sonic points in scalar conservation laws, and for supersonic flows in the Euler equations. Near sonic points, or for subsonic flows, flux splitting is necessarily somewhat artificial. Even when a physical flux splitting exists, there may be compelling numerical reasons to use something else. In summary, flux splitting may yield successful numerical methods even when based on fantasy relationships between fluxes and waves. This section concerns flux splitting for scalar conservation laws. Section 18.2 describes flux splitting for the Euler equations.

Example 13.6 Design a first-order upwind method for the linear advection equation using flux splitting.

Solution The unique physical flux splitting is $f(u) = au$, $f^+(u) = \max(0, a)u$, and $f^-(u) = \min(0, a)u$. Then a split flux form of the linear advection equation is as follows

$$\frac{\partial u}{\partial t} + \frac{\partial}{\partial x}(\max(0, a)u) + \frac{\partial}{\partial x}(\min(0, a)u) = 0.$$

A backward-space approximation gives:

$$\frac{\partial}{\partial x}(\max(0, a)u) \approx \max(0, a)\frac{u_i^n - u_{i-1}^n}{\Delta x},$$

and a forward-space approximation gives

$$\frac{\partial}{\partial x}(\min(0, a)u) \approx \min(0, a)\frac{u_{i+1}^n - u_i^n}{\Delta x}.$$

Combining these with a forward-time approximation yields

$$u_i^{n+1} = u_i^n - \lambda \max(0, a)\left(u_i^n - u_{i-1}^n\right) - \lambda \min(0, a)\left(u_{i+1}^n - u_i^n\right).$$

Equivalently, in conservation form

$$u_i^{n+1} = u_i^n - \lambda a \begin{cases} u_i^n - u_{i-1}^n & \text{for } a > 0 \\ u_{i+1}^n - u_i^n & \text{for } a < 0 \end{cases};$$

which is exactly the same as the method found in Example 13.5.

Example 13.7 Design a first-order upwind method for Burgers' equation using flux splitting.

Solution The unique physical flux splitting is $f(u) = u^2/2$, $f^+(u) = \max(0, u)$ $u/2$, and $f^-(u) = \min(0, u)u/2$, as illustrated in Figure 13.6. Then a split flux form of Burgers' equation is as follows:

$$\frac{\partial u}{\partial t} + \frac{1}{2}\frac{\partial}{\partial x}(\max(0, u)u) + \frac{1}{2}\frac{\partial}{\partial x}(\min(0, u)u) = 0.$$

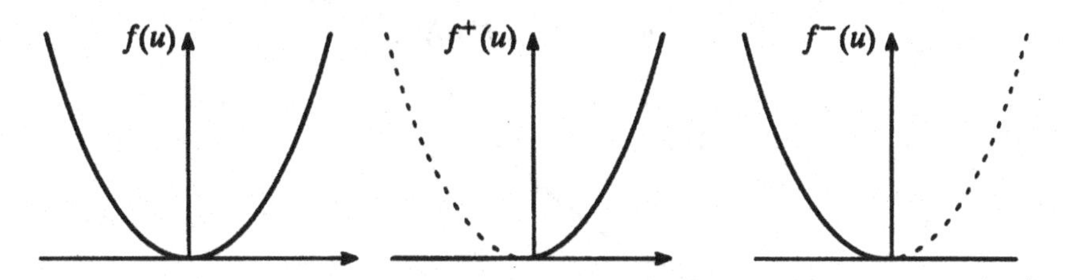

Figure 13.6 The "physical" flux splitting for Burgers' equation.

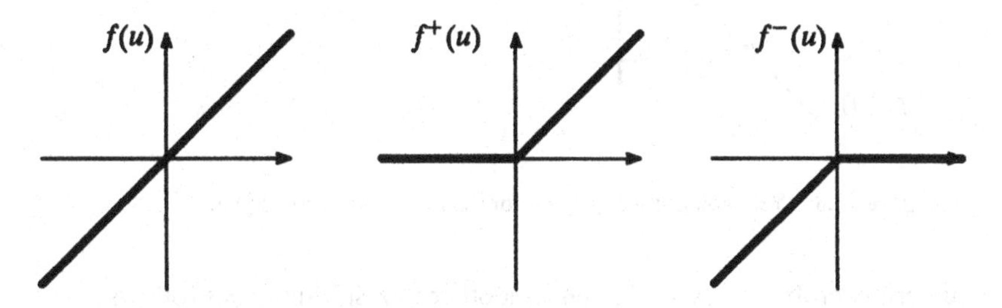

Figure 13.7 Another flux splitting for the linear advection equation.

A backward-space approximation gives

$$\frac{\partial}{\partial x}(\max(0, u)u) \approx \frac{\max(0, u_i^n)u_i^n - \max(0, u_{i-1}^n)u_{i-1}^n}{\Delta x}$$

and a forward-space approximation gives

$$\frac{\partial}{\partial x}(\min(0, u)u) \approx \frac{\min(0, u_{i+1}^n)u_{i+1}^n - \max(0, u_i^n)u_i^n}{\Delta x}.$$

Combining these with a forward-time approximation yields

$$\begin{aligned} u_i^{n+1} = u_i^n &- \frac{\lambda}{2}\left(\max(0, u_i^n)u_i^n - \max(0, u_{i-1}^n)u_{i-1}^n\right) \\ &- \frac{\lambda}{2}\left(\min(0, u_{i+1}^n)u_{i+1}^n - \min(0, u_i^n)u_i^n\right). \end{aligned}$$

Example 13.8 In the last two examples, the split fluxes $f^+(u)$ and $f^-(u)$ were completely smooth, continuous, and physical. Is this always true?

Solution No – split fluxes are not necessarily smooth, continuous, or physical. For example, consider the linear advection equation. For $a > 0$, let $f^+(u) = a\max(0, u)$ and $f^-(u) = a\min(0, u)$, as illustrated in Figure 13.7. Unlike the flux splittings in the last two examples, this flux splitting is *not* physical. For example, $f^- = f$ and $f^+ = 0$ for $u < 0$. In other words, this flux splitting falsely attributes all of the flux to left-running waves for $u < 0$, even though all of the flux is actually due to right-running waves since $a > 0$. As another issue, these split fluxes have discontinuous first derivatives at $u = 0$, which may

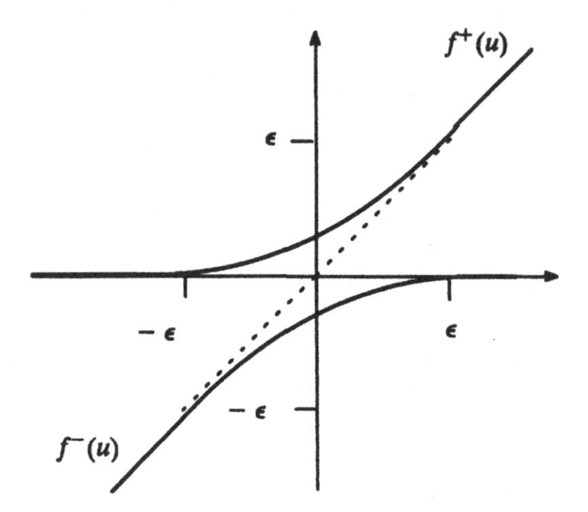

Figure 13.8 Yet another flux splitting for the linear advection equation.

cause numerical problems. A similar but smoother flux splitting is as follows:

$$f^+(u) = \begin{cases} 0 & u \leq -\epsilon, \\ \epsilon\left(\frac{u+\epsilon}{2\epsilon}\right)^2 & -\epsilon < u < \epsilon, \\ u & u \geq \epsilon, \end{cases}$$

$$f^-(u) = \begin{cases} u & u \leq -\epsilon, \\ -\epsilon\left(\frac{u-\epsilon}{2\epsilon}\right)^2 & -\epsilon < u < \epsilon, \\ 0 & u \geq \epsilon, \end{cases}$$

where ϵ is any small number. This flux splitting is illustrated in Figure 13.8. These split fluxes have continuous first derivatives but discontinuous second derivatives at $u = \pm\epsilon$.

In many cases, it may be difficult to find a continuous flux splitting, much less a flux splitting with continuous derivatives. For example, consider the flux function illustrated in Figure 13.9. The sonic points are where the wind changes direction. At sonic points, $f(u)$ has a maximum or minimum, that is, $a(u) = f'(u) = 0$. Sonic points are a natural place to switch between $f^+(u)$ and $f^-(u)$, and, in particular, sonic point splittings are the unique physical flux splittings. However, sonic point splittings have jump discontinuities at sonic points unless $f(u) = 0$ at sonic points. The jumps experienced by the unique physical flux splitting means that the physical flux splitting is not always the best choice numerically speaking. Even when the physical flux splitting is continuous at sonic points, the first derivatives may not be, which can cause numerical problems, again meaning that physical flux splitting is not always the best choice numerically speaking.

13.4.1 *Flux Split Form*

This subsection concerns a general flux splitting form for numerical approximations of scalar conservation laws. Assume that $\partial f^+/\partial x$ is discretized with a leftward bias,

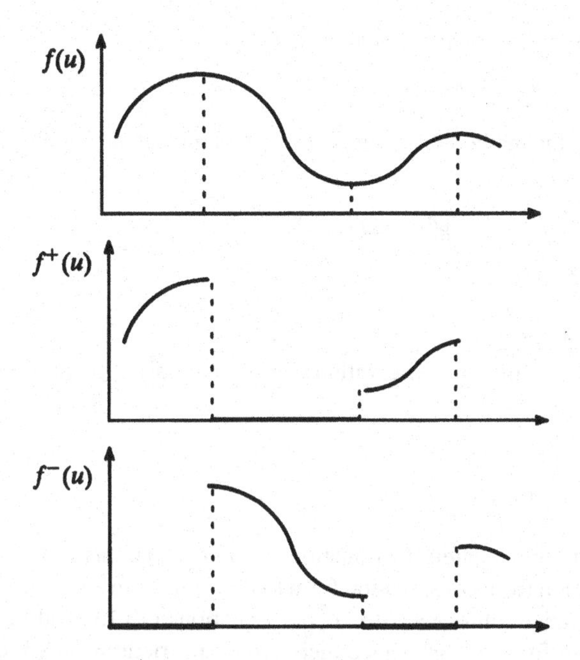

Figure 13.9 A nonsmooth flux splitting for a typical scalar conservation law.

so that the approximation at $x = x_i$ is centered on or biased towards $x = x_{i-1/2}$. More specifically, assume that

$$\frac{\partial f^+}{\partial x} \approx \frac{\Delta \hat{f}^+_{i-1/2}}{\Delta x} \tag{13.9}$$

for some $\Delta \hat{f}^+_{i-1/2}$. Similarly, assume that $\partial f^-/\partial x$ is discretized with a rightward bias, so that its approximation at $x = mx_i$ is centered on or biased towards $x = x_{i+1/2}$. More specifically, assume that

$$\frac{\partial f^-}{\partial x} \approx \frac{\Delta \hat{f}^-_{i+1/2}}{\Delta x} \tag{13.10}$$

for some $\Delta \hat{f}^-_{i+1/2}$. Finally, use the forward-Euler time discretization to obtain

$$u_i^{n+1} = u_i^n - \lambda\left(\Delta \hat{f}^+_{i-1/2} + \Delta \hat{f}^-_{i+1/2}\right). \tag{13.11}$$

This is called the *flux split form* of a numerical approximation.

Any explicit forward-time approximation can be written in flux split form, whether or not the approximation was derived from a flux split form of the governing equations. Thus, putting aside physical flux splitting considerations such as Equations (13.6) and (13.7), any explicit conservative numerical method can be written in form (13.11). Of course, without conditions (13.6) and (13.7), a flux split form (13.11) says literally nothing about the connection between fluxes and waves. However, it can still be useful for other things, such as nonlinear stability analysis.

A method in flux split form is conservative if and only if

$$\Delta\hat{f}^{+}_{i+1/2} + \Delta\hat{f}^{-}_{i+1/2} = g^n_{i+1} - g^n_i \tag{13.12}$$

for some g^n_i. The proof follows the method of Example 11.9. In particular, add and subtract λg^n_i in Equation (13.11) to get

$$u^{n+1}_i = u^n_i - \lambda\left(\Delta\hat{f}^{+}_{i-1/2} + g^n_i - g^n_i + \Delta\hat{f}^{-}_{i+1/2}\right).$$

Compare with the conservation form

$$u^{n+1}_i = u^n_i - \lambda\left(\hat{f}^{n}_{i+1/2} - \hat{f}^{n}_{i-1/2}\right).$$

Then the flux split form can be written in conservation form if and only if there exists g^n_i such that

$$\hat{f}^{n}_{i+1/2} = \Delta\hat{f}^{-}_{i+1/2} + g^n_i, \tag{13.13}$$

$$\hat{f}^{n}_{i-1/2} = -\Delta\hat{f}^{+}_{i-1/2} + g^n_i. \tag{13.14}$$

Then $\hat{f}^{n}_{i+1/2} = \hat{f}^{n}_{(i-1/2)+1}$ leads immediately to Equation (13.12). Equations (13.13) and (13.14) allow easy transformation from conservation form to flux split form and vice versa. In fact, for any given conservative numerical flux $\hat{f}^{n}_{+1/2}$, Equations (13.13) and (13.14) yield a different flux split form for every g^n_i. Since there are no restrictions on g^n_i, every conservative method has *infinitely* many flux split forms.

Example 13.9 Write the first-order upwind method for Burgers' equation found in Example 13.7 in flux split form and conservation form.

Solution Example 13.7 naturally yields the flux split form as follows:

$$\Delta\hat{f}^{+}_{i-1/2} = \frac{1}{2}\left(\max\left(0, u^n_i\right)u^n_i - \max\left(0, u^n_{i-1}\right)u^n_{i-1}\right),$$

$$\Delta\hat{f}^{-}_{i+1/2} = \frac{1}{2}\left(\min\left(0, u^n_{i+1}\right)u^n_{i+1} - \min\left(0, u^n_i\right)u^n_i\right).$$

To write the method in conservation form, use the following identities:

$$|x| = \max(0, x) - \min(0, x), \tag{13.15}$$

and

$$x = \max(0, x) + \min(0, x) \tag{13.16}$$

for any number x. Then, as the reader can show, Equation (13.12) has the following unique solution:

$$g^n_{i+1} = \frac{1}{2}\left(u^n_{i+1}\right)^2, \qquad g^n_i = \frac{1}{2}\left(u^n_i\right)^2.$$

After some algebra, Equation (13.13) yields

$$\hat{f}^{n}_{i+1/2} = \frac{1}{2}\left(\min\left(0, u^n_{i+1}\right)u^n_{i+1} + \max\left(0, u^n_i\right)u^n_i\right).$$

This expression can also be derived using the approach illustrated in Example 11.10.

13.4.2 *Introduction to Flux Reconstruction*

As mentioned earlier, many numerical methods combine flux splitting and flux averaging. In this regard, Shu and Osher (1988) suggested a particularly interesting general approach, based on the method of lines. From Subsection 11.2.1, recall that the spatial discretization in the method of lines constructs $\hat{f}^n_{s,i+1/2}$ such that

$$\frac{\partial f}{\partial x}(u(x_i, t^n)) \approx \frac{\hat{f}^n_{s,i+1/2} - \hat{f}^n_{s,i-1/2}}{\Delta x}.$$

Rather than interpolating or extrapolating between two stable and well-established conservative numerical fluxes to find $\hat{f}^n_{s,i+1/2}$, as in flux averaged methods, Shu and Osher construct the semidiscrete conservative numerical flux function starting from scratch. In particular, suppose that $\hat{f}^n_{s,i+1/2} = \hat{f}^n_s(x_{i+1/2})$, that is, suppose $\hat{f}^n_{s,i+1/2}$ depends on space directly, rather than only indirectly through u^n_i. Then the spatial discretization step constructs $\hat{f}^n_{s,i+1/2}$ such that

$$\frac{\partial f}{\partial x}(u(x_i, t^n)) \approx \frac{\hat{f}^n_s(x_{i+1/2}) - \hat{f}^n_s(x_{i-1/2})}{\Delta x}.$$

But, by the fundamental theorem of calculus, this is equivalent to

$$f(u(x_i, t^n)) \approx \frac{1}{\Delta x}\int_{x_{i-1/2}}^{x_{i+1/2}} \hat{f}^n_s(x)\,dx. \tag{13.17}$$

Then, in this approach, the spatial discretization step reconstructs $\hat{f}^n_s(x)$ from cell-integral averages $f(u(x_i, t^n)) \approx f(u^n_i)$. But this is just reconstruction via the primitive function, as described in Section 9.3. Reconstruction via the primitive function based on essentially nonoscillatory (ENO) reconstruction, as seen in Chapter 9, yields a rich variety of semidiscrete fluxes, some associated with standard methods and some completely new. From one point of view, such flux reconstructions naturally yield adaptive flux averages of fixed-stenciled methods, as seen in Section 21.4.

ENO flux reconstruction avoids shocks but does not introduce upwinding. In other words, ENO reconstruction considers only the size of divided differences during stencil selection, as discussed in Chapter 9, and thus may not choose any upwind points, in violation of the CFL condition. So we must use flux splitting to introduce the minimal amount of upwinding required by the CFL condition. In particular, reconstruct $\hat{f}^+_s(x)$ from $f^+(u^n_i)$ and $f^+(u^n_{i-1})$ plus other points as determined by ENO reconstruction techniques, and reconstruct $\hat{f}^-_s(x)$ from $f^-(u^n_i)$ and $f^-(u^n_{i+1})$ plus other points as determined by ENO reconstruction techniques. Then

$$\hat{f}^n_{s,i+1/2} = \hat{f}^+_s(x_{i+1/2}) + \hat{f}^-_s(x_{i+1/2}). \tag{13.18}$$

13.5 Introduction to Wave Speed Splitting

The last section concerned flux splitting. This section concerns a closely related type of splitting called *wave speed splitting*. Flux splitting uses the governing equations in conservation form and tends to yield conservative numerical approximations. In contrast wave speed splitting uses the governing equations in nonconservation form and tends to yield nonconservative approximations. Thus, in most cases, flux splitting is preferred over wave speed splitting except when the flux function has the special property $\mathbf{f} = (d\mathbf{f}/d\mathbf{u})\mathbf{u}$.

As discussed below, this special property makes flux splitting and wave speed splitting essentially equivalent. Among other applications, wave speed splitting leads to an interesting wave speed split form often used in nonlinear stability analysis.

For scalar conservation laws, suppose that the wave speed is split as follows:

$$a(u) = a^+(u) + a^-(u), \tag{13.19}$$

where

$$a^+ \geq 0, \qquad a^- \leq 0. \tag{13.20}$$

Then the scalar conservation law can be written as

$$\frac{\partial u}{\partial t} + a^+\frac{\partial u}{\partial x} + a^-\frac{\partial u}{\partial x} = 0, \tag{13.21}$$

which is called a *wave speed split form* of the scalar conservation law. Then $a^+\partial u/\partial x$ can be discretized conservatively using at least one point on the left, and $a^-\partial u/\partial x$ can be discretized conservatively using at least one point on the right, thus obtaining conservation and the CFL condition. Notice that $a^+\partial u/\partial x$ and $a^-\partial u/\partial x$ are not in conservation form, which makes it more difficult, but certainly not impossible, to obtain a conservative discretization.

Now consider vector conservation laws such as the Euler equations. Suppose that the flux Jacobian matrix $A = d\mathbf{f}/d\mathbf{u}$ is split as follows:

$$A(\mathbf{u}) = A^+(\mathbf{u}) + A^-(\mathbf{u}), \tag{13.22}$$

where the characteristic values of A^+ are nonnegative and the characteristic values of A^- are nonpositive. In standard matrix notation,

$$A^+ \geq 0, \qquad A^- \leq 0. \tag{13.23}$$

The matrices A^+ and A^- are usually obtained by splitting the characteristic values of A, which represent wave speeds, into positive and negative parts. Then the vector conservation law can be written as

$$\frac{\partial \mathbf{u}}{\partial t} + A^+\frac{\partial \mathbf{u}}{\partial x} + A^-\frac{\partial \mathbf{u}}{\partial x} = 0. \tag{13.24}$$

This is called a *wave speed split form* of the vector conservation law. Then $A^+\partial\mathbf{u}/\partial x$ can be discretized conservatively using at least one point on the left, and $A^-\partial\mathbf{u}/\partial x$ can be discretized conservatively using at least one point on the right, thus obtaining conservation and the CFL condition.

The flux vector in the Euler equations has a special relationship with its Jacobian matrix. In particular,

$$\mathbf{f} = A\mathbf{u}. \tag{13.25}$$

Then the flux function $\mathbf{f}$ in the Euler equations is called a *homogenous function of order one*. As some alert readers may recall, Problem 2.2 introduced homogeneous flux functions of order one. By Equation (13.25), every wave speed splitting $A^\pm$ of the Euler equations leads immediately to the flux vector splitting $\mathbf{f}^\pm = A^\pm\mathbf{u}$, although this flux vector splitting may or may not satisfy condition (13.6).

Example 13.10 Design a first-order upwind method for the linear advection equation using wave speed splitting.

Solution The unique physical wave speed splitting is $a^+ = \max(0, a)$ and $a^- = \min(0, a)$. For right-running waves with $a > 0$, $a^+ = a$ and $a^- = 0$. In other words, this physical wave speed splitting properly attributes everything to right-running waves. Similarly, for left-running waves with $a < 0$, $a^- = a$ and $a^+ = 0$. In other words, this physical wave speed splitting properly attributes everything to left-running waves. Written in terms of this wave speed splitting, the linear advection equation becomes

$$\frac{\partial u}{\partial t} + \max(0, a)\frac{\partial u}{\partial x} + \min(0, a)\frac{\partial u}{\partial x} = 0.$$

Then a backward-space approximation gives

$$\max(0, a)\frac{\partial u}{\partial x} \approx \max(0, a)\frac{u_i^n - u_{i-1}^n}{\Delta x},$$

and a forward-space approximation gives

$$\min(0, a)\frac{\partial u}{\partial x} \approx \min(0, a)\frac{u_{i+1}^n - u_i^n}{\Delta x}.$$

Combining these with a forward-time approximation yields

$$u_i^{n+1} = u_i^n - \lambda \max(0, a)\left(u_i^n - u_{i-1}^n\right) - \lambda \min(0, a)\left(u_{i+1}^n - u_i^n\right),$$

which is the same as the method found in Examples 13.5 and 13.6.

As it turns out, there is no real difference between this example and Example 13.6. The linear advection equation has the special property $f(x) = (df/du)u = au$. Thus, just like the flux function for the Euler equations, the flux function for the linear advection equation is a homogenous function of order one. Although there are many homogenous *vector* functions of order one, the only homogenous *scalar* functions of order one are linear functions. Thus, among scalar conservation laws, only the linear advection equation has a homogenous flux function of order one – Burgers' equation and all other scalar conservation laws do not have this special property. Since $f(u) = au$, any wave speed splitting $a^\pm$ implies a flux splitting $f^\pm = a^\pm u$. In particular, the wave speed splitting $a^+ = \max(0, a)$ and $a^- = \min(0, a)$ used in this example immediately implies the flux splitting $f^+(u) = \max(0, a)u$ and $f^-(u) = \min(0, a)u$ used in Example 13.6. Vice versa, any flux splitting $f^\pm$ implies a wave speed splitting $a^\pm = f^\pm/u$.

13.5.1 *Wave Speed Split Form*

This section introduces a general wave speed split form for numerical methods often used in nonlinear stability analysis, as seen in Section 16.4. Assume that $a^+\partial u/\partial x$ is discretized with a leftward bias, so that its approximation at $x = x_i$ is centered on or biased towards $x = x_{i-1/2}$. More specifically, assume

$$a^+\frac{\partial u}{\partial x} \approx a_{i-1/2}^+\frac{u_i^n - u_{i-1}^n}{\Delta x}. \tag{13.26}$$

Of course, the first-order upwind approximation $(u_i^n - u_{i-1}^n)/\Delta x$ to $\partial u/\partial x$ can be replaced by any number of other approximations. Similarly, assume that $a^- \partial u/\partial x$ is discretized with a rightward bias, so that its approximation at $x = x_i$ is centered on or biased towards $x = x_{i+1/2}$. More specifically, assume

$$a^- \frac{\partial u}{\partial x} \approx a_{i+1/2}^- \frac{u_{i+1}^n - u_i^n}{\Delta x}. \tag{13.27}$$

Again, the first-order upwind approximation $(u_{i+1}^n - u_i^n)/\Delta x$ to $\partial u/\partial x$ can be replaced by any number of other approximations. Finally, use the forward-time discretization to obtain

$$u_i^{n+1} = u_i^n - \lambda a_{i+1/2}^- \left(u_{i+1}^n - u_i^n\right) - \lambda a_{i-1/2}^+ \left(u_i^n - u_{i-1}^n\right), \tag{13.28}$$

which is called a *wave speed split form* of the numerical approximation. Any explicit conservative numerical method can be written in wave speed split form, whether or not it was derived using wave speed splitting. In other words, putting aside physical splitting considerations such as Equations (13.19) and (13.20), any explicit conservative numerical method can be written in form (13.28). Of course, without conditions (13.19) and (13.20), a wave speed split form (13.28) of the numerical approximation says literally nothing about the connections between fluxes and waves.

There is a simple relationship between the flux split form seen in Subsection 13.4.1 and the wave speed split form seen in this subsection. In particular,

$$\Delta \hat{f}_{i+1/2}^{\pm} = a_{i+1/2}^{\pm} \left(u_{i+1}^n - u_i^n\right). \tag{13.29}$$

In other words, wave speed split form simply factors the split fluxes $\Delta f_{i+1/2}^{\pm}$ into coefficients $a_{i+1/2}^{\pm}$ times first differences $u_{i+1}^n - u_i^n$. This factorization is natural and convenient when the split fluxes happen to include factors $u_{i+1}^n - u_i^n$. Otherwise, wave speed split form still works, but with a caveat. Suppose that $\Delta f_{i+1/2}^{\pm}$ does not include a factor $u_{i+1}^n - u_i^n$ and, furthermore, suppose that $\Delta f_{i+1/2}^{\pm} \neq 0$ when $u_{i+1}^n - u_i^n = 0$. Then the coefficients $a_{i+1/2}^{\pm}$ must be *infinite* since only infinity times zero yields anything other than zero. In fact, for most methods with more than two or three points in their stencils, all possible wave speed splitting forms involve infinite coefficients. Even with the niggling issue of infinite coefficients, wave speed split form is still more useful than flux split form. This is not to say that wave speed splitting is better than flux splitting (in fact, quite the opposite) but only to say that wave speed split *form* is better than flux split *form* in certain applications, especially nonlinear stability analysis.

By Equations (13.12) and (13.29), a method in wave speed split form is conservative if and only if

$$\left(a_{i+1/2}^+ + a_{i+1/2}^-\right)\left(u_{i+1}^n - u_i^n\right) = g_{i+1}^n - g_i^n \tag{13.30}$$

for some flux function g_i^n. Similarly, Equations (13.13), (13.14), and (13.29) imply

$$\hat{f}_{i+1/2}^n = a_{i+1/2}^- \left(u_{i+1}^n - u_i^n\right) + g_i^n, \tag{13.31}$$

$$\hat{f}_{i-1/2}^n = -a_{i-1/2}^+ \left(u_i^n - u_{i+1}^n\right) + g_i^n. \tag{13.32}$$

Equations (13.31) and (13.32) allow easy transformation from conservation form to wave speed split form and vice versa. In fact, for any given conservative numerical flux $\hat{f}_{i+1/2}^n$, Equations (13.31) and (13.32) yield a different wave speed split form for every different g_i^n. Since there are no restrictions on g_i^n, every conservative method has infinitely many wave speed split forms. By the way, if the coefficients $a^{\pm}_{i+1/2}$ are finite, and if $\hat{f}^n_{i+1/2}$ is consistent with $f(u)$, then Equations (13.31) and (13.32) imply that g^n_i is consistent with $f(u)$.

Example 13.11 Assuming that g^n_i is consistent with $f(u)$, the most obvious choice for the splitting function g^n_i is

$$g^n_i = f(u^n_i). \tag{13.33}$$

Then the conservation condition (13.30) for scalar conservation laws implies

$$a^+_{i+1/2} + a^-_{i+1/2} = a^n_{i+1/2}, \tag{13.34}$$

where, as usual, $a^n_{i+1/2}$ is defined as follows:

$$a^n_{i+1/2} = \begin{cases} \dfrac{f(u^n_{i+1}) - f(u^n_i)}{u^n_{i+1} - u^n_i} & u^n_{i+1} \neq u^n_i, \\ f'(u^n_i) & u^n_{i+1} = u^n_i. \end{cases} \tag{13.35}$$

In this example, the numerical wave speed splitting $a^n_{i+1/2} = a^+_{i+1/2} + a^-_{i+1/2}$ directly reflects the physical wave speed splitting $a(u) = a^+(u) + a^-(u)$.

The above notation for the wave speed split form is the standard notation when wave speed splitting is used to derive new methods. Unfortunately, the above notation is not the standard notation when existing methods are rewritten in wave speed split form as a preliminary step in nonlinear stability analysis. Instead, the nonlinear stability literature uses the following notation for wave speed split form:

$$\blacklozenge \qquad u^{n+1}_i = u^n_i + C^+_{i+1/2}(u^n_{i+1} - u^n_i) - C^-_{i-1/2}(u^n_i - u^n_{i-1}), \tag{13.36}$$

which follows Harten (1983). Comparing Equations (13.28) and (13.36), we see that Harten's wave speed split form notation is related to the earlier wave speed split form notation through

$$C^+_{i+1/2} = -\lambda a^-_{i+1/2}, \qquad C^-_{i+1/2} = \lambda a^+_{i+1/2}. \tag{13.37}$$

If a method is derived using wave speed splitting, and not just written in wave speed split form, then Equations (13.19) and (13.20) become

$$\lambda a(u) = C^+(u) - C^-(u), \tag{13.38}$$

$$C^+(u) \geq 0, \qquad C^-(u) \geq 0. \tag{13.39}$$

Also, conservation condition (13.30) becomes

$$(C^-_{i+1/2} - C^+_{i+1/2})(u^n_{i+1} - u^n_i) = \lambda(g^n_{i+1} - g^n_i). \tag{13.40}$$

Similarly, Equations (13.31) and (13.32) become

$$\lambda \hat{f}^n_{i+1/2} = -C^+_{i+1/2}(u^n_{i+1} - u^n_i) + \lambda g^n_i, \tag{13.41}$$

$$\lambda \hat{f}^n_{i-1/2} = -C^-_{i-1/2}(u^n_i - u^n_{i-1}) + \lambda g^n_i. \tag{13.42}$$

Unfortunately, the coefficients $C^{\pm}_{i+1/2}$ are often infinite when $u^n_{i+1} = u^n_i$. If the coefficients $C^{\pm}_{i+1/2}$ are always finite, then $u^{n+1}_i = u^n_i$ whenever $u^n_{i-1} = u^n_i = u^n_{i+1}$. Many methods with two- or three-point stencils satisfy this condition, but many methods with wider stencils do not. Rephrasing this observation in the reverse sense, any method that allows $u^{n+1}_i \neq u^n_i$ when $u^n_{i-1} = u^n_i = u^n_{i+1}$ cannot have uniformly finite coefficients $C^{\pm}_{i+1/2}$. Remember that there are infinitely many wave speed split forms, depending on the choice of g^n_i. For a given numerical method, of the infinitely many possible wave speed split forms, *at most* one has finite coefficients $C^{\pm}_{i+1/2}$. If the coefficients $C^{\pm}_{i+1/2}$ are always finite, and if $\hat{f}^n_{i+1/2}$ is consistent with $f(u)$, then Equations (13.41) and (13.42) imply that g^n_i is consistent with $f(u)$.

Example 13.12 Find two wave speed splitting forms for FTCS for scalar conservation laws. If possible, one of the forms should have finite $C^{\pm}_{i+1/2}$.

Solution Recall that the conservative numerical flux of FTCS is

$$\hat{f}^n_{i+1/2} = \frac{f(u^n_{i+1}) + f(u^n_i)}{2}.$$

Suppose that $g^n_i = f(u^n_i)$. Then Equations (13.41) and (13.42) yield

$$\lambda\frac{f(u^n_{i+1}) + f(u^n_i)}{2} = -C^+_{i+1/2}(u^n_{i+1} - u^n_i) + \lambda f(u^n_i),$$
$$\lambda\frac{f(u^n_i) + f(u^n_{i-1})}{2} = -C^-_{i-1/2}(u^n_i - u^n_{i-1}) + \lambda f(u^n_i).$$

Solving for $C^+_{i+1/2}$ and $C^-_{i-1/2}$ yields

$$C^+_{i+1/2} = -\frac{\lambda}{2}a^n_{i+1/2},$$
$$C^-_{i-1/2} = \frac{\lambda}{2}a^n_{i-1/2},$$

where $a^n_{i+1/2}$ is defined by Equation (13.35). Notice that $C^+_{i+1/2}$ and $C^-_{i-1/2}$ are always finite, assuming only that the physical wave speed $a(u)$ is always finite. Similarly, to chose another arbitrary example, if $g^n_i = f(u^n_i) + (u^n_{i+1} - u^n_i)/\lambda$ then

$$C^+_{i+1/2} = -\frac{\lambda}{2}a^n_{i+1/2} + 1,$$
$$C^-_{i-1/2} = \frac{\lambda}{2}a^n_{i-1/2} + \frac{u^n_{i+1} - u^n_i}{u^n_{i-1} - u^n_i}.$$

Notice that $C^+_{i+1/2}$ is always finite, but $C^-_{i-1/2}$ is infinite if $u^n_i = u^n_{i-1}$.

Example 13.13 Find three flux derivative splitting forms for FTFS for scalar conservation laws. If possible, one of the forms should have finite $C^+_{i+1/2}$.

Solution Recall that the conservative numerical flux of FTFS is

$$\hat{f}^n_{i+1/2} = f(u^n_{i+1}).$$

Suppose that $g_i^n = f(u_i^n)$. Then

$$C^+_{i+1/2} = -\lambda a_{i+1/2},$$
$$C^-_{i-1/2} = 0.$$

Alternatively, splitting FTFS about $g_i^n = \frac{1}{2}(f(u_{i+1}^n) + f(u_i^n))$ yields

$$C^+_{i+1/2} = -\frac{1}{2}\lambda a^n_{i+1/2},$$
$$C^-_{i-1/2} = \frac{1}{2}\lambda a^n_{i+1/2}\frac{u^n_{i+1} - u^n_i}{u^n_i - u^n_{i-1}}.$$

Finally, splitting FTFS about $g_i^n = f(u_{i+1}^n)$ yields

$$C^+_{i+1/2} = 0,$$
$$C^-_{i-1/2} = \lambda a^n_{i+1/2}\frac{u^n_{i+1} - u^n_i}{u^n_i - u^n_{i-1}}.$$

Notice that the first form has bounded coefficients $C^+_{i+1/2}$ and $C^-_{i-1/2}$, whereas the last two forms have bounded $C^+_{i+1/2}$ and unbounded $C^-_{i-1/2}$. Although the examples in this section always found at least one wave speed split form with finite coefficients, this is only because the stencils are so narrow.

13.6 Introduction to Reconstruction–Evolution Methods

The last two sections have concerned splitting techniques for finite-difference methods. Although splitting techniques also work for finite-volume methods, finite-volume methods more often use a different approach towards the same ends. Consider the following two-step finite-volume design approach:

(1) Spatial Reconstruction Reconstruct the solution $u(x, t^n)$ using, for example, the ENO techniques of Chapter 9. This step need not account for the upwind direction. Also, this step may involve solution averaging and slope limiting, analogous to flux averaging and flux limiting seen in finite-difference methods. Put another way, finite-volume methods form u and then $f(u)$, while finite-difference methods form $f(u)$ directly.

(2) Temporal Evolution Evolve the solution from time level n to $n+1$ using characteristics or some other technique. In particular, approximate $u(x_{i+1/2}, t)$ for $t^n \leq t \leq t^{n+1}$. Then

$$\hat{f}^n_{i+1/2} = \frac{1}{\Delta t}\int_{t^n}^{t^{n+1}} f(u(x_{i+1/2}, t))\, dt.$$

Any reasonable approximation based on waves and characteristics naturally introduces the minimal amount of upwinding required by the CFL condition.

This procedure is illustrated in Figure 13.10. Methods derived using this procedure are sometimes called *reconstruction–evolution methods*, a term coined by Harten, Engquist,

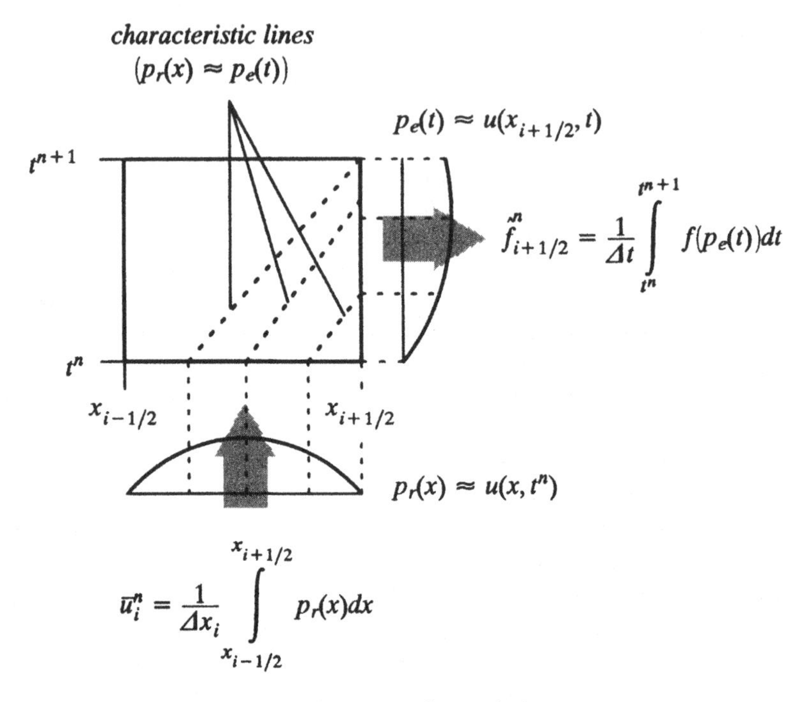

Figure 13.10 An illustration of reconstruction–evolution.

Osher, and Chakravarthy (1987). Alternatively, they are sometimes called *Godunov-type methods*, after the earliest such first-order accurate method, especially when the time evolution uses the exact Riemann solver. They are sometimes called *MUSCL-type methods*, after the earliest such second-order accurate method, discussed in Chapter 23. They are sometimes called *flux difference splitting methods*, to stress their relationship with flux vector splitting, especially when the time evolution uses a linearized Riemann solver such as Roe's approximate Riemann solver, as explained in Section 18.3.2. This book will mainly use the term reconstruction–evolution.

Reconstruction–evolution methods are extremely physical and elegant. However, in practice, reconstruction–evolution methods may be unduly expensive. After expending a huge amount of effort to reconstruct and evolve the solution, most of the detailed information is simply thrown away by the integral averaging at the end of the time evolution. Then the process starts all over again, with the reconstruction step attempting to reconstruct information just recently tossed aside. Of course, the polynomial reconstruction could be retained from time step to time step, and updated directly in terms of the polynomial coefficients rather than indirectly via cell-integral averages and reconstruction, but this is getting into finite-element territory. Other options will be explored in Section 23.1.

Example 13.14 Find a first-order reconstruction–evolution method for the linear advection equation. Use piecewise-constant reconstruction and exact evolution.

Solution Suppose that the reconstruction is $p_r(x) \approx u(x, t^n)$. For piecewise-constant reconstruction, let $p_r(x) = \bar{u}_i^n$ on cell $[x_{i-1/2}, x_{i+1/2}]$. Suppose that the evolution is $p_{e,i+1/2}(t) \approx u(x_{i+1/2}, t)$. For the linear advection equation, the exact solution implies $u(x,t) = u(x - a(t - t^n), t^n)$ or

$$p_{e,i+1/2}(t) = p_r(x_{i+1/2} - a(t - t^n)) = \begin{cases} \bar{u}_i^n & \text{for } 0 \le \lambda a \le 1, \\ \bar{u}_{i+1}^n & \text{for } -1 \le \lambda a \le 0. \end{cases}$$

Then

$$\hat{f}_{i+1/2}^n = \frac{1}{\Delta t}\int_{t^n}^{t^{n+1}} f(p_{e,i+1/2}(t))\,dt = \frac{1}{\Delta t}\int_{t^n}^{t^{n+1}} a p_{e,i+1/2}(t)\,dt = \begin{cases} a\bar{u}_i^n & \text{for } 0 \le \lambda a \le 1, \\ a\bar{u}_{i+1}^n & \text{for } -1 \le \lambda a \le 0. \end{cases}$$

The resulting method can be written as follows:

$$\bar{u}_i^{n+1} = \bar{u}_i^n - \lambda a \begin{cases} \bar{u}_i^n - \bar{u}_{i-1}^n & \text{for } a > 0, \\ \bar{u}_{i+1}^n - \bar{u}_i^n & \text{for } a < 0. \end{cases}$$

The same method was found in Examples 13.5, 13.6, and 13.10. Of course, this method is a finite-volume one whereas the earlier methods were finite difference. However, since this method is only first-order accurate, there is no need to distinguish finite-volume from finite-difference methods.

Why do completely different derivation techniques keep yielding the same numerical approximation to the linear advection equation? Mainly, because the linear advection equation is linear and linearity tends to wipe out the differences that would otherwise exist.

Example 13.15 Rederive the method found in the last example. This time, use a variant of reconstruction–evolution, in which the evolution step approximates $u(x, t^{n+1})$ rather than $u(x_{i+1/2}, t)$. Form the cell-integral averages of $u(x, t^{n+1})$ to approximate $\bar{u}_i^{n+1}$.

Solution Suppose that the reconstruction is $p_r(x) \approx u(x, t^n)$. For piecewise-constant reconstruction, let $p_r(x) = \bar{u}_i^n$ on cell $[x_{i-1/2}, x_{i+1/2}]$. Suppose that the temporal evolution is $p_e^{n+1}(x) \approx u(x, t^{n+1})$. For the linear advection equation, the exact solution implies $u(x,t) = u(x - a(t - t^n), t^n)$ or

$$p_e^{n+1}(x) = p_r(x_{i+1/2} - a\Delta t).$$

The desired alternative derivation of the linear first-order upwind method is illustrated in Figure 13.11. For $a > 0$, the mathematical details are as follows:

$$\bar{u}_i^{n+1} = \frac{1}{\Delta x}\int_{x_{i-1/2}}^{x_{i+1/2}} p_e^{n+1}(x)\,dx = \frac{1}{\Delta x}\int_{x_{i-1/2}}^{x_{i+1/2}} p_r(x - a\Delta t)\,dx$$
$$= \frac{1}{\Delta x}\left[(\Delta x - a\Delta t)\bar{u}_i^n + a\Delta t\bar{u}_{i-1}^n\right] = \bar{u}_i^n - \lambda a\left(\bar{u}_i^n - \bar{u}_{i-1}^n\right).$$

A similar calculation for $a < 0$ verifies that this is indeed the first-order upwind method for the linear advection equation. In the last example, the solution was time evolved just at cell edges $x_{i+1/2}$. In this example, the *entire* solution was time evolved. The approach

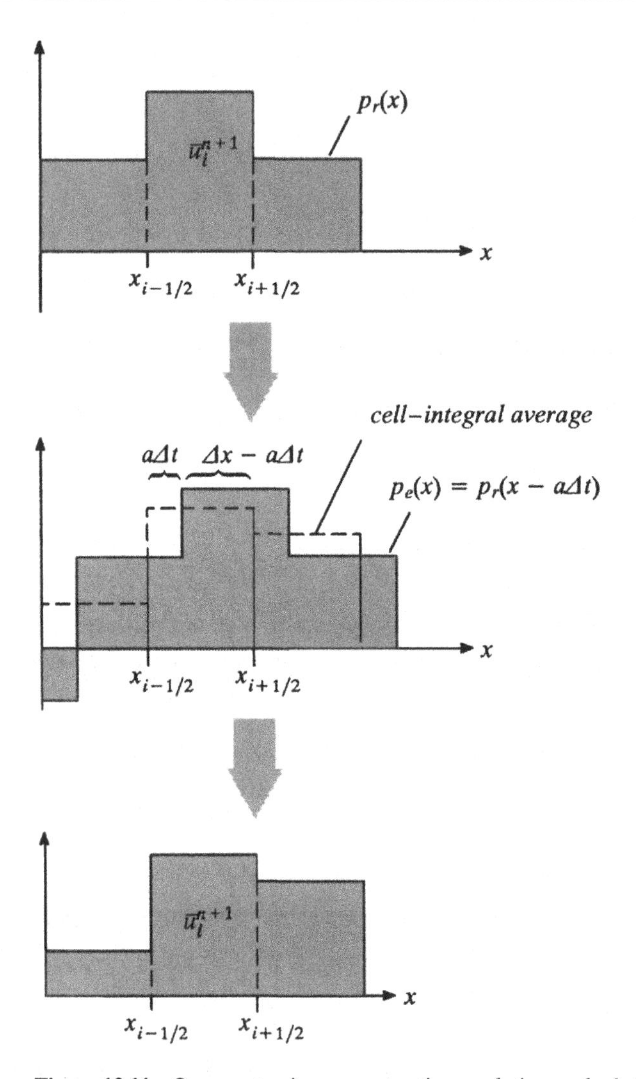

Figure 13.11 One way to view reconstruction–evolution methods for the linear advection equation.

in this example is easily illustrated, and the two approaches are entirely equivalent when the time evolution is exact, as for linear equations. However, the approach in this example does not naturally yield conservative numerical fluxes $\hat{f}^n_{i+1/2}$ and, in fact, will give rise to nonconservative methods in many cases unless great care is exercised.

The time-evolution step in the preceding examples was trivial, since the linear advection equation has a simple exact solution. But what about nonlinear conservation laws, which lack simple exact solutions? First, consider piecewise-constant reconstruction. Piecewise-constant reconstruction gives rise to a Riemann problem at each cell edge, where the

Riemann problem has an exact solution, as seen in Chapter 5. Reconstruction–evolution methods often use approximate Riemann solvers that deviate from the true Riemann solver in surprising ways without changing the numerical solution. In fact, in some cases, nonphysical deviations can actually improve numerical accuracy. Thus, just like flux splitting, reconstruction–evolution methods allow and even benefit from nonphysicality in their wave description. This is discussed further in Sections 17.3 and 18.3.

Now consider higher-order reconstructions, such as piecewise-linear, piecewise-quadratic, or some other higher-order piecewise polynomial reconstruction. Unfortunately, the jump at each cell edge in the piecewise-linear or higher-order piecewise-polynomial reconstruction gives rise to a problem that lacks an exact solution. However, the exact solution can be approximated using Riemann solvers. In fact, just as before with first-order methods, a well-constructed but artificial approximate wave solution may actually improve results compared with the unattainable exact physical solution. This is discussed in Chapter 23.

References

Boris, J. P., and Book, D. L. 1973. "Flux-Corrected Transport I. SHASTA, a Fluid Transport Algorithm that Works," *Journal of Computational Physics*, 11: 38–69.

Courant, R., Isaacson, E., and Rees, M. 1952. "On the Solution of Nonlinear Hyperbolic Differential Equations by Finite Differences," *Communications on Pure and Applied Mathematics*, 5: 243–255.

Harten, A. 1983. "High Resolution Schemes for Hyperbolic Conservation Laws," *Journal of Computational Physics*, 49: 357–393.

Harten, A., Engquist, B., Osher S., and Chakravarthy, S. R. 1987. "Uniformly High Order Accurate Essentially Non-Oscillatory Schemes, III," *Journal of Computational Physics*, 71: 231–303.

Harten, A., and Zwas, G. 1972. "Self-Adjusting Hybrid Schemes for Shock Computations," *Journal of Computational Physics*, 9: 568–583.

Shu, C.-W., and Osher, S. 1988. "Efficient Implementation of Essentially Non-Oscillatory Shock-Capturing Schemes," *Journal of Computational Physics*, 77: 439–471.

Steger, J. L., and Warming, R. F. 1979. "Flux Vector Splitting of the Inviscid Gasdynamic Equations with Application to Finite-Difference Methods," *Journal of Computational Physics*, 40: 263–293.

Van Leer, B. 1974. "Towards the Ultimate Conservative Difference Scheme. II. Monotonicity and Conservation Combined in a Second-Order Scheme," *Journal of Computational Physics*, 14: 361–370 .

Problems

13.1 Consider the following numerical method:

$$u_i^{n+1} = u_i^n - \frac{\lambda}{2}\left(3f\left(u_i^n\right) - 4f\left(u_{i-1}^n\right) + f\left(u_{i-2}^n\right)\right).$$

(a) Is this method upwind, downwind, or centered?

(b) Write the method in flux split form in two different ways.

(c) Write the method in wave speed split form in two different ways. If possible, find a wave speed split form with finite coefficients.

13.2 Consider the following numerical method:

$$u_i^{n+1} = u_i^n + \frac{\lambda}{12}\left(f\left(u_{i+2}^n\right) - 8f\left(u_{i+1}^n\right) + 8f\left(u_{i-1}^n\right) - f\left(u_{i-2}^n\right)\right).$$

(a) Is this method upwind, downwind, or centered?
(b) Write the method in flux split form in two different ways.
(c) Write the method in wave speed split form in two different ways. If possible, find a wave speed split form with finite coefficients.

13.3 Consider a scalar conservation law with flux function $f(u)$. In each case, suggest a first-order upwind method based on flux splitting.
(a) $f(u) = u^3/3$
(b) $f(u) = u^4/4$
(c) $f(u) = u^2 - 2u + 1$

13.4 Consider the scalar flux function $f(u) = \frac{1}{3}u^3 - u^2 + u$. Suggest a flux splitting. If possible, choose the flux splitting to be continuous.

13.5 Consider wave speed split forms, as described in Section 13.5. Argue briefly that, for a given numerical method, there is at most one wave speed split form whose coefficients are always finite.

CHAPTER 14

Artificial Viscosity

14.0 Introduction

It sometimes makes sense to divide numerical approximations into first differences and second differences. For example, Equation (10.17) can be written as

$$\frac{df}{dx}(x_{i-1}) = \frac{f(x_i) - f(x_{i-1})}{\Delta x} - \frac{f(x_{i+1}) - 2f(x_i) + f(x_{i-1})}{2\Delta x} + O(\Delta x^2).$$

From one point of view, second differences naturally appear as part of approximations to first derivatives, as in the preceding expression. From another point of view, first differences approximate first derivatives and second differences approximate second derivatives. In our applications, we will often associate first differences with the first derivatives in the Euler equations and scalar conservation laws and second differences with *viscous terms* in the Navier–Stokes equations. Then the second differences are called *artificial viscosity*. The term "artificial" reminds us that the second differences in question may not have much to do with physical viscosity. For example, artificial viscosity almost never involves the true coefficient of viscosity. In fact, the true viscosity is typically small in gasdynamics, outside of thin boundary layers coating solid surfaces; this justifies dropping the viscous terms in the first place and means that second differences that truly approximated viscous terms would not have much effect in most of the flow. At best, artificial viscosity has viscous-like effects, but the form and amount of artificial viscosity are chosen on a purely numerical basis.

Many first-derivative approximations also involve higher-order differences; fourth, sixth, and other even-order differences are generally also called artificial viscosity, while third, fifth, and other odd-order differences are generally called *artificial dispersion*. These higher-order differences are discussed in the next chapter. This chapter concerns only first- and second-order differences.

14.1 Physical Viscosity

To help introduce artificial viscosity and its relationship to numerical stability, this section concerns physical viscosity and its relationship to physical stability. The *Navier–Stokes equations* govern viscous flows. Making standard assumptions such as Stokes' hypothesis and Fourier's law for heat transfer, the conservation form of the Navier–Stokes equations for one-dimensional compressible flows is

$$\frac{\partial \rho}{\partial t} + \frac{\partial(\rho u)}{\partial x} = 0, \tag{14.1a}$$

$$\frac{\partial(\rho u)}{\partial t} + \frac{\partial}{\partial x}(\rho u^2 + p) = \frac{\partial}{\partial x}\left(\frac{4}{3}\mu\frac{\partial u}{\partial x}\right), \tag{14.1b}$$

$$\frac{\partial(\rho e_T)}{\partial t} + \frac{\partial}{\partial x}(\rho u e_T + pu) = \frac{\partial}{\partial x}\left(\frac{4}{3}\mu u\frac{\partial u}{\partial x} + k\frac{\partial T}{\partial x}\right), \tag{14.1c}$$

where $\mu \geq 0$ is the *coefficient of viscosity* and $k \geq 0$ is the *coefficient of thermal conductivity*. Notice that the left-hand side of the Navier–Stokes equations is exactly the same as the left–hand side of the Euler equations. However, where the Euler equations have zeros on the right-hand side, the Navier–Stokes equations have second derivative terms involving μ and k. The two terms involving μ are viscosity terms, where viscosity in fluids is the same thing as friction in solids, and the term involving k is a heat conduction term.

Before discussing the relationship between physical viscosity and physical stability, we must first define physical stability. In general, a physical system is *stable* if the system returns to its original state after any small disturbance. A physical system is *neutrally stable* if small disturbances cause proportionately small but permanent disturbances. Finally, a physical system is *unstable* if small disturbances lead to large disturbances. A classic example is a ball on a surface. A ball in a valley is stable – a small disturbance causes the ball to oscillate inside the valley, running up one side of the valley and down the other, until friction damps the motion, and the ball returns to rest. A ball on a perfectly flat surface is neutrally stable – a small disturbance causes the ball to move a small distance. Finally, a ball on a peaked surface is unstable – any small disturbance causes the ball to run down the sides of the peak, gaining speed as it releases its gravitational potential energy, until it comes to rest in some stable, neutrally stable, or even unstable position. As you might expect, friction has a stabilizing effect on the ball, although no amount of friction can stabilize a ball balanced on top of a sharp peak.

As another simple example of stability, consider a driven damped pendulum. This simple system allows all sorts of strange and complicated behaviors, making it a favorite example in the study of dynamical systems and chaos. Whereas large amounts of damping always stabilize the pendulum's motion, or even prevent the pendulum from moving, small amounts of damping lead to unstable and chaotic behaviors not found in an ideal frictionless pendulum. See Tabor (1989) for more information on the stability of the driven damped pendulum.

Now we turn to the most relevant example – fluid flows. Like any physical system, fluid flows may be stable, neutrally stable, or unstable. Stable flows damp out small disturbances, while unstable flows amplify them. For example, small disturbances in unstable laminar flows lead to oscillatory waves, called *Tollmien–Schlichting* waves, whose amplitudes grow downstream and with time, ultimately creating first bursts of turbulence and then continuous turbulence, characterized by highly unstable small-scale features. Examples of unstable laminar flows include jets, wakes, shear layers, and boundary layers with adverse pressure gradients.

A classic stability analysis technique called *small perturbation analysis* leads to *Rayleigh's equation* in the case of incompressible laminar inviscid flow, to the *Orr–Sommerfeld equation* in the case of incompressible laminar viscous flow, and to more complicated equations for compressible flows. Small perturbation analysis shows that viscosity has a dramatic effect on the stability of fluid flows, just as in other physical systems. In most cases, viscosity tends to stabilize fluid flows but, like the driven damped pendulum, sometimes small increases in viscosity may actually destabilize fluid flows. See White (1991) for a more thorough introduction to the stability of fluid flows. This completes our necessarily brief introduction to physical viscosity and stability.

14.2 Artificial Viscosity Form

Chapter 11 discussed conservation form, and Chapter 13 discussed flux split forms and wave speed split forms. This section introduces a fourth form, artificial viscosity form, composed of FTCS plus second-order artificial viscosity. Like the flux and wave speed split forms, artificial viscosity form can aid stability analyses.

The Euler equations omit viscosity; however, discretization generally reintroduces viscosity or, more precisely, second-difference terms that have viscous-like effects. Second differences that arise naturally as part of first-derivative approximation are called *implicit artificial viscosity*. Second differences purposely added to first-derivative approximations are called *explicit artificial viscosity*. In this context, the term implicit means "hidden and probably unintended" while explicit means "out in the open and intentional." Hopefully, these usages of "implicit" and "explicit" will not cause too much confusion with the earlier usages. This section concerns explicit (second meaning) artificial viscosity in explicit (first meaning) finite-difference approximations.

The viscous terms in the Navier–Stokes equations (14.1) have the following general form:

$$\frac{\partial}{\partial x}\left(\epsilon \frac{\partial U}{\partial x}\right),$$

where $\epsilon = \mu, \mu u, k$ and $U = u, T$. Now consider a scalar conservation law,

$$\frac{\partial u}{\partial t} + \frac{\partial f(u)}{\partial x} = 0.$$

Add a viscous-like term to the right-hand side as follows:

$$\frac{\partial u}{\partial t} + \frac{\partial f(u)}{\partial x} = \frac{\partial}{\partial x}\left(\epsilon(u) \frac{\partial u}{\partial x}\right), \tag{14.2}$$

where $\epsilon \geq 0$. Remember that the viscous-like term is added for numerical reasons and generally has no relationship to physical viscosity. As a result, the discretization of the viscous term need not be accurate, provided only that the discretized viscosity term has desirable numerical effects. A conservative forward-time central-space discretization of Equation (14.2) yields

$$\frac{u_i^{n+1} - u_i^n}{\Delta t} + \frac{f(u_{i+1}^n) - f(u_{i-1}^n)}{2\Delta x} = \frac{\epsilon_{i+1/2}^n \frac{u_{i+1}^n - u_i^n}{\Delta x} - \epsilon_{i-1/2}^n \frac{u_i^n - u_{i-1}^n}{\Delta x}}{\Delta x}.$$

For convenience, absorb the factor $2/\Delta x$ into $\epsilon_{i+1/2}^n$. Then rearranging terms yields the following:

$$\blacklozenge \quad \begin{aligned} u_i^{n+1} = u_i^n &- \frac{\lambda}{2}\left(f(u_{i+1}^n) - f(u_{i-1}^n)\right) \\ &+ \frac{\lambda}{2}\left(\epsilon_{i+1/2}^n (u_{i+1}^n - u_i^n) - \epsilon_{i-1/2}^n (u_i^n - u_{i-1}^n)\right). \end{aligned} \tag{14.3}$$

This is known as an *artificial viscosity form*, the terms involving $\epsilon_{i+1/2}^n$ are called *second-order explicit artificial viscosity*, and $\epsilon_{i+1/2}^n$ is called the *coefficient of second-order explicit artificial viscosity*. This artificial viscosity form applies equally to scalar and vector conservation laws. For vector conservation laws, the coefficient of viscosity $\epsilon_{i+1/2}^n$ can be either a scalar or a matrix. Although useful for purposes of introduction, few numerical methods actually derive from equations such as (14.2). Instead, artificial viscosity arises naturally as part of the first-derivative approximations, or gets added after the fact, so that the coefficient of artificial viscosity $\epsilon_{i+1/2}^n$ generally has no relationship to any $\epsilon(u)$.

Equation (14.3) involves FTCS. Of course, FTCS can be replaced by any other method. For example, suppose the viscous term in Equation (14.2) is discretized as before, while the flux term is discretized using FTFS. Then one obtains

$$u_i^{n+1} = u_i^n - \lambda\left(f\left(u_{i+1}^n\right) - f\left(u_i^n\right)\right)$$
$$+ \frac{\lambda}{2}\left(\epsilon_{i+1/2}^n\left(u_{i+1}^n - u_i^n\right) - \epsilon_{i-1/2}^n\left(u_i^n - u_{i-1}^n\right)\right).$$

Besides changing the discretization of the flux term, one could also change the discretization of the viscous term. Despite these myriad of possibilities, this book reserves the phrase "artificial viscosity form" specifically for the traditional equation (14.3).

Artificial viscosity form is related to conservation form by

$$\blacklozenge \qquad \hat{f}_{i+1/2}^n = \frac{1}{2}\left[f\left(u_{i+1}^n\right) + f\left(u_i^n\right) - \epsilon_{i+1/2}^n\left(u_{i+1}^n - u_i^n\right)\right]. \tag{14.4}$$

Equation (14.4) allows easy transformation from conservation form to artificial viscosity form and vice versa. In fact, for scalar conservation laws, every $\hat{f}_{i+1/2}^n$ yields a unique $\epsilon_{i+1/2}^n$ and vice versa. For vector conservation laws, Equation (14.4) may have no solution if $\epsilon_{i+1/2}^n$ is a scalar or infinitely many solutions if $\epsilon_{i+1/2}^n$ is a matrix. The solution for matrix coefficients $\epsilon_{i+1/2}^n$ proceeds as in Section 5.3.

From one point of view, artificial viscosity form is just a variation on the flux-correction form seen in Section 13.3. As seen in Equation (13.5), the flux-correction form is

$$\hat{f}_{i+1/2}^n = \hat{f}_{i+1/2}^{(1)} + \hat{f}_{i+1/2}^{(C)},$$

where $\hat{f}_{i+1/2}^{(1)}$ is the conservative numerical flux of some reference method and $\hat{f}_{i+1/2}^{(C)}$ is a flux correction to the reference. Compare this last equation with Equation (14.4) to find the following:

$$\hat{f}_{i+1/2}^{(1)} = \frac{1}{2}\left(f\left(u_{i+1}^n\right) + f\left(u_i^n\right)\right) = \hat{f}_{i+1/2}^{FTCS},$$
$$\hat{f}_{i+1/2}^{(C)} = -\frac{1}{2}\epsilon_{i+1/2}^n\left(u_{i+1}^n - u_i^n\right).$$

Thus artificial viscosity form is FTCS plus a flux correction, where the flux correction is factored into a coefficient $\epsilon_{i+1/2}^n$ times a first difference $u_{i+1}^n - u_i^n$. Oftentimes, when a method is written as a flux correction to FTCS, the flux correction includes first-difference factors $u_{i+1}^n - u_i^n$, making the artificial viscosity form especially convenient and natural. Unfortunately, the artificial viscosity form generally involves infinite coefficients when the flux correction does not include first-difference factors $u_{i+1}^n - u_i^n$. In particular, suppose that $u_{i+1}^n - u_i^n = 0$. Then if $\epsilon_{i+1/2}^n$ is finite, the flux correction is zero and the method equals FTCS or, more specifically, $\hat{f}_{i+1/2}^n = f(u_i^n) = f(u_{i+1}^n)$. Otherwise, if the method does not equal FTCS then $\epsilon_{i+1/2}^n$ is infinite, which is usually the case when the method's stencil contains more than two or three points. Recall that a similar discussion took place in Section 13.5, where the first differences appearing in the wave speed split form also lead to infinite coefficients.

Artificial viscosity forms sometimes suggest alterations and improvements. For example, in one interpretation, the implicit artificial viscosity in FTCS is too small, making FTCS unstable, and adding explicit second-order artificial viscosity with a positive coefficient has a smoothing and stabilizing effect. In other cases, a numerical method may have too much artificial viscosity, causing smearing or even instability. In this case, adding explicit artificial viscosity with a negative coefficient partially cancels the implicit artificial viscosity, resulting in a sharper, more detailed, and even more stable solution. Second-order artificial viscosity with a positive coefficient is sometimes called *artificial dissipation*; second-order artificial viscosity with a negative coefficient is sometimes also called *artificial antidissipation*.

Example 14.1 Write FTFS and FTBS in artificial viscosity form for both vector and scalar conservation laws. Discuss the relationship between stability and the sign of the coefficient of artificial viscosity.

Solution First consider FTFS. By Equation (14.4),

$$f(u^n_{i+1}) = \frac{1}{2}\left[f(u^n_{i+1}) + f(u^n_i) - \epsilon^n_{i+1/2}(u^n_{i+1} - u^n_i)\right].$$

For scalar conservation laws, solve for $\epsilon^n_{i+1/2}$ to find

$$\epsilon^n_{i+1/2} = -a^n_{i+1/2},$$

where as usual

$$a^n_{i+1/2} = \begin{cases} \dfrac{f(u^n_{i+1}) - f(u^n_i)}{u^n_{i+1} - u^n_i} & u^n_{i+1} \neq u^n_i, \\ f'(u^n_i) & u^n_{i+1} = u^n_i. \end{cases}$$

Similarly, FTBS yields

$$\epsilon^n_{i+1/2} = a^n_{i+1/2}.$$

Notice that both coefficients of artificial viscosity are finite. Also notice that FTBS and FTFS are stable when their coefficients of artificial viscosity are positive and unstable when their coefficients of artificial viscosity are negative. Then, from this point of view, a sufficient amount of artificial dissipation can successfully stabilize FTCS.

For vector conservation laws, the previous expressions apply except that $a^n_{i+1/2}$ is replaced by the Roe-average Jacobian matrix $A^n_{i+1/2}$ or some other secant plane slope matrix, as described in Section 5.3. Such artificial viscosity successfully stabilizes the method only when all of the characteristic values of $A^n_{i+1/2}$ are positive and sufficiently large.

There are a number of potential pitfalls with the concept of artificial viscosity. As mentioned before, artificial viscosity may give a false impression of physicality (remember that it is called *artificial* viscosity for a reason). Furthermore, many people treat artificial viscosity as an afterthought. A typical statement along these lines is as follows: "Standard artificial viscosity was added to the numerical method to reduce spurious oscillations." However, taking the broadest view of artificial viscosity, any method can be written as any other method plus artificial viscosity. Thus, adding artificial viscosity fundamentally alters a method; changing the second differences has almost as much effect as changing the first differences. In some cases, it is more convenient to design or analyze the coefficient of artificial viscosity $\epsilon^n_{i+1/2}$ rather than the conservative fluxes $\hat{f}^n_{i+1/2}$, but the one leads to the other, and the design or analysis problem is never trivial either way.

References

Tabor, M. 1989. *Chaos and Integrability in Nonlinear Dynamics: An Introduction*, New York: Wiley.
White, F. M. 1991. *Viscous Fluid Flow*, 2nd ed., New York: McGraw-Hill, Chapter 5.

Problems

14.1 Consider the following numerical approximation to a scalar conservation law:

$$u_i^{n+1} = u_i^n - \frac{\lambda}{2}\left(3f\left(u_i^n\right) - 4f\left(u_{i-1}^n\right) + f\left(u_{i-2}^n\right)\right).$$

(a) Write this method in standard artificial viscosity form. That is, write the method as FTCS plus second-order artificial viscosity.

(b) Write this method as FTBS plus second-order artificial viscosity. Does the method have more or less artificial viscosity than FTBS? In particular, for $a > 0$, show that the method has more artificial viscosity than FTBS at maxima and minima and less artificial viscosity than FTBS in monotone regions.

14.2 Consider the following numerical approximation to a scalar conservation law:

$$u_i^{n+1} = u_i^n + \frac{\lambda}{12}\left(f\left(u_{i+2}^n\right) - 8f\left(u_{i+1}^n\right) + 8f\left(u_{i-1}^n\right) - f\left(u_{i-2}^n\right)\right).$$

Write this method in standard artificial viscosity form.

14.3 This problem concerns the relationship between artificial viscosity form and wave speed split form.

(a) Find an expression for $\epsilon_{i+1/2}^n$ in terms of $C_{i+1/2}^+$, $C_{i+1/2}^-$, and g_i^n. Specifically, suppose a conservative numerical method is written in wave speed split form as follows:

$$u_i^{n+1} = u_i^n + C_{i+1/2}^+\left(u_{i+1}^n - u_i^n\right) - C_{i-1/2}^-\left(u_i^n - u_{i-1}^n\right)$$

such that

$$\left(C_{i+1/2}^- - C_{i+1/2}^+\right)\left(u_{i+1}^n - u_i^n\right) = \lambda\left(g_{i+1}^n - g_i^n\right).$$

Then show that the method can be written in the following form:

$$u_i^{n+1} = u_i^n - \frac{\lambda}{2}\left(g_{i+1}^n - g_{i-1}^n\right) + \frac{1}{2}\lambda\epsilon_{i+1/2}^n\left(u_{i+1}^n - u_i^n\right) - \frac{1}{2}\lambda\epsilon_{i-1/2}^n\left(u_i^n - u_{i-1}^n\right),$$

where

$$\lambda\epsilon_{i+1/2}^n = C_{i+1/2}^+ + C_{i+1/2}^-.$$

If $g_i^n = f(u_i^n)$ then

$$u_i^{n+1} = u_i^n - \frac{\lambda}{2}\left(f\left(u_{i+1}^n\right) - f\left(u_{i-1}^n\right)\right) + \frac{1}{2}\lambda\epsilon_{i+1/2}^n\left(u_{i+1}^n - u_i^n\right) - \frac{1}{2}\lambda\epsilon_{i-1/2}^n\left(u_i^n - u_{i-1}^n\right),$$

which is the standard artificial viscosity form.

(b) Reverse the results found in part (a) to find expressions for $C_{i+1/2}^+$ and $C_{i+1/2}^-$ in terms of $\epsilon_{i+1/2}^n$ and g_i^n.

14.4 By examining conservative numerical flux, argue that second-order artificial viscosity transfers conserved quantities from cells that have more to cells that have less. Like Robin Hood, it takes from the rich and gives to the poor.

CHAPTER 15

Linear Stability

15.0 Introduction

This chapter concerns linear stability, while Chapter 16 will treat nonlinear stability. We begin with a general introduction to linear and nonlinear stability. Unfortunately, there are many different definitions for numerical stability, most of which differ from the definition of physical stability seen in Section 14.1. In particular, unlike physical instability, numerical instability does not necessarily imply sensitivity to small disturbances. Four common definitions of numerical stability are as follows:

- ***Unbounded Growth*** A method is *unstable* if the error grows to infinity as time goes to infinity. In some definitions, the error is required to grow at a certain minimum rate (e.g., exponentially or algebraically). A method that is not unstable is *stable*.
- ***Convergence*** A method is *stable* if it converges as $\Delta x \to 0$ and $\Delta t \to 0$, assuming only a few basic conditions such as consistency, conservation, and well-posed initial and boundary conditions. Otherwise, a method is *unstable*. In some definitions, the solution is required not only to converge but also to converge to the solution that satisfies the entropy condition. Somewhat surprisingly, the "unbounded growth" definition and the "convergence" definition are closely related, as discussed in Sections 15.4 and 16.11.
- ***Physical*** A method is *unstable* if it exhibits significant errors created by interactions between various time and space approximations and, in particular, any errors that start small and grow with time. In other words, instability is any significant error beyond that found in the individual component approximations such as the forward-time approximation or the central-space approximation. Conversely, a method is *stable* when it exhibits only small errors beyond those caused by flaws in the individual component approximations. This definition of numerical instability is the closest in spirit to the definition of physical instability, described in Section 14.1, in that it involves space-time interactions and growth in time. However, unlike true physical instability, this definition of numerical instability does not necessarily require sensitivity to small disturbances.
- ***Broad*** A method is *unstable* if it exhibits large errors that increase with time, or any local errors significantly larger than the average error, most especially *oscillatory errors* and expansion shocks. Otherwise, a method is *stable*.

Regardless of definition, spurious oscillations and overshoots are the most common symptoms of numerical instability. Thus, as a unifying principle, *this book will discuss stability mainly in terms of spurious oscillations and overshoots*. This principle follows the "broad" definition of stability. For example, this principle concerns itself with any significant oscillations, not just unbounded oscillations, contradicting the "unbounded growth" definition of instability. Furthermore, this principle concerns itself with oscillations at ordinary values of Δx and Δt, rather than oscillations or other types of errors in the limit $\Delta x \to 0$ and $\Delta t \to 0$, contradicting the "convergence" definition of stability. Finally, this principle concerns oscillations regardless of source; oscillations may originate either in the individual space and time approximations or in the interactions between various space and time approximations, contradicting the "physical" definition of instability.

In one natural interpretation, regardless of the definition of stability, stability conditions commonly amount to conditions restricting oscillations. The "unbounded growth" and "convergence" definitions of stability tend to place relatively weak restrictions on oscillations, whereas the "broad" definition tends to place relatively strong restrictions on oscillations. But, in any event, stability analysis most often amounts to oscillation analysis. To distinguish the oscillation aspect of stability from other aspects, some references use terms such as *oscillation control* or *extremum control*; this book mostly uses such terms as synonyms for stability.

Although oscillations are the primary symptom of numerical instability, they are also the primary symptom of physical instability, as seen in Section 14.1. By definition, unstable physical systems amplify small disturbances as time progresses, including those introduced by numerical approximations. Then every numerical approximation experiences large errors: The best approximations experience errors characteristic of the physical instability; lesser approximations introduce additional forms of error and instability. How can one distinguish between physical and numerical oscillations without an exact solution? Fortunately, this difficult question does not arise in our applications since, as seen Section 4.11, the unsteady one-dimensional Euler equations and scalar conservation laws are essentially *nonoscillatory*. In particular, scalar conservation laws neither allow new extrema nor allow existing extrema to grow, so that any oscillations except those found in the initial conditions are purely numerical in origin.

This observation greatly simplifies stability analysis for scalar conservation laws and the one-dimensional Euler equations. Of course, scalar conservation laws and the one-dimensional Euler equations are ultimately just first steps towards the multidimensional Euler and Navier–Stokes equations, and any numerical methods based on specific properties of the one-dimensional Euler equations may fail in the extension to the multidimensional Euler and Navier–Stokes equations. In particular, although scalar conservation laws and the one-dimensional Euler equations are always nonoscillatory as described in Section 4.11, the multidimensional Euler and Navier–Stokes equations may be unstable and highly oscillatory as discussed in Section 14.1. However, having said this, by discretizing the nonoscillatory wave/flux terms separately from the oscillatory viscous terms, many numerical methods based on specific properties of scalar conservation laws and the one-dimensional Euler equations extend surprisingly well to the multidimensional Euler and Navier–Stokes equations and, in fact, routinely and dramatically outperform methods that are not based on specific properties of scalar conservation laws and the one-dimensional Euler equations.

Numerical errors often mimic physical behaviors – numerical instability mimics physical instability, numerical dissipation mimics physical dissipation, numerical dispersion mimics physical dispersion, and so forth. Distinguishing the physical from the numerical requires a reliable reference standard. One possible reference is experimental results. However, experimental results sometimes contain artifacts of the specific testing conditions, such as disturbances to the flow due to test probes, stings, wall effects in wind tunnels, and so forth. Such artifacts may be difficult to identify and may indicate errors in numerical methods where none exist. Fortunately, again, the limited scope of this book mostly sidesteps these troubling issues.

Unstable numerical methods are amazingly common. For example, FTCS is highly unstable by any definition, except for a few special and useless cases such as $u_i^0 = const.$ For another historical example, in 1910, Richardson devised one of the first finite-difference methods, a simple linear central-time central-space method approximating the heat equation. Richardson's method is highly unstable, although this was not immediately recognized because the cost of hand calculations prevented anyone from taking more than a few time steps. Instability generally takes time to grow, so that even unstable methods may look fine for the first little while. Sometimes numerical instability has a physical explanation such as, for example, a violation of the CFL condition. Other times there are no specific identifiable factors, either physical or numerical. This is certainly the case for FTCS. Rather than attempting to identify the root causes of instability, this book targets only the symptoms: spurious oscillations and overshoots.

This completes the general introduction to stability. The rest of this chapter specifically concerns stability for linear methods. An equation or numerical approximation is *linear* if any linear combination of solutions is also a solution. If the governing equations are nonlinear, then any numerical approximation should also be nonlinear. However, the reverse is not necessarily true: Many numerical methods are nonlinear even when the governing equations are linear. Such numerical methods are called *inherently nonlinear*. All of the methods seen in this book so far are linear when applied to the linear advection equation or a linear system of equations but nonlinear when applied to Burgers' equation or the Euler equations. In other words, none of the methods seen so far are inherently nonlinear. However, most of the modern methods seen in Part V *are* inherently nonlinear.

This chapter concerns stability for linear methods. Although linearity substantially simplifies stability analysis, *never assume that the conclusions of linear stability analysis apply to nonlinear methods*. At best, nonlinear methods may be approximately locally linearized, so that the results of linear analyses apply approximately and locally. However, many modern methods, especially inherently nonlinear methods, do not allow for local linearizations. As a result, linear stability analysis has increasingly lost ground to nonlinear stability analysis, to the extent that linear stability analysis rarely appears in research papers written after the early 1980s.

15.1 von Neumann Analysis

This section describes the most popular type of linear stability analysis, known as *von Neumann* or *Fourier series analysis*. As described in the introduction to Chapter 8 of Hirsch (1988), von Neumann developed this technique during World War II at Los Alamos National Labs in New Mexico. The technique was classified until published by others in the open literature just after the war. Although this book will discuss von Neumann analysis specifically in the context of gasdynamics, it is a general technique that applies to all sorts of linear finite-difference approximations.

Since unstable solutions typically oscillate, it makes sense to express the solution as a Fourier series or, in other words, as a sum of oscillatory trigonometric functions, following the philosophy espoused in Example 6.5. Many other subjects use the same strategy: Acoustics, vibrations, and so forth also use Fourier series because periodic and oscillatory behaviors are also expected. By Equation (7.30), the Fourier series for the solution $u(x, t^n)$

on any spatial domain $[a, b]$ is

$$u(x, t^n) = a_0^n + \sum_{m=1}^{\infty} a_m^n \cos\left(2\pi m \frac{x-a}{b-a}\right) + \sum_{m=1}^{\infty} b_m^n \sin\left(2\pi m \frac{x-a}{b-a}\right), \quad (15.1)$$

where a_m^n and b_m^n are the Fourier series coefficients for $u(x, t^n)$, not to be confused with the limits a and b of the spatial domain. For a discrete solution, sample the Fourier series to obtain

$$u(x_i, t^n) = a_0^n + \sum_{m=1}^{\infty} \left(a_m^n \cos 2\pi m \frac{x_i - a}{b-a} + b_m^n \sin 2\pi m \frac{x_i - a}{b-a}\right).$$

Assume evenly space samples $x_{i+1} - x_i = \Delta x = const.$ Also assume that the first and last samples are the endpoints of the domain (i.e., $x_0 = a$ and $x_N = b$). Then $x_i - a = i\Delta x$ and $b - a = N\Delta x$. The sampled Fourier series then becomes

$$u(x_i, t^n) = a_0^n + \sum_{m=1}^{\infty} \left(a_m^n \cos 2\pi m \frac{i}{N} + b_m^n \sin 2\pi m \frac{i}{N}\right).$$

Unfortunately, samples cannot support an infinite range of frequencies and wavelengths. Instead, samples can only support wavelengths of $2\Delta x$ or longer, as discussed in Section 8.2. Truncate the sampled Fourier series accordingly to find

$$u(x_i, t^n) \approx a_0^n + \sum_{m=1}^{N/2} \left(a_m^n \cos \frac{2\pi m i}{N} + b_m^n \sin \frac{2\pi m i}{N}\right), \quad (15.2)$$

which is called a *discrete Fourier series.* Although this expression allows $2\Delta x$- and $3\Delta x$-waves, the Nyquist sampling theorem says that these waves do not contain useful information about the solution, as discussed in Section 8.2. An equivalent complex expression for a discrete Fourier series is as follows:

$$u(x_i, t^n) \approx \sum_{m=-N/2}^{N/2} C_m^n \exp\left(\frac{2\pi I m i}{N}\right), \quad (15.3)$$

where $I = \sqrt{-1}$. (A more common notation for $\sqrt{-1}$ is i or j; however, these can easily be confused with the spatial indices i and j.) The complex form simplifies the algebra seen later in the chapter. From elementary complex analysis, *Euler's formula* is

♦ $$e^{I\theta} = \cos\theta + I\sin\theta. \quad (15.4)$$

Euler's formula yields the following relationships between the coefficients in the real and complex Fourier series:

$$C_0 = a_0, \quad (15.5a)$$

$$C_m = \frac{a_m - Ib_m}{2}, \quad (15.5b)$$

$$C_{-m} = \frac{a_m + Ib_m}{2}. \quad (15.5c)$$

In conclusion, any numerical solution can be written in the form of a discrete Fourier series, preferably using complex notation, assuming only that $\Delta x = const.$ and that u_i^n is periodic or, in other words, $u_0^n = u_N^n$.

Example 15.1 Write the solution of FTCS for the linear advection equation in terms of Fourier series coefficients C_m^n. Use these expressions to show that FTCS for the linear advection equation "blows up" in time.

Solution FTCS for the linear advection equation is

$$u_i^{n+1} - u_i^n + \lambda a \frac{u_{i+1}^n - u_{i-1}^n}{2} = 0.$$

Let

$$u_i^n = \sum_{m=-N/2}^{N/2} C_m^n \exp\left(\frac{2\pi Imi}{N}\right).$$

Then

$$u_i^{n+1} = \sum_{m=-N/2}^{N/2} C_m^{n+1} \exp\left(\frac{2\pi Imi}{N}\right)$$

and

$$\begin{aligned} u_{i\pm1}^n &= \sum_{m=-N/2}^{N/2} C_m^n \exp\left(\frac{2\pi Imi(i \pm 1)}{N}\right) \\ &= \sum_{m=-N/2}^{N/2} C_m^n \exp\left(\pm\frac{2\pi Im}{N}\right) \exp\left(\frac{2\pi Imi}{N}\right). \end{aligned}$$

Then FTCS for the linear advection equation becomes

$$\sum_{m=-N/2}^{N/2} \left[C_m^{n+1} - C_m^n + \frac{\lambda a}{2} C_m^n \exp\left(\frac{2\pi Im}{N}\right) - \frac{\lambda a}{2} C_m^n \exp\left(-\frac{2\pi Im}{N}\right)\right] \exp\left(\frac{2\pi Imi}{N}\right) = 0.$$

Euler's relation, Equation (15.4), implies

$$e^{I\theta} - e^{-I\theta} = 2I \sin\theta.$$

Assuming $C_m^n \neq 0$, then FTCS for the linear advection equation becomes

$$\sum_{m=-N/2}^{N/2} C_m^n \left[\frac{C_m^{n+1}}{C_m^n} - 1 + I\lambda a \sin\frac{2\pi m}{N}\right] \exp\left(\frac{2\pi Imi}{N}\right) = 0.$$

This last expression is a linear combination of trigonometric functions $\exp(2\pi Imi/N)$, where $m = -N/2, \ldots, N/2$. Since these trigonometric functions are linearly independent, the coefficients in the linear combination must equal zero. Then

$$\frac{C_m^{n+1}}{C_m^n} = 1 - I\lambda a \sin\frac{2\pi m}{N} \tag{15.6}$$

for all m. This last equation is the desired expression for FTCS written in terms of Fourier series coefficients. Again, notice that all we have done so far is to rewrite FTCS.

Let us use this last expression to prove that FTCS blows up. First,

$$\left|\frac{C_m^{n+1}}{C_m^n}\right|^2 = \left|1 - I\lambda a \sin\frac{2\pi m}{N}\right|^2 = 1 + (\lambda a)^2 \sin^2\left(\frac{2\pi m}{N}\right) \geq 1, \tag{15.7}$$

where

$$|\operatorname{Re} z + I \operatorname{Im} z|^2 = (\operatorname{Re} z)^2 + (\operatorname{Im} z)^2$$

for any complex number z. Then

$$|C_m^{n+1}| > |C_m^n|$$

for all $m \neq 0$. Thus, except for C_0^n, all of the Fourier series coefficients increase in magnitude as time increases. Thus FTCS for the linear advection equation blows up in time.

The preceding example illustrates a general technique that can be used to analyze the stability of any linear method. However, rather than take the long way around each time, the preceding example suggests several simplifications. First of all, from now on, the principle of superposition will be invoked immediately. Thus, instead of using the entire Fourier series, use only one term in the series:

$$u_i^n = C_m^n \exp\left(\frac{2\pi Imi}{N}\right), \tag{15.8}$$

where $m = -N/2, \ldots, N/2$. As a second simplification, consider the following convenient notation:

♦ $$\phi_m = \frac{2\pi m}{N}. \tag{15.9}$$

As a third simplification, the solution of a linear equation can always be written in terms of the ratio C_m^{n+1}/C_m^n. For example, Equation (15.6) expresses FTCS in terms of C_m^{n+1}/C_m^n. Furthermore, and rather remarkably, the ratio C_m^{n+1}/C_m^n does not depend on n! In other words, *each of the Fourier series coefficients changes by exactly the same factor at every time step*. This strange property is a direct consequence of linearity. Then, as a convenient notation, let

♦ $$G_m = \frac{C_m^{n+1}}{C_m^n}, \tag{15.10}$$

where G_m does not have an n superscript, since G_m does not depend on the time level n. The ratio G_m is sometimes called the *amplification factor*. Then Equation (15.8) becomes

$$u_i^n = \frac{C_m^n}{C_m^{n-1}} \cdots \frac{C_m^2}{C_m^1}\frac{C_m^1}{C_m^0} C_m^0 e^{I\phi_m i}$$

or

$$u_i^n = G \cdot G \cdot \cdots \cdot G \cdot C_m^0 e^{I\phi_m i}$$

or

$$u_i^n = G_m^n C_m^0 e^{I\phi_m i}, \tag{15.11}$$

where G_m^n means "G_m to the power n" as opposed to "G_m at time level n" (unlike the other quantities in this chapter). As a fourth simplification, assume that $C_m^0 = 1$ in Equation (15.11); any other value except 0 would do just as well and would not affect the final results, but 1 is the most convenient choice. As a final simplification, drop the subscript m in Equation (15.11) to obtain the following final result:

♦ $$u_i^n = G^n e^{I\phi i}, \tag{15.12}$$

where G is now understood to be a function of ϕ. The dependencies on N and m have thus been eliminated in favor of a dependence on ϕ.

This section began assuming nothing more than that the solution could be expressed as a discrete Fourier series, in keeping with the "broad" or "oscillatory" definition of stability. However, the preceding results indicate that the amplitudes of each sinusoid in the discrete Fourier series can only have a very narrow range of behaviors, which inspires a definition for the stability of linear methods along the lines of the "unbounded growth" definition. In particular, suppose a linear method satisfies the CFL condition. Then, by the usual linear definitions, the linear method is *linearly stable* if $|G| < 1$ for all $-\pi \leq \phi \leq \pi$. Similarly, the linear method is *neutrally linearly stable* if $|G| \leq 1$ for all $-\pi \leq \phi \leq \pi$ and $|G| = 1$ for some $-\pi \leq \phi \leq \pi$. Finally, the linear method is *linearly unstable* if $|G| > 1$ for some $-\pi \leq \phi \leq \pi$. These simple definitions of stable, unstable, and neutrally stable are direct consequences of the simple behaviors of linear methods. In particular, for linear methods, the amplitudes of the terms in the discrete Fourier series either grow like $|G^n|$, shrink like $|G^n|$, or remain completely the same.

Von Neumann analysis or *Fourier series analysis* refers to any linear analysis based on Equation (15.12). As the first priority, von Neumann analysis is used to determine linear stability, as indicated by the magnitude of the amplification factor $|G| = \sqrt{\text{Re}(G)^2 + \text{Im}(G)^2}$. After establishing linear stability, von Neumann analysis may also examine the phase $\tan^{-1}(\text{Im}(G)/\text{Re}(G))$. Whereas the magnitude indicates the amplitude of each term in the Fourier series, the phase indicates the speed of each term in the Fourier series. In a perfect method, all sinusoids would travel at the same wave speed a. However, in real methods, different frequencies travel at different speeds. This causes some frequencies to lag and some frequencies to lead. In the worst case, sinusoids with different frequencies may become completely separated, leading to large spatial oscillations. To introduce some common terminology, frequency-dependent wave speeds cause *dispersion*. Dispersion can occur physically, but this discussion concerns purely numerical dispersion. Although dispersion never leads linear methods to blow up, large oscillations caused by dispersion may still be considered a form of instability, under the "broad" definition of instability, although it is certainly not instability under the "unbounded growth" definition most often used in linear analysis. Von Neumann analysis commonly investigates both amplitude and phase, and thus von Neumann analysis commonly investigates both the "unbounded growth" and "broad" definitions of instability, although this chapter is mainly concerned with the "unbounded growth" definition.

As one way to view Fourier series analysis, imagine striking a series of tones on a xylophone. The amplitude C_m^0 indicates how loud the note begins; $C_m^0 = 0$ corresponds to skipping a note. As the mallet strikes each plate on the xylophone, the tone either dies out, preferably slowly, or resonates and becomes increasingly louder. If one strikes all of the notes simultaneously or in quick succession, the method is said to have low dispersion if far away listeners hear all of the tones simultaneously or in quick succession and high dispersion if far away listeners hear some tones first, well before the others, somewhat like the Doppler effect.

Example 15.2 Use von Neumann analysis to determine the stability of FTFS for the linear advection equation.

Solution FTFS for the linear advection equation is

$$u_i^{n+1} = u_i^n - \lambda a\left(u_{i+1}^n - u_i^n\right).$$

Let $u_i^n = G^n e^{I\phi i}$. Then FTFS for the linear advection equation becomes

$$G^{n+1}e^{I\phi i} = G^n e^{I\phi i} - \lambda a G^n e^{I\phi(i+1)} + \lambda a G^n e^{I\phi i}.$$

Divide by $G^n e^{I\phi i}$ and apply Euler's relation to obtain

$$G = 1 - \lambda a e^{I\phi} + \lambda a = 1 + \lambda a - \lambda a \cos\phi - I\lambda a \sin\phi.$$

Then

$$|G|^2 = (1 + \lambda a - \lambda a \cos\phi)^2 + (\lambda a \sin\phi)^2.$$

After some algebraic manipulation, this becomes

$$|G|^2 = 1 + 2\lambda a(1 + \lambda a)(1 - \cos\phi).$$

Since $1 - \cos\phi \geq 0$ for all $-\pi \leq \phi \leq \pi$, $|G| \leq 1$ if and only if

$$2\lambda a(1 + \lambda a) \leq 0,$$

or equivalently,

$$-1 \leq \lambda a \leq 0.$$

In conclusion, FTFS is linearly stable if $-1 \leq \lambda a \leq 0$. Notice that, in this case, the linear stability condition is the same as the CFL condition.

Example 15.3 Use von Neumann analysis to determine the stability of BTCS for the linear advection equation.

Solution BTCS for the linear advection equation is

$$u_i^{n+1} = u_i^n - \lambda a \frac{u_{i+1}^{n+1} - u_{i-1}^{n+1}}{2}.$$

Let $u_i^n = G^n e^{I\phi i}$. Then BTCS for the linear advection equation becomes

$$G^{n+1} e^{I\phi i} = G^n e^{I\phi i} - \frac{\lambda a}{2} G^{n+1} e^{I\phi(i+1)} + \frac{\lambda a}{2} G^{n+1} e^{I\phi(i-1)}.$$

Divide by $G^n e^{I\phi i}$ to obtain

$$G = 1 - \frac{\lambda a}{2} G e^{I\phi} + \frac{\lambda a}{2} G e^{-I\phi} = 1 - I\lambda a G \sin\phi.$$

Solve for G to find

$$G = \frac{1}{1 + I\lambda a \sin\phi}.$$

Then

$$|G|^2 = GG^* = \frac{1}{1 + I\lambda a \sin\phi} \frac{1}{1 - I\lambda a \sin\phi} = \frac{1}{1 + (\lambda a \sin\phi)^2},$$

where the asterisk superscript indicates complex conjugation. By definition:

$$(\text{Re}\, z + I \text{Im}\, z)^* = \text{Re}\, z - I \text{Im}\, z$$

for any complex number z. Then $|G| \leq 1$ for all $-\pi \leq \phi \leq \pi$ and for all λa. Then BTCS is *unconditionally stable.*

Example 15.4 Show that any method written in artificial viscosity form (14.3) is linearly stable if $(\lambda a)^2 \leq \lambda\epsilon \leq 1$. This indicates that instability may be caused by either too little or too much artificial viscosity. Assume that both ϵ and a are constant.

Solution For constant ϵ and a, the artificial viscosity form is

$$u_i^{n+1} = u_i^n - \frac{\lambda a}{2}\left(u_{i+1}^n - u_{i-1}^n\right) + \frac{\lambda\epsilon}{2}\left(u_{i+1}^n - 2u_i^n + u_{i-1}^n\right),$$

which allows only methods whose stencils contain the three points u_{i-1}^n, u_i^n, and u_{i+1}^n. This includes FTBS, FTFS, and FTCS applied to the linear advection equation. Let $u_i^n = G^n e^{I\phi i}$. Then

$$\begin{aligned} G &= 1 - \frac{\lambda a}{2}(e^{I\phi} - e^{-I\phi}) + \frac{\lambda\epsilon}{2}(e^{I\phi} - 2 + e^{-I\phi}) \\ &= 1 - \lambda\epsilon + \lambda\epsilon \cos\phi - I\lambda a \sin\phi \end{aligned}$$

and

$$|G|^2 = (1 - \lambda\epsilon + \lambda\epsilon\cos\phi)^2 + (\lambda a \sin\phi)^2.$$

The rest of this example is an exercise in freshman calculus. The maximum of a smooth function on a closed domain occurs at the endpoints of the domain or where the first derivative equals zero and the second derivative is less than or equal to zero. In this case, the maximum of $|G|^2$ occurs either at $\phi = 0$ or the endpoints $\pm\pi$. Notice that $|G|^2 = 1$ when $\phi = 0$ regardless of ϵ or a. Then the method is unstable unless all values near $\phi = 0$ are less than or equal to one, which implies that $\phi = 0$ must be a local maximum regardless of whether or not it is a global maximum. But if $\phi = 0$ is a local maximum then the second derivative at $\phi = 0$ must be less than or equal to zero, which is true if $(\lambda a)^2 \leq \lambda\epsilon$. Also, a simple calculation shows that $|G|^2 \leq 1$ at $\phi = \pm\pi$ if $0 \leq \lambda\epsilon \leq 1$. Combining these two conditions, we see that $(\lambda a)^2 \leq \lambda\epsilon \leq 1$, implying that all candidate maxima are less than or equal to one, and thus $|G|^2 \leq 1$; therefore the method is linearly stable.

Unfortunately, von Neumann stability analysis does not work for nonlinear methods. In particular, the principle of superposition no longer applies, so that different frequencies interact nonlinearly, making von Neumann analysis practically impossible. However, suppose that the method is *locally linearized*. For example, for a scalar conservation law with nonlinear flux function $f(u)$, FTFS is

$$u_i^{n+1} = u_i^n - \lambda\big(f\left(u_{i+1}^n\right) - f\left(u_i^n\right)\big).$$

This method can be locally linearized by replacing the true nonlinear flux function $f(u)$ by the locally linearized flux function $a_{i+1/2}^n u + const.$, where the slope $a_{i+1/2}^n$ is some average value of the wave speed $f'(u) = a(u)$ for u between u_i^n and u_{i+1}^n, such as, for example, $a_{i+1/2}^n = (a(u_{i+1}^n) + a(u_i^n))/2$. After this local linearization, FTFS becomes

$$u_i^{n+1} = u_i^n - \lambda a_{i+1/2}^n\big(u_{i+1}^n - u_i^n\big).$$

Von Neumann analysis proceeds as before with the result $-1 \leq \lambda a_{i+1/2}^n \leq 0$. Unfortunately, in general, locally linearized stability is neither necessary nor sufficient for true nonlinear stability. Nonlinear stability is considered in the next chapter.

Von Neumann analysis also applies to linear systems of equations. However, linear scalar analysis usually provides just as much information as linear systems analysis, since the scalar characteristic equations composing a linear system do not interact, and thus von Neumann analysis is rarely conducted on systems. See Section 3.6.2 in Anderson, Tannehill, and Pletcher (1984) for more details.

15.2 Alternatives to von Neumann Analysis

Besides linearity, von Neumann analysis requires constant grid spacing Δx and periodic solutions. However, there are other linear stability analyses that allow nonconstant grid spacings and nonperiodic solutions. For example, the *energy method* for linear stability analysis requires that the 2-norm error $\|e\|_2^2 = \sum(u(x_i, t^n) - u_i^n)^2$ should not increase with n. Unfortunately, this is usually extremely difficult to prove, even for relatively simple methods. A more practical alternative is the *matrix method* for linear stability analysis. Any linear finite-difference method can be written as follows:

$$u_i^{n+1} = \sum_{j=-K}^{K} a_{ij} u_{i+j}^n. \tag{15.13}$$

An equivalent vector–matrix formulation is

$$\mathbf{u}^{n+1} = A\mathbf{u}^n, \tag{15.14}$$

where

$$\mathbf{u}^n = \left[u_0^n | u_1^n | \cdots | u_N^n\right]^T.$$

The vector $\mathbf{u}^n$ is called the *vector of samples*. Do not confuse the vector of samples with the vector of conserved quantities. The matrix A is composed of the coefficients a_{ij}. For example, in Example 11.4, BTCS was written as $A^{-1}\mathbf{u}^{n+1} = \mathbf{u}^n$. Invert to find $\mathbf{u}^{n+1} = A\mathbf{u}^n$. Then

$$\begin{aligned}
\mathbf{u}^1 &= A\mathbf{u}^0, \\
\mathbf{u}^2 &= A\mathbf{u}^1 = A^2\mathbf{u}^0, \\
&\vdots \\
\mathbf{u}^n &= A\mathbf{u}^{n-1} = A^2\mathbf{u}^{n-2} = \cdots = A^n\mathbf{u}^0.
\end{aligned}$$

Suppose that $Q^{-1}AQ = \Lambda$ where Λ is a diagonal matrix of the characteristic values λ_i of A, as seen in Section 3.2. Then $Q^{-1}A^nQ = \Lambda^n$ where Λ^n is a diagonal matrix of λ_i^n, where all the ns in these expressions are exponents and not time indices. Then

$$Q^{-1}\mathbf{u}^n = \Lambda^n Q^{-1}\mathbf{u}^0.$$

Notice that λ_i^n either grows rapidly with n if $\lambda_i > 1$ or shrinks rapidly with n if $\lambda_i < 1$. Then Λ^n grows rapidly with n if $\lambda_i > 1$ for any i and shrinks rapidly with n if $\lambda_i < 1$ for all i. Let the *spectral radius* $\rho(A)$ be the largest characteristic value of A in absolute value. Assume that the method satisfies the CFL condition. Then using "unbounded growth" definitions as in the last section, a linear method is *linearly stable* if $\rho(A) < 1$, *neutrally linearly stable* if $\rho(A) = 1$, and *linearly unstable* if $\rho(A) > 1$. Unfortunately, finding the spectral radius of A may be difficult. Also notice that, unlike von Neumann analysis, this analysis does not say anything about dispersion and the resulting spurious oscillations, which is an important part of linear stability analysis under the "broad" definition. On the positive side, matrix stability analysis works in cases where Fourier series stability analysis does not. In particular, matrix stability analysis can register the effects of different types of boundary conditions, the number of grid points, and the variable spacing between grid points, all of which radically affect stability, whereas von Neumann analysis is restricted to periodic boundaries and constant grid spacings, as mentioned at the beginning of the chapter. For more on the energy method, the matrix method, and other alternatives to Fourier series analysis see, for example, Chapter 10 of Hirsch (1988). For more on linear stability in the presence of nonperiodic boundaries, see Section 19.1.

15.3 Modified Equations

Finite-difference methods solve linear advection equations *approximately*, but they solve modified linear advection equations *exactly*. In other words, modified equations choose the questions to fit the answers, just like the television quiz show Jeopardy. Modified equations are best explained by example.

Example 15.5 Find the modified equation for FTCS applied to the linear advection equation.

Solution FTCS for the linear advection equation is

$$\frac{u_i^{n+1} - u_i^n}{\Delta t} + a\frac{u_{i+1}^n - u_{i-1}^n}{2\Delta x} = 0.$$

By Taylor series we have

$$\frac{u_i^{n+1} - u_i^n}{\Delta t} = \frac{\partial u}{\partial t}(x_i, t^n) + \frac{\Delta t}{2}\frac{\partial^2 u}{\partial t^2}(x_i, t^n) + \frac{\Delta t^2}{6}\frac{\partial^3 u}{\partial t^3}(x_i, t^n) + O(\Delta t^3)$$

and

$$\frac{u_{i+1}^n - u_{i-1}^n}{2\Delta x} = \frac{\partial u}{\partial x}(x_i, t^n) + \frac{\Delta x^2}{6}\frac{\partial^3 u}{\partial x^3}(x_i, t^n) + O(\Delta x^4).$$

Then FTCS solves the following partial differential equation *exactly*:

$$\frac{\partial u}{\partial t} + a\frac{\partial u}{\partial x} = -\frac{a\Delta x^2}{6}\frac{\partial^3 u}{\partial x^3} - \frac{\Delta t}{2}\frac{\partial^2 u}{\partial t^2} - \frac{\Delta t^2}{6}\frac{\partial^3 u}{\partial t^3} + O(\Delta x^4) + O(\Delta t^3). \tag{15.15}$$

This is called a *modified equation.*

To yield useful information, the time derivatives on the right hand of the modified equation must be replaced by spatial derivatives. In particular, it will be shown that

$$\frac{\partial^2 u}{\partial t^2} = a^2\frac{\partial^2 u}{\partial x^2} + a^3\Delta t\frac{\partial^3 u}{\partial x^3} + O(\Delta x^2) + O(\Delta t^2), \tag{15.16}$$

$$\frac{\partial^3 u}{\partial t^3} = -a^3\frac{\partial^3 u}{\partial x^3} + O(\Delta x^2) + O(\Delta t). \tag{15.17}$$

The proof involves repeatedly differentiating the modified equation and substituting the results back into the modified equation or derivatives of the modified equation. This process is sometimes called the *Cauchy–Kowalewski* procedure. To start with, take the partial derivatives of the modified equation with respect to x and t:

$$\frac{\partial^2 u}{\partial t^2} + a\frac{\partial^2 u}{\partial t\partial x} = -\frac{a\Delta x^2}{6}\frac{\partial^4 u}{\partial t\partial x^3} - \frac{\Delta t}{2}\frac{\partial^3 u}{\partial t^3} - \frac{\Delta t^2}{6}\frac{\partial^4 u}{\partial t^4} + O(\Delta x^4) + O(\Delta t^3), \tag{15.18}$$

$$\frac{\partial^2 u}{\partial x\partial t} + a\frac{\partial^2 u}{\partial x^2} = -\frac{a\Delta x^2}{6}\frac{\partial^4 u}{\partial x^4} - \frac{\Delta t}{2}\frac{\partial^3 u}{\partial x\partial t^2} - \frac{\Delta t^2}{6}\frac{\partial^4 u}{\partial x\partial t^3} + O(\Delta x^4) + O(\Delta t^3). \tag{15.19}$$

Multiply Equation (15.19) by a and subtract from Equation (15.18) to obtain

$$\frac{\partial^2 u}{\partial t^2} = a^2\frac{\partial^2 u}{\partial x^2} - \frac{\Delta t}{2}\frac{\partial^3 u}{\partial t^3} + \frac{a\Delta t}{2}\frac{\partial^3 u}{\partial x\partial t^2} + O(\Delta x^2) + O(\Delta t^2), \tag{15.20}$$

where the several higher-order terms are dropped, since they will not be needed. In Equation (15.20), there is one pure x derivative, one pure t derivative, and one mixed x–t derivative. To eliminate the mixed derivative, take the partial derivative of Equation (15.20) with respect to x to obtain

$$\frac{\partial^3 u}{\partial x\partial t^2} = a^2\frac{\partial^3 u}{\partial x^3} + O(\Delta t) + O(\Delta x^2), \tag{15.21}$$

where the higher-order terms are again dropped, since they will not be needed. Substitute Equation (15.21) into Equation (15.20) to get:

$$\frac{\partial^2 u}{\partial t^2} = a^2 \frac{\partial^2 u}{\partial x^2} - \frac{\Delta t}{2}\frac{\partial^3 u}{\partial t^3} + \frac{a^3 \Delta t}{2}\frac{\partial^3 u}{\partial x^3} + O(\Delta x^2) + O(\Delta t^2). \tag{15.22}$$

In Equation (15.22), there is one pure t derivative and two pure x derivatives. To eliminate the t derivative, take the second partial derivative of the modified Equation (15.15) with respect to t to obtain

$$\frac{\partial^3 u}{\partial t^3} = -a \frac{\partial}{\partial x}\left(\frac{\partial^2 u}{\partial t^2}\right) - \frac{a\Delta x^2}{6}\frac{\partial^4 u}{\partial t^2 \partial x^3} - \frac{\Delta t}{2}\frac{\partial^4 u}{\partial t^4} - \frac{\Delta t^2}{6}\frac{\partial^5 u}{\partial t^5} + O(\Delta x^4) + O(\Delta t^3). \tag{15.23}$$

Finally, substitute Equation (15.22) into Equation (15.23) to obtain Equation (15.17), and substitute (15.17) into (15.22) to obtain Equation (15.16).

Now substitute Equations (15.16) and (15.17) into the modified Equation (15.15) to obtain the final result

$$\frac{\partial u}{\partial t} + a\frac{\partial u}{\partial x} = -\frac{a^2 \Delta t}{2}\frac{\partial^2 u}{\partial x^2} - \frac{a\Delta x^2}{6}(1 - 2(\lambda a)^2)\frac{\partial^3 u}{\partial x^3} + O(\Delta t^3) + O(\Delta x^3) + O(\Delta x^2 \Delta t). \tag{15.24}$$

The right-hand side of the modified equation is written entirely in terms of x derivatives. By the discussion of Section 14.2, the second derivative on the right-hand side is second-order implicit artificial viscosity. Since the coefficient of the second-order artificial viscosity is negative, rather than positive, the artificial viscosity is destabilizing rather than stabilizing, or *antidissipative*. The third derivative on the right-hand side is third-order implicit *artificial dispersion*. Whereas artificial dissipation affects the amplitudes of the sinusoids in a Fourier series representation, artificial dispersion affects the speed. Notice that artificial dispersion vanishes for the special CFL numbers $\lambda a = \pm 1/\sqrt{2}$. The artificial antidissipation never vanishes except in the limit $\Delta t \to 0$.

Dissipation reduces the amplitudes of sinusoids in a Fourier series. The following derivatives are generally dissipative:

$$+\frac{\partial^2 u}{\partial x^2}, -\frac{\partial^4 u}{\partial x^4}, +\frac{\partial^6 u}{\partial x^6}, -\frac{\partial^8 u}{\partial x^8}, \ldots.$$

By contrast, antidissipation increases the amplitudes of sinusoids in a Fourier series. The following derivatives are generally antidissipative:

$$-\frac{\partial^2 u}{\partial x^2}, +\frac{\partial^4 u}{\partial x^4}, -\frac{\partial^6 u}{\partial x^6}, +\frac{\partial^8 u}{\partial x^8}, \ldots.$$

Finally, dispersion affects the speed of sinusoids in a discrete Fourier series, causing them to lead or lag depending on the sign of the coefficient of dispersion and the order of the dispersive derivative. The following derivatives are generally dispersive:

$$\pm\frac{\partial^3 u}{\partial x^3}, \pm\frac{\partial^5 u}{\partial x^5}, \pm\frac{\partial^7 u}{\partial x^7}, \pm\frac{\partial^9 u}{\partial x^9}, \ldots.$$

Dispersion can cause different frequencies to separate, resulting in oscillations. Dispersion is especially dramatic near jump discontinuities, where the full range of frequencies occur simultaneously in significant proportions. Explicit artificial dissipation (also known as explicit artificial viscosity) can control or cancel implicit artificial dissipation. Similarly, explicit artificial dispersion can control or cancel implicit artificial dispersion. Artificial dissipation takes the form of even-order differences approximating even-order derivatives. Similarly, artificial dispersion takes the form of odd-order differences approximating odd-order derivatives.

As mentioned at the end of Section 12.1, stability tends to decrease as formal accuracy increases. Modified equations help to justify this observation. Looking at the right-hand side of a modified equation, suppose that the first term is an even derivative, the second term is an odd derivative, the third term is an even derivative, and so forth. Assume that the first term is dissipative and that the method is consequently stable. The first term is the largest source of error, and thus it must be reduced or eliminated to increase the formal order of accuracy. However, if the first term is reduced, the second term will dominate the error. The second term is dispersive and thus the error tends to become oscillatory, especially near jump discontinuities. Furthermore, as the first dissipative term decreases, the scheme's stability becomes heavily influenced by the third term, which may be antidissipative, so that the scheme is unstable. In conclusion, large dissipative terms may be preferable to smaller antidissipative or dispersive terms. Put another way, increasing formal order of accuracy may decrease stability and thus may decrease the overall accuracy.

The right-hand side of a modified equation is sometimes called *local truncation error*. This definition of the local truncation error is closely related to the definition of the local truncation error seen in Equation (11.53). By either definition, a numerical method is called *consistent* if the local truncation error goes to zero as $\Delta x \to 0$ and $\Delta t \to 0$, as discussed in Subsection 11.2.2. Notice that this is generally not true if the solution is discontinuous, since the derivatives on the right-hand side of the modified equation will be infinite at jump discontinuities – in other words, numerical methods are generally *not* consistent for discontinuous solutions. Unfortunately, modified equations are mainly useful for linear methods. The modified equations for nonlinear methods contain nonlinear terms such as $(\partial u/\partial x)^2$ and $(\partial u/\partial x)(\partial u/\partial t)$ which have unpredictable effects, unlike the purely dispersive or purely dissipative derivatives seen here.

15.4 Convergence and Linear Stability

With some minimal sensible assumptions, linear stability as defined above guarantees convergence to the true solution as $\Delta x \to 0$ and $\Delta t \to 0$, at least for smooth solutions. Assume that the linear approximate equation is consistent with the linear advection equation, in the sense that the local truncation error as defined in Subsection 11.2.2 or as in the previous section goes to zero as $\Delta x \to 0$ and $\Delta t \to 0$. Also, assume that the linear advection equation has *well-posed* boundary and initial conditions, as discussed in Sections 3.1 and 3.2. Then *a consistent linear finite-difference approximation to a well-posed linear problem is convergent if and only if it is linearly stable*. Hence, for linear methods, the "unbounded increase" notion of stability is equivalent to the "convergence" notion of stability. This is known as the *Lax Equivalence Theorem*. Recall that, by one definition, the local truncation error is the error incurred after a single time step. Then a method is consistent if the error incurred after a single time step always goes to zero in the limit of an infinitesimal time step. The Lax Equivalence Theorem says that if the error goes to zero in the limit of a single infinitesimal time step, then the error goes to zero in the limit of infinitely many infinitesimal time steps if and only if the numerical method is linearly stable. Like so many of the stability properties seen in this chapter, the Lax Equivalence Theorem is a special property of linear methods.

As an often overlooked point, for most numerical methods, the truncation error does not go to zero in the presence of jump discontinuities. In other words, most numerical approximations are not consistent for nonsmooth solutions! Then, in essence, *the Lax Equivalence Theorem only applies to smooth solutions – linear stability and convergence are distinct for nonsmooth solutions.* Skeptical readers should look ahead to the Lax–Wendroff method, seen in Section 17.2. The Lax–Wendroff method is linearly stable when applied to the linear advection equation. Also, the Lax–Wendroff method is consistent with the linear advection equation for smooth solutions. Then, as expected by the Lax Equivalence Theorem, the Lax–Wendroff method converges to the true solution as $\Delta x \to 0$ and $\Delta t \to 0$ for smooth solutions. However, for solutions with jump discontinuities, the Lax–Wendroff method does not converge; for example, in the linear advection of a square wave, oscillations originating on the jump discontinuities spread slowly as time increases, as seen in Figure 17.9.

References

Anderson, D. A., Tannehill, J. C., and Pletcher, R. H. 1984. *Computational Fluid Mechanics and Heat Transfer*, New York: Hemisphere, Sections 3.3 and 3.6.

Hirsch, C. 1988. *Numerical Computation of Internal and External Flows, Volume 1: Fundamentals of Numerical Discretization*, Chichester: Wiley, Chapters 8, 9, and 10.

Problems

15.1 Apply von Neumann analysis to FTBS for the linear advection equation.
(a) Show that $G = 1 - \lambda a + \lambda a\, e^{I\phi}$.
(b) Based on the results of part (a), show that FTBS is linearly stable if $0 \leq \lambda a \leq 1$ and unstable otherwise. How does the linear stability condition compare with the CFL condition?

15.2 Apply von Neumann stability to CTCS for the linear advection equation.
(a) Show that $G = -I\lambda a \sin\phi \pm \sqrt{1 - (\lambda a)^2 \sin^2\phi}$.
(b) Based on the results of part (a), show that CTCS is neutrally linearly stable if $|\lambda a| \leq 1$ and unstable otherwise. How does the linear stability condition compare with the CFL condition?

15.3 Apply von Neumann stability to CTBS for the linear advection equation.
(a) Show that $G = -z \pm \sqrt{z^2 + 1}$, where $z = \lambda a(1 - e^{-I\phi})$. Hint: The quadratic formula applies to complex numbers.
(b) Based on the results of part (a), show that CTBS is unconditionally linearly unstable. As a hint, multiply the two solutions found in part (a). Use this product to show that one of the two solutions found in part (a) is always greater than or equal to one, while the other is always less than or equal to one.

15.4 Apply von Neumann stability to BTBS for the linear advection equation.
(a) Show that $G = 1/(1 + \lambda a - \lambda a e^{-I\phi})$.

(b) Based on the results of part (a), show that BTBS is linearly stable for $\lambda a \geq 0$ or $\lambda a \leq -1$. How does the linear stability condition compare with the CFL condition? In particular, what do you think of the negative CFL numbers allowed by the linear stability analysis?

15.5 Apply von Neumann analysis to the following method:

$$u_i^{n+1} = u_i^n - \frac{\lambda a}{12}\left(u_{i+2}^n - 8u_{i+1}^n + 8u_{i-1}^n - u_{i-2}^n\right).$$

(a) Show that $G = 1 - \frac{\lambda a}{3}(4 - \cos\phi)\sin\phi I$.

(b) Based on the results of part (a), show that this method is unconditionally linearly unstable.

(c) This method is centered, just like FTCS. Does the wider centered stencil make the method more or less stable relative to FTCS? In other words, is $|G|^2$ for this method greater than or less than $|G|^2$ for FTCS, as given by Equation (15.7)?

15.6 Consider the following numerical method:

$$u_{i+1}^n = u_i^n - \lambda\left(\hat{f}_{i+1/2}^n - \hat{f}_{i-1/2}^n\right),$$

where

$$\hat{f}_{i+1/2}^n = \frac{1}{2}\left(f\left(u_{i+1}^n\right) + f\left(u_i^n\right)\right) + \frac{1}{2}\delta\left|a_{i+1/2}^n\right|\left(u_{i+2}^n - 3u_{i+1}^n + 3u_i^n - u_{i-1}^n\right)$$

and where $\delta = const.$ and $a_{i+1/2}^n$ is any average wave speed. Notice that this method is central differences plus fourth-order artificial viscosity; this method will be used in Section 22.3. Perform a von Neumann stability analysis. Write $|G|^2$ in the simplest possible form and plot $|G|^2$ versus $-\pi \leq \phi \leq \pi$ for

(a) $\lambda a = 0.8$, $\delta = 1/2$, (b) $\lambda a = 0.8$, $\delta = 1/4$,
(c) $\lambda a = 0.4$, $\delta = 1/2$, (d) $\lambda a = 0.4$, $\delta = 1/4$.

15.7 Consider the following numerical method:

$$u_{i+1}^n = u_i^n - \lambda\left(\hat{f}_{i+1/2}^n - \hat{f}_{i-1/2}^n\right),$$

where

$$\hat{f}_{i+1/2}^n = \frac{1}{2}\left(f\left(u_{i+1}^n\right) + f\left(u_i^n\right)\right) - \frac{1}{2}\left|a_{i+1/2}^n\right|\left[\frac{1}{2}\left(u_{i+1}^n - u_i^n\right) - \frac{1}{2}\delta\left(u_{i+2}^n - 3u_{i+1}^n + 3u_i^n - u_{i-1}^n\right)\right]$$

and where $\delta = const.$ and $a_{i+1/2}^n$ is any average wave speed. This method is central differences plus second- and fourth-order artificial viscosity; a similar method will be used in Section 22.3. Perform a von Neumann stability analysis. Write $|G|^2$ in the simplest possible form and plot $|G|^2$ versus $-\pi \leq \phi \leq \pi$ for

(a) $\lambda a = 0.8$, $\delta = 1/2$, (b) $\lambda a = 0.8$, $\delta = 1/4$,
(c) $\lambda a = 0.4$, $\delta = 1/2$, (d) $\lambda a = 0.4$, $\delta = 1/4$.

15.8 Write FTFS for the linear advection equation in artificial viscosity form. Use the results of Example 15.4 to find linear stability bounds on the CFL condition. Do your results agree with those of Example 15.2?

15.9 Consider the Boris and Book first-order upwind method, as described in Problem 11.7. Show that this method is linearly stable for $\lambda|a| \leq \sqrt{3}/2 \approx 0.866$. You may wish to use the results of Example 15.4.

15.10 Consider the following finite-difference approximation:

$$u_i^{n+1} = u_i^n - \lambda u_i^n \left(u_i^n - u_{i-1}^n\right).$$

What does von Neumann stability analysis say about this method? Be careful.

15.11 Show that FTCS for the linear advection equation is not sensitive to small disturbances, despite the fact that it is highly unstable. More specifically, consider the following linear advection problem on a periodic domain $[-1, 1]$:

$$\frac{\partial u}{\partial t} + \frac{\partial u}{\partial x} = 0,$$

$$u(x,0) = \begin{cases} 1 & |x| \le 1/3, \\ 0 & |x| > 1/3. \end{cases}$$

(a) Approximate $u(x, 2)$ using FTCS with 20 cells and $\lambda = \Delta t/\Delta x = 0.8$. Verify that your results agree with Figure 11.6.

(b) Perturb your initial conditions using a $2\Delta x$-wave with an amplitude of 0.1. That is, alternatively add and subtract 0.1 from the samples in the initial conditions. Make sure that the initial conditions remain periodic. Approximate $u(x, 2)$ using FTCS with 20 cells and $\lambda = \Delta t/\Delta x = 0.8$. Compare the results with those of part (a). The initial conditions differ by 10% in the ∞-norm. Do the results for $u(x, 2)$ differ by more or less than 10% in the ∞-norm? That is, have the relative differences grown or shrunk?

(c) Repeat part (b), except this time use a $10\Delta x$-wave. Is the relative difference in $u(x, 2)$ sensitive to the frequency of the disturbance in the initial conditions?

15.12 Consider the leapfrog method applied to the linear advection equation.

(a) Show that the first term on the right-hand side of the modified equation is

$$\frac{\partial u}{\partial t} + a\frac{\partial u}{\partial x} = \frac{a\Delta x^2}{6}((\lambda a)^2 - 1)\frac{\partial^3 u}{\partial t^3} + \cdots.$$

What does this indicate about the dissipation, dispersion, and stability of the leapfrog method? Are there any special CFL numbers where the first term on the right-hand side vanishes?

(b) Show that the first two terms on the right-hand side of the modified equation are

$$\frac{\partial u}{\partial t} + a\frac{\partial u}{\partial x} = \frac{a\Delta x^2}{6}((\lambda a)^2 - 1)\frac{\partial^3 u}{\partial t^3} - \frac{a\Delta x^4}{120}(9(\lambda a)^4 - 10(\lambda a)^2 + 1)\frac{\partial^5 u}{\partial t^5} + \cdots.$$

What do these two terms indicate about the dissipation, dispersion, and stability of the leapfrog method? Are there any special CFL numbers where either or both of the first two terms on the right-hand side vanish?

CHAPTER 16

Nonlinear Stability

16.0 Introduction

The last chapter concerned linear stability. This chapter concerns nonlinear stability. Linear stability theory is classic, dating to the late 1940s. The study of nonlinear stability is far newer. Starting with papers by Boris and Book (1973) and Van Leer (1974), nonlinear stability theory developed over roughly the next fifteen years. Although nonlinear stability theory may someday undergo major revision, no significant new developments have appeared in the literature since the late 1980s. Thus after a period of intensive development, nonlinear stability theory has plateaued, at least temporarily.

To keep the discussion within reasonable bounds, this chapter concerns only explicit forward-time finite-difference approximations. Furthermore, this chapter concerns mainly one-dimensional scalar conservation laws on infinite spatial domains. As far as more realistic scenarios go, the Euler equations are discussed briefly in Section 16.12, multidimensions are discussed briefly at the end of this introduction, and solid and far-field boundaries are discussed briefly in Section 19.1. Unlike solid and far-field boundaries, the periodic boundaries in Chapter 15 and the infinite boundaries in this chapter do not pose any difficult stability issues.

The last chapter began with a general introduction to stability, both linear and nonlinear. Any impatient readers who skipped the last chapter should go back and read Section 15.0. While linear and nonlinear stability share the same broad philosophical principles, especially the emphasis on spurious oscillations, the details are completely different. Thus, except for its introduction, the last chapter is not a prerequisite for this chapter. One of the more important nonlinear stability conditions relies heavily on the wave speed split form described in Section 13.5. Readers who skipped this section should go back and read it. Also, most of the nonlinear stability conditions discussed in this chapter rely heavily on the principles of waveform preservation, waveform destruction, and waveform creation, as described in Section 4.11. Again, any readers who skipped this section should go back and read it.

Perhaps the best known nonlinear stability condition is the *total variation diminishing* (TVD) condition, suggested by Harten in 1983. People have several common misconceptions about TVD. First, the term "TVD" commonly refers to three distinct nonlinear stability conditions – the actual TVD condition and two other stronger conditions, which imply TVD. Second, some people believe that TVD completely eliminates all spurious oscillations for all Δx and Δt. It does not. In fact, the actual TVD condition may allow large spurious oscillations; of the two other conditions commonly called TVD conditions, only the strongest one truly eliminates spurious oscillations, and then only for scalar model problems on infinite spatial domains away from sonic points. Third, although the term TVD is widely used, few outside of the mathematics community recognize that TVD refers to stability.

This chapter will only address nonlinear stability conditions; it will not discuss how numerical methods actually achieve nonlinear stability. Suffice it to say that *TVD methods* typically have two ingredients: first, they involve flux averaging as introduced in Section 13.3; second, they carefully exploit the freedom of flux averaging to enforce nonlinear stability conditions. A few TVD methods, such as first-order upwind methods, do not need to use flux averaging – they achieve nonlinear stability by good fortune rather than by design.

As discussed in Section 15.0, this book defines stability mainly in terms of spurious oscillations and overshoots. The last chapter described spurious oscillations in linear methods using Fourier series. However, the Euler equations and most scalar conservation laws are highly nonlinear. Numerical approximations should contain at least as much nonlinearity as the governing equations. Most modern numerical methods add their own nonlinearity, making them highly nonlinear even for linear governing equations. Without the principle of superposition, different frequencies in a discrete Fourier series interact nonlinearly, making any sort of von Neumann analysis intractable. Although nonlinear stability analyses still focus on spurious oscillations and overshoots, they must take a completely different tack.

To further expound on the connections between linear stability analysis as seen in the last chapter and nonlinear stability analysis as seen in this chapter, recall that linear stability analysis naturally uses the "unbounded growth" definition of stability. In other words, linear stability analysis requires only that the solution should not "blow up" or, more specifically, that each component in the Fourier series representation of the solution should not increase to infinity. Nonlinear stability analysis may use a similar notion – while it never attempts to decompose the solution into sinusoidal components, as in a Fourier series, nonlinear stability conditions can require that the overall amount of oscillation, as measured by the total variation, remains bounded; this is known as the *total variation bounded* (*TVB*) condition. In linear stability analysis, the special properties of linear equations imply that if the solution does not blow up then it must monotonically shrink or stay the same. More specifically, if each component in the Fourier series representation of the solution remains bounded, then each component in the Fourier series representation either shrinks by the same amount or remains exactly the same at each time step. Nonlinear stability analysis may use a similar notion – while it never attempts to decompose the solution into sinusoidal components, as in a Fourier series, it can require that the overall amount of oscillation, as measured by the total variation, either shrinks or remains exactly that same at each time step; this is known as the total variation diminishing (TVD) condition. Not "blowing up" and "shrinking" are equivalent notions for linear equations, at least when viewed in terms of Fourier series coefficients. However, these notions are distinct when viewed in terms of total variation, for either linear or especially nonlinear equations. In particular, TVD implies TVB but not necessarily vice versa.

After determining whether or not the solution blows up, as a secondary goal, von Neumann stability analysis investigates spurious oscillations caused by phase errors rather than amplitude errors in the Fourier series representation. Phase errors cause dispersion and spurious oscillations, a form of instability if you care to view it that way, although dispersion never causes the solution to blow up. Nonlinear stability analysis may use similar notions – it can investigate any sort of spurious oscillation, regardless of whether or not it causes the solution to blow up, and regardless of how it affects the total size of the oscillations as measured by the total variation. Indeed, in most cases, it is impossible to enforce TVB or TVD directly. Instead, most nonlinear stability analyses actually use much stronger conditions that address the individual maxima and minima found in spurious oscillations, rather than the overall growth of the spurious oscillations as measured by the total variation.

These conditions include the positivity condition and especially the upwind range condition, both of which imply TVB and TVD.

As we have said, rather than fitting maxima and minima into regular periodic oscillations, as in Fourier series, nonlinear stability analysis focuses on the individual maxima and minima and sums thereof. In fact, maxima and minima are the most important features of any function. Here and throughout the chapter, the terms "maxima" and "minima" include "suprema" and "infema," such as horizontal asymptotes. As discussed following Equation (6.8), there is a technical distinction between the values that a function actually attains versus the values that a function draws arbitrarily close to. However, it is extremely awkward to say "maxima, minima, suprema, and infema." Thus, this chapter will take the liberty of saying "maxima and minima" or simply "extrema" to cover all four terms. To continue, as any freshman calculus student knows, a function can be plotted roughly knowing only its maxima and minima. Of course, roots, inflection points, randomly chosen samples, and so on are also extremely helpful, but not as helpful as maxima and minima. Without maxima and minima, or at least estimates thereof, any graph of the function will not look much like it should. Sometimes one knows that a function has a certain parametric form. For example, sometimes one knows that the function is an Nth-order polynomial or an Nth-order trigonometric series, as discussed in Chapter 8. In such cases, almost any sort of further information allows you to determine the constants in the parametric form and reconstruct the function perfectly, including the maxima and minima. However, the functions in gasdynamics do not generally assume any predictable parametric forms. Thus, for the functions in gasdynamics, or for any arbitrary function, the most important pieces of information are the maxima and minima.

Von Neumann analysis as seen in the last chapter applies to any linear method. However, the nonlinear stability conditions in this chapter exploit properties that are, to varying degrees, specific to scalar conservation laws and the one-dimensional Euler equations. The maxima and minima in the solutions to scalar conservation laws and the one-dimensional Euler equations behave in simple, predictable, and exploitable fashions. In particular, according to the discussion of Section 4.11, any waveform-preserving portions of the solution preserve maxima and minima. Of course, maxima and minima may change positions, but their values never change, except near shocks. In contrast, waveform-destroying portions of the solutions, near shocks, may reduce or eliminate maxima and increase or eliminate minima. Besides waveform preservation and waveform destruction, one must also consider waveform creation. For scalar conservation laws, waveform creation never creates new maxima or minima – created waveforms are purely monotone increasing or monotone decreasing. *In summary, for scalar conservation laws, existing maxima either stay the same or decrease near shocks, existing minima either stay the same or increase near shocks, and no new maxima and minima are ever created.* This is called the *range diminishing* property. The same conclusions hold for the characteristic variables in the one-dimensional Euler equations, except that new maxima and minima may appear in the waveforms created by jump discontinuities in the initial conditions, intersections between jump discontinuities, and reflections of jump discontinuities from solid surfaces. This proves more annoying in theory than in practice.

Many of the nonlinear stability conditions seen in this chapter stem directly from the range diminishing property and thus indirectly from the properties of waveform preservation, waveform destruction, and waveform creation as found in scalar conservation laws and the one-dimensional Euler equations. In particular, the range diminishing property implies TVD and TVB. However, since TVD and TVB are much weaker properties than the range diminishing property, they may apply in circumstances where the range diminishing condition does not; remember that scalar conservation laws and the one-dimensional Euler equations are only model problems for more realistic equations, and we would like to find principles that apply generally, not just in model problems, making the TVD and TVB conditions more widely usable than range diminishing.

In one interpretation, the CFL condition is a necessary nonlinear stability condition, as discussed in Chapter 12. Unlike most of the nonlinear stability conditions described in this chapter, the CFL condition has no direct relationship to spurious maxima and minima. The nonlinear stability conditions in this chapter often do not imply the CFL condition and, in fact, generally have no direct connection to the CFL condition, since they concern maxima and minima whereas the CFL condition does not. Any nonlinear stability condition that does not imply the CFL condition is certainly not, by itself, a sufficient stability condition. Thus most of the nonlinear stability conditions in this chapter are not, in and of themselves, sufficient for nonlinear stability. Furthermore, they may not, in isolation, be necessary for nonlinear stability.

Nonlinear stability is especially vital at shocks and contacts, which tend to create large spurious oscillations in otherwise stable and monotone solutions. Of course, conservation is also vital at shocks and contacts, in order to locate such features correctly, as discussed in Chapter 11. In one sense, conservation addresses the phase (location) of the solution, while nonlinear stability addresses the complementary issue of the amplitude (shape) of the solution. This naturally leads us to consider the combined effects of nonlinear stability conditions and conservation. In fact, there is an established body of theory describing the effects of conservation combined with nonlinear stability conditions in the convergence limit $\Delta x \to 0$ and $\Delta t \to 0$, as discussed briefly in Section 16.11. Unfortunately, there are far fewer results on the combined effects of conservation and nonlinear stability for ordinary values of Δx and Δt. As a rough rule of thumb, for ordinary values of Δx and Δt, conservation accentuates both the positive and negative effects of nonlinear stability conditions. For example, if the stability condition reduces the order of accuracy, the combination of stability and conservation may further reduce the order of accuracy.

Unfortunately, no one has yet discovered the perfect nonlinear stability condition. If they had, this chapter would be considerably shorter: it would describe the perfect condition and be over. Instead, this chapter considers *nine* imperfect conditions, each with their own strengths and weakness. Some of these conditions, especially the weakest, are useful mostly in the limit $\Delta x \to 0$ and $\Delta t \to 0$; on the positive side, these weaker conditions may apply to all sorts of equations, not just to scalar conservation laws or to the one-dimensional Euler equations. Others of these conditions limit the numerical solution in entirely unphysical ways. For example, some of the stronger conditions cause first- or second-order *clipping errors* at extrema, whereas the strongest limit the order of accuracy to one throughout the entire solution.

In summary, although the stronger nonlinear stability conditions restrict the formal order of accuracy, especially at extrema, they also may effectively reduce or completely eliminate spurious oscillations and overshoots; by contrast, the weaker nonlinear stability conditions place little or no restrictions on the formal order of accuracy at extrema or elsewhere, but they may allow large spurious oscillations and overshoots. To some extent, there is thus a trade-off between two types of errors, and one must choose between oscillatory errors in monotone regions or clipping errors at extrema.

This book mainly concerns one-dimensional flows. However, before continuing, let us say a few words about nonlinear stability in multidimensions. Unfortunately, in general, there are still many unresolved issues in the theory of multidimensional gasdynamics.

Fortunately, putting aside theory, there are established heuristics, usually heavily based on one-dimensional concepts, that work well enough in practice. Nonlinear stability is no exception. Recall that the one-dimensional stability analysis, as seen in this book, uses spurious oscillations as a fundamental unifying principle. As one primary obstacle to multidimensional stability analysis, a function can oscillate in multidimensions without creating maxima or minima. For example, imagine a wavy surface, like a corrugated tin roof. If the ridges of the surface are kept perfectly horizontal, then every oscillatory wave in the surface is either a maximum or a minimum. However, if the surface is tilted, then only the ridges on the highest and lowest ends of the surface represent true maxima and minima. This poses a dilemma: should we attempt to control all oscillations, or just those oscillations that cause maxima and minima? Goodman and LeVeque (1985) proposed a multidimensional definition of TVD that accounts for all of the oscillations and not just extrema; unfortunately, they found that even weak limits on such oscillations limited the order of accuracy to one. Laney and Caughey (1991a, 1991b, 1991c) proposed another multidimensional definition of TVD that only controls true multidimensional extrema; although such definitions do not restrict the order of accuracy away from extrema, they might allow large multidimensional oscillations of the sort that do not cause extrema.

As the other primary obstacle to nonlinear stability theory in multidimensions, the Euler equations are no longer nonoscillatory in multidimensions, at least in many regions of flow; in other words, the multidimensional characteristic variables do not necessarily share the nonoscillatory properties of the one-dimensional characteristic variables, as exploited in this chapter. Even if they were completely nonoscillatory, there is always the issue of how to access multidimensional characteristic variables to enforce nonoscillatory nonlinear stability conditions. The classic one-dimensional characteristic access techniques – flux splitting, wave speed splitting, and Riemann solvers – do not extend in any obvious truly multidimensional way. Thus at this point, nonlinear stability theory in multidimensions remains a work in progress. However, as always, even when the theory is lacking, there are practical procedures that seem to work well enough. Thus even though the nonlinear analysis seen in this chapter is strictly one dimensional in theory, it still yields major improvements in multidimensional methods in practice.

Many of the results in this chapter require mathematical proofs. To avoid disrupting the flow of the discussion, several of the longer proofs are relegated to Section 16.13 at the end.

16.1 Monotonicity Preservation

The solutions of scalar conservation laws on infinite spatial domains are *monotonicity preserving*, meaning that if the initial conditions $u(x, 0)$ are monotone increasing, the solution $u(x, t)$ is monotone increasing for all time; and if the initial conditions are monotone decreasing, the solution is monotone decreasing for all time. As with many of the nonlinear stability conditions in this chapter, monotonicity preservation is a consequence of waveform preservation, waveform destruction, and waveform creation, as described in Section 4.11.

Suppose that a numerical approximation inherits monotonicity preservation. Then, if the initial conditions are monotone, monotonicity-preserving methods do not allow spurious oscillations and, in this sense, are stable. Monotonicity preservation was first suggested by Godunov (1959). However, overall, monotonicity preservation is a poor stability condition for the following reasons:

- Monotonicity preservation does not address the stability of nonmonotone solutions.

- Most attempts at enforcing monotonicity preservation end up enforcing much stronger nonlinear stability conditions, such as the positivity condition discussed in Section 16.4.
- Monotonicity preservation does not allow even small benign oscillations in monotone solutions. Small oscillations do not necessarily indicate serious instability, and attempts to purge all oscillatory errors, however small, may cause much larger nonoscillatory errors.

These results imply that monotonicity preservation is too weak in some situations and too strong in others. As one specific example of the effects of monotonicity preservation, consider linear methods. (Nonlinear stability conditions such as monotonicity preservation apply to linear as well as nonlinear methods. Of course, the reverse is not true – linear stability conditions do not apply to nonlinear methods, except possibly locally and approximately.) As it turns out, *linear monotonicity-preserving methods are first-order accurate, at best*. This is called *Godunov's theorem*. Unfortunately, first-order accuracy is not enough for most practical computations; second-order accuracy is usually minimal. Because of this, Godunov's theorem is often used to justify *inherently nonlinear* methods, which are nonlinear even when the governing equations are linear. However, Godunov's theorem is possibly better used as an argument against monotonicity preservation – a more relaxed policy towards minor oscillations would reduce or avoid the order-of-accuracy sacrifice. But whatever its drawbacks, monotonicity preservation is an established element of modern nonlinear stability theory. In fact, all but two of the nonlinear stability conditions considered in the remainder of the chapter imply monotonicity preservation.

16.2 Total Variation Diminishing (TVD)

As its greatest drawback, monotonicity preservation fails to address nonmonotone solutions. This section concerns the *total variation diminishing* (*TVD*) condition, first proposed by Harten (1983). In essence, TVD is the smallest possible step beyond monotonicity preservation. In other words, TVD is the weakest possible condition that implies monotonicity preservation and yet addresses the stability of both monotone and nonmonotone solutions.

Total variation is a classic concept in real and functional analysis, as seen in any number of mathematical textbooks; for example, see Royden (1968). The most general definition of total variation covers all sorts of strange and arbitrary functions. However, for the sorts of functions seen in computational gasdynamics, the *total variation* of the exact solution may be defined as follows:

$$TV(u(\cdot, t)) = \sup_{\text{all possible } x_i} \sum_{i=-\infty}^{\infty} |u(x_{i+1}, t) - u(x_i, t)|, \tag{16.1}$$

where this "sup" notation means that the supremum is taken over all possible sets of samples taken from the function's infinite domain. As in Equation (6.8), most readers should mentally replace "sup" by "max." This chapter will hereafter take the liberty of including suprema under the heading of maxima and infema under the heading of minima.

To better understand the meaning of total variation, consider the following result:

♦ *The total variation of a function on an infinite domain is a sum of extrema, maxima counted positively and minima counted negatively. The two infinite boundaries are always extrema and they both count once; every other extrema counts twice.* (16.2)

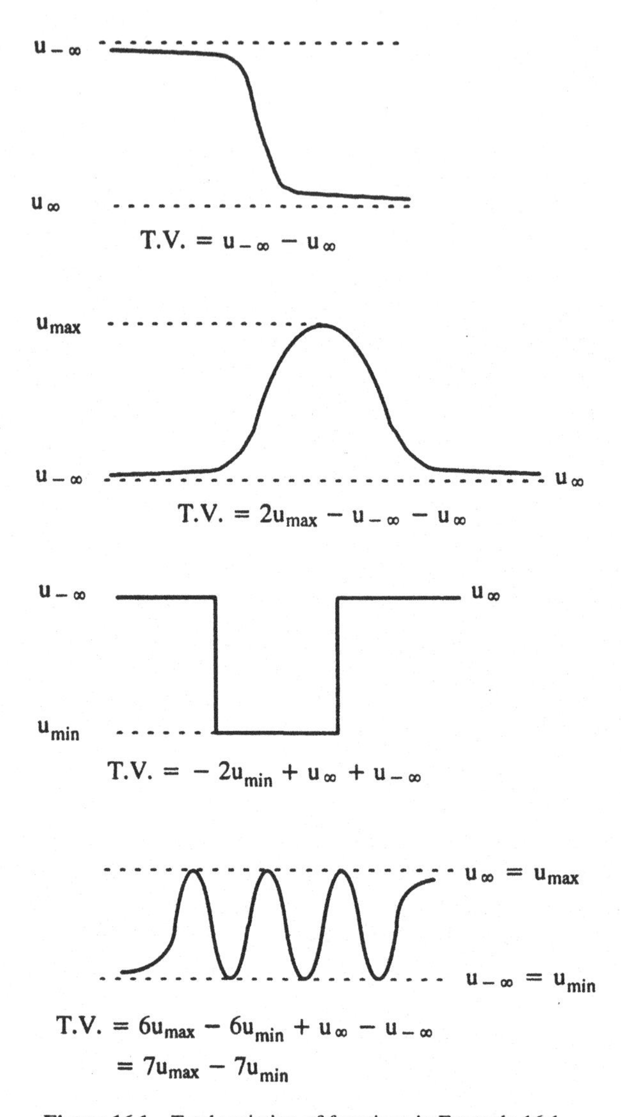

Figure 16.1 Total variation of functions in Example 16.1.

This result was first reported in the computational gasdynamics literature by Laney and Caughey (1991a). It is proven in Section 16.13. Intuitively, (16.2) says that the total variation measures the overall amount of oscillation in a function.

Example 16.1 Using (16.2), the total variation of any function is easily determined by inspecting its graph, as illustrated in Figure 16.1.

What causes the total variation to increase? Obviously, by (16.2), only maxima and minima affect the total variation. A new local maximum cannot occur unless a new local minimum also occurs, that is, maxima and minima always come in pairs. The reader can

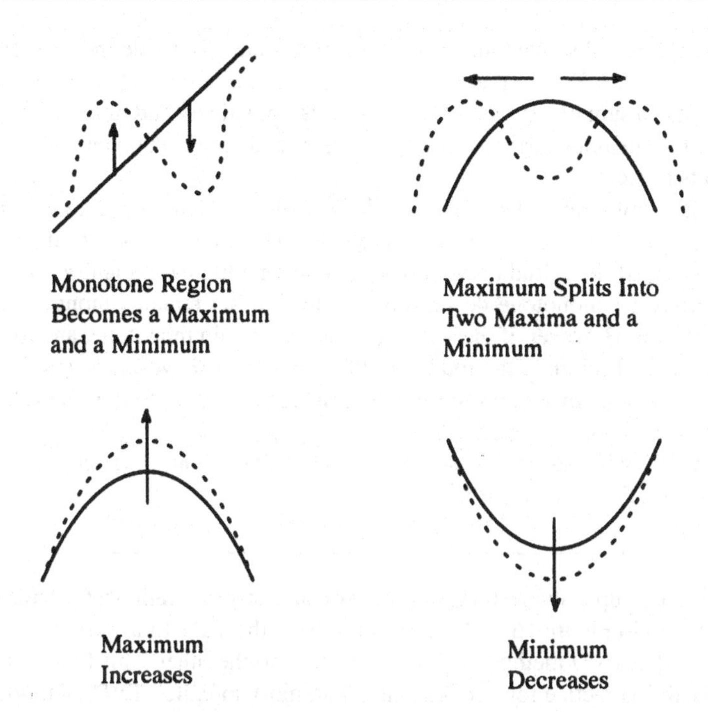

Figure 16.2 Things that cause total variation to increase.

easily verify this by sketching a curve and then attempting to create a new local maximum without also creating a new local minimum; even allowing jump discontinuities, this is an exercise in frustration. By (16.2), any new local maximum–minimum pair always increases the total variation. Furthermore, the total variation increases if an existing local maximum increases or an existing local minimum decreases. Figure 16.2 illustrates all of the things that can cause total variation to increase.

The discussion of waveform preservation, creation, and destruction in Section 4.11 implies that maxima do not increase, minima do not decrease, and no new maxima or minima are created in solutions to scalar conservation laws. In other words, none of the things that cause total variation to increase can occur in the solutions of scalar conservations. Then solutions of scalar conservation laws are *total variation diminishing* (*TVD*) and we can write

$$TV(u(\cdot, t_2)) \leq TV(u(\cdot, t_1)) \tag{16.3}$$

for all $t_2 \geq t_1$. In the original description, Harten (1983) used the term *total variation nonincreasing* (*TVNI*). However, every subsequent paper has used the term "diminishing" as a synonym for "nonincreasing."

Suppose that a numerical approximation inherits the total variation diminishing property. That is, suppose that

$$TV(u^{n+1}) \leq TV(u^n) \tag{16.4}$$

for all n where

$$TV(u^n) = \sum_{i=-\infty}^{\infty} \left|u_{i+1}^n - u_i^n\right|. \tag{16.5}$$

Fortunately, (16.2) applies to discrete functions as well as to continuously defined functions. In other words, Equation (16.5) says that the total variation of a numerical approximation on an infinite domain is a sum of extrema, maxima counted positively and minima counted negatively. The two infinite boundaries are always extrema and they both count once; every other extrema counts twice.

As promised at the beginning of the section, *TVD implies monotonicity preservation.* To see this, suppose that the initial conditions are monotone. The total variation of the initial conditions is $u_{\infty} - u_{-\infty}$ if the initial conditions are monotone increasing and $u_{-\infty} - u_{\infty}$ if the initial conditions are monotone decreasing. If the solution remains monotone, the total variation is constant. However, if the solution does not remain monotone, and instead develops new maxima and minima, the total variation increases. However, of course, the total variation cannot increase in a total variation diminishing method, and thus the solution must remain monotone.

As shown in the following examples, sometimes the TVD condition is too weak and sometimes the TVD condition is too strong.

Example 16.2 This example shows that TVD allows spurious oscillations. Consider the two solutions illustrated in Figure 16.3. Both solutions have the same total variation of 12. Thus, in theory at least, a TVD method could evolve one into the other. This kind of thing does not happen much in practice for two reasons. First, most so-called TVD methods actually satisfy much stronger nonlinear stability conditions, which prevent such oscillations. Second, most TVD methods are also conservative, and the combination tends to prevent oscillations that TVD alone would allow. In short, it is not TVD itself that prevents oscillations in practice, but other conditions that imply or supplement TVD.

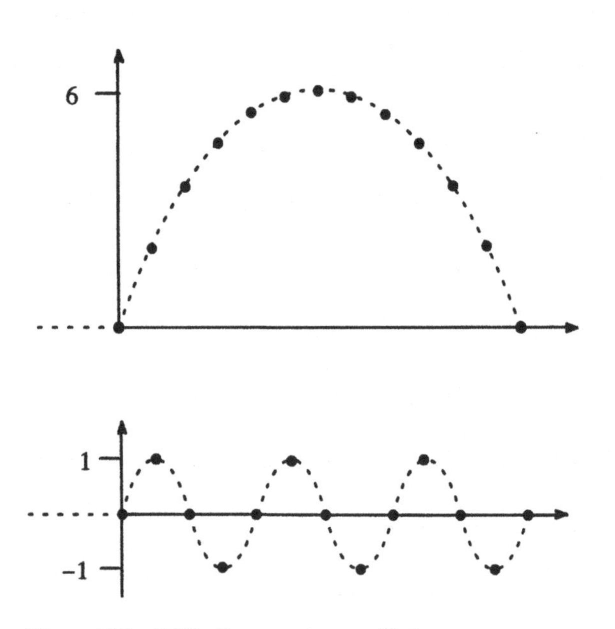

Figure 16.3 TVD allows spurious oscillations.

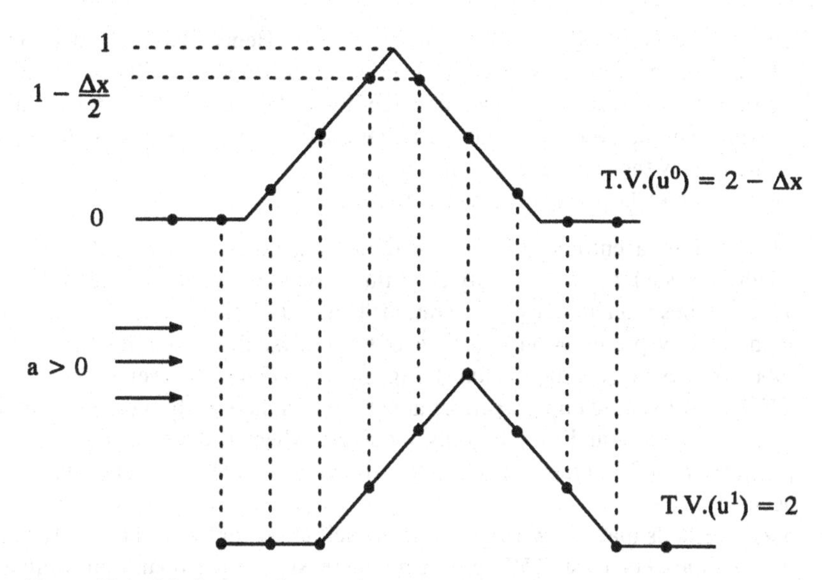

Figure 16.4 Maxima may increase in samples of the true solution; thus the total variation of the samples may increase.

Example 16.3 This example shows how TVD affects accuracy at extrema. Consider the linear advection of a triangular shape, as shown in Figure 16.4. In the figure, the total variation should increase by Δx between time steps, but this increase is disallowed by TVD. In actuality, the total variation of the numerical approximation should both decrease *and increase* as the alignment between the grid and the exact solution changes. Preventing maxima from increasing, or minima from decreasing, causes an error called *clipping*. In this example, TVD forces an $O(\Delta x)$ clipping error at the nonsmooth maximum. Clipping error is $O(\Delta x^2)$ for most smooth extrema, as discussed in the next section.

In theory, TVD does not always require clipping. For example, the local maximum could increase, provided that a local maximum decreased somewhere else, or a local minimum increased somewhere else, or a local maximum–minimum pair disappeared somewhere else. In theory, even when some clipping occurs, it might be less severe than first- or second-order, for the same reasons. However, in practice, TVD methods typically clip every time, reducing the order of accuracy at extrema to two or less, as measured by equations such as (11.52). This is partly because most so-called TVD methods actually satisfy much stronger nonlinear stability conditions than TVD. Unlike TVD, these stronger nonlinear stability conditions imply second order or greater clipping errors at every extrema, as described in later sections. Even if you could devise a method that was only TVD, and did not satisfy any stronger nonlinear stability condition, most TVD methods are also conservative and the combination tends to exaggerate the effects of TVD.

We have just seen some of the drawbacks with TVD. Let us now look at one advantage. Oscillations always add to the total variation, and thus oscillations cannot grow indefinitely without violating the TVD condition. If the number of oscillations increases, the size of the oscillations must eventually decrease, and vice versa. Hence, assuming that the total variation of the initial conditions is finite, TVD prevents unbounded oscillatory growth and thus TVD eliminates the worst-case type of instability. In fact, taking the total variation to be a rough estimate of the overall amount of oscillation, the total amount of oscillation must either decrease or stay the same at every time step.

The properties of TVD are summarized as follows:

- Most attempts at enforcing TVD end up enforcing much stronger nonlinear stability conditions, such as the positivity condition discussed in Section 16.4.
- TVD implies monotonicity preservation. This is desirable in circumstances where monotonicity preservation is too weak but undesirable in circumstances where monotonicity preservation is too strong, since TVD may be even stronger.
- TVD tends to cause clipping errors at extrema. In theory, clipping need not occur at every extrema and may be only moderate where it does occur. However, in practice, most TVD methods clip all extrema to between first- and second-order accuracy.
- TVD methods may allow large spurious oscillations in theory but rarely in practice. At the very least, TVD puts a nonincreasing upper bound on oscillations, eliminating worst-case "unbounded growth" instability.

Regardless of its drawbacks, TVD is a nearly unavoidable element of current nonlinear stability theory. As it turns out, all but two of the stability conditions seen in the remainder of this chapter imply TVD.

16.3 Range Diminishing

TVD allows large spurious oscillations, at least under certain circumstances. This section concerns a stronger condition which completely eliminates spurious oscillations. The exact solutions of scalar conservation laws on infinite spatial domains have the following property:

♦ *Maxima do not increase, minima do not decrease, and no new extrema are created with time.* (16.6)

As described in the introduction to the chapter, this property is a direct consequence of waveform preservation, waveform destruction, and monotone waveform creation. It implies that both the global and local range of the exact solution continuously contract in time. More specifically, the local range is preserved during waveform preservation or creation, and the local range is reduced during waveform destruction at shocks. Thus property (16.6) is called the *range diminishing* or *range reducing* property. Range diminishing is illustrated in Figure 16.5. In particular, Figure 16.5 shows that the range may diminish either globally or locally: in Figure 16.5a, the global maximum decreases, and thus the global range also decreases; in Figure 16.5b, a local minimum is eliminated, which decreases the range locally, near the center of the solution, but not globally.

Range diminishing was first proposed by Boris and Book (1973) and later clarified by Harten (1983). The reader should be aware that range diminishing is not a standard term. However, it makes sense to introduce a new term, rather than saying "maxima do not increase, minima do not decrease, and there are no new maxima or minima" each time, as in the existing literature.

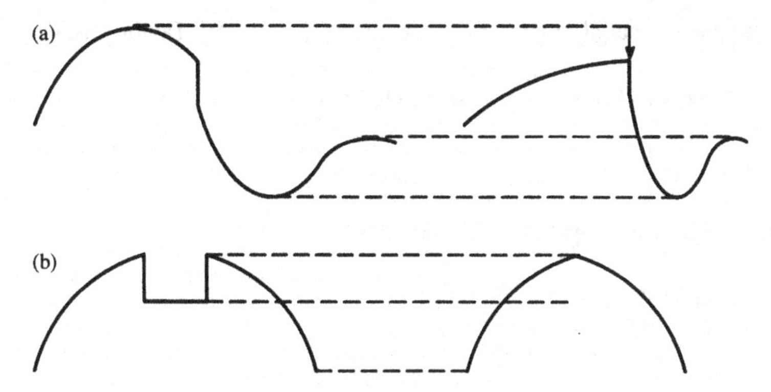

Figure 16.5 (a) An illustration of global range reduction. (b) An illustration of local range reduction.

Suppose that a numerical approximation inherits the range diminishing property. In other words, in the discrete approximation, suppose that maxima do not increase, minima do not decrease, and that no new maxima and minima are created. Then the numerical approximation is obviously free of any spurious oscillations and overshoots. Unfortunately, range diminishment exacts a toll on accuracy. To see this, notice that in the perfect finite-difference method

$$u_i^n = u(x_i, t^n).$$

Hence, the perfect finite-difference approximation yields samples of the exact solution. However, the local and global ranges of the true samples both decrease *and increase* in time, unless a sample always falls exactly on the crest of every maxima and minima in the true solution. For example, suppose that the closest sample falls just to the right of a right-running global maximum in the exact solution. Then this sample is a local maximum among samples, but yet it should increase as the maximum in the true solution moves to the right. This is illustrated in Figure 16.4. For another example, suppose a maximum in the true solution falls between two samples (the maximum falls between the cracks, so to speak) such that the samples are monotone or constant. At some later time, one of the samples may fall directly on top of the local maximum and thus should create a new maximum in the samples. Actually, the biggest problem in this case is the sample spacing, which must be reduced to reliably capture such fine details. An *adequate* sampling of the exact solution would not produce new maxima or minima.

Because of all this, range reduction in numerical methods prevents certain legitimate physical behaviors. The same can be said of monotonicity preservation and TVD, and for exactly the same reasons. Physically, the numerical solution should allow small increases in existing maxima and small decreases in existing minima. The numerical solution should also allow small new maxima and minima, if not for physical reasons then for numerical reasons, since eliminating all new maxima and minima may create larger errors than it prevents, especially in linear methods, by Godunov's theorem.

Like TVD, range reduction implies clipping at extrema. Unlike TVD, range reduction clips every extrema, causing at least second-order errors. Thus

♦ *Range diminishing methods are formally second-order accurate at extrema, at best.* (16.7)

To prove this, suppose a spatial maximum occurs at (x_{max}, t_{max}). By Taylor series

$$u(x,t) = u(x_{max}, t_{max}) + \frac{\partial u}{\partial x}(x_{max}, t_{max})(x - x_{max}) + \frac{1}{2}\frac{\partial^2 u}{\partial x^2}(x_{max}, t_{max})(x - x_{max})^2 + \cdots.$$

But if (x_{max}, t_{max}) is a smooth spatial maximum then

$$\frac{\partial u}{\partial x}(x_{max}, t_{max}) = 0$$

and

$$\frac{\partial^2 u}{\partial x^2}(x_{max}, t_{max}) \leq 0,$$

as proven in any elementary calculus book. Then

$$u(x, t_{max}) = u(x_{max}, t_{max}) - O(\Delta x^2)$$

for all $x_{max} - O(\Delta x) \leq x \leq x_{max} + O(\Delta x)$. In the best case, a sample falls exactly on x_{max}. But, in general, the nearest sample is $O(\Delta x)$ away from x_{max} and the true maximum $u(x_{max}, t_{max})$ is represented by a sample $u(x_{max}, t_{max}) - O(\Delta x^2)$. This is illustrated in Figure 16.6. If the samples ever once miss the maximum, then range reduction never allows the samples to increase, making the second-order reduction permanent. In Figure 16.6, clipping appears to neatly trim the top of the maximum, leaving a flat plateau. In practice, clipping need not act quite so crisply but, in any event, clipping limits accuracy at most smooth extrema to second order.

All statements about formal order of accuracy contain a certain degree of ambiguity, as discussed in Section 6.3 and Subsection 11.2.2. Equation (16.7) is no exception. For example, the proof of (16.7) assumed two continuous derivatives. If the solution does not have the required continuous derivatives, the formal order of accuracy will decrease, as in Example 16.3. Furthermore, the proof of (16.7) assumes that the second derivative is strictly less than zero. If the second derivative equals zero, the formal order of accuracy will increase. Finally, (16.7) does not account for space–time interactions, instability, or any sort

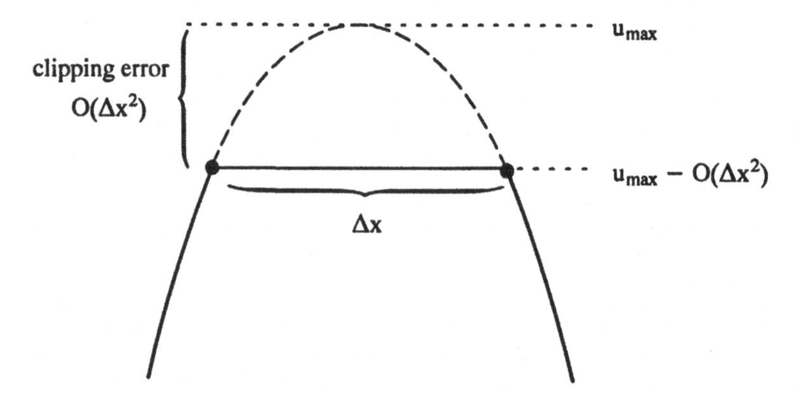

Figure 16.6 Clipping error is formally second order or greater at smooth extrema.

of numerical error except for clipping. Because of these additional sources of error, most range reducing methods do not achieve full second-order accuracy at extrema, but instead produce a fractional pointwise order of accuracy at smooth extrema somewhere between one and two.

As it turns out, *range diminishing implies TVD*, since range diminishing specifically disallows all behaviors that increase total variation, such as those illustrated in Figure 16.2. The properties of range diminishing are summarized as follows:

- Most attempts at enforcing range diminishing end up enforcing stronger nonlinear stability conditions, such as the upwind range condition of Section 16.5. However, there have been some attempts to enforce the range diminishing condition directly and precisely. For example, see Laney and Caughey (1991b, 1991c) or Coray and Koebbe (1993).
- Range diminishing implies TVD. This is desirable in circumstances where TVD is too weak. However, this is undesirable in circumstance where TVD is too strong, since range diminishing is even stronger.
- Range diminishing always causes clipping error at extrema. Clipping limits the formal pointwise order of accuracy at extrema to two.
- Range diminishing completely eliminates any spurious oscillations, overshoots, or extrema.

This chapter has now described monotonicity preservation, TVD, and range diminishing conditions. Unfortunately, all three of these conditions are difficult to prove or enforce directly. The next two sections concern conditions that are relatively easy to prove or enforce, and thus these next two conditions are among the most popular and practical nonlinear stability conditions.

16.4 Positivity

This section concerns a nonlinear stability condition based on the wave speed split form discussed in Section 13.5. Recall that the wave speed split form is

$$u_i^{n+1} = u_i^n + C_{i+1/2}^+(u_{i+1}^n - u_i^n) - C_{i-1/2}^-(u_i^n - u_{i-1}^n). \tag{13.36}$$

Recall that methods derived using wave speed splitting naturally arrive in wave speed split form where $C_{i+1/2}^{\pm} \geq 0$; however, any method can be written in a wave speed split form, in infinitely many different ways, provided one can tolerate infinite coefficients $C_{i+1/2}^{\pm}$ when $u_{i+1}^n - u_i^n = 0$. But suppose that a numerical method can be written in a wave speed split form with finite nonnegative coefficients such that

♦
$$\begin{aligned} &C_{i+1/2}^+ \geq 0, \qquad C_{i+1/2}^- \geq 0, \\ &C_{i+1/2}^+ + C_{i+1/2}^- \leq 1 \end{aligned} \tag{16.8}$$

for all i. This is called the *positivity condition*. Positivity was first suggested by Harten (1983). The reader should be warned that the term "positivity" has many meanings in the literature. Instead of a positivity condition, in the existing literature, Equation (16.8) is most often called a TVD condition. Obviously this leads to confusion with the true TVD condition, Equation (16.3), which justifies the introduction of a distinct term. Although positivity implies TVD, as discussed below, positivity is certainly not the same thing as TVD. Unlike previous nonlinear stability conditions, the positivity condition does not derive from any specific physical attributes of the exact solution, except possibly $C^{\pm}_{i+1/2} \geq 0$, which is a physical splitting condition, if you care to interpret it that way, as seen in Equation (13.39).

A numerical method can be split in infinitely many different ways, as seen in Section 13.5. A method that does not satisfy Equation (16.8) when written in one wave speed split form may yet satisfy Equation (16.8) when written in some other wave speed split form. However, in general, at most one splitting will yield finite coefficients $C^{\pm}_{i+1/2}$ and thus at most one splitting even has the potential to satisfy Equation (16.8).

The positivity condition, Equation (16.8), does not look much like a nonlinear stability condition, in the sense that it does not have any obvious connections to spurious oscillations. However, the positivity condition lies between two nonlinear stability conditions. In particular, *positivity implies TVD*, as shown by Harten (1983); the reader may wish to attempt this half-page proof as an exercise. Also, *the upwind range condition implies positivity*, as discussed in the next section and as proven in Section 16.13. Its intimate association with admitted nonlinear stability conditions makes positivity a nonlinear stability condition.

One of the common myths about positivity is that it eliminates spurious oscillations and overshoots. Although positivity limits spurious oscillations and overshoots, just like all nonlinear stability conditions, it does not necessarily eliminate spurious oscillations and overshoots. A good example is the Lax–Friedrichs method. The Lax–Friedrichs method is positive and yet may develop large spurious oscillations in as little as one time step, as discussed in Section 17.1. Unfortunately, the myth about positivity has led to the myth that the Lax–Friedrichs method does not allow spurious oscillations and overshoots. In fact, in some quarters, Lax–Friedrichs has an unwarranted reputation as the ultimate nonoscillatory method because it satisfies not only positivity but also every nonlinear stability condition seen in this chapter, except for the range diminishing and upwind range conditions. However, this reflects more on the weaknesses of current nonlinear stability theory than on the strengths of the Lax–Friedrichs method.

Like range reduction, positivity usually restricts accuracy at extrema to between first and second order. However, positivity by itself does not limit the order of accuracy at extrema but instead the *combination* of positivity and conservation limits the order of accuracy at extrema to between first and second order in a "clipping-like" way; see Subsection 3.3.2 of Laney and Caughey (1991c) for a discussion.

Example 16.4 Show that FTCS is not positive.

Solution Example 13.12 described wave speed split forms for FTCS. Every method can be written in, at most, one wave speed form with finite coefficients. For FTCS, that form is as follows:

$$C^{+}_{i+1/2} = -\frac{\lambda}{2}a^{n}_{i+1/2},$$

$$C^{-}_{i-1/2} = \frac{\lambda}{2}a^{n}_{i-1/2},$$

where as usual

$$a^n_{i+1/2} = \begin{cases} \frac{f(u^n_{i+1})-f(u^n_i)}{u^n_{i+1}-u^n_i} & \text{if } u^n_{i+1} \neq u^n_i, \\ a(u) = f'(u) & \text{if } u^n_{i+1} = u^n_i = u. \end{cases}$$

Since $C^+_{i+1/2} = -C^-_{i+1/2}$, the coefficients $C^+_{i+1/2}$ and $C^-_{i+1/2}$ cannot both be nonnegative. Thus this form violates $C^\pm_{i+1/2} \geq 0$. Every other wave speed split form of FTCS has infinite coefficients – either $C^+_{i+1/2}$ or $C^-_{i+1/2}$ or both are infinite when $u^n_i = u^n_{i+1}$. Thus every other wave speed split form violates either $C^\pm_{i+1/2} \geq 0$ or $C^+_{i+1/2} + C^-_{i+1/2} \leq 1$ or both. Thus FTCS does not satisfy positivity. This is consistent with the known instability of FTCS.

Example 16.5 Show that FTFS is positive for $-1 \leq \lambda a^n_{i+1/2} \leq 0$.

Solution As shown in Example 13.13, FTFS can be written in wave speed split form with the following finite coefficients:

$$C^+_{i+1/2} = -\lambda a_{i+1/2},$$
$$C^-_{i-1/2} = 0.$$

Then FTFS is positive if $0 \leq C^+_{i+1/2} \leq 1$ or $-1 \leq \lambda a^n_{i+1/2} \leq 0$. Notice that, in this case, the positivity condition is equivalent to the CFL condition.

Example 16.6 Positivity is sometimes defined in terms of the artificial viscosity form seen in Section 14.2. Show that a method is positive if

$$\left|\lambda a^n_{i+1/2}\right| \leq \lambda \epsilon^n_{i+1/2} \leq 1, \tag{16.9}$$

where as usual

$$a^n_{i+1/2} = \begin{cases} \frac{f(u^n_{i+1})-f(u^n_i)}{(u^n_{i+1})-u^n_i} & \text{if } u^n_{i+1} \neq u^n_i, \\ a(u) = f'(u) & \text{if } u^n_{i+1} = u^n_i = u. \end{cases}$$

Solution The artificial viscosity form is

$$u^{n+1}_i = u^n_i - \frac{\lambda}{2}\left(f\left(u^n_{i+1}\right) - f\left(u^n_{i-1}\right)\right)$$
$$- \frac{\lambda}{2}\left(\epsilon^n_{i+1/2}\left(u^n_{i+1} - u^n_i\right) + \epsilon^n_{i-1/2}\left(u^n_i - u^n_{i-1}\right)\right). \tag{14.3}$$

This can be rewritten in wave speed split form as follows:

$$u^{n+1}_i = u^n_i - \frac{\lambda}{2}\left(\epsilon^n_{i+1/2} + a^n_{i+1/2}\right)\left(u^n_{i+1} - u^n_i\right)$$
$$- \frac{\lambda}{2}\left(\epsilon^n_{i-1/2} - a^n_{i-1/2}\right)\left(u^n_i - u^n_{i-1}\right). \tag{16.10}$$

This implies

$$C^+_{i+1/2} = \frac{\lambda}{2}\left(\epsilon^n_{i+1/2} + a^n_{i+1/2}\right),$$
$$C^-_{i+1/2} = \frac{\lambda}{2}\left(\epsilon^n_{i+1/2} - a^n_{i+1/2}\right) \tag{16.11}$$

or

$$C^+_{i+1/2} + C^-_{i+1/2} = \lambda\epsilon^n_{i+1/2},$$
$$C^+_{i+1/2} - C^-_{i+1/2} = \lambda a^n_{i+1/2}. \tag{16.12}$$

By Equation (16.11), notice that $\epsilon^n_{i+1/2} \geq |a^n_{i+1/2}|$ implies $C^+_{i+1/2} \geq 0$ and $C^-_{i+1/2} \geq 0$. By Equation (16.12), notice that $\lambda\epsilon^n_{i+1/2} \leq 1$ implies $C^+_{i+1/2}+C^-_{i+1/2} \leq 1$. Putting $\epsilon^n_{i+1/2} \geq |a^n_{i+1/2}|$ and $\lambda\epsilon^n_{i+1/2} \leq 1$ together, we have $|\lambda a^n_{i+1/2}| \leq \lambda\epsilon^n_{i+1/2} \leq 1$, which implies positivity.

Unfortunately, $|\lambda a^n_{i+1/2}| \leq \lambda\epsilon^n_{i+1/2} \leq 1$ only allows first-order accurate methods, as shown by Osher (1984). However, in the artificial viscosity form, suppose that FTCS is replaced by some other reference method, written in wave speed split form as follows:

$$u^{n+1}_i = u^n_i + D^+_{i+1/2}\left(u^n_{i+1} - u^n_i\right) - D^-_{i-1/2}\left(u^n_i - u^n_{i-1}\right).$$

Then this reference method plus second-order artificial viscosity is

$$u^{n+1}_i = u^n_i + \left(D^+_{i+1/2} + \frac{\lambda}{2}\epsilon^n_{i+1/2}\right)\left(u^n_{i+1} - u^n_i\right) - \left(D^-_{i-1/2} + \frac{\lambda}{2}\epsilon^n_{i-1/2}\right)\left(u^n_i - u^n_{i-1}\right)$$

or

$$u^{n+1}_i = u^n_i + C^+_{i+1/2}\left(u^n_{i+1} - u^n_i\right) - C^-_{i-1/2}\left(u^n_i - u^n_{i-1}\right),$$

where

$$C^+_{i+1/2} = D^+_{i+1/2} + \frac{\lambda}{2}\epsilon^n_{i+1/2},$$
$$C^-_{i+1/2} = D^-_{i+1/2} + \frac{\lambda}{2}\epsilon^n_{i+1/2}.$$

Increasing the artificial viscosity $\epsilon^n_{i+1/2}$ increases both $C^+_{i+1/2}$ and $C^-_{i+1/2}$. Thus $C^+_{i+1/2} \geq 0$ and $C^-_{i+1/2} \geq 0$ for sufficiently large $\epsilon^n_{i+1/2}$. However, if $\epsilon^n_{i+1/2}$ is too large then $C^+_{i+1/2} + C^-_{i+1/2} \leq 1$ is violated. Hence, according to the positivity condition, artificial viscosity is stabilizing in small amounts and destabilizing in large amounts. Whereas positivity implies first-order accuracy for FTCS plus artificial viscosity, it allows arbitrary orders of accuracy for other reference methods plus artificial viscosity, at least away from extrema, especially for reference methods with stencils wider than three points.

The properties of positivity are summarized as follows:

- Positivity is relatively easy to prove and enforce. This makes positivity one of the most widely cited nonlinear stability conditions. However, while many papers discuss positivity, the vast majority of them actually enforce the upwind range condition, which is a special case of positivity, as discussed in the next section.

- Positivity implies TVD. This is desirable in circumstances where TVD is too weak. However, this is undesirable in circumstances where TVD is too strong, since positivity is even stronger. Most so-called TVD methods are actually positive.
- Positivity causes a "clipping-like" error at extrema. Positivity and conservation generally limit the formal order of accuracy at extrema to between first and second order.
- Positivity allows large spurious oscillations and overshoots.

16.5 Upwind Range Condition

Genetics tells us that children inherit their properties from their parents, grandparents, and more distant ancestors. Similarly, $u(x, t^{n+1})$ inherits its properties from its "ancestors" such as $u(x, t^n)$. In particular, by following wavefronts or, in other words, by tracing characteristics, we see that $u(x_i, t^{n+1})$ inherits its value from an upwind ancestor such as $u(x_{i-1}, t^n)$. If all of the upwind ancestors fall within a certain range, as defined by a maximum and a minimum, then $u(x_i, t^{n+1})$ must also lie within this same range. Written mathematically and precisely, the exact solution of a scalar conservation law satisfies the condition

$$\min_{x_{i-1} \le x \le x_i} u(x, t^n) \le u(x_i, t^{n+1}) \le \max_{x_{i-1} \le x \le x_i} u(x, t^n) \tag{16.13a}$$

for $0 \le \lambda a(x_i, t^{n+1}) \le 1$. Also,

$$\min_{x_i \le x \le x_{i+1}} u(x, t^n) \le u(x_i, t^{n+1}) \le \max_{x_i \le x \le x_{i+1}} u(x, t^n) \tag{16.13b}$$

for $-1 \le \lambda a(x_i, t^{n+1}) \le 0$. This is called the *upwind range* property. The upwind range property is illustrated in Figure 16.7.

Unfortunately, in a numerical approximation, a child cannot know about all of its possible ancestors – it can only know about the ancestors that live at discrete neighboring sample points. But notice that

$$\min_{x_{i-1} \le x \le x_i} u(x, t^n) \le \min(u(x_i, t^n), u(x_{i-1}, t^n)),$$
$$\min_{x_i \le x \le x_{i+1}} u(x, t^n) \le \min(u(x_i, t^n), u(x_{i+1}, t^n))$$

and

$$\max(u(x_i, t^n), u(x_{i-1}, t^n)) \le \max_{x_{i-1} \le x \le x_i} u(x, t^n),$$
$$\max(u(x_i, t^n), u(x_{i+1}, t^n)) \le \max_{x_i \le x \le x_{i+1}} u(x, t^n).$$

Then Equation (16.13a) is true if

$$\min(u(x_i, t^n), u(x_{i-1}, t^n)) \le u(x_i, t^{n+1}) \le \max(u(x_i, t^n), u(x_{i-1}, t^n)) \tag{16.14a}$$

for $0 \le \lambda a(x_i, t^{n+1}) \le 1$. Also, Equation (16.13b) is true if

$$\min(u(x_i, t^n), u(x_{i+1}, t^n)) \le u(x_i, t^{n+1}) \le \max(u(x_i, t^n), u(x_{i+1}, t^n)) \tag{16.14b}$$

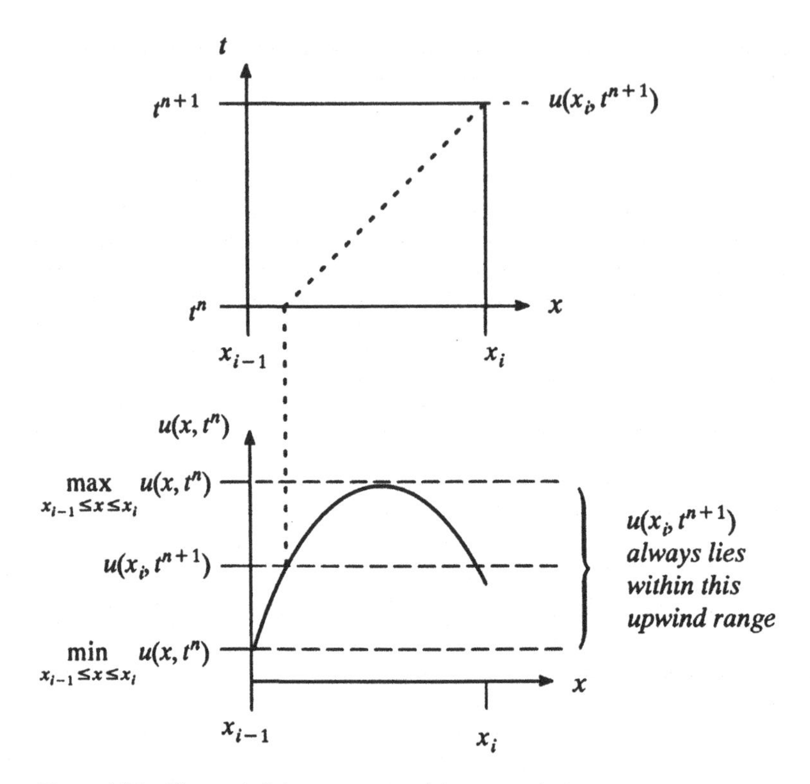

Figure 16.7 The upwind range property of the exact solution.

for $-1 \leq \lambda a(x_i, t^{n+1}) \leq 0$. This is a simplified version of the upwind range property. The simplified version is equivalent to the original version if and only if the solution is monotone increasing or monotone decreasing. Unfortunately, at maxima and minima, the simplified version unphysically restricts the solution, capping maxima too low and minima too high.

Suppose that a numerical approximation has the simplified version of the upwind range property. In other words, suppose that

♦ $$\min\left(u_i^n, u_{i-1}^n\right) \leq u_i^{n+1} \leq \max\left(u_i^n, u_{i-1}^n\right), \qquad 0 \leq \lambda a\left(u_i^{n+1}\right) \leq 1, \tag{16.15a}$$

♦ $$\min\left(u_i^n, u_{i+1}^n\right) \leq u_i^{n+1} \leq \max\left(u_i^n, u_{i+1}^n\right), \qquad -1 \leq \lambda a\left(u_i^{n+1}\right) \leq 0. \tag{16.15b}$$

This is called the *upwind range* or *upwind compartment condition*. The upwind range condition was first suggested by Boris and Book (1973) and Van Leer (1974). It was further developed by Sweby (1984) and others. The reader should be aware that "upwind range" is not a standard term. In fact, in the existing literature, Equation (16.15) is most often called a TVD condition. Obviously, this leads to confusion with the true TVD condition, Equation (16.3), which justifies the introduction of a distinct term.

Let us now consider some of the issues surrounding the upwind range condition. First, the upwind range condition allows values of the solution to travel at most one grid point per time step, either to the left or to the right. Then the upwind range condition only makes sense when $\lambda|a| \leq 1$. Second, the upwind range condition involves $\lambda a(u_i^{n+1})$. At time level n, by definition, explicit methods cannot depend on $\lambda a(u_i^{n+1})$. However, the unknown $\lambda a(u_i^{n+1})$

has the same sign as the known $\lambda a(u_i^n)$ unless the waves change direction during the time step. The wind can only change directions at *sonic points* where $a(u) = 0$. Thus, in practice, it may be difficult to enforce the upwind range condition at sonic points. Third, *the upwind range condition implies the range diminishing condition, except possibly at sonic points*, as shown in Section 16.13. Fourth and finally, also as shown in Section 16.13, an explicit method has the upwind range condition if and only if it can be written in wave speed split form where

$$\blacklozenge \quad 0 \le C^-_{i+1/2} \le 1 \qquad C^+_{i+1/2} = 0, \qquad 0 \le \lambda a\left(u_i^{n+1}\right) \le 1, \tag{16.16a}$$

$$\blacklozenge \quad 0 \le C^+_{i+1/2} \le 1 \qquad C^-_{i+1/2} = 0, \qquad -1 \le \lambda a\left(u_i^{n+1}\right) \le 0 \tag{16.16b}$$

for all i. Then *the upwind range condition implies positivity*. As a possibly more convenient expression, for conservative methods, Equation (16.16) is equivalent to

$$0 \le \lambda \frac{\hat{f}^n_{i+1/2} - \hat{f}^n_{i-1/2}}{u_i^n - u_{i-1}^n} \le 1, \qquad 0 \le \lambda a\left(u_i^{n+1}\right) \le 1, \tag{16.17a}$$

$$-1 \le \lambda \frac{\hat{f}^n_{i+1/2} - \hat{f}^n_{i-1/2}}{u_{i+1}^n - u_i^n} \le 0, \qquad -1 \le \lambda a\left(u_i^{n+1}\right) \le 0. \tag{16.17b}$$

Example 16.7 Show that FTFS satisfies the upwind range condition if $-1 \le \lambda a^n_{i+1/2} \le 0$.

Solution As shown in Example 13.13, FTFS can be written in wave speed split form with the following coefficients:

$$C^+_{i+1/2} = -\lambda a^n_{i+1/2},$$
$$C^-_{i-1/2} = 0.$$

Then FTFS has the upwind range condition if $0 \le C^+_{i+1/2} \le 1$. This is true if and only if $-1 \le \lambda a^n_{i+1/2} \le 0$.

The properties of the upwind range condition are summarized as follows:

- The upwind range condition is relatively easy to prove and enforce, except possibly at sonic points. This makes the upwind range condition one of the most popular nonlinear stability conditions.
- The upwind range condition implies positivity. In fact, it can be seen as a special case of the positivity condition.
- The upwind range condition implies the range diminishing condition, except possibly at sonic points. Then the upwind range condition always causes second-order clipping error at extrema, except possibly at sonic points, just like the range diminishing condition. Also, the upwind range condition completely eliminates spurious oscillations and overshoots, just like the range diminishing condition, except possibly at sonic points.
- The upwind range condition assumes $\lambda|a| \le 1$.

16.6 Total Variation Bounded (TVB)

All of the nonlinear stability conditions considered so far are too strong, at least under certain circumstances. In particular, all of the nonlinear stability conditions considered so far imply monotonicity preservation which is sometimes too strong, as argued in Section 16.1. This section and the next consider some weaker stability conditions; in particular, the conditions in this section and the next do not imply monotonicity preservation. Equation (16.3) implies that exact solutions to scalar conservation laws have the following property:

$$TV(u(\cdot, t)) \leq M \leq \infty \tag{16.18}$$

for all t and for some constant M. For example, M could be the total variation of the initial conditions. Then the exact solutions of scalar conservation laws are said to be *total variation bounded* (*TVB*) or to have *bounded variation* (*BV*). Like total variation, bounded variation is a classic concept in real and functional analysis, although it typically has a slightly different meaning than that seen here; see Royden (1968). Suppose that a numerical approximation inherits the total variation bounded property. In other words, suppose that

$$TV(u^n) \leq M \leq \infty \tag{16.19}$$

for all n and for some constant M. TVB is easily the weakest nonlinear stability condition seen in this chapter. In particular, TVB allows large oscillations provided only that spurious oscillations do not grow unboundedly large as time increases. Thus TVB simply ensures that a method does not blow up, at least not in an oscillatory fashion.

Of all the stability conditions in this chapter, along with TVD, TVB is perhaps the most similar to the linear stability conditions studied in the last chapter. Specifically, both linear stability conditions and TVB prevent unbounded oscillatory growth. For linear methods, if the solution does not blow up then it must either shrink or stay the same size. Thus, to say that a linear method does not blow up is actually quite a strong statement. Nonlinear methods, however, have a much richer variety of behaviors. Thus, when you say that a nonlinear method does not blow up or, equivalently, when you say that a nonlinear method is TVB, the method could still have other provocative behaviors. For example, in theory, the error could start small and grow in time, provided the error eventually stopped growing or reached a horizontal asymptote, however large that asymptote might be.

Historically, in the computational gasdynamics literature, conditions bounding total variation appeared years before TVD, in the context of convergence proofs. When the limitations of TVD (especially clipping) became better known, Shu (1987) officially suggested TVB as a substitute for TVD. Certainly, TVB does not place any unphysical restrictions on the solution like TVD does, except possibly when combined with other properties such as conservation. TVB does place restrictions on the solution in the limit $\Delta x \to 0$ and $\Delta t \to 0$, but even then TVB is only really helpful in combination with other conditions such as conservation. Shu (1987) suggested a method which he called a TVB method, as mentioned in Section 21.4. Obviously this method's success cannot be explained solely by the fact that it is TVB.

TVB follows the "unbounded growth" definition of stability. There is a subtle but important variant on TVB often used in the "convergence" definition of stability. In particular, suppose that the total variation is bounded by a constant M for all times t, as with ordinary TVB, *and also for all sufficiently small* Δx *and* Δt. LeVeque (1992) calls this the *total variation stability* condition to distinguish it from the ordinary TVB condition. Notice that total variation stability implies TVB but not vice versa. In fact, many common numerical methods are TVB but not total variation stable. For example, the Lax–Wendroff method seen in Section 17.2 is TVB but not total variation stable, at least for discontinuous solutions. In particular, for given initial conditions, the total variation is bounded for all t for any fixed Δx and Δt, but the total variation grows without bound for discontinuous solutions as $\Delta x \to 0$ and $\Delta t \to 0$ for fixed t.

The properties of TVB are summarized as follows:

- Most attempts at TVB end up enforcing much stronger nonlinear stability conditions, such as the positivity condition discussed in Section 16.4.
- TVB is the weakest nonlinear stability condition seen in this chapter. In particular, every other stability condition in this chapter implies TVB, except for monotonicity preservation, but TVB does not imply any of the other stability conditions in this chapter.
- TVB does not force errors such as clipping, unlike most of the other nonlinear stability conditions in this chapter.
- TVB tolerates large spurious oscillations, provided only that the oscillations remain bounded. Although TVB puts an upper bound on oscillations, the upper bound can be as large as you like.
- A variant of TVB, sometimes called total variation stability, is useful in convergence proofs, as discussed in Section 16.11.
- As the weakest nonlinear stability condition seen in this chapter, TVB is the most likely condition to apply to other equations in addition to scalar conservation laws and the one-dimensional Euler equations.

16.7 Essentially Nonoscillatory (ENO)

Equation (16.3) obviously implies

$$TV(u(\cdot, t_2)) \leq TV(u(\cdot, t_1)) + O(\Delta x^r) \tag{16.20}$$

for all $t_2 \geq t_1$, where $O(\Delta x^r)$ refers to any positive rth-order term. Then exact solutions of scalar conservation on infinite domains are called *essentially nonoscillatory* (*ENO*). Suppose that a numerical approximation inherits the essentially nonoscillatory property. In other words, suppose

$$TV(u^m) \leq TV(u^n) + O(\Delta x^r) \tag{16.21}$$

for all $m \geq n$ and for some arbitrary $r > 0$. In particular, Equation (16.21) implies

$$TV(u^{n+1}) \leq TV(u^n) + O(\Delta x^r). \tag{16.22}$$

The ENO condition was suggested by Harten, Engquist, Osher, and Chakravarthy (1987). Notice that TVD implies ENO, which in turn implies TVB. Also, ENO becomes TVD in the limit $\Delta x \to 0$. The constant r can equal the formal order of accuracy of the method. Better yet, r can equal two. Then ENO allows second-order increases in the total variation of the numerical approximation and, in particular, second-order increases in maxima and second-order decreases in minima, which eliminates the need for clipping. Unfortunately, ENO theoretically allows large oscillations and overshoots, more so than TVD but less so than

TVB, since oscillations and overshoots can grow by an rth-order amount at every time step. Furthermore, like TVD and TVB, it is extremely difficult to show directly that a numerical method is ENO. Of course, TVD, range reducing, positivity, and the upwind range condition all imply ENO. But if a method satisfies a stronger nonlinear stability condition then one should say so, and not pretend that it only satisfies some weaker nonlinear stability condition. In fact, while ENO methods (Sections 21.4 and 23.5) based on ENO reconstructions exist (Chapter 9) there is no rigorous proof that these methods have ENO stability. In other words, it is important to distinguish between ENO stability and ENO methods, just as it is important to distinguish between TVD stability and TVD methods, or TVB stability and TVB methods.

16.8 Contraction

This section considers one of the more obscure nonlinear stability conditions. Consider two exact solutions $u(x,t)$ and $v(x,t)$ of the same scalar conservation law on the same unbounded domain. The solutions are different only because their initial conditions are different. As shown by Lax (1973), *any two solutions always draw closer together as measured in the 1-norm.* That is,

$$\|u(\cdot,t_2) - v(\cdot,t_2)\|_1 \leq \|u(\cdot,t_1) - v(\cdot,t_1)\|_1 \tag{16.23}$$

for all $t_2 \geq t_1$. This is called the *contraction* property. Suppose a numerical approximation inherits contraction. That is, for any two numerical approximations, suppose that

$$\|u^{n+1} - v^{n+1}\|_1 \leq \|u^n - v^n\|_1. \tag{16.24}$$

Numerical contraction is difficult to directly verify or enforce. The contraction condition implies TVD and the monotone condition seen in the next section implies contraction. Unlike previous nonlinear stability conditions, the contraction condition appears to have no direct link to spurious oscillations. Contraction is an occasionally useful property in mathematical proofs, especially convergence proofs. See Lax (1973) and Section 15.6 of LeVeque (1992) for more details.

16.9 Monotone Methods

Consider two exact solutions $u(x,t)$ and $v(x,t)$ of the same scalar conservation law on the same unbounded domain. The solutions are different only because their initial conditions are different. In particular, suppose that the initial conditions satisfy $u(x,0) \geq v(x,0)$ for all x. Then $u(x,t) \geq v(x,t)$ for all x and t. This is called the *monotone* property. Suppose that a numerical approximation inherits the monotone property. In other words, suppose $u_i^0 \geq v_i^0$ for all i implies $u_i^n \geq v_i^n$ for all i and n. It can be shown that an explicit forward-time method is monotone if and only if u_i^{n+1} increases or stays the same when any value in its stencil $(u_{i-K_1}^n, \ldots, u_{i+K_2}^n)$ increases. Be careful not to confuse monotone *methods* with monotone increasing or decreasing *solutions* or with monotonicity-preserving methods. The monotone condition was first proposed by Harten, Hyman, and Lax (1976). The monotone condition limits the order of accuracy to one, which strongly limits the practical appeal of monotone methods. The main application for the monotone condition is convergence proofs. Like contraction, the monotone condition appears to have no direct link to spurious oscillations, except that it implies TVD. See Crandall and Majda (1980) or Section 15.7 of LeVeque (1992) for more details on monotone methods.

16.10 A Summary of Nonlinear Stability Conditions

Figure 16.8 summarizes seven of the nine nonlinear stability conditions seen in this chapter. The contraction and monotone conditions are omitted from the figure since they are rarely seen outside of the mathematical literature. In Figure 16.8, the stability conditions are arranged with the strongest conditions on the top and the weakest conditions on the bottom. A line connecting two conditions means that the upper condition implies the lower condition. Going from bottom to top in Figure 16.8, the strength of the stability conditions increases, both in terms of positive effects in reducing spurious oscillations and overshoots and in terms of negative effects in reducing the order of accuracy. More specifically, as the strength of the stability conditions increase, spurious overshoots and oscillations decrease, while formal accuracy decreases at extrema or even globally.

Confusingly, as mentioned earlier, the term TVD commonly refers to positivity, the upwind range condition, or even the range reducing condition. In other words, in current parlance, the term TVD is almost synonymous with nonlinear stability. Since the term TVD has been used so indiscriminantly, the vocabulary of nonlinear stability is underdeveloped in the standard literature; this book has introduced new terms to address the situation.

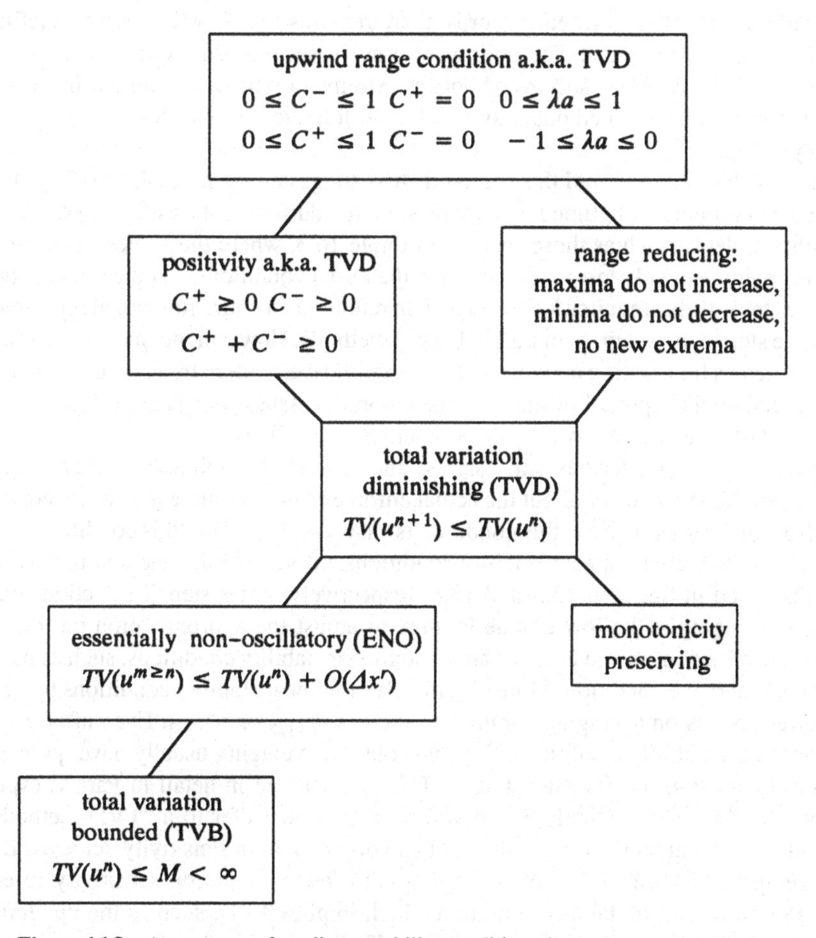

Figure 16.8 A summary of nonlinear stability conditions for scalar conservation laws.

Except for the positivity condition, all of the nonlinear stability conditions seen in this chapter are properties of the exact solution of scalar conservation laws. However, paradoxically, these physical properties become unphysical when imposed on finite-difference approximations, since the maxima and minima of the samples of the numerical solution behave differently than the maxima and minima of the exact solution. One way around this problem is to apply nonlinear stability conditions to some continuously defined *functional* representation of the solution, such as a piecewise-polynomial reconstruction. Maxima and minima in functional representations behave like maxima and minima in the exact functional solution, since there is no longer any concern about alignment between the samples and extrema. Finite-element methods use piecewise-polynomials as their base representation. However, this chapter is confined to finite-difference methods. For finite-difference methods, a functional solution must be reconstructed from samples. However, lacking knowledge of the true maxima and minima, any maximum in the reconstruction that exceeds the corresponding maximum in the samples could very well represent an overshoot, and any minimum in the reconstruction that falls below the corresponding minimum in the samples could very well represent an undershoot. Thus, the idea of applying stability conditions to a functional representation rather than samples does not really change the situation for finite-difference methods. In other words, there remains a trade-off between stability and formal accuracy at extrema. The only way to avoid second-order clipping at extrema is to allow occasional under- and overshoots at extrema, a type of instability, if you care to view it that way. Many methods play it safe and tolerate clipping for the sake of robust stability.

One possible way to avoid this trade-off, first suggested by Zalesak (1979), is to store maxima and minima from time step to time step, in addition to the ordinary samples. This prevents problems, such as those seen in Example 16.3, where the values of maxima and minima are immediately lost when sampling the initial conditions or, otherwise, in the first few time steps. It also requires less storage than retaining an entire functional representation from time step to time step, as in finite-element methods. However, no one has yet devised a practical means for tracking maxima and minima, and the modern trend is surely away from such special bookkeeping. For more on these sorts of issues, see Zalesak (1979); Harten, Engquist, Osher, and Chakravarthy (1987); and Sanders (1988).

Until now, this chapter has only defined and described nonlinear stability conditions, without providing any clues about the actual enforcement of nonlinear stability conditions, aside from assertions such as "this condition is easy to enforce" or "this condition is hard to enforce." So how are nonlinear stability conditions enforced? Flux and solution averaging were discussed in Sections 13.3 and 13.6, respectively. After stencil selection, flux and solution averaging still allow a wide latitude to adjust the approximation on the chosen stencil. This freedom can be used to satisfy nonlinear stability conditions, such as positivity or the upwind range condition. More specifically, nonlinear stability conditions place upper and lower bounds on averaging parameters such as $\theta^n_{i+1/2}$ or $\phi^n_{i+1/2}$. Even after enforcing such nonlinear stability conditions, flux and solution averaging usually have quite a bit of freedom to spare to use for other things. This is discussed in detail in Part V, especially Chapter 20. Thus, for example, when you hear someone refer to a "TVD method" they generally mean a numerical method that (1) involves solution sensitivity achieved through flux averaging or, rarely, solution averaging (2) uses that solution sensitivity to enforce some sort of nonlinear stability condition which implies TVD, such as the upwind range condition, at least for certain model problems (3) limits the order of accuracy at extrema, usually to between first and second order; and (4) came after the invention of the term

TVD, circa 1983. Therefore, when applied to specific methods, terms such as TVD, TVB, and ENO usually imply much more than nonlinear stability.

To keep the discussion within reasonable bounds, this chapter has only concerned explicit finite-difference methods for scalar conservation laws on unbounded domains. Let us very briefly consider the effects of lifting these restrictions. First consider implicit methods. Except for the positivity condition, all of the nonlinear stability conditions in this chapter are properties of the exact solution and thus apply equally well to explicit and implicit methods; see Harten (1984) and Problem 16.8 for a simple extension of positivity to a class of implicit methods. Second, consider finite-volume methods. The maxima and the minima of cell-integral averages behave differently than the maxima and minima of samples or, for that matter, the maxima and minima of the exact solution. Unfortunately, there have been few attempts to account for the subtle but important second-order differences between samples and cell-integral averages. See Section 2.7 of Laney and Caughey (1991c) for a short discussion. Third, consider boundary conditions. As with linear stability, solid and far-field boundary conditions have profound effects on nonlinear stability; this is discussed briefly in Section 19.1. In practice, most boundary treatments attempt to make the boundary method as physical as possible and hope for the best as far as numerics and stability. Unlike solid and far-field boundaries, periodic boundaries behave very much like the infinite boundaries studied in this chapter; see Problem 16.3.

16.11 Stability and Convergence

By the Lax Equivalence Theorem seen in Section 15.4, the "unbounded growth" and "convergence" definitions of stability and instability are essentially equivalent for linear methods, at least as for smooth solutions. Unfortunately, nonlinearity disturbs this simple relationship; although nonlinear "unbounded growth" stability is necessary for "convergence" stability, they are no longer exactly equivalent. Suppose that a numerical method for a scalar conservation law on an infinite domain is conservative, is consistent for smooth solutions (it need not be consistent for nonsmooth solutions), and has well-posed initial conditions. Also, for given initial conditions, suppose that the total variation is bounded for all n and for all sufficiently small Δx and Δt. In other words, suppose that the solution is total variation stable. Then the numerical approximation converges to an exact weak solution in the 1-norm as $\Delta x \to 0$ and $\Delta t \to 0$. For a mathematically rigorous statement and proof of this result, see Theorem 15.2 of LeVeque (1992). Unlike the Lax Equivalence Theorem, this result applies both to smooth and nonsmooth solutions.

Unfortunately, converged solutions may not satisfy the entropy conditions discussed in Chapter 4. In other words, although total variation stability implies convergence to an exact solution, it does not necessarily imply convergence to *the* exact solution. Entropy conditions are not an issue for linear problems, which have only a single weak solution. However, entropy conditions are a major issue for nonlinear problems when the solution might contain shocks. The monotone condition seen in Section 16.9 ensures convergence to the entropy solution; see, for example, Theorem 15.7 in LeVeque (1992). However, the monotone condition also restricts the order of accuracy to first order, limiting its usefulness in most practical applications. In general, for higher-order accurate methods, entropy conditions must be proven on a case by case basis, exploiting the specific details of each numerical method, in addition to general properties such as positivity and conservation. Such proofs tend to be extremely long, complicated, and mathematical and are well beyond the scope of this book. For some examples, see Osher and Chakravarthy (1984) or Osher and Tadmor

(1988). Of course, just because the solution satisfies entropy conditions in the limit $\Delta x \to 0$ and $\Delta t \to 0$ does not mean that the solution satisfies entropy conditions for ordinary values of Δx and Δt, as discussed in Section 4.10.

Because of the Lax Equivalence Theorem, many people equate stability and convergence, even for nonlinear methods. Whereas most nonlinear stability conditions imply convergence, most nonlinear stability conditions imply much more than just convergence. By most definitions of stability, the main goal of stability is to prevent large errors for ordinary values of Δx and Δt, rather than to ensure zero errors in the limit $\Delta x \to 0$ and $\Delta t \to 0$. Hence, by most definitions of stability, convergence is simply a fringe benefit of nonlinear stability. However, for those who equate stability with convergence, the TVB and total variation stability conditions are the most important nonlinear stability conditions, and the other nonlinear stability conditions seen in this chapter are interesting only because they imply the TVB or total variation stability conditions and are easier to prove than TVB or total variation stability directly. Similarly, if stability is taken to mean convergence to the solution that satisfies entropy conditions, then the TVB or total variation stability conditions are the most important stability conditions, in the sense that they ensure convergence, while the other stronger nonlinear stability conditions in this chapter are introduced primarily to help encourage convergence to the solution that satisfies the entropy condition.

Although desirable, perfection in the limit $\Delta x \to 0$ and $\Delta t \to 0$ may not imply much about the accuracy of the solution for ordinary values of Δx and Δt. For example, as seen in Chapter 7, Legendre polynomial series, Chebyshev series, and Fourier series all converge in the 1-norm and 2-norm as $\Delta x \to 0$, but the maximum error or ∞-norm error is large for any finite Δx in the presence of jump discontinuities, due to Gibbs oscillations. The most helpful convergence results take the form of a strict upper bound on the maximum error that decreases rapidly as $\Delta x \to 0$ and $\Delta t \to 0$. Unfortunately, such convergence results are rare in computational gasdynamics.

16.12 The Euler Equations

This section concerns nonlinear stability for the Euler equations. In theory, the nonlinear stability analysis seen in the rest of this chapter only applies if the Euler equations share the relevant nonlinear stability properties of scalar conservation laws. Of course, the weaker the nonlinear stability property, the more likely it is that the Euler equations will satisfy it. For example, most equations satisfy TVB, since this only requires that solutions do not blow up. On the other hand, only a few equations satisfy the range diminishing or upwind range conditions since, for starters, these only make sense for pure wavelike solutions. Surprisingly, the nonlinear stability conditions described in this chapter can often be used even when the governing equations do not have the relevant nonlinear stability traits. For example, in the Navier–Stokes equations, the inviscid "wavelike" terms may be discretized in a way that enforces nonlinear stability conditions, whereas the viscous "nonwavelike" terms can be discretized separately in a way that does not enforce nonlinear stability conditions. Historically, the potential usefulness of nonlinear stability conditions for equations that do not share the relevant nonlinear properties of scalar conservation laws was noted along with one of the very first nonlinear stability conditions – see Figure 5 in Boris and Book (1973) and the associated discussion.

Fortunately, the characteristic variables of the one-dimensional Euler equations share most of the nonlinear stability properties of scalar conservation laws – including monotonicity preservation, TVD, range diminishing, the upwind range condition, TVB, and ENO – except possibly for waveforms created by jump discontinuity intersections, waveforms created by reflections of jump discontinuities from solid boundaries, and waveforms created by jump discontinuities in the initial conditions. For example, the solution to the Riemann problem seen in Example 5.1 creates a new minimum in the entropy, which is a characteristic variable. Notice that, in general, the primitive variables, the conservative variables, or any other set of variables besides the characteristic variables of the Euler equations most certainly do not share most of the nonlinear stability properties of scalar conservation laws, regardless of whether the solution is waveform creating or not.

Although the total variation of any single characteristic variable may increase during waveform creation, the total variation summed over all three families of characteristic variables cannot increase. Unfortunately, a numerical method that shares this property may allow large spurious oscillations and overshoots, even more so than TVD methods for scalar conservation laws. As an alternative, the characteristic variables of the Euler equations are TVB, measured either separately or summed together, although this property allows even larger spurious oscillations and overshoots than TVD. Even if the characteristic variables of the Euler equations always had the same nonlinear stability properties as scalar conservation laws, few numerical methods can actually access characteristic variables directly. Instead, most numerical methods access characteristic variables indirectly via flux vector splitting or Riemann solvers, as discussed in Chapter 13. Without direct access to characteristics, it may be difficult to enforce the nonlinear stability conditions discussed in this chapter even in those regions where they apply.

Certainly there are sticky theoretical obstacles to the extension of nonlinear stability conditions from scalar conservation laws to the Euler equations, the Navier–Stokes equations, and other equations. However, there are a number of more or less successful practical techniques for enforcing nonlinear stability conditions. While these practical techniques do not always rigorously ensure nonlinear stability, they do well enough for most purposes, as judged by the numerical results. In fact, in practice, almost all modern shock-capturing methods for the Euler and Navier–Stokes equations have been heavily influenced by the nonlinear stability concepts described in this chapter. Specific techniques will be studied in Part V.

16.13 Proofs

This section contains four proofs deferred from other sections in the chapter. To begin with, consider Equation (16.2) from Section 16.2. First, let us show that the total variation of a *discrete* function equals a signed sum of extrema, maxima counted positively and minima counted negatively, interior extrema counted twice, and boundary extrema counted once. Rewrite Equation (16.5) in terms of each u_i^n separately rather that in terms of the differences $u_{i+1}^n - u_i^n$. To find the coefficient of u_i^n, consider the two terms in the total variation sum that involve u_i^n as follows:

$$
\begin{aligned}
&\left|u_i^n - u_{i-1}^n\right| + \left|u_{i+1}^n - u_i^n\right| \\
&\quad = \begin{cases}
u_{i+1}^n - u_{i-1}^n & \text{if } u_i^n - u_{i-1}^n \geq 0 \quad \text{and} \quad u_{i+1}^n - u_i^n \geq 0, \\
u_{i-1}^n - u_{i+1}^n & \text{if } u_i^n - u_{i-1}^n \leq 0 \quad \text{and} \quad u_{i+1}^n - u_i^n \leq 0, \\
2u_i^n - u_{i+1}^n - u_{i-1}^n & \text{if } u_i^n - u_{i-1}^n > 0 \quad \text{and} \quad u_{i+1}^n - u_i^n < 0, \\
-2u_i^n + u_{i+1}^n + u_{i-1}^n & \text{if } u_i^n - u_{i-1}^n < 0 \quad \text{and} \quad u_{i+1}^n - u_i^n > 0.
\end{cases}
\end{aligned}
$$

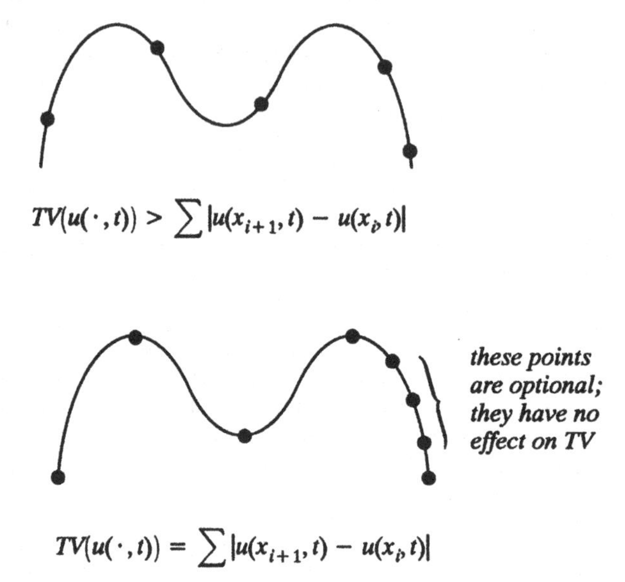

Figure 16.9 Maximizing the total variation of the samples of a continuously defined function.

Thus the coefficient of u_i^n is 2 if u_i^n is a maximum, -2 if u_i^n is a minimum, and 0 otherwise. Boundary terms only occur once in the sum and thus can only contribute once rather than twice to the sum. For a somewhat more rigorous proof, see Laney and Caughey (1991a).

Now let us prove that the total variation of a *continuously defined* function equals a signed sum of extrema, maxima counted positively and minima counted negatively, interior extrema counted twice, and boundary extrema counted once. Recall definition (16.1). For any grid

$$\sum_{i=-\infty}^{\infty} |u(x_{i+1}, t) - u(x_i, t)|$$

equals a signed sum of sample extrema, with sample maxima counted positively and sample minima counted negatively, interior sample extrema counted twice, and boundary sample extrema counted once, as proven in the last paragraph. The samples $u(x_i, t)$ may have fewer maxima and minima than $u(x, t)$. However, for any extremum in the samples $u(x_i, t)$, there is a corresponding extremum in $u(x, t)$. In particular, if a sample $u(x_i, t)$ is a local sample maximum then a corresponding local maximum exists in $u(x, t)$ which is greater than or equal to $u(x_i, t)$; similarly if $u(x_i, t)$ is a local sample minimum then a corresponding local minimum exists in $u(x, t)$ which is less than or equal to $u(x_i, t)$. Now choose a set of samples that maximizes the total variation. In particular, suppose that one sample falls on the crest of each and every local maxima and minima of $u(x, t)$ and that there is one sample at each of the infinite boundaries. There can be any number of other samples as well, but these other samples do not affect the total variation. This is illustrated in Figure 16.9. Then the maxima of the samples are all as large as possible, the minima are all as small as possible, and the samples have as many maxima and minima as possible. This implies that

$$\sum_{i=-\infty}^{\infty} |u(x_{i+1}, t) - u(x_i, t)|$$

is as large as possible and thus

$$TV(u(\cdot,t)) = \sum_{i=-\infty}^{\infty} |u(x_{i+1},t) - u(x_i,t)|.$$

But, as shown above, the total variation of the samples equals a signed sum of extrema, maxima counted positively and minima counted negatively, interior sample extrema counted twice, and boundary sample extrema counted once. By construction, the maxima and minima of the samples $u(x_i, t)$ are exactly the same as the maxima and minima of $u(x, t)$, and thus the result for the total variation of the samples $u(x_i, t)$ carries over to $u(x, t)$. This completes the proof.

As asserted in Section 16.5, an explicit method has the upwind range condition if and only if it can be written in wave speed split form such that

$$0 \le C^-_{i+1/2} \le 1, \qquad C^+_{i+1/2} = 0, \qquad 0 \le \lambda a\left(u_i^{n+1}\right) \le 1, \tag{16.16a}$$

$$0 \le C^+_{i+1/2} \le 1, \qquad C^-_{i+1/2} = 0, \qquad -1 \le \lambda a\left(u_i^{n+1}\right) \le 0 \tag{16.16b}$$

for all i. The following is a proof of Equation (16.16a). The proof of Equation (16.16b) is similar. First, let us show that (16.16a) implies the upwind range condition. Suppose a method can be written as follows:

$$u_i^{n+1} = u_i^n - C^-_{i-1/2}\left(u_i^n - u_{i-1}^n\right),$$

where $0 \le C^-_{i-1/2} \le 1$. Letting $C^-_{i-1/2} = 0$ gives $u_i^{n+1} = u_i^n$, while letting $C^-_{i-1/2} = 1$ gives $u_i^{n+1} = u_{i-1}^n$. Then $0 < C^-_{i-1/2} < 1$ gives a value for u_i^{n+1} between u_i^n and u_{i-1}^n. In other words, $0 \le C^-_{i-1/2} \le 1$ implies

$$\min\left(u_i^n, u_{i-1}^n\right) \le u_i^{n+1} \le \max\left(u_i^n, u_{i-1}^n\right),$$

which is the upwind range condition. Going in reverse, let us show that the upwind range condition implies (16.16a). If a method is upwind range and $0 \le \lambda a(u_i^{n+1}) \le 1$ then

$$\min\left(u_i^n, u_{i-1}^n\right) \le u_i^{n+1} \le \max\left(u_i^n, u_{i-1}^n\right).$$

Any explicit forward-time method can be written uniquely in the following wave speed split form:

$$u_i^{n+1} = u_i^n - C^-_{i-1/2}\left(u_i^n - u_{i-1}^n\right).$$

To see this, notice that any forward-time method can be written as

$$u_i^{n+1} = u_i^n - H_i^n,$$

where H_i may represent conservative flux differences, FTCS plus artificial viscosity, or any other form you care to name. Then

$$C^-_{i-1/2} = H_i^n / \left(u_i^n - u_{i-1}^n\right).$$

The upwind range condition implies

$$\min\left(u_i^n, u_{i-1}^n\right) \le u_i^n - C^-_{i-1/2}\left(u_i^n - u_{i-1}^n\right) \le \max\left(u_i^n, u_{i-1}^n\right)$$

or

$$\min\left(0, u_{i-1}^n - u_i^n\right) \le -C^-_{i-1/2}\left(u_i^n - u_{i-1}^n\right) \le \max\left(0, u_{i-1}^n - u_i^n\right)$$

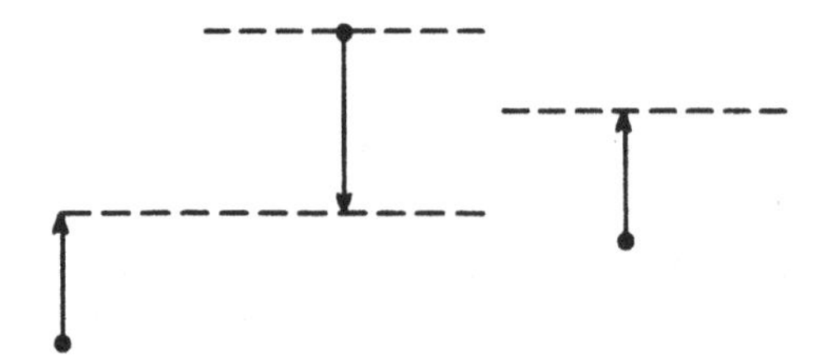

Figure 16.10 Trapped inside a compartment, the maximum can move to an adjacent point but cannot increase.

or

$$\min(0, u_i^n - u_{i-1}^n) \le C_{i-1/2}^-(u_{i-1}^n - u_i^n) \le \max(0, u_i^n - u_{i-1}^n).$$

If $u_i^n - u_{i-1}^n \ge 0$ this says

$$0 \le C_{i-1/2}^-(u_{i-1}^n - u_i^n) \le u_i^n - u_{i-1}^n$$

or

$$0 \le C_{i-1/2}^- \le 1.$$

If $u_i^n - u_{i-1}^n \le 0$ this says

$$u_i^n - u_{i-1}^n \le C_{i-1/2}^-(u_{i-1}^n - u_i^n) \le 0$$

or

$$0 \le C_{i-1/2}^- \le 1.$$

Thus the upwind range condition implies that a method can be written in wave speed split form such that $0 \le C_{i-1/2}^- \le 1$ and $C_{i+1/2}^+ = 0$. This completes the proof of Equation (16.16a).

As the last proof in this section, let us show that the upwind range condition implies range diminishing, except possibly at sonic points, as asserted in Section 16.5. A *compartment* is any region whose upper and lower bounds lie between u_i^n and u_{i-1}^n and between u_i^n and u_{i+1}^n. For example, the upwind range condition traps each u_i^{n+1} inside compartments whose upper and lower bounds are u_i^n and its upwind neighbor, either u_{i-1}^n or u_{i+1}^n. Let us show that a solution is range diminishing provided that u_i^{n+1} is trapped inside a compartment, regardless of how the compartment is constructed. In monotone regions, the compartments are disjoint, and there is no way for two points to cross one another, and thus there is no way to create a new maximum and minimum. A compartment enclosing a maximum has its upper edge exactly aligned with the maximum, which prevents the maximum from increasing. Also, in this case, the compartment may overlap at most one neighboring compartment, which allows the maximum to move to a neighboring compartment if appropriate. A compartment condition at a maximum is illustrated in Figure 16.10. Similar reasoning applies at minima. Thus compartments imply range reduction. The upwind range condition implies compartments except possibly at sonic points. This completes the proof.

References

Boris, J. P., and Book, D. L. 1973. "Flux-Corrected Transport I. SHASTA, a Fluid Transport Algorithm that Works," *Journal of Computational Physics*, 11: 38–69.

Coray, C., and Koebbe, J. 1993. "Accuracy Optimized Methods for Constrained Numerical Solutions of Hyperbolic Conservation Laws," *Journal of Computational Physics*, 109: 115–132.

Crandall, M. G., and Majda, A. 1980. "Monotone Difference Approximations for Scalar Conservation Laws," *Mathematics of Computation*, 34: 1–21.

Godunov, S. K. 1959. "A Difference Scheme for Numerical Computation of Discontinuous Solutions of Hydrodynamics Equations," *Math. Sbornik*, 47: 271–306.

Goodman, J. B., and LeVeque, R. J. 1985. "On the Accuracy of Stable Schemes for 2D Scalar Conservation Law," *Mathematics of Computation*, 45: 15–21.

Harten, A. 1983. "High Resolution Schemes for Hyperbolic Conservation Laws," *Journal of Computational Physics*, 49: 357–393.

Harten, A. 1984. "On a Class of High Resolution Total-Variation-Stable Finite-Difference Schemes," *SIAM Journal on Numerical Analysis*, 21: 1–23.

Harten, A., Engquist, B., Osher, S., and Chakravarthy, S. R. 1987. "Uniformly High Order Accurate Essentially Non-Oscillatory Schemes, III," *Journal of Computational Physics*, 71: 231–303.

Harten, A., Hyman, J. M., and Lax, P. D. 1976. "On Finite-Difference Approximations and Entropy Conditions for Shocks," *Communications on Pure and Applied Mathematics*, 29: 297–322.

Jameson, A. 1995. "Positive Schemes and Shock Modelling for Compressible Flows," *International Journal for Numerical Methods in Fluids*, 20: 743–776.

Laney, C. B., and Caughey, D. A. 1991a. "Extremum Control II: Semidiscrete Approximations to Conservation Laws," *AIAA Paper 91-0632* (unpublished).

Laney, C. B., and Caughey, D. A. 1991b. "Extremum Control III: Fully-Discrete Approximations to Conservation Laws," *AIAA Paper 91-1534* (unpublished).

Laney, C. B., and Caughey, D. A. 1991c. "Monotonicity and Overshoot Conditions for Numerical Approximations to Conservation Laws," Sibley School of Mechanical and Aerospace Engineering, Fluid Dynamics Program, Cornell University, *Report FDA-91-10* (unpublished).

Lax, P. D. 1973. *Hyperbolic Systems of Conservation Laws and the Mathematical Theory of Shock Waves*, Regional Conference Series in Applied Mathematics, 11, SIAM.

LeVeque, R. J. 1992. *Numerical Methods for Conservation Laws*, 2nd ed., Basel: Birkhäuser-Verlag, Chapter 15.

Osher, S. 1984. "Riemann Solvers, the Entropy Condition, and Difference Approximations," *SIAM Journal on Numerical Analysis*, 21: 217–235.

Osher, S., and Chakravarthy, S. R. 1984. "High Resolution Schemes and the Entropy Condition," *SIAM Journal on Numerical Analysis*, 21: 955–984.

Osher, S., and Tadmor, E. 1988. "On the Convergence of Difference Approximations to Scalar Conservation Laws," *Mathematics of Computation*, 50: 19–51.

Royden, H. L. 1968. *Real Analysis*, 2nd ed., New York: MacMillan.

Sanders, R. 1988. "A Third-Order Accurate Variation Nonexpansive Difference Scheme for Single Nonlinear Conservation Laws," *Mathematics of Computation*, 51: 535–558.

Shu, C.-W. 1987. "TVB Uniformly High-Order Accurate Schemes for Conservation Laws," *Mathematics of Computation*, 49: 105–121.

Sweby, P. K. 1984. "High Resolution Schemes Using Flux Limiters for Hyperbolic Conservation Laws," *SIAM Journal on Numerical Analysis*, 21: 985–1011.

Van Leer, B. 1974. "Towards the Ultimate Conservative Difference Scheme. II. Monotonicity and Conservation Combined in a Second-Order Scheme," *Journal of Computational Physics*, 14: 361–370.

Zalesak, S. T. 1979. "Fully Multidimensional Flux-Corrected Transport Algorithms for Fluids," *Journal of Computational Physics*, 31: 335–362.

Problems

16.1 Under what circumstances will FTBS have the following nonlinear stability conditions?

(a) monotonicity preservation (b) total variation diminishing
(c) range diminishing (d) positivity
(e) upwind range (f) total variation bounded
(g) essentially nonoscillatory (h) contracting
(i) monotone (j) total variation stable

16.2 Consider the following method:

$$u_i^{n+1} = u_i^n + \frac{\lambda}{12}\left(f\left(u_{i+2}^n\right) - 8f\left(u_{i+1}^n\right) + 8f\left(u_{i-1}^n\right) - f\left(u_{i-2}^n\right)\right).$$

(a) Using Godunov's theorem, argue that this method cannot be monotonicity preserving when applied to the linear advection equation.
(b) Find the coefficient of second-order artificial viscosity $\epsilon_{i+1/2}^n$. Does this method satisfy the nonlinear stability condition $|\lambda a_{i+1}^n| \leq \lambda \epsilon_{i+1/2}^n \leq 1$ found in Example 16.5? If so, what are the constraints on the CFL number? Explain carefully.
(c) Does this method satisfy the positivity condition? If so, what are the constraints on the CFL number?

16.3 This chapter described nonlinear stability on infinite domains. This problem concerns nonlinear stability on periodic domains.
(a) Argue that monotonicity preservation does not make sense on periodic domains.
(b) Express the total variation of a continuously defined and discretely defined function on a periodic domain as a linear combination of maxima and minima.

16.4 Consider the following function:

$$u(x) = 1 + \frac{-2x + 5}{3x^2 + 4}$$

on an infinite domain $-\infty < x < \infty$.
(a) Find the total variation of the continuously defined function $u(x)$.
(b) Find the total variation of the samples $u(i)$ where i is an integer between $-\infty$ and ∞. In other words, find the total variation of the discretely defined function $\{\ldots, u(-2), u(-1), u(0), u(1), u(2), \ldots\}$. Explain your reasoning carefully. Why does your answer differ from that found in part (a)?
(c) Find samples x_i that maximize the total variation of $u(x_i)$.

16.5 This problem concerns the range diminishing condition. A function is range diminishing if its range diminishes both globally *and locally* in time. If the global range of a function increases, then the local range must also increase somewhere, and then the function is not range reducing. In contrast, if the global range of a function decreases, the local range may decrease *or increase*. The whole concept of range reduction only makes sense if the function has wavelike behavior, so that individual pieces of the function can be tracked through time, along wavefronts, to determine whether the local range is increasing or decreasing in time. If the range of each local piece is decreasing, then the function is range reducing. With this background in mind, are the following functions range reducing for $t > 0$? If not, are the following functions TVD or TVB?
(a) $u(x,t) = e^{-(x-at)^2-t}$ where $a = const.$

(b) $u(x, t) = \cos(k_a x - \omega_a t) - \cos(k_b x - \omega_b t)$ where k_a, k_b, ω_a, and ω_b are constants.
(c) $u(x, t) = \frac{x^2-1}{x^2+1} \sin t$.

16.6 In 1979, Zalesak suggested the following nonlinear stability condition:
$\min\left(u_{i-1}^n, u_i^n, u_{i+1}^n\right) \leq u_i^{n+1} \leq \min\left(u_{i-1}^n, u_i^n, u_{i+1}^n\right)$.
(a) Argue that this nonlinear stability condition does not allow existing maxima to increase or existing minima to decrease.
(b) Argue that this nonlinear condition allows large new maxima and minima. In other words, argue that this nonlinear stability condition allows large spurious oscillations.

16.7 While the main body of this chapter concerns only explicit methods, this problem and the next concern implicit methods. Consider an implicit backward-time conservative approximation in the following form:

$$u_i^{n+1} = u_i^n - \lambda\left(\hat{f}_{i+1/2}^{n+1} - \hat{f}_{i-1/2}^{n+1}\right),$$

where $\hat{f}_{i\pm1/2}^{n+1}$ are functions of the solution at time level $n + 1$.
(a) Suppose that this method can be written in implicit wave speed split form as follows:

$$u_i^{n+1} = u_i^n + C_{i+1/2}^{+}\left(u_{i+1}^{n+1} - u_i^{n+1}\right) - C_{i-1/2}^{-}\left(u_i^{n+1} - u_{i-1}^{n+1}\right).$$

Show that the positivity condition $C_{i+1/2}^{+} \geq 0$ and $C_{i+1/2}^{-} \geq 0$ for all i implies TVD.
(b) Suppose that this method can be written in implicit artificial viscosity form as follows:

$$u_i^{n+1} = u_i^n - \frac{\lambda}{2}\left(f\left(u_{i+1}^{n+1}\right) - f\left(u_{i-1}^{n+1}\right)\right)$$
$$+ \frac{\lambda}{2}\left(\epsilon_{i+1/2}^{n+1}\left(u_{i+1}^{n+1} - u_i^{n+1}\right) - \epsilon_{i-1/2}^{n+1}\left(u_i^{n+1} - u_{i-1}^{n+1}\right)\right).$$

Show that the positivity condition from part (a) is satisfied if $|a_{i+1/2}^n| \leq \epsilon_{i+1/2}^n$.

16.8 An explicit forward-time conservative approximation is as follows:

$$u_i^{n+1} = u_i^n - \lambda\left(\hat{f}_{i+1/2}^{n} - \hat{f}_{i-1/2}^{n}\right).$$

Also, an implicit backward-time conservative approximation is as follows:

$$u_i^{n+1} = u_i^n - \lambda\left(\hat{f}_{i+1/2}^{n+1} - \hat{f}_{i-1/2}^{n+1}\right).$$

Consider a convex linear combination of the explicit forward-time approximation and the implicit backward-time approximations as follows:

$$u_i^{n+1} = u_i^n - \lambda(1-\theta)\left(\hat{f}_{i+1/2}^{n} - \hat{f}_{i-1/2}^{n}\right) - \lambda\theta\left(\hat{f}_{i+1/2}^{n+1} - \hat{f}_{i-1/2}^{n+1}\right),$$

where $0 \leq \theta \leq 1$.
(a) Suppose that this method can be written in a wave speed split form as follows:

$$u_i^{n+1} = u_i^n + (1-\theta)\left(C_{i+1/2}^{+}\right)^n\left(u_{i+1}^n - u_i^n\right) - (1-\theta)\left(C_{i-1/2}^{-}\right)^n\left(u_i^n - u_{i-1}^n\right)$$
$$+ \theta\left(C_{i+1/2}^{+}\right)^{n+1}\left(u_{i+1}^{n+1} - u_i^{n+1}\right) - \theta\left(C_{i-1/2}^{-}\right)^{n+1}\left(u_i^{n+1} - u_{i-1}^{n+1}\right).$$

Show that the positivity condition $(C_{i+1/2}^{+})^n \geq 0$,

$(C_{i+1/2}^{-})^n \geq 0$, $\left(C_{i+1/2}^{+}\right)^n + \left(C_{i+1/2}^{+}\right)^n \leq 1/(1-\theta)$

for all i and n implies the TVD condition.

(b) Suppose that this method can be written in implicit artificial viscosity form as follows:

$$
\begin{aligned}
u_i^{n+1} = u_i^n &- \frac{\lambda}{2}(1-\theta)\left(f\left(u_{i+1}^n\right) - f\left(u_{i-1}^n\right)\right) - \frac{\lambda}{2}\theta\left(f\left(u_{i+1}^{n+1}\right) - f\left(u_{i-1}^{n+1}\right)\right) \\
&+ \frac{\lambda}{2}(1-\theta)\left(\epsilon_{i+1/2}^n\left(u_{i+1}^n - u_i^n\right) - \epsilon_{i-1/2}^n\left(u_i^n - u_{i-1}^n\right)\right) \\
&+ \frac{\lambda}{2}\theta\left(\epsilon_{i+1/2}^{n+1}\left(u_{i+1}^{n+1} - u_i^{n+1}\right) - \epsilon_{i-1/2}^{n+1}\left(u_i^{n+1} - u_{i-1}^{n+1}\right)\right).
\end{aligned}
$$

Show that the positivity condition from part (a) is satisfied if

$\left|a_{i+1/2}^n\right| \leq \epsilon_{i+1/2}^n \leq 1/(1-\theta)$ for all i and n.

Part IV

Basic Methods of Computational Gasdynamics

Most of the simple methods described in Part III have almost no potential for practical applications, due mainly to stability problems. This fourth part of the book concerns methods at the next higher level of sophistication, here referred to as the *first generation* of numerical methods for computational gasdynamics. However sophisticated they may be in other ways, by definition, first-generation methods do not use flux averaging, slope averaging, or other forms of solution sensitivity, except possibly upwinding. As a result, first-generation methods experience a sharp trade-off between accuracy and stability: They can model shocks well but then experience low accuracy in smooth regions; or they can model smooth regions well but then experience poor stability near shocks in the form of spurious oscillations and overshoots. The next part of the book, Part V, describes solution-sensitive second- and third-generation numerical methods, which reduce this trade-off by doing one thing in smooth regions and another at shocks. First-generation methods prove useful in undemanding applications and, more importantly, are the basic building blocks of second- and third-generation methods.

Chapter 17 describes numerical methods for scalar conservation laws. Chapter 18 extends those methods to the Euler equations using flux vector splitting and Riemann solvers. Chapter 19 concerns solid and far-field boundary treatments, a crucial topic avoided until now by using periodic boundaries as in Chapter 15 or infinite boundaries as in Chapter 16.

CHAPTER 17

Basic Numerical Methods for Scalar Conservation Laws

17.0 Introduction

This chapter surveys simple classic methods for scalar conservation laws. As seen in Part III, the six basic design techniques used in computational gasdynamics are as follows:

(1) flux averaging (Section 13.3),
(2) flux and wave speed splitting (Sections 13.4 and 13.5),
(3) numerical integration and numerical differentiation (Chapter 10),
(4) Cauchy–Kowalewski (Section 15.3),
(5) method of lines (Subsection 11.2.1), and
(6) reconstruction–evolution (Section 13.6).

The numerical methods found in this chapter use all of these derivational techniques, albeit in the simplest possible ways. To begin with, Section 17.1 derives the Lax–Friedrichs method using numerical differentiation and the method of lines. Section 17.2 derives the Lax–Wendroff method using the Cauchy–Kowalewski technique and numerical differentiation. Section 17.3 derives three first-order upwind methods using reconstruction–evolution and numerical integration. Section 17.4 derives the Beam–Warming second-order upwind method using the Cauchy–Kowalewski technique, numerical differentiation, and flux splitting. Finally, Section 17.5 derives Fromm's method using a simple fixed flux average of the Lax–Wendroff method and the Beam–Warming second-order upwind method.

To help evaluate the numerical methods found in this chapter, consider the following twelve-point checklist based on the material found in Part III:

(1) Artificial viscosity (Chapter 14, Sections 15.1, 15.3, and 16.4).
(2) CFL condition (Chapter 12).
(3) Conservation (Chapter 11).
(4) Consistency (Sections 11.1, 15.3, 15.4 and 16.11).
(5) Convergence (Sections 15.4 and 16.11).
(6) Explicit versus implicit (Sections 11.1 and 12.1).
(7) Finite volume versus finite difference (Chapter 11).
(8) Linear stability (Section 11.1, Chapter 15).
(9) Linear versus nonlinear (Chapters 15 and 16).
(10) Nonlinear stability (Sections 11.1 and 12.1, Chapter 16).
(11) Order of accuracy (Sections 11.1 and 11.2).
(12) Upwinding and stencil selection (Chapter 13).

These theoretical specifications are no substitute for actual performance testing. To put each method through its paces, the rest of this introduction constructs a series of five

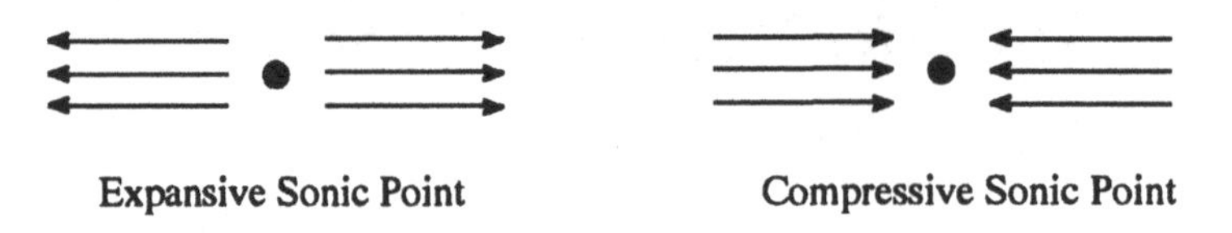

Figure 17.1 Expansive and compressive sonic points.

numerical test problems designed to cover the range of solution features, especially solution features that commonly cause numerical problems.

What kinds of solution features cause numerical problems? To begin with, many numerical methods have difficulties when a wave speed equals zero. In three dimensions, wave speeds equal zero along *sonic surfaces*; in two dimensions, wave speeds equal zero along *sonic lines*; and in one dimension, wave speeds equal zero at *sonic points*. In other words, u^* is a sonic point if $f'(u^*) = a(u^*) = 0$. Sonic points usually signal a change in the wind direction. Sonic points are *expansive* if the wave direction switches from left to right and *compressive* if the wave direction switches from right to left, as illustrated in Figure 17.1. Expansive sonic points typically occur inside sonic expansion fans, which contain one stationary characteristic separating left- from right-running characteristics. Compressive sonic points typically occur inside stationary or slowly moving shocks.

Section 4.5 introduced convex flux functions. Convex flux functions have at most one sonic point. For example, Burgers' equation has a convex flux function $f(u) = u^2/2$ with a unique sonic point $u^* = 0$. By contrast, the nonconvex flux $f(u) = \cos \pi u$ has infinitely many sonic points $u^* = 0, \pm 1, \pm 2, \pm 3, \ldots$.

Unless specific steps are taken, many numerical methods produce significant errors near sonic points, especially expansive sonic points, which often cause spurious expansion shocks. Upwind methods are forced to give sonic points special consideration, since the upwind direction changes at sonic points. Although not required, centered methods should also treat sonic points differently from other points, or suffer the consequences.

Besides sonic points, shock waves, contact discontinuities, expansion fans, and other nonsmooth flow features are major stumbling blocks for many numerical methods. Typical symptoms include oscillations, overshoots, and a smearing that spreads the discontinuities over a region of from two to seven cells. As far as a shock smearing goes, the physical compression and numerical expansion reach an equilibrium after only a few time steps, effectively halting any further smearing. Contact discontinuities do not have any physical compression, and thus numerical smearing increases progressively with the number of time steps, as if the contact were an expansion fan. Not surprisingly, contact smearing decreases as the order of accuracy increases. Specifically, Harten (1978) estimated that, unless specific steps are taken, numerical methods will smear contact discontinuities over a region proportional to $n^{1/(r+1)}$ where n is the number of time steps and r is the order of accuracy. In the same paper, Harten described the most popular measure to combat contact smearing, known as *artificial compression*; this technique is discussed in Section 21.3. For reconstruction–evolution methods, the other alternative is subcell resolution, as described in Section 9.4. The first-generation methods seen in this chapter incorporate neither artificial compression nor subcell resolution and are all subject to contact smearing according to their order of accuracy. As far as expansions are concerned, expansions typically have corners (discontinuities in the first derivative of the solution) at their head and tail. Many numerical methods may experience spurious overshoots, oscillations, and smearing at the heads and tails, albeit to a lesser extent than at shocks and contacts. Furthermore, some numerical methods partially or completely replace sonic expansions by expansion shocks.

While recognizing the importance of shocks, contacts, expansion fans, and sonic points, let us not forget about ordinary smooth solutions. Although far less challenging than discontinuous solutions, most numerical methods exhibit obvious flaws even on completely smooth solutions, at least for large enough Δx, Δt, and t. For example, for smooth solutions of scalar conservation laws, the exact solution preserves the local and global range, as discussed in Sections 4.11 and Chapter 16, especially Section 16.3. By contrast, most stable numerical approximations continuously erode the solution, dramatically reducing the local and global range. Besides amplitude and dissipative errors, most numerical methods also experience phase errors, causing the approximate solution to lead or lag the true solution, and dispersive errors, causing spurious oscillations as different frequencies separate; see section 15.3. Unlike shocks, contacts, expansions, and sonic points, the formal order of accuracy is a generally reliable indicator of the size of phase, dispersive, and amplitude errors for smooth solutions, assuming only that the numerical method is reasonably stable.

The following five test cases involve all of the flow features identified above – shocks, contacts, expansion fans, sonic points, and smooth regions. To avoid the complicating influence of boundary conditions, all five test cases involve a periodic domain $[-1, 1]$. In other words, $u(-1, t) = u(1, t)$ for all t and, furthermore, $u(x - 1, t) = u(x + 1, t)$ for all x and t.

Test Case 1 Find $u(x, 30)$ where

$$\frac{\partial u}{\partial t} + \frac{\partial u}{\partial x} = 0,$$
$$u(x, 0) = -\sin(\pi x),$$

which is linear advection of one period of a sinusoid. Use 40 evenly spaced grid points and $\lambda = \Delta t/\Delta x = 0.8$. At $t = 30$, the initial conditions have traveled around the periodic domain $[-1, 1]$ exactly 15 times so that $u(x, 30) = u(x, 0)$. This test case has a completely smooth exact solution with no sonic points. This test case illustrates mainly phase and amplitude errors but not dispersion; since there is only one frequency in the exact solution, there is no way for different frequencies to separate as happens in dispersion.

Test Case 2 Find $u(x, 4)$ where

$$\frac{\partial u}{\partial t} + \frac{\partial u}{\partial x} = 0,$$
$$u(x, 0) = \begin{cases} 1 & \text{for } |x| < \frac{1}{3}, \\ 0 & \text{for } \frac{1}{3} < |x| \leq 1, \end{cases}$$

which is linear advection of a square wave. Use 40 evenly spaced grid points and $\lambda = \Delta t/\Delta x = 0.8$. At $t = 4$, the initial conditions have traveled around the periodic domain $[-1, 1]$ exactly twice so that $u(x, 4) = u(x, 0)$. The two jump discontinuities in the solution correspond to contact discontinuities. This test case illustrates progressive contact smearing and dispersion.

Test Case 3 Find $u(x,4)$ and $u(x,40)$ where

$$\frac{\partial u}{\partial t}+\frac{\partial u}{\partial x}=0,$$

$$u(x,0)=\begin{cases}1 & \text{for } |x|<\frac{1}{3},\\ 0 & \text{for } \frac{1}{3}<|x|\le 1,\end{cases}$$

which is linear advection of a square wave. Use 600 evenly spaced grid points and $\lambda = \Delta t/\,\Delta x = 0.8$. This example illustrates convergence as $\Delta x \to 0$ and $\Delta t \to 0$. It also illustrates how dissipation, dispersion, and other numerical artifacts accumulate with large times for discontinuous solutions.

Test Case 4 Find $u(x,0.6)$ where

$$\frac{\partial u}{\partial t}+\frac{\partial}{\partial x}\left(\frac{1}{2}u^2\right)=0,$$

$$u(x,0)=\begin{cases}1 & \text{for } |x|<\frac{1}{3},\\ 0 & \text{for } \frac{1}{3}<|x|\le 1.\end{cases}$$

Use 40 evenly spaced grid points and $\lambda = \Delta t/\Delta x = 0.8$. The scalar conservation law is Burgers' equation and the initial condition is a square wave. This problem was solved in Example 4.3. To review briefly, the jump from zero to one at $x=-1/3$ creates an expansion fan, while the jump from one to zero at $x=1/3$ creates a shock. The unique sonic point for Burgers' equation is $u^*=0$. Although the exact solution never crosses the sonic point, a numerical method with spurious overshoots or oscillations may cross the sonic point once or even several times, causing a dramatic error if the numerical method has sonic point problems.

Test Case 5 Find $u(x,0.3)$ where

$$\frac{\partial u}{\partial t}+\frac{\partial}{\partial x}\left(\frac{1}{2}u^2\right)=0,$$

$$u(x,0)=\begin{cases}1 & \text{for } |x|<\frac{1}{3},\\ -1 & \text{for } \frac{1}{3}<|x|\le 1.\end{cases}$$

Use 40 evenly spaced grid points and $\lambda = \Delta t/\Delta x = 0.8$. The scalar conservation law is Burgers' equation and the initial condition is a square wave. This problem was solved in Example 4.4. To review briefly, the jump from minus one to one at $x=-1/3$ creates a sonic expansion fan, while the jump from one to minus one at $x=1/3$ creates a steady shock. Notice that both the expansion fan and the shock symmetrically span the sonic point $u^*=0$.

17.1 Lax–Friedrichs Method

This section concerns the Lax–Friedrichs method, discovered in 1954. To derive the Lax–Friedrichs method, first consider FTCS:

$$u_i^{n+1}=u_i^n-\frac{\lambda}{2}\left(f\left(u_{i+1}^n\right)-f\left(u_{i-1}^n\right)\right). \tag{11.16}$$

As seen repeatedly in Part III, FTCS is unconditionally unstable. However, suppose u_i^n is replaced by $(u_{i+1}^n + u_{i-1}^n)/2$. With this modification, FTCS becomes the *Lax–Friedrichs method*:

$$u_i^{n+1} = \frac{1}{2}(u_{i+1}^n + u_{i-1}^n) - \frac{\lambda}{2}(f(u_{i+1}^n) - f(u_{i-1}^n)). \tag{17.1}$$

In conservation form, introduced in Chapter 11, the Lax–Friedrichs method is

$$u_i^{n+1} = u_i^n - \lambda(\hat{f}_{i+1/2}^n - \hat{f}_{i-1/2}^n),$$

where

$$\hat{f}_{i+1/2}^n = \frac{1}{2}(f(u_{i+1}^n) + f(u_i^n)) - \frac{1}{2\lambda}(u_{i+1}^n - u_i^n). \tag{17.2}$$

In artificial viscosity form, introduced in Chapter 14, the Lax–Friedrichs method is

$$u_i^{n+1} = u_i^n - \frac{\lambda}{2}(f(u_{i+1}^n) - f(u_{i-1}^n)) + \frac{\lambda}{2}(\epsilon_{i+1/2}^n(u_{i+1}^n - u_i^n) - \epsilon_{i-1/2}^n(u_i^n - u_{i-1}^n)),$$

where

$$\epsilon_{i+1/2}^n = \frac{1}{\lambda}. \tag{17.3}$$

In wave speed split form, introduced in Subsection 13.5.1, the Lax–Friedrichs method is

$$u_i^{n+1} = u_i^n + C_{i+1/2}^+(u_{i+1}^n - u_i^n) - C_{i-1/2}^-(u_i^n - u_{i-1}^n),$$

where

$$C_{i+1/2}^+ = \frac{1}{2}(1 - \lambda a_{i+1/2}^n), \tag{17.4a}$$

$$C_{i+1/2}^- = \frac{1}{2}(1 + \lambda a_{i+1/2}^n). \tag{17.4b}$$

Of course, as always, there are infinitely many other wave speed split forms. However, this is the only wave speed split form with finite coefficients – in this sense, this is the unique *natural* wave speed split form.

The above derivation paints Lax–Friedrichs as a minor modification of FTCS. However, despite the seemingly small size of the modification, Lax–Friedrichs and FTCS end up completely different.

The twelve-point checklist for the Lax–Friedrichs method is as follows:

(1) The Lax–Friedrichs method has the maximum amount of artificial viscosity allowed by the linear stability condition $(\lambda a)^2 \le \lambda\epsilon \le 1$ found in Example 15.4. Furthermore, it has the maximum amount of artificial viscosity allowed by the nonlinear stability condition $|\lambda a_{i+1/2}^n| \le \lambda\epsilon_{i+1/2}^n \le 1$ found in Example 16.6.
(2) CFL condition: $|\lambda a(u)| \le 1$.
(3) Conservative.
(4) Consistent.
(5) Converges to the correct solution as $\Delta x \to 0$ and $\Delta t \to 0$ provided that the CFL condition is satisfied.
(6) Explicit.
(7) Finite difference.

(8) Linearly stable provided that the CFL condition is satisfied.
(9) Linear when applied to the linear advection equation. Nonlinear when applied to any nonlinear scalar conservation law.
(10) Consider the nonlinear stability conditions seen in Chapter 16. The Lax–Friedrichs method satisfies the monotonicity preservation, TVD, positivity, TVB, ENO, contraction, and monotone conditions provided that the CFL condition is satisfied. It does not satisfy the range diminishing or the upwind range condition and, in fact, the Lax–Friedrichs method allows large spurious oscillations, especially for small λ and large Δx, as seen in Test Case 2 and Problem 17.4 below.
(11) First-order accurate in time and space. This is true of any method that satisfies $|\lambda a^n_{i+1/2}| \leq \lambda \epsilon^n_{i+1/2} \leq 1$, as mentioned in Example 16.6.
(12) Centered.

As an additional property, not listed above, steady-state solutions of the Lax–Friedrichs method depend on the time-step Δt. If a steady-state solution exists, then $u^{n+1}_i - u^n_i \to 0$ as $n \to \infty$. If $u^{n+1}_i = u^n_i$, the Lax–Friedrichs method becomes

$$\lambda\left(f\left(u^n_{i+1}\right) - f\left(u^n_{i-1}\right)\right) = u^n_{i+1} - 2u^n_i + u^n_{i-1}.$$

The solution to this equation clearly depends on $\lambda = \Delta t/\Delta x$ and thus on Δt. This steady-state dependency on Δt is highly unphysical – the steady-state solution should depend only on the grid, through Δx, and not on the time steps Δt taken to obtain the steady-state solution. The behavior of the Lax–Friedrichs method is illustrated using the five standard test cases defined in Section 17.0.

Test Case 1 As seen in Figure 17.2, the sinusoid experiences devastating dissipation. The sinusoid's shape is well preserved, and the phase error is relatively small, but the amplitude is only about 3.6% of what it should be.

Test Case 2 As seen in Figure 17.3, the solution exhibits large "odd–even" oscillations. In other words, the solution exhibits oscillations with the shortest possible wavelength of $2\Delta x$. Notice that these large spurious oscillations occur despite the nonlinear

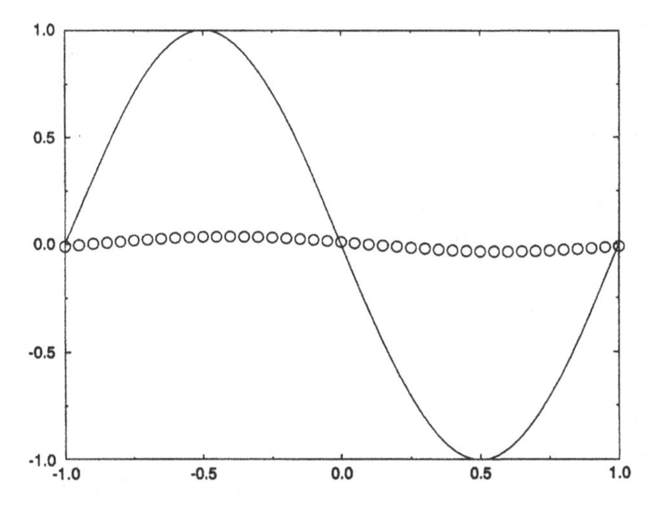

Figure 17.2 Lax–Friedrichs method for Test Case 1.

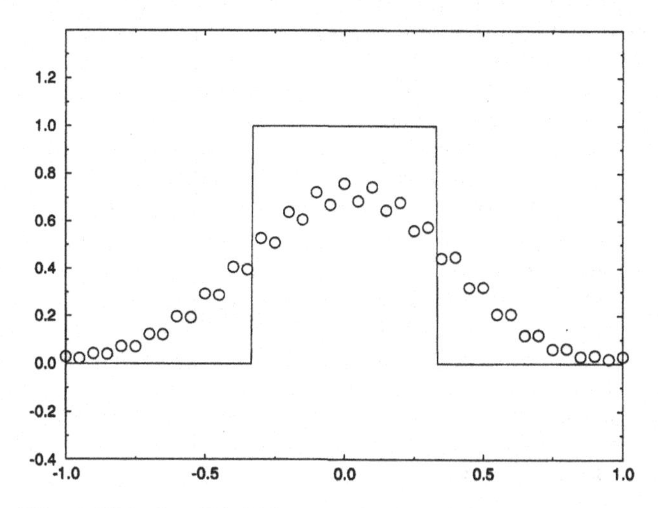

Figure 17.3 Lax–Friedrichs method for Test Case 2.

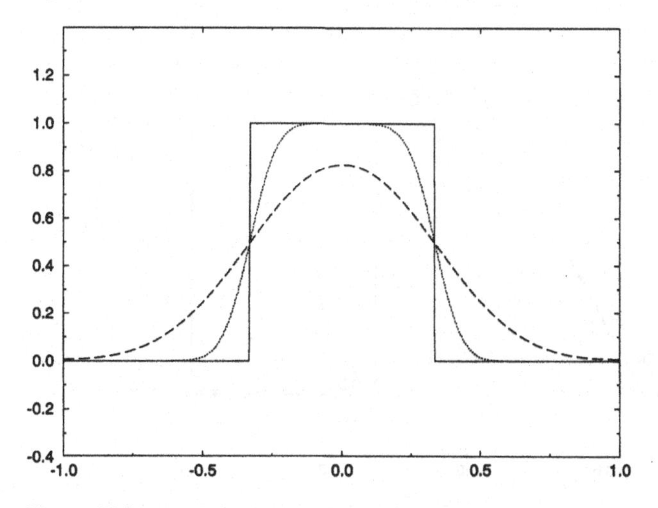

Figure 17.4 Lax–Friedrichs method for Test Case 3.

stability properties of the Lax–Friedrichs method listed above. As one explanation, the Lax–Friedrichs method has too much artificial viscosity which, just like too little artificial viscosity, leads to instability and spurious oscillations. As another explanation, these oscillations are caused by odd–even decoupling, much as in FTCS or the leapfrog method; see the discussion in Subsection 11.1.3. However, suppose the artificial viscosity were moderately reduced, or suppose a nonuniform grid were used, which would reduce or eliminate odd–even decoupling, as seen in Subsection 10.1.2. In this case, there would still be too much artificial viscosity, and thus there would still be spurious oscillations in the solution. In addition to the spurious oscillations, the contacts are extremely smeared and the peak of the square wave has been reduced by about 25%. On the positive side, the solution is reasonably symmetric and properly located.

Test Case 3 In Figure 17.4, the dotted line represents the Lax–Friedrichs approximation to $u(x, 4)$, the long dashed line represents the Lax–Friedrichs approximation

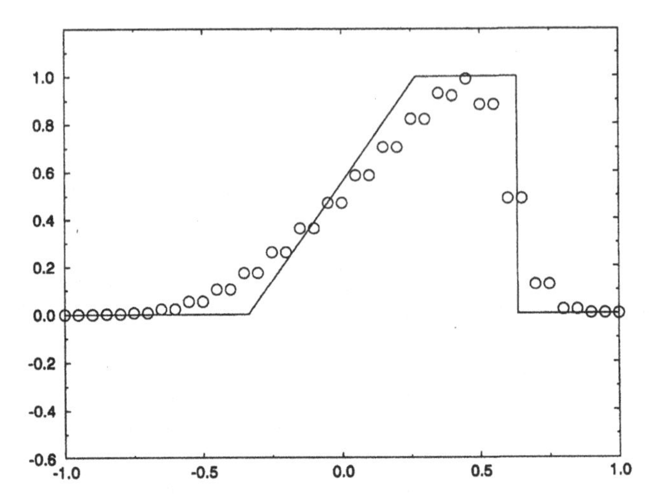

Figure 17.5 Lax–Friedrichs method for Test Case 4.

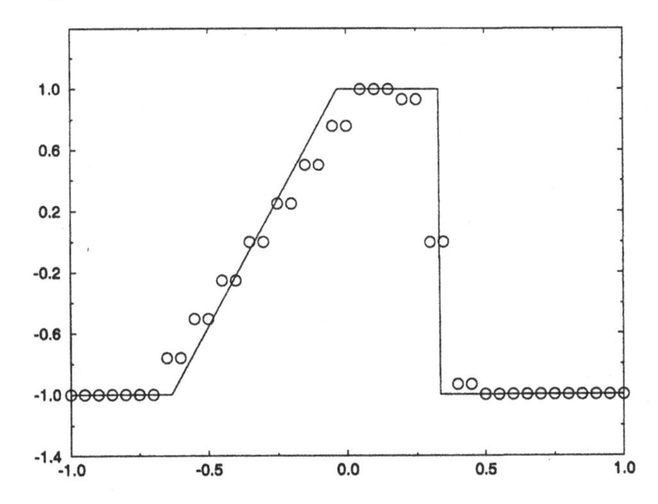

Figure 17.6 Lax–Friedrichs method for Test Case 5.

to $u(x, 40)$, and the solid line represents the exact solution for $u(x, 4)$ or $u(x, 40)$. Clearly, increasing the number of grid points, and decreasing Δx and Δt, dramatically improves the approximation for $u(x, 4)$, as compared with Figure 17.3. There are still some very tiny, nearly invisible, odd–even plateaus in the solution. Thus the slight roughness seen in the plot is not a printer artifact but a genuine solution behavior.

Test Cases 4 and 5 As seen in Figures 17.5 and 17.6, the Lax–Friedrichs solution contains a number of strange odd–even two-point plateaus.

17.2 Lax–Wendroff Method

The Lax–Friedrichs method has the greatest amount of artificial viscosity allowed by the linear stability condition $(\lambda a)^2 \leq \lambda\epsilon \leq 1$ found in Example 15.4. This section concerns the method at the other extreme – the Lax–Wendroff method has the *least* amount of artificial viscosity allowed by the linear stability condition $(\lambda a)^2 \leq \lambda\epsilon \leq 1$. To derive the Lax–Wendroff method, consider a Taylor series for $u(x, t + \Delta t)$:

$$u(x, t + \Delta t) = u(x, t) + \Delta t \frac{\partial u}{\partial t}(x, t) + \frac{\Delta t^2}{2} \frac{\partial^2 u}{\partial t^2}(x, t) + O(\Delta t^3).$$

The t derivatives can be transformed to x derivatives using the governing equation. This is called the *Lax–Wendroff* or *Cauchy–Kowalewski* technique, as introduced in Section 15.3, where it was used to convert t derivatives to x derivatives on the right-hand sides of modified equations. In this case, the governing equation is

$$\frac{\partial u}{\partial t} + \frac{\partial f(u)}{\partial x} = 0,$$

which implies

$$\frac{\partial u}{\partial t} = -\frac{\partial f(u)}{\partial x} = -\frac{\partial f(u)}{\partial u}\frac{\partial u}{\partial x} = -a(u)\frac{\partial u}{\partial x}$$

and

$$\begin{aligned}\frac{\partial^2 u}{\partial t^2} &= \frac{\partial}{\partial t}\left(\frac{\partial u}{\partial t}\right) = \frac{\partial}{\partial t}\left(-\frac{\partial f(u)}{\partial x}\right) = -\frac{\partial}{\partial x}\left(\frac{\partial f(u)}{\partial t}\right) \\ &= -\frac{\partial}{\partial x}\left(a(u)\frac{\partial u}{\partial t}\right) = \frac{\partial}{\partial x}\left(a(u)\frac{\partial f(u)}{\partial x}\right).\end{aligned}$$

Substitute the preceding expressions for $\partial u/\partial t$ and $\partial^2 u/\partial t^2$ into the Taylor series for $u(x, t + \Delta t)$ to obtain

$$u(x, t + \Delta t) = u(x, t) - \Delta t \frac{\partial f}{\partial x}(x, t) + \frac{\Delta t^2}{2}\frac{\partial}{\partial x}\left(a(u)\frac{\partial f}{\partial x}\right)(x, t) + O(\Delta t^3).$$

This expression is discretized using central differences as follows:

$$\begin{aligned}u_i^{n+1} &= u_i^n - \Delta t \frac{f(u_{i+1}^n) - f(u_{i-1}^n)}{2\Delta x} \\ &\quad + \frac{\Delta t^2}{2} \frac{a_{i+1/2}^n \frac{f(u_{i+1}^n) - f(u_i^n)}{\Delta x} - a_{i-1/2}^n \frac{f(u_i^n) - f(u_{i-1}^n)}{\Delta x}}{\Delta x}\end{aligned}$$

or

$$\begin{aligned}◆ \quad u_i^{n+1} &= u_i^n - \frac{\lambda}{2}\left(f(u_{i+1}^n) - f(u_{i-1}^n)\right) \\ &\quad + \frac{\lambda^2}{2}\left[a_{i+1/2}^n\left(f(u_{i+1}^n) - f(u_i^n)\right) - a_{i-1/2}^n\left(f(u_i^n) - f(u_{i-1}^n)\right)\right], \qquad (17.5)\end{aligned}$$

which is called the *Lax–Wendroff method*. The Lax–Wendroff method was discovered in 1960.

As mentioned before, the preceding derivation of the Lax–Wendroff method illustrates a general technique called the Lax–Wendroff or the Cauchy–Kowalewski technique. Let us say a few words about the general Cauchy–Kowalewski technique before continuing. Suppose that $u(x,t)$ is expressed as a two-dimensional Taylor series about (x_i, t^n) as follows:

$$u(x,t) \approx \sum_{j=0}^{N} \sum_{k=0}^{j} \frac{\partial^j u(x_i,t^n)}{\partial x^k \partial t^{j-k}} \frac{(x-x_i)^k}{k!} \frac{(t-t^n)^{j-k}}{(j-k)!}. \tag{17.6}$$

Using the governing equation, all of the temporal derivatives in the preceding expression can be exchanged for spatial derivatives as follows:

$$\frac{\partial u}{\partial t} = -\frac{df}{du}\frac{\partial u}{\partial x},$$

$$\frac{\partial^2 u}{\partial x \partial t} = -\frac{df}{du}\frac{\partial^2 u}{\partial x^2} - \frac{d^2 f}{du^2}\left(\frac{\partial u}{\partial x}\right)^2,$$

$$\frac{\partial^2 u}{\partial t^2} = \left(\frac{\partial f}{\partial u}\right)^2 \frac{\partial^2 u}{\partial x^2} + 2\frac{d^2 f}{du^2}\frac{df}{du}\left(\frac{\partial u}{\partial x}\right)^2,$$

$$\frac{\partial^3 u}{\partial x^2 \partial t} = -\frac{d^3 f}{du^3}\left(\frac{\partial u}{\partial x}\right)^3 - 3\frac{d^2 f}{du^2}\frac{\partial^2 u}{\partial x^2}\frac{\partial u}{\partial x} - \frac{df}{du}\frac{\partial^3 u}{\partial x^3},$$

and so forth. Then, any value of $u(x,t)$ at time t can be approximated to any order of accuracy using pure spatial discretization of pure spatial derivatives. The Cauchy–Kowalewski technique will be used again to the derive the Beam–Warming second-order upwind method in Section 17.4 and periodically throughout the rest of the book.

In the Lax–Wendroff method, the average wave speed $a^n_{i+1/2}$ may be defined in a variety of different ways. For example,

$$a^n_{i+1/2} = a\left(\frac{u^n_{i+1} + u^n_i}{2}\right) \tag{17.7}$$

or

$$a^n_{i+1/2} = \frac{a(u^n_{i+1}) + a(u^n_i)}{2}. \tag{17.8}$$

In other words, the Lax–Wendroff method is actually an entire class of methods differing only in the choice of the average wave speed $a^n_{i+1/2}$. This book will use the following definition:

$$a^n_{i+1/2} = \begin{cases} \frac{f(u^n_{i+1}) - f(u^n_i)}{u^n_{i+1} - u^n_i} & \text{for } u^n_i \neq u^n_{i+1}, \\ a(u^n_i) & \text{for } u^n_i = u^n_{i+1}. \end{cases} \tag{17.9}$$

This is not the best choice for $a^n_{i+1/2}$ if the Lax–Wendroff method is used on a stand-alone basis. However, this choice makes the theoretical analysis simpler, and it also makes the Lax–Wendroff method well suited for adaptive combinations with other numerical methods, to form second- and third-generation methods. The reader may recall a similar discussion about averages from Section 5.4 – in particular, Equation (17.7) is the same as Equation (5.62), Equation (17.8) is the same as Equation (5.64), and Equation (17.9) is the same as the Roe average seen in Section 5.3.

In conservation form, the Lax–Wendroff method is

$$u_i^{n+1} = u_i^n - \lambda\left(\hat{f}^n_{i+1/2} - \hat{f}^n_{i-1/2}\right),$$

where

$$\blacklozenge \quad \hat{f}^n_{i+1/2} = \frac{1}{2}\left(f(u^n_{i+1}) + f(u^n_i)\right) - \frac{1}{2}\lambda a^n_{i+1/2}\left(f(u^n_{i+1}) - f(u^n_i)\right)$$
$$= \frac{1}{2}\left(f(u^n_{i+1}) + f(u^n_i)\right) - \frac{1}{2}\lambda\left(a^n_{i+1/2}\right)^2\left(u^n_{i+1} - u^n_i\right). \tag{17.10}$$

In artificial viscosity form, the Lax–Wendroff method is

$$u_i^{n+1} = u_i^n - \frac{\lambda}{2}\left(f(u^n_{i+1}) - f(u^n_{i-1})\right) + \frac{\lambda}{2}\left(\epsilon^n_{i+1/2}(u^n_{i+1} - u^n_i) - \epsilon^n_{i-1/2}(u^n_i - u^n_{i-1})\right),$$

where

$$\epsilon^n_{i+1/2} = \lambda\left(a^n_{i+1/2}\right)^2. \tag{17.11}$$

In wave speed split form, the Lax–Wendroff method is

$$u_i^{n+1} = u_i^n + C^+_{i+1/2}(u^n_{i+1} - u^n_i) - C^-_{i-1/2}(u^n_i - u^n_{i-1}),$$

where

$$C^+_{i+1/2} = -\frac{1}{2}\lambda a^n_{i+1/2}\left(1 - \lambda a^n_{i+1/2}\right), \tag{17.12a}$$

$$C^-_{i+1/2} = \frac{1}{2}\lambda a^n_{i+1/2}\left(1 + \lambda a^n_{i+1/2}\right). \tag{17.12b}$$

Of course, as always, there are infinitely many other wave speed split forms. However, this is the only wave speed split form with finite coefficients – in this sense, this is the unique *natural* wave speed split form. The twelve-point checklist for the Lax–Wendroff method is as follows:

(1) The Lax–Wendroff method has the minimum amount of artificial viscosity allowed by the linear stability condition $(\lambda a)^2 \leq \lambda\epsilon \leq 1$ found in Example 15.4.
(2) CFL condition: $|\lambda a(u)| \leq 1$.
(3) Conservative.
(4) Consistent.
(5) The Lax–Wendroff method may not converge on nonsmooth solutions as $\Delta x \to 0$ and $\Delta t \to 0$. Even if it does converge, the converged solution may not satisfy entropy conditions. In particular, the Lax–Wendroff method allows expansion shocks, at least with $a^n_{i+1/2}$ defined as in Equation (17.9).
(6) Explicit.
(7) Finite difference.
(8) Linearly stable provided that the CFL condition is satisfied.
(9) Linear when applied to the linear advection equation. Nonlinear when applied to any nonlinear scalar conservation law.
(10) Consider the nonlinear stability conditions seen in Chapter 16. Although there is no rigorous proof, numerical tests show that the Lax–Wendroff method is TVB. However, although the total variation is bounded for any fixed Δx and Δt, the

total variation may go to infinity in the limit $\Delta x \to 0$ and $\Delta t \to 0$. Thus the Lax–Wendroff method is always TVB but *not* always total variation stable; see the discussion in Section 16.11. Furthermore, by Godunov's theorem seen in Section 16.1, the Lax–Wendroff method is not monotonicity preserving. Then the Lax–Wendroff method cannot satisfy any condition that implies monotonicity preservation including TVD, positivity, range reduction, the upwind range condition, contraction, or the monotone condition.

(11) Formally second-order accurate in time and space.

(12) Centered.

As an additional property, not listed above, steady-state solutions of the Lax–Wendroff method depend on the time step Δt, since the conservative numerical flux depends on λ and thus on Δt. The behavior of the Lax–Wendroff method is illustrated using the five standard test cases defined in Section 17.0.

Test Case 1 As seen in Figure 17.7, the sinusoid's shape and amplitude are well captured. The only visible error is a slight lagging phase error.

Test Case 2 As seen in Figure 17.8, the solution overshoots and undershoots by about 20%. As one explanation, the Lax–Wendroff method has insufficient artificial viscosity, which leads to instability and spurious oscillations. On the positive side, the solution is apparently properly located, and thus the overall phase error is slight. Also, the Lax–Wendroff method smears the contacts far less than the Lax–Friedrichs method seen in Figure 17.3. This is to be expected given the increased order of accuracy.

Test Case 3 In Figure 17.9, the dotted line represents the Lax–Wendroff approximation to $u(x, 4)$, the long dashed line represents the Lax–Wendroff approximation to $u(x, 40)$, and the solid line represents the exact solution for $u(x, 4)$ or $u(x, 40)$. Comparing Figures 17.8 and 17.9, we see that increasing the number of grid points improves

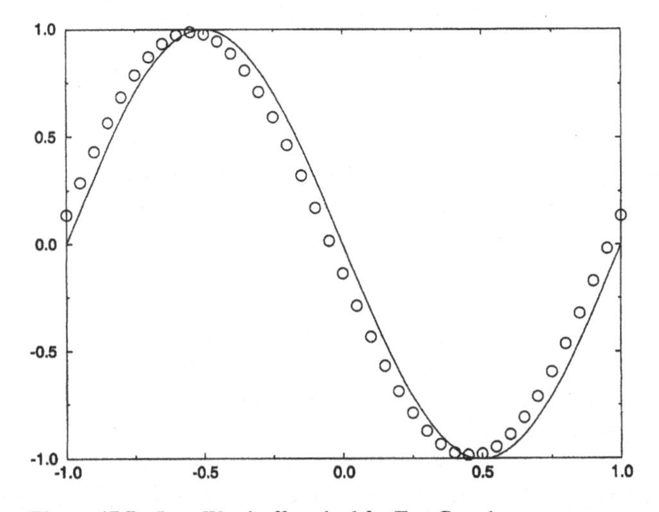

Figure 17.7 Lax–Wendroff method for Test Case 1.

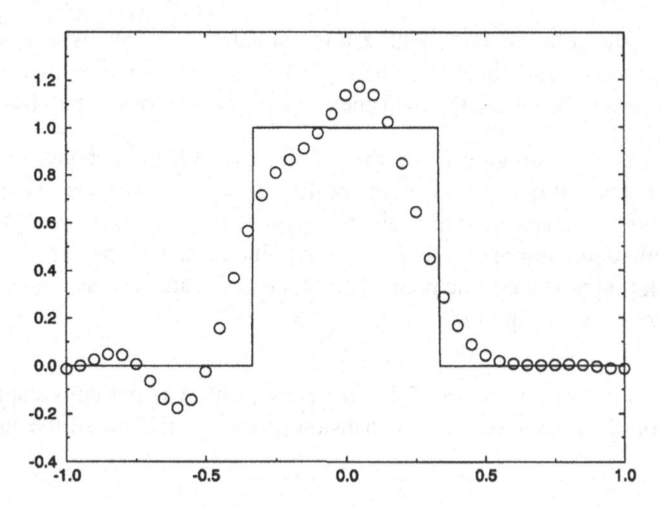

Figure 17.8 Lax–Wendroff method for Test Case 2.

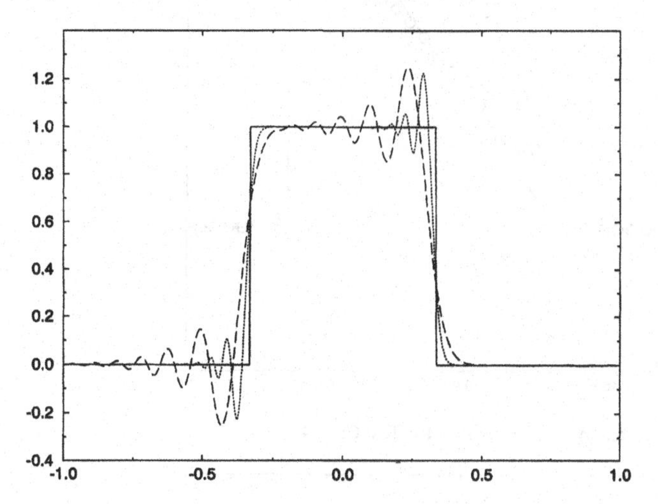

Figure 17.9 Lax–Wendroff method for Test Case 3.

the solution for $u(x, 4)$ in some ways but makes it worse in other ways. In particular, increasing the number of grid points creates large ringing oscillations to the left of the jump discontinuities. In fact, these ringing oscillations grow to infinity in the limit $\Delta x \to 0$ and $\Delta t \to 0$. Comparing the solutions for $u(x, 4)$ and $u(x, 40)$, as seen in Figure 17.9, we see that increasing the final time also increases the ringing oscillations. However, for fixed Δx and Δt, the total variation created by the jump must eventually stop growing as $t \to \infty$. To see this, from von Neumann analysis, recall that the amplitudes of the sinusoids in a discrete Fourier series representation of the solution cannot increase. Thus the oscillations in the solution are created entirely by dispersion, that is, by frequency-dependent propagation speeds. But dispersion can only create a limited amount of oscillation for a fixed Δx. Once dispersion creates the maximum amount of oscillation possible, by fully separating various frequencies, the only way for the total variation to increase is for the amplitude of the oscillations to increase, which is impossible by the results of linear stability analysis.

As one way to view it, the solution starts with a finite amount of energy, which can be redistributed but whose total cannot increase. Once all of the energy is redistributed from the jumps into the spurious oscillations, the total energy in the oscillations cannot increase.

Test Case 4 As seen in Figure 17.10, the solution overshoots by about 12% near the shock and undershoots and overshoots by about 40% near the tail of the expansion. Notice that the undershoot overlaps the sonic point $u^* = 0$ and, as seen in the next test case, the Lax–Wendroff method has severe problems at sonic points, which helps to explain the large tearing in the solution near the sonic point. The shock is smeared across two or three grid points but is otherwise well captured.

Test Case 5 As seen in Figure 17.11, the steady shock is perfectly captured. However, the expansion fan is captured as an expansion shock, in violation of the entropy

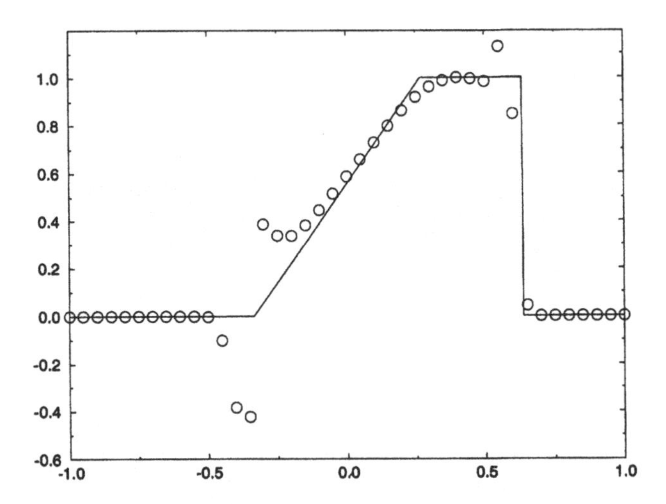

Figure 17.10 Lax–Wendroff method for Test Case 4.

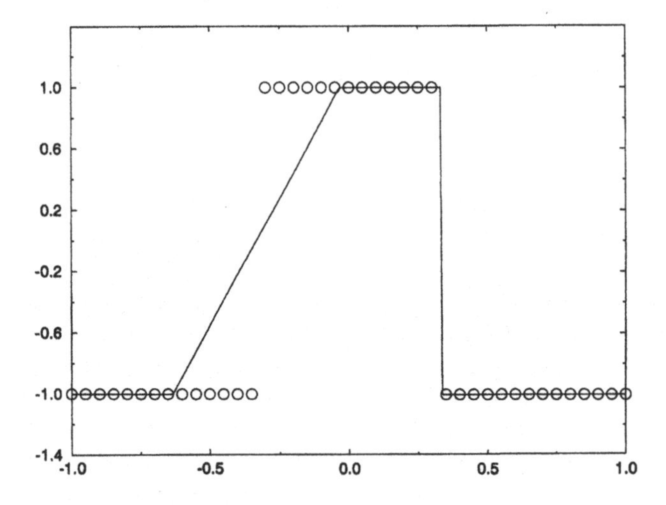

Figure 17.11 Lax–Wendroff method for Test Case 5.

conditions. In fact, the Lax–Wendroff method fails to alter its initial conditions in any way, which is perfectly correct at the shock and a complete disaster at the expansion. To be fair, the performance in this test case can be greatly improved by using a different definition for $a^n_{i+1/2}$ rather than the "Roe average" definition of Equation (17.9).

17.3 First-Order Upwind Methods

The Lax–Friedrichs method has too much artificial viscosity. The Lax–Wendroff method does not have enough artificial viscosity. This section concerns methods somewhere in the middle. First-order upwind methods were introduced in Chapter 13. In particular, a first-order upwind method for the linear advection equation, sometimes called CIR, was derived using flux averaging in Example 13.5, flux splitting in Example 13.6, wave speed splitting in Example 13.10, and reconstruction–evolution in Example 13.14. Whereas the derivation technique did not matter much for linear equations, which lack sonic points, the derivation technique is critical for nonlinear equations, which do contain sonic points. This section derives first-order upwind methods for nonlinear scalar conservation laws using reconstruction–evolution. Since reconstruction–evolution is central to this section, the reader should carefully review the introduction to reconstruction–evolution seen in Section 13.6 before proceeding. Also, the reader should review Section 5.6, which concerns the Riemann problem for scalar conservation laws. Although this section concerns first-order upwind methods based on reconstruction–evolution, first-order upwind methods can also be found using flux splitting or wave speed splitting, as described later in Subsection 18.2.5.

Suppose the reconstruction is piecewise-constant. Then each cell edge gives rise to a Riemann problem, as illustrated in Figure 17.12. Then the exact evolution of the piecewise-constant reconstruction yields

$$u_i^{n+1} = u_i^n - \lambda\left(\hat{f}^n_{i+1/2} - \hat{f}^n_{i-1/2}\right),$$

where

$$\hat{f}^n_{i+1/2} = \frac{1}{\Delta t}\int_{t^n}^{t^{n+1}} f(u_{\text{RIEMANN}}(x_{i+1/2}, t))\, dt,$$

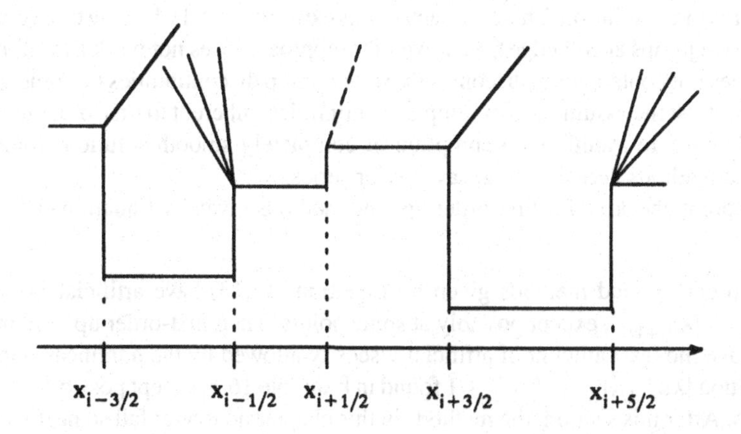

Figure 17.12 Each cell edge in a piecewise-constant reconstruction gives rise to a Riemann problem.

where $u_{\text{RIEMANN}}(x_{i+1/2}, t)$ is a real or approximate solution to the Riemann problem centered on the cell edge $x_{i+1/2}$. Since the solution to the Riemann problem is self-similar or, in other words, strictly a function of $(x - x_{i+1/2})/t$, $u_{\text{RIEMANN}}(x_{i+1/2}, t)$ is constant for all time since $(x_{i+1/2} - x_{i+1/2})/t = 0$. Then

$$\blacklozenge \qquad \hat{f}^n_{i+1/2} = f(u_{\text{RIEMANN}}(x_{i+1/2}, t)), \qquad (17.13)$$

where $f(u_{\text{RIEMANN}}(x_{i+1/2}, t))$ is found using any exact or approximate Riemann solver, such as those seen in Section 5.6, and where t can be any arbitrary time greater than t^n. In this section, the term *first-order upwind method* refers to any method based on Equation (17.13).

Equation (17.13) assumes that waves from different cell edges do not interact or, at least, that any interactions do not affect the solution at the cell edges. If $\lambda|a(u)| \leq 1/2$ then waves travel at most one-half cell per time step – then waves originating at one cell edge cannot interact with waves originating from any other cell edge during a single time step. If $\lambda|a(u)| \leq 1$ then waves travel at most one cell per time step – then waves can interact, but the interactions cannot reach the cell edges during a single time step. In fact, if $\lambda|a(u)| \leq 1$, waves from adjacent cell edges can interact only if there is a compressive sonic point inside the cell, so that the waves at the left cell edge are right-running whereas the waves from the right cell edge are left-running. Notice that, by coincidence, the noninteraction condition $\lambda|a(u)| \leq 1$ is also the CFL condition.

If the wave speeds are always positive, the waves originating from cell edge $x_{i+1/2}$ all travel to the right, and then Equation (17.13) always yields $f^n_{i+1/2} = f(u^n_i)$, which is FTBS. Similarly, if the wave speeds are always negative, the waves originating from cell edge $x_{i+1/2}$ all travel to the left, and then Equation (17.13) always yields $f^n_{i+1/2} = f(u^n_{i+1})$, which is FTFS. This is true for any reasonable real or approximate Riemann used in Equation (17.13). Thus *all first-order upwind methods given by Equation* (17.13) *are the same – either FTBS or FTFS – except near sonic points, where the wave speed changes sign.*

In general, first-order upwind methods have excellent shock-capturing abilities. As one explanation, this is because first-order upwind methods use the minimum possible stencil allowed by the CFL condition; this explanation was pursued earlier in Chapter 13. As another explanation, this is because first-order upwind methods assume that a jump discontinuity exists at every cell edge, and then they evolve the cell-edge discontinuities in time using an exact or approximate solution. This approach works extremely well when the true solution actually does have jumps at cell edges. However, this approach does not work as well if the true solution does not contain jump discontinuities, or if jump discontinuities lie somewhere inside a cell. Contrast the assumption of jump discontinuities, inherent to first-order upwind methods, with the more traditional assumption of completely smooth solutions found in finite-difference methods derived using, say, Taylor series.

The twelve-point checklist for first-order upwind methods given by Equation (17.13) is as follows:

(1) First-order upwind methods given by Equation (17.13) have artificial viscosity $\epsilon^n_{i+1/2} = |\lambda a^n_{i+1/2}|$, except possibly at sonic points. Then first-order upwind methods have the least amount of artificial viscosity allowed by the nonlinear stability condition $|\lambda a^n_{i+1/2}| \leq \lambda\epsilon^n_{i+1/2} \leq 1$ found in Example 16.6, except possibly at sonic points. After this section, the methods in this chapter no longer fall so neatly at the extremes of a stability condition. Furthermore, $\epsilon^n_{i+1/2}$ will no longer be a simple function of $a^n_{i+1/2}$.

(2) CFL condition: $|\lambda a(u)| \leq 1$.
(3) Conservative.
(4) Consistent.
(5) The first-order upwind methods given by Equation (17.13) converge as $\Delta x \to 0$ and $\Delta t \to 0$ provided that the CFL condition is satisfied. Whether the converged solution satisfies the entropy condition depends on how the first-order upwind method treats sonic points. At sonic points, the more artificial viscosity the better, within reason.
(6) Explicit.
(7) Finite volume.
(8) Linearly stable provided that the CFL condition is satisfied.
(9) The first-order upwind methods given by Equation (17.13) are linear when applied to the linear advection equation. In fact, all first-order upwind methods given by Equation (17.13) are the same when applied to the linear advection equation – they equal either FTBS for $a > 0$ or FTFS for $a < 0$. First-order upwind methods are nonlinear when applied to any nonlinear scalar conservation law. The behavior at sonic points in a nonlinear scalar conservation law is the only thing that distinguishes the first-order upwind methods given by Equation (17.13).
(10) The first-order upwind methods given by Equation (17.13) satisfy all of the nonlinear stability conditions considered in Chapter 16. As always, this assumes that the CFL condition is satisfied and may not be true at sonic points. In particular, first-order upwind methods satisfy monotonicity preservation, TVD, positivity, range reduction, the upwind range condition, TVB, ENO, contraction, and the monotone condition, except possibly at sonic points. Thus first-order upwind methods do not allow spurious oscillations or overshoots, except possibly at sonic points.
(11) Formally first-order accurate in time and space, except possibly at sonic points.
(12) Upwind.

Note that steady-state solutions of first-order upwind methods do *not* depend on Δt, unlike all of the other methods in this chapter. The following subsections consider some specific first-order upwind methods.

17.3.1 *Godunov's First-Order Upwind Method*

The exact Riemann solver for a scalar conservation law is given by Equation (5.66). In Equation (5.66), replace u_L by u_i^n, replace u_R by u_{i+1}^n, replace $t = 0$ by $t = t^n$, and replace $x = 0$ by $x = x_{i+1/2}$. Substitute into Equation (17.13) to find

$$\blacklozenge \qquad \hat{f}^n_{i+1/2} = \begin{cases} \min\limits_{u_i^n \leq u \leq u_{i+1}^n} f(u) & \text{if } u_i^n < u_{i+1}^n, \\ \max\limits_{u_i^n \geq u \geq u_{i+1}^n} f(u) & \text{if } u_i^n > u_{i+1}^n. \end{cases} \tag{17.14a}$$

This is called *Godunov's first-order upwind method.* Godunov's first-order upwind method was discovered in 1959. As shown in any calculus book, the maximum of a function on a closed interval occurs either at the endpoints of the interval or where the derivative of the function equals zero. The derivative of the flux function is the wave speed, and the wave speed is zero at sonic points. Then Equation (17.14a) is equivalent to the following:

$$\hat{f}^n_{i+1/2} = \begin{cases} \min\left(f\left(u^n_i\right), f\left(u^n_{i+1}\right), f(u^*)\right) & \text{if } u^n_i < u^n_{i+1}, \\ \max\left(f\left(u^n_i\right), f\left(u^n_{i+1}\right), f(u^*)\right) & \text{if } u^n_i > u^n_{i+1}, \end{cases} \tag{17.14b}$$

where u^* refers to any and all sonic points between u^n_i and u^n_{i+1}. In artificial viscosity form, Godunov's first-order upwind method is

$$u^{n+1}_i = u^n_i - \frac{\lambda}{2}\left(f\left(u^n_{i+1}\right) - f\left(u^n_{i-1}\right)\right) + \frac{\lambda}{2}\left(\epsilon^n_{i+1/2}\left(u^n_{i+1} - u^n_i\right) - \epsilon^n_{i-1/2}\left(u^n_i - u^n_{i-1}\right)\right),$$

where

$$\epsilon^n_{i+1/2} = \max_{\substack{u \text{ between} \\ u_i \text{ and } u_{i+1}}} \left(\frac{f\left(u^n_{i+1}\right) - 2f(u) + f\left(u^n_i\right)}{u^n_{i+1} - u^n_i}\right)$$

$$= \max\left(\left|a^n_{i+1/2}\right|, \frac{f\left(u^n_{i+1}\right) - 2f(u^*) + f\left(u^n_i\right)}{u^n_{i+1} - u^n_i}\right) \tag{17.15}$$

and where $a^n_{i+1/2}$ is given by Equation (17.9). In wave speed split form, Godunov's first-order upwind method is

$$u^{n+1}_i = u^n_i + C^+_{i+1/2}\left(u^n_{i+1} - u^n_i\right) - C^-_{i-1/2}\left(u^n_i - u^n_{i-1}\right),$$

where

$$C^+_{i+1/2} = -\lambda \min_{\substack{u \text{ between} \\ u_i \text{ and } u_{i+1}}} \left(\frac{f(u) - f\left(u^n_i\right)}{u^n_{i+1} - u^n_i}\right)$$

$$= -\min\left(0, \lambda a^n_{i+1/2}, \lambda \frac{f(u^*) - f\left(u^n_i\right)}{u^n_{i+1} - u^n_i}\right) \tag{17.16a}$$

and

$$C^-_{i+1/2} = \lambda \max_{\substack{u \text{ between} \\ u_i \text{ and } u_{i+1}}} \left(\frac{f\left(u^n_{i+1}\right) - f(u)}{u^n_{i+1} - u^n_i}\right)$$

$$= \max\left(0, \lambda a^n_{i+1/2}, \lambda \frac{f\left(u^n_{i+1}\right) - f(u^*)}{u^n_{i+1} - u^n_i}\right). \tag{17.16b}$$

Of course, as always, there are infinitely many other wave speed split forms. However, this is the only wave speed split form with finite coefficients – in this sense, this is the unique *natural* wave speed split form.

Godunov's first-order upwind method satisfies the nonlinear stability condition $|\lambda a^n_{i+1/2}| \leq \lambda \epsilon^n_{i+1/2} \leq 1$ found in Example 16.6 provided that the CFL condition is satisfied, except possibly at expansive sonic points. At expansive sonic points u^*, Godunov's method satisfies $|\lambda a^n_{i+1/2}| \leq \lambda \epsilon^n_{i+1/2} \leq 1$ if the CFL condition is satisfied and if

$$\lambda \frac{f\left(u^n_{i+1}\right) - 2f(u^*) + f\left(u^n_i\right)}{u^n_{i+1} - u^n_i} \leq 1. \tag{17.17}$$

It can be shown that this is true if

$$\max_{\substack{u \text{ between} \\ u_i \text{ and } u_{i+1}}} \lambda f''(u)\left(u_{i+1}^n - u_i^n\right) \leq 2. \tag{17.18}$$

Whereas the CFL condition limits $f'(u)$, this last stability condition limits $f''(u)$. Although far less common than limits on the first derivative, stability limits on the second derivative crop up from time to time. The behavior of Godunov's first-order upwind method is illustrated using the five standard test cases defined in Section 17.0.

Test Case 1 As seen in Figure 17.13, the sinusoid's shape is well preserved, and its phase is nearly correct, but its amplitude has been reduced by about 82%. In this case, Godunov's first-order upwind method equals FTBS.

Test Case 2 As seen in Figure 17.14, the contacts are extremely smeared and the square wave's peak has been reduced by about 10%. On the positive side, the solution

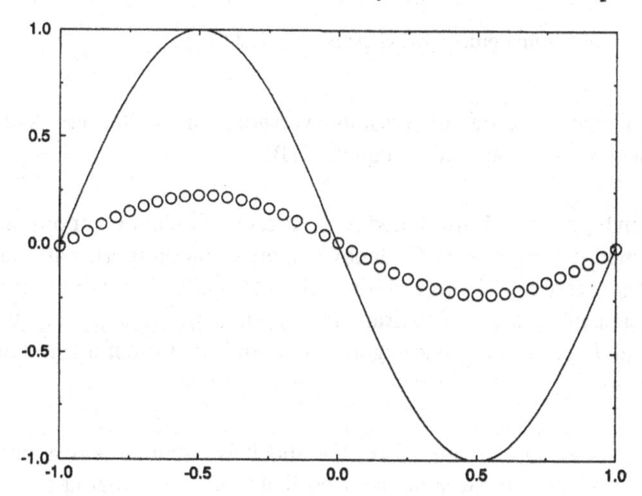

Figure 17.13 Godunov's first-order upwind method for Test Case 1.

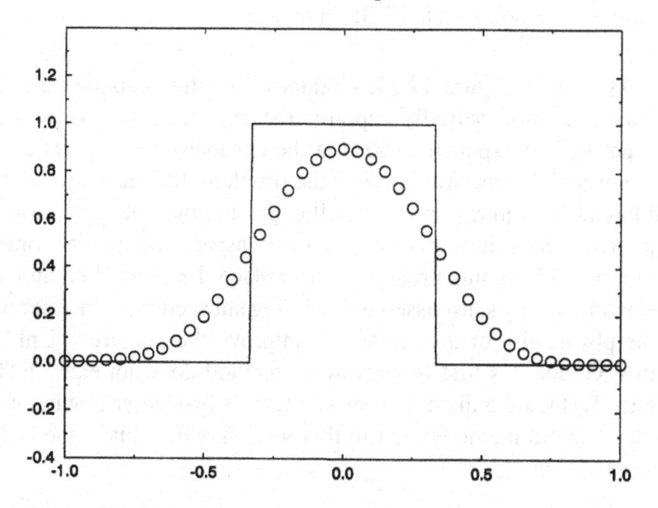

Figure 17.14 Godunov's first-order upwind method for Test Case 2.

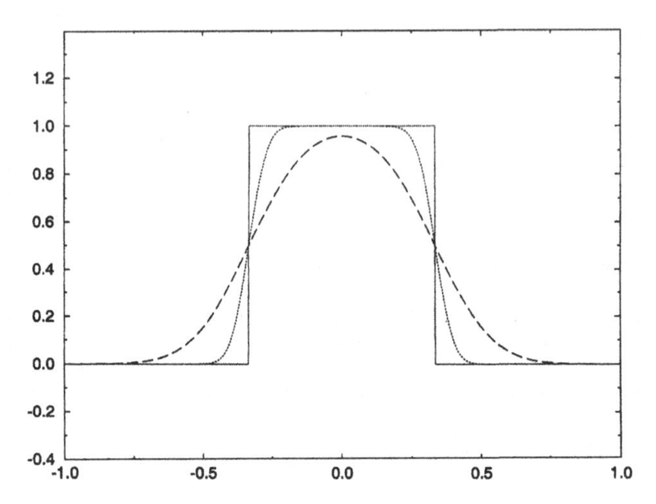

Figure 17.15 Godunov's first-order upwind method for Test Case 3.

is symmetric, properly located, and free of spurious overshoots or oscillations. As in Test Case 1, Godunov's first-order upwind method equals FTBS.

Test Case 3 In Figure 17.15, the dotted line represents Godunov's approximation to $u(x, 4)$, the long dashed line represents Godunov's approximation to $u(x, 40)$, and the solid line represents the exact solution for $u(x, 4)$ or $u(x, 40)$. Clearly, increasing the number of grid points, and decreasing Δx and Δt, dramatically improve the approximation for $u(x, 4)$, as compared with Figure 17.14. Once again, Godunov's first-order upwind method equals FTBS in this test case.

Test Case 4 As seen in Figure 17.16, the shock is captured across only two grid points and without any spurious overshoots or oscillations. The corner at the head of the expansion fan has been slightly rounded off. As with all of the preceding test cases, Godunov's first-order upwind method equals FTBS in this case.

Test Case 5 As seen in Figure 17.17, Godunov's method captures the steady shock perfectly. Godunov's method partially captures the expansion fan. However, the expansion fan contains an $O(\Delta x)$ expansion shock at the expansive sonic point found in the center of the expansion fan. As one way to view the problem, the reconstruction used by Godunov's method has $O(\Delta x)$ jumps at every cell edge, and the time evolution tends to retain the $O(\Delta x)$ jumps if the solution contains an expansive sonic point. Somewhat surprisingly, the best way to address this problem is to replace the *exact* Riemann solver by an *approximate* Riemann solver, as discussed in following subsections – in other words, one needs to reduce the physicality of the method to improve the numerics. Unlike the preceding four test cases, Godunov's first-order upwind method does not equal FTBS in this test case. In fact, only Test Case 5 distinguishes Godunov's first-order upwind method from the other first-order upwind methods seen in this section, since this is the only test case to involve sonic points.

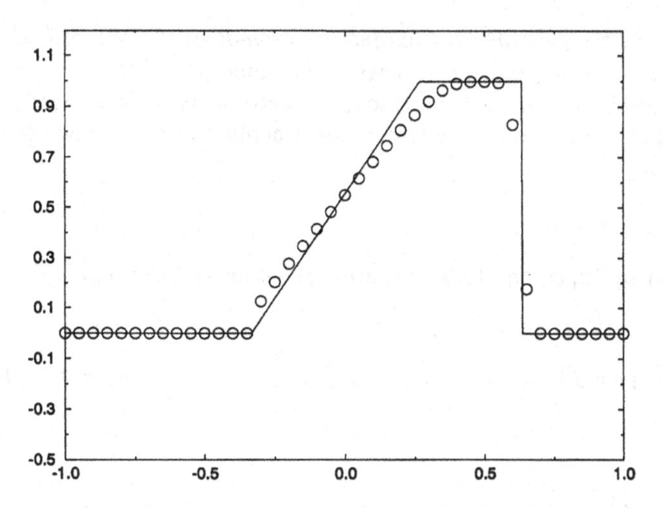

Figure 17.16 Godunov's first-order upwind method for Test Case 4.

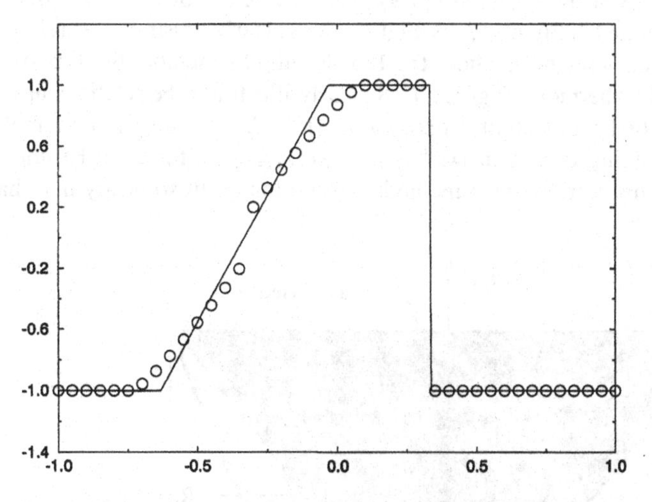

Figure 17.17 Godunov's first-order upwind method for Test Case 5.

17.3.2 *Roe's First-Order Upwind Method*

Roe's approximate Riemann solver is given by Equation (5.69). In Equation (5.69), replace u_L by u_i^n, replace u_R by u_{i+1}^n, replace $t = 0$ by $t = t^n$, and replace $x = 0$ by $x = x_{i+1/2}$. Substitute into Equation (17.13) to find

$$\hat{f}^n_{i+1/2} = \begin{cases} \min\left(f\left(u_i^n\right), f\left(u_{i+1}^n\right)\right) & \text{if } u_i^n < u_{i+1}^n, \\ \max\left(f\left(u_i^n\right), f\left(u_{i+1}^n\right)\right) & \text{if } u_i^n > u_{i+1}^n. \end{cases} \tag{17.19}$$

This is called *Roe's first-order upwind method*. Roe's first-order upwind method for the Euler equations was discovered in 1981. However, the scalar version was well-known before this. For example, the version for the linear advection equation is sometimes called the *Courant–Isaacson–Rees (CIR) method*; then the version for nonlinear scalar conservation laws is sometimes called the *generalized Courant–Isaacson–Rees (GCIR) method.* Also, the version for nonlinear scalar conservation laws is sometimes called the *Murman–Cole scheme*; see Murman and Cole (1971). One way to think of Roe's first-order upwind method is that it equals Godunov's first-order upwind method applied to the following locally linearized flux function:

$$f_l(u) = a^n_{i+1/2}u + f\left(u^n_i\right), \tag{17.20}$$

where $a^n_{i+1/2}$ is given by Equation (17.9). In artificial viscosity form, Roe's first-order upwind method is

$$u^{n+1}_i = u^n_i - \frac{\lambda}{2}\left(f\left(u^n_{i+1}\right) - f\left(u^n_{i-1}\right)\right) + \frac{\lambda}{2}\left(\epsilon^n_{i+1/2}\left(u^n_{i+1} - u^n_i\right) - \epsilon^n_{i-1/2}\left(u^n_i - u^n_{i-1}\right)\right),$$

where

$$\epsilon^n_{i+1/2} = \left|a^n_{i+1/2}\right| \tag{17.21}$$

and where $a^n_{i+1/2}$ is given by Equation (17.9). The coefficient of artificial viscosity for Roe's method is illustrated in Figure 17.18. Figure 17.18 clearly illustrates the relationships among Roe's first-order upwind method, the Lax–Friedrichs method, the Lax–Wendroff method, and FTCS. Furthermore, Figure 17.18 clearly illustrates the relationships among these four methods, the linear stability condition $(\lambda a)^2 \leq \lambda\epsilon \leq 1$ found in Example 15.4, and the nonlinear stability condition $|\lambda a^n_{i+1/2}| \leq \lambda\epsilon^n_{i+1/2} \leq 1$ found in Example 16.6. Unfortunately, after this section, most methods will no longer fit so neatly into diagrams

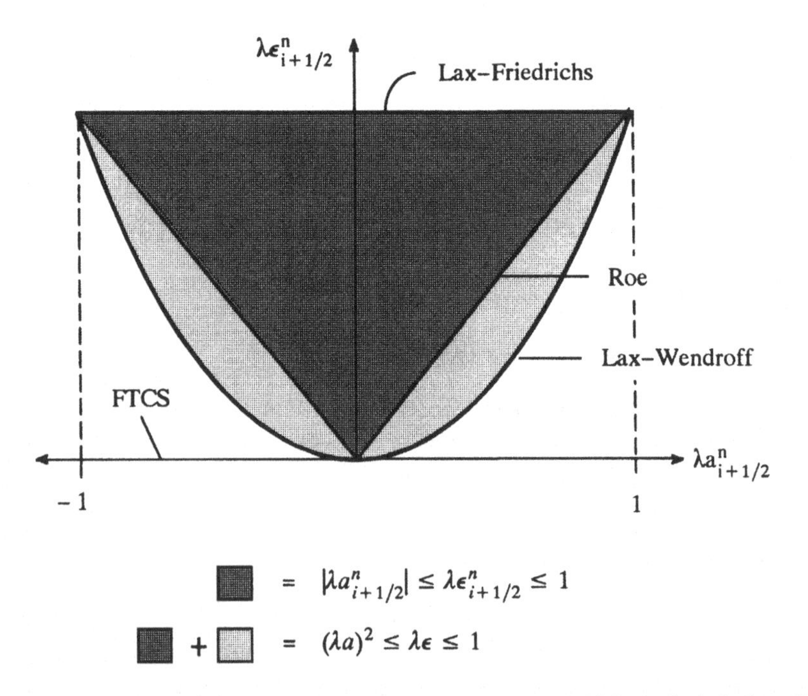

Figure 17.18 The artificial viscosity of FTCS, the Lax–Friedrichs method, the Lax–Wendroff method, and Roe's first-order upwind method.

such as Figure 17.18. In wave speed splitting form, Roe's first-order upwind method is

$$u_i^{n+1} = u_i^n + C_{i+1/2}^+(u_{i+1}^n - u_i^n) - C_{i-1/2}^-(u_i^n - u_{i-1}^n),$$

where

$$C_{i+1/2}^+ = -\lambda \min\left(0, a_{i+1/2}^n\right), \tag{17.22a}$$

$$C_{i+1/2}^- = \lambda \max\left(0, a_{i+1/2}^n\right). \tag{17.22b}$$

Of course, as always, there are infinitely many other wave speed split forms. However, this is the only wave speed split form with finite coefficients – in this sense, this is the unique *natural* wave speed split form. Roe's first-order upwind method can also be written as follows:

$$u_i^{n+1} = \begin{cases} u_i^n - \lambda\left(f\left(u_i^n\right) - f\left(u_{i-1}^n\right)\right) & \text{if } a_{i+1/2}^n \geq 0 \text{ and } a_{i-1/2}^n \geq 0, \\ u_i^n - \lambda\left(f\left(u_{i+1}^n\right) - f\left(u_i^n\right)\right) & \text{if } a_{i+1/2}^n \leq 0 \text{ and } a_{i-1/2}^n \leq 0, \\ u_i^n & \text{if } a_{i+1/2}^n \geq 0 \text{ and } a_{i-1/2}^n \leq 0, \\ u_i^n - \lambda\left(f\left(u_{i+1}^n\right) - f\left(u_{i-1}^n\right)\right) & \text{if } a_{i+1/2}^n \leq 0 \text{ and } a_{i-1/2}^n \geq 0. \end{cases} \tag{17.23}$$

Comparing Equations (17.14) and (17.19), we notice that Roe's first-order upwind method differs from Godunov's first-order upwind method only at sonic points, just as expected. More specifically, *Roe's first-order method differs from Godunov's first-order upwind method only at expansive sonic points.* Comparing Equations (17.15) and (17.21), we see that Roe's first-order upwind method has less artificial viscosity than Godunov's first-order upwind method at expansive sonic points. The reduced artificial viscosity reduces the accuracy of Roe's method at expansive sonic points. On the positive side, Roe's first-order upwind method is simpler and cheaper than Godunov's first-order upwind method.

The behavior of Roe's first-order upwind method is illustrated using the five standard test cases defined in Section 17.0.

Test Cases 1 through 4 Roe's first-order upwind method is identical to Godunov's first-order method and to FTBS, since these tests cases all lack expansive sonic points. See Figures 17.13 to 17.16 and the associated discussion.

Test Case 5 As seen in Figure 17.19, the steady shock is captured perfectly. Unfortunately, like the Lax–Wendroff method, Roe's method fails to alter the initial conditions in any way, which is a total disaster at the expansion. As one way to view the situation, Roe's approximate Riemann solver cannot capture the finite spread of expansion fans, and this defect carries over to Roe's first-order upwind method. As another way to view the situation, Godunov's first-order method has somewhat inadequate artificial viscosity at expansive sonic points, whereas Roe's first-order upwind method has drastically inadequate artificial viscosity at expansive sonic points. In particular, at the expansive sonic point in this test case, by Equation (17.21), Roe's first-order upwind method has

$$\epsilon = \frac{f(1) - f(-1)}{1 - (-1)} = 0,$$

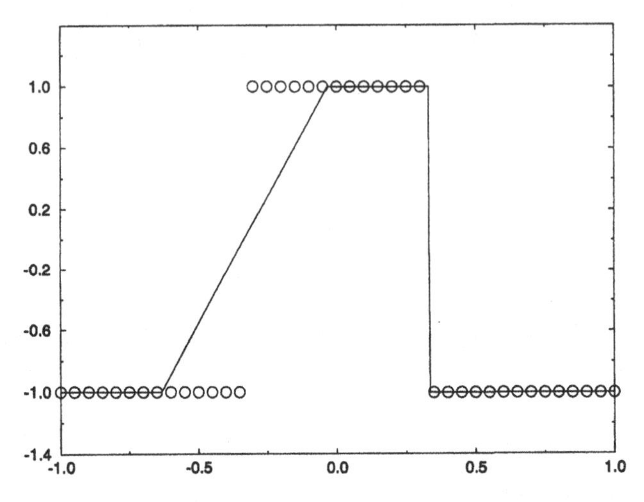

Figure 17.19 Roe's first-order upwind method for Test Case 5.

which makes Roe's first-order upwind method the same as FTCS. By Equation (17.15) Godunov's first-order upwind method has

$$\epsilon = \frac{f(1) + f(-1)}{1 - (-1)} = \frac{1}{2},$$

which is a little small but better than nothing. Although Roe's and Godunov's first-order upwind methods have too little artificial viscosity at expansive sonic points, too much artificial viscosity at expansive sonic points can also cause problems. In particular, too little artificial viscosity may cause a spurious vertical jump, whereas too much may cause a spurious horizontal jump. Clearly, expansive sonic points are delicate things.

17.3.3 *Harten's First-Order Upwind Method*

Whereas Roe's approximate Riemann solver uses locally linear approximations to the flux function, this section uses locally quadratic approximations to the flux function. A linear approximation to the flux is not very accurate at the maxima and minima of the flux function (i.e., sonic points) because a linear function is always monotone increasing or decreasing. Unlike linear functions, quadratic functions allow extrema, and thus a quadratic approximation should be much more successful for modeling the extrema of the flux function. Suppose that the left and right states to the Riemann problem are $u_L = u_i^n$ and $u_R = u_{i+1}^n$, respectively. Also, suppose that the quadratic approximation to the flux function passes through $(u_i^n, f(u_i^n))$ and $(u_{i+1}^n, f(u_{i+1}^n))$. This leaves one degree of freedom in the quadratic which can be used any way you like. For example, this degree of freedom can be used to model the true flux function. The closer the quadratic resembles the true flux function, the more the resulting first-order upwind method resembles Godunov's first-order upwind method, but, unfortunately, Godunov's first-order upwind method allows $O(\Delta x)$ jumps at expansive sonic points. Thus the free parameter in the quadratic should be chosen to model the true flux function everywhere *except* near expansive sonic points, where it should be chosen to address the special numerical needs of expansive sonic points. It is exciting to realize that an approximate Riemann solver may actually yield better numerical results than the exact Riemann solver, at least near expansive sonic points. In other words, errors in the solution to the Riemann problem may either increase *or decrease* errors in first-order upwind approximations based on Riemann solvers.

Suppose that the locally quadratic approximation to the flux function is written as follows:

$$f_{q,i+1/2}(u) = \frac{\delta^n_{i+1/2}}{u^n_i - u^n_{i+1}}(u - u^n_{i+1})(u - u^n_i) + a^n_{i+1/2}(u - u^n_i) + f(u^n_i), \tag{17.24}$$

where $\delta^n_{i+1/2}$ is a free parameter and $a^n_{i+1/2}$ is defined by Equation (17.9). Notice that the quadratic in Equation (17.24) is in Newton form, as defined in Subsection 8.1.2. Among other virtues, the Newton form separates the linear term $a^n_{i+1/2}(u - u^n_i) + f(u^n_i)$ used in the last subsection from the new quadratic term $\delta^n_{i+1/2}(u - u^n_{i+1})(u - u^n_i)/(u^n_{i+1} - u^n_i)$ added in this subsection. The factor $1/(u^n_{i+1} - u^n_i)$ in the new quadratic term is for convenience and for consistency with the literature. By Equations (5.66) and (17.13), the exact solution to this approximate locally quadratic Riemann problem yields the following first-order upwind method:

$$\hat{f}^n_{i+1/2} = \begin{cases} \min\limits_{u^n_i \le u \le u^n_{i+1}} f_{q,i+1/2}(u) & \text{if } u^n_i < u^n_{i+1}, \\ \max\limits_{u^n_i \ge u \ge u^n_{i+1}} f_{q,i+1/2}(u) & \text{if } u^n_i > u^n_{i+1}, \end{cases} \tag{17.25}$$

which is called *Harten's first-order upwind method.* As shown in Problem 17.10 Harten's first-order upwind method can be written as

$$\hat{f}^n_{i+1/2} = \frac{1}{2}\left(f(u^n_{i+1}) + f(u^n_i)\right) - \frac{1}{2}\epsilon^n_{i+1/2}(u^{n+1}_i - u^n_i),$$

where

$$\blacklozenge \qquad \epsilon_{i+1/2} = \begin{cases} \dfrac{\left(a^n_{i+1/2}\right)^2 + \left(\delta^n_{i+1/2}\right)^2}{2\delta^n_{i+1/2}} & \left|a^n_{i+1/2}\right| < \delta^n_{i+1/2}, \\ \left|a^n_{i+1/2}\right| & \left|a^n_{i+1/2}\right| > \delta^n_{i+1/2}, \end{cases} \tag{17.26}$$

which is the expression given in Harten's original 1983 paper. The coefficient of artificial viscosity of Harten's first-order upwind method is illustrated in Figure 17.20. Comparing

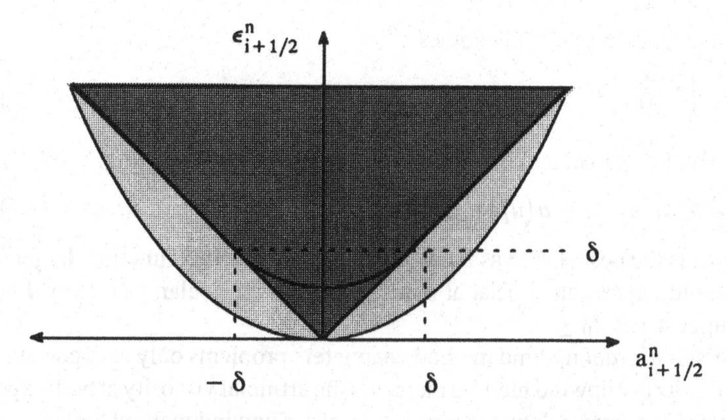

Figure 17.20 The artificial viscosity of Harten's first-order upwind method.

Figures 17.18 and 17.20, for small $\delta^n_{i+1/2}$, we notice that Harten's first-order upwind method is exactly the same as Roe's first-order upwind method except for the increased artificial viscosity near sonic points, which may reduce or eliminate entropy-condition-violating expansion shocks. Harten's first-order upwind method is sometimes called an *entropy fix* of Roe's first-order upwind method. Of course, every first-order upwind method in this section differs from Roe's first-order upwind methods only near sonic points. In this sense, every first-order upwind method in this section is some sort of "entropy fix" of Roe's first-order upwind method.

Notice that the quadratic in Equation (17.24) equals an interpolation quadratic if

$$\frac{\delta^n_{i+1/2}}{u^n_{i+1} - u^n_i} = f[u_{i-1}, u_i, u_{i+1}] = \frac{1}{2}\frac{d^2 f(\xi)}{du^2}$$

or if

$$\frac{\delta^n_{i+1/2}}{u^n_{i+1} - u^n_i} = f[u_i, u_{i+1}, u_{i+2}] = \frac{1}{2}\frac{d^2 f(\xi')}{du^2},$$

where ξ is some local value of u. Either of these interpolation quadratics equals the true flux function to third-order accuracy. Alternatively, by matching terms in a Taylor series, the quadratic in Equation (17.24) equals the true flux function to third-order accuracy if

$$\frac{\delta^n_{i+1/2}}{u^n_{i+1} - u^n_i} = \frac{1}{2}\frac{d^2 f\left(u^n_i\right)}{du^2},$$

as shown in Problem 17.10. More generally, Equation (17.24) equals the true flux function to third-order accuracy if

$$\frac{\delta^n_{i+1/2}}{u^n_{i+1} - u^n_i} = \frac{1}{2}(f'')^n_{i+1/2}, \tag{17.27}$$

where $(f'')^n_{i+1/2}$ is any average value of $f''(u)$ for u between u^n_i and u^n_{i+1}. For example, Van Leer, Lee, and Powell (1989) suggested the following "Roe average" second derivative:

$$(f'')^n_{i+1/2} = \begin{cases} \frac{f'\left(u^n_{i+1}\right) - f'\left(u^n_i\right)}{u^n_{i+1} - u^n_i} & u^n_i \neq u^n_{i+1}, \\ f''\left(u^n_i\right) & u^n_i = u^n_{i+1}. \end{cases} \tag{17.28}$$

With this average, Equation (17.27) becomes

$$\delta^n_{i+1/2} = \frac{1}{2}\left(a\left(u^n_{i+1}\right) - a\left(u^n_i\right)\right). \tag{17.29}$$

More generally, Van Leer, Lee, and Powell (1989) suggested the following:

$$\delta^n_{i+1/2} = \delta_0\left(a\left(u^n_{i+1}\right) - a\left(u^n_i\right)\right). \tag{17.30}$$

Although $\delta_0 = 1/2$ is the best choice as far as modeling the true flux function, larger values are required to avoid expansion shocks at sonic points; in particular, Van Leer, Lee, and Powell (1989) suggest $1 \leq \delta_0 \leq 2$.

Although Roe's first-order upwind method encounters problems only at expansive sonic points, Harten's first-order upwind method increases the artificial viscosity at both expansive and compressive sonic points, relative to Roe's first-order upwind method, unless $\delta = 0$ at compressive sonic points. Roe's first-order upwind method performs well at compressive sonic points, and Harten's "entropy fix" only degrades performance in this instance. This problem is easily addressed by using a linear approximation at compressive sonic points, as in Roe's first-order upwind method, and a quadratic approximation at expansive sonic points, as in Harten's first-order upwind method. This yields a first-order upwind method with the following coefficient of artificial viscosity:

$$\epsilon^n_{i+1/2} = \begin{cases} \dfrac{\left(a^n_{i+1/2}\right)^2 + \left(\delta^n_{i+1/2}\right)^2}{2\delta^n_{i+1/2}} & \begin{array}{l} \left|a^n_{i+1/2}\right| < \delta^n_{i+1/2}, \\ a\left(u^n_i\right) \le 0, a\left(u^n_{i+1}\right) \ge 0, \end{array} \\ \left|a^n_{i+1/2}\right| & \text{otherwise.} \end{cases} \tag{17.31}$$

This coefficient was first suggested by Van Leer, Lee, and Powell (1989). Having said all of this, in practice, the most common approach is to set $\delta^n_{i+1/2}$ equal to some small constant rather than attempting any solution sensitivity of the sort seen in Equations (17.27), (17.30), or (17.31).

The behavior of Harten's first-order upwind method is illustrated using the five standard test cases defined in Section 17.0.

Test Cases 1 through 4 Harten's first-order upwind method is identical to Roe's method and Godunov's first-order upwind method. See Figures 17.13 to 17.16 and the associated discussion.

Test Case 5 As seen in Figure 17.21, using Equations (17.30) and (17.31) with $\delta_0 = 1$, Harten's first-order upwind method captures the expansion fan without any expansion shock. The main defect in the expansion fan is a slight rounding off of the corners at the head and tail. Away from the expansion fan, Harten's first-order method is just the same as Roe's and Godunov's first-order upwind method. In particular, it captures the steady shock perfectly.

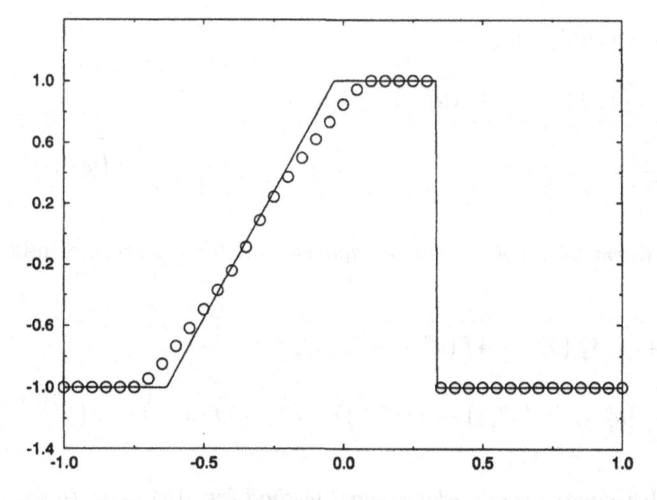

Figure 17.21 Harten's first-order upwind method for Test Case 5.

17.4 Beam–Warming Second-Order Upwind Method

The first-order upwind methods seen in the last section are accurate and stable at shocks. Unfortunately, first-order accuracy in smooth regions is too low for most practical computations. This leads to the following question: do higher-order accurate upwind methods in general share the stability and shock capturing abilities of their first-order brethren? If so, higher-order accurate upwind methods will be the methods of choice for shock capturing. This section concerns a specific second-order accurate upwind method proposed by Warming and Beam (1976). The Beam–Warming second-order upwind method exemplifies some of the general virtues and limitations of higher-order upwind methods.

The derivation of the Beam–Warming second-order upwind method is similar to the derivation of the Lax–Wendroff method seen in Section 17.2. To begin with, consider the following Taylor series:

$$u(x, t + \Delta t) = u(x, t) - \Delta t \frac{\partial f}{\partial x}(x, t) + \frac{\Delta t^2}{2} \frac{\partial}{\partial x}\left(a(u)\frac{\partial f}{\partial x}\right)(x, t) + O(\Delta t^3),$$

where the governing equation has been used to replace time derivatives by space derivatives. Assume $a(u) > 0$. Use Equation (10.18) to discretize the first derivative:

$$\frac{\partial f}{\partial x}(x_i, t^n) = \frac{3f(u_i^n) - 4f(u_{i-1}^n) + f(u_{i-2}^n)}{2\Delta x} + O(\Delta x^2).$$

Also let

$$\frac{\partial}{\partial x}\left(a(u)\frac{\partial f}{\partial x}\right)(x_i, t^n) = \frac{\partial}{\partial x}\left(a(u)\frac{\partial f}{\partial x}\right)(x_{i-1}, t^n) + O(\Delta x),$$

where

$$\frac{\partial}{\partial x}\left(a(u)\frac{\partial f}{\partial x}\right)(x_{i-1}, t^n)$$
$$= \frac{a_{i-1/2}^n(f(u_i^n) - f(u_{i-1}^n)) - a_{i-3/2}^n(f(u_{i-1}^n) - f(u_{i-2}^n))}{\Delta x^2} + O(\Delta x^2).$$

The resulting method is as follows:

$$\blacklozenge \quad u_i^{n+1} = u_i^n - \frac{\lambda}{2}(3f(u_i^n) - 4f(u_{i-1}^n) + f(u_{i-2}^n))$$
$$+ \frac{\lambda^2}{2}\left[a_{i-1/2}^n(f(u_i^n) - f(u_{i-1}^n)) - a_{i-3/2}^n(f(u_{i-1}^n) - f(u_{i-2}^n))\right], \tag{17.32a}$$

which is the *Beam–Warming second-order upwind method* for $a(u) > 0$. Similarly, for $a(u) < 0$,

$$u_i^{n+1} = u_i^n + \frac{\lambda}{2}(3f(u_i^n) - 4f(u_{i+1}^n) + f(u_{i+2}^n))$$
$$+ \frac{\lambda^2}{2}\left[a_{i+3/2}^n(f(u_{i+2}^n) - f(u_{i+1}^n)) - a_{i+1/2}^n(f(u_{i+1}^n) - f(u_i^n))\right], \tag{17.32b}$$

which is the *Beam–Warming second-order upwind method* for $a(u) < 0$. In the Beam–Warming method, the average wave speed $a_{i+1/2}^n$ may be defined in a variety of different ways. In other words, the Beam–Warming method is actually an entire *class* of methods differing in the choice of $a^n_{i+1/2}$, just like the Lax–Wendroff method. As usual, this book will use definition (17.9).

The above expressions for the Beam–Warming second-order method are fine except at sonic points, where $a(u)$ changes sign. There are at least three possible ways to define the method at sonic points: flux averaging, flux splitting, and reconstruction–evolution. In 1976, at the time of the original paper, neither flux splitting nor higher-order accurate reconstruction–evolution had been invented yet. Thus the original paper used flux averaging, such that the Beam–Warming second-order upwind method was used in smooth regions away from sonic points, and a first-order upwind method was used at shocks and at sonic points. However, this approach is a little too advanced for now; flux averaging will not be considered in detail until Part V. So now consider reconstruction–evolution. Reconstruction–evolution easily yields first-order accurate methods with piecewise-constant reconstructions, as seen in the last section, since an exact or approximate solution to the Riemann problem easily yields the time evolution. However, reconstruction–evolution requires a great deal more effort for second- and higher-order accurate methods, which use piecewise-linear and higher-order polynomial reconstructions, because these spatial reconstructions lack any obvious time evolution. Thus the reconstruction–evolution approach is also a little too advanced for the time being. This leaves flux splitting, first introduced in Section 13.4. In particular, suppose that $f(u) = f^+(u) + f^-(u)$, where $df^+/du \leq 0$ and $df^-/du \leq 0$. Then $\partial f/\partial x = \partial f^+/\partial x + \partial f^-/\partial x$. Use the Beam–Warming method for $a > 0$ to discretize $\partial f^+/\partial x$ and use the Beam–Warming method for $a < 0$ to discretize $\partial f^-/\partial x$. The resulting method is as follows:

$$\begin{aligned}
u_i^{n+1} = u_i^n &- \frac{\lambda}{2}\left(3f^+(u_i^n) - 4f^+(u_{i-1}^n) + f^+(u_{i-2}^n)\right) \\
&+ \frac{\lambda}{2}\left(3f^-(u_i^n) - 4f^-(u_{i+1}^n) + f^-(u_{i+2}^n)\right) \\
&+ \frac{\lambda^2}{2}\left[a^+_{i-1/2}\left(f^+(u_i^n) - f^+(u_{i-1}^n)\right) - a^+_{i-3/2}\left(f^+(u_{i-1}^n) - f^+(u_{i-2}^n)\right)\right] \\
&+ \frac{\lambda^2}{2}\left[a^-_{i+3/2}\left(f^-(u_{i+2}^n) - f^-(u_{i+1}^n)\right) - a^-_{i+1/2}\left(f^-(u_{i+1}^n) - f^-(u_i^n)\right)\right],
\end{aligned} \tag{17.33}$$

where

$$a^{\pm}_{i+1/2} = \begin{cases} \dfrac{(f^{\pm})'(u_{i+1}^n) - (f^{\pm})'(u_i^n)}{u_{i+1}^n - u_i^n} & u_i^n \neq u_{i+1}^n, \\ (f^{\pm})''(u_i^n) & u_i^n = u_{i+1}^n. \end{cases} \tag{17.34}$$

To keep things simple, the rest of the section concerns the Beam–Warning second-order upwind method for $a(u) > 0$. In conservation form, the Beam–Warming second-order upwind method for $a(u) > 0$ is

$$u_i^{n+1} = u_i^n - \lambda\left(\hat{f}^n_{i+1/2} - \hat{f}^n_{i-1/2}\right),$$

where

$$\begin{aligned}
\hat{f}^n_{i+1/2} &= \frac{1}{2}\left(3f(u_i^n) - f(u_{i-1}^n)\right) - \frac{\lambda}{2}a^n_{i-1/2}\left(f(u_i^n) - f(u_{i-1}^n)\right) \\
&= \frac{1}{2}\left(3f(u_i^n) - f(u_{i-1}^n)\right) - \frac{\lambda}{2}\left(a^n_{i-1/2}\right)^2\left(u_i^n - u_{i-1}^n\right).
\end{aligned} \tag{17.35}$$

In artificial viscosity form, the Beam–Warming second-order upwind method for $a(u) > 0$ is

$$u_i^{n+1} = u_i - \frac{\lambda}{2}\left(f\left(u_{i+1}^n\right) - f\left(u_{i-1}^n\right)\right) + \frac{\lambda}{2}\left(\epsilon_{i+1/2}^n\left(u_{i+1}^n - u_i^n\right) - \epsilon_{i-1/2}^n\left(u_i^n - u_{i-1}^n\right)\right),$$

where

$$\epsilon_{i+1/2}^n = a_{i+1/2}^n - a_{i-1/2}^n\left(1 - \lambda a_{i-1/2}^n\right) r_i^n \tag{17.36}$$

and where

$$r_i^n = \frac{u_i^n - u_{i-1}^n}{u_{i+1}^n - u_i^n}. \tag{17.37}$$

Notice that $r_i^n \geq 0$ if the solution is monotone increasing or decreasing, whereas $r_i^n \leq 0$ if the solution has a maximum or minimum. Figure 17.22 illustrates the coefficient of artificial viscosity for the Beam–Warming method applied to the linear advection equation. Unlike the previous methods in this chapter, $\lambda\epsilon$ is not a simple function of λa. Instead, $\lambda\epsilon$ depends on both λa and the ratio r_i^n. Depending on r_i^n, the coefficient of artificial viscosity of the Beam–Warming second-order upwind method may equal that of Roe's first-order upwind method, or the Lax–Wendroff method, or a variety of other unnamed methods. The complicated and possibly infinite coefficient of artificial viscosity is a sad but common consequence of straying outside the stencil $(u_{i-1}^n, u_i^n, u_{i+1}^n)$. In wave speed split form, the

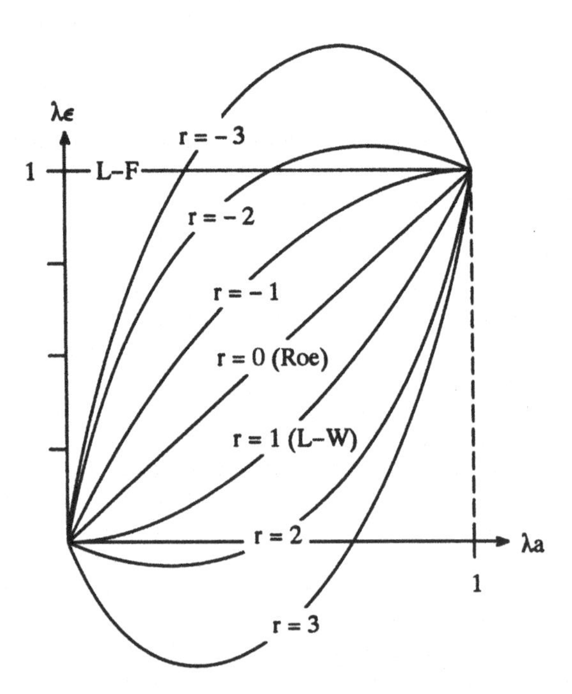

Figure 17.22 Artificial viscosity for the Beam–Warming second-order upwind method applied to the linear advection equation.

Beam–Warming second-order upwind method for $a(u) > 0$ is

$$u_i^{n+1} = u_i^n + C_{i+1/2}^{+}(u_{i+1}^n - u_i^n) - C_{i-1/2}^{-}(u_i^n - u_{i-1}^n),$$

where

$$C_{i+1/2}^{+} = \frac{1}{2}\lambda a_{i-1/2}^n r_i^n (\lambda a_{i-1/2}^n - 3), \tag{17.38a}$$

$$C_{i+1/2}^{-} = \frac{1}{2}\lambda a_{i-1/2}^n r_i^n (\lambda a_{i-1/2}^n - 1). \tag{17.38b}$$

Of course, as always, there are infinitely many other wave speed split forms. Unlike the previous methods in this chapter, the coefficients $C_{i+1/2}^{+}$ in Equation (17.38) are infinite when the ratio r_i^n is infinite. In fact, there is *no* wave speed split form of the Beam–Warming second-order upwind method with finite coefficients. This is another sad but common consequence of straying outside the stencil $(u_{i-1}^n, u_i^n, u_{i+1}^n)$.

The twelve-point checklist for the Beam–Warming second-order upwind method is as follows:

(1) The coefficient of artificial viscosity may be large or infinite, depending on the ratio $r_i^n = (u_i^n - u_{i-1}^n)/(u_{i+1}^n - u_i^n)$ and the CFL number.
(2) CFL condition: $|\lambda a(u)| \leq 2$. This is the first time we have seen a CFL limit larger than 1. This is a major benefit of using a wider stencil.
(3) Conservative.
(4) Consistent.
(5) The Beam–Warming second-order upwind method may not converge on nonsmooth solutions as $\Delta x \to 0$ and $\Delta t \to 0$. Even if it does converge, the converged solution may not satisfy entropy conditions. In particular, the Beam–Warming second-order upwind method may allow expansion shocks, depending on the sonic point treatment.
(6) Explicit.
(7) Finite difference.
(8) Linearly stable provided that the CFL condition is satisfied.
(9) Linear when applied to the linear advection equation. Nonlinear when applied to any nonlinear scalar conservation law.
(10) Consider the nonlinear stability conditions seen in Chapter 16. Although there is no rigorous proof, numerical tests show that the Beam–Warming second-order upwind method is TVB. However, although the total variation is bounded for any fixed Δx and Δt, the total variation may equal infinity in the limit $\Delta x \to 0$ and $\Delta t \to 0$. In other words, the Beam–Warming second-order upwind method is always TVB but *not* always total variation stable; see the discussion in Section 16.11. Furthermore, by Godunov's theorem seen in Section 16.1, the Beam–Warming second-order method is not monotonicity preserving. Then the Beam–Warming second-order upwind method cannot satisfy any condition that implies monotonicity preservation including TVD, positivity, range reduction, the upwind range condition, contraction, or the monotone condition.
(11) Formally second-order accurate in time and space, except possibly at sonic points.
(12) Upwind.

As an additional property, not listed above, steady-state solutions of the Beam–Warming second-order upwind method depend on the time step Δt. The behavior of the Beam–Warming second-order upwind method is illustrated using the five standard test cases defined in Section 17.0.

Test Case 1 As seen in Figure 17.23, the sinusoid's shape and amplitude are well captured. The only visible error is a slight leading phase error.

Test Case 2 As seen in Figure 17.24, the numerical solution oscillates about the true solution causing errors of up to 25%. On the whole, this solution is distinctly worse than the Lax–Wendroff solution seen in Figure 17.8. The relatively poor performance of the Beam–Warming second-order upwind method in Test Case 2 could have been predicted by examining the artificial viscosity. As seen in Figure 17.22, for large $|r_i^n|$, the coefficient

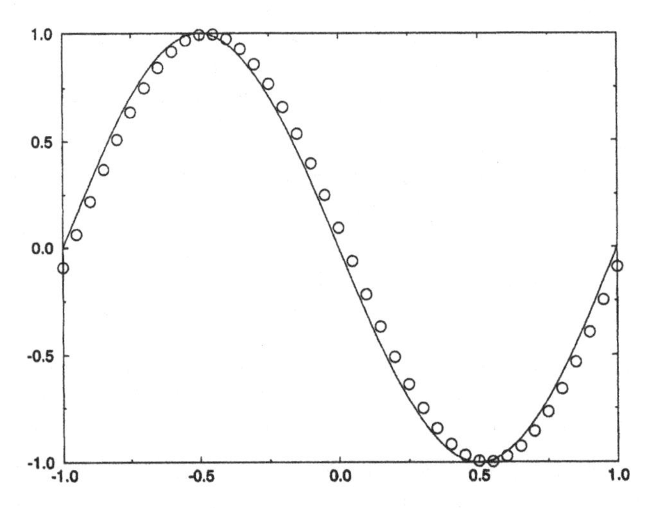

Figure 17.23 Beam–Warming second-order upwind method for Test Case 1.

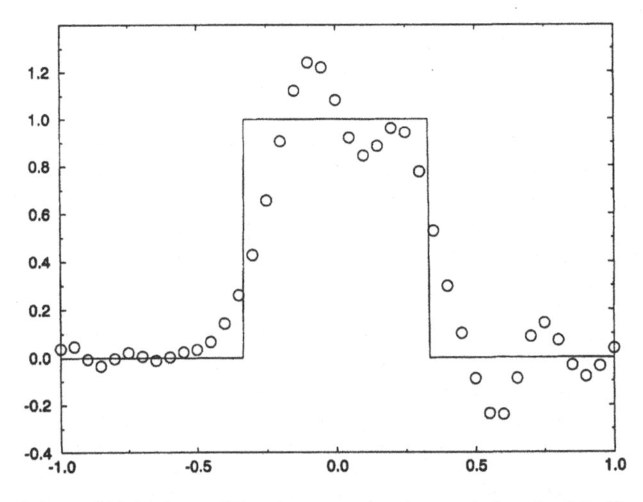

Figure 17.24 Beam–Warming second-order upwind method for Test Case 2.

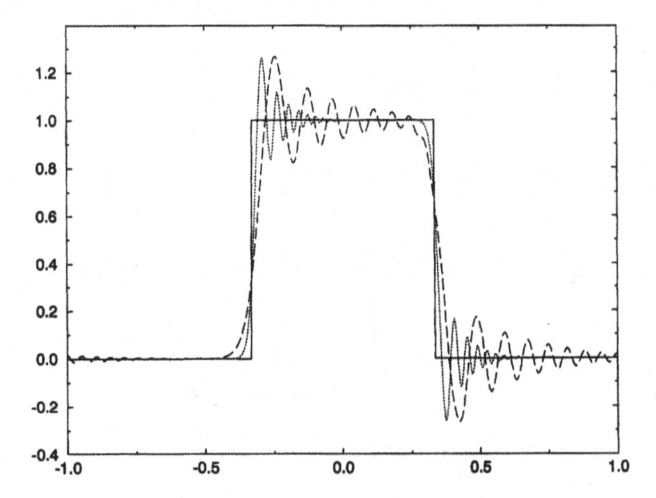

Figure 17.25 Beam–Warming second-order upwind method for Test Case 3.

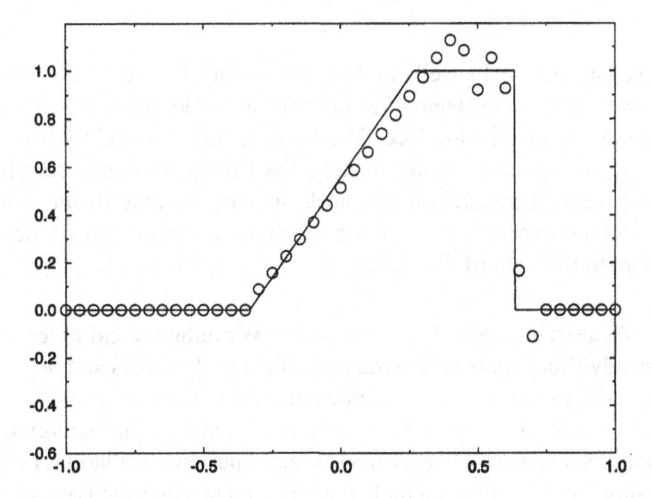

Figure 17.26 Beam–Warming second-order upwind method for Test Case 4.

of artificial viscosity may be either way too large or way too small, either of which leads to instability and spurious oscillations.

Test Case 3 In Figure 17.25, the dotted line represents the Beam–Warming approximation to $u(x, 4)$, the long dashed line represents the Beam–Warming approximation to $u(x, 40)$, and the solid line represents the exact solution for $u(x, 4)$ and $u(x, 40)$. The spurious oscillations are similar to those found in the Lax–Wendroff method, except that they are far more severe and lie on opposite sides of the jump discontinuities.

Test Case 4 This test case and the next use the natural flux splitting $f^{+}(u) = \max(0, u)u/2$ and $f^{-}(u) = \max(0, u)u/2$ for Burgers' equation, as described in Example 13.7. As seen in Figure 17.26, the Beam–Warming second-order upwind method

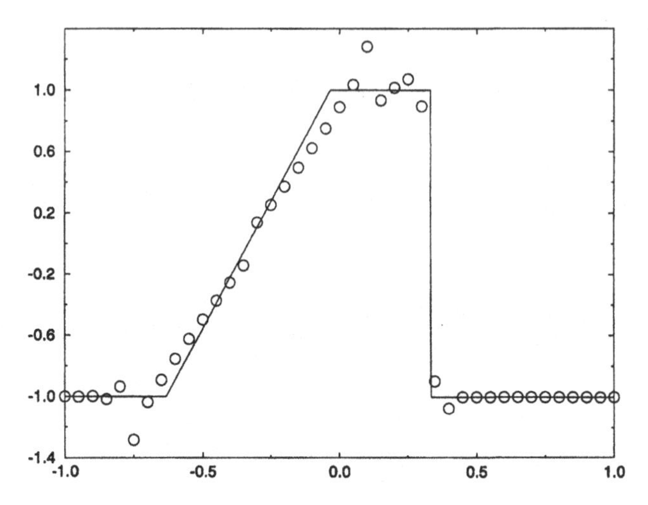

Figure 17.27 Beam–Warming second-order upwind method for Test Case 5.

captures the expansion fan extremely well. In fact, comparing Figure 17.26 with Figures 17.10 and 17.16, we see that the Beam–Warming second-order upwind method captures the expansion fan far better than the Lax–Wendroff method and slightly better than Godunov's first-order upwind method. Unfortunately, the Beam–Warming second-order upwind method is not quite so successful on the shock. In particular, the Beam–Warming second-order upwind method experiences large $2\Delta x$ oscillations to the left of the shock and a smaller undershoot to the right of the shock.

Test Case 5 As seen in Figure 17.27, the Beam–Warming second-order upwind method captures the steady shock quite well, with only slight undershoots and overshoots. Unfortunately, the Beam–Warming second-order upwind method has more difficulties with the sonic expansion fan. First of all, it experiences a small but significant spurious expansion shock near the sonic point. Second of all, it experiences large spurious oscillations to the left and right of the expansion fan. Still, this is a far better performance than the Lax–Wendroff method or Roe's first-order upwind method, which do not affect the initial conditions in any way.

We can now answer the question posed at the beginning of the section: unlike first-order upwind methods, higher-order upwind methods may be extremely oscillatory. In fact, higher-order upwind methods may be even more oscillatory than centered methods, as is the case comparing the Beam–Warming second-order upwind method with the Lax–Wendroff method. Of course, this conclusion is completely consistent with the discussion of upwind methods found in Section 13.1. In their original paper, Warming and Beam (1976) recognized the oscillatory nature of their second-order upwind method; they proposed a blending that used a first-order upwind method at shocks and sonic points and their second-order upwind method in smooth regions, as mentioned earlier. As a final note, there is an implicit method commonly referred to as the Beam–Warming method, and this method should not be confused with the explicit method discussed here.

17.5 Fromm's Method

This section concerns Fromm's method, discovered in 1968. Flux averaging was introduced in Section 13.3, and Fromm's method is best seen as a simple average of the Lax–Wendroff method and the Beam–Warming second-order upwind method. For an average of two methods to make sense, the two methods should have complementary properties. For example, one method might have too much artificial viscosity, while the other has too little. As another example, one method might have the best stencil for smooth regions, while the other method might have the best stencil for shocks. As a third example, one method may have negative artificial dispersion, while the other has positive artificial dispersion.

To start this section, let us show that the Lax–Wendroff method and the Beam–Warming second-order upwind method have complementary properties, at least when applied to the linear advection equation. Modified equations are an extremely powerful analysis tool for linear methods, as described in Section 15.3. The modified equation for the Lax–Wendroff method applied to the linear advection equation is

$$\frac{\partial u}{\partial t} + a\frac{\partial u}{\partial x} = -\frac{a\Delta x^2}{6}(1-(\lambda a)^2)\frac{\partial^3 u}{\partial x^3} - \frac{a\Delta x^3}{8}\lambda a(1-(\lambda a)^2)\frac{\partial^4 u}{\partial x^4} + \cdots. \tag{17.39}$$

The modified equation for the Beam–Warming second-order upwind method applied to the linear advection equation for $a > 0$ is

$$\frac{\partial u}{\partial t} + a\frac{\partial u}{\partial x} = \frac{a\Delta x^2}{6}(1-\lambda a)(2-\lambda a)\frac{\partial^3 u}{\partial x^3} - \frac{a\Delta x^3}{8}(1-\lambda a)^2(2-\lambda a)\frac{\partial^4 u}{\partial x^4} + \cdots. \tag{17.40}$$

Notice that the coefficients of the third-derivative terms in Equations (17.39) and (17.40) have opposite signs for $0 \leq \lambda a \leq 1$. This suggests that an average of the Lax–Wendroff method and the Beam–Warming second-order upwind method will have less dispersive error than either method separately, at least when applied to the linear advection equation. In particular, the sum of Equations (17.39) and (17.40) divided by two yields

$$\begin{aligned}\frac{\partial u}{\partial t} + a\frac{\partial u}{\partial x} &= \frac{a\Delta x^2}{6}(1-\lambda a)\left(\frac{1}{2}-\lambda a\right)\frac{\partial^3 u}{\partial x^3} \\ &\quad - \frac{a\Delta x^3}{8}(1-\lambda a)((\lambda a)^2 - \lambda a + 1)\frac{\partial^4 u}{\partial x^4} + \cdots.\end{aligned} \tag{17.41}$$

As the reader can easily verify, the coefficient of the third derivative in Equation (17.41) is substantially less in absolute value than the ones in Equation (17.39) or (17.40) for $0 \leq \lambda a \leq 1$.

Let us average the Lax–Wendroff and Beam–Warming second-order upwind methods in artificial viscosity form. Instead of using FTCS as the reference method, as in the standard artificial viscosity form, it is more convenient here to use FTBS as the reference. The Lax–Wendroff method may be written as FTBS plus second-order artificial viscosity as follows:

$$\begin{aligned}u_i^{n+1} = u_i^n - \lambda\left(f\left(u_i^n\right) - f\left(u_{i-1}^n\right)\right) &- \frac{1}{2}\lambda a_{i+1/2}^n\left(1-\lambda a_{i+1/2}^n\right)\left(u_{i+1}^n - u_i^n\right) \\ &+ \frac{1}{2}\lambda a_{i-1/2}^n\left(1-\lambda a_{i-1/2}^n\right)\left(u_i^n - u_{i-1}^n\right).\end{aligned} \tag{17.42}$$

The Beam–Warming second-order upwind method for $a(u) > 0$ may be written as FTBS plus second-order artificial viscosity as follows:

$$u_i^{n+1} = u_i^n - \lambda\left(f\left(u_i^n\right) - f\left(u_{i-1}^n\right)\right) - \frac{1}{2}\lambda a_{i-1/2}^n\left(1 - \lambda a_{i-1/2}^n\right)\left(u_i^n - u_{i-1}^n\right) + \frac{1}{2}\lambda a_{i-3/2}^n\left(1 - \lambda a_{i-3/2}^n\right)\left(u_{i-1}^n - u_{i-2}^n\right). \tag{17.43}$$

The sum of Equations (17.42) and (17.43) divided by two yields the following:

♦
$$u_i^{n+1} = u_i^n - \lambda\left(f\left(u_i^n\right) - f\left(u_{i-1}^n\right)\right) - \frac{1}{4}\lambda a_{i+1/2}^n\left(1 - \lambda a_{i+1/2}^n\right)\left(u_{i+1}^n - u_i^n\right) + \frac{1}{4}\lambda a_{i-3/2}^n\left(1 - \lambda a_{i-3/2}^n\right)\left(u_{i-1}^n - u_{i-2}^n\right), \tag{17.44}$$

which is *Fromm's method* for $a(u) > 0$. Fromm's method for $a(u) < 0$ is left as an exercise for the reader. Also, Fromm's method can be extended to any $a(u)$, regardless of sign, using flux averaging, reconstruction–evolution, or flux splitting, just as in the last section; the details are omitted. Finally, expressions for the conservation form, the artificial viscosity form, and wave speed split forms are left as exercises. The extremely simple fixed arithmetic average in Fromm's method is fairly successful, which indicates that more sophisticated solution-sensitive averages might be wildly successful. We shall return to this idea in Part V; in particular, see Section 20.1. The standard twelve-point checklist for Fromm's method is as follows:

(1) The coefficient of artificial viscosity may be large or infinite, depending on the ratio $r_i^n = (u_i^n - u_{i-1}^n)/(u_{i+1}^n - u_i^n)$ and the CFL number.
(2) CFL condition $|\lambda a(u)| \leq 2$.
(3) Conservative.
(4) Consistent.
(5) Fromm's method may not converge on nonsmooth solutions as $\Delta x \to 0$ and $\Delta t \to 0$. Even if it does converge, the converged solution may not satisfy entropy conditions. In particular, Fromm's method allows expansion shocks, depending on the sonic point treatment.
(6) Explicit.
(7) Finite difference.
(8) Linearly stable provided that $|\lambda a| \leq 1$. Unlike the other methods in this chapter, the linear stability condition is more restrictive than the CFL condition. The linear stability condition is inherited from the Lax–Wendroff method, whereas the CFL condition is inherited from the Beam–Warming second-order upwind method.
(9) Linear when applied to the linear advection equation. Nonlinear when applied to any nonlinear scalar conservation law.
(10) Consider the nonlinear stability conditions seen in Chapter 16. Although there is no rigorous proof, numerical tests show that Fromm's method is TVB for $|\lambda a(u)| \leq 1$. However, although the total variation is bounded for any fixed Δx and Δt, the total variation may go to infinity in the limit $\Delta x \to 0$ and $\Delta t \to 0$. In other words, Fromm's method is always TVB but *not* always total variation stable; see the discussion in Section 16.11. Furthermore, by Godunov's theorem seen in Section 16.1, Fromm's method is not monotonicity preserving. Then Fromm's method cannot

satisfy any condition that implies monotonicity preservation including TVD, positivity, range reduction, the upwind range condition, contraction, or the monotone condition.

(11) Formally second-order accurate in time and space.

(12) Upwind.

As an additional property, not listed above, steady-state solutions of Fromm's method depend on the time step Δt. The behavior of Fromm's method is illustrated using the five standard test cases defined in Section 17.0.

Test Case 1 As seen in Figure 17.28, Fromm's method yields the best results on this test case of any method seen in this chapter. Whereas the Lax–Wendroff method had a slight lagging error as seen in Figure 17.7, and the Beam–Warming second-order upwind method had a slight leading error as seen in Figure 17.23, Fromm's method is nearly indistinguishable from the true solution in Figure 17.28.

Test Case 2 As seen in Figure 17.29, the numerical solution oscillates about the true solution, but the maximum error is only about 7.5%. The contact discontinuities have been smeared out over about five grid cells, which is the least smearing seen so far. In fact, overall, Fromm's method does better than any other method seen in this chapter on this test case. It should be noted that the linear advection equation is perhaps an unfair choice, since Fromm's method was specifically designed to have a low error when applied to the linear advection equation. Fromm's method may not do quite so well when applied to a nonlinear equation.

Test Case 3 In Figure 17.30, the dotted line represents Fromm's approximation to $u(x, 4)$, the long dashed line represents Fromm's approximation to $u(x, 40)$, and the solid line represents the exact solution for $u(x, 4)$ and $u(x, 40)$. The spurious oscillations are similar to those found in the Lax–Wendroff and Beam–Warming second-order upwind methods, except that they are far less severe.

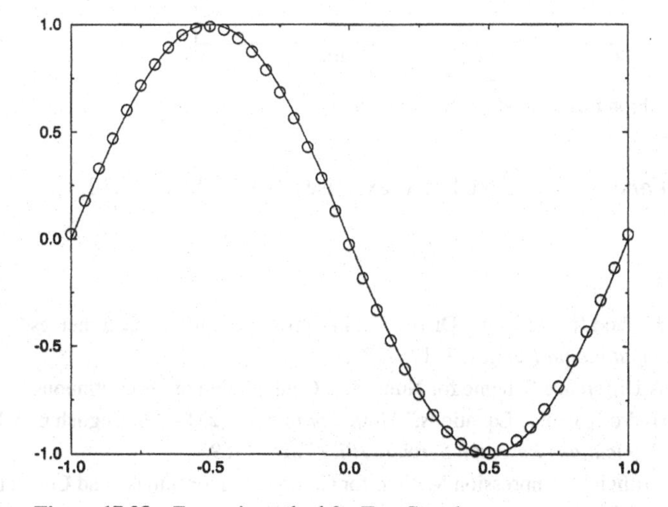

Figure 17.28 Fromm's method for Test Case 1.

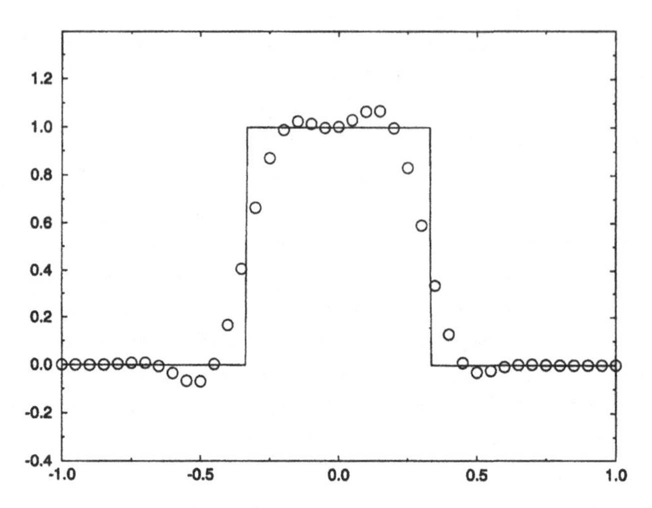

Figure 17.29 Fromm's method for Test Case 2.

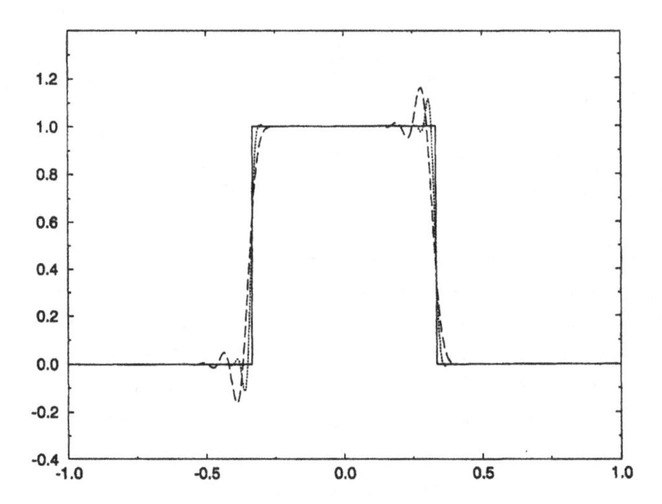

Figure 17.30 Fromm's method for Test Case 3.

Test Cases 4 and 5 These are left as exercises. See Problem 17.12.

References

Fromm, J. E. 1968. "A Method for Reducing Dispersion in Convective Difference Schemes," *Journal of Computational Physics*, 3: 176–189.

Godunov, S. K. 1959. "A Difference Scheme for Numerical Computation of Discontinuous Solutions of Hydrodynamics Equations," *Math. Sbornik*, 47: 271–306. English translation in *U.S. Joint Publications Research Service, JPRS 7226*, 1969.

Harten, A. 1978. "The Artificial Compression Method for Computation of Shocks and Contact Discontinuities: III. Self-Adjusting Hybrid Schemes," *Mathematics of Computation*, 32: 363–389.

Harten, A. 1983. "High Resolution Schemes for Hyperbolic Conservation Laws," *Journal of Computational Physics*, 49: 357–393.

Lax, P. D. 1954. "Weak Solutions of Nonlinear Hyperbolic Equations and Their Numerical Computation," *Communications on Pure and Applied Mathematics*, 7: 159–193.

Lax, P. D., and Wendroff, B. 1960. "Systems of Conservation Laws," *Communications on Pure and Applied Mathematics*, 13: 217–237.

Murman, E. M., and Cole, J. D. 1971. "Calculations of Plane Unsteady Transonic Flow," *AIAA Journal*, 9: 114–121.

Roe, P. L. 1981. "Approximate Riemann Solvers, Parameter Vectors, and Difference Schemes," *Journal of Computational Physics*, 43: 357–372.

Van Leer, B., Lee, W.-T., and Powell, K. G. 1989. "Sonic-Point Capturing," *AIAA Paper 89-1945* (unpublished).

Warming, R. F., and Beam, R. M. 1976. "Upwind Second-Order Difference Schemes and Applications in Aerodynamic Flows," *AIAA Journal*, 14: 1241–1249.

Problems

17.1 In general, linearity allows for simplifications. In particular, the expressions for the numerical methods found in this chapter may be simplified when the methods are applied to the linear advection equation. The simplified forms often improve computational efficiency and minimize round-off errors. Thus the simplified form is often preferred for actual programming. Unfortunately, the simplified linear forms look quite a bit different from the general forms for nonlinear scalar conservation laws. This can be confusing, in that some sources only provide the simplified expressions for linear advection equations, making it hard to recognize the method.

(a) Show that the Lax–Friedrichs method applied to the linear advection equation can be written as follows:

$$u_i^{n+1} = \frac{1}{2}(1 - \lambda a)u_{i+1}^n + \frac{1}{2}(1 + \lambda a)u_{i-1}^n.$$

(b) Show that the Lax–Wendroff method applied to the linear advection equation can be written as follows:

$$u_i^{n+1} = \frac{1}{2}\lambda a(\lambda a - 1)u_{i+1}^n - (\lambda a - 1)(\lambda a + 1)u_i^n + \frac{1}{2}\lambda a(\lambda a + 1)u_{i-1}^n.$$

(c) Show that the Beam–Warming second-order method applied to the linear advection equation with $a > 0$ can be written as follows:

$$u_i^{n+1} = \frac{1}{2}(\lambda a - 1)(\lambda a - 2)u_i^n - \lambda a(\lambda a - 2)u_{i-1}^n + \frac{1}{2}\lambda a(\lambda a - 1)u_{i-2}^n.$$

(d) Show that Fromm's method applied to the linear advection equation with $a > 0$ can be written as follows:

$$u_i^{n+1} = u_i^n - \frac{1}{4}\lambda a(1 - \lambda a)\left(u_{i+1}^n - u_i^n\right) - \lambda a\left(u_i^n - u_{i-1}^n\right) + \frac{1}{4}\lambda a(1 - \lambda a)\left(u_{i-1}^n - u_{i-2}^n\right).$$

(e) Argue that the approximations found in parts (a)–(d) are all *perfect* when $\lambda a = 1$.

17.2 Under ordinary circumstances, Fromm's method is second-order accurate. Show that Fromm's method is third-order accurate when applied to the linear advection equation with $\lambda a = 1/2$. You may prove this result either analytically or with numerical results.

17.3 In Section 17.1, the Lax–Friedrichs method is derived from FTCS by replacing u_i^n by $(u_{i+1}^n + u_{i-1}^n)/2$. Remember that FTCS is formally second-order accurate in space. Also notice that u_i^n and $(u_{i+1}^n + u_{i-1}^n)/2$ are the same to within second-order spatial accuracy. Does this imply that the Lax–Friedrichs method is formally second-order accurate in space? If so, then why is the Lax–Friedrichs method usually considered a first-order accurate method? Your answer should be relatively brief and intuitive. You may wish to review the discussion on order of accuracy found in Subsection 11.2.2.

17.4 Consider the Lax–Friedrichs method for the linear advection equation with the following initial conditions:

$$u_0^0 = 0.4, \quad u_1^0 = 0.5, \quad u_2^0 = 0.6, \quad u_3^0 = 1.0, \quad u_4^0 = 0.6, \quad u_5^0 = 0.5, \quad u_6^0 = 0.4.$$

Notice that the initial conditions contain a single maximum at u_3^0.

(a) Find u_1^1, u_2^1, u_3^1, u_4^1, and u_5^1 using the CFL number $\lambda a = 0.25$. Notice that the solution develops large oscillations in a single time step. In particular, u_2^1 and u_4^1 are maxima, while u_3^1 is a minimum.

(b) For the given initial conditions, show that the Lax–Friedrichs method for the linear advection equation develops spurious oscillations in one time step for $0 < \lambda a < 0.6$ but does not develop spurious oscillations in one time step for $0.6 \leq \lambda a \leq 1$. You may prove this result either analytically or with numerical results.

(c) The oscillations in the Lax–Friedrichs method and in other situations are often attributed to odd–even decoupling. To eliminate odd–even decoupling, reduce the coefficient of artificial viscosity by 10%, that is, let $\epsilon = 0.9/\lambda$, and repeat part (a). The resulting method depends on both even and odd points. Show that although reducing the artificial viscosity reduces the spurious oscillations, the method still develops spurious oscillations after a single time step. In other words, spurious oscillations in the Lax–Friedrichs method cannot be completely attributed to odd–even decoupling.

17.5 (a) Consider the Lax–Friedrichs method. What are the advantages and disadvantages of reducing its coefficient of artificial viscosity by a constant amount?

(b) Consider the Lax–Wendroff method. What are the advantages and disadvantages of increasing its coefficient of artificial viscosity by a constant amount?

17.6 Consider the scalar flux functions illustrated below.

(a) Sketch the flux functions. In your sketches, label all sonic points and state whether they are expansive or compressive.

(b) Find the conservative numerical flux $f_{i+1/2}^n$ of Godunov's and Roe's first-order upwind method.

17.7 Write Equation (17.33) for the Beam–Warming second-order upwind method in conservation form.

17.8 As seen in Figure 17.11, the Lax–Wendroff method using the "Roe average" definition of $a_{i+1/2}^n$ seen in Equation (17.9) allows a large expansion shock in Test Case 5. In this problem, we shall see how alternative definitions of $a_{i+1/2}^n$ affect expansive sonic point capturing.

(a) Repeat Test Case 5 using definition (17.7) for $a_{i+1/2}^n$.

(b) Repeat Test Case 5 using definition (17.8) for $a_{i+1/2}^n$.

17.9 Use Equation (17.18) to prove that Godunov's first-order upwind method applied to Burgers' equation *always* satisfies the nonlinear stability condition $|\lambda a_{i+1/2}^n| \leq \lambda \epsilon_{i+1/2}^n \leq 1$ found in Example 16.6, provided that the CFL condition is satisfied, *regardless of expansive sonic points*.

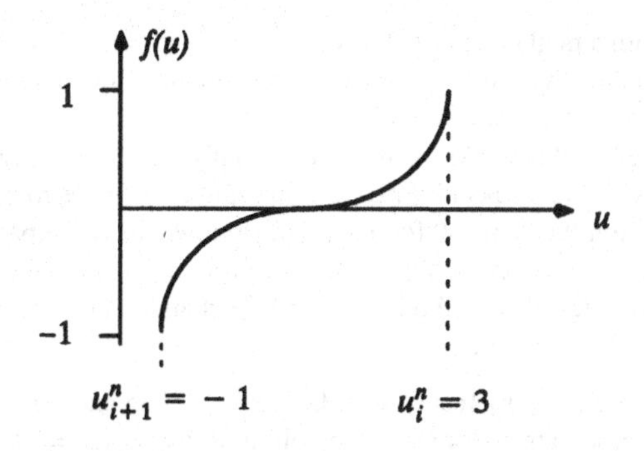

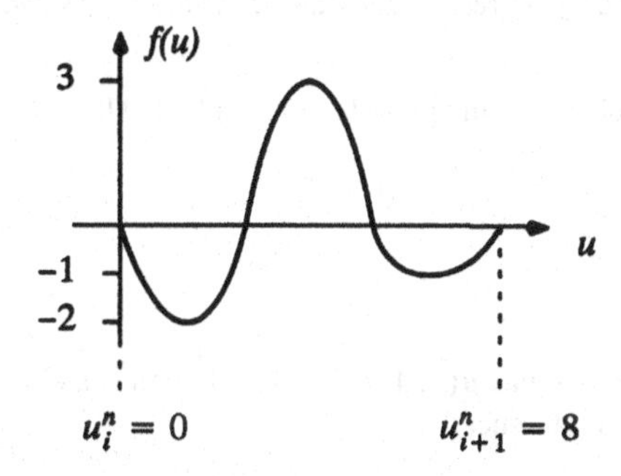

Problem 17.6.

17.10 Consider the following quadratic approximation to $f(u)$:

$$f_{q,i+1/2}(u) = \frac{\delta}{u^n_{i+1} - u^n_i}\left(u - u^n_i\right)^2 - \left(\delta - a^n_{i+1/2}\right)\left(u - u^n_i\right) + f\left(u^n_i\right).$$

(a) Show that this is the same quadratic as Equation (17.24), except that this has been written in Taylor series form rather than Newton form, which you may find more convenient in this problem.
(b) Verify that $f_{q,i+1/2}(u^n_i) = f(u^n_i)$ and $f_{q,i+1/2}(u^n_{i+1}) = f(u^n_{i+1})$.
(c) Assuming $\delta \neq 0$, the quadratic $f_{q,i+1/2}$ always has a global extremum. Show that the global extremum of $f_{q,i+1/2}$ occurs at u^* where

$$u^* - u^n_i = \frac{1}{2\delta}\left(\delta - a^n_{i+1/2}\right)\left(u^n_{i+1} - u^n_i\right).$$

(d) When is the extremum found in part (c) a maximum? When is the extremum found in part (c) a minimum? When does the extremum found in part (c) fall between u^n_i and u^n_{i+1}?
(e) Using the result from parts (a)–(d), prove that Equation (17.25) implies Equation (17.26).
(f) Show that $f_{q,i+1/2}(u) = f(u)$ if $f(u) = au$ and $\delta = 0$.
(g) Show that $f_{q,i+1/2}(u) = f(u)$ if $f(u) = u^2/2$ and $\delta = (u^n_{i+1} - u^n_i)/2$.
(h) Consider any flux function $f(u)$. Matching terms in a Taylor series for $f(u)$ centered on u^n_i, show that $f_{q,i+1/2}(u) = f(u) + O(u^n_{i+1} - u^n_i)^3$ if $\delta = f''(u^n_i)(u^n_{i+1} - u^n_i)/2$.

17.11 Write Fromm's method for $a(u) > 0$ in wave speed split form using $g^n_i = (f(u^n_i) + f(u^n_{i-1}))/2$.

17.12 (a) Find an expression for Fromm's method for $a(u) < 0$.

(b) Find a version of Fromm's method that works for any $a(u)$ by averaging Equations (17.5) and (17.33). Test this method on Test Cases 4 and 5.

(c) Find a version of Fromm's method that works for any $a(u)$ by applying (1) the expression for $a(u) > 0$ found in Section 17.5 to the positive part of a flux split form of the governing equations and (2) the expression for $a(u) < 0$ found in part (a) to the negative part of a flux split form of the governing equations. Is this version of Fromm's method the same as that found in part (b)? If not, test this method on Test Cases 4 and 5, and say how it compares to the method found in part (b).

17.13 Apply Harten's first-order upwind method to Test Case 5. Use $\Delta x = 0.025$ and $\Delta t = 0.02$ and evolve the solution for 30 time steps. Compare the solutions for $\delta^n_{i+1/2} = 0, 0.01, 0.025, 0.05, 0.1, 0.2, 0.5$, and 0.8. How do your results compare with Godunov's and Roe's first-order upwind methods?

17.14 Consider the following initial value problem on the periodic domain $[-1, 1]$:

$$\frac{\partial u}{\partial t} + \frac{\partial u}{\partial x} = 0,$$

$$u(x, 0) = \begin{cases} 1 - 3|x| & |x| \leq 1/3, \\ 0 & |x| \geq 1/3. \end{cases}$$

For each of the following methods, approximate $u(x, 4)$ using 40 grid points and $\lambda = 0.8$. Make sure to plot the exact solution as a reference.

(a) the Lax–Friedrichs method
(b) the Lax–Wendroff method
(c) FTBS
(d) the Beam–Warming second-order upwind method
(e) Fromm's method

CHAPTER 18

Basic Numerical Methods for the Euler Equations

18.0 Introduction

This chapter concerns numerical methods for the Euler equations. Rather than starting from scratch, this chapter mainly converts the numerical methods for scalar conservation laws seen in the last chapter into numerical methods for the Euler equations. The approaches described here can be divided into two categories – flux approaches and wave approaches. Wave approaches can be further subdivided into two categories – flux vector splitting approaches and reconstruction–evolution approaches. Whereas flux approaches consider only fluxes, wave approaches model both fluxes and waves, and especially the interactions between various families of waves, using either flux vector splitting or Riemann solvers, which makes them more physical and accurate but also more expensive and complicated. Before continuing, the reader may wish to review pertinent material from previous chapters. In particular, in preparation for flux vector splitting, discussed in Section 18.2, the reader should review the introduction to flux splitting and wave speed splitting found in Sections 13.4 and 13.5, respectively. Furthermore, in preparation for reconstruction–evolution methods, which use real or approximate Riemann solvers, the reader should review Chapter 5 and the introduction to reconstruction–evolution methods found in Section 13.6. And, of course, the reader should not even consider reading this chapter until carefully ingesting the last chapter.

This chapter converts all of the numerical methods for scalar conservation laws seen in the last chapter, except for Fromm's method, into numerical methods for the vector Euler equations in one if not several ways. For most numerical methods, the vector versions retain most of the properties of the scalar versions. However, while flaws in the scalar versions carry over to vector versions, the vector versions may introduce additional flaws, due to the interaction among various conserved quantities and various families of waves, and can otherwise change certain properties of the scalar methods. Vector and scalar versions generally share properties such as conservation, explicit versus implicit, finite volume versus finite difference, linear stability, linearity, and formal order of accuracy. Vector and scalar versions sometimes do not share other properties such as artificial viscosity, convergence, and nonlinear stability. For example, the vector version will usually have a matrix coefficient of artificial viscosity, generally not uniquely defined, which may or may not resemble the scalar coefficient of artificial viscosity. As another example, convergence proofs for scalar conservation laws may not carry over to vector conservation laws such as the Euler equations. Fortunately, the CFL condition extends rather easily from scalar to vector versions. For example, if the scalar method has the CFL condition $\lambda|a(u)| \leq 1$ then the vector method has the CFL condition $\lambda\rho(A(\mathbf{u})) \leq 1$, where $\rho(A)$ is the largest wave speed in absolute value.

Now let us consider two standard test problems. The Riemann problem for the Euler equations has one of the few exact solutions of the Euler equations, as discussed in Section 5.1. Thus, not by coincidence, both of the standard test problems used in this chapter are Riemann problems. These two test problems were first suggested by Sod (1978) and are sometimes referred to as *Sod's test problems*. These two test problems are classics and appear often in modern research papers such as, for example, Zha and Bilgen (1993). Before stating the test problems, a short background discussion is in order. In any Euler calculation, one must choose between fixing the maximum CFL number or fixing the time step. In the most relevant example, the solution to the Riemann problem generally generates new and faster waves than those found in the initial conditions, causing the CFL number to increase rapidly for the first few time steps, assuming fixed time steps, or causing the time step to decrease rapidly for the first few time steps, assuming fixed CFL numbers. In our examples, the time step is fixed rather than the CFL number, which allows for more meaningful comparisons between methods. If the CFL number is fixed, and if a method overestimates the wave speeds, then it must take smaller time steps and more time steps than other methods to reach the final time and is thus, in this sense, double penalized. Conversely, if a method underestimates the wave speeds, then it can take larger time steps and fewer time steps than other methods and is thus, in this sense, rewarded for its mistake. Of course, when fixing the time step, one must ensure that any overshoots in the wave speeds do not violate the CFL condition. Although the results in this chapter use fixed time steps, under ordinary practical circumstances it usually makes more sense to take the largest time steps possible, consistent with accuracy and stability, which usually means fixing the CFL number.

Besides CFL-condition-violating overshoots in the wave speeds, another potential problem is undershoots in quantities like pressure and density. Although these quantities are physically positive, undershoots may drive them numerically negative, causing the method to fail when, for example, computing the speed of sound according to $a = \sqrt{\gamma p/\rho}$. It is thus important to constantly monitor quantities such as pressure, energy, and density to ensure that they remain positive. If they do become negative, some possible responses include presmoothing the initial conditions, adding artificial viscosity, or replacing the negative quantities by zero, as in $p = \max(0, p)$.

What follows are the two standard test cases used in this chapter. In the following descriptions, any properties of the exact solution, such as the maximum wave speeds, come from the exact solution found as described in Section 5.1.

Test Case 1 Find the pressure, velocity, speed of sound, density, entropy, and Mach number at $t = 0.01$ s where

$$\mathbf{w}(x, 0) = \begin{cases} \mathbf{w}_L & x < 0 \\ \mathbf{w}_R & x \geq 0, \end{cases}$$

and where

$$\mathbf{w}_L = \begin{bmatrix} \rho_L \\ u_L \\ p_L \end{bmatrix} = \begin{bmatrix} 1\,\text{kg/m}^3 \\ 0\,\text{m/s} \\ 100{,}000\,\text{N/m}^2 \end{bmatrix},$$

$$\mathbf{w}_R = \begin{bmatrix} \rho_R \\ u_R \\ p_R \end{bmatrix} = \begin{bmatrix} 0.125\,\text{kg/m}^3 \\ 0\,\text{m/s} \\ 10{,}000\,\text{N/m}^2 \end{bmatrix}.$$

Use $N = 50$ points on the domain $[-10\,\text{m}, 10\,\text{m}]$ – then $\Delta x = (20\,\text{m})/50 = 0.4$ m. Also use an initial CFL number of 0.4. In the initial conditions, the maximum wave speed is 374.17 m/s. The fixed time step is thus $\Delta t = 0.4 \times 0.4\,\text{m}/(374.17\,\text{m/s}) = 4.276 \times 10^{-4}$ s. Then $\lambda = \Delta t/\Delta x = 1.069 \times 10^{-3}$ s/m. Then the final time $t = 0.01$ s is reached in approximately 23 time steps. For all $t > 0$, in the exact solution, the maximum wave speed is 693.03 m/s which corresponds to a CFL number of $(1.069 \times 10^{-3}\,\text{s/m})(693.03\,\text{m/s}) = 0.7408$. Different numerical methods will use slightly different CFL numbers, depending on how well they approximate the maximum wave speeds. For example, a highly dissipative method like the Lax–Friedrichs method will have a CFL number of 0.69 after the first few time steps, whereas a method prone to overshoots and undershoots like MacCormack's version of the Lax–Wendroff method will have a CFL number near 0.87.

Test Case 2 Find the pressure, velocity, speed of sound, density, entropy, and Mach number at $t = 0.01$ s where

$$\mathbf{w}(x,0) = \begin{cases} \mathbf{w}_L & x < 0 \\ \mathbf{w}_R & x \geq 0, \end{cases}$$

and where:

$$\mathbf{w}_L = \begin{bmatrix} \rho_L \\ u_L \\ p_L \end{bmatrix} = \begin{bmatrix} 1\ \text{kg/m}^3 \\ 0\ \text{m/s} \\ 100{,}000\ \text{N/m}^2 \end{bmatrix},$$

$$\mathbf{w}_R = \begin{bmatrix} \rho_R \\ u_R \\ p_R \end{bmatrix} = \begin{bmatrix} 0.010\ \text{kg/m}^3 \\ 0\ \text{m/s} \\ 1{,}000\ \text{N/m}^2 \end{bmatrix}.$$

Use $N = 50$ points on the domain $[-10\ \text{m}, 15\ \text{m}]$ – then $\Delta x = (25\,\text{m})/50 = 0.5$ m. Also use an initial CFL number of 0.3. In the initial conditions, the maximum wave speed is 374.17 m/s. The fixed time step is thus $\Delta t = 0.3 \times 0.5\ \text{m}/(374.17\,\text{m/s}) = 4.01 \times 10^{-4}$ s. Then $\lambda = \Delta t/\Delta x = 8.02 \times 10^{-4}$ s/m. Then the final time $t = 0.01$ s is reached in approximately 25 time steps. For all $t > 0$, in the exact solution, the maximum wave speed is 1,138.7 m/s, which corresponds to a CFL number of $(8.02 \times 10^{-3}\ \text{s/m})\ (1138.7\ \text{m/s}) = 0.91$. Different numerical methods will achieve different final CFL numbers, depending on how well they approximate the maximum wave speeds.

As far as Riemann problems go, there is nothing very special or exotic about Test Case 1. It does not contain any particularly strong shocks, contacts, or expansions. It does not contain any sonic points, although undershoots in the Mach number may cause the solution to cross the sonic point in the entropy waves $u = 0$ and overshoots in the Mach number may cause the solution to cross the sonic point in the acoustic waves $u - a = 0$. About the only thing that makes Test Case 1 unusually challenging is that the shock and contact are close together. Even at the final time, when their separation is greatest, the shock and contact are separated by only 2.5 m, which corresponds to only about five grid cells. It is next to impossible for most numerical methods to capture both a shock and a contact in five grid cells or less, unless they incorporate special features like artificial compression or subcell resolution, which the methods in this chapter do not.

Test Case 2 is much more challenging than Test Case 1. First, the shock and contact are separated by only five grid cells or less. Second, the expansion fan contains an expansive sonic point $u - a = 0$ at $x = 0$, which causes small expansion shocks to form in many numerical methods. Third, the shock contains a compressive sonic point $u + a = 0$, although this does not tend to cause most numerical methods any particular problems. Finally, the initial conditions of Test Case 2 involve the unusually large pressure ratio of 100. However, as it turns out, the large pressure ratio in the initial conditions does not translate into a large pressure ratio across the shock – in fact, the pressure ratio across the shock is only 6.392, which is not particularly strong, and is only about twice as strong as the pressure ratio across the shock in Test Case 1. In other words, most of the large pressure ratio is taken up in the smooth expansion fan, which makes the large pressure ratio easier to handle. Still, many numerical methods have a hard time getting started on Test Case 2, although they tend to settle down after the first few times steps, after the large initial jump discontinuity has had a chance to spread out.

Even though many sources use Sod's test cases, it is still possible to choose the exact circumstances to flatter or embarrass any given method. For example, a method such as the Lax–Friedrichs method looks relatively convincing for small Δx but quite alarming for large Δx, as seen in Subsection 18.1.1. Besides the parameter Δx, most methods are also sensitive to the CFL number, so that a large CFL number may favor some methods, while a small CFL number may favor others. In fact, not surprisingly, most of the original research papers include mainly flattering results, making a fair assessment based on the original papers difficult. In this chapter, to be as even handed as possible, the circumstances for Test Cases 1 and 2 were chosen to make *every* method look just slightly dismal. For example, as mentioned above, both test cases allow at most five or six grid points between the shock and contact and between the contact and the tail of the expansion fan, which is simply not enough space for most methods. As a result, most methods will show a good deal of smearing, undershoot, and overshoot in these regions, which would lessen considerably if the number of grid points were increased. Although we could have chosen to use more grid points, or in other ways tailored the circumstances to show each method in a better light, this would lead to more subtle differences between methods, and our goal is to show differences, possibly even to exaggerate the differences, to help stress the relative strengths and weaknesses of each method.

Sod's test cases and, indeed, all Riemann problems will tend to flatter first-order accurate methods. The solution to the Riemann problem consists mainly of flat regions punctuated by sharp transitions such as shocks and contacts. First-order accurate methods tend to shine under such circumstances, by design, whereas second- and higher-order accurate methods tend to develop large spurious oscillations, even more so than for scalar conservation laws, due to the interactions between the several conserved variables and several families of waves. Second- and higher-order accuracy has the potential to improve the interior of the expansion fan and to reduce the rate of smearing of the contact, but these benefits do not generally outweigh the detriments in the case of the Riemann problem. In fact, ideally, one should turn off the second- and higher-order terms near trouble spots like shocks, but this chapter concerns only first-generation methods and thus, by definition, the higher-order methods in this chapter are committed to leaving their higher-order terms on all of the time, appropriately or not. To be completely fair, this chapter should include test cases that favor second- and higher-order methods over first-order methods, especially problems with completely smooth solutions, but the Riemann problem has an exact solution whereas other smoother test cases do not. Thus this chapter will stick to the two Riemann problem test cases defined above.

18.1 Flux Approach

This section describes the simplest approach for converting numerical methods for scalar conservation laws into numerical methods for the Euler equations. The technique described considers only flux – it makes no efforts to model waves. The numerical differentiation formulae derived in Section 10.1 and the numerical integration formulae derived in Section 10.2 apply equally to scalars and vectors. For example, the central difference formula given by Equation (10.16) is

$$\frac{df}{dx}(x_i) = \frac{f(x_{i+1}) - f(x_{i-1})}{2\Delta x} + O(\Delta x^2)$$

for scalar functions and

$$\frac{d\mathbf{f}}{dx}(x_i) = \frac{\mathbf{f}(x_{i+1}) - \mathbf{f}(x_{i-1})}{2\Delta x} + O(\Delta x^2)$$

for vector functions. Many of the numerical methods seen in the book are based directly on the integration and differentiation formulae of Sections 10.1 and 10.2. For such numerical methods, simply replace the scalar u by the vector $\mathbf{u}$, the scalar f by the vector $\mathbf{f}$, and the scalar derivative $a = df/du$ by the Jacobian matrix $A = d\mathbf{f}/d\mathbf{u}$.

18.1.1 *Lax–Friedrichs Method*

The Lax–Friedrichs method for scalar conservation laws was described in Section 17.1. In Equation (17.1), replace u by $\mathbf{u}$ and f by $\mathbf{f}$. Then the Lax–Friedrichs method for the Euler equations is

$$\blacklozenge \qquad \mathbf{u}_i^{n+1} = \frac{1}{2}(\mathbf{u}_{i+1}^n + \mathbf{u}_{i-1}^n) - \frac{\lambda}{2}(\mathbf{f}(\mathbf{u}_{i+1}^n) - \mathbf{f}(\mathbf{u}_{i-1}^n)). \qquad (18.1)$$

Nothing could be easier than that. The behavior of the Lax–Friedrichs method is illustrated using the two tests cases defined in Section 18.0. As seen in Figures 18.1 and 18.2, the Lax–Friedrichs method exhibits considerable smearing and dissipation, as well as a number of two-point odd–even plateaus. Of course, the Lax–Friedrichs method for scalar conservation laws experiences similar difficulties, which the Lax–Friedrichs method for the Euler equations has simply inherited. On the positive side, the Lax–Friedrichs method handles the expansive sonic point in Test Case 2 with aplomb and, in general, has no more difficulties with Test Case 2 than with Test Case 1.

18.1.2 *Lax–Wendroff Methods*

The Lax–Wendroff method for scalar conservation laws was described in Section 17.2. In Equation (17.5), replace u by $\mathbf{u}$, f by $\mathbf{f}$, and $a_{i+1/2}^n$ by $A_{i+1/2}^n$. Then the Lax–Wendroff method for the Euler equations is

$$\begin{aligned}\mathbf{u}_i^{n+1} &= \mathbf{u}_i^n - \frac{\lambda}{2}(\mathbf{f}(\mathbf{u}_{i+1}^n) - \mathbf{f}(\mathbf{u}_{i-1}^n)) \\ &\quad + \frac{\lambda^2}{2}\left[A_{i+1/2}^n(\mathbf{f}(\mathbf{u}_{i+1}^n) - \mathbf{f}(\mathbf{u}_i^n)) - A_{i-1/2}^n(\mathbf{f}(\mathbf{u}_i^n) - \mathbf{f}(\mathbf{u}_{i-1}^n))\right]. \qquad (18.2)\end{aligned}$$

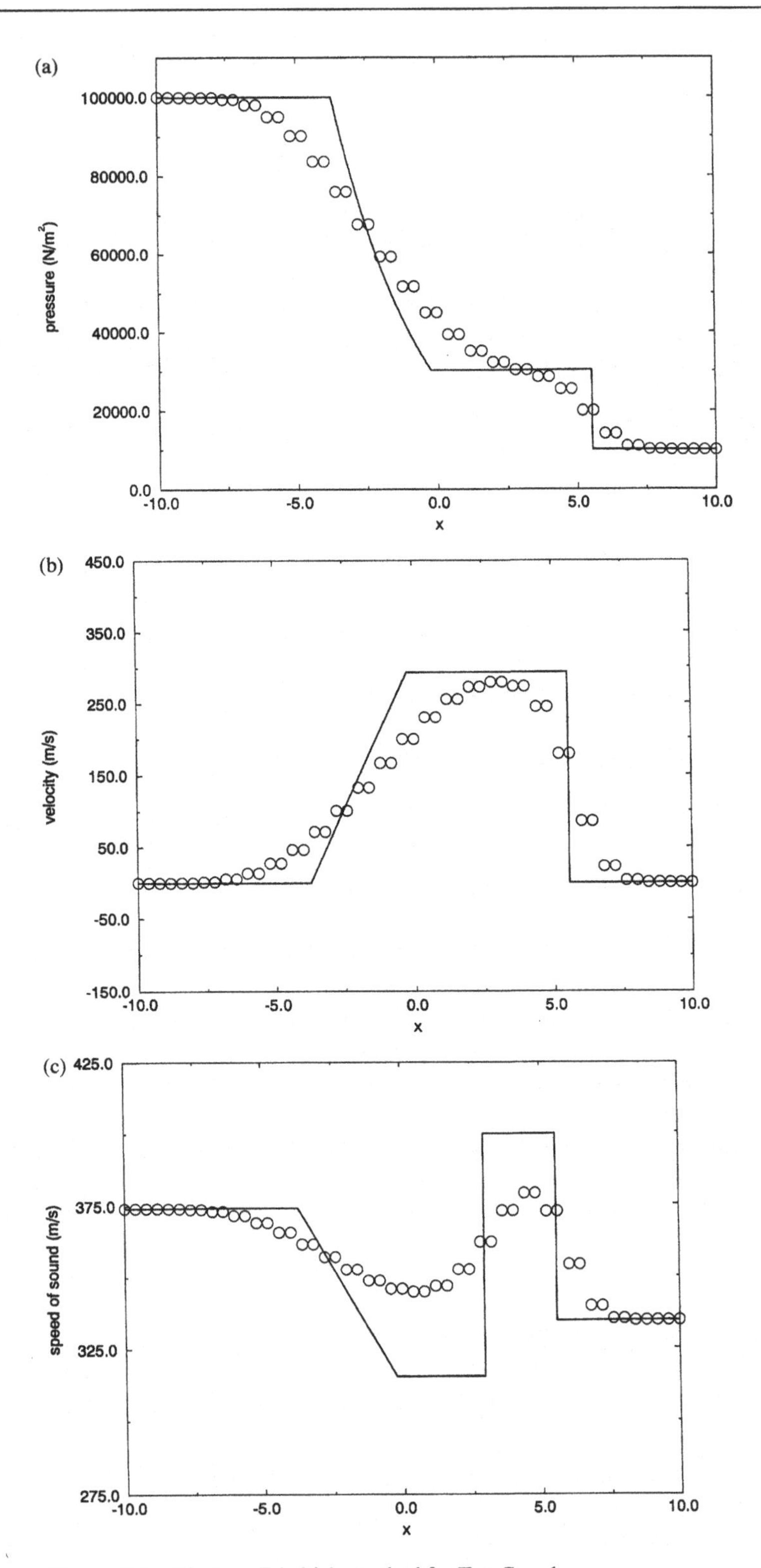

Figure 18.1 The Lax–Friedrichs method for Test Case 1.

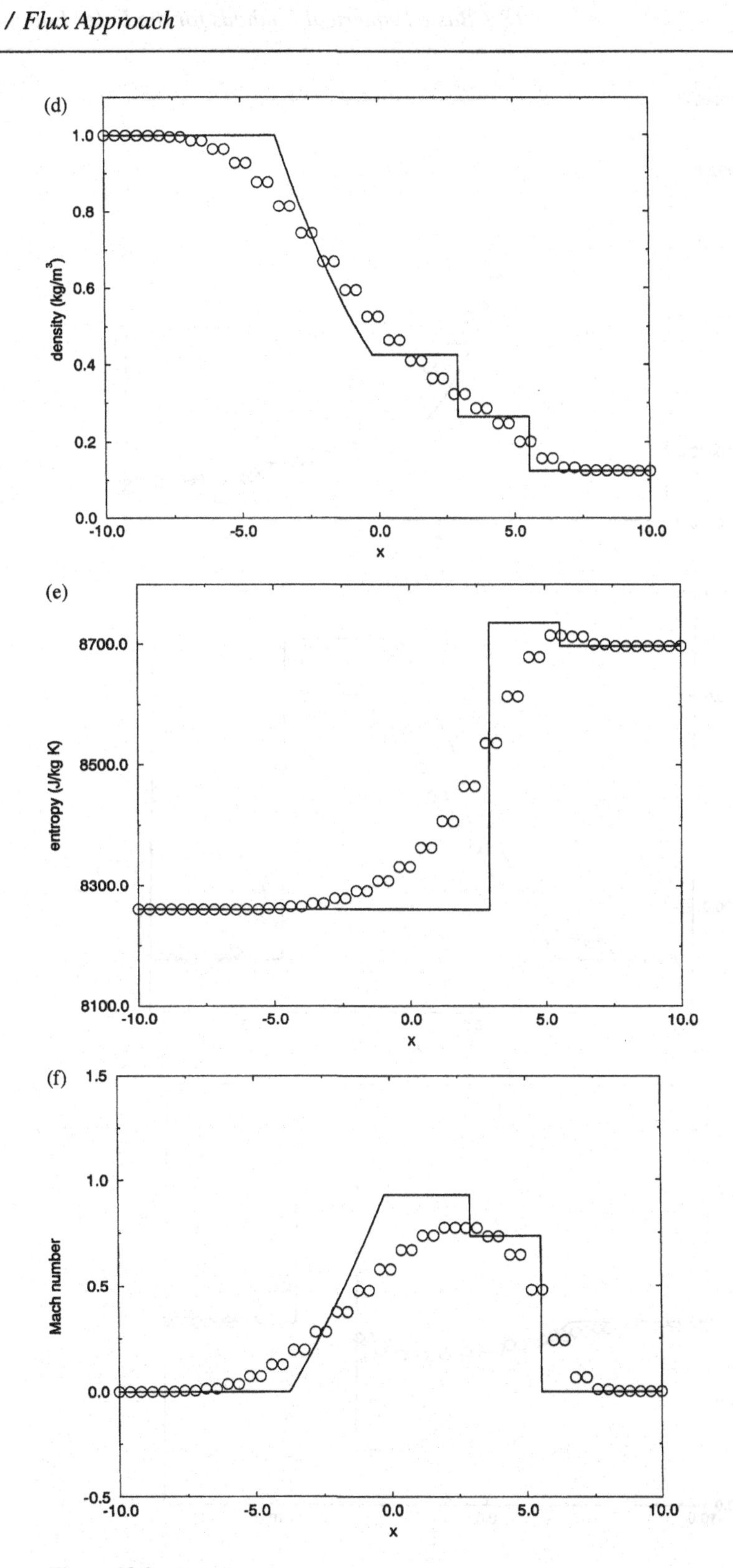

Figure 18.1 *(cont.)*

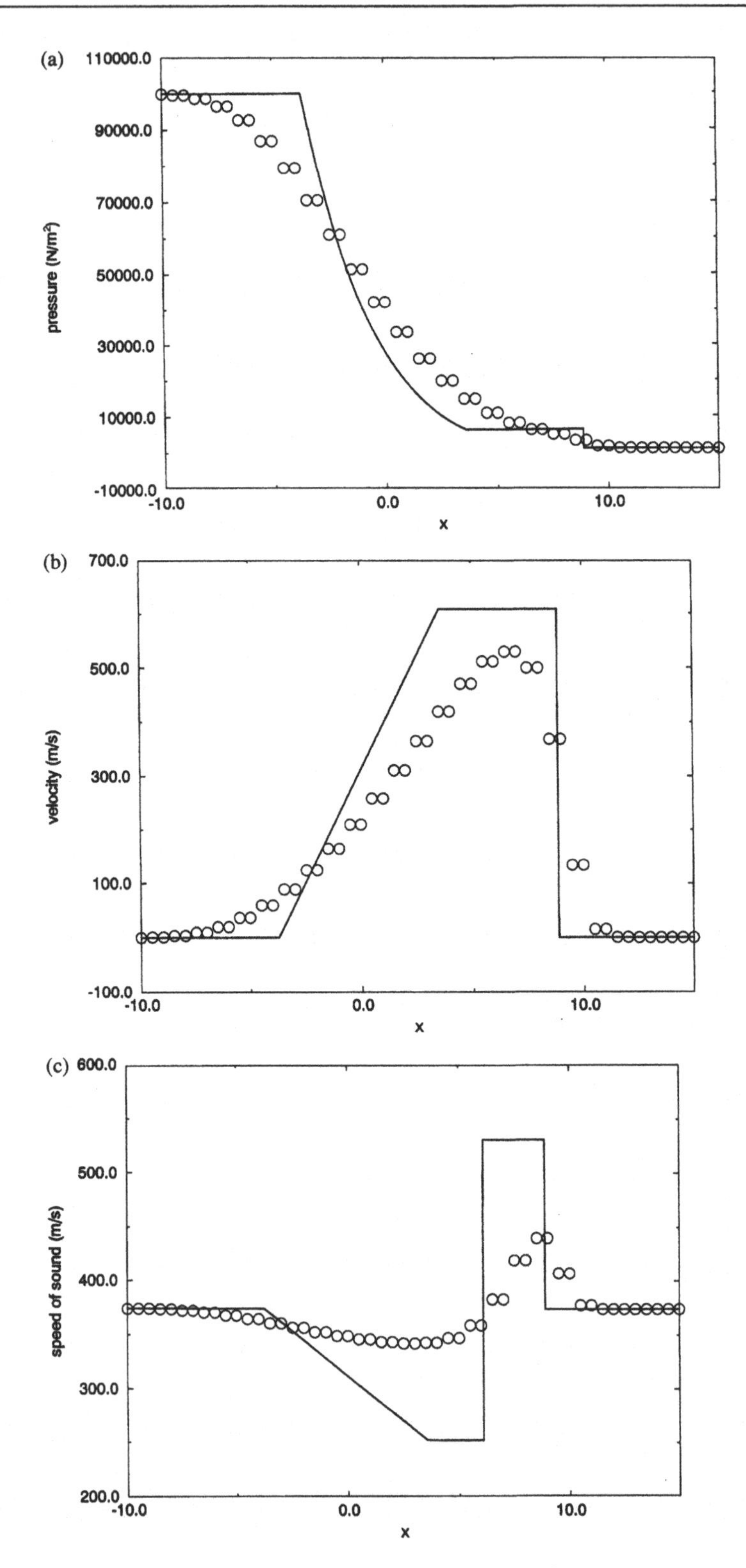

Figure 18.2 The Lax–Friedrichs method for Test Case 2.

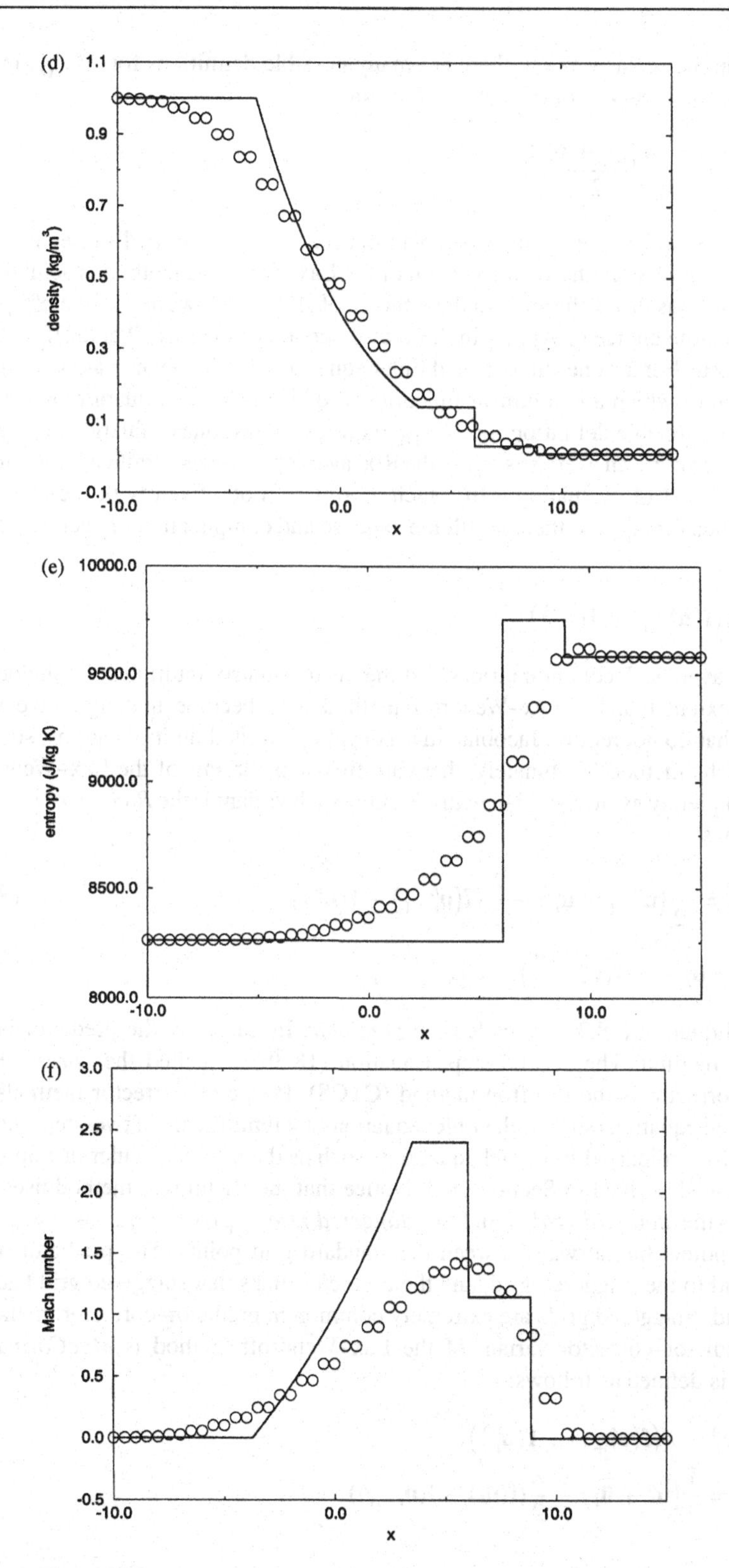

Figure 18.2 (*cont.*)

Like $a^n_{i+1/2}$ in the scalar version, there are many possible definitions for $A^n_{i+1/2}$ in the vector version of the Lax–Wendroff method. For example,

$$A^n_{i+1/2} = A\left(\frac{\mathbf{u}^n_{i+1} + \mathbf{u}^n_i}{2}\right).$$

Another possibility is the Roe-average Jacobian matrix $A^n_{i+1/2}$ described in Section 5.3. Thus, like the scalar version, the vector version of the Lax–Wendroff method is actually an entire class of methods that differ only in their choice of $A^n_{i+1/2}$. However, unlike $a^n_{i+1/2}$ in the scalar version, the choice of $A^n_{i+1/2}$ in the vector version of the Lax–Wendroff method greatly affects costs. For the one-dimensional Euler equations, the Jacobian matrix $A(\mathbf{u})$ has nine entries, each of which are nonlinear functions of $\mathbf{u}$. Thus every evaluation of $A(\mathbf{u})$ is expensive. Any reasonable definition for $A^n_{i+1/2}$ requires at least one evaluation of $A(\mathbf{u})$ if not several. Furthermore, an average such as the Roe average involves significant additional costs besides the cost of evaluating $A(\mathbf{u})$, such as square roots. Even with the cheapest possible definitions for $A^n_{i+1/2}$, there is still the expense and complication of vector–matrix multiples such as

$$A^n_{i+1/2}\left(\mathbf{f}(\mathbf{u}^n_{i+1}) - \mathbf{f}(\mathbf{u}^n_i)\right).$$

Forming the average Jacobian matrices and the vector–matrix multiples are major expenses, to the extent that the Lax–Wendroff method may become uncompetitive with other methods that do not require Jacobian matrix evaluations and multiplications, such as the Lax–Friedrichs method. Fortunately, there are two-step variants of the Lax–Wendroff method that completely avoid Jacobian matrices. One such variant is the *Richtmyer method*, defined as follows:

$$\blacklozenge \quad \mathbf{u}^{n+1/2}_{i+1/2} = \frac{1}{2}\left(\mathbf{u}^n_{i+1} + \mathbf{u}^n_i\right) - \frac{\lambda}{2}\left(\mathbf{f}(\mathbf{u}^n_{i+1}) - \mathbf{f}(\mathbf{u}^n_i)\right), \tag{18.3a}$$

$$\blacklozenge \quad \mathbf{u}^{n+1}_i = \mathbf{u}^n_i - \lambda\left(\mathbf{f}\left(\mathbf{u}^{n+1/2}_{i+1/2}\right) - \mathbf{f}\left(\mathbf{u}^{n+1/2}_{i-1/2}\right)\right). \tag{18.3b}$$

The first step, Equation (18.3a), is called the *predictor*. In this case, the predictor is the Lax–Friedrichs method. The second step, Equation (18.3b), is called the *corrector*. In this case, the corrector is the leapfrog method (CTCS). Predictor–corrector methods for partial differential equations such as the Euler equations are reminiscent of two-step Runge–Kutta methods for ordinary differential equations, such as the modified Euler or improved Euler methods, as described in Section 10.3. Notice that the Richtmyer method uses two different grids – the *standard grid* x_i and the *staggered grid* $x_{i+1/2} = (x_{i+1} + x_i)/2$. The staggered grid points lie halfway between the standard grid points. The predictor maps the standard grid to the staggered grid, and the corrector maps that staggered grid back to the standard grid. Staggered grids are extremely common in predictor–corrector methods.

Another predictor–corrector variant of the Lax–Wendroff method is *MacCormack's method*, which is defined as follows:

$$\bar{\mathbf{u}}_i = \mathbf{u}^n_i - \lambda\left(\mathbf{f}(\mathbf{u}^n_{i+1}) - \mathbf{f}(\mathbf{u}^n_i)\right),$$

$$\mathbf{u}^{n+1}_i = \frac{1}{2}\left(\mathbf{u}^n_i + \bar{\mathbf{u}}_i\right) - \frac{\lambda}{2}(\mathbf{f}(\bar{\mathbf{u}}_i) - \mathbf{f}(\bar{\mathbf{u}}_{i-1})).$$

This looks a bit like the Richtmyer method with the predictor and corrector reversed. Equivalently, MacCormack's method can be written as

$$\blacklozenge \qquad \bar{\mathbf{u}}_i = \mathbf{u}_i^n - \lambda\left(\mathbf{f}(\mathbf{u}_{i+1}^n) - \mathbf{f}(\mathbf{u}_i^n)\right), \tag{18.4a}$$

$$\blacklozenge \qquad \bar{\bar{\mathbf{u}}}_i = \bar{\mathbf{u}}_i - \lambda(\mathbf{f}(\bar{\mathbf{u}}_i) - \mathbf{f}(\bar{\mathbf{u}}_{i-1})), \tag{18.4b}$$

$$\blacklozenge \qquad \mathbf{u}_i^{n+1} = \frac{1}{2}\left(\mathbf{u}_i^n + \bar{\bar{\mathbf{u}}}_i\right). \tag{18.4c}$$

In this form, the predictor, Equation (18.4a), is FTFS. The corrector, Equation (18.4b), is FTBS. The predictor and corrector in MacCormack's method can be reversed as follows:

$$\blacklozenge \qquad \bar{\mathbf{u}}_i = \mathbf{u}_i^n - \lambda\left(\mathbf{f}(\mathbf{u}_i^n) - \mathbf{f}(\mathbf{u}_{i-1}^n)\right), \tag{18.5a}$$

$$\blacklozenge \qquad \bar{\bar{\mathbf{u}}}_i = \bar{\mathbf{u}}_i - \lambda(\mathbf{f}(\bar{\mathbf{u}}_{i+1}) - \mathbf{f}(\bar{\mathbf{u}}_i)), \tag{18.5b}$$

$$\blacklozenge \qquad \mathbf{u}_i^{n+1} = \frac{1}{2}\left(\mathbf{u}_i^n + \bar{\bar{\mathbf{u}}}_i\right). \tag{18.5c}$$

This method is again called MacCormack's method. Left-running waves are better captured by the first version of MacCormack's method, whereas right-running waves are better captured by the second version. To avoid favoring either left- or right-running waves, the two versions of MacCormack's method are often combined, reversing the order of FTBS and FTFS after every time step. Notice that, as promised, none of the above two-step predictor–corrector methods involves Jacobian matrices. It is rather exciting to realize that two unstable methods, applied in succession, can result in a completely stable method. In the case of MacCormack's method, the predictor or corrector or both are always unconditionally unstable, and yet the sequence is completely stable provided only that the CFL condition $\lambda\rho(A) \leq 1$ is satisfied.

The above two-step methods seem quite different from the original one-step Lax–Wendroff method and quite different from each other. In fact, they *are* different, certainly different enough to qualify as distinct methods. However, if the flux function is linear, then the Lax–Wendroff method, the Richtmyer method, and the two versions of MacCormack's method are identical. Furthermore, if the flux function is nonlinear, then the Lax–Wendroff method, the Richtmyer method, and the two versions of MacCormack's method are equal to within second-order accuracy. Of course, any second-order accurate method is equal to any other second-order accurate method to within second-order accuracy. But, although they may have differences, MacCormack's method, the Richtmyer method, and the original Lax–Wendroff method have distinctive similarities both in theory and practice that set them apart from other methods. Historically, MacCormack's method is usually preferred over other Lax–Wendroff variants, although the differences are not necessarily major nor consistently in favor of one variant over another. Even with the sophisticated methods available today, many people still turn to MacCormack's method first, especially in the absence of shocks, mainly because of its inviting simplicity and efficiency.

Lax–Wendroff methods for the Euler equations were never meant to be used *au natural*, at least not for solutions containing strong shocks or contacts. In fact, almost every practical Lax–Wendroff code adds artificial viscosity of some variety. The most traditional choice is the *von Neumann–Richtmyer artificial viscosity*, described in Subsection 17.3.2 of Hirsch (1990). A more modern approach is to add some sort of adaptive artificial viscosity, perhaps designed to enforce nonlinear stability conditions, as discussed in Part V. However, to keep things as simple as possible, this section will use constant-coefficient second-order artificial viscosity. For example, the Richtmyer method with this sort of artificial viscosity becomes

$$\mathbf{u}_{i+1/2}^{n+1/2} = \frac{1}{2}(\mathbf{u}_{i+1}^{n} + \mathbf{u}_{i}^{n}) - \frac{\lambda}{2}(\mathbf{f}(\mathbf{u}_{i+1}^{n}) - \mathbf{f}(\mathbf{u}_{i}^{n})),$$
$$\mathbf{u}_{i}^{n+1} = \mathbf{u}_{i}^{n} - \lambda\left(\mathbf{f}(\mathbf{u}_{i+1/2}^{n+1/2}) - \mathbf{f}(\mathbf{u}_{i-1/2}^{n+1/2})\right) + \epsilon\left(\mathbf{u}_{i+1}^{n} - 2\mathbf{u}_{i}^{n} + \mathbf{u}_{i-1}^{n}\right).$$

The behaviors of MacCormack's method and the Richtmyer method are illustrated using the two tests cases defined in Section 18.0. Version (18.4) of MacCormack's method is used, since it seems marginally superior to either version (18.5) or to switching back and forth between the two versions. The results for MacCormack's method applied to Test Case 1 are shown in Figure 18.3, and the results for the Richtmyer method applied to Test Case 1 are shown in Figure 18.4. Without artificial viscosity, neither MacCormack's method nor the Richtmyer method can complete the first time step without overshooting and creating negative pressures. Thus, for Test Case 1, both methods take two time steps with a constant coefficient of second-order artificial viscosity $\epsilon = 0.02$. After the first two time steps have smoothed the initial conditions, both methods are able to continue without artificial viscosity. However, although neither method fails catastrophically, they both experience severe overshoots and oscillations without continued use of artificial viscosity, especially at the tail of the expansion fan. To show the benefits of artificial viscosity, it was left on throughout Test Case 2, and the results are shown in Figures 18.5 and 18.6. Even though Test Case 2 is substantially more difficult than Test Case 1, the results are substantially better due to the continued use of artificial viscosity. Notice that the Richtmyer method performs noticeably better than MacCormack's method. Specifically, besides the size of the spurious oscillations, MacCormack's method has a strong expansion shock at the expansive sonic point, while the Richtmyer method does not. In all fairness, however, the results could be evened out by feeding MacCormack's method more artificial viscosity than the Richtmyer method.

18.2 Wave Approach I: Flux Vector Splitting

This section concerns flux vector splitting methods for the Euler equations. So far, the book has only considered flux splitting for scalar conservation laws, especially the linear advection equation and Burgers' equation, as seen in Section 13.4. Unfortunately, things are much more complicated for the Euler equations. Although there are only a few plausible flux splittings for the linear advection equation and Burgers' equation, there are any number of plausible flux vector splittings for the Euler equations, many of which are quite involved. This section will consider four possible flux vector splittings: Steger–Warming, Van Leer, Liou–Steffen, and Zha–Bilgen. This section discusses these four splittings in historical order although, rather perversely, the most difficult flux splittings were discovered first, whereas the simplest flux splittings came later.

Let us review flux vector splitting. Suppose the flux vector can be written as follows:

$$\mathbf{f}(\mathbf{u}) = \mathbf{f}^{+}(\mathbf{u}) + \mathbf{f}^{-}(\mathbf{u}), \tag{18.6}$$

where the characteristic values of $d\mathbf{f}^{+}/d\mathbf{u}$ are all nonnegative and the characteristic values of $d\mathbf{f}^{-}/d\mathbf{u}$ are all nonpositive. In standard notation,

$$\frac{d\mathbf{f}^{+}}{d\mathbf{u}} \geq 0, \qquad \frac{d\mathbf{f}^{-}}{d\mathbf{u}} \leq 0. \tag{18.7}$$

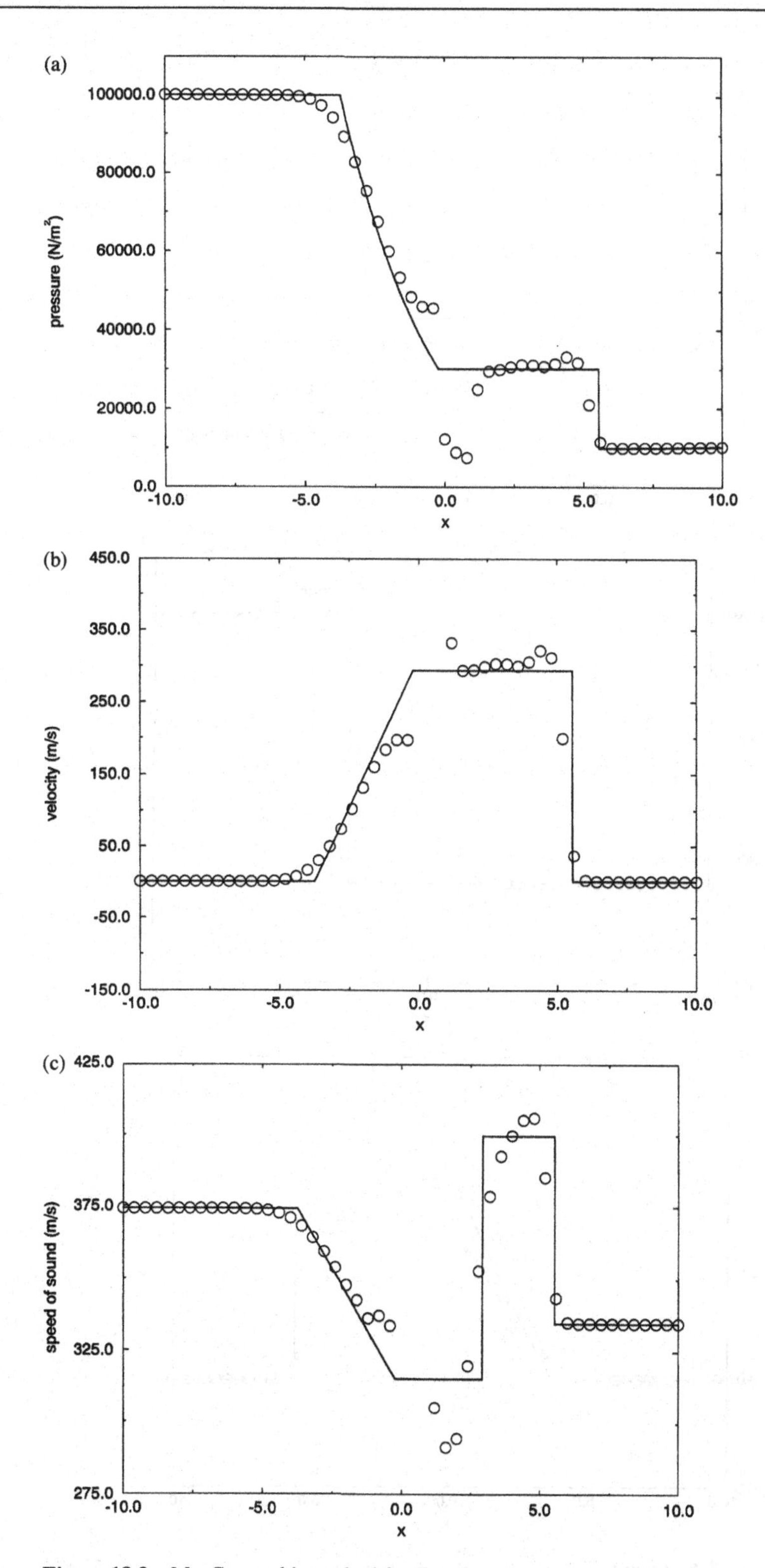

Figure 18.3 MacCormack's method for Test Case 1.

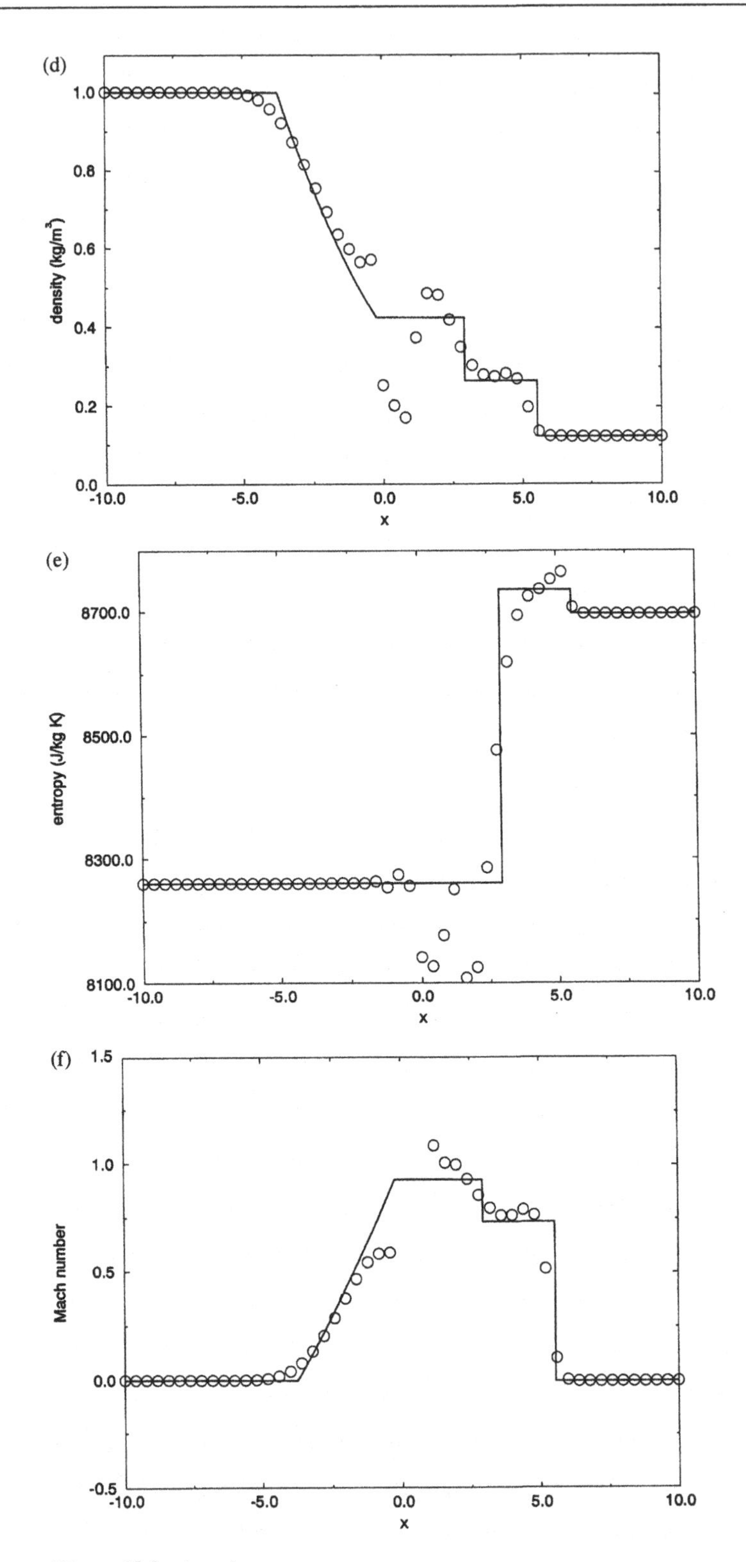

Figure 18.3 *(cont.)*

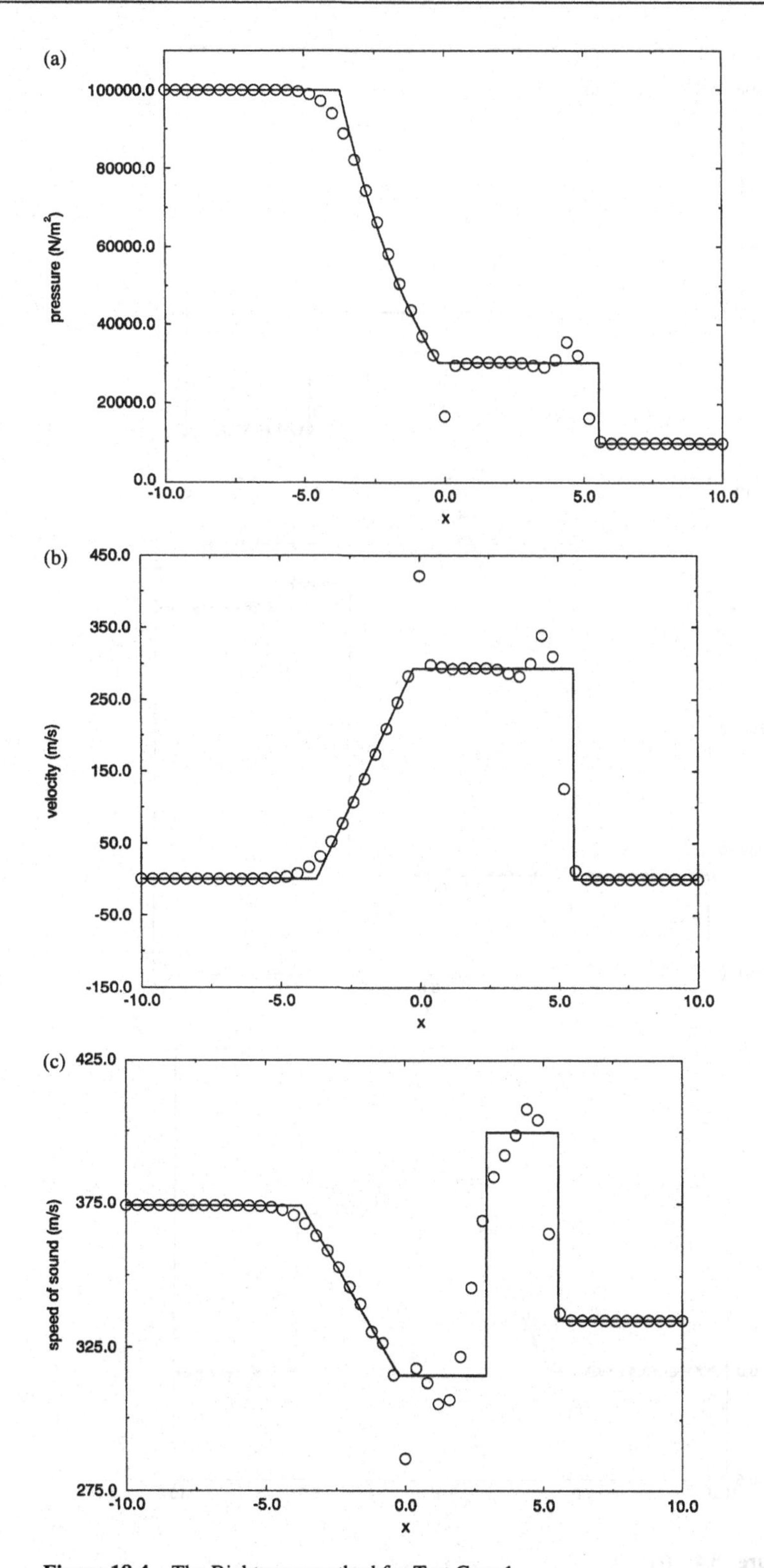

Figure 18.4 The Richtmyer method for Test Case 1.

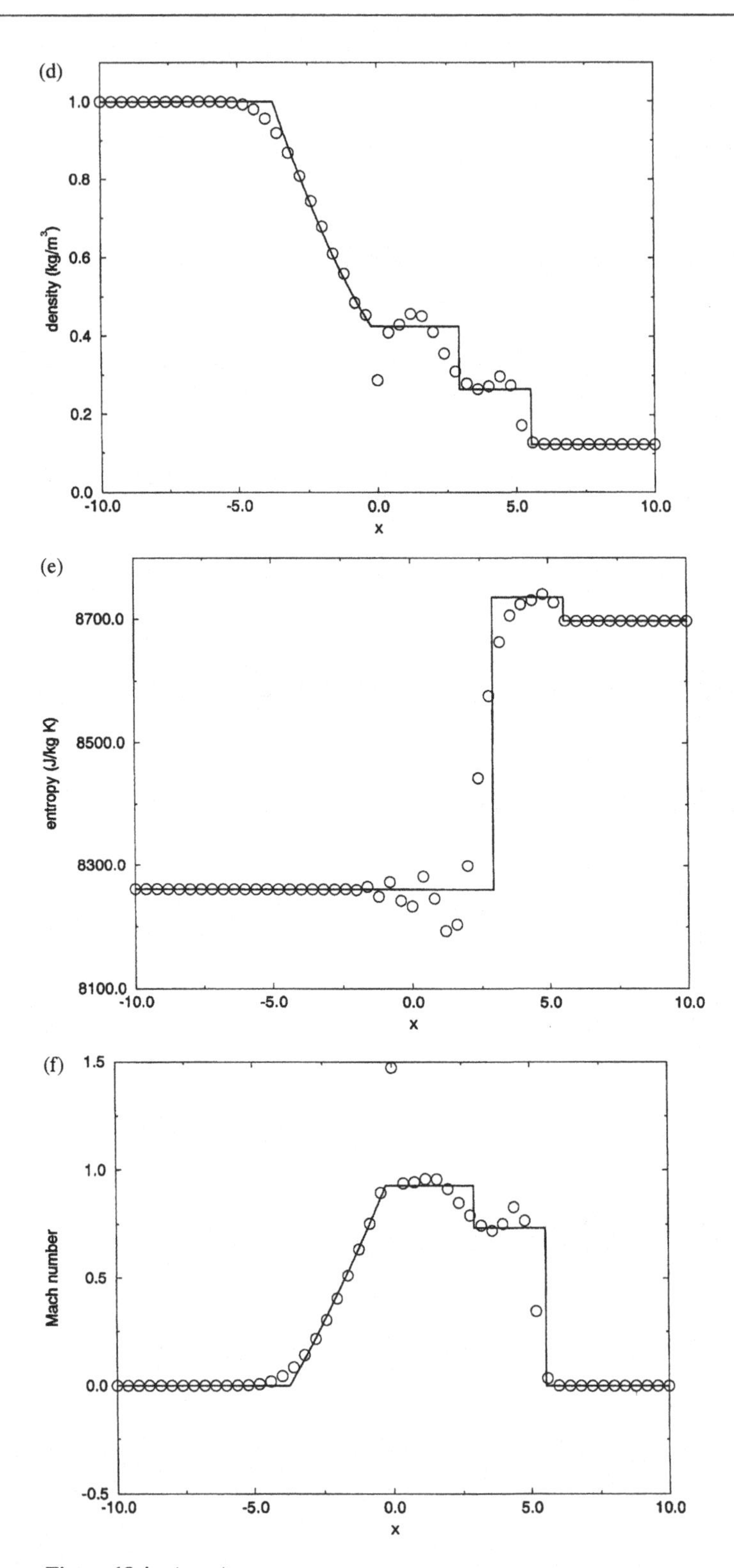

Figure 18.4 *(cont.)*

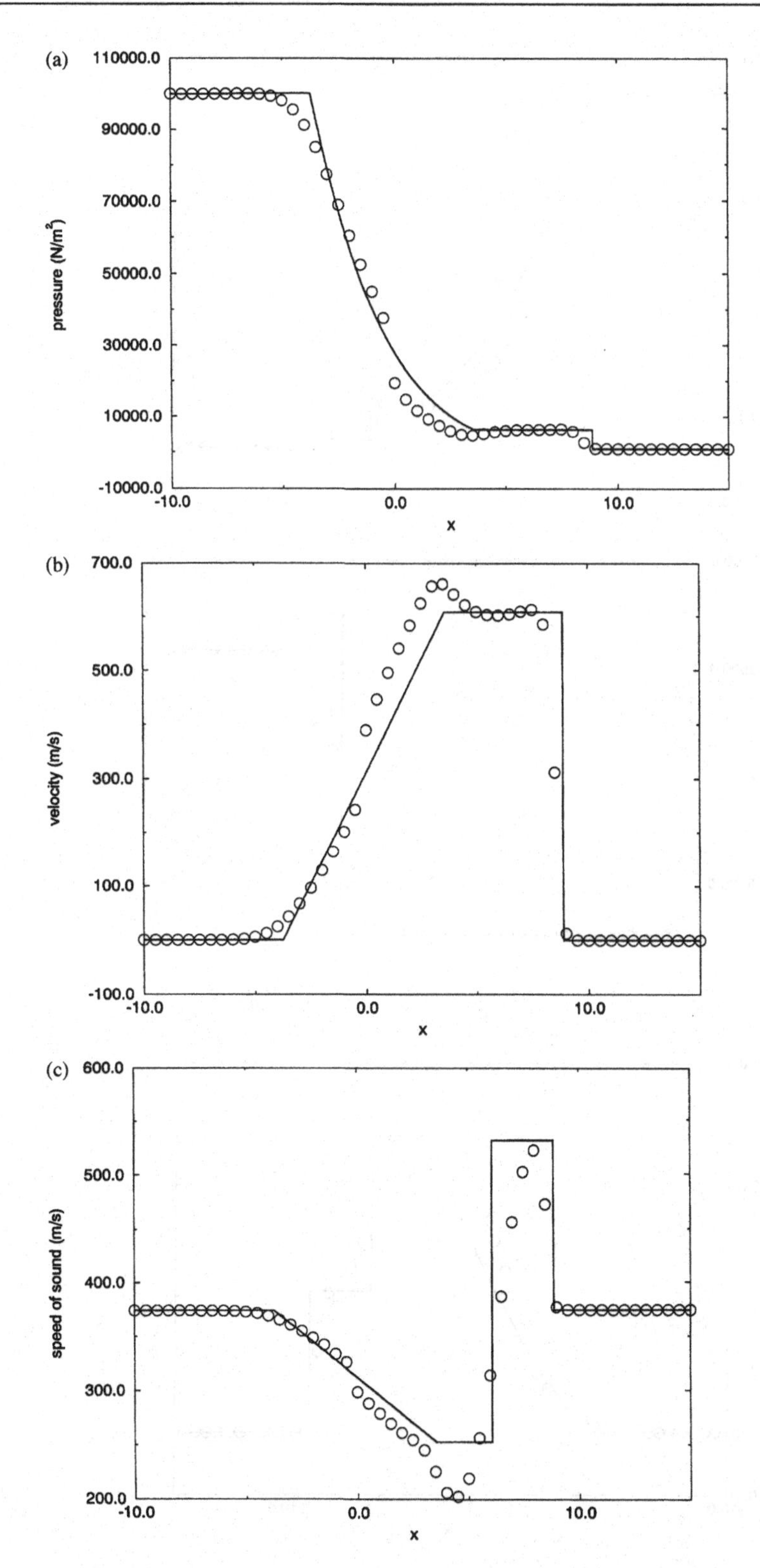

Figure 18.5 MacCormack's method with artificial viscosity for Test Case 2.

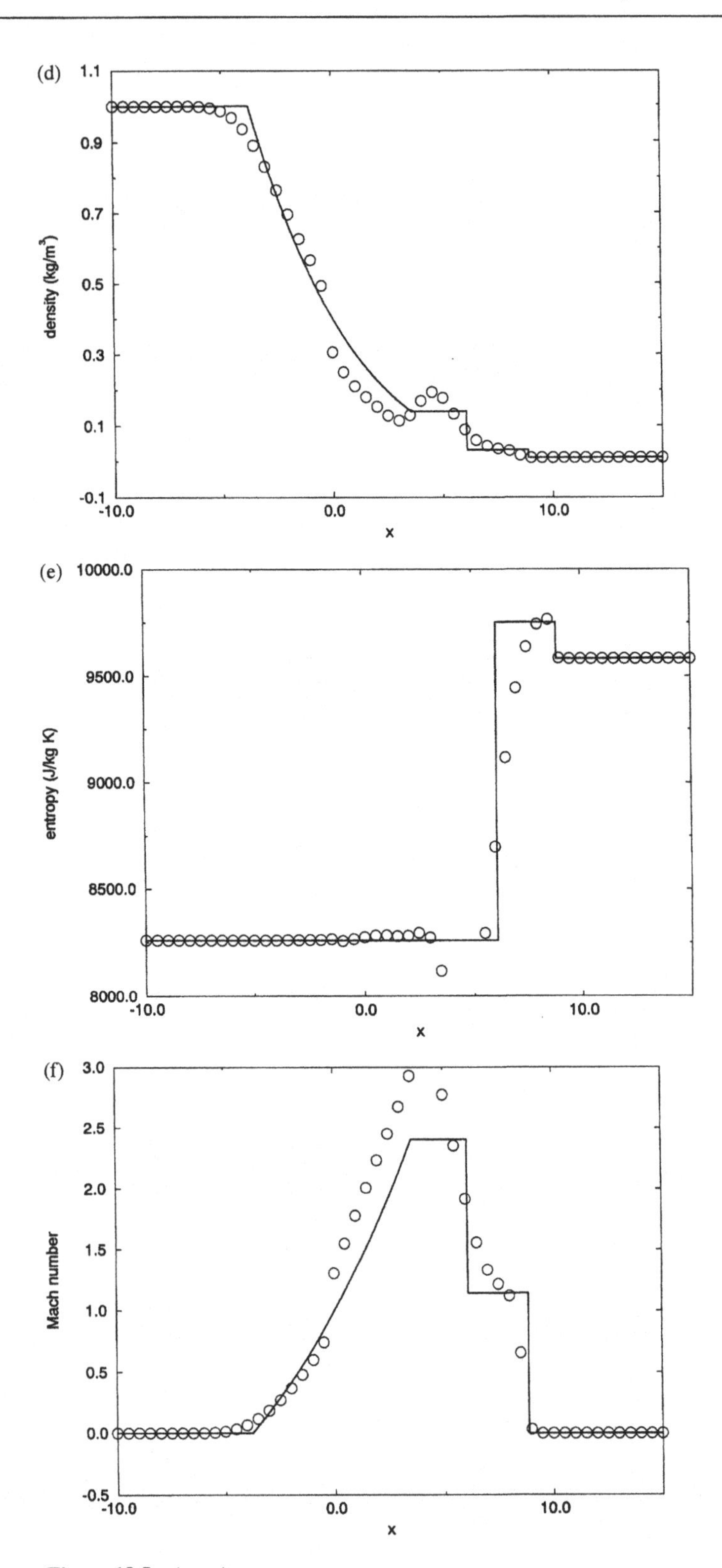

Figure 18.5 *(cont.)*

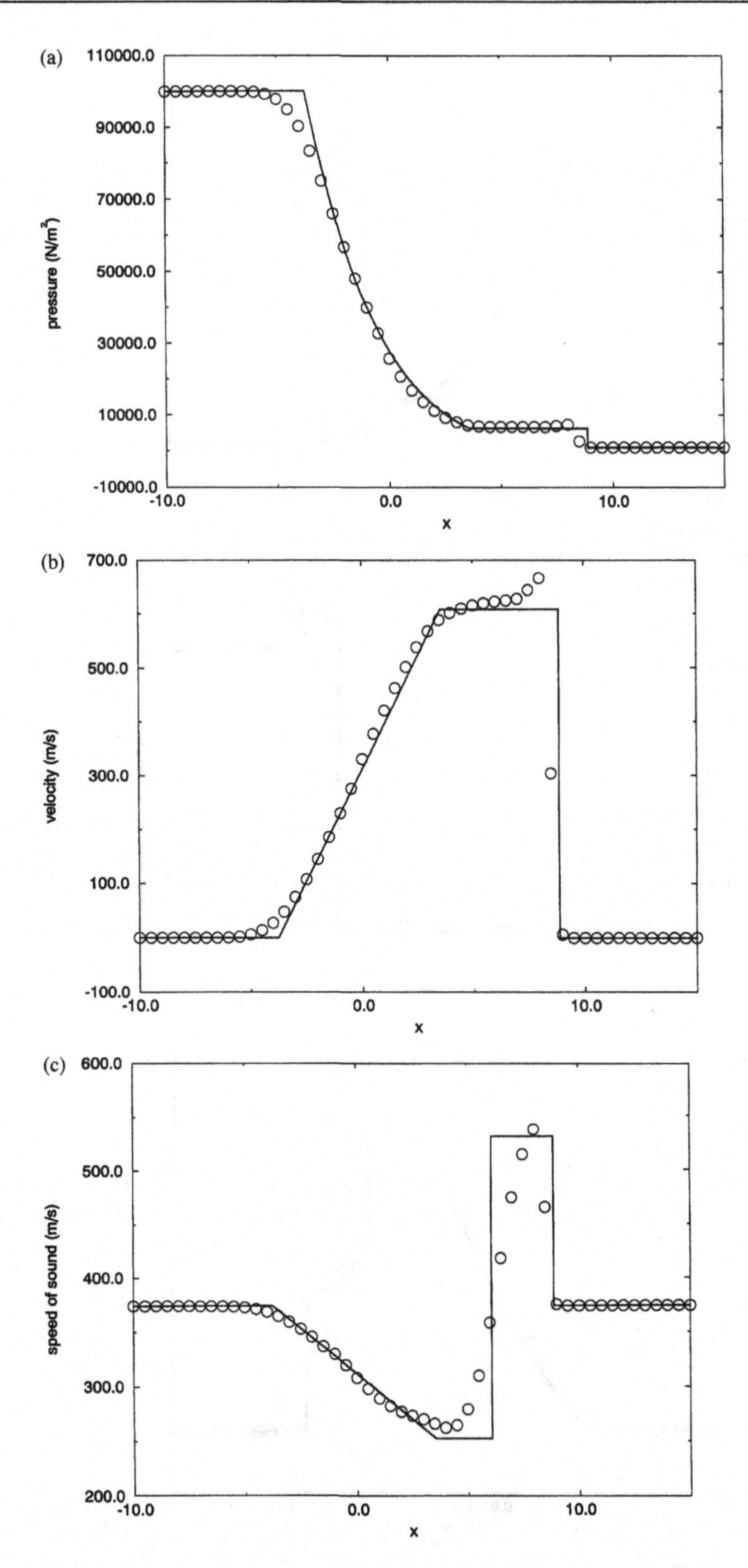

Figure 18.6 The Richtmyer method with artificial viscosity for Test Case 2.

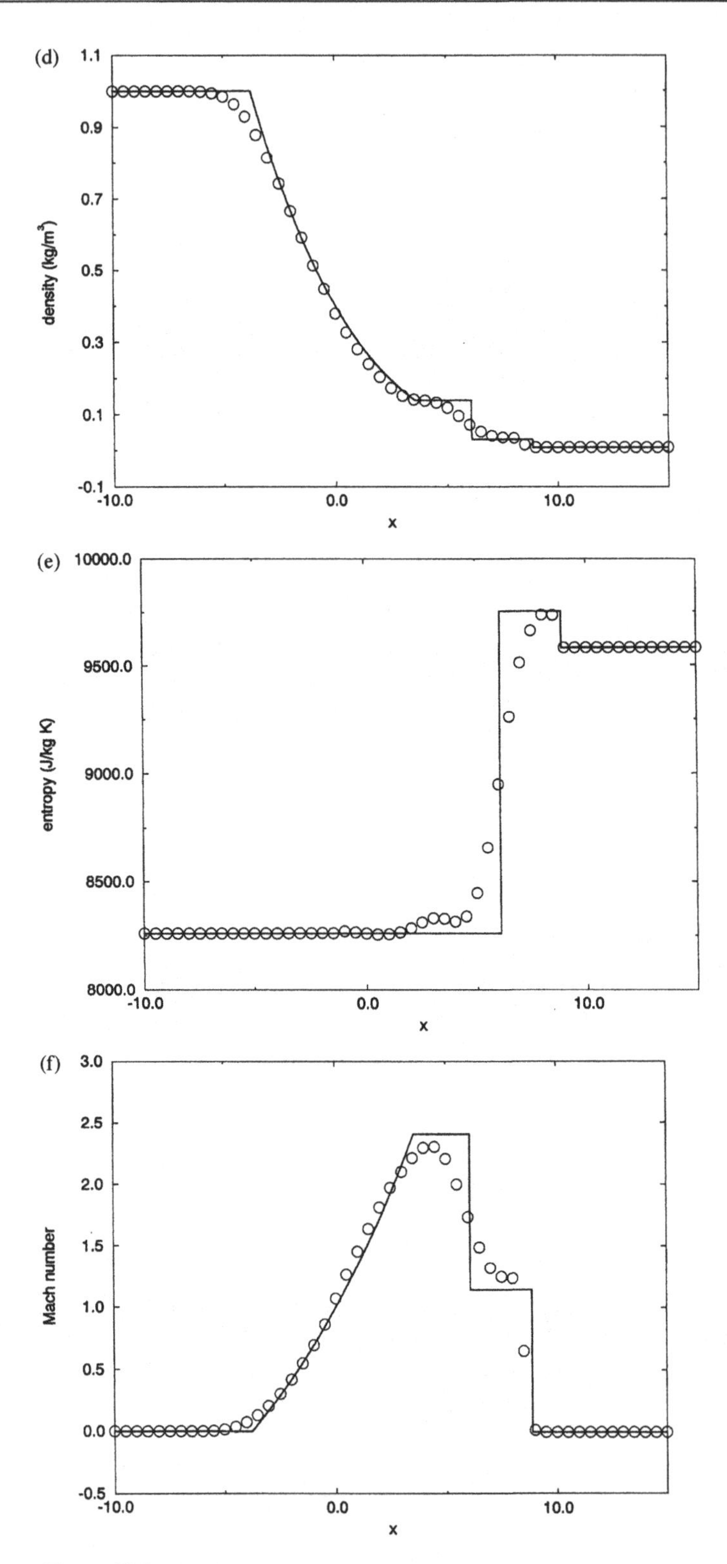

Figure 18.6 *(cont.)*

This is called *flux vector splitting*. Then the vector conservation law can be written as

$$\frac{\partial \mathbf{u}}{\partial t} + \frac{\partial \mathbf{f}^+}{\partial x} + \frac{\partial \mathbf{f}^-}{\partial x} = 0. \tag{18.8}$$

Then $\partial \mathbf{f}^+/\partial x$ can be discretized using FTBS or some other such method, and $\partial \mathbf{f}^-/\partial x$ can be discretized using FTFS or some other method. The resulting method will be completely stable for both left- and right-running waves.

As a related concept, consider wave speed splitting, as introduced in Section 13.5. To review, suppose the flux Jacobian matrix can be written as follows:

$$A(\mathbf{u}) = A^+(\mathbf{u}) + A^-(\mathbf{u}), \tag{13.22}$$

where the characteristic values of A^+ are nonnegative and the characteristic values of A^- are nonpositive. In standard notation,

$$A^+ \geq 0, \qquad A^- \leq 0. \tag{13.23}$$

Then the vector conservation law can be written as

$$\frac{\partial \mathbf{u}}{\partial t} + A^+\frac{\partial \mathbf{u}}{\partial x} + A^-\frac{\partial \mathbf{u}}{\partial x} = 0. \tag{13.24}$$

The matrices A^+ and A^- are usually obtained by splitting the characteristic values of A (a.k.a. the wave speeds) into positive and negative parts. This is called *wave speed splitting*. Specifically, let the wave speeds be λ_i. The wave speeds are split as follows:

$$\lambda_i = \lambda_i^+ + \lambda_i^-, \tag{18.9}$$

where

$$\lambda_i^+ \geq 0, \qquad \lambda_i^- \leq 0. \tag{18.10}$$

Let us express wave speed splitting in matrix notation. Let Λ be a diagonal matrix of λ_i, Λ^+ be a diagonal matrix of λ_i^+, and let Λ^- be a diagonal matrix of λ_i^-. Then Λ is split as follows:

$$\Lambda = \Lambda^+ + \Lambda^-, \tag{18.11}$$

where

$$\Lambda^+ \geq 0, \qquad \Lambda^- \leq 0. \tag{18.12}$$

The wave speeds Λ and the Jacobian matrix A are related by

$$Q_A^{-1} A Q_A = \Lambda, \tag{3.22}$$

where

$$Q_A = \begin{bmatrix} 1 & \frac{\rho}{2a} & -\frac{\rho}{2a} \\ u & \frac{\rho}{2a}(u+a) & -\frac{\rho}{2a}(u-a) \\ \frac{u^2}{2} & \frac{\rho}{2a}\left(\frac{u^2}{2} + \frac{a^2}{\gamma-1} + au\right) & -\frac{\rho}{2a}\left(\frac{u^2}{2} + \frac{a^2}{\gamma-1} - au\right) \end{bmatrix}, \tag{3.23}$$

$$Q_A^{-1} = \frac{\gamma - 1}{\rho a} \begin{bmatrix} \frac{\rho}{a}\left(-\frac{u^2}{2} + \frac{a^2}{\gamma-1}\right) & \frac{\rho}{a}u & -\frac{\rho}{a} \\ \frac{u^2}{2} - \frac{au}{\gamma-1} & -u + \frac{a}{\gamma-1} & 1 \\ -\frac{u^2}{2} - \frac{au}{\gamma-1} & u + \frac{a}{\gamma-1} & -1 \end{bmatrix}. \tag{3.24}$$

Then

$$A = A^+ + A^-, \tag{18.13}$$

where

$$A^+ = Q_A^{-1}\Lambda^+ Q_A \geq 0, \qquad A^- = Q_A^{-1}\Lambda^- Q_A \leq 0 \tag{18.14}$$

and where, in this standard notation, the matrix inequalities refer to the characteristic values of the matrix.

In general, flux vector splitting and wave speed splitting are distinct. However, the Euler equations have a special property: The flux vector is a homogenous function of order one. In other words,

$$\mathbf{f}(\mathbf{u}) = \frac{d\mathbf{f}}{d\mathbf{u}}\mathbf{u} = A\mathbf{u}. \tag{18.15}$$

Then the wave speed splitting $A^\pm = Q_A^{-1}\Lambda^\pm Q_A$ leads immediately to the following flux vector splitting:

$$\mathbf{f}^+ = A^+\mathbf{u} = Q_A^{-1}\Lambda^+ Q_A\mathbf{u}, \tag{18.16a}$$

$$\mathbf{f}^- = A^-\mathbf{u} = Q_A^{-1}\Lambda^- Q_A\mathbf{u}. \tag{18.16b}$$

A tedious calculation shows that

$$\mathbf{f}^\pm = \frac{1}{\gamma} Q_A \begin{bmatrix} (\gamma - 1)\rho\lambda_1^\pm \\ a\lambda_2^\pm \\ -a\lambda_3^\pm \end{bmatrix} \tag{18.17a}$$

or

♦ $$\mathbf{f}^\pm = \frac{\gamma - 1}{\gamma}\rho\lambda_1^\pm \begin{bmatrix} 1 \\ u \\ \frac{1}{2}u^2 \end{bmatrix} + \frac{\rho}{2\gamma}\lambda_2^\pm \begin{bmatrix} 1 \\ u + a \\ \frac{u^2}{2} + \frac{a^2}{\gamma-1} + au \end{bmatrix} + \frac{\rho}{2\gamma}\lambda_3^\pm \begin{bmatrix} 1 \\ u - a \\ \frac{u^2}{2} + \frac{a^2}{\gamma-1} - au \end{bmatrix}. \tag{18.17b}$$

This result was first obtain by Steger and Warming (1981). Conversely, a flux vector splitting $\mathbf{f}^\pm$ leads immediately to the following wave speed splitting:

$$\begin{bmatrix} (\gamma - 1)\rho\lambda_1^\pm \\ a\lambda_2^\pm \\ -a\lambda_3^\pm \end{bmatrix} = \gamma Q_A^{-1}\mathbf{f}^\pm \tag{18.18a}$$

or

$$\blacklozenge \quad \begin{bmatrix} \lambda_1^{\pm} \\ \lambda_2^{\pm} \\ \lambda_3^{\pm} \end{bmatrix} = \frac{\gamma}{\rho a^2} \begin{bmatrix} -\frac{u^2}{2} + \frac{a^2}{\gamma-1} \\ \frac{\gamma-1}{2}u^2 - au \\ \frac{\gamma-1}{2}u^2 + au \end{bmatrix} f_1^{\pm} + \frac{\gamma}{\rho a^2} \begin{bmatrix} u \\ -(\gamma-1)u + a \\ -(\gamma-1)u - a \end{bmatrix} f_2^{\pm} + \frac{\gamma}{\rho a^2} \begin{bmatrix} -1 \\ \gamma - 1 \\ \gamma - 1 \end{bmatrix} f_3^{\pm}. \tag{18.18b}$$

To summarize, by Equations (18.17) and (18.18) *flux vector splitting and wave speed splitting are essentially equivalent for the Euler equations* or for any conservation law with a homogeneous flux function of order one. Unfortunately, a wave speed splitting converted to a flux vector splitting using Equation (18.17) may not satisfy the flux vector splitting condition (18.7) and, conversely, a flux vector splitting converted to a wave speed splitting using Equation (18.18) may not satisfy the wave speed splitting condition (13.23). In general, it is best to convert all wave speed splittings to flux vector splittings prior to discretization, since discretizations of conservative flux vector splitting forms such as Equation (13.8) are more likely to be conservative than discretizations of nonconservative wave speed splitting forms such as Equation (13.24).

18.2.1 *Steger–Warming Flux Vector Splitting*

Steger and Warming (1981) proposed the first flux vector splitting, based on wave speed splitting. One possible wave speed splitting is as follows:

$$\blacklozenge \quad \lambda_i^+ = \max(0, \lambda_i) = \frac{1}{2}(\lambda_i + |\lambda_i|), \tag{18.19a}$$

$$\blacklozenge \quad \lambda_i^- = \min(0, \lambda_i) = \frac{1}{2}(\lambda_i - |\lambda_i|). \tag{18.19b}$$

Convert this wave speed splitting to flux vector splitting using Equation (18.17). Then for $M \leq -1$:

$$\begin{aligned} \mathbf{f}^+ &= \mathbf{0}, \\ \mathbf{f}^- &= \mathbf{f}, \end{aligned} \tag{18.20a}$$

for $-1 < M \leq 0$:

$$\begin{aligned} \mathbf{f}^+ &= \frac{\rho}{2\gamma}(u+a) \begin{bmatrix} 1 \\ u + a \\ \frac{u^2}{2} + \frac{a^2}{\gamma-1} + au \end{bmatrix}, \\ \mathbf{f}^- &= \frac{\gamma - 1}{\gamma}\rho u \begin{bmatrix} 1 \\ u \\ \frac{1}{2}u^2 \end{bmatrix} + \frac{\rho}{2\gamma}(u-a) \begin{bmatrix} 1 \\ u - a \\ \frac{u^2}{2} + \frac{a^2}{\gamma-1} - au \end{bmatrix}, \end{aligned} \tag{18.20b}$$

for $0 < M \leq 1$:

$$\mathbf{f}^{+} = \frac{\gamma - 1}{\gamma} \rho u \begin{bmatrix} 1 \\ u \\ \frac{1}{2}u^2 \end{bmatrix} + \frac{\rho}{2\gamma}(u + a) \begin{bmatrix} 1 \\ u + a \\ \frac{u^2}{2} + \frac{a^2}{\gamma - 1} + au \end{bmatrix},$$

$$\mathbf{f}^{-} = \frac{\rho}{2\gamma}(u - a) \begin{bmatrix} 1 \\ u - a \\ \frac{u^2}{2} + \frac{a^2}{\gamma - 1} - au \end{bmatrix}, \tag{18.20c}$$

and for $M > 1$:

$$\mathbf{f}^{+} = \mathbf{f}, \qquad \mathbf{f}^{-} = \mathbf{0}, \tag{18.20d}$$

where $M = u/a$ is the Mach number. Then, for supersonic right-running flows, all of the waves are right-running, and the flux vector splitting (18.20d) correctly attributes all of the flux to right-running waves and none to left-running waves. Similarly, for supersonic left-running flows, all of the waves are left-running, and the flux vector splitting (18.20a) correctly attributes all of the flux to left-running waves and none to right-running waves. For subsonic flows, waves are both left- and right-running, and the flux vector splitting (18.20b) or (18.20c) correctly attributes some flux to left-running waves and some flux to right-running waves, although there is no reason to believe that it gets the exact physical proportions correct.

Sonic points in the entropy waves occur for $M = 0$, whereas sonic points in the acoustic waves occur for $M = \pm 1$. Unfortunately, wave speed splitting (18.19) may experience difficulties at these sonic points, due to the discontinuity in the first derivative of the split wave speeds at these sonic points. A simple "entropy fix" that rounds out the corners is

♦ $$\lambda_i^{\pm} = \frac{1}{2}\left(\lambda_i \pm \sqrt{\lambda_i^2 + \delta^2}\right), \tag{18.21}$$

where δ is a small user-adjustable parameter. All of the numerical results given later in this section use the simple wave speed splitting (18.19) rather than this "entropy-fixed" wave speed splitting.

18.2.2 *Van Leer Flux Vector Splitting*

Van Leer (1982) suggested a second type of flux vector splitting. Unlike the Steger–Warming flux vector splitting, Van Leer's flux vector splitting is not based on wave speed splitting. As seen in Section 13.4, sonic points are natural flux splitting points; at the very least, sonic points require special consideration to avoid numerical problems. For the Euler equations, the Mach number M indicates sonic points; in particular, sonic points occur when $M = u/a = 0, \pm 1$. Thus, to help address sonic points, Van Leer bases his flux vector splitting on Mach number splitting. The flux vector for the Euler equations can be written in terms of the Mach number as follows:

$$\mathbf{f} = \begin{bmatrix} f_1 \\ f_2 \\ f_3 \end{bmatrix} = \begin{bmatrix} \rho a M \\ \frac{\rho a^2}{\gamma}(\gamma M^2 + 1) \\ \rho a^3 M\left(\frac{1}{2}M^2 + \frac{1}{\gamma-1}\right) \end{bmatrix}. \tag{18.22}$$

Notice that the mass flux f_1 depends linearly on M. Then the mass flux can be split much like the flux function for the linear advection equation, as seen in Example 13.8. In particular, the linear function M can be split into two quadratic functions as follows:

♦ $$M^+ = \begin{cases} 0 & M \leq -1, \\ \left(\frac{M+1}{2}\right)^2 & -1 < M < 1, \\ M & M \geq 1. \end{cases} \tag{18.23a}$$

♦ $$M^- = \begin{cases} M & M \leq -1, \\ -\left(\frac{M-1}{2}\right)^2 & -1 < M < 1, \\ 0 & M \geq 1. \end{cases} \tag{18.23b}$$

Obviously $M^+ + M^- = M$ for $|M| \geq 1$. The same is true for $|M| < 1$, which is easily proven as follows:

$$M^+ + M^- = \frac{M^2 + 2M + 1}{4} - \frac{M^2 - 2M + 1}{4} = M.$$

Notice that $M^\pm$ and its first derivative are continuous. Unfortunately, the second derivative is highly discontinuous at the sonic points $M = \pm 1$. This could be prevented by splitting the linear flux into two cubic pieces – the higher the order of the polynomials in the splitting, the more degrees of freedom there are to ensure continuity in the derivatives. Mach number splitting (18.23) implies the following mass flux splitting:

$$f_1^\pm = \rho a M^\pm. \tag{18.24}$$

The momentum flux f_2 depends on $\gamma M^2 + 1$. By the same principles as before, this quadratic is split into two cubics, where the cubics ensure continuity of the split momentum flux and its first derivative. Omitting the details, the result is

$$(\gamma M^2 + 1)^+ = \begin{cases} 0 & M \leq -1, \\ \left(\frac{M+1}{2}\right)^2((\gamma - 1)M + 2) & -1 < M < 1, \\ \gamma M^2 + 1 & M \geq 1, \end{cases} \tag{18.25a}$$

$$(\gamma M^2 + 1)^- = \begin{cases} \gamma M^2 + 1 & M \leq -1, \\ -\left(\frac{M-1}{2}\right)^2((\gamma - 1)M - 2) & -1 < M < 1, \\ 0 & M \geq 1. \end{cases} \tag{18.25b}$$

This implies the following momentum flux splitting:

$$f_2^\pm = \frac{\rho a^2}{\gamma}(\gamma M^2 + 1)^\pm \tag{18.26}$$

or

$$f_2^+ = \begin{cases} 0 & M \le -1, \\ \frac{1}{\gamma} f_1^+((\gamma-1)u + 2a) & -1 < M \le 1, \\ f_2 & M > 1, \end{cases} \tag{18.27a}$$

$$f_2^- = \begin{cases} f_2 & M \le -1, \\ \frac{1}{\gamma} f_1^-((\gamma-1)u - 2a) & -1 < M \le 1, \\ 0 & M > 1. \end{cases} \tag{18.27b}$$

For future reference, remember that the momentum flux includes a pressure term, by standard convention; then the above momentum flux splitting implicitly involves a pressure splitting. In particular, as the reader can show, Van Leer splits the pressure as follows:

$$p^+ = p \begin{cases} 0 & M \le -1, \\ \left(\frac{M+1}{2}\right)^2 (2 - M) & -1 < M < 1, \\ 1 & M \ge 1, \end{cases} \tag{18.28a}$$

$$p^- = p \begin{cases} 1 & M \le -1, \\ -\left(\frac{M-1}{2}\right)^2 (2 + M) & -1 < M < 1, \\ 0 & M \ge 1. \end{cases} \tag{18.28b}$$

Finally, the energy flux depends on the cubic $M(\frac{1}{2}M^2 + \frac{1}{\gamma-1})$, which is split into two quartics. Omitting the rather complicated algebra, the final result is

$$f_3^+ = \begin{cases} 0 & M \le -1, \\ \frac{1}{2(\gamma+1)(\gamma-1)} f_1^+((\gamma-1)u + 2a)^2 & -1 < M \le 1, \\ f_3 & M > 1, \end{cases} \tag{18.29a}$$

$$f_3^+ = \begin{cases} f_3 & M \le -1, \\ \frac{1}{2(\gamma+1)(\gamma-1)} f_1^-((\gamma-1)u - 2a)^2 & -1 < M \le 1, \\ 0 & M > 1. \end{cases} \tag{18.29b}$$

To summarize, from Equations (18.24), (18.27), and (18.29), notice that

$$\blacklozenge \quad \mathbf{f}^\pm = \pm\frac{\rho a}{4}(M \pm 1)^2 \begin{bmatrix} 1 \\ \frac{(\gamma-1)u \pm 2a}{\gamma} \\ \frac{((\gamma-1)u \pm 2a)^2}{2(\gamma+1)(\gamma-1)} \end{bmatrix} \tag{18.30}$$

for $|M| < 1$. Otherwise, $\mathbf{f}^+ = \mathbf{f}$ and $\mathbf{f}^- = \mathbf{0}$ for $M \ge 1$, and $\mathbf{f}^- = \mathbf{f}$ and $\mathbf{f}^+ = \mathbf{0}$ for $M \le 1$. Like the Steger–Warming flux vector splitting, Van Leer's flux vector splitting correctly attributes all of the flux to right-running waves for right-running supersonic flow and all of the flux to left-running waves for left-running supersonic flow. Furthermore, in the original paper, Van Leer proves that his splitting satisfies $d\mathbf{f}^+/d\mathbf{u} \ge 0$ and $d\mathbf{f}^-/d\mathbf{u} \le 0$, as required by Equation (13.7).

Van Leer's flux vector splitting may be converted to wave speed splitting using Equation (18.18)

$$\lambda_1^+ = \frac{\gamma}{\rho a^2}\left[\left(-\frac{u^2}{2} + \frac{a^2}{\gamma - 1}\right) f_1^+ + u f_2^+ - f_3^+\right].$$

Then by Equation (18.30)

$$\lambda_1^+ = \frac{\gamma f_1^+}{\rho a^2}\left[-\frac{u^2}{2} + \frac{a^2}{\gamma - 1} + \frac{u}{\gamma}((\gamma - 1)u + 2a) - \frac{(\gamma - 1)^2 u^2 + 4(\gamma - 1)au + 4a^2}{2(\gamma + 1)(\gamma - 1)}\right].$$

Combine terms to find

$$\lambda_1^+ = \frac{\gamma f_1^+}{\rho a^2}\left[\left(-\frac{1}{2} + \frac{\gamma - 1}{\gamma} - \frac{\gamma - 1}{2(\gamma + 1)}\right) u^2 \right.$$
$$\left. + 2\left(\frac{1}{\gamma} - \frac{1}{\gamma + 1}\right) au + \frac{1}{\gamma - 1}\left(1 - \frac{2}{\gamma + 1}\right) a^2\right].$$

Simplify the coefficients of u^2, au, and a^2 to find

$$\lambda_1^+ = \frac{\gamma f_1^+}{\rho a^2}\left(-\frac{1}{\gamma(\gamma + 1)} u^2 + \frac{2}{\gamma(\gamma + 1)} ua + \frac{1}{\gamma + 1} a^2\right)$$

or, equivalently,

$$\lambda_1^+ = \frac{f_1^+}{\rho(\gamma + 1)}(-M^2 + 2M + \gamma).$$

Proceed similarly for the other eigenvalues to find the following final results, as given in Subsection 20.2.3 of Hirsch (1990):

$$\lambda_1^+ = \frac{a}{4}(M + 1)^2\left[1 + \frac{(M - 1)^2}{\gamma + 1}\right],$$
$$\lambda_2^+ = \frac{a}{4}(M + 1)^2\left[3 - M + \frac{\gamma - 1}{\gamma + 1}(M - 1)^2\right], \tag{18.31a}$$
$$\lambda_3^+ = \frac{a}{2}(M + 1)^2 \frac{M - 1}{\gamma + 1}\left[1 + \frac{\gamma - 1}{2} M\right]$$

and

$$\lambda_1^-(M) = -\lambda_1^+(-M),$$
$$\lambda_2^-(M) = -\lambda_3^+(-M), \tag{18.31b}$$
$$\lambda_3^-(M) = -\lambda_2^+(-M).$$

The Van Leer flux vector splitting was generalized and extended to real gases by Liou, Van Leer, and Shuen (1990) and to chemically reacting flows by Shuen, Liou, and Van Leer (1990). This completes our discussion of the Van Leer flux vector splitting.

18.2.3 *Liou–Steffen Flux Vector Splitting*

The Beam–Warming and Van Leer flux vector splittings are both relatively complicated and expensive. As an alternative, Liou and Steffen (1993) suggested a far simpler flux vector splitting. The Liou–Steffen flux vector splitting uses Van Leer's approach to split the Mach number and pressure appearing in the momentum flux but, unlike Van Leer's flux vector splitting, leaves the rest of the variables in the flux vector alone. The Liou–Steffen flux vector splitting method distinguishes between true momentum flux and pressure force terms, much as in Equation (2.20). Specifically, the flux vector is written as follows:

$$\mathbf{f} = \begin{bmatrix} \rho u \\ \rho u^2 \\ \rho h_T u \end{bmatrix} + \begin{bmatrix} 0 \\ p \\ 0 \end{bmatrix}$$

or

$$\mathbf{f} = M \begin{bmatrix} \rho a \\ \rho u a \\ \rho h_T a \end{bmatrix} + \begin{bmatrix} 0 \\ p \\ 0 \end{bmatrix}. \tag{18.32}$$

The Mach number and the pressure appearing in the second vector are split just as in Van Leer's flux vector splitting. Specifically,

$$M = M^+ + M^-$$

and

$$p = p^+ + p^-,$$

where $M^\pm$ are defined by Equation (18.23) and $p^\pm$ are defined by Equation (18.28). As an alternative to Van Leer's pressure splitting, Liou and Steffen suggested the following simplified pressure splitting:

♦ $$p^+ = p \begin{cases} 0 & M \le -1, \\ \frac{1}{2}(1+M) & -1 < M < 1, \\ 1 & M \ge 1, \end{cases} \tag{18.33a}$$

♦ $$p^- = p \begin{cases} 1 & M \le -1, \\ \frac{1}{2}(1-M) & -1 < M < 1, \\ 0 & M \ge 1, \end{cases} \tag{18.33b}$$

which seems to yield results almost as good as Van Leer's pressure splitting. Interestingly enough, Liou and Steffen split the Mach number a second time as follows:

$$M = \max(0, M^+ + M^-) + \min(0, M^+ + M^-).$$

Then the Liou–Steffen flux vector splitting with "double Mach number splitting" is

$$\mathbf{f}^+ = \max(0, M^+ + M^-) \begin{bmatrix} \rho a \\ \rho u a \\ \rho h_T a \end{bmatrix} + \begin{bmatrix} 0 \\ p^+ \\ 0 \end{bmatrix}, \tag{18.34a}$$

$$\mathbf{f}^- = \min(0, M^+ + M^-) \begin{bmatrix} \rho a \\ \rho u a \\ \rho h_T a \end{bmatrix} + \begin{bmatrix} 0 \\ p^- \\ 0 \end{bmatrix}. \tag{18.34b}$$

18.2.4 *Zha–Bilgen Flux Vector Splitting*

Inspired by Liou and Steffen, Zha and Bilgen (1993) suggested another simple flux vector splitting. Whereas Liou and Steffen separated true flux from pressure terms only in conservation of momentum, Zha and Bilgen separate pressure terms from true flux in both conservation of momentum and conservation of energy, which seems sensible. In particular, Zha and Bilgen write the flux vector as

$$\mathbf{f} = u \begin{bmatrix} \rho \\ \rho u \\ \rho e_T \end{bmatrix} + \begin{bmatrix} 0 \\ p \\ pu \end{bmatrix} \tag{18.35}$$

much like Equation (2.20). Then the Zha–Bilgen flux vector splitting is

♦ $$\mathbf{f}^+ = \max(0, u) \begin{bmatrix} \rho \\ \rho u \\ \rho e_T \end{bmatrix} + \begin{bmatrix} 0 \\ p^+ \\ (pu)^+ \end{bmatrix}, \tag{18.36a}$$

♦ $$\mathbf{f}^- = \min(0, u) \begin{bmatrix} \rho \\ \rho u \\ \rho e_T \end{bmatrix} + \begin{bmatrix} 0 \\ p^- \\ (pu)^- \end{bmatrix}, \tag{18.36b}$$

where $p^\pm$ are defined by Equation (18.33) and $(pu)^\pm$ are defined as follows:

♦ $$(pu)^+ = p \begin{cases} 0 & M \le -1, \\ \frac{1}{2}(u + a) & -1 < M < 1, \\ u & M \ge 1, \end{cases} \tag{18.37a}$$

♦ $$(pu)^- = p \begin{cases} u & M \le -1, \\ \frac{1}{2}(u - a) & -1 < M < 1, \\ 0 & M \ge 1. \end{cases} \tag{18.37b}$$

Notice that the Zha–Bilgen splitting completely eliminates any sort of Van Leer splitting, making theirs the simplest flux vector splitting yet.

18.2.5 *First-Order Upwind Methods*

The flux vector splittings described in the last four subsections are only the first step. The next step is to discretize $\partial \mathbf{f}^+/\partial x$ and $\partial \mathbf{f}^-/\partial x$. Consider the Euler equations written in a split flux form:

$$\frac{\partial \mathbf{u}}{\partial t} + \frac{\partial \mathbf{f}^+}{\partial x} + \frac{\partial \mathbf{f}^-}{\partial x} = 0.$$

Discretize $\partial \mathbf{f}^+/\partial x$ using FTBS and discretize $\partial \mathbf{f}^-/\partial x$ using FTFS to obtain the following *flux split first-order upwind method*:

$$\blacklozenge \qquad \mathbf{u}_i^{n+1} = \mathbf{u}_i^n - \lambda\left(\mathbf{f}^+(\mathbf{u}_i^n) - \mathbf{f}^+(\mathbf{u}_{i-1}^n) + \mathbf{f}^-(\mathbf{u}_{i+1}^n) - \mathbf{f}^-(\mathbf{u}_i^n)\right). \tag{18.38}$$

In conservation form, the flux split first-order upwind method is

$$\hat{\mathbf{f}}_{i+1/2}^n = \mathbf{f}^+(\mathbf{u}_i) + \mathbf{f}^-(\mathbf{u}_{i+1}). \tag{18.39}$$

Equation (18.39) can be written in the following interesting form:

$$\hat{\mathbf{f}}_{i+1/2}^n = \frac{1}{2}\left(\mathbf{f}(u_{i+1}^n) + \mathbf{f}(u_i^n)\right) - \left(A_{i+1/2}^+ - A_{i+1/2}^-\right)\left(\mathbf{u}_{i+1}^n - \mathbf{u}_i^n\right), \tag{18.40}$$

where

$$\mathbf{f}^+(\mathbf{u}_{i+1}^n) - \mathbf{f}^+(\mathbf{u}_i^n) = A_{i+1/2}^+(\mathbf{u}_{i+1}^n - \mathbf{u}_i^n), \tag{18.41a}$$

$$\mathbf{f}^-(\mathbf{u}_{i+1}^n) - \mathbf{f}^-(\mathbf{u}_i^n) = A_{i+1/2}^-(\mathbf{u}_{i+1}^n - \mathbf{u}_i^n). \tag{18.41b}$$

Equation (18.40) has the form of central differences plus second-order artificial viscosity, where the matrix coefficient of artificial viscosity is $A_{i+1/2}^+ - A_{i+1/2}^-$. Equation (18.41) defines the matrices $A_{i+1/2}^\pm$ in much the same way as the Roe-average matrix was defined in Section 5.3. In fact, Equation (18.41) yields an *infinite* number of solutions for the matrices $A_{i+1/2}^\pm$, as discussed in Section 5.3. Of course, you should not use the artificial viscosity form seen in Equation (18.40) in computations, since it involves two expensive-to-compute matrices $A_{i+1/2}^\pm$. However, the artificial viscosity form is useful in later analysis, especially for describing the connections between flux vector splitting methods and reconstruction–evolution methods.

Equation (18.39) used with the Steger–Warming flux vector splitting is called the *Steger–Warming flux split first-order upwind method*; Equation (18.39) used with Van Leer's flux vector splitting is called *Van Leer's flux split first-order upwind method*; and Equation (18.39) used with the Zha–Bilgen flux vector splitting is called the *Zha–Bilgen flux split first-order upwind method.* Liou and Steffen (1993) suggested a nonstandard first-order upwind method based on Liou–Steffen flux vector splitting. In particular,

$$\blacklozenge \qquad \hat{\mathbf{f}}_{i+1/2}^n = \max\left(0, M_i^+ + M_{i+1}^-\right)\begin{bmatrix}\rho_i^n a_i^n \\ \rho_i^n u_i^n a_i^n \\ \rho_i^n (h_\mathrm{T})_i^n a_i^n\end{bmatrix} + \begin{bmatrix}0 \\ p_i^+ \\ 0\end{bmatrix}$$

$$+ \min\left(0, M_i^+ + M_{i+1}^-\right)\begin{bmatrix}\rho_{i+1}^n a_{i+1}^n \\ \rho_{i+1}^n u_{i+1}^n a_{i+1}^n \\ \rho_{i+1}^n (h_\mathrm{T})_{i+1}^n a_{i+1}^n\end{bmatrix} + \begin{bmatrix}0 \\ p_{i+1}^- \\ 0\end{bmatrix}. \tag{18.42}$$

Liou and Steffen called this method the *advection upstream splitting method* (*AUSM*) although, as usual, this book will refer to it by author and type – the *Liou–Steffen flux split first-order upwind method* – rather than by acronym. The Liou–Steffen flux vector splitting was designed specifically for use in the Liou–Steffen first-order upwind method. Similarly, the Zha–Bilgen flux vector splitting was designed specifically for use in the Zha–Bilgen first-order upwind method. By contrast, the Steger–Warming and Van Leer flux vector splittings were designed as general purpose, and they have been used successfully with a variety of different discretizations down through the years. In the original paper, Liou and Steffen (1993) indicate that their first-order upwind method obtains results comparable to other first-order flux vector splitting methods and flux difference splitting methods, including Roe's first-order upwind method as discussed in Subsection 18.3.2, and with substantially reduced costs. Zha and Bilgen (1993) make the same claim for their first-order upwind method, and they provide interesting comparisons between the Zha–Bilgen first-order upwind method, the Liou–Steffen first-order upwind method, Van Leer's first-order upwind method, and Roe's first-order upwind method. As another variation on flux vector splitting, Radespiel and Kroll (1995) suggested an adaptive combination of the Van Leer and Liou–Steffen first-order upwind methods, with the Liou–Steffen first-order upwind method splitting used in smooth regions and Van Leer flux first-order upwind method used near shocks.

The performance of flux split first-order upwind methods is illustrated by the two standard tests cases defined in Section 18.0. For Test Case 1, the results of the Steger–Warming flux split first-order upwind method are shown in Figure 18.7, the results of the Van Leer flux split first-order upwind method are shown in Figure 18.8, the results of the Liou–Steffen flux split first-order upwind method are shown in Figure 18.9, and the results of the Zha–Bilgen flux split first-order upwind method are shown in Figure 18.10. The results are all reasonably similar, although the Steger–Warming splitting is decidedly worse than the others. In this test case, the simple Liou–Steffen and Zha–Bilgen flux vector splittings acquit themselves admirably and are, if anything, just slightly better than the more complicated Van Leer flux vector splitting. For Test Case 2, the results of the Steger-Warming flux split first-order upwind method are shown in Figure 18.11, the results of the Van Leer flux split first-order upwind method are shown in Figure 18.12, the results of the Liou–Steffen flux split first-order upwind method are shown in Figure 18.13, and the results of the Zha–Bilgen flux split first-order upwind method are shown in Figure 18.14. In Test Case 2, to get them started, both the Liou–Steffen and Zha–Bilgen flux split first-order upwind methods needed a small amount of fixed-coefficient second-order artificial viscosity for the first two time steps. In particular, the Liou–Steffen flux split first-order upwind method uses $\epsilon = 0.025$ and Zha–Bilgen flux split first-order upwind method uses $\epsilon = 0.02$ for the first two time steps, which are near the minimum possible values, after which the artificial viscosity is turned completely off. Although the two simplest flux splitting methods may need a little bit of help getting started, the final results are just as good as those of the two more complicated flux splitting methods. In fact, all four methods perform about the same, although Van Leer's flux splitting leads to a slightly larger expansion shock at the expansive sonic point, while the Zha–Bilgen flux splitting leads to a slight drop in the entropy between the contact and expansion. Still, when it comes to first-order upwind methods, there is not too much to choose from between the four flux vector splittings. The real differences show up in the next subsection, where flux vector splitting is used in a second-order upwind method.

18.2.6 *Beam–Warming Second-Order Upwind Method*

The Beam–Warming second-order upwind method for scalar conservation laws was described in Section 17.4. This section concerns a version for the Euler equations

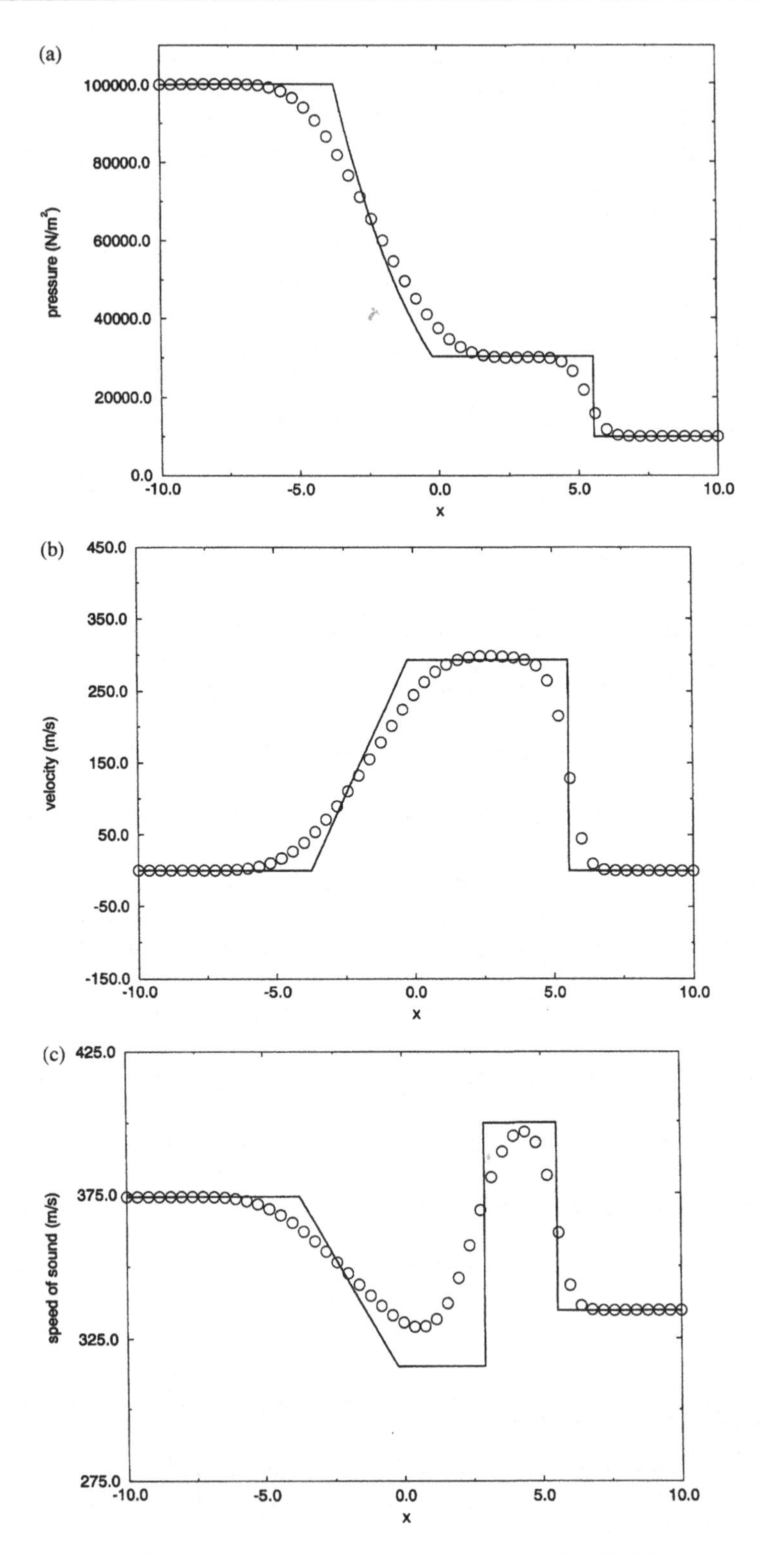

Figure 18.7 Steger–Warming flux split first-order upwind method for Test Case 1.

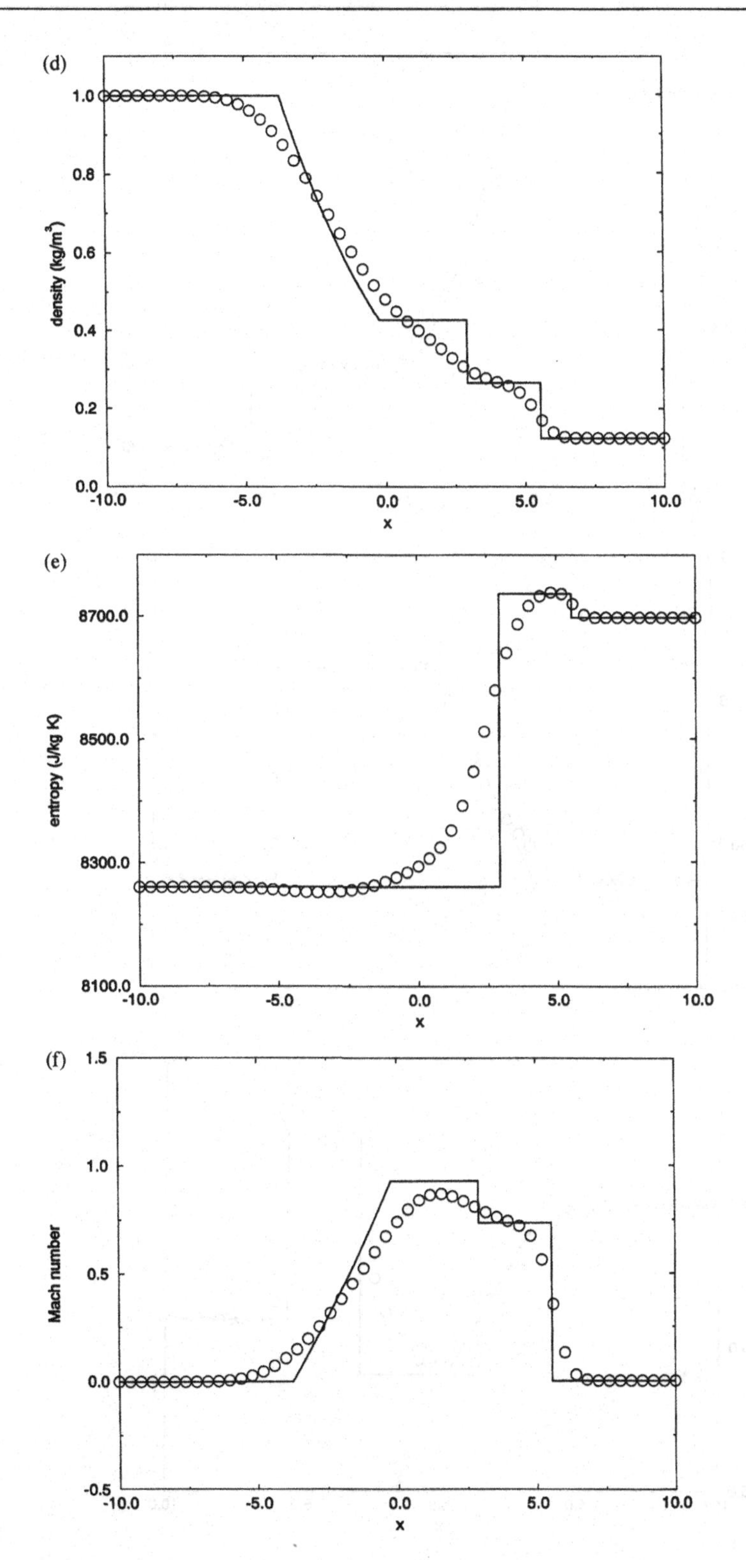

Figure 18.7 *(cont.)*

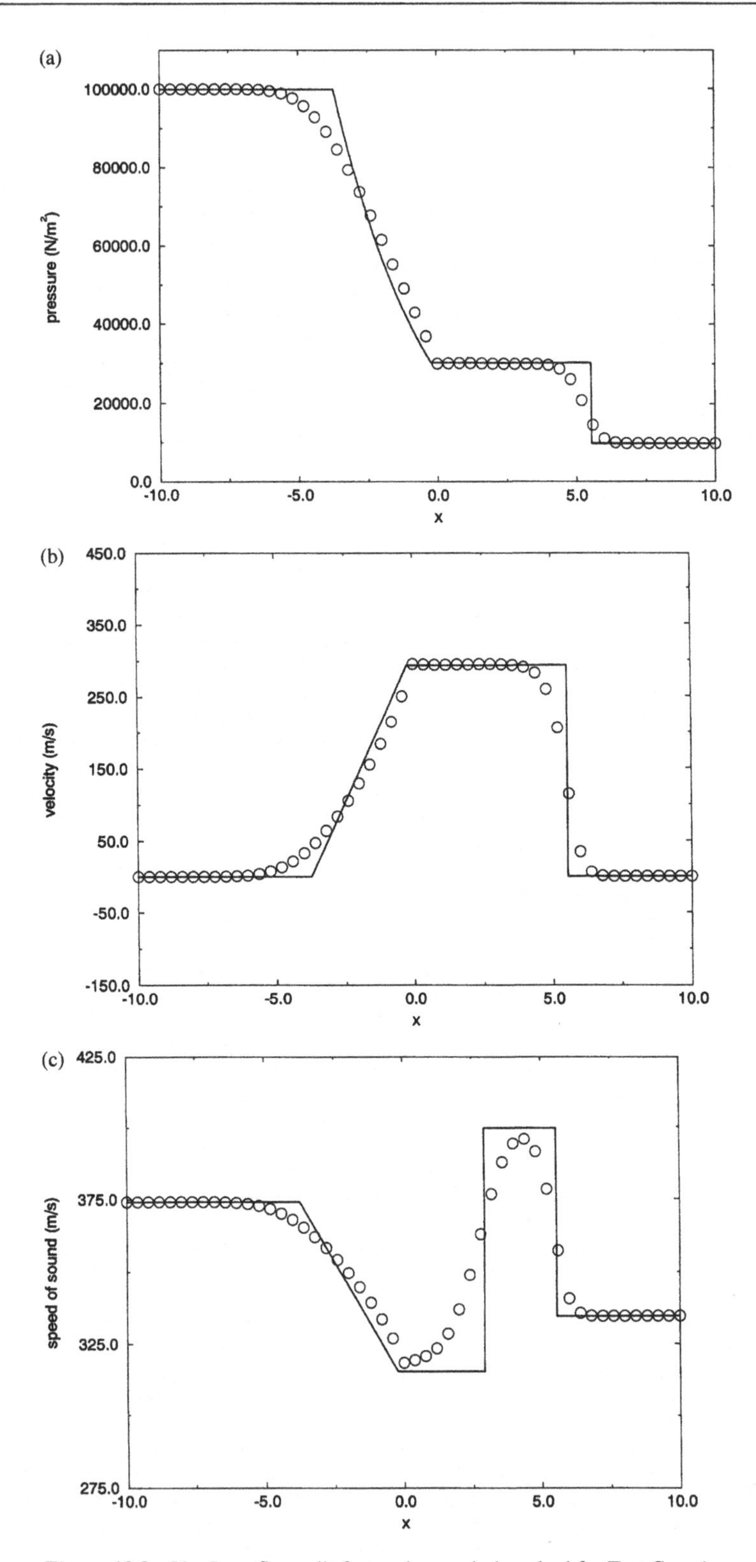

Figure 18.8 Van Leer flux split first-order upwind method for Test Case 1.

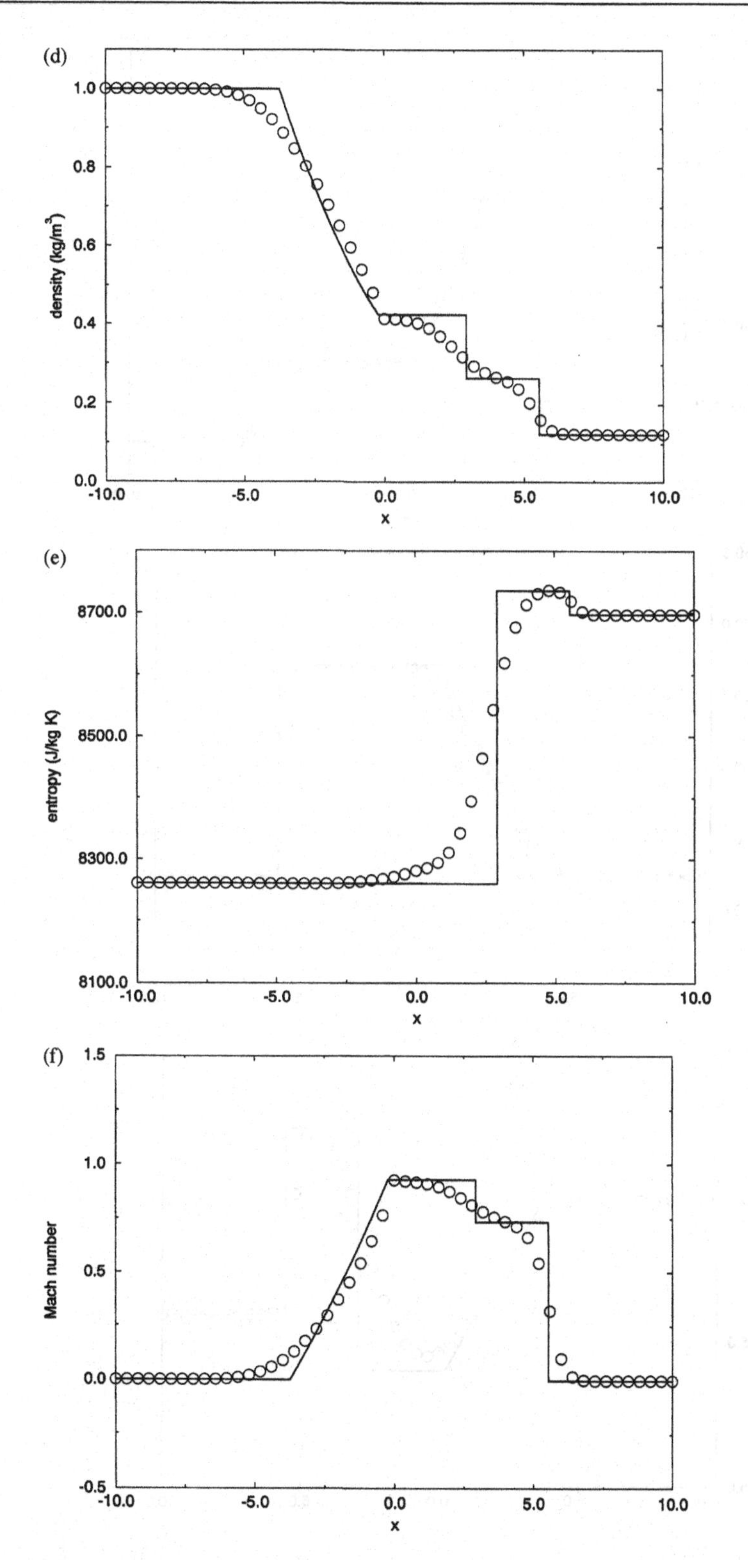

Figure 18.8 *(cont.)*

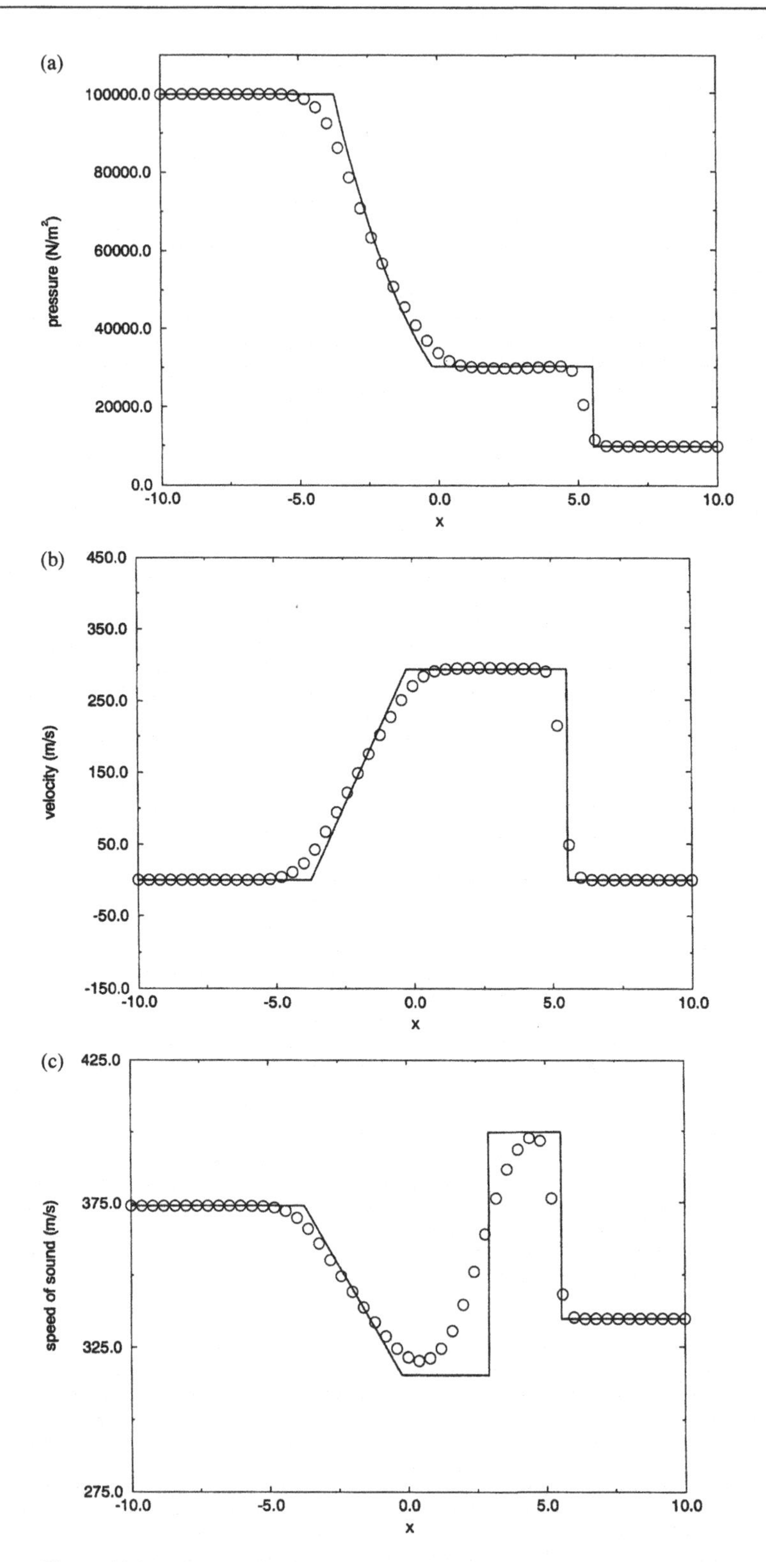

Figure 18.9 Liou–Steffen flux split first-order upwind method for Test Case 1.

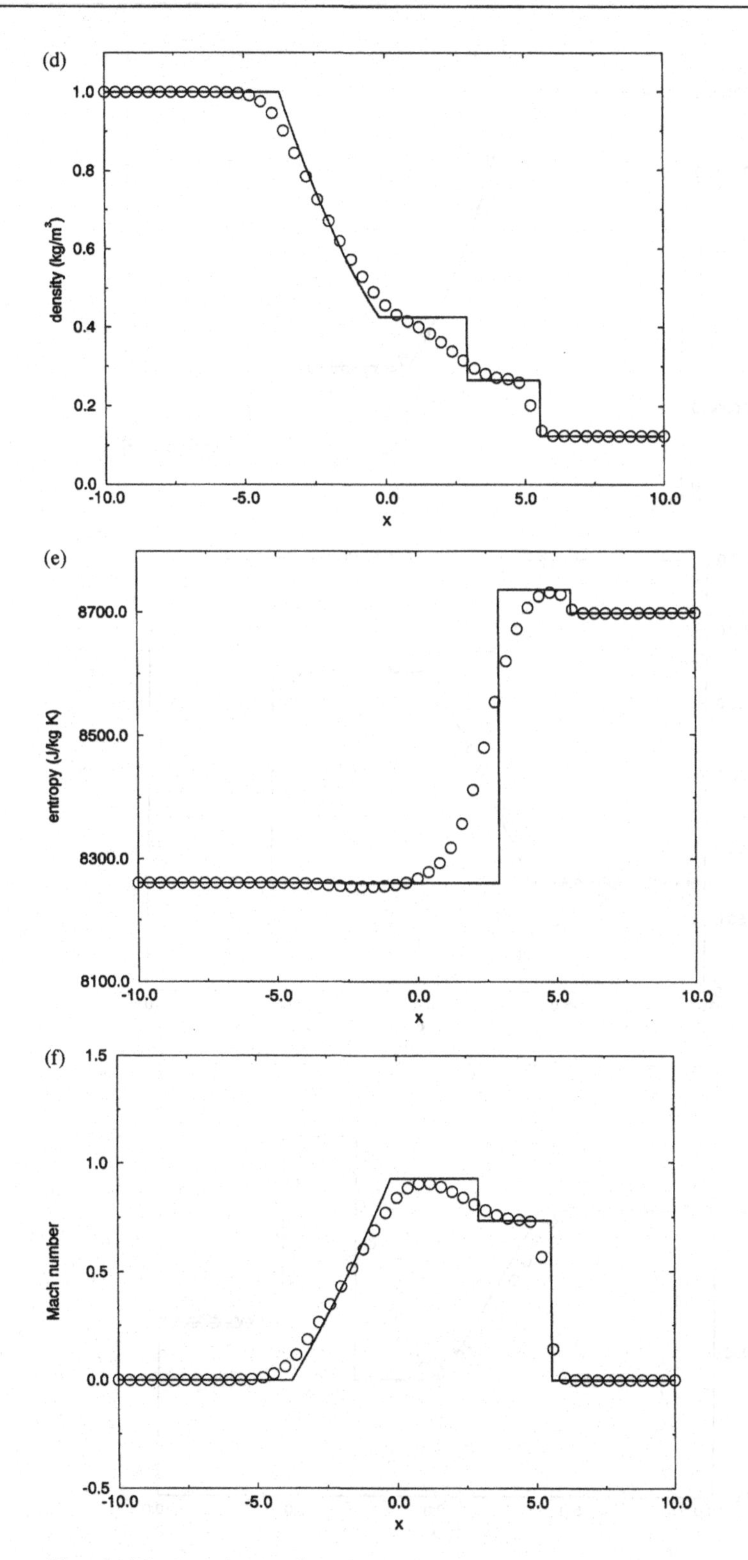

Figure 18.9 *(cont.)*

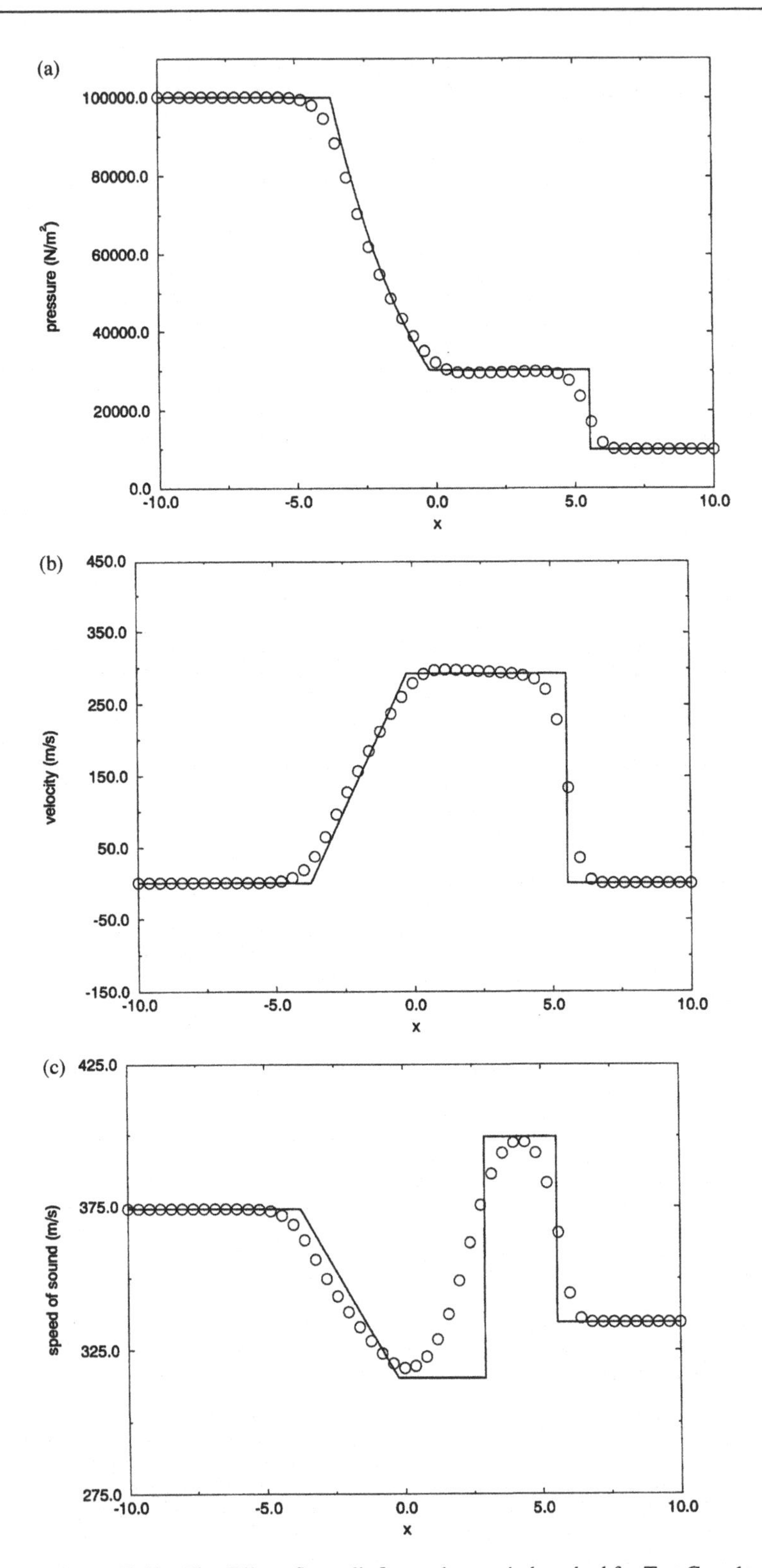

Figure 18.10 Zha–Bilgen flux split first-order upwind method for Test Case 1.

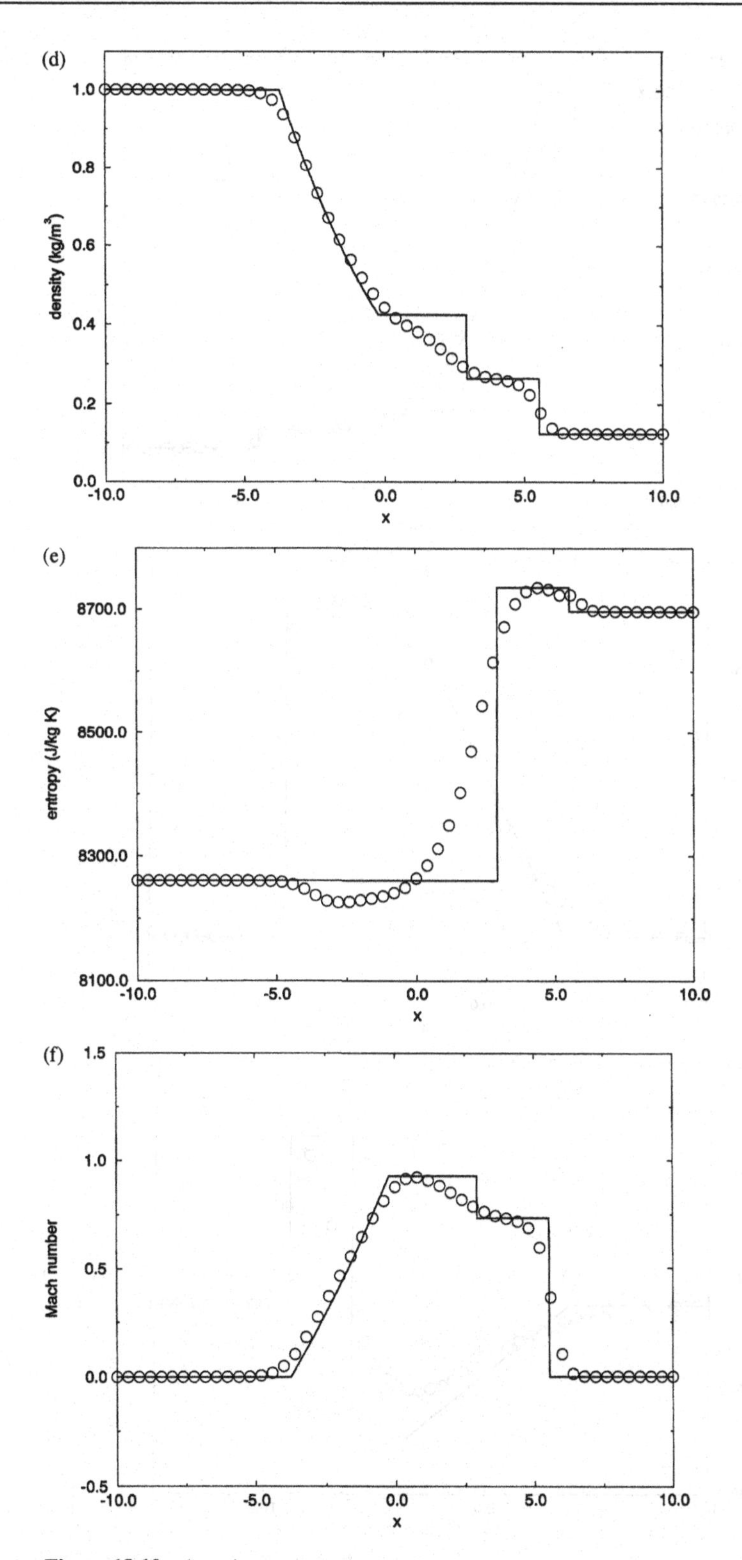

Figure 18.10 (*cont.*)

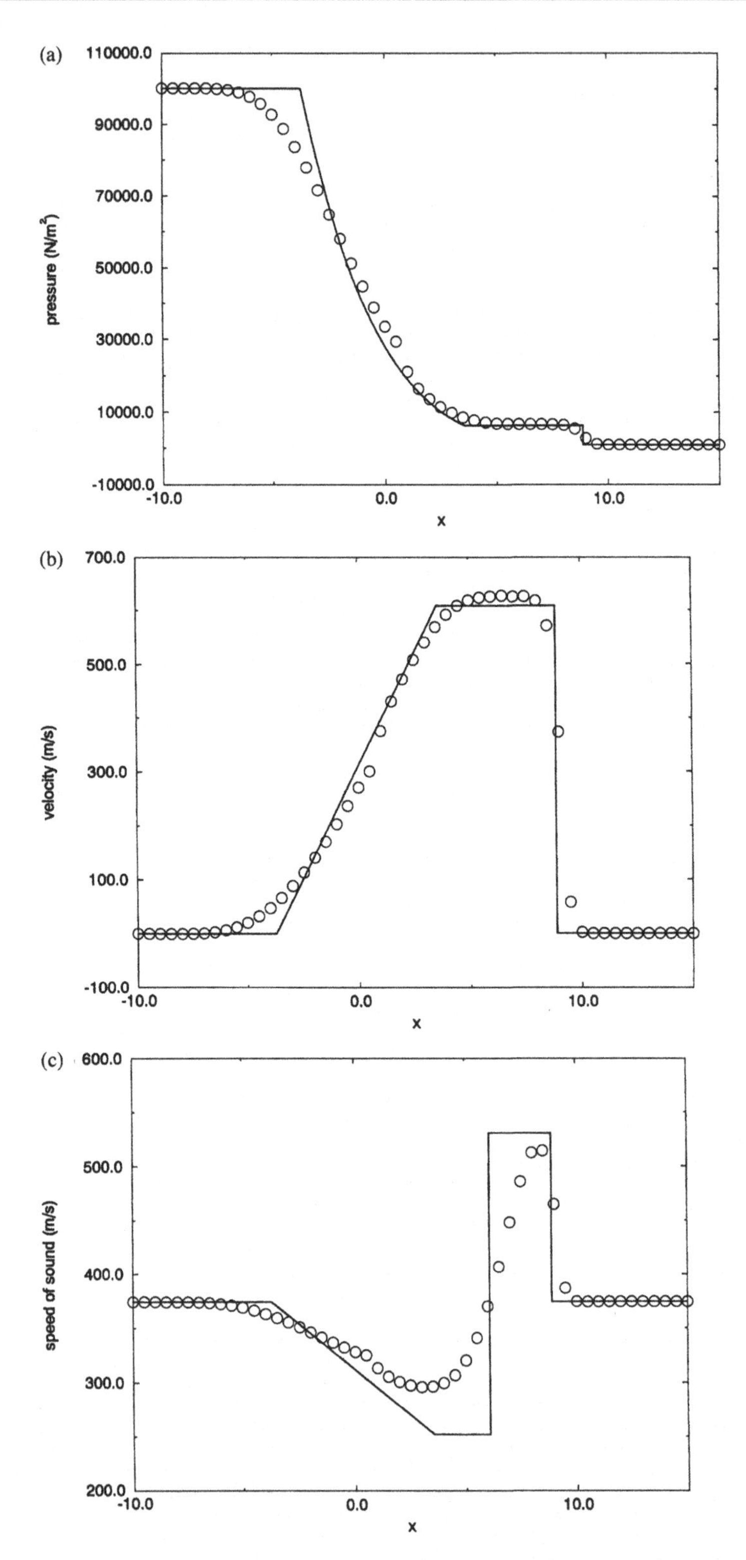

Figure 18.11 Steger–Warming flux split first-order upwind method for Test Case 2.

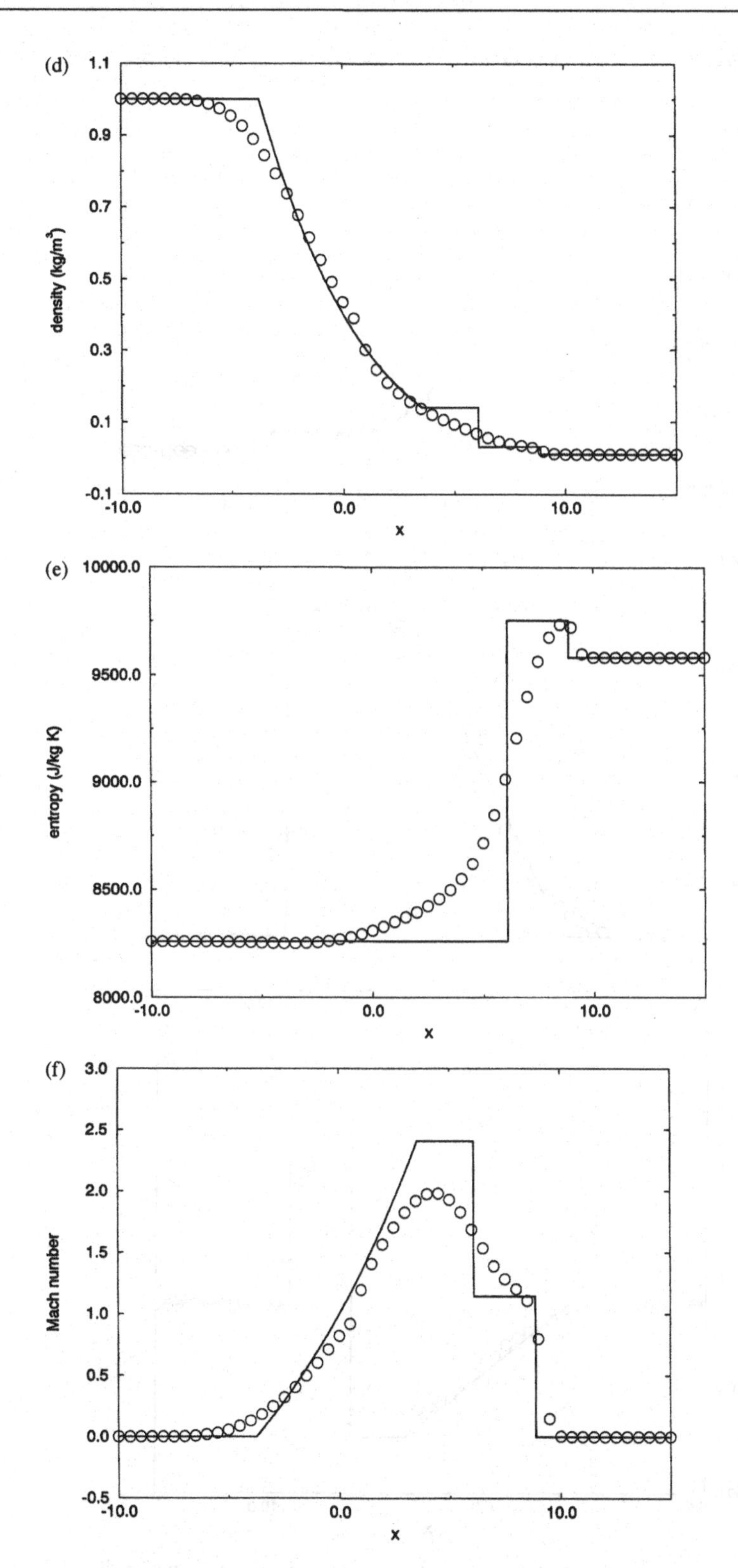

Figure 18.11 *(cont.)*

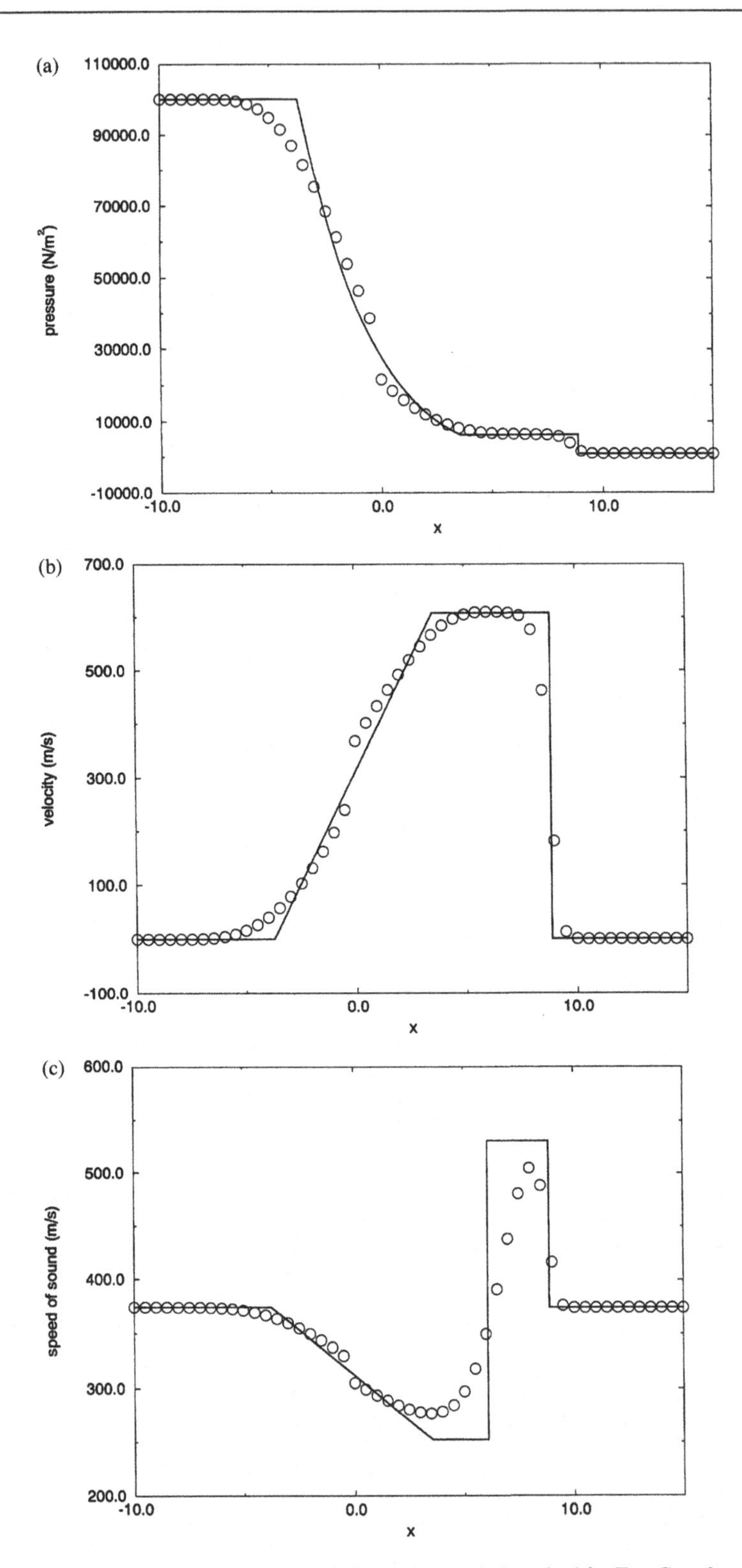

Figure 18.12 Van Leer flux split first-order upwind method for Test Case 2.

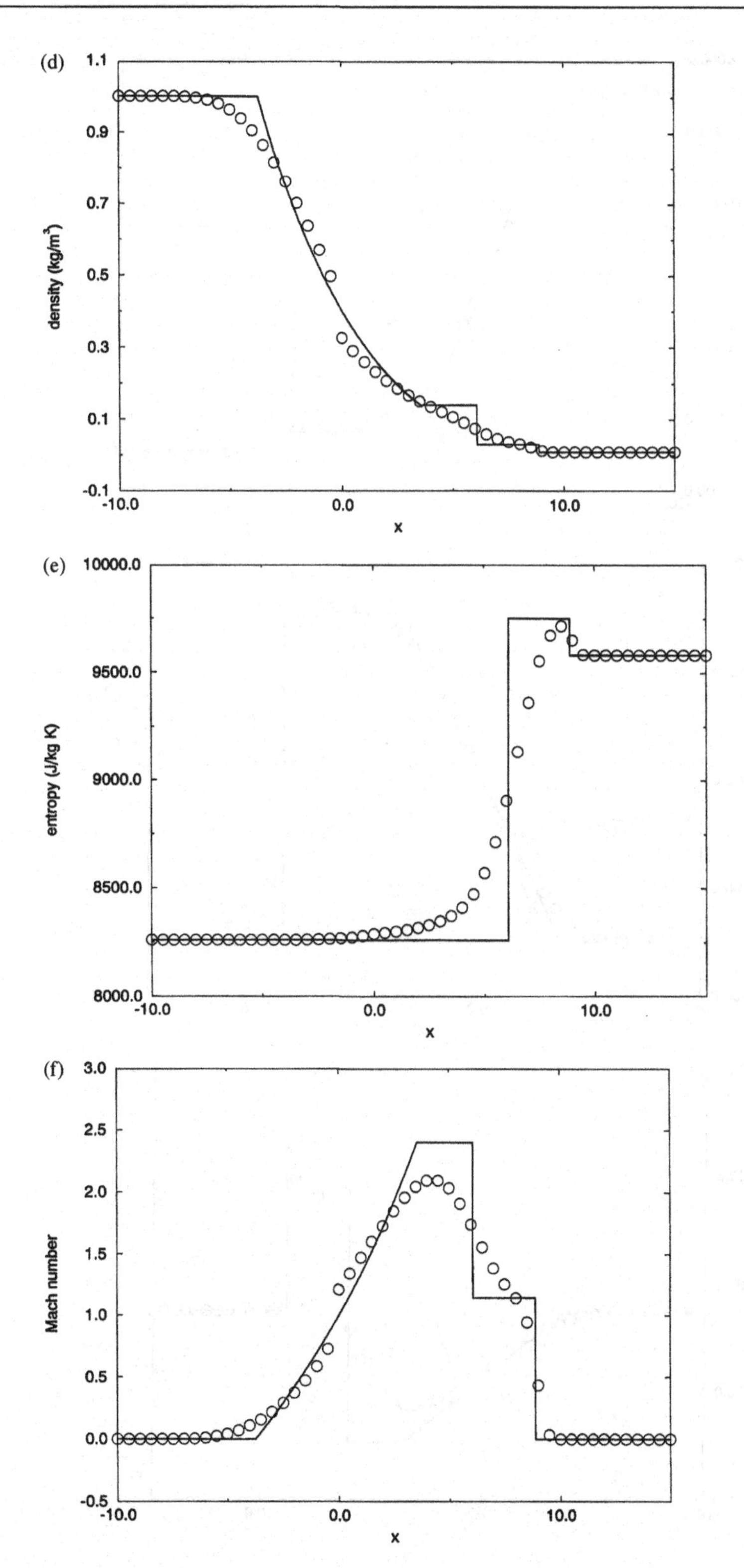

Figure 18.12 *(cont.)*

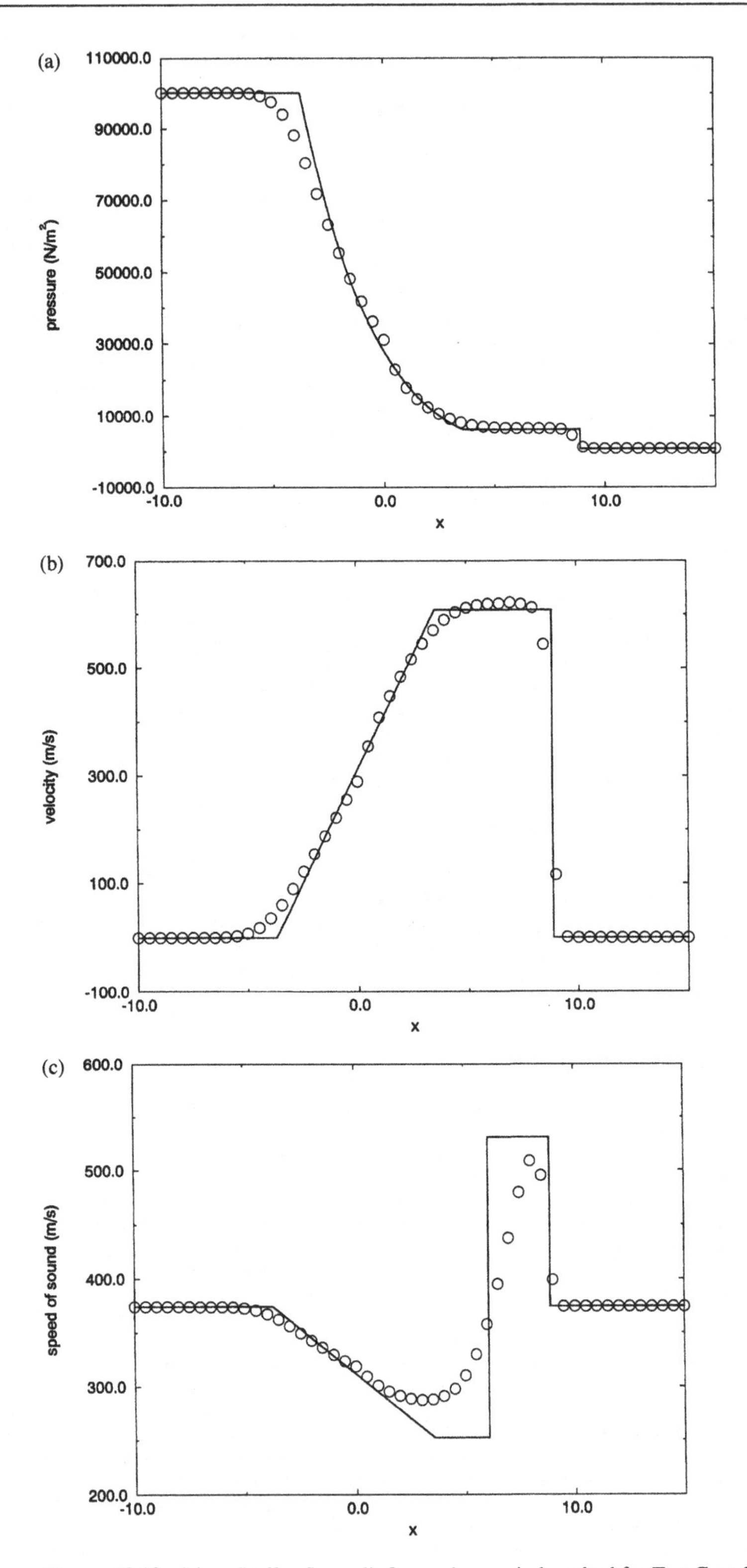

Figure 18.13 Liou–Steffen flux split first-order upwind method for Test Case 2.

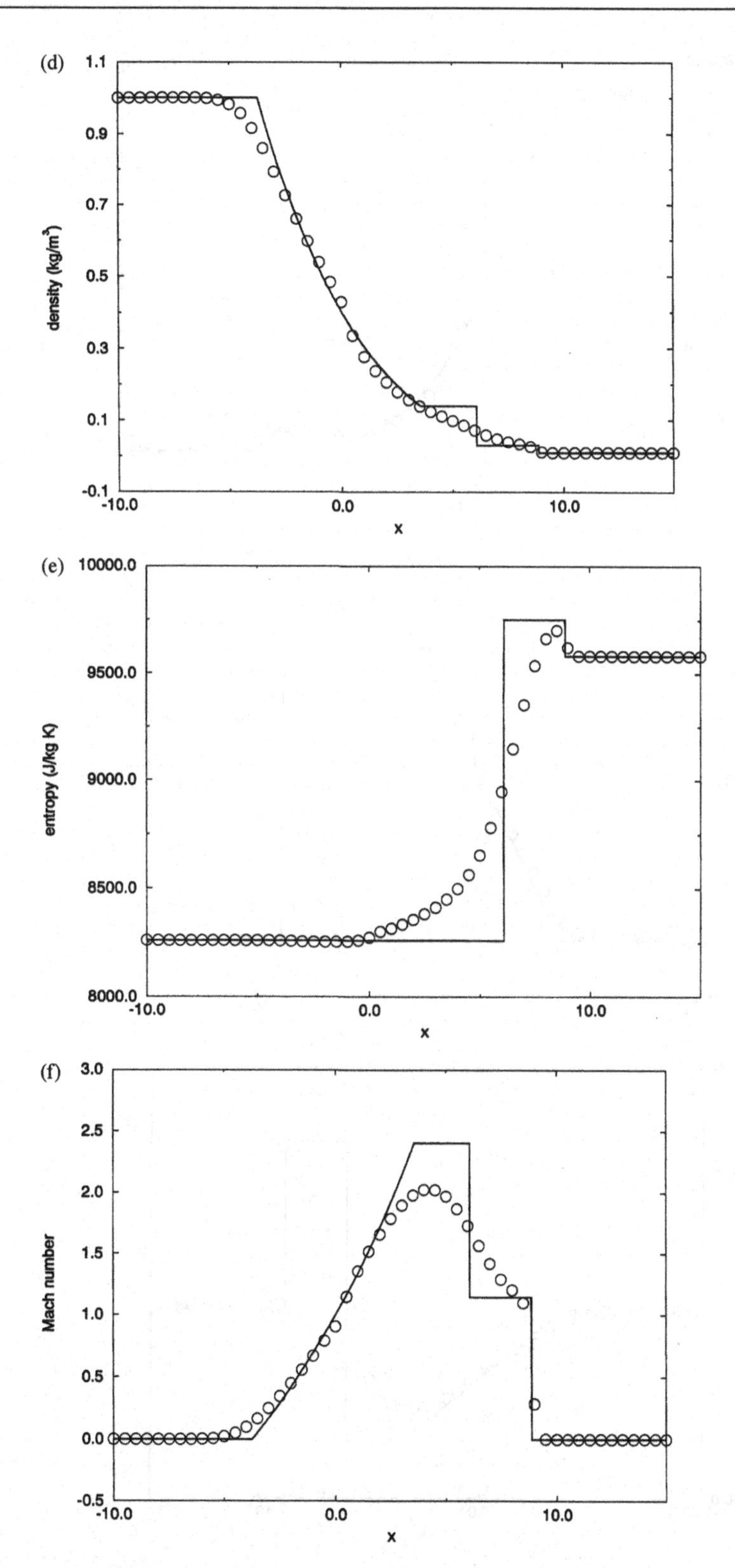

Figure 18.13 *(cont.)*

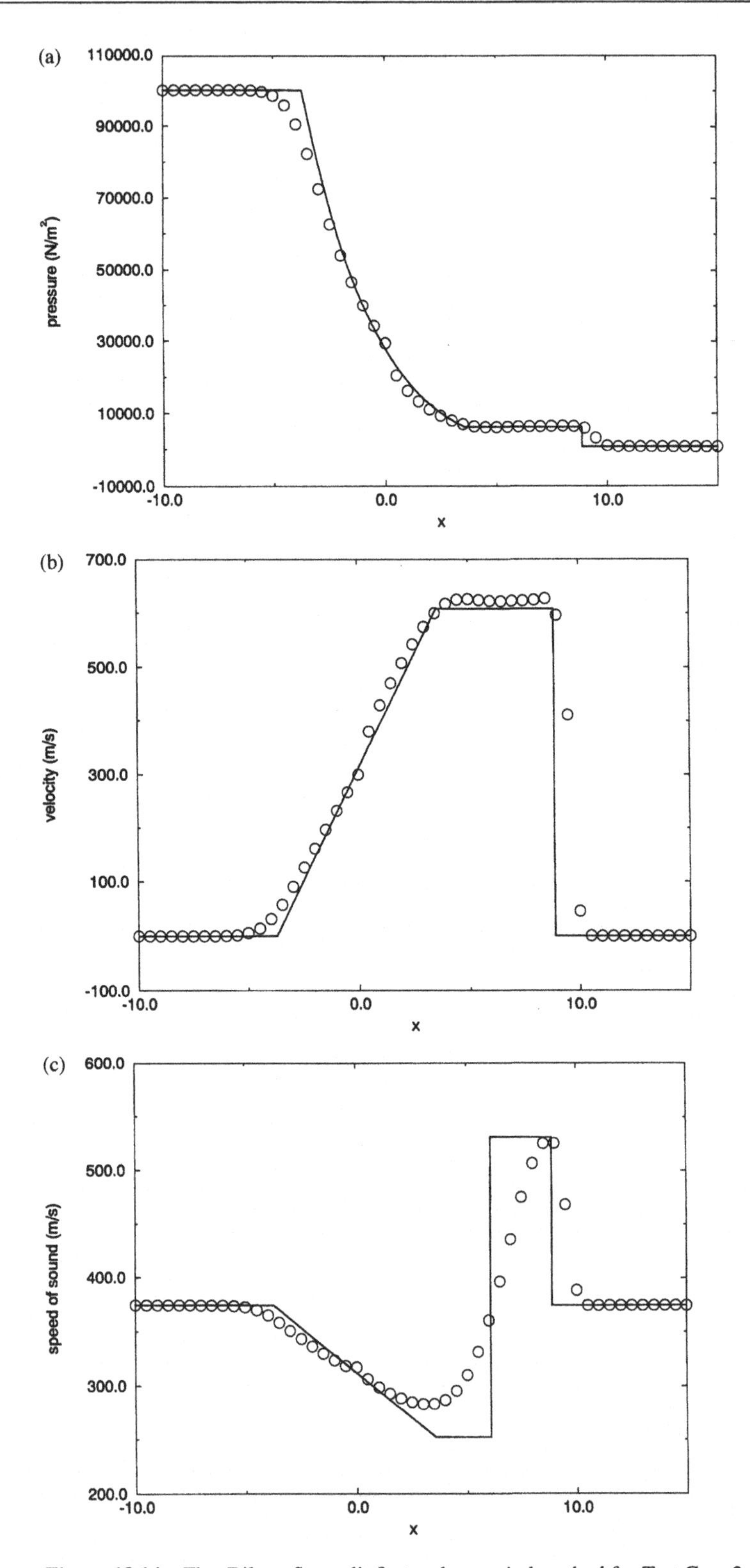

Figure 18.14 Zha–Bilgen flux split first-order upwind method for Test Case 2.

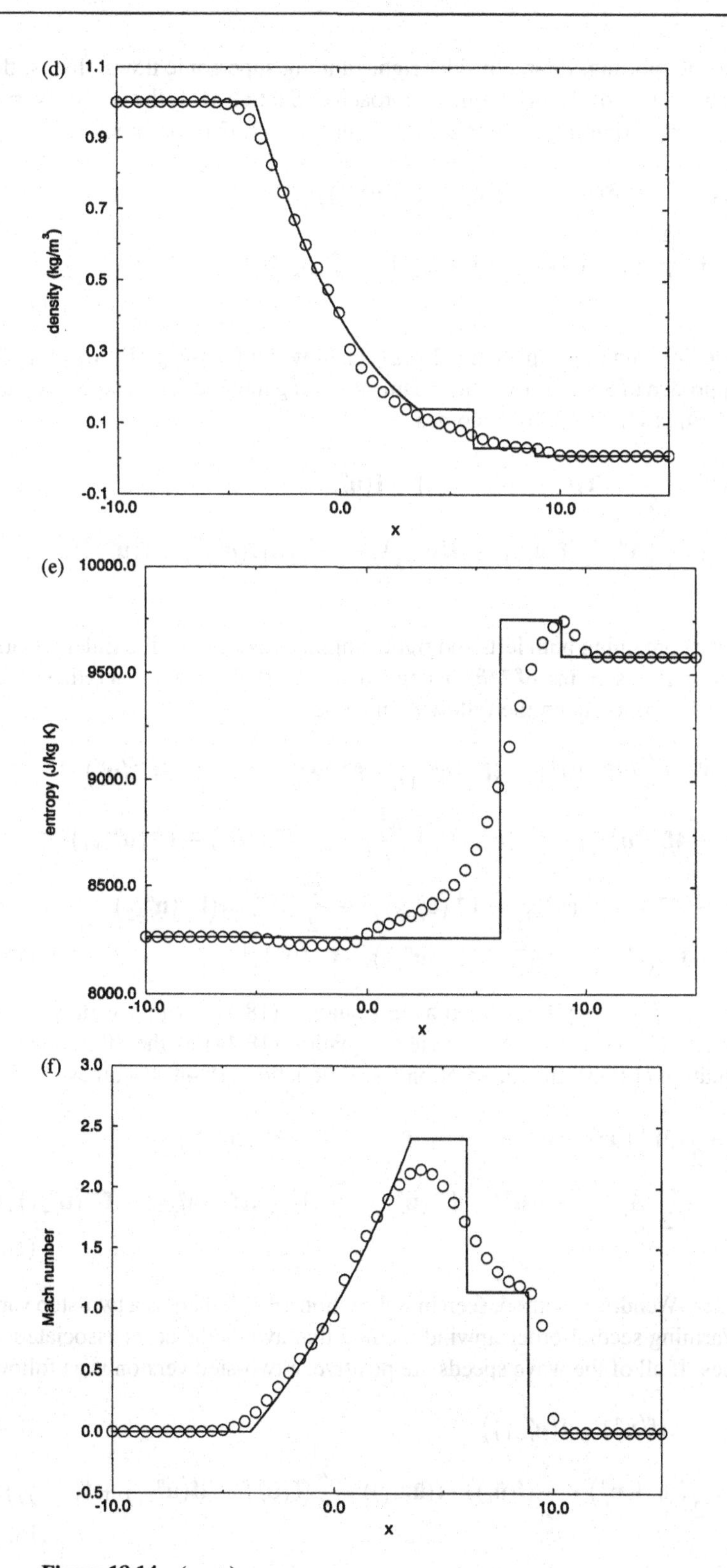

Figure 18.14 *(cont.)*

based on flux vector splitting. First, consider right-running supersonic flows, that is, flows with all-positive wave speeds. Using the flux approach of Section 18.1, the Beam–Warming method for scalar conservation laws with $a > 0$, Equation (17.32a), becomes

$$\begin{aligned}\mathbf{u}_i^{n+1} = &\mathbf{u}_i^n - \frac{\lambda}{2}\left(3\mathbf{f}(\mathbf{u}_i^n) - 4\mathbf{f}(\mathbf{u}_{i-1}^n) + \mathbf{f}(\mathbf{u}_{i-2}^n)\right)\\ &+ \frac{\lambda^2}{2}\left[A_{i-1/2}^n\left(\mathbf{f}(\mathbf{u}_i^n) - \mathbf{f}(\mathbf{u}_{i-1}^n)\right) - A_{i-3/2}^n\left(\mathbf{f}(\mathbf{u}_{i-1}^n) - \mathbf{f}(\mathbf{u}_{i-2}^n)\right)\right].\end{aligned} \tag{18.43a}$$

Similarly, consider left-running supersonic flows (i.e., flows with all-negative wave speeds). Using the flux approach of Section 18.1, the Beam–Warming method for scalar conservation laws with $a < 0$, Equation (17.32b), becomes

$$\begin{aligned}\mathbf{u}_i^{n+1} = &\mathbf{u}_i^n - \frac{\lambda}{2}\left(-3\mathbf{f}(\mathbf{u}_i^n) + 4\mathbf{f}(\mathbf{u}_{i+1}^n) - \mathbf{f}(\mathbf{u}_{i+2}^n)\right)\\ &+ \frac{\lambda^2}{2}\left[A_{i+3/2}^n\left(\mathbf{f}(\mathbf{u}_{i+2}^n) - \mathbf{f}(\mathbf{u}_{i+1}^n)\right) - A_{i+1/2}^n\left(\mathbf{f}(\mathbf{u}_{i+1}^n) - \mathbf{f}(\mathbf{u}_i^n)\right)\right].\end{aligned} \tag{18.43b}$$

For subsonic flows containing both left- and right-running waves, write the Euler equations in flux split form, and discretize $\partial\mathbf{f}^+/\partial x$ using Equation (18.43a) and discretize $\partial\mathbf{f}^-/\partial x$ using Equation (18.43b) to obtain the following method:

$$\begin{aligned}\mathbf{u}_i^{n+1} = &\mathbf{u}_i^n - \frac{\lambda}{2}\left(3\mathbf{f}^+(\mathbf{u}_i^n) - 4\mathbf{f}^+(\mathbf{u}_{i-1}^n) + \mathbf{f}^+(\mathbf{u}_{i-2}^n)\right) - \frac{\lambda}{2}\left(-3\mathbf{f}^-(\mathbf{u}_i^n)\right.\\ &\left.+ 4\mathbf{f}^-(\mathbf{u}_{i+1}^n) - \mathbf{f}^-(\mathbf{u}_{i+2}^n)\right) + \frac{\lambda^2}{2}\left[A_{i-1/2}^+\left(\mathbf{f}^+(\mathbf{u}_i^n) - \mathbf{f}^+(\mathbf{u}_{i-1}^n)\right)\right.\\ &\left.- A_{i-3/2}^+\left(\mathbf{f}^+(\mathbf{u}_{i-1}^n) - \mathbf{f}^+(\mathbf{u}_{i-2}^n)\right)\right] + \frac{\lambda^2}{2}\left[A_{i+3/2}^-\left(\mathbf{f}^-(\mathbf{u}_{i+2}^n)\right.\right.\\ &\left.\left.- \mathbf{f}^-(\mathbf{u}_{i+1}^n)\right) - A_{i+1/2}^-\left(\mathbf{f}^-(\mathbf{u}_{i+1}^n) - \mathbf{f}^-(\mathbf{u}_i^n)\right)\right],\end{aligned} \tag{18.44}$$

where the matrices $A_{i+1/2}^{\pm}$ may be defined as in Equation (18.41) or in a variety of other ways. Alternatively, if you wish, you can view Equation (18.44) as the "flux approach" extension of Equation (17.33). In conservation form, Equation (18.44) becomes

$$\begin{aligned}\hat{\mathbf{f}}_{i+1/2}^n = &\frac{1}{2}\left(3\mathbf{f}^+(\mathbf{u}_i^n) - \mathbf{f}^+(\mathbf{u}_{i-1}^n) + 3\mathbf{f}^-(\mathbf{u}_{i+1}^n) - \mathbf{f}^-(\mathbf{u}_{i+2}^n)\right)\\ &- \frac{\lambda}{2}\left[A_{i-1/2}^+\left(\mathbf{f}^+(\mathbf{u}_i^n) - \mathbf{f}^+(\mathbf{u}_{i-1}^n)\right) + A_{i+3/2}^-\left(\mathbf{f}^-(\mathbf{u}_{i+2}^n) - \mathbf{f}^-(\mathbf{u}_{i+1}^n)\right)\right].\end{aligned} \tag{18.45}$$

As with the Lax–Wendroff method, seen in Subsection 18.1.2, there is a two-step variant of the Beam–Warming second-order upwind method that avoids the costs associated with Jacobian matrices. If all of the wave speeds are positive, a two-step version is as follows:

$$\bar{\mathbf{u}}_i = \mathbf{u}_i^n - \lambda\left(\mathbf{f}(\mathbf{u}_i^n) - \mathbf{f}(\mathbf{u}_{i-1}^n)\right), \tag{18.46a}$$

$$\mathbf{u}_i^{n+1} = \frac{1}{2}\left(\bar{\mathbf{u}}_i + \mathbf{u}_i^n\right) - \frac{\lambda}{2}(\mathbf{f}(\bar{\mathbf{u}}_i) - \mathbf{f}(\bar{\mathbf{u}}_{i-1})) - \frac{\lambda}{2}\left(\mathbf{f}(\mathbf{u}_i^n) - 2\mathbf{f}(\mathbf{u}_{i-1}^n) + \mathbf{f}(\mathbf{u}_{i-2}^n)\right). \tag{18.46b}$$

If all of the wave speeds are negative, a two-step version is as follows:

$$\bar{\mathbf{u}}_i = \mathbf{u}_i^n - \lambda\left(\mathbf{f}\left(\mathbf{u}_{i+1}^n\right) - \mathbf{f}\left(\mathbf{u}_i^n\right)\right), \tag{18.47a}$$

$$\mathbf{u}_i^{n+1} = \frac{1}{2}\left(\bar{\mathbf{u}}_i + \mathbf{u}_i^n\right) - \frac{\lambda}{2}(\mathbf{f}(\bar{\mathbf{u}}_{i+1}) - \mathbf{f}(\bar{\mathbf{u}}_i)) + \frac{\lambda}{2}\left(\mathbf{f}\left(\mathbf{u}_{i+2}^n\right) - 2\mathbf{f}\left(\mathbf{u}_{i+1}^n\right) + \mathbf{f}\left(\mathbf{u}_i^n\right)\right). \tag{18.47b}$$

For flows with both positive and negative wave speeds, a two-step version of the Beam–Warming second-order upwind method is as follows:

♦ $$\bar{\mathbf{u}}_i = \mathbf{u}_i^n - \lambda\left(\mathbf{f}^+\left(\mathbf{u}_i^n\right) - \mathbf{f}^+\left(\mathbf{u}_{i-1}^n\right) + \mathbf{f}^-\left(\mathbf{u}_{i+1}^n\right) - \mathbf{f}^-\left(\mathbf{u}_i^n\right)\right), \tag{18.48a}$$

♦ $$\begin{aligned}\mathbf{u}_i^{n+1} = {} & \frac{1}{2}\left(\bar{\mathbf{u}}_i + \mathbf{u}_i^n\right) - \frac{\lambda}{2}\left(\mathbf{f}^+(\bar{\mathbf{u}}_i) - \mathbf{f}^+(\bar{\mathbf{u}}_{i-1})\right) - \frac{\lambda}{2}(\mathbf{f}^-(\bar{\mathbf{u}}_{i+1}) - \mathbf{f}^-(\bar{\mathbf{u}}_i)) \\ & - \frac{\lambda}{2}\left(\mathbf{f}^+\left(\mathbf{u}_i^n\right) - 2\mathbf{f}^+\left(\mathbf{u}_{i-1}^n\right) + \mathbf{f}^+\left(\mathbf{u}_{i-2}^n\right)\right) \\ & + \frac{\lambda}{2}\left(\mathbf{f}^-\left(\mathbf{u}_{i+2}^n\right) - 2\mathbf{f}^-\left(\mathbf{u}_{i+1}^n\right) + \mathbf{f}^-\left(\mathbf{u}_i^n\right)\right).\end{aligned} \tag{18.48b}$$

As an exercise, the reader can show that the two-step version, Equation (18.48), is equivalent to the one-step version, Equation (18.44), for linear systems of equations.

Method (18.48) is actually an entire class of methods, depending on the choice of the flux vector splitting. This section will illustrate the performance of the Beam–Warming second-order upwind method with Steger–Warming, Van Leer, and Zha–Bilgen flux vector splitting using the two standard test cases defined in Section 18.0. For Test Case 1, the results of the Beam–Warming second-order upwind method based on the Steger–Warming flux splitting are shown in Figure 18.15, the results of the Beam–Warming second-order upwind method based on the Van Leer flux splitting are shown in Figure 18.16, and the results of the Beam–Warming second-order upwind method based on the Zha–Bilgen flux splitting are shown in Figure 18.17. In this test case, all three versions of the Beam–Warming second-order upwind method experience large oscillations and overshoots at the head of the expansion and, especially, to the right of the shock. The Van Leer flux splitting is marginally superior to the Steger–Warming flux splitting which is, in turn, noticeably superior to the Zha–Bilgen flux splitting.

All three versions of the Beam–Warming second-order upwind method require extensive coddling and coercion to complete Test Case 2. First of all, they require constant-coefficient second-order artificial viscosity with $\epsilon = 0.05$ throughout the calculation. The second-order artificial viscosity has a three-point stencil, while the main Beam–Warming second-order upwind method has a five-point stencil. Unfortunately, the Beam–Warming second-order upwind method runs into trouble two points away from the initial jump discontinuity during the first time step, while the second-order artificial viscosity can only see one point away; since the second-order artificial viscosity sees only a constant solution one point to the right and one point to the left, the second difference is zero, and thus the artificial viscosity remains zero. In other words, second-order artificial viscosity is powerless to help smooth over the troubles caused by the initial jump discontinuity. Fourth-order artificial viscosity has a wider stencil and thus can at least affect the difficulty caused by the jump in the initial conditions. Sadly, and this is typical of fourth-order artificial viscosity at jumps, it only makes things worse. Thus, since one cannot rely on artificial

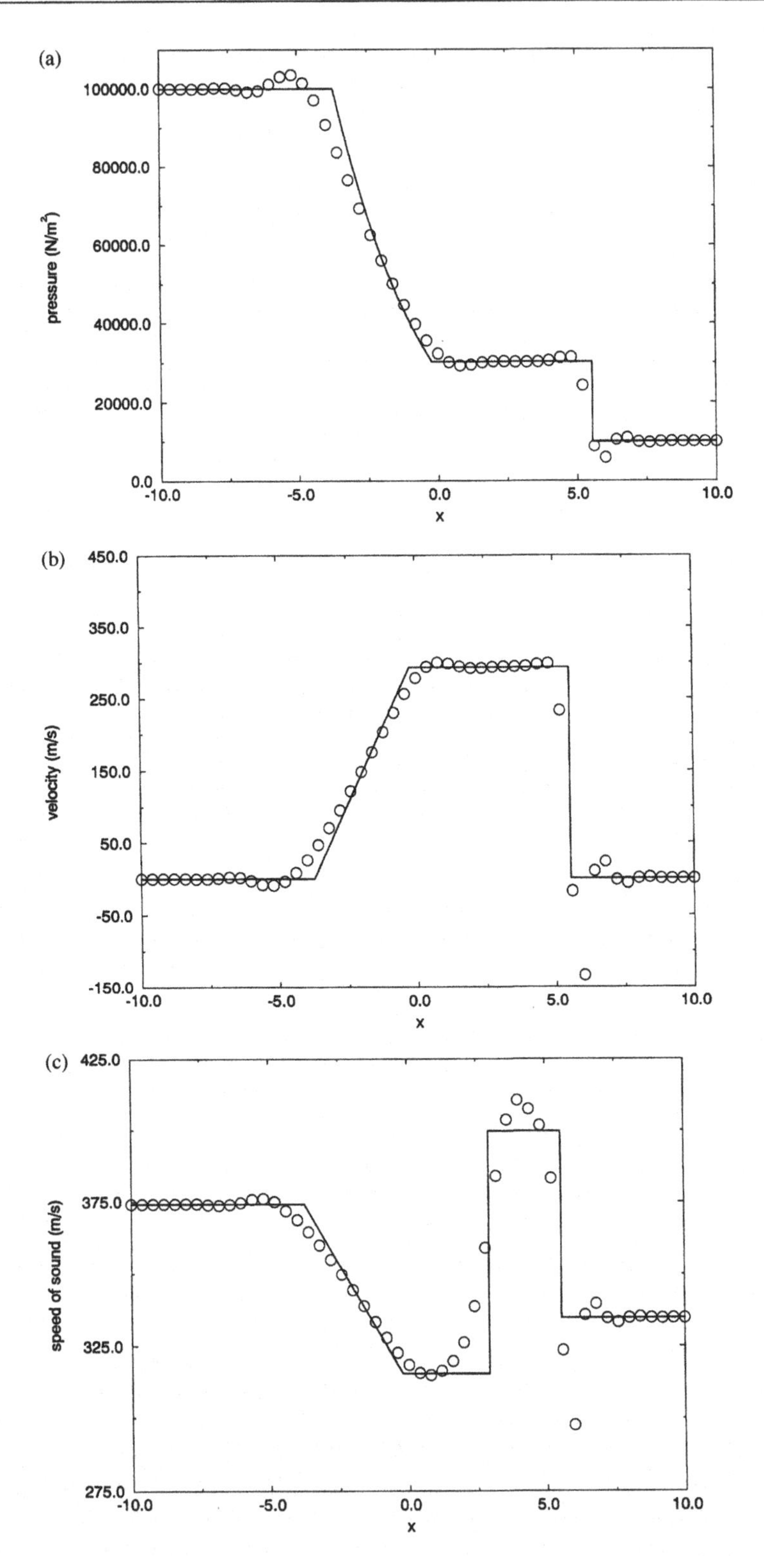

Figure 18.15 Beam–Warming second-order upwind method based on Steger–Warming flux vector splitting for Test Case 1.

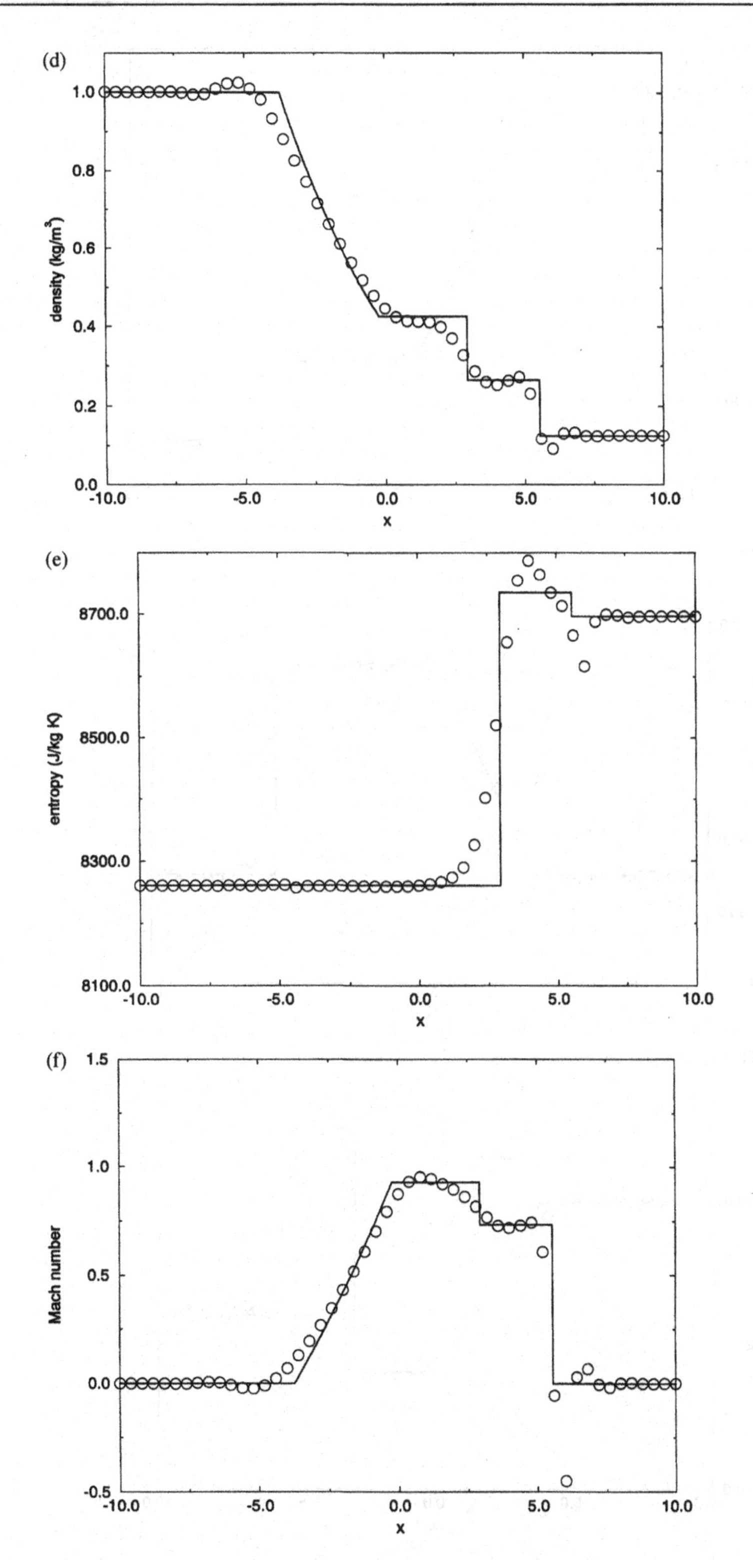

Figure 18.15 *(cont.)*

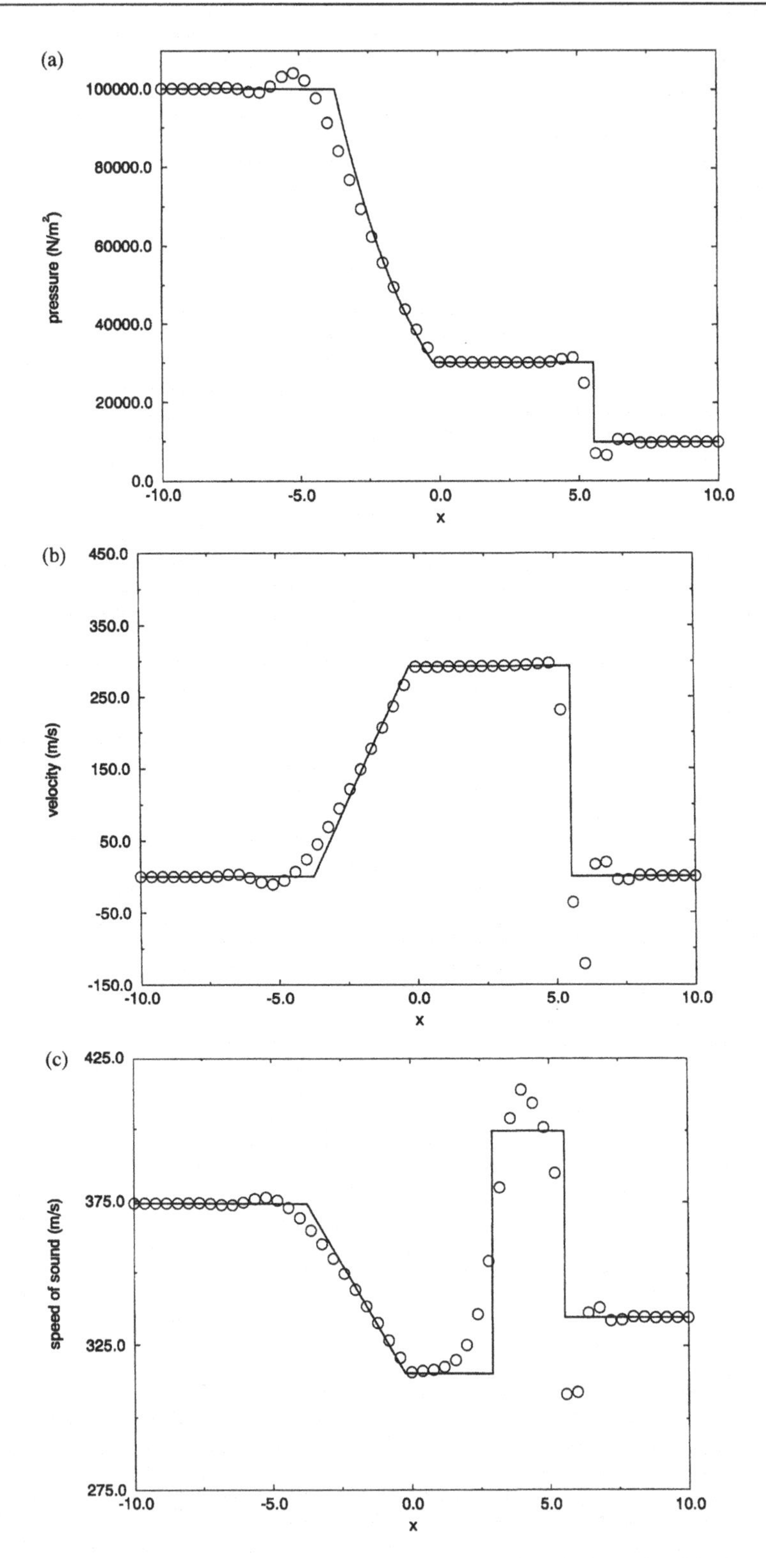

Figure 18.16 Beam–Warming second-order upwind method based on Van Leer flux vector splitting for Test Case 1.

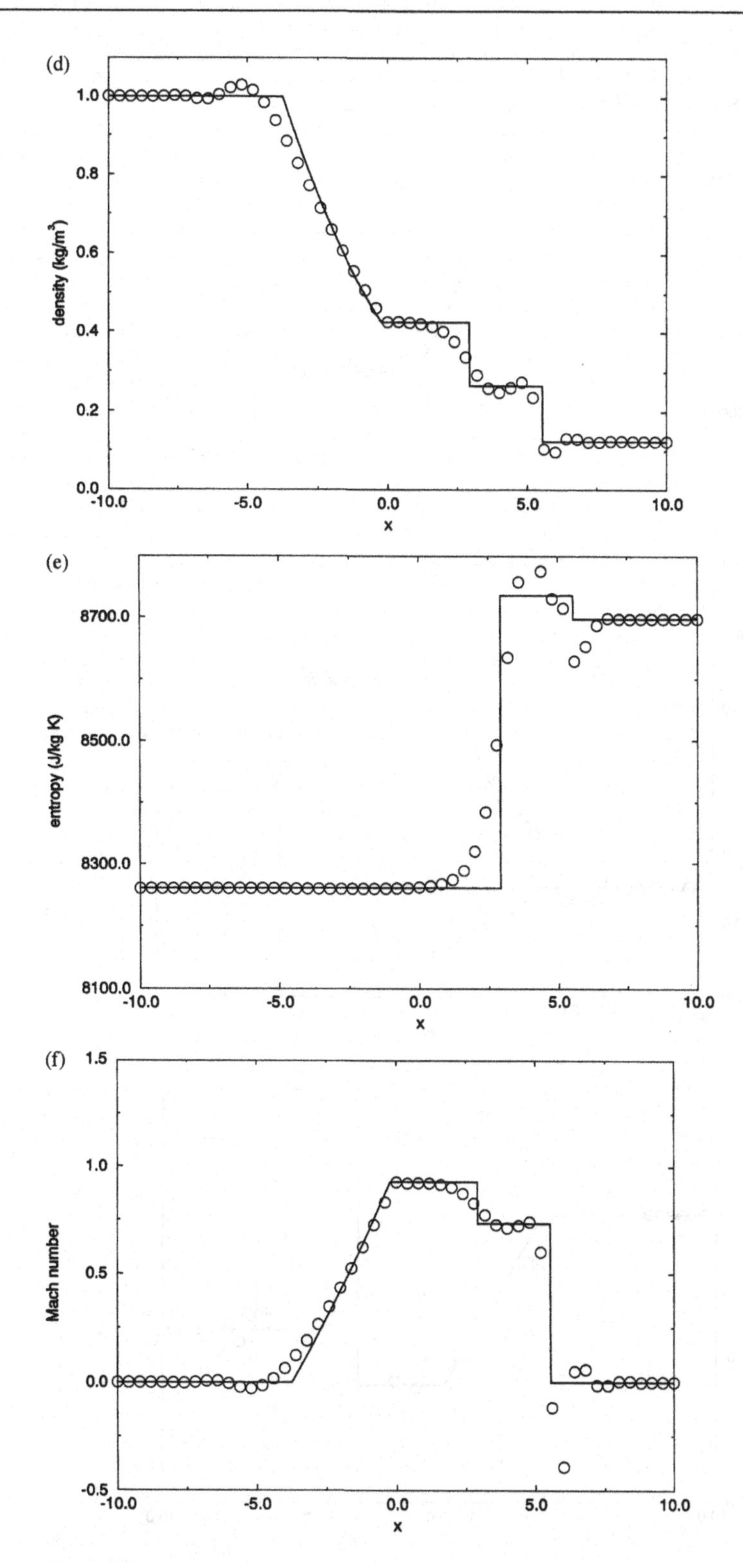

Figure 18.16 *(cont.)*

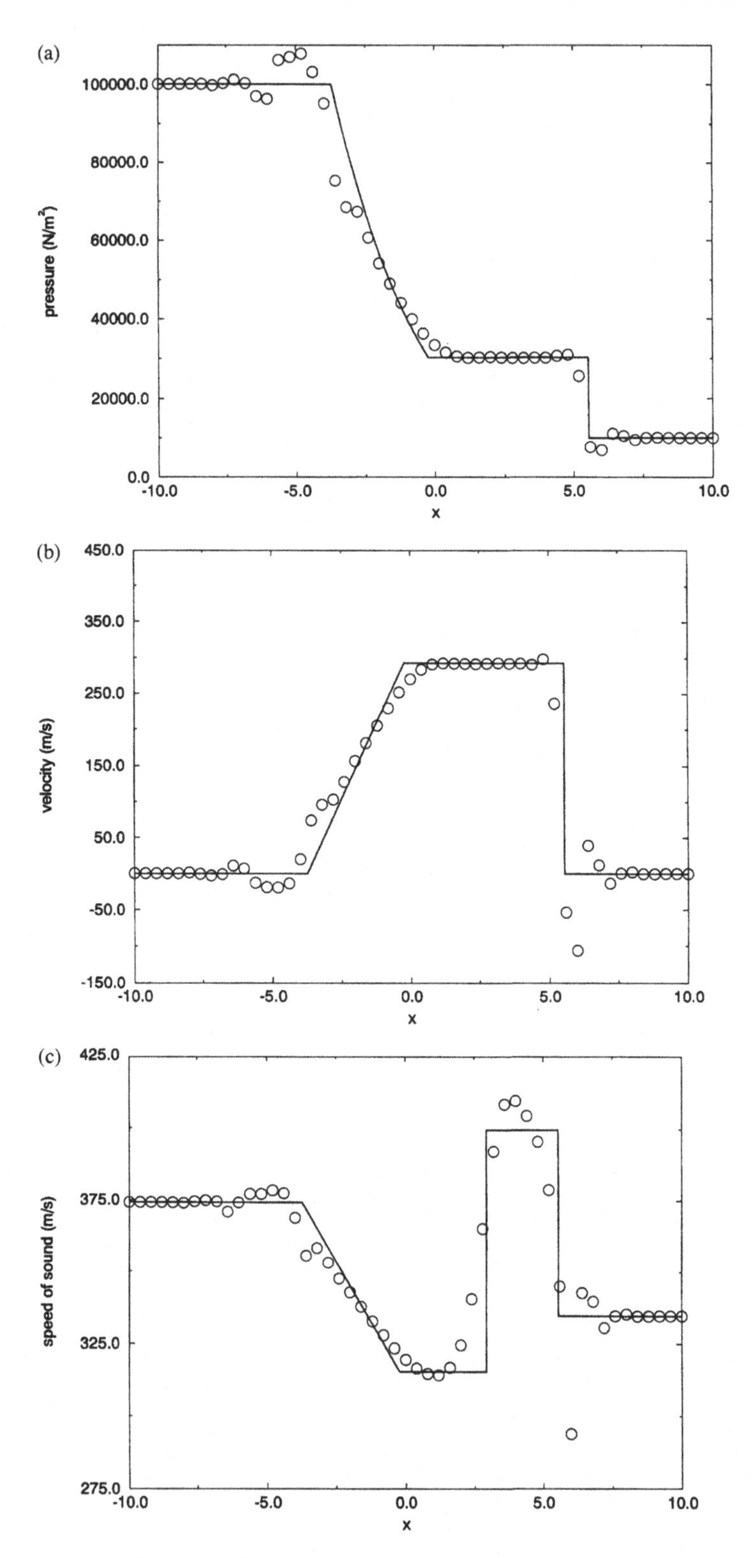

Figure 18.17 Beam–Warming second-order upwind method based on Zha–Bilgen flux vector splitting for Test Case 1.

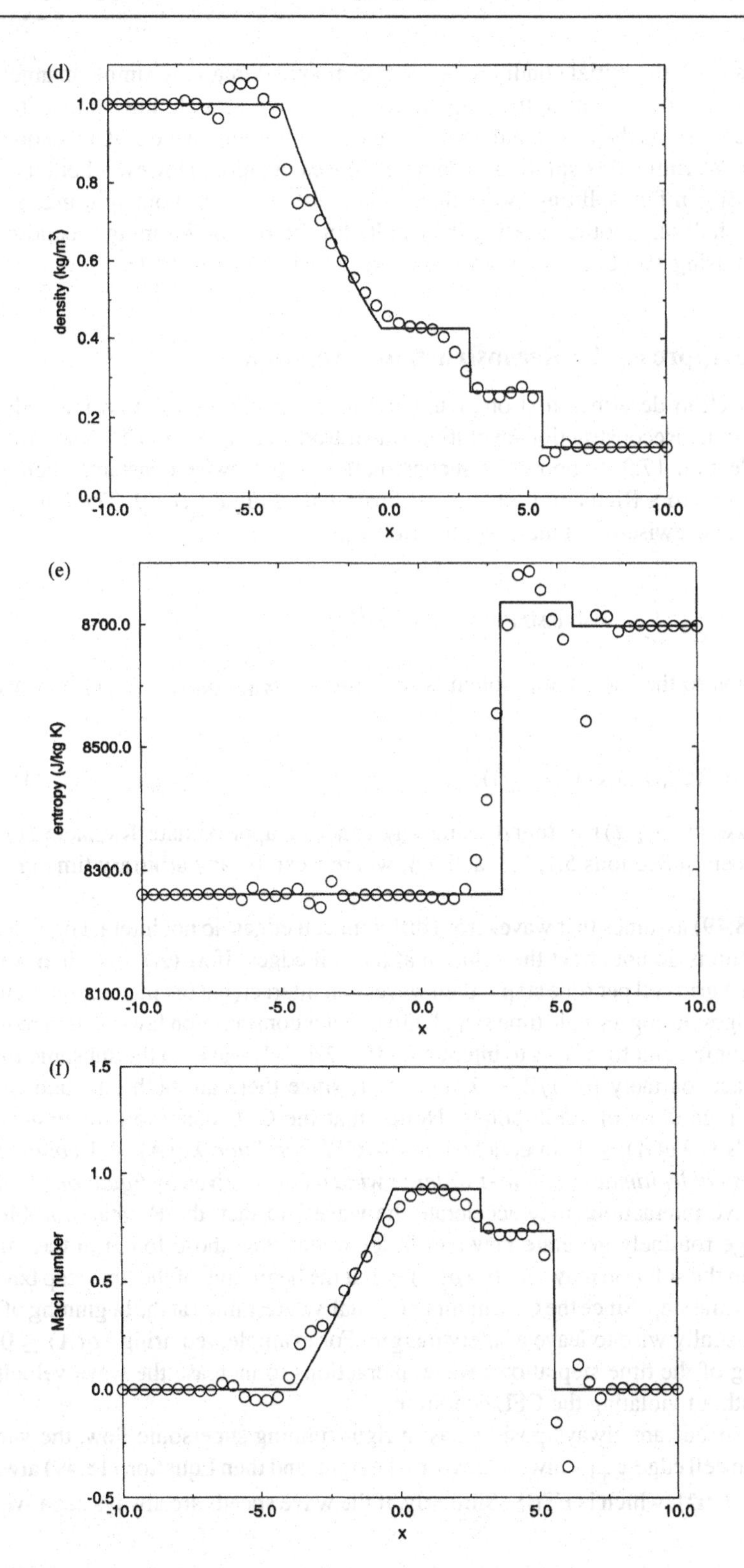

Figure 18.17 *(cont.)*

viscosity in this case, the initial conditions were presmoothed in a very simple manner, by replacing the three center points in the jump discontinuity by a straight line connecting the left and right states. Even then, artificial viscosity and presmoothing were still not enough to coax the Steger–Warming flux splitting to complete the calculation. However, both the Van Leer and Zha–Bilgen flux splittings were fine. In fact, the results of both splittings are so similar that we shall only bother plotting the results for the Beam–Warming second-order upwind method using Van Leer flux vector splitting, which is Figure 18.18.

18.3 Wave Approach II: Reconstruction–Evolution

This section describes first-order upwind reconstruction–evolution methods for the Euler equations. Reconstruction–evolution was introduced in Section 13.6 and further developed in Section 17.3. Suppose the reconstruction is piecewise-constant. Then each cell edge gives rise to a Riemann problem, as illustrated in Figure 18.19. Then the exact evolution of the piecewise-constant reconstruction yields

$$\hat{\mathbf{f}}^n_{i+1/2} = \frac{1}{\Delta t}\int_{t^n}^{t^{n+1}} \mathbf{f}(\mathbf{u}_{\text{RIEMANN}}(x_{i+1/2}, t))\,dt.$$

Since the solution to the Riemann problem is self-similar, $\mathbf{u}_{\text{RIEMANN}}(x_{i+1/2}, t)$ is constant for all time. Then

$$\blacklozenge \qquad \hat{\mathbf{f}}^n_{i+1/2} = \mathbf{f}(\mathbf{u}_{\text{RIEMANN}}(x_{i+1/2}, t)), \tag{18.49}$$

where $\mathbf{f}(\mathbf{u}_{\text{RIEMANN}}(x_{i+1/2}, t))$ is found using any exact or approximate Riemann solver, such as those seen in Sections 5.1, 5.3, and 5.4, where t can be any arbitrary time greater than t^n.

Equation (18.49) assumes that waves from different cell edges do not interact or, at least, that any interactions do not affect the solution at the cell edges. If $\lambda\rho(A) \leq 1$, then waves travel at most one grid cell per time step – then waves can interact, but the interactions cannot reach the cell edges during a single time step. Unlike scalar conservation laws, which require a compressive sonic point for waves to interact for $0 \leq \lambda a \leq 1$, waves in the subsonic Euler equations interact routinely for $1/2 < \lambda\rho(A) \leq 1$, since there are both left- and right-running waves regardless of sonic points. Notice that the CFL condition for first-order upwind methods is $\lambda\rho(A) \leq 1$. In conclusion, *the CFL condition $\lambda\rho(A) \leq 1$ completely ensures the physical legitimacy of all first-order upwind methods given by Equation (18.49).* By the way, wave interactions may accelerate the waves. In fact, the Riemann problems at each cell edge routinely generate new and faster waves than those found in the initial conditions. Then the solution may satisfy $\lambda\rho(A) \leq 1$ at the beginning of the time step but not at the end of the time step. Since the CFL number is usually determined at the beginning of the time step, it is usually wise to leave a safety margin. For example, requiring $\lambda\rho(A) \leq 0.95$ at the beginning of the time step allows wave interactions to increase the wave velocities by up to 5% without violating the CFL condition.

If the wave speeds are always positive, as in right-running supersonic flow, the waves originating from cell edge $x_{i+1/2}$ always travel to the right, and then Equation (18.49) always yields $\hat{f}^n_{i+1/2} = \mathbf{f}(\mathbf{u}^n_i)$, which is FTBS. Similarly, if the wave speeds are always negative, as

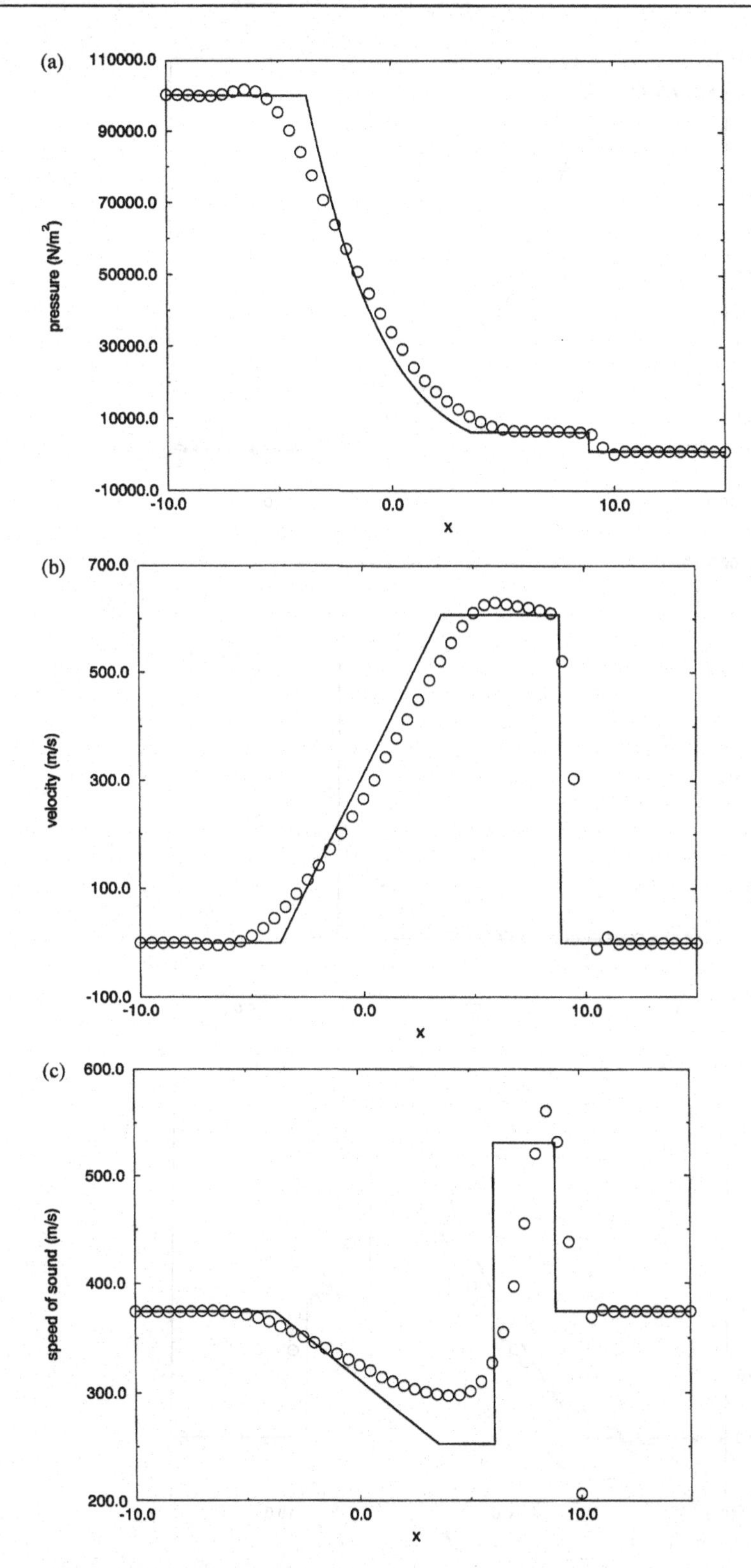

Figure 18.18 Beam–Warming second-order upwind method based on Van Leer flux vector splitting with artificial viscosity for Test Case 2.

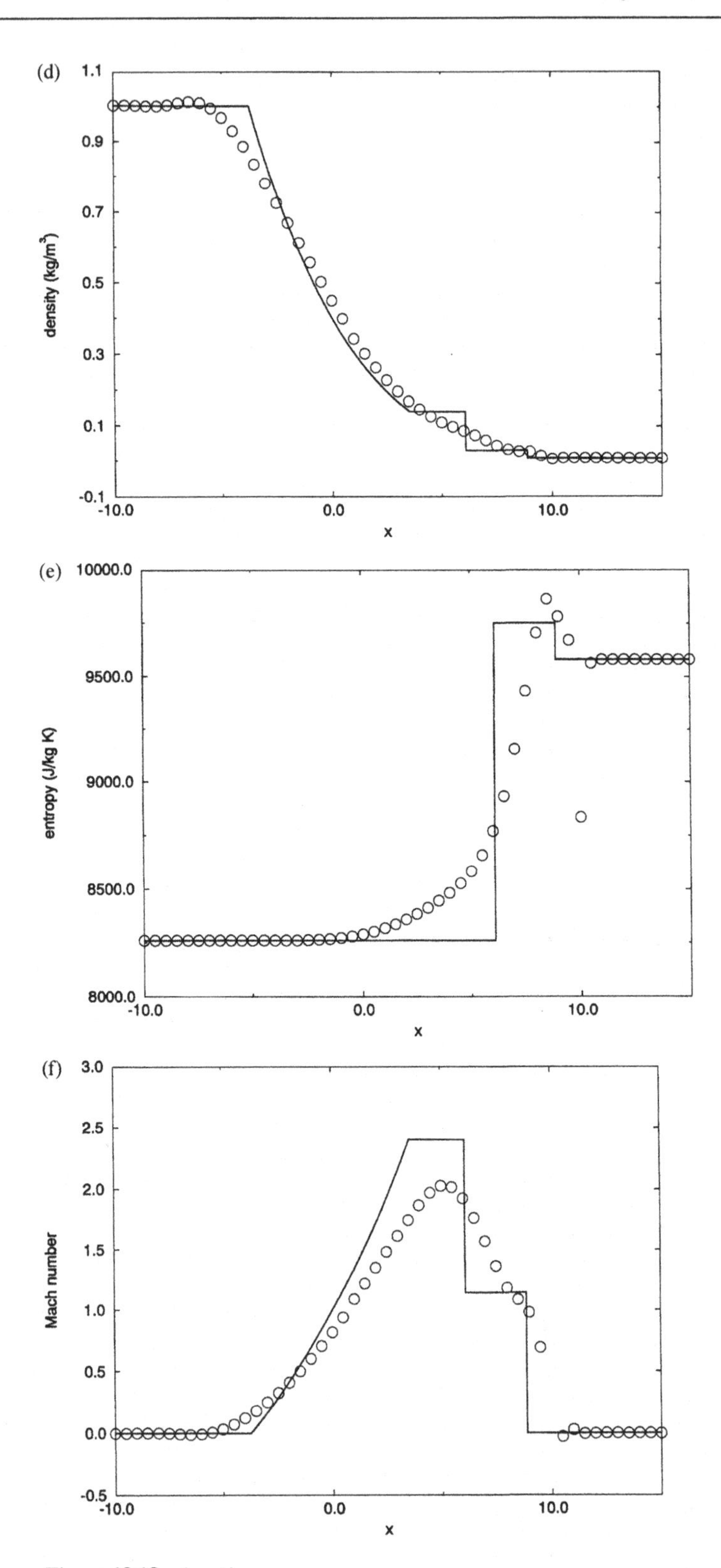

Figure 18.18 *(cont.)*

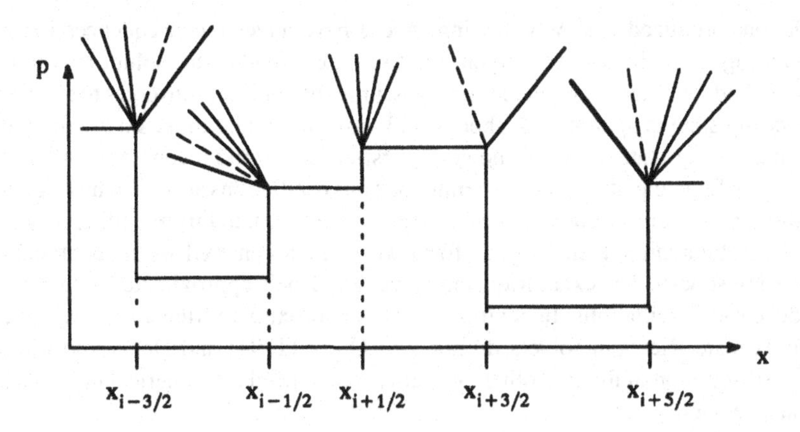

Figure 18.19 Each cell edge in a piecewise-constant reconstruction gives rise to a Riemann problem.

in left-running supersonic flow, the waves originating from cell edge $x_{i+1/2}$ always travel to the left, and then Equation (18.49) always yields $\hat{\mathbf{f}}^n_{i+1/2} = \mathbf{f}(\mathbf{u}^n_{i+1})$, which is FTFS. This is true for any reasonable real or approximate Riemann solver used in Equation (18.49). In other words, *all first-order upwind methods given by Equation (18.49) are the same – either FTBS or FTFS – for supersonic flows.* Notice that the same thing is true of the first-order upwind methods based on flux vector splitting, as seen in Subsection 18.2.5.

As usual, most of the properties of first-order upwind methods for scalar conservation laws carry over to first-order upwind methods for the Euler equations. For example, first-order upwind methods for the Euler equations have excellent shock-capturing abilities, for much the same reasons as first-order upwind methods for scalar conservation laws, as described in Section 17.3. For another example, first-order upwind methods for the Euler equations may experience difficulties at expansive sonic points, just like first-order upwind methods for scalar conservation laws. In particular, depending on the sonic point treatment, first-order upwind methods for the Euler equations may allow large expansion shocks at expansive sonic points. In short, anything that can happen in first-order upwind methods for scalar conservation laws can also happen in first-order upwind methods for the Euler equations. However, the reverse is not necessarily true – some things can happen in first-order upwind methods for the Euler equations that could never happen in first-order upwind methods for scalar conservation laws. For example, whereas the Riemann problem for scalar conservation laws is monotonicity preserving, the Riemann problem for the Euler equations is not. This lack of monotonicity preservation may undercut some of the nonlinear stability properties of the method. For example, unlike scalar first-order upwind methods, vector first-order upwind methods sometimes produce spurious oscillations, especially at steady and slowly moving shocks. These oscillations were first observed by Colella and Woodward (1984). They were further described by Ben-Artzi and Falcowitz (1984), Roberts (1990), Lin (1995), Jin and Liu (1996), and others. Since steady and slowly moving shocks span compressive sonic points, first-order upwind reconstruction–evolution methods for the Euler equations may have problems at both expansive and compressive sonic points, whereas first-order upwind reconstruction–evolution methods for scalar conservation laws only have problems at expansive sonic points.

Oscillations produced at slowly moving shocks have adverse consequences for steady shock capturing. To elaborate, a common way to capture steady-state solutions is to run an unsteady method for a large number of time steps, until the solution converges to steady state. Then, as steady state is approached, shocks will move more and more slowly until attaining their final steady-state positions; however, first-order upwind reconstruction–evolution methods for the Euler equations may generate spurious oscillations at slowly moving shocks, which substantially hamper convergence to steady state. Thus, in this regard, upwind methods based on Riemann solvers may perform worse than centered methods at steady or slowly moving shocks. The exact Riemann solver and Roe's approximate Riemann solver may produce such oscillations. In contrast, Osher's approximate Riemann solver and one-wave approximate Riemann solvers do not. See Quirk (1994) and Donat and Marquina (1996) for a discussion of this and other problems experienced by numerical methods based on Riemann solvers.

18.3.1 *Godunov's First-Order Upwind Method*

Consider the exact solution to the Riemann problem for the Euler equations, as discussed in Section 5.1. In the results of Section 5.1, replace $\mathbf{u}_L$ by $\mathbf{u}_i^n$, replace $\mathbf{u}_R$ by $\mathbf{u}_{i+1}^n$, replace $t=0$ by $t=t^n$, and replace $x=0$ by $x=x_{i+1/2}$. Then the vector version of Godunov's method can be summarized as follows:

$$\hat{\mathbf{f}}_{i+1/2}^n = \mathbf{f}(\mathbf{u}_{\mathrm{RIEMANN}}(x_{i+1/2}, t)),$$

where $\mathbf{u}_{\mathrm{RIEMANN}}(x_{i+1/2}, t)$ is the exact solution to the Riemann problem for the Euler equations. Unfortunately, because there is no explicit expression for the solution of the Riemann problem for the Euler equations, there is no explicit expression for $\hat{f}_{i+1/2}^n$. The performance of Godunov's first-order upwind method is illustrated using the two standard test cases defined in Section 18.0. The results for Test Case 1 are shown in Figure 18.20 and the results for Test Case 2 are shown in Figure 18.21.

18.3.2 *Roe's First-Order Upwind Method*

Consider Roe's approximate Riemann solver for the Euler equations, as discussed in Section 5.3. In the results of Section 5.3, replace $\mathbf{u}_L$ by $\mathbf{u}_i^n$, replace $\mathbf{u}_R$ by $\mathbf{u}_{i+1}^n$, replace A_{RL} by $A_{i+1/2}^n$, replace $t = 0$ by $t = t^n$, and replace $x = 0$ by $x = x_{i+1/2}$. Then the vector version Roe's method can be summarized as follows:

$$\hat{\mathbf{f}}_{i+1/2}^n = \mathbf{f}(\mathbf{u}_{\mathrm{RIEMANN}}(x_{i+1/2}, t)).$$

Unlike Godunov's first-order upwind method, there are some nice analytical expressions for the conservative numerical flux of Roe's first-order upwind method. Assume that $A_{i+1/2}^n$ has characteristic values $(\lambda_{i+1/2}^n)_j$, right characteristic vectors $(\mathbf{r}_{i+1/2}^n)_j$, and left characteristic vectors $(\mathbf{l}_{i+1/2}^n)_j$. Also let $(\Delta v_{i+1/2}^n)_j = (\mathbf{l}_{i+1/2}^n)_j \cdot (\mathbf{u}_{i+1}^n - \mathbf{u}_i)$. Section 5.3 provides the full details on how to compute $(\lambda_{i+1/2}^n)_j$, $(\mathbf{r}_{i+1/2}^n)_j$, $(\mathbf{l}_{i+1/2}^n)_j$, and $(\Delta v_{i+1/2}^n)_j$. By Equation (5.56),

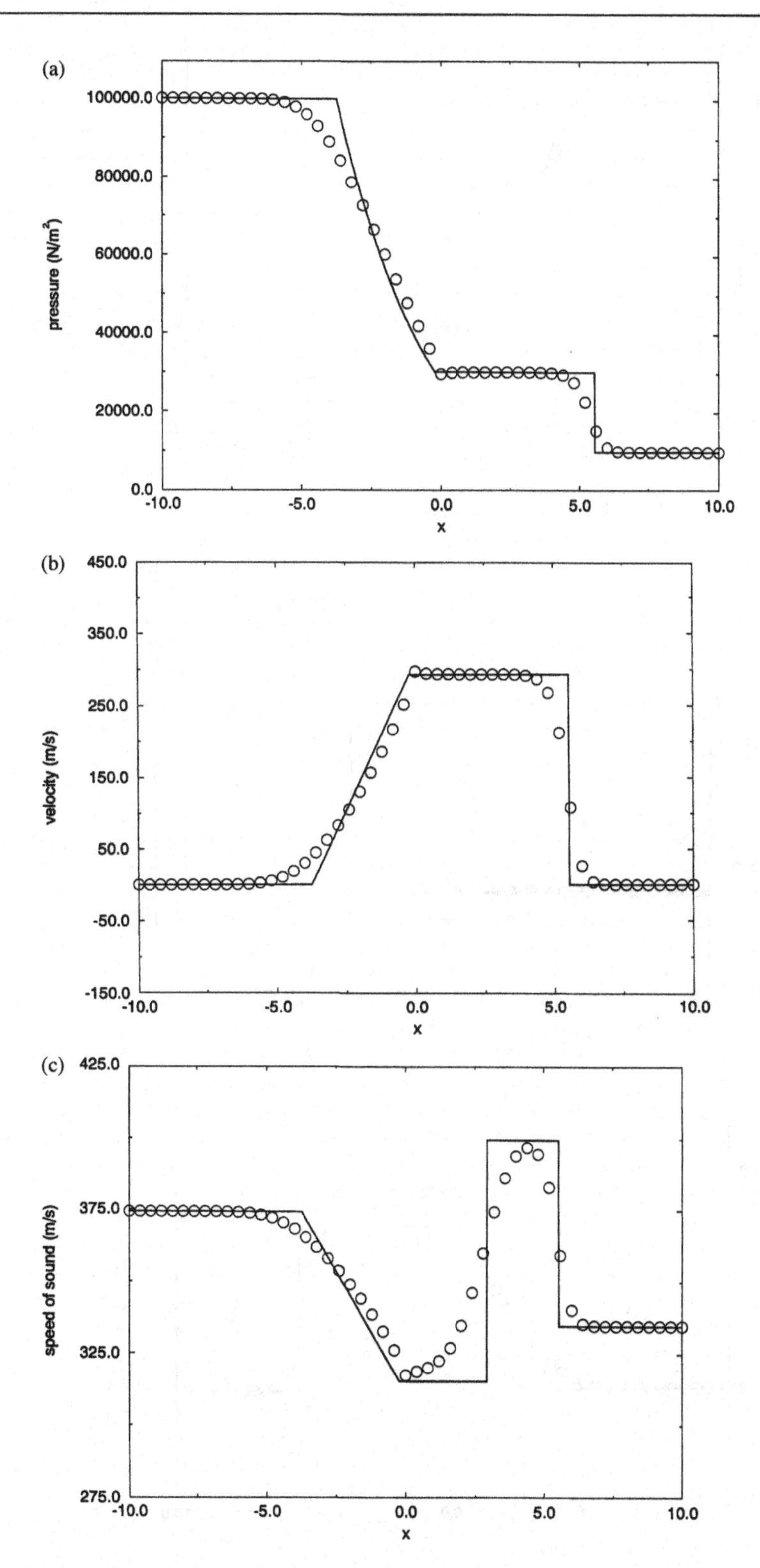

Figure 18.20 Godunov's first-order upwind method for Test Case 1.

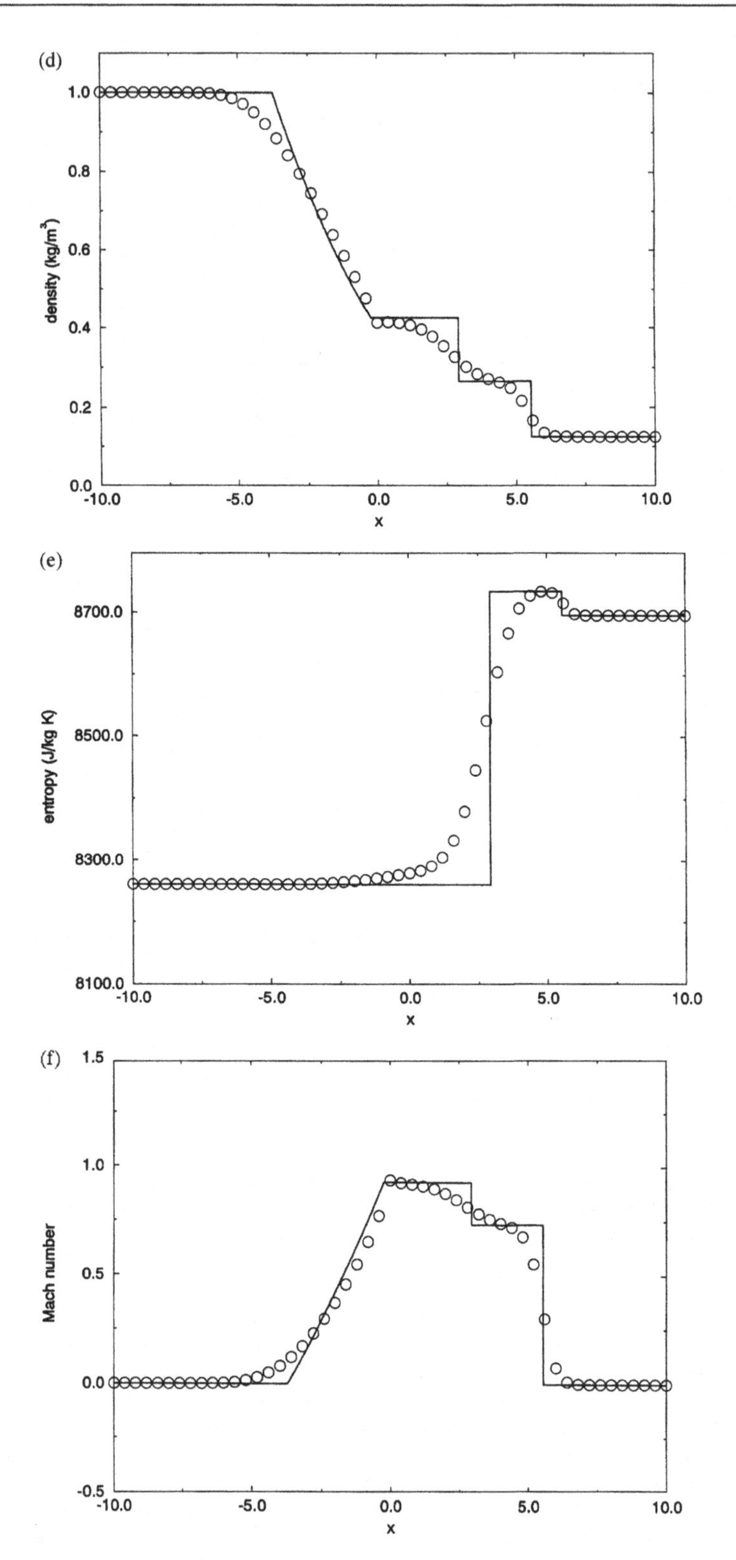

Figure 18.20 *(cont.)*

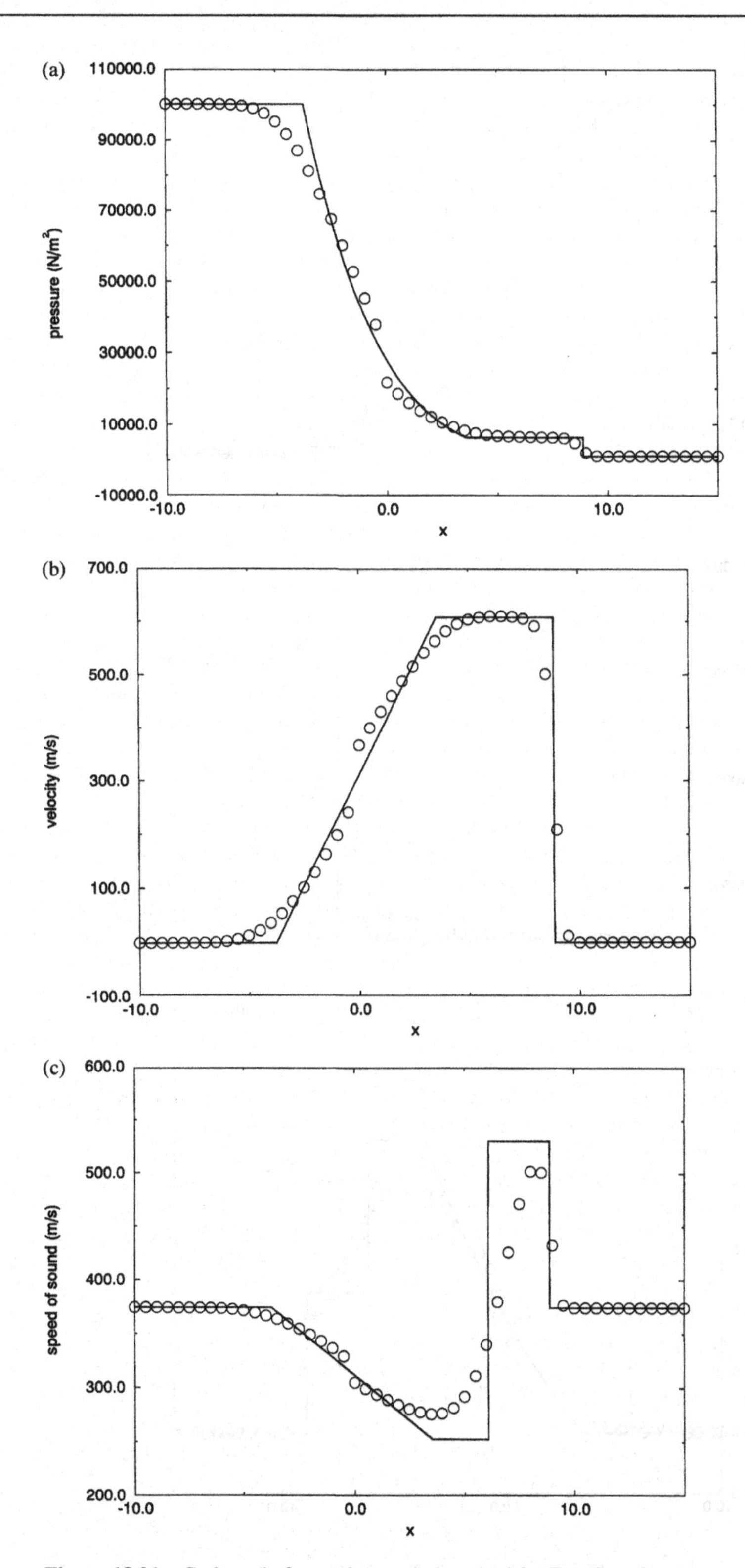

Figure 18.21 Godunov's first-order upwind method for Test Case 2.

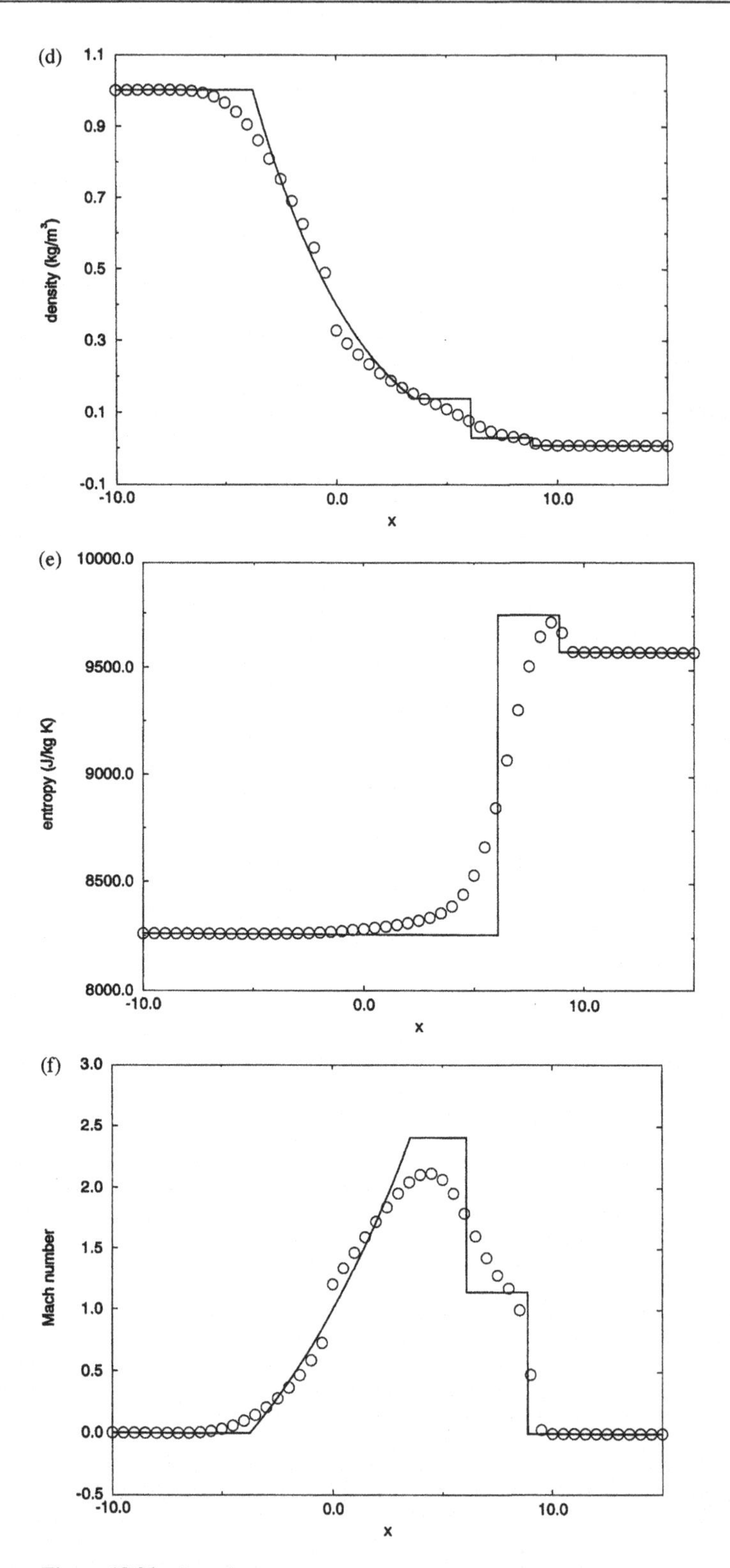

Figure 18.21 (*cont.*)

the conservative numerical flux of Roe's method can be written in three different ways:

- $$\hat{\mathbf{f}}^n_{i+1/2} = \mathbf{f}(\mathbf{u}^n_i) + \sum_{j=1}^{3} (\mathbf{r}^n_{i+1/2})_j \min\left[0, (\lambda^n_{i+1/2})_j\right] (\Delta v^n_{i+1/2})_j, \tag{18.50a}$$
- $$\hat{\mathbf{f}}^n_{i+1/2} = \mathbf{f}(\mathbf{u}^n_{i+1}) - \sum_{j=1}^{3} (\mathbf{r}^n_{i+1/2})_j \max\left[0, (\lambda^n_{i+1/2})_j\right] (\Delta v^n_{i+1/2})_j, \tag{18.50b}$$
- $$\hat{\mathbf{f}}^n_{i+1/2} = \frac{1}{2}(\mathbf{f}(\mathbf{u}^n_{i+1}) + \mathbf{f}(\mathbf{u}^n_i)) - \frac{1}{2}\sum_{j=1}^{3} (\mathbf{r}^n_{i+1/2})_j |\lambda^n_{i+1/2}|_j (\Delta v^n_{i+1/2})_j. \tag{18.50c}$$

Equivalently, Equation (5.27) yields the following matrix expressions:

- $$\hat{\mathbf{f}}^n_{i+1/2} = \mathbf{f}(\mathbf{u}^n_i) + A^-_{i+1/2}(\mathbf{u}^n_{i+1} - \mathbf{u}^n_i), \tag{18.51a}$$
- $$\hat{\mathbf{f}}^n_{i+1/2} = \mathbf{f}(\mathbf{u}^n_{i+1}) - A^+_{i+1/2}(\mathbf{u}^n_{i+1} - \mathbf{u}^n_i), \tag{18.51b}$$
- $$\hat{\mathbf{f}}^n_{i+1/2} = \frac{1}{2}(\mathbf{f}(\mathbf{u}^n_{i+1}) + \mathbf{f}(\mathbf{u}^n_i)) - \frac{1}{2}|A^n_{i+1/2}|(\mathbf{u}^n_{i+1} - \mathbf{u}^n_i), \tag{18.51c}$$

where $A^\pm_{i+1/2}$ and $|A^n_{i+1/2}|$ are defined by Equations (5.22), (5.23), and (5.24). Despite the seeming differences, the last six equations are all identical. The first three expressions are the best for computations, while the last three expressions are the best for analysis. Notice that Equation (18.51c) is FTCS plus second-order artificial viscosity, where the coefficient of artificial viscosity is $\epsilon^n_{i+1/2} = |A^n_{i+1/2}|$.

From Chapter 5, recall that the Roe-average matrix $A^n_{i+1/2}$ must satisfy

$$\mathbf{f}(\mathbf{u}^n_{i+1}) - \mathbf{f}(\mathbf{u}^n_i) = A^n_{i+1/2}(\mathbf{u}^n_{i+1} - \mathbf{u}^n_i). \tag{18.52}$$

Also recall that

$$A^+_{i+1/2} + A^-_{i+1/2} = A^n_{i+1/2}, \tag{5.25}$$

$$A^+_{i+1/2} - A^-_{i+1/2} = |A^n_{i+1/2}|. \tag{5.26}$$

Then Equation (18.52) becomes

$$\mathbf{f}(\mathbf{u}^n_{i+1}) - \mathbf{f}(\mathbf{u}^n_i) = (A^+_{i+1/2} + A^-_{i+1/2})(\mathbf{u}^n_{i+1} - \mathbf{u}^n_i). \tag{18.53}$$

Then Roe's first-order upwind method splits the flux difference $\mathbf{f}(\mathbf{u}^n_{i+1}) - \mathbf{f}(\mathbf{u}^n_i)$ into positive and negative parts $A^\pm_{i+1/2}(\mathbf{u}^n_{i+1} - \mathbf{u}^n_i)$. As a result, Roe's first-order upwind method is sometimes called a *flux difference splitting method.* In fact, there are infinitely many solutions for $A^n_{i+1/2}$ in Equation (18.52) in addition to the Roe-average matrix, as discussed in Chapter 5. For each one of these solutions, there are infinitely many matrix splittings $A^\pm_{i+1/2}$ besides the ones seen in Equations (5.22) and (5.23). In other words, there are infinitely many flux difference splitting methods. The term "flux difference splitting" stresses the parallels with flux vector splitting; unfortunately, these two terms are also easily confused, and the reader should take care to distinguish between them. To help the reader keep separate flux difference splitting and flux vector splitting, the terms "difference" and "vector" will be italicized in the following two paragraphs.

Let us compare flux *difference* splitting with flux *vector* splitting. Notice that the flux *difference* splitting method seen in Equation (18.51c) can be written as

$$\hat{\mathbf{f}}^n_{i+1/2} = \frac{1}{2}\left(f\left(\mathbf{u}^n_{i+1}\right) + f\left(\mathbf{u}^n_i\right)\right) - \frac{1}{2}\left(A^+_{i+1/2} - A^-_{i+1/2}\right)\left(\mathbf{u}^n_{i+1} - \mathbf{u}^n_i\right), \tag{18.54}$$

which is identical to the flux *vector* splitting method seen in Equation (18.40) except, of course, that the matrices $A^{\pm}_{i+1/2}$ may be defined in completely different ways. In particular, in flux *difference* splitting, the matrices $A^{\pm}_{i+1/2}$ are defined by Equations (18.53) and (5.25). By contrast, in flux vector splitting, the matrices $A^{\pm}_{i+1/2}$ are defined as follows:

$$\mathbf{f}^+\left(\mathbf{u}^n_{i+1}\right) - \mathbf{f}^+\left(\mathbf{u}^n_i\right) = A^+_{i+1/2}\left(\mathbf{u}^n_{i+1} - \mathbf{u}^n_i\right), \tag{18.41a}$$

$$\mathbf{f}^-\left(\mathbf{u}^n_{i+1}\right) - \mathbf{f}^-\left(\mathbf{u}^n_i\right) = A^-_{i+1/2}\left(\mathbf{u}^n_{i+1} - \mathbf{u}^n_i\right). \tag{18.41b}$$

Notice that all of the solutions of the flux *vector* splitting equation (18.41) are also solutions of the flux *difference* splitting equation (18.53) assuming that $A^+_{i+1/2} + A^-_{i+1/2} = A^n_{i+1/2}$. To see this, simply add Equations (18.41a) and (18.41b) to obtain Equation (18.53), using the fact that $\mathbf{f}^+ + \mathbf{f}^- = \mathbf{f}$. Unfortunately, the reverse is not true – the solutions of the flux *difference* splitting equation (18.53) are not necessarily solutions of the flux vector splitting equation (18.41). Thus, given $A^{\pm}_{i+1/2}$ from Equation (18.53), there may or may not be any flux functions $\mathbf{f}^{\pm}$ that satisfy Equation (18.41). In particular, there is no flux *vector* splitting method corresponding to Roe's flux *difference* splitting method. In conclusion, for the first-order upwind methods described in this chapter, *flux vector splitting is a subset of flux difference splitting*. All first-order upwind flux *vector* splitting methods can be considered as first-order upwind flux *difference* splitting methods but not vice versa, and, in particular, Roe's first-order upwind method is a flux *difference* splitting method but not a flux *vector* splitting method.

Flux *vector* splitting methods are naturally more efficient than flux *difference* splitting methods, since they involve only vectors rather than matrices. Although flux *vector* splitting methods can be written in matrix forms, such as Equation (18.41), they are most naturally written in purely vector forms. On the other hand, in general, flux *difference* splitting methods naturally involve matrices, or at least the sort of sums required in matrix–vector products such as those seen as in Equation (18.50). Since flux *vector* splitting uses vectors while flux *difference* splitting uses matrices, flux *vector* splitting is naturally less costly than flux *difference* splitting, much like MacCormack's method is less costly than the original Lax–Wendroff method, as discussed in Section 18.1.2. Of course, neither flux *vector* splitting nor flux *difference* splitting is any where near as cheap as "flux approach" methods such as Lax–Friedrichs and Lax–Wendroff methods, except possibly for very simple flux *vector* splitting methods such as those based on Zha–Bilgen splitting.

Roe's first-order upwind method yields results almost identical to those of Godunov's first-order upwind method on the standard test cases defined in Section 18.0. This is a bit surprising, given the often huge differences between the solutions yielded by Roe's approximate Riemann solver and the exact Riemann solver, as seen in Chapter 5. Once again, large nonphysical deviations can have little effect, or even positive effects, on numerical results. Since the results are so similar to Godunov's method, we shall not bother to plot them, and the reader is referred instead to Figures 18.20 and 18.21.

18.3.3 *Harten's First-Order Upwind Method*

Subsection 17.3.3 derived Harten's first-order upwind method for scalar conservation laws using a locally quadratic approximation to the flux function. Alternatively, Harten's first-order upwind method for scalar conservation laws may be formally derived by replacing $|a^n_{i+1/2}|$ with $\psi(a^n_{i+1/2})$ in Roe's first-order upwind method for scalar conservation laws, where

$$\blacklozenge \qquad \psi(x) = \begin{cases} \dfrac{x^2+\delta^2}{2\delta} & |x| < \delta, \\ |x| & |x| \geq \delta, \end{cases} \tag{18.55}$$

assuming that $\delta^n_{i+1/2} = \delta = const.$ Similarly, Harten's first-order upwind method for the Euler equations is formally derived by replacing $|\lambda^n_{i+1/2}|$ by $\psi(\lambda^n_{i+1/2})$ in Equation (18.50c). Thus Harten's first-order upwind method for the Euler equations is

$$\blacklozenge \qquad \hat{\mathbf{f}}^n_{i+1/2} = \frac{1}{2}\left(\mathbf{f}(\mathbf{u}^n_{i+1}) + \mathbf{f}(\mathbf{u}^n_i)\right) - \frac{1}{2}\sum_{j=1}^{3}\left(\mathbf{r}^n_{i+1/2}\right)_j \psi\left(\lambda^n_{i+1/2}\right)_j \left(\Delta v^n_{i+1/2}\right)_j, \tag{18.56}$$

where $(\lambda^n_{i+1/2})_j$, $(\mathbf{r}^n_{i+1/2})_j$, and $(\Delta v^n_{i+1/2})_j$ are exactly the same as in Roe's first-order upwind method for the Euler equations. Harten's first-order upwind method is also called Harten's *entropy fixed* version of Roe's first-order upwind method. Like Roe's first-order upwind method, Harten's first-order upwind method is a flux difference splitting method, as described in the last subsection. The performance of Harten's first-order upwind method is illustrated using the two standard test cases defined in Section 18.0. The results for Test Case 1 are essentially identical to those of Godunov's and Roe's first-order upwind methods; see Figure 18.20. The results for Test Case 2 are shown in Figure 18.22. After some trial and error, the value $\delta = 200$ m/s was found to be the minimum value which adequately eliminated the expansion shock seen in Godunov's and Roe's first-order upwind methods. Although 200 may seem like a large number, remember that terms like "large" and "small" are relative, and 200 is a moderately small number compared with the wave speeds and other flow properties found in Test Case 2; for example, the wave speeds in the exact solution run well over 1,000 m/s. Notice that Harten's entropy fix only affects the sonic point in Test Case 2, while the rest of the solution is left essentially untouched. Overall, Harten's first-order upwind method delivers the best results of any method in this chapter.

18.3.4 *First-Order Upwind Method Based on One-Wave Solver*

Although cheaper than Godunov's first-order upwind method, Roe's and Harten's first-order upwind methods are still relatively expensive. So consider the linear one-wave solver discussed in Section 5.4. In the results of Section 5.4, replace $\mathbf{u}_L$ by $\mathbf{u}^n_i$, replace $\mathbf{u}_R$ by $\mathbf{u}^n_{i+1}$, replace A_{RL} by $A^n_{i+1/2}$, replace $t = 0$ by $t = t^n$, and replace $x = 0$ by $x = x_{i+1/2}$. In artificial viscosity form, the resulting method is as follows:

$$\blacklozenge \qquad \hat{\mathbf{f}}^n_{i+1/2} = \frac{1}{2}\left(\mathbf{f}(\mathbf{u}^n_{i+1}) + \mathbf{f}(\mathbf{u}^n_i)\right) - \frac{1}{2}\rho\left(A^n_{i+1/2}\right)\left(\mathbf{u}^n_{i+1} - \mathbf{u}^n_i\right), \tag{18.57}$$

where $A^n_{i+1/2}$ is any average Jacobian matrix and $\rho(A^n_{i+1/2})$ is the spectral radius of $A^n_{i+1/2}$. The spectral radius is the largest characteristic value of $A^n_{i+1/2}$ in absolute value (i.e., the larger of $u^n_{i+1/2} + a^n_{i+1/2}$ and $u^n_{i+1/2} - a^n_{i+1/2}$ in absolute value). Comparing Equation (18.57)

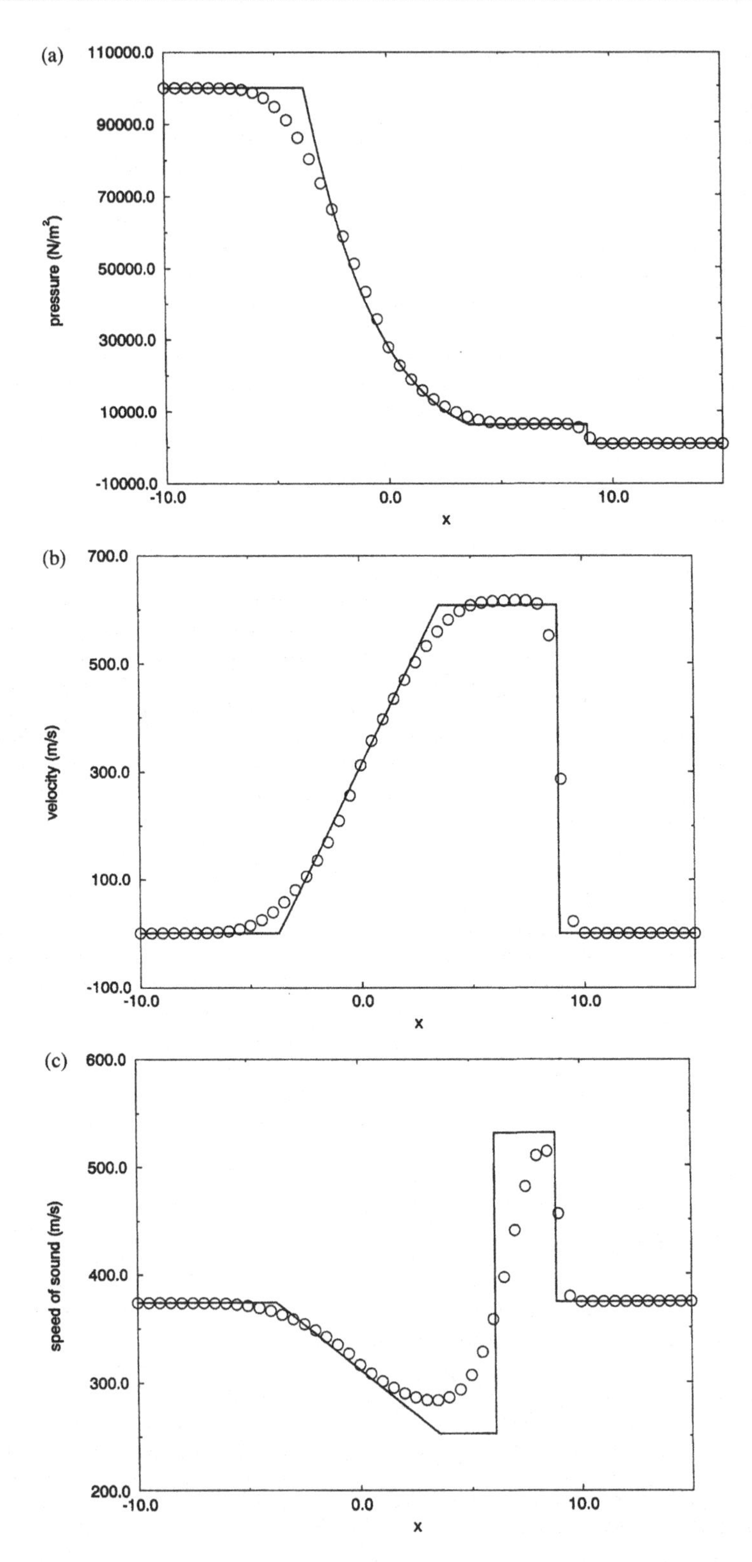

Figure 18.22 Harten's first-order upwind method for Test Case 2.

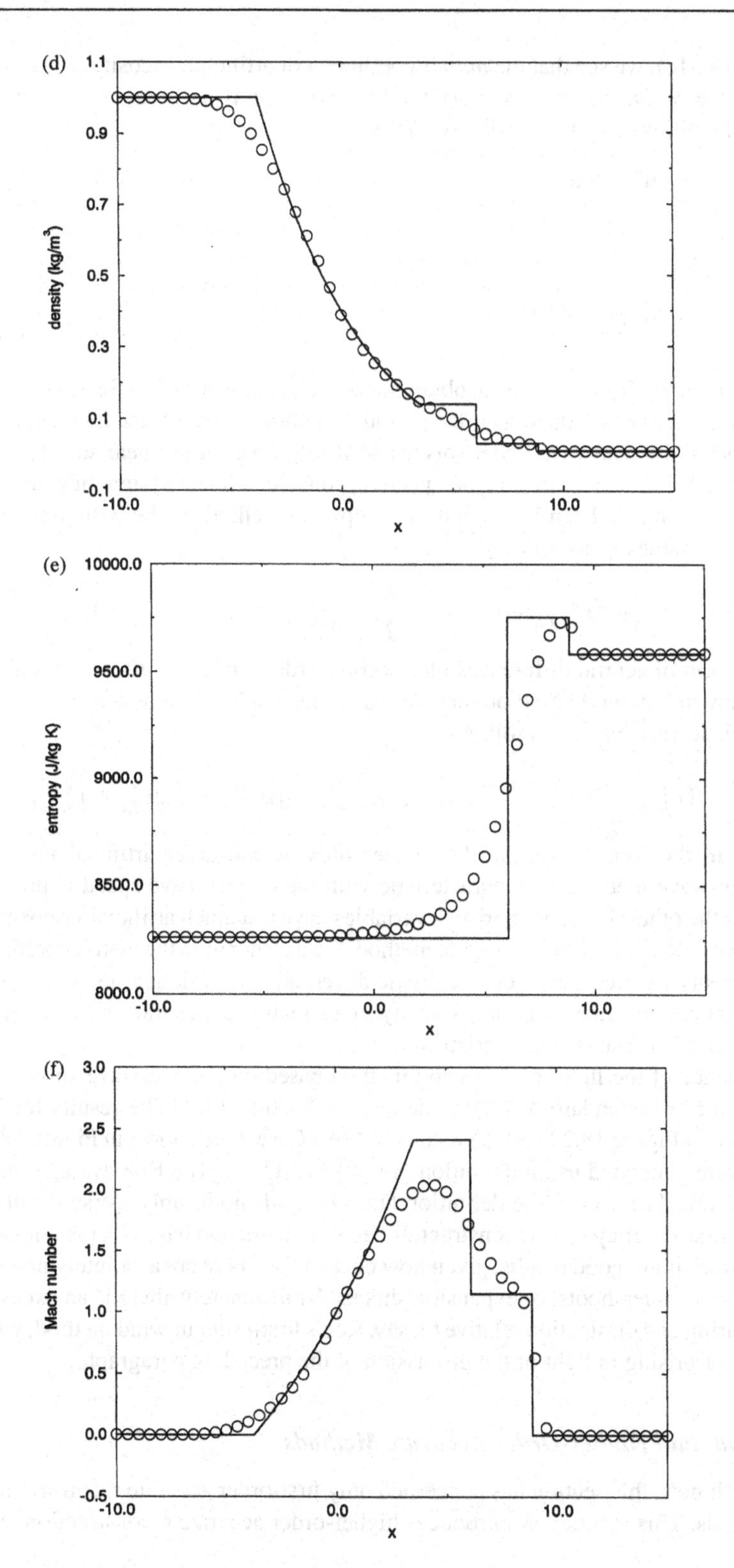

Figure 18.22 *(cont.)*

with Equation (18.51c), we see that the *matrix* coefficient of artificial viscosity $|A^n_{i+1/2}|$ has been traded for the *scalar* coefficient of artificial viscosity $\rho(A^n_{i+1/2})$. As usual, there are many possible definitions for $A^n_{i+1/2}$. For example,

$$A^n_{i+1/2} = A\left(\frac{\mathbf{u}^n_{i+1} + \mathbf{u}^n_i}{2}\right). \tag{18.58}$$

For another example,

$$A^n_{i+1/2} = \frac{A\left(\mathbf{u}^n_{i+1}\right) + A\left(\mathbf{u}^n_i\right)}{2}. \tag{18.59}$$

A third possibility is the Roe-average Jacobian matrix $A^n_{i+1/2}$ described in Section 5.3.

The differences between Equation (18.57) and Equation (18.51c) are much like the differences between the one-wave linear solver and Roe's three-wave linear solver, as described in Chapter 5. For one interesting perspective, consider a linear system of equations, as described in Section 5.2. Then Roe's first-order upwind method can be written in terms of characteristic variables v_j as follows:

$$v^{n+1}_{j,i} = v^n_{j,i} - \frac{\lambda}{2}a_j\left(v^n_{j,i+1} - v^n_{j,i-1}\right) + \frac{\lambda}{2}|a_j|\left(v^n_{j,i+1} - 2v^n_{j,i} + v^n_{j,i-1}\right),$$

which is in the form of central differences plus second-order artificial viscosity. Similarly, the first-order upwind method based on the one-wave linear solver can be written in terms of the characteristic variables v_j as follows:

$$v^{n+1}_{j,i} = v^n_{j,i} - \frac{\lambda}{2}a_j\left(v^n_{j,i+1} - v^n_{j,i-1}\right) + \frac{\lambda}{2}\max(|a_1|, |a_2|, |a_3|)\left(v^n_{j,i+1} - 2v^n_{j,i} + v^n_{j,i-1}\right),$$

which is again in the form of central differences plus second-order artificial viscosity. Then, in the one-wave method, the characteristic with the largest wave speed is properly treated, whereas the other two characteristics variables have too much artificial viscosity, as measured relative to Roe's first-order upwind method. Put another way, the matrix coefficient of artificial viscosity $|A|$ treats each characteristic differently, according to its wave speed, whereas the scalar coefficient of artificial viscosity $\rho(A)$ treats every characteristic the same, like the "worst case" or fastest characteristic.

The performance of the first-order upwind method based on the one-wave solver is illustrated using the two standard test cases defined in Section 18.0. The results for Test Case 1 are shown in Figure 18.23 and the results for Test Case 2 are shown in Figure 18.24. These results were generated using definition (18.58) for $A^n_{i+1/2}$. The Roe-average matrix yields nearly identical results, while definition (18.59) yields noticeably worse results. In some ways, the first-order upwind reconstruction–evolution method based on the one-wave solver yields surprisingly good results, given how cheap it is. There are absolutely no oscillations, overshoots, undershoots, or expansion shocks. Unfortunately, there is an excessive amount of smearing and dissipation relative to, say, Roe's first-order upwind method, which is certainly not surprising in light of the discussion of the preceding paragraph.

18.3.5 *Second- and Higher-Order Accurate Methods*

Up until now, this section has concerned only first-order accurate reconstruction–evolution methods. This subsection introduces higher-order accurate reconstruction–evol-

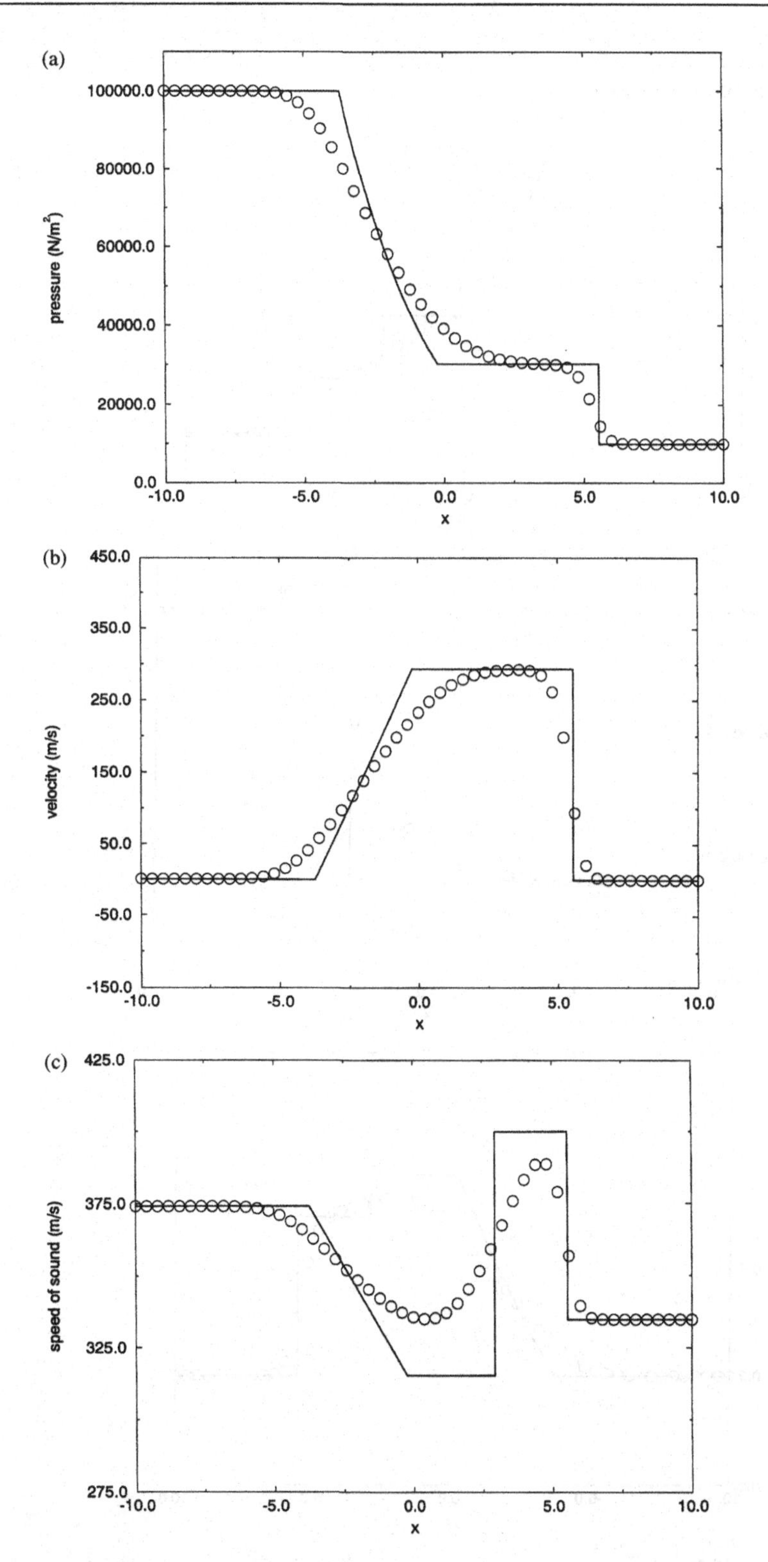

Figure 18.23 First-order upwind reconstruction–evolution method based on the one-wave linear Riemann solver for Test Case 1.

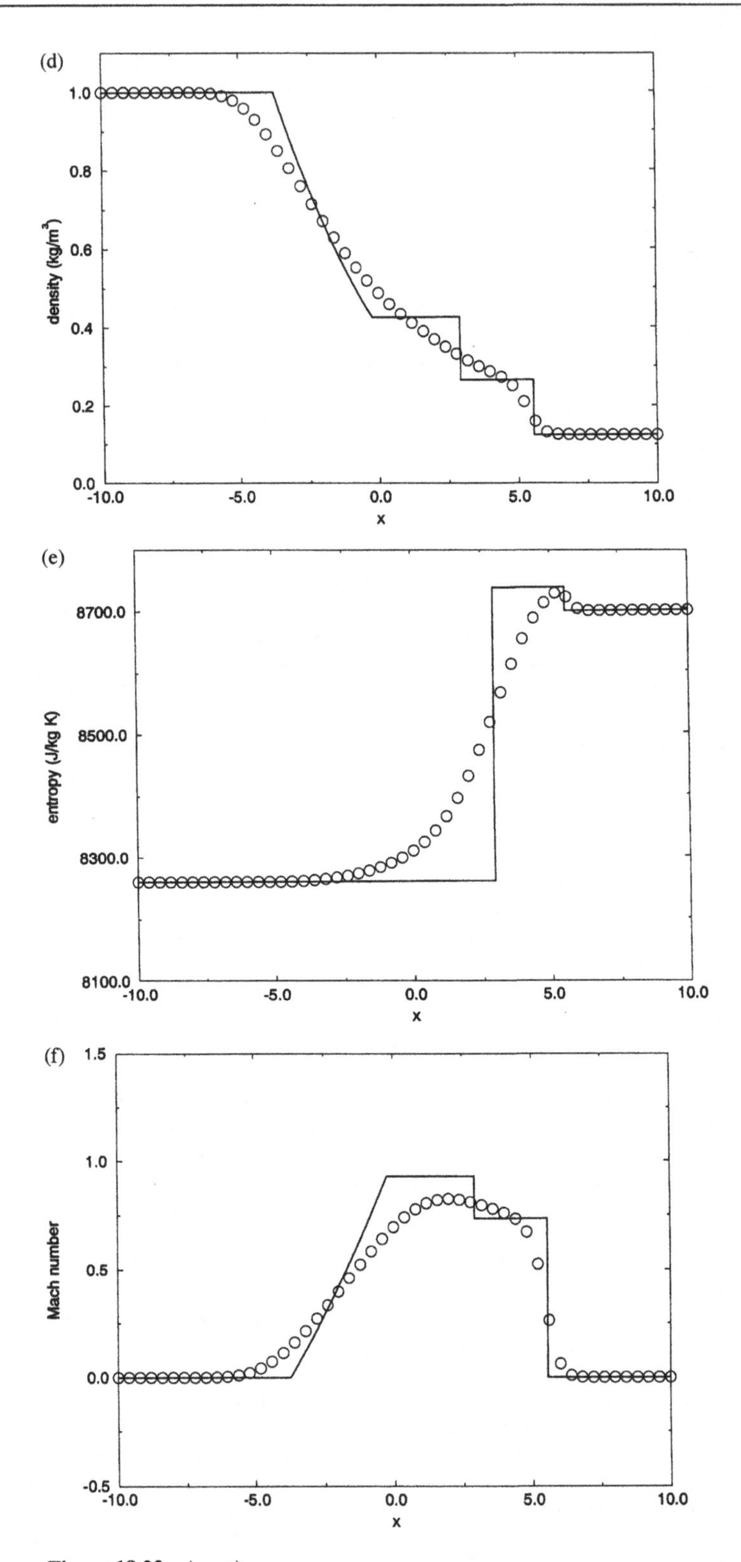

Figure 18.23 *(cont.)*

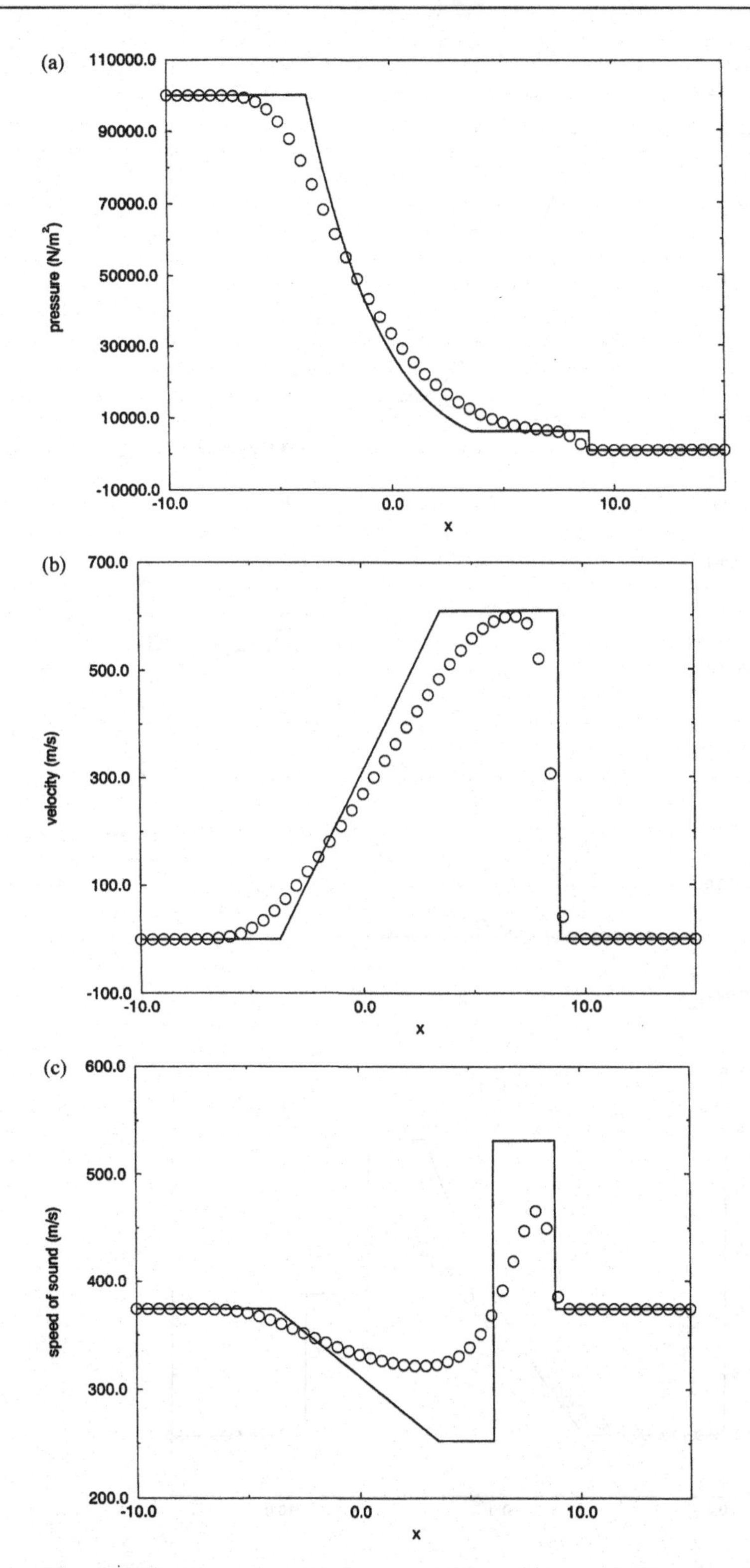

Figure 18.24 First-order upwind reconstruction–evolution method based on the one-wave linear Riemann solver for Test Case 1.

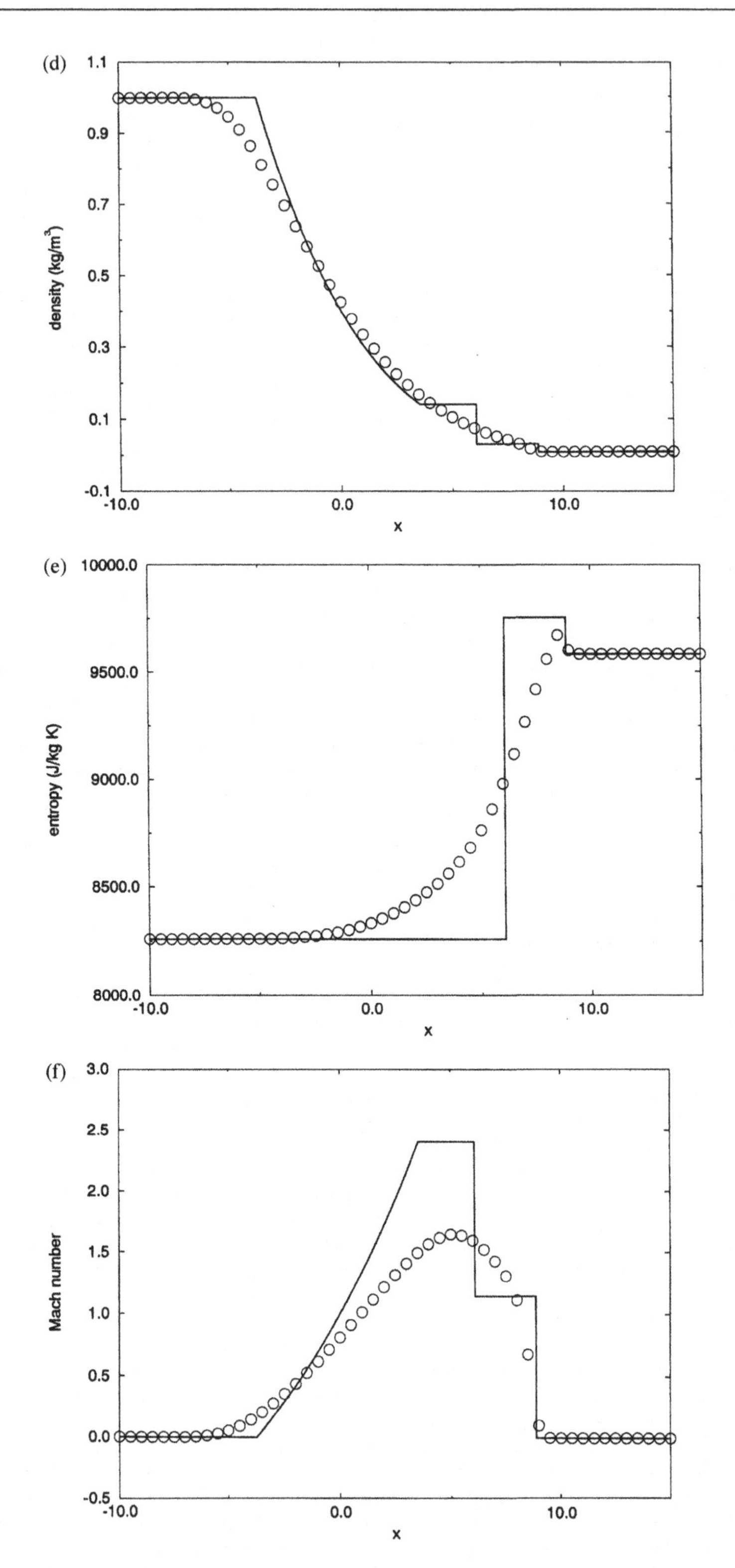

Figure 18.24 (*cont.*)

ution methods. This section is only an introduction – specific methods will have to wait until Part V, and especially Chapter 23. To begin with, consider higher-order accurate methods based on the exact Riemann solver, that is, higher-order accurate versions of Godunov's method. The spatial reconstruction is relatively straightforward: Use any piecewise-linear, piecewise-quadratic, or higher-order piecewise-polynomial reconstruction, such as those discussed in Chapter 9. Now consider the temporal evolution. Recall that

$$\hat{\mathbf{f}}^n_{i+1/2} \approx \frac{1}{\Delta t}\int_{t^n}^{t^{n+1}} \mathbf{f}(\mathbf{u}(x_{i+1/2}, t))\, dt.$$

As the first step, approximate the time-integral average using any of the numerical integration formulae seen in Section 10.2. For example,

$$\frac{1}{\Delta t}\int_{t^n}^{t^{n+1}} \mathbf{f}(\mathbf{u}(x_{i+1/2}, t))\, dt = \mathbf{f}(\mathbf{u}(x_{i+1/2}, t^{n+1/2})) + O(\Delta t^2).$$

For another example,

$$\frac{1}{\Delta t}\int_{t^n}^{t^{n+1}} \mathbf{f}(\mathbf{u}(x_{i+1/2}, t))\, dt = \frac{1}{2}\mathbf{f}(\mathbf{u}(x_{i+1/2}, t^{n+1})) + \frac{1}{2}\mathbf{f}(\mathbf{u}(x_{i+1/2}, t^n)) + O(\Delta t^2).$$

As the second step, find a Taylor series about $t = t^n$. For example,

$$\mathbf{f}(\mathbf{u}(x_{i+1/2}, t)) = \mathbf{f}(\mathbf{u}(x_{i+1/2}, t^n)) + \frac{\partial \mathbf{f}}{\partial t}(\mathbf{u}(x_{i+1/2}, t^n))(t - t^n) + O(t - t^n)^2.$$

As the third step, exchange time derivatives in the Taylor series for space derivatives using Cauchy–Kowalewski. For example, by conservation of momentum, the time derivative of the momentum flux can be written as

$$\frac{\partial}{\partial t}(\rho u) = -\rho u \frac{\partial u}{\partial x} + u \frac{\partial}{\partial x}(\rho u) - \frac{\partial p}{\partial x}.$$

As the fourth step, differentiate the spatial reconstruction at time level n to approximate the spatial derivatives at $(x_{i+1/2}, t^n)$. In general, the reconstruction and/or its derivatives contain jump discontinuities at the cell edges $x = x_{i+1/2}$. Let $\mathbf{u}_{i+1/2,L}(t)$ and $\mathbf{u}_{i+1/2,R}(t)$ be the left- and right-hand limits of the approximation to $u(x_{i+1/2}, t)$ found in the preceding four steps. As the fifth and final step, average $\mathbf{u}_{i+1/2,L}(t)$ and $\mathbf{u}_{i+1/2,R}(t)$. In particular, use $\mathbf{u}_{i+1/2,L}(t)$ and $\mathbf{u}_{i+1/2,R}(t)$ as the left- and right-hand states, respectively, in the exact solution to the Riemann problem. Although other averages could be used besides the Riemann solver, only the Riemann solver average yields the exact solution in the case of piecewise-constant reconstructions. Because this procedure is based on Taylor series and differential forms of the Euler equations, it does not apply at shocks. In fact, at shocks, it is absolutely vital to eliminate all of the higher-order terms and return to first-order piecewise-constant reconstruction, since the higher-order terms computed using the preceding procedure will actually decrease rather than increase accuracy. For specific examples of this technique, see Chapter 23 and especially Section 23.5.

Now consider higher-order accurate methods based on locally linearized approximate Riemann solvers such as, for example, Roe's approximate Riemann solver or the linearized one-wave solver. To help introduce this idea, consider a linear system of equations:

$$\frac{\partial \mathbf{u}}{\partial t} + A\frac{\partial \mathbf{u}}{\partial x} = 0,$$

where the matrix A is constant. Assume that $A = Q\Lambda Q^{-1}$, where Q is a matrix whose columns are right characteristic vectors, Q^{-1} is a matrix whose rows are left characteristic vectors, and Λ is a diagonal matrix of characteristic values. Define the characteristic variables as $\mathbf{v} = Q^{-1}\mathbf{u}$. In characteristic form, the linearized governing equations can be written as

$$\frac{\partial \mathbf{v}}{\partial t} + \Lambda \frac{\partial \mathbf{v}}{\partial x} = 0.$$

This is a set of independent linear advection equations. Each equation in the system has the following form:

$$\frac{\partial v_j}{\partial t} + \lambda_j \frac{\partial v_j}{\partial x} = 0.$$

Consider any conservative numerical approximation to the linear advection equation

$$\left(v_i^{n+1}\right)_j = \left(v_i^n\right)_j - \lambda\left[\left(\hat{f}_{i+1/2}^n\right)_j - \left(\hat{f}_{i-1/2}^n\right)_j\right],$$

where

$$\begin{aligned}\left(\hat{f}_{i+1/2}^n\right)_j &= \hat{f}_j\left[\left(v_{i-K_1+1}^n\right)_j, \ldots, \left(v_{i-K_2}^n\right)_j\right] \\ &= \hat{f}_j\left[\left(Q^{-1}\mathbf{u}_{i-K_1+1}^n\right)_j, \ldots, \left(Q^{-1}\mathbf{u}_{i-K_2}^n\right)_j\right].\end{aligned}$$

Equivalently, in vector form, we have

$$\mathbf{v}_i^{n+1} = \mathbf{v}_i^n - \lambda\left(\hat{\mathbf{f}}_{i+1/2}^n - \hat{\mathbf{f}}_{i-1/2}^n\right),$$

where

$$\hat{\mathbf{f}}_{i+1/2}^n = \hat{\mathbf{f}}\left[Q^{-1}\mathbf{u}_{i-K_1+1}^n, \ldots, Q^{-1}\mathbf{u}_{i-K_2}\right].$$

Multiply the vector characteristic form by Q to recover the conserved variables

$$\begin{aligned}\mathbf{u}_i^{n+1} &= \mathbf{u}_i^n - \lambda\left(Q\hat{\mathbf{f}}_{i+1/2}^n - Q\hat{\mathbf{f}}_{i-1/2}^n\right), \\ Q\hat{\mathbf{f}}_{i+1/2}^n &= Q\hat{\mathbf{f}}\left[Q^{-1}\mathbf{u}_{i-K_1+1}^n, \ldots, Q^{-1}\mathbf{u}_{i-K_2}\right].\end{aligned}$$

Now suppose that the true governing equations are nonlinear but that they have been replaced by locally linearized governing equations such as the following:

$$\frac{\partial \mathbf{u}}{\partial t} + A_{i+1/2}\frac{\partial \mathbf{u}}{\partial x} = 0.$$

As before, multiply by $Q_{i+1/2}^{-1}$ to transform to locally linearized characteristics, apply any scalar method to each locally linearized characteristic equation, and multiply by $Q_{i+1/2}$ to transform back to conservation form. Unfortunately, in most cases, things are not quite this simple: Rather than staying with one single locally linearized Jacobian matrix $Q_{i+1/2}$, the locally linearized Jacobian matrix changes with position. Specifically,

$$\mathbf{u}_i^{n+1} = \mathbf{u}_i^n - \lambda\left(Q_{i+1/2}\hat{\mathbf{f}}_{i+1/2}^n - Q_{i-1/2}\hat{\mathbf{f}}_{i-1/2}^n\right),$$

where the spatial index on Q matches that of $\hat{\mathbf{f}}$. By choosing the index on Q appropriately, terms such as $f(u_i^n)$ in the scalar method become $\mathbf{f}(\mathbf{u}_i^n)$ in the vector method. Also, factors such as $a_{i+1/2}^n(u_{i+1}^n - u_i^n)$ in the scalar method become

$$\begin{aligned}Q_{i+1/2}\Lambda_{i+1/2}\left(\mathbf{v}_{i+1}^n - \mathbf{v}_i^n\right) &= Q_{i+1/2}\Lambda_{i+1/2}Q_{i+1/2}^{-1}\left(\mathbf{u}_{i+1}^n - \mathbf{u}_i^n\right) \\ &= A_{i+1/2}^n\left(\mathbf{u}_{i+1}^n - \mathbf{u}_i^n\right)\end{aligned}$$

in the vector method. For more specifics on this technique see, for example, Subsections 20.5.2 and 21.3.2. This all becomes extremely simple in the case of the linearized one-wave solvers, where $Q_{i+1/2} = Q^{-1}_{i+1/2} = I$. For example, any factors such as $a^n_{i+1/2}(u^n_{i+1/2} - u^n_i)$ in the scalar method become $\rho(A^n_{i+1/2})(\mathbf{u}^n_{i+1} - \mathbf{u}^n_i)$ in the vector method. For more specifics on this technique see, for example, Subsections 20.4.2 and 22.3.2.

References

Ben-Artzi, M., and Falcovitz, J. 1984. "A Second-Order Godunov-Type Scheme for Compressible Fluid Dynamics," *Journal of Computational Physics*, 55: 1–32.

Colella, P., and Woodward, P. R. 1984. "The Piecewise Parabolic Method (PPM) for Gas-Dynamical Simulations," *Journal of Computational Physics*, 54: 174–201.

Donat, R., and Marquina, A. 1996. "Capturing Shock Reflections: An Improved Flux Formula," *Journal of Computational Physics*, 125: 42–58.

Hirsch, C. 1990. *Numerical Computation of Internal and External Flows. Volume 2: Computational Methods for Inviscid and Viscous Flows*, Chichester: Wiley.

Jin, S., and Liu, J.-G. 1996. "The Effects of Numerical Viscosities. I. Slowly Moving Shocks," *Journal of Computational Physics*, 126: 373–389.

Lin, H.-C. 1995. "Dissipation Additions to Flux-Difference Splitting," *Journal of Computational Physics*, 117: 20–27.

Liou, M.-S., and Steffen, C. J. 1993. "A New Flux Splitting Scheme," *Journal of Computational Physics*, 107: 23–39.

Liou, M.-S., Van Leer, B., and Shuen, J.-S. 1990. "Splitting of Inviscid Fluxes for Real Gases," *Journal of Computational Physics*, 87: 1–24.

MacCormack, R. W. 1969. "The Effect of Viscosity in Hypervelocity Impact Cratering," *AIAA Paper 69–0354* (unpublished).

Quirk, J. J. 1994. "A Contribution to the Great Riemann Solver Debate," *International Journal for Numerical Methods in Fluids*, 18: 555–574.

Radespiel, R., and Kroll, N. 1995. "Accurate Flux Vector Splitting for Shocks and Shear Layers," *Journal of Computational Physics*, 121: 66–78.

Roberts, T. W. 1990. "The Behavior of Flux Difference Splitting Schemes Near Slowly Moving Shock Waves," *Journal of Computational Physics*, 90: 141–160.

Roe, P. L. 1981. "Approximate Riemann Solvers, Parameter Vectors, and Difference Schemes," *Journal of Computational Physics*, 43: 357–372.

Shuen, J.-S., Liou, M.-S., and Van Leer, B. 1990. "Inviscid Flux-Splitting Algorithms for Real Gases with Non-Equilibrium Chemistry," *Journal of Computational Physics*, 90: 371–395.

Sod, G. A. 1978. "A Survey of Several Finite-Differences Methods for Systems of Nonlinear Hyperbolic Conservation Laws," *Journal of Computational Physics*, 27: 1–31.

Steger, J. L., and Warming, R. F. 1981. "Flux Vector Splitting of the Inviscid Gasdynamic Equations with Applications to Finite-Difference Methods," *Journal of Computational Physics*, 40: 263–293.

Van Leer, B. 1982. "Flux-Vector Splitting for the Euler Equations." In *Lecture Notes in Physics, Volume 170, Eighth International Conference of Numerical Methods in Fluid Dynamics*, ed. E. Krause, pp. 507–512, Berlin: Springer-Verlag.

Zha, G.-C., and Bilgen, E. 1993. "Numerical Solutions of Euler Equations by Using a New Flux Vector Splitting Scheme," *International Journal for Numerical Methods in Fluids*, 17: 115–144.

Problems

18.1 From the last chapter, recall that steady-state solutions of the Lax–Wendroff and Beam–Warming second-order upwind methods both depend on the time step Δt.

(a) Do the steady-state solutions of the two-step Richtmyer method for the Euler equations depend on Δt?

(b) Do the steady-state solutions of two-step MacCormack's method for the Euler equations depend on Δt?

(c) Do the steady-state solutions of two-step Beam–Warming second-order upwind method for the Euler equations depend on Δt?

18.2 (a) Find an expression for the conservative numerical flux of the two-step MacCormack's method.

(b) Find an expression for the conservative numerical flux of the two-step Richtmyer method.

(c) Find an expression for the conservative numerical flux of the two-step Beam–Warming method.

18.3 (a) Show that all of the Lax–Wendroff methods seen in Subsection 18.1.2 are equivalent when applied to linear systems of equations. In other words, show that the original Lax–Wendroff method, Equation (18.2), the Richtmyer method, Equation (18.3), and both versions of MacCormack's method, Equations (18.4) and (18.5), are the same when $\mathbf{f} = A\mathbf{u}$ where $A = const.$

(b) Show that one-step and two-step Beam–Warming second-order upwind methods seen in Subsection 18.2.6 are equivalent when applied to linear systems of equations. In other words, show that Equations (18.44) and (18.48) are the same when $\mathbf{f} = A\mathbf{u}$ where $A = const.$

18.4 In Section 18.1, the Lax–Friedrichs and Lax–Wendroff methods were extended from scalar conservation laws to the Euler equations using the flux approach. However, we could just as well have used a wave approach or, more specifically, flux vector splitting, since techniques like flux vector splitting can be used for both upwind *or centered* methods. Let $\mathbf{f}^+ + \mathbf{f}^- = \mathbf{f}$ be any flux vector splitting. Define $A^{\pm}_{i+1/2}$ as follows:

$$\mathbf{f}^+\left(u^n_{i+1}\right) - \mathbf{f}^+\left(u^n_i\right) = A^+_{i+1/2}\left(u^n_{i+1} - u^n_i\right),$$
$$\mathbf{f}^-\left(u^n_{i+1}\right) - \mathbf{f}^-\left(u^n_i\right) = A^-_{i+1/2}\left(u^n_{i+1} - u^n_i\right).$$

Then write $\mathbf{f}^+ + \mathbf{f}^-$ wherever f appears in the method for the scalar conservation law, and write $A^+_{i+1/2} + A^-_{i+1/2}$ wherever $a^n_{i+1/2}$ appears in the method for the scalar conservation law.

(a) Find a version of the Lax–Friedrichs method for the Euler equations using flux vector splitting. When is this version of the Lax–Friedrichs method equivalent to the "flux approach" version given by Equation (18.1)?

(b) Find a version of the Lax–Wendroff method for the Euler equations using flux vector splitting. When is this version of the Lax–Wendroff method equivalent to the "flux approach" version given by Equation (18.2)?

18.5 (a) Write the Steger–Warming flux vector splitting $\mathbf{f}^{\pm}$ in terms of the conserved quantities per unit volume $\mathbf{u} = (\rho, \rho u, \rho e_T)^T$.

(b) Use the expression from part (a) to find $d\mathbf{f}^{\pm}/d\mathbf{u}$.

(c) Does the Steger–Warming flux vector splitting satisfy condition (13.7)? In other words, is $d\mathbf{f}^+/d\mathbf{u} \geq 0$ and $d\mathbf{f}^-/d\mathbf{u} \leq 0$?

18.6 (a) Prove that Van Leer's flux vector splitting satisfies $d\mathbf{f}^+/d\mathbf{u} \geq 0$ and $d\mathbf{f}^-/d\mathbf{u} \leq 0$.
(b) Does the Liou–Steffen flux vector splitting satisfy $d\mathbf{f}^+/d\mathbf{u} \geq 0$ and $d\mathbf{f}^-/d\mathbf{u} \leq 0$?
(c) Does the Zha–Bilgen flux vector splitting satisfy $d\mathbf{f}^+/d\mathbf{u} \geq 0$ and $d\mathbf{f}^-/d\mathbf{u} \leq 0$?

18.7 (a) Convert the Van Leer flux vector splitting to a wave speed splitting using Equation (18.18). In other words, prove Equation (18.31).
(b) Convert the Zha–Bilgen flux vector splitting to a wave speed splitting using Equation (18.18).

18.8 As mentioned in Section 18.2, Radespiel and Kroll (1995) suggested an adaptive combination of Van Leer and Liou–Steffen first-order upwind methods, with the Liou–Steffen first-order upwind method splitting used in smooth regions and Van Leer flux first-order upwind method used near shocks. Briefly explain why this might be a good idea.

18.9 This problem concerns forms of flux split first-order upwind methods that are something like artificial viscosity forms. Recall that for any x:

$$\max(0, x) = \frac{x + |x|}{2}, \quad \min(0, x) = \frac{x - |x|}{2}.$$

In each case, after finding the desired expression, state whether there would be any advantage to using the expression in practical calculations.
(a) Write the Liou–Steffen first-order upwind method as follows:

$$\hat{\mathbf{f}}^n_{i+1/2} = \frac{M_i^+ + M_{i+1}^-}{2}\left(\begin{bmatrix} \rho^n_{i+1}a^n_{i+1} \\ \rho^n_{i+1}u^n_{i+1}a^n_{i+1} \\ \rho^n_{i+1}(h_T)^n_{i+1}a^n_{i+1} \end{bmatrix} + \begin{bmatrix} \rho^n_i a^n_i \\ \rho^n_i u^n_i a^n_i \\ \rho^n_i (h_T)^n_i a^n_i \end{bmatrix}\right) - \frac{|M_i^+ + M_{i+1}^+|}{2}$$
$$\times\left(\begin{bmatrix} \rho^n_{i+1}a^n_{i+1} \\ \rho^n_{i+1}u^n_{i+1}a^n_{i+1} \\ \rho^n_{i+1}(h_T)^n_{i+1}a^n_{i+1} \end{bmatrix} - \begin{bmatrix} \rho^n_i a^n_i \\ \rho^n_i u^n_i a^n_i \\ \rho^n_i (h_T)^n_i a^n_i \end{bmatrix}\right) + (p_i^+ + p_{i+1}^-)\begin{bmatrix} 0 \\ 1 \\ 0 \end{bmatrix}.$$

(b) Write the Zha–Bilgen first-order upwind method as follows:

$$\hat{\mathbf{f}}^n_{i+1/2} = \frac{1}{2}\left(\mathbf{f}(\mathbf{u}^n_{i+1}) + \mathbf{f}(\mathbf{u}^n_i)\right) - \frac{1}{2}\left(|\mathbf{u}^n_{i+1}|\begin{bmatrix} \rho^n_{i+1} \\ \rho^n_{i+1}u^n_{i+1} \\ \rho^n_{i+1}(e_T)^n_{i+1} \end{bmatrix} - |u^n_i|\begin{bmatrix} \rho^n_i \\ \rho^n_i u^n_i \\ \rho^n_i (e_T)^n_i \end{bmatrix}\right)$$
$$-\frac{1}{2}\left(\begin{bmatrix} 0 \\ p^n_{i+1}M^n_{i+1} \\ p^n_{i+1}a^n_{i+1} \end{bmatrix} - \begin{bmatrix} 0 \\ p^n_i M^n_i \\ p^n_i a^n_i \end{bmatrix}\right).$$

18.10 Consider Harten's "entropy fix" given by Equation (18.56). Show that an equivalent expression for the function ψ is

$$\psi(x) = \max\left[|x|, \min\left(\delta, \frac{x^2 + \delta^2}{2\delta}\right)\right].$$

What advantages might this expression have in practical programming terms, if any?

CHAPTER 19

Boundary Treatments

19.0 Introduction

This chapter concerns boundary treatments. Before now, the book has avoided boundary treatments by using either infinite boundaries or periodic boundaries, but one cannot remain innocent forever. Boundary conditions and governing equations have equal importance, despite the fact that most sources, including this one, spend most of their time focused on the governing equations. For the same governing equations, boundary conditions distinguish flow over a plane from flow over a train from flow over a space shuttle from any other sort of flow. In practice, numerical boundary treatments often consume a large percentage of computational gasdynamics codes, both in terms of the number of lines of code and in terms of the development effort.

This chapter concerns two types of boundaries – *solid* and *far-field boundaries*. Solid boundaries are also known as *surface*, *wall*, *rigid*, or *impermeable boundaries*; naturally enough, solid boundaries occur at the surfaces of solid objects. This chapter considers only stationary solid boundaries. Far-field boundaries are also known as *open*, *artificial*, *permeable*, or *remote boundaries*. Far-field boundaries limit the computational domain to a reasonable finite size; the true boundaries may be extremely far away, or even infinitely far away, at least conceptually. In general, when the physical domain is very large or infinite, then the farther away the numerical far-field boundary is, the more accurate but also the more costly the numerical approximation will be. Far-field boundaries are divided into *inflow* boundaries, where fluid enters the computational domain, and *outflow* boundaries, where fluid exits the computational domain. In multidimensions, far-field boundaries may also be *streamlines*, across which fluid neither exits nor enters. Solid boundaries are purely physical whereas far-field boundaries are purely numerical. Solid boundaries reflect existing waves whereas far-field boundaries both absorb existing waves and emit new ones. There are other types of boundaries, including *permeable* or *porous* boundaries, but these will not be discussed here.

Because numerical boundary treatments are so much of an art, even more so than the interior, there is often no way to predict in advance when the numerical results will be especially sensitive to boundary treatments and, if so, which boundary treatments will yield the best results. Some computer codes include several different far-field boundary treatments, including boundary treatments based on constant, linear, and quadratic extrapolation. Then, if the code exhibits mysterious instabilities, or a drop in accuracy, or a failure to converge as $t \to \infty$ or as $\Delta x \to 0$ and $\Delta t \to 0$, the user can easily switch to different far-field boundary treatments, sometimes with dramatic improvements.

There are two basic types of grids, depending on the alignment between the grid and the boundaries, as illustrated in Figure 19.1. In type 1 grids, the boundary is aligned with a cell edge; type 1 grids are the most common type of grid for finite-volume methods. In type 2 grids, the boundary is aligned with a cell center; type 2 grids are the most common type of grid for finite-difference methods.

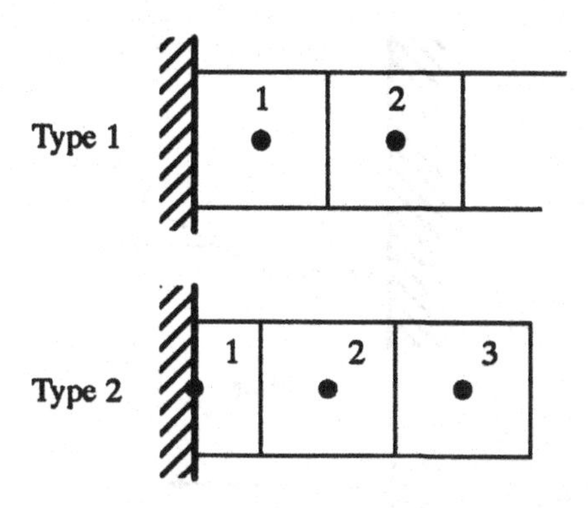

Figure 19.1 The two basic types of grid alignments.

Near boundaries of either sort, numerical methods usually require values from outside the computational domain. For example, recall that the Lax–Wendroff method uses a centered three-point stencil. Then the Lax–Wendroff method at point 1 in Figure 19.1 requires the value of the solution at the nonexistent point 0. There are two ways of dealing with this problem: *Either change the method or change the computational domain.* In other words, the interior method may be changed to a special boundary method, which is completely or partially biased away from the boundary, such as FTFS at a left-hand boundary, and thus does not require points from outside the computational domain; alternatively, the computational domain may be changed, by introducing cells outside of the computational domain, called *ghost* cells. Either approach, or a combination of the two, combined with the physical boundary conditions constitutes a numerical *boundary treatment.* Note the distinction between numerical boundary *treatments* and physical boundary *conditions*. Boundary conditions, which include the no-penetration boundary condition and, in viscous flows, the no-slip boundary condition, only partially specify boundary flows. For example, the no-penetration condition requires $u = 0$ at stationary solid boundaries, so that fluid neither enters nor exits through the solid, but leaves all other flow properties free. Boundary treatments reconstruct and model the flow properties at the boundary using both boundary conditions and the interior flow. Unfortunately, many sources use the terms "boundary treatments" and "boundary conditions" synonymously.

As you might expect, boundary treatments affect the accuracy, order of accuracy, and stability of numerical methods at the boundaries. Figure 19.2 illustrates two examples of instability and inaccuracy at boundaries caused by faulty boundary treatments. However, besides the boundaries, boundary treatments may also profoundly affect the accuracy, order of accuracy, and stability of numerical methods in the interior. Section 19.1 briefly discusses the effects of boundary treatments on local and global stability. As far as order of accuracy, a rather remarkable result, due to Gustafsson (1975), says that the formal order of accuracy in the interior may be higher than the formal order of accuracy on the boundaries. More specifically, zeroth-order accurate boundary treatments allow first-order accuracy in the interior; first-order accurate boundary treatments allow second-order accuracy in the interior; and so forth. Gustafsson's result seems less surprising in light of the fact that order of accuracy can drop to first order or less near shocks, contacts, and other trouble spots in the interior and yet rebound to *any* order of accuracy elsewhere. Unfortunately, in many applications, the boundary is the primary reason for the calculation. For example, the lift of an airfoil is

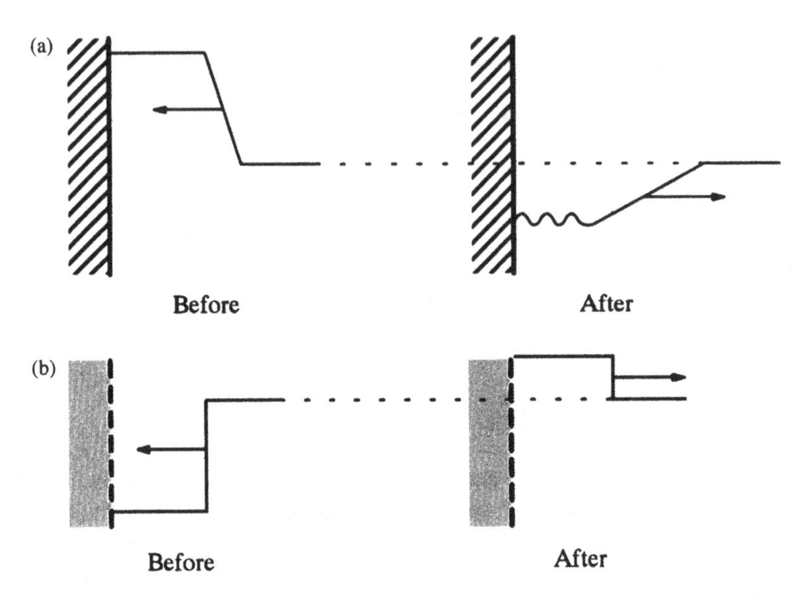

Figure 19.2 Some typical local problems caused by faulty boundary treatments. (a) Small spurious oscillations in the wake of an expansion wave reflected from a solid boundary. (b) Small spurious reflection of a shock wave from a far-field boundary.

completely determined by the pressure on the solid surface of the airfoil. In this case, it is of little comfort knowing that a low accuracy on the solid surfaces rebounds in the interior. Furthermore, at least in some cases, recent work appears to show that any waves passing through a low order-of-accuracy region, either through a shock or a low order-of-accuracy boundary treatment, retain the low order of accuracy *permanently*; for example, Casper and Carpenter (1995) found that numerical results downstream of a sound–shock interaction in unsteady flow achieved only first-order accuracy for small enough Δx.

This chapter mainly concerns boundary treatments for unsteady explicit numerical methods. In many cases, steady-state flows are computed in the limit as time goes to infinity, in which case, of course, unsteady boundary treatments also work for steady-state flows. However, for steady-state flows, the far-field boundary treatments are especially critical, since unsteady waves exiting the far-field constitute a major mechanism for convergence to steady state – the numerical method "blows out" the unsteadiness through the far-field boundaries – and any reflections from the far-field boundaries may slow or even prevent steady-state convergence. In the worst case, unsteady waves bounce back and forth across the numerical domain, unable to leave, creating standing waves that prevent full convergence. For a discussion of steady-state boundary treatments, see Rudy and Strikwerda (1980, 1981), Jameson, Schmidt, and Turkel (1981), Bayliss and Turkel (1982), Mazaheri and Roe (1991), Saxer and Giles (1991), Karni (1991), and Ferm (1995).

The same boundary treatments can be used for explicit and implicit numerical methods. However, the choice of boundary treatments greatly impacts on the structure of the implicit equations and thus the cost of solving the implicit equations. For example, depending on the method, assuming periodic boundaries might yield a periodic tridiagonal system of equations, as in Example 11.4, whereas other types of boundaries yield much more complicated systems of implicit equations, depending on the boundary treatment. For a discussion of boundary treatments for implicit methods, see Yee, Beam, and Warming (1982), Pulliam (1982), and Chakravarthy (1983).

To keep things simple, this chapter considers only formally zero-, first-, and second-order accurate boundary treatments. Then there is no need to distinguish boundary treatments for finite-difference and finite-volume methods, since finite-difference and finite-volume methods are the same to within second-order accuracy, at least in one dimension, as discussed in Chapter 11. This chapter will not include numerical results, but readers are encouraged to experiment for themselves, using any of the boundary treatments found in this chapter with any of the numerical methods for the Euler equations found in the last chapter. One nice test case replaces the infinite boundaries in the Riemann problem by a solid wall on the low-pressure side and a far-field boundary on the high-pressure side. To get a sense of the exact solution, it may help to read a discussion of one-dimensional wave reflections, as found in any gasdynamics text; see, for example, Figure 45 in Courant and Friedrichs (1948) and the associated discussion.

19.1 Stability

This section very briefly discusses the effects of boundary treatments on stability. However stable and successful a numerical method might be with infinite or periodic boundaries, improper solid and far-field boundary treatments may destabilize the method either locally or globally and may even cause the entire numerical method to blow up. For example, some numerical methods exhibit small oscillations near solid boundaries, depending on the solid boundary treatment and the solution near the solid boundaries, even when the solution is completely nonoscillatory and stable elsewhere.

With regard to linear stability, remember that ordinary von Neumann analysis does not apply except at periodic boundaries. Kreiss (1966, 1968, 1970) and Gustafsson, Kreiss, and Sundstrom (1972) proposed a modification of von Neumann analysis to determine the effects of boundary treatments on linear stability. For somewhat more modern information on Kreiss stability analysis and related approaches, see Gottlieb and Turkel (1978); Beam, Warming, and Yee (1982); and Higdon (1986). As an alternative to Kreiss' variant on von Neumann analysis, Section 15.2 of this book describes matrix stability analysis. Among other uses, matrix stability analysis can determine the effects of boundary treatments on linear stability; for more information see, for example, Appendix A of Roe (1989) and Gustafsson (1982).

With regard to nonlinear stability analysis, recall that the CFL condition is necessary for nonlinear stability. The CFL condition requires that the numerical domain of dependence contain the true domain of dependence, as seen in Chapter 12. At solid boundaries, the true domain of dependence is entirely to the right of a left-hand wall and entirely to the left of a right-hand wall. Similarly, the numerical domain of dependence must be entirely to the right of a left-hand wall and entirely to the left of a right-hand wall. Thus the CFL condition is automatically satisfied at solid boundaries. By contrast, if any waves enter through far-field boundaries, the CFL condition is automatically *violated* at far-field boundaries! To see this, notice that if any waves enter the far-field boundary, then the physical domain of dependence lies partly or fully outside of the numerical domain. But the numerical domain of dependence is always, of course, completely contained within the numerical domain, so that the numerical domain of dependence cannot contain the physical domain of dependence. As the only possible response, to satisfy the CFL condition, the numerical method must somehow know something about any waves entering through the far-field boundaries, as discussed later in Section 19.3. The amount and type of outside information specified at the far-field boundaries determine whether the exact problem is well-posed and whether the numerical approximation is stable and accurate. Recall that stability requires well-posed problems, and well-posed problems require proper boundary and initial conditions. The Kreiss papers cited above discuss conditions for well-posed problems. Also see the discussion of well-posed problems in Sections 3.1 and 3.2.

Besides the CFL condition, most of the nonlinear stability theory studied in Chapter 16 does not apply at boundaries, especially solid boundaries. The nonlinear stability conditions of Chapter 16 derive from specific behaviors of extrema in the solution. Unfortunately, solid boundaries change the behavior of extrema in the solution. For example, a maximum may reflect from a solid boundary as a minimum and vice versa, as seen in Figure 19.2. Except for some of the weaker conditions such as TVD, TVB, and ENO, the nonlinear stability conditions of Chapter 16 will not allow such behaviors and thus do not apply at solid boundaries. As a rare example of a nonlinear stability analysis at boundaries, see Shu (1987).

19.2 Solid Boundaries

Physically, solid surfaces in inviscid one-dimensional flow have but one property: $u = 0$ on stationary solid surfaces. This is known as the *no-penetration boundary condition.* Then, in a numerical method, a solid boundary treatment should enforce the no-penetration boundary condition without restricting the flow in any other way. For ease of presentation, this section and the next will only consider left-hand solid surfaces, positioned at $x = x_L$. The extension to right-hand boundaries should be obvious.

The *method of images* is a well-known technique for solving linear equations, such as Laplace's equation, in the presence of solid boundaries. A closely related technique applies to the Euler equations. In particular, a solid surface may be replaced by an *image* or *ghost* flow. The only restriction on the ghost flow is that it must ensure the no-penetration boundary condition $u(x_L, t) = 0$ along the interface between the real and ghost flows. If the no-penetration boundary condition holds on the boundary between the real and ghost flows, the real flow neither knows nor cares what happens beyond the boundary. The best technique for constructing a ghost flow is *reflection* or *imaging*. In reflection, scalar quantities reflect symmetrically onto the ghost region, whereas vector quantities such as velocity reflect with a change in sign, just as if the solid surface were a mirror. Thus, in terms of primitive variables, the solid surface is replaced by a ghost flow where

$$\rho(x_L - x, t) = \rho(x_L + x, t), \tag{19.1a}$$

$$u(x_L - x, t) = -u(x_L + x, t), \tag{19.1b}$$

$$p(x_L - x, t) = p(x_L + x, t). \tag{19.1c}$$

Since zero is the only number that equals its own negative, Equation (19.1b) implies $u(x_L, t) = 0$. In other words, Equation (19.1b) ensures the no-penetration boundary condition, assuming only that the velocity is continuous. Except for the no-penetration boundary condition, Equation (19.1) does not affect the true flow in any way – in fact, Equation (19.1) ensures the no-penetration boundary condition, nothing more and nothing less, just as it should. Although Equation (19.1) does not restrict the physical flow, it certainly restricts the ghost flow. In particular, it ensures that the density, velocity, and pressure are continuous

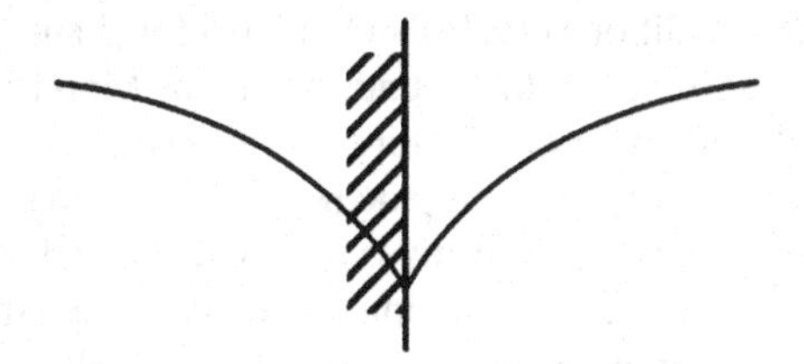

Figure 19.3 Typical cusped flow variable plotted across the boundary between real and ghost flows.

across the boundary between the real and ghost flows. While all this is not absolutely necessary, jump discontinuities at the boundary would cause numerical problems, much like shocks and contacts in the interior do.

Before proceeding, let us make three observations. First, the ghost flow has been defined in terms of primitive variables. However, it could just as well have been defined in terms of conservative variables, characteristic variables, or any other set of variables. For example, reflecting the conservative variables yields

$$\begin{aligned}
\rho(x_L - x, t) &= \rho(x_L + x, t), \\
\rho u(x_L - x, t) &= -\rho u(x_L + x, t), \\
\rho e_T(x_L - x, t) &= \rho e_T(x_L + x, t),
\end{aligned}$$

which is entirely equivalent to Equation (19.1). Second, although simple reflection ensures the no-penetration boundary condition and continuity across the boundary, other ghost flows might just as well accomplish the same thing; however, to date, no other means of constructing ghost flows have caught on. Third, although the flow properties are continuous across the boundary between the real and ghost flows, the first derivatives of the flow properties are not. In particular, flow variables may be cusped across the boundary, as illustrated in Figure (19.3). The solid boundary is a characteristic – a stationary solid boundary that moves with the characteristic speed $u = 0$ – both before and after the introduction of the ghost flow. This explains how cusping can occur at the boundary between real and ghost flows; in general, jump discontinuities in the first derivative, such as cusps, may occur across characteristic surfaces, as mentioned in Chapter 3.

Every inviscid flow may be described entirely in terms of waves and characteristics, as seen in Chapter 3. Then, as one common interpretation of ghost flows based on reflection, every wave in the real flow has an equal and opposite wave in the ghost flow. For example, a shock wave located 10 meters to the right of a solid wall traveling at −50 m/s has a ghost shock 10 meters to the left of the wall traveling at 50 m/s. Eventually, the real shock meets the ghost shock, producing the same result in the real flow as if the real shock reflected from the real solid wall. By the way, in this example, notice that the shock reflection problem is identical to the shock intersection problem, which is, in turn, identical to the Riemann problem, as discussed in Chapter 5.

Although widely used, Moretti (1969) roundly condemned ghost cell reflection techniques early on, making them seem like a guilty pleasure. For example, at the start of their paper, Dadone and Grossman (1994) say "the pioneering work of Moretti focused on this issue [solid boundary treatments] and described the inadequacies of reflection techniques. . . ." Despite this recognition of Moretti's criticism, Dadone and Grossman's paper goes on to consider reflection techniques exclusively. Let us consider Moretti's specific criticisms – Moretti claimed that reflection conditions (19.1a) and (19.1c) for ρ and p imply that the first derivatives of ρ and p are zero across the boundary and, as Moretti says, "these are redundant conditions and are physically wrong." Moretti's criticism consists of two distinct parts: first, he claims that the derivatives of ρ and p do not equal zero in the true flow; and, second, he claims that reflection makes the derivatives of ρ and p equal to zero in the numerical approximation. Let us examine these two claims more closely. First, as it turns out, in one-dimensional flow, the first spatial derivative of pressure actually *is* zero at the wall. To prove this, by Equation (2.35) the conservation of momentum in primitive variable form is as follows:

$$\frac{\partial u}{\partial t} + u\frac{\partial u}{\partial x} + \frac{1}{\rho}\frac{\partial p}{\partial x} = 0.$$

But $u(x_L, t) = 0$ for all t, and thus $\frac{\partial u}{\partial t}(x_L, t) = 0$ at solid boundaries. Then conservation of momentum becomes

$$\frac{\partial p}{\partial x}(x_L, t) = 0 \tag{19.2}$$

and the spatial derivative of pressure is zero at solid boundaries, as promised. If the entropy is constant near the wall, as in homentropic flow, then all other flow properties, including density and internal energy, will also have zero spatial derivatives at the wall. However, if the entropy is not constant near the wall, then the derivatives of density and other flow properties need not be zero at the wall. In multidimensions, the pressure derivative normal to the wall is proportional to the wall's curvature, and the normal derivatives of other flow properties will also be nonzero. Thus, in general, Moretti is correct in saying that the first derivatives of ρ and p are nonzero at solid boundaries, especially for multidimensional flows. This brings us to the second prong of Moretti's criticism: Does reflection wrongly impose zero derivatives on ρ and p at the boundaries? If the density is symmetric about the boundary, then the first derivative of density must be antisymmetric. In other words, the symmetry condition

$$\rho(x_L - x, t) = \rho(x_L + x, t)$$

implies the following antisymmetry condition:

$$\frac{\partial \rho}{\partial x}(x_L - x, t) = -\frac{\partial \rho}{\partial x}(x_L + x, t).$$

In particular,

$$\frac{\partial \rho}{\partial x}(x_L, t) = -\frac{\partial \rho}{\partial x}(x_L, t).$$

If $\partial \rho / \partial x$ is continuous then this implies that

$$\frac{\partial \rho}{\partial x}(x_L, t) = 0,$$

in agreement with Moretti. However, this conclusion assumes that $\partial \rho / \partial x$ is continuous. But, as seen in Figure 19.3, $\partial \rho / \partial x$ is not necessarily continuous, but instead generally contains a jump discontinuity at the wall. Then, contrary to Moretti's assertion, first spatial derivatives can have *any* value near the boundary, including but not limited to zero. Therefore, contrary to Moretti's assertion, reflection places absolutely no restrictions on the physical flow except, of course, for the no-penetration boundary condition, exactly as it should.

The ghost flow technique simply rephrases the true problem, which still leaves open the question of how to approximate the true problem. Consider an explicit numerical method whose stencil contains K_1 points to the left. Then the numerical method requires at least K_1 ghost cells at each left-hand solid boundary. In other words, although the ghost flow is semi-infinite in the true problem, the numerical approximation only requires a relatively small number of ghost cells. For example, suppose $K_1 = 2$. For type 1 grids, the numerical approximations in the required ghost cells are defined as follows:

$$\begin{aligned} \rho_0^n &= \rho_1^n, & \rho_{-1}^n &= \rho_2^n, \\ u_0^n &= -u_1^n, & u_{-1}^n &= -u_2^n, \\ p_0^n &= p_1^n, & p_{-1}^n &= p_2^n. \end{aligned}$$

Similarly, for type 2 grids, the numerical approximations in the required ghost cells are defined as follows:

$$\begin{aligned} \rho_0^n &= \rho_2^n, & \rho_{-1}^n &= \rho_3^n, \\ u_0^n &= -u_2^n, & u_{-1}^n &= -u_3^n, \\ p_0^n &= p_2^n, & p_{-1}^n &= p_3^n. \end{aligned}$$

Although the true derivatives may be nonzero near solid walls, certain numerical approximations might say otherwise. For example, on a type 1 grid, a first-order forward-space approximation says

$$\frac{\partial \rho}{\partial x} \approx \frac{\rho_1^n - \rho_0^n}{\Delta x} = 0.$$

Similarly, a third-order accurate approximation says

$$\frac{\partial \rho}{\partial x} \approx \frac{\rho_2^n - 3\rho_1^n + 3\rho_0^n - \rho_{-1}^n}{\Delta x} = 0,$$

which is illustrated in Figure 19.4. In Figure 19.4, the reader should recall the relationships between numerical differentiation and interpolation, as described in Chapter 10. Specifically, the spatial derivative of the density is approximately equal to the spatial derivative of the interpolation polynomial passing through the density samples, which, in Figure 19.4,

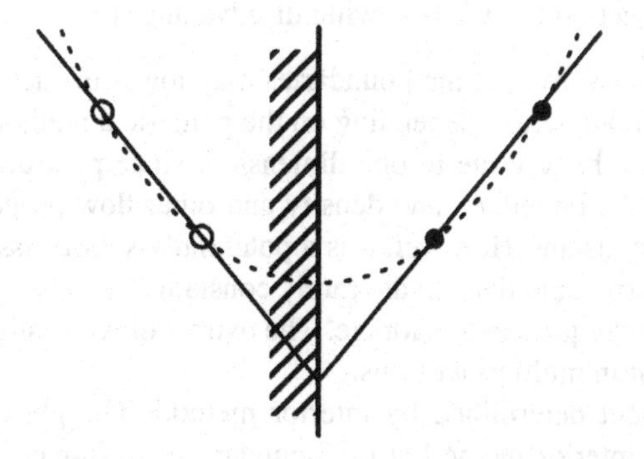

Figure 19.4 Interpolation across boundary yields zero numerical derivative at boundary.

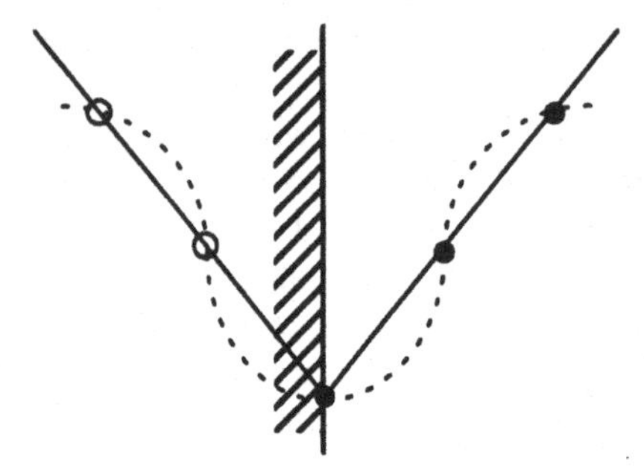

Figure 19.5 Interpolation across boundary cusp may experience spurious oscillations.

equals zero at the boundary or, in other words, the slope of the interpolation polynomial is zero at the boundary. To the extent that a numerical method senses the cusp at the boundary, the cusp may cause numerical problems, just like any jump in the first derivative, especially spurious overshoots and oscillations, such as those seen in Figure 19.5. Of course, oscillations and instabilities caused by boundary cusps may be combatted in exactly the same way as oscillations and instabilities caused by shocks, contacts, and the corners at the heads and tails of expansion fans, using the solution-sensitive techniques introduced in Section 13.3 and developed intensively in Part V.

Although numerical methods must use points on both sides of the boundary, any post processing of the final results at the boundaries need not. For example, in analyzing the final results, one could use only the black dots in Figures 19.4 and 19.5, resulting in much improved approximations for the flow variables and their derivatives at the wall. Two common sorts of postprocessing include (1) plotting and (2) averaging the pressure over a solid surface to determine lift. It is an especially good idea to extrapolate from the interior to the solid boundaries when the interior solution has a greater accuracy or order of accuracy than the solid boundary solution. For more details, see Dadone and Grossman (1994).

To summarize, the ghost cell technique based on reflection has the following advantages:

- Simple and elegant.
- No need to alter the method near boundaries, saving the expense and complication of determining whether the method is near a boundary and, if so, changing the method accordingly.

The ghost cell technique based on reflection has the following disadvantages:

- Cusping. Jumps in the first derivatives at the boundaries may lower the accuracy and order of accuracy at the boundaries, depending on the numerical method and the type of grid. This is not a huge issue in one dimension, since pressure and velocity are constant across the boundary, and density and other flow properties are often constant or nearly constant. However, it is a potentially serious issue in multidimensions, where pressure and density are rarely constant.
- Extra storage. The ghost cells require extra storage. The extra storage is slight in one dimension but significant in multidimensions.
- Quality of boundary treatment determined by interior method. The ghost cell technique uses the ordinary interior method at the boundary. Unfortunately, the interior method is usually not specifically designed for boundaries, and thus it

may not perform well at boundaries. In particular, the interior method may not accurately enforce the no-penetration boundary condition, or it may be unusually sensitive to cusping. The only way around this problem is to specially modify the method at solid boundaries, but this sacrifices some of the simplicity of the ghost cell approach. If the numerical method accurately enforces the no-penetration boundary condition, then it should ensure

$$\hat{\mathbf{f}}^n_{1/2} = \begin{bmatrix} 0 \\ p^n_{1/2} \\ 0 \end{bmatrix}$$

at type 1 boundaries and

$$\mathbf{u}^n_1 = \begin{bmatrix} \rho^n_1 \\ 0 \\ \rho e^n_{T,1} \end{bmatrix}$$

at type 2 boundaries, as discussed below.

Now let us examine the alternative to the ghost cell technique. The ghost cell technique alters the boundaries, creating additional imaginary cells as required. The alternative is to alter the method, so that it no longer requires values outside of the boundaries. More specifically, in the altered method, the conservative numerical fluxes $\hat{\mathbf{f}}^n_{1/2}, \hat{\mathbf{f}}^n_{3/2}, \hat{\mathbf{f}}^n_{5/2}, \ldots$ should only depend on $\mathbf{u}^n_1, \mathbf{u}^n_2, \mathbf{u}^n_3, \ldots$. For example, $\hat{\mathbf{f}}^n_{1/2}$ might be FTFS:

$$\hat{\mathbf{f}}^n_{1/2} = \mathbf{f}(\mathbf{u}^n_1)$$

whereas $\hat{\mathbf{f}}^n_{3/2}$ might be the Lax–Wendroff method:

$$\hat{\mathbf{f}}^n_{3/2} = \frac{1}{2}(\mathbf{f}(\mathbf{u}^n_2) + \mathbf{f}(\mathbf{u}^n_1)) - \frac{\lambda}{2}A^n_{3/2}(\mathbf{f}(\mathbf{u}^n_2) - \mathbf{f}(\mathbf{u}^n_1)).$$

Then

$$\mathbf{u}^{n+1}_1 = \mathbf{u}^n_1 - \lambda(\hat{\mathbf{f}}^n_{3/2} - \hat{\mathbf{f}}^n_{1/2})$$

does not require points from outside the physical domain. The further the method is from the left-hand boundary, the further the stencil can extend to the left, until the standard interior method can be used once again. One possible concern is that the numerical boundary method is often a downwind method. For instance, in the preceding example, FTFS is downwind when the wave speeds are positive. In the interior of the flow, downwind methods such as FTFS ordinarily cause enormous instability, since they violate the CFL condition. However, at solid boundaries, such downwind methods do not violate the CFL condition, as explained in Section 19.1, and thus do *not* necessarily cause instability. Clearly, intuition developed in the interior may prove false at solid boundaries.

Unfortunately, altering or replacing the interior method near the solid boundaries does not, in and of itself, ensure the no-penetration boundary condition. For example, using FTFS to compute $\hat{\mathbf{f}}^n_{1/2}$ will not effectively ensure the no-penetration boundary condition. Of course, this is nothing new – as mentioned before, the ghost cell approach also may not effectively enforce the no-penetration boundary condition. Thus, in either approach, it sometimes makes sense to explicitly and exactly enforce the no-penetration boundary condition. The remainder of this section is devoted to some possible strategies.

First, consider a type 1 grid. For finite-volume methods:

$$\hat{\mathbf{f}}^n_{1/2} \approx \frac{1}{\Delta t}\int_{t^n}^{t^{n+1}} \mathbf{f}(\mathbf{u}(x_{1/2}, t))\, dt$$

as seen in Chapter 11. Since finite-volume methods are the same as finite-difference methods to within second-order accuracy, the conservative numerical flux of finite-difference methods should also satisfy this same equation, to within second-order accuracy. By the no-penetration boundary condition, $u = 0$ at $x_{1/2} = x_L$ for a stationary solid boundary. As seen in Equation (2.19), all of the elements in the flux vector $\mathbf{f}$ include a factor of u, except for the pressure in the second element. Then

$$\mathbf{f}(\mathbf{u}(x_{1/2}, t)) = \begin{bmatrix} 0 \\ p(x_{1/2}, t) \\ 0 \end{bmatrix}. \tag{19.3}$$

Thus, to approximate $\hat{f}^n_{1/2}$ while enforcing the no-penetration boundary condition, *on a type 1 grid, approximate the pressure at the wall, and set all of the other flux terms to zero.* The simplest option for determining the pressure at the wall is pressure extrapolation. For example,

$$p(x_{1/2}, t) \approx p^n_1 + O(\Delta x^2) + O(t - t^n). \tag{19.4}$$

In general, this constant extrapolation is only first-order accurate in space. However, for one-dimensional flows, the first derivative of pressure is zero at solid boundaries, by Equation (19.2), making constant extrapolation formally second-order accurate in space. As another example, linear extrapolation yields

$$p(x_{1/2}, t) \approx \frac{3}{2}p^n_1 - \frac{1}{2}p^n_2 + O(\Delta x^2) + O(t - t^n). \tag{19.5}$$

The usual accuracy trade-offs between shocks and smooth flows apply to these pressure extrapolation formulae. For completely smooth flows, linear extrapolation should be superior to constant extrapolation, at least after setting aside stability considerations; however, as a shock nears the wall, constant extrapolation may become superior to linear extrapolation.

Now consider type 2 grids, where the cell center rather than the cell edge is aligned with the solid boundary – then one must consider the solution $\mathbf{u}(x_1, t) = \mathbf{u}(x_L, t)$ rather than the flux $\mathbf{f}(\mathbf{u}(x_{1/2}, t)) = \mathbf{f}(\mathbf{u}(x_L, t))$. By the no-penetration boundary condition, the velocity is zero at $x_1 = x_L$. Then $\mathbf{u}^n_1$ must have the following form:

$$\mathbf{u}^n_1 = \begin{bmatrix} \rho^n_1 \\ 0 \\ \rho e^n_{T,1} \end{bmatrix} \tag{19.6}$$

for all n. How can the second element of $\mathbf{u}^n_1$ be kept equal to zero? For a conservative method, the second element of $\mathbf{u}^n_1$ changes as follows:

$$(\mathbf{u}^{n+1}_1)_2 = (\mathbf{u}^n_1)_2 - \lambda\left[(\hat{\mathbf{f}}^n_{3/2})_2 - (\hat{\mathbf{f}}^n_{1/2})_2\right].$$

But then $(\mathbf{u}^{n+1}_1)_2 = (\mathbf{u}^n_1)_2 = 0$ implies

$$(\hat{\mathbf{f}}^n_{1/2})_2 = (\hat{\mathbf{f}}^n_{3/2})_2.$$

In other words, the second element of the conservative numerical flux vector $\hat{\mathbf{f}}^n_{1/2}$ must be equal to the second element of the conservative numerical flux vector $\hat{\mathbf{f}}^n_{3/2}$ to avoid changing the second element of $\mathbf{u}^n_1$ from its proper value of zero. So long as this is true, the no-penetration boundary condition is exactly true on a type 2 grid. Enforcing the no-penetration boundary condition on type 2 grids is an essentially trivial task compared with type 1 grids. Also, type 2 grids have a solution sample placed right on the solid boundary, as opposed to type 1 grids, where the closest sample is one-half cell away, giving type 2 grids a natural accuracy advantage over type 1 grids on solid surfaces. Both of these observations favor type 2 grids at solid boundaries; however, other factors may conspire to favor type 1 grids in practice, especially when using ghost cell techniques.

Before ending this section, let us say a few words about shock reflections from solid walls. Donat and Marquina (1996) document some of the common problems found in this situation. First, many numerical methods exhibit a spike near the wall in the wake of reflected shocks, whose size depends only slightly on parameters such as the CFL number. For example, in one instance, Roe's first-order upwind method, usually one of the least oscillatory methods, exhibits a 10% spike near the solid wall during shock reflection. Unfortunately, higher-order methods based on Roe's first-order upwind method tend to create even greater spikes. In the boundary spike, the density is too low, while the temperature and specific internal energy are too high. Because the temperature is too high, this spiking problem is sometimes called *overheating*. Besides overheating during shock reflection, as a second problem, Donat and Marquina (1996) report that some numerical methods do not capture the shock speed correctly near the wall; however, unlike overheating, this problem sometimes stems from incorrect conservation enforcement near the wall. In particular, some boundary treatments may not properly enforce the no-penetration condition, but may instead allow spurious flux to pass through the wall. Notice that a stationary solid wall is always a sonic point of the 0-characteristic family. Donat and Marquina (1996) show that adding extra dissipation to the 0-characteristic family improves shock reflection; furthermore, they show that adding extra artificial viscosity near sonic points, at the wall or elsewhere, has other beneficial effects, including reducing spurious expansion shocks and reducing the spurious oscillations often found in the wake of slowly moving shocks in methods such as Roe's first-order upwind method. As a word of caution, other nonsmooth waves including expansion waves, with kinks at their heads and tails, may suffer from problems much like shocks during reflection. However, if you think shocks and other nonsmooth waves cause difficulties at solid boundaries, just wait until you see what they do at far-field boundaries. In fact, most far-field boundaries will not pass shocks as they should but will largely reflect them.

19.3 Far-Field Boundaries

We now turn to the tricky topic of far-field boundary treatments. Stationary solid boundaries should simply reflect waves from the interior; they should neither emit nor absorb waves. By contrast, far-field boundaries should allow waves to travel freely in and out. Thus far-field boundary treatments must specify incoming waves and prevent reflections of outgoing waves. Of course, the interior has no information about incoming waves. Although outgoing waves may influence incoming waves to some extent, incoming waves basically carry information from the exterior. Thus far-field boundary treatments must be told about the exterior. Unfortunately, there is usually only sketchy information about the exterior solution – truly complete information would require explicitly modeling the exterior, or at least some part of the exterior, meaning that part or all of the exterior, in this sense, becomes part of the interior or, in other words, part of the computational domain. Unfortunately, costs increase proportionally to the size of the interior. Thus, far-field boundaries impose a trade-off between efficiency and accuracy; the smaller the interior and the larger the exterior, the lower the costs, and the less accurate the information about the exterior; the larger the interior and the smaller the exterior, the higher the costs, and the more accurate the information about the exterior. In general, even with a very large interior, and a very small exterior, there is still uncertainty about the far-field flow, and thus every numerical approximation has to tolerate at least some ambiguity about the far field. This ambiguity makes far-field boundaries much more difficult to deal with than solid boundaries. Indeed, for every research paper concerning solid boundary treatments, there are half a dozen or more concerning far-field boundary treatments, especially *subsonic* far-field boundary treatments. If the exterior flow is not known exactly, then the trick is to feed the far-field boundaries the right type and amount of perhaps slightly wrong exterior information. This must be done in such a way as not to blatantly contradict other information, such as information about the interior flow determined by the interior numerical method, which is itself probably slightly wrong, or information about the exterior flow given at other far-field boundaries.

There are three steps in designing a far-field boundary treatment:

- First, decide how many flow variables to specify – zero, one, two, or three.
- Second, decide which flow variables to specify.
- Third, decide what values to assign to the chosen flow variables, e.g., steady uniform free-stream values. This is the hardest step.

This section considers each step in turn, starting with the first. Not every far-field boundary requires complete information about the exterior flow. For example, a supersonic outflow boundary requires absolutely no information about the exterior flow, since no information about the exterior can propagate upstream to influence the interior. In general, the amount of information required is given by the following table:

Number of quantities required at a *supersonic inflow* boundary	Number of quantities required at a *subsonic inflow* boundary	Number of quantities required at a *subsonic outflow* boundary	Number of quantities required at a *supersonic outflow* boundary
3	2	1	0

One example is illustrated in Figure 19.6. Notice that the characteristics are drawn as straight lines, as opposed to curves, which is consistent with steady uniform flow (far-field flow is often approximately steady and uniform, as discussed later). If you are absolutely sure that your information is correct, you could specify the entire flow in the far field, rather than just the part of the flow required by the table. However, even if your information is 100% correct, it may conflict with the slightly incorrect information found in the interior, as determined by the numerical approximation, leading to numerical problems. In other words, ignoring the above table leads to duplicative information at best, and contradictory information at worst, which possibly leads to numerically ill-posed problems and instability.

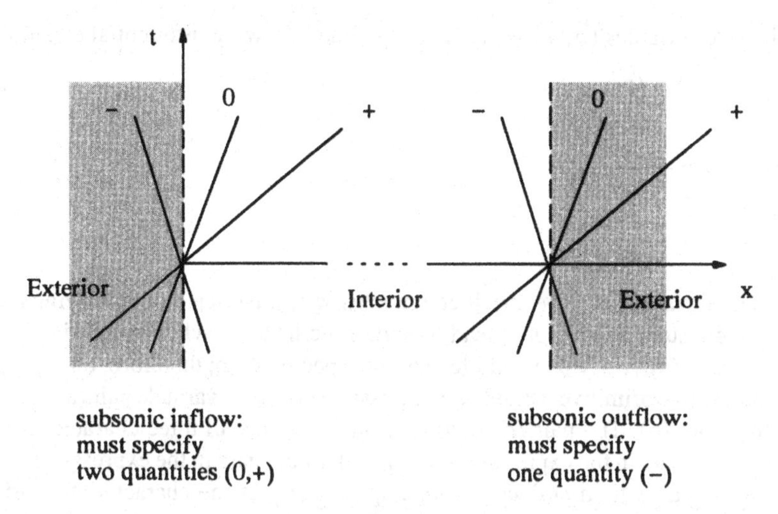

Figure 19.6 An illustration of the number of flow quantities required at far-field boundaries.

The above table specifies the number of flow properties required at far-field boundaries. However, the table does not say exactly which flow properties to specify at far-field boundaries. One can specify primitive variables, characteristic variables, conservative variables, or some combination thereof. The information carried by an incoming characteristic is, of course, a characteristic variable. From this point of view, characteristic variables are the most natural variables to specify at far-field boundaries. However, in general, characteristic variables are defined only by differential equations, which lack analytical solutions, entropy being the primary exception. Furthermore, in trying to numerically model experimental results (i.e., real life) it is difficult or impossible to directly measure most characteristic variables in the far field or elsewhere. Instead, in general, characteristic variables must be inferred from directly measurable variables. As the reader will recall from Chapter 2, the primitive variables – density, velocity, and pressure – are directly measurable. Thus, from this point of view, primitive variables are the most natural variables to specify at far-field boundaries and, indeed, in most cases, far-field boundary conditions are stated in terms of primitive variables. For example, at a subsonic outflow, the single specified variable would typically be pressure, such as the easily measured atmospheric pressure, rather than the characteristic variable v_-.

In general, primitive variables or other types of variables may be used instead of characteristic variables provided that the specified variables uniquely determine the incoming characteristic variables, when combined with interior flow information, but do not in any way restrict or specify the outgoing characteristic variable. The problem is *underspecified* unless the given variables specify incoming characteristic variables, and the problem is *overspecified* if the given variables determine outgoing characteristic variables. In fact, under- or overspecification makes the problem *ill-posed*, which is a classic cause of instability, as discussed several times now in this book, starting with Section 3.1.

Let us discuss how to choose variables to avoid under- and overspecification at far-field boundaries. As an example, consider a subsonic inflow from the left. There are two incoming characteristics, which require specification of two characteristic variables or, alternatively, two primitive variables. Recall that the characteristic variables (v_0, v_+, v_-)

and the primitive variables (ρ, u, p) are related by the following differential equations:

$$dv_0 = d\rho - \frac{dp}{a^2}, \tag{3.20a}$$

$$dv_+ = du + \frac{dp}{\rho a}, \tag{3.20b}$$

$$dv_- = du - \frac{dp}{\rho a}. \tag{3.20c}$$

The characteristic variables to be specified are (dv_0, dv_+). Notice that specifying a single primitive variable such as dp would lead to underspecification, while specifying all three primitive variables $(d\rho, du, dp)$ would lead to overspecification; therefore, the only possibility is to specify two primitive variables. The possible primitive variable pairs are $(d\rho, du)$, $(d\rho, dp)$, and (du, dp). The pair (du, dp) *partially* specifies all three characteristic variables, since $d\rho$ appears in the expression for dv_0 and du appears in the expressions for dv_+ and dv_-, but the pair $(d\rho, du)$ does not *fully* specify any of the characteristic variables. However, remember that dv_- is known from the interior flow, and then $(d\rho, du, dv_-)$ completely specifies (dv_0, dv_+) as required. Similarly, $(d\rho, dp)$ *fully* specifies dv_0, and *partially* specifies dv_+ and dv_-; however, again, dv_- is known from the interior flow, and $(d\rho, dp, dv_-)$ completely specifies (dv_0, dv_+) as required. The ringer is (du, dp), which partially specifies dv_0 and *completely* specifies dv_+ and dv_-. But specification of dv_- is wrong, since this should be determined by the interior flow. *To summarize, for subsonic inflow, one can specify density and pressure, or density and velocity, but not pressure and velocity.* While this is just one example, there is no potential for conflict when all, none, or only one variable must be specified. In one-dimensional flow, subsonic inflow is the only case with a potential for conflict, since it is the only case where two characteristic variables must be specified. Supersonic far-field boundaries require either all or nothing – they either require everything about the far-field flow or nothing about the far-field flow. Since all or nothing cannot cause conflict, supersonic boundaries are relatively simple – it does not matter which set of variables you choose to specify. Thus most papers on far-field boundary treatments concern subsonic flows, whether they say so explicitly or not.

The differential equations seen above somewhat cloud the relationships between characteristic variables and other variables. Of course, one can certainly specify dv_- (a typical choice is $dv_- = 0$ which corresponds to $v_- = const.$), but it feels better to talk about v_- and its relationships to p and u, rather than dv_- and its relationships to dp and du. In fact, in certain cases, one can dispense with the differential equations entirely. For example, for homentropic flow, the simple wave relationships are as follows:

$$s = const.,$$

$$v_+ = u + \frac{2a}{\gamma - 1},$$

$$v_- = u - \frac{2a}{\gamma - 1},$$

as seen in Section 3.4, where $a^2 = \gamma p/\rho$ and $s = c_v \ln p - c_p \ln \rho + const.$ Besides the benefit of algebraic relations, constant entropy means that there is only one variable to specify at the boundaries. Unfortunately, although the flow often enters the domain homentropically, unsteady shocks and other nonisentropic flow regimes may cause the entropy to vary, so that the flow exits the computational domain nonhomentropically. In other words, homentropic flow approximations may work better at inflow than at outflow boundaries.

We have now seen how to choose which variables to specify at the inflow and outflow to avoid duplication or conflict at the inflow or outflow. The only other worry, as far as choosing which variables to specify in the far field, is that the inflow conditions may duplicate or contradict information given at the outflow or, vice versa, the outflow conditions may duplicate or contradict information given at the inflow. For instance, suppose the inflow and outflow both have the same subsonic conditions and that the density and pressure are specified at inflow. Then velocity should be specified at the outflow; if the density or pressure were specified at the outflow, this would constitute a repetition of the information given at the inflow, and there would only be a total of two conditions specified between the inflow and outflow, rather than the required three. This issue was discussed by Wornum and Hafez (1984).

Once we have decided on a set of variables, we must still decide what values to give them. This is the trickiest part. In most cases, the only given information is the free-stream flow. Free-stream flow properties are generally known from experimental measurements, or they can be approximated by the values found in standard tables, such as tables of the standard atmosphere. By definition, the *free-stream flow* is the flow infinitely far away from any solid bodies, such as airfoils or solid walls. Put another way, the free-stream flow is the flow that would occur if all of the solid bodies disturbing the flow were removed. In some cases, the free-stream flow changes with time or space. However, for simplicity, this section will assume that the free-stream flow is both temporally *steady* and spatially *uniform*. Conceptually, the free stream is attained infinitely far away from the disturbing solid bodies; thus, in standard notation, the free-stream flow properties are given as p_∞, ρ_∞, a_∞, and so on. Of course, the numerical far-field field boundary is not at infinity, but at some finite location. Then the free-stream and far-field flow properties are generally different. However, if the far-field boundary is far enough away, then the far-field properties may approximately equal the free-stream properties. For example, if you wish to specify pressure and density at a far-field subsonic inlet, and the far-field subsonic inlet is far enough away, then you might set the far-field subsonic inlet pressure and density equal to the free-stream pressure and density.

Although it is an almost universal practice, if only by necessity, fixing far-field values equal to free-stream values raises many serious objections. Indeed, any outgoing waves violate the assumption that the free-stream conditions approximately equal the far-field conditions. For example, suppose that an expansion fan passes through the far field; the expansion fan changes the values of the far-field pressure, density, velocity, and so forth, so that they no longer equal their free-stream values. In this case, you should certainly not set the far-field pressure, density, velocity, and so forth equal to the free-stream pressure, density, velocity, and so forth.

What are the practical ramifications of mistakenly fixing the far-field flow conditions equal to the free-stream flow conditions? Well, most importantly, any outgoing waves will partially reflect from the far field, especially shock waves, as seen in Figure 19.2. Videos can dramatically illustrate the effects of reflections from far-field boundary treatments. For example, Mazaheri and Roe (1991) made a video showing outgoing waves continuously and spuriously reflecting from numerical far-field boundaries.

As another difficulty, an accumulation of very small errors over a very large far-field surface may result in large overall errors in certain crucial integrated quantities. Specifically, as the far field is moved farther and farther away, the far-field flow approaches free-stream flow; however, in multidimensions, the size of the far field also increases, so that an integral or sum of flow properties across the far field may not decrease as the far-field location increases. For example, as seen in any fluid dynamics or gasdynamics text, *circulation* is the line integral of the component of velocity parallel to a two-dimensional closed curve. For subsonic inviscid flow over an airfoil, the circulation around the airfoil should have the same value, regardless of the curve chosen to measure the circulation, provided only that the curve encircles but does not intersect the airfoil. In the limit as the far field becomes infinitely far away from the airfoil, the circulation measured across the far field equals an infinitesimal difference from the free-stream integrated over an infinite curve, yielding a finite nonzero quantity. By setting the values on the far-field surface exactly equal to free-stream values, this balance is completely thrown off, resulting in a zero value for the circulation. This may result in incorrect numerical values for the airfoil lift, which is intimately related to circulation by results such as the Kutta–Joukowski theorem, as explained in Thomas and Salas (1986) and Roe (1989).

Under ordinary circumstances, one must chose extremely distant far-field boundaries to justify fixing far-field flow conditions equal to free-stream conditions. If the far-field boundaries are drawn far enough away, then outgoing waves will be heavily dissipated by artificial viscosity by the time they reach the far field. Rather than rely on artificial viscosity, it is also possible to construct special wave-absorbing buffer zones to damp outgoing waves before they can reach the far field. However, as mentioned before, any waves that actually reach the far-field boundaries may cause difficulties. In airfoil simulations, Thomas and Salas (1986) estimate that the far field should be 50 chord lengths (50 times the length of the airfoil) away to achieve acceptable results for airfoil lift. Such distant far-field boundaries require either very large cells between the main region of interest and the far field or a very large number of cells. Very large cells in the far field require an abrupt increase in cell sizes as the far field is approached, and abrupt changes may dramatically decrease overall accuracy and possibly lead to spurious reflections from the zones where cell sizes change. On the other hand, very large numbers of more uniformly sized cells require extra expense, most of which is wasted calculating nearly uniform flows distant from the flow region of interest. In short, assuming that far-field flow properties equal free-stream flow properties may create an unacceptable trade-off between accuracy and efficiency.

Fortunately, there are several alternatives to assuming that primitive or conservative variables equal their free-stream values at far-field boundaries. As a group, these alternatives are commonly known as *nonreflecting* boundary treatments. They have also been called *radiation*, *absorbing*, *silent*, *transmitting*, *transparent*, and *one-way* boundary treatments.

Many nonreflecting boundary treatments are based on linearized approximations to the Euler equations. Thus, as an aside, let us briefly discuss the linearized Euler equations. Suppose that the far-field flow is nearly but not exactly equal to the steady uniform free-stream flow, which is usually the case, assuming that the far-field boundary is chosen sufficiently far enough away. Then one can linearize the Euler equations at the far-field boundaries about the steady uniform free-stream flow. Of course, it is also possible to linearize the Euler equations about the steady uniform free-stream flow in the interior, but only if the interior does not differ much from the free stream, which assumes thin or slender solid bodies, at the very least. Fortunately, if the far-field boundaries are drawn far enough a way from the main solid bodies, it is appropriate to linearize the governing equations in the far field even when it is not appropriate to linearize the governing equations in the interior. Thus, at the far-field boundaries, assume the following:

$$\rho = \rho_\infty + \rho', \tag{19.7a}$$

$$u = u_\infty + u', \tag{19.7b}$$

$$p = p_\infty + p', \tag{19.7c}$$

where $(\rho_\infty, u_\infty, p_\infty)$ are steady uniform free-stream values and (ρ', u', p') are small *perturbations* from the steady uniform free stream. After substituting Equation (19.7) into the Euler equations, keep any single primed quantities, but drop any terms that are products of primed quantities, on the assumption that a small quantity is much larger than a small quantity times a small quantity. This yields the following approximate system of linear partial differential equations:

$$\frac{\partial \mathbf{w}}{\partial t} + C_\infty \frac{\partial \mathbf{w}}{\partial x} = 0, \tag{19.8}$$

where:

$$C_\infty = \begin{bmatrix} u_\infty & \rho_\infty & 0 \\ 0 & u_\infty & \frac{1}{\rho_\infty} \\ 0 & \rho_\infty a_\infty^2 & u_\infty \end{bmatrix}. \tag{19.9}$$

In the far field, the Euler equations and, for that matter, the Navier–Stokes equations approximately equal the linearized Euler equations. So now let us discuss nonreflecting boundary treatments for the linearized Euler equations. The two basic approaches are as follows: (1) Model the outgoing waves to prevent their reflection using some sort of asymptotic analysis or, (2) set the characteristic variables or their derivatives equal to constants. Let us briefly consider the wave modeling approach first; although this section will discuss wave models, it will not discuss the details of how to discretize these waves models, nor how to integrate the discretized far-field wave models with the interior method. The outgoing waves typically satisfy a governing differential equation, separate from that satisfied by incoming waves or by the total combination of incoming and outgoing waves. In other words, although the outgoing waves satisfy the linearized Euler equations, they also satisfy a governing differential equation that differs from the linearized Euler equations. In many cases, the governing differential equation for outgoing waves has an exact analytical solution, giving a general functional form for the outgoing waves; some people choose to use the governing equation and some people choose to use the general functional form for the outgoing waves. As one simple approach, for subsonic inflow, Bayliss and Turkel (1982) transform the linearized Euler equations into the wave equation for pressure; this then allows them to exploit the extensive base of mathematical and numerical techniques for modeling waves in the wave equation. Notice that a single equation, in this case a single wave equation for pressure, is enough for subsonic inlets, which only allows a single outgoing family of waves.

Bayliss and Turkel (1982) and Engquist and Majda (1977) discuss various ways of modeling the outgoing waves in the wave equation. In the far field, to a first approximation, outgoing waves are often planar and one dimensional. From Section 3.0, recall that the solutions of the one-dimensional wave equation for pressure have the following form:

$$p = p_1(x - at) + p_2(x + at).$$

This solution is a superposition of two wave solutions – right-running waves with wavefronts $x - at = const.$ and left-running waves with wavefronts $x + at = const.$ At at a left-hand boundary, the outgoing waves are the left-running waves. In other words, a functional form for the outgoing waves is

$$p(x + at). \tag{19.10}$$

Alternatively, in addition to the wave equation, the outgoing waves also satisfy the following linear advection equation:

$$\frac{\partial p}{\partial t} - a\frac{\partial p}{\partial x} = 0. \tag{19.11}$$

For two-dimensional flows, many times, the outgoing waves are approximately cylindrical in the far field. In this case, a functional form for the outgoing waves is

$$\frac{p(r - at)}{\sqrt{r}}. \tag{19.12}$$

Alternatively, the outgoing waves satisfy the following differential equation:

$$\frac{\partial p}{\partial t} + a\frac{\partial p}{\partial r} + \frac{a}{2r}p = 0. \tag{19.13}$$

Similarly, in three-dimensional flows, many times, the outgoing waves are approximately spherical in the far field; the expressions in this case are omitted. In general, the waves in the far field are neither exactly planar, cylindrical, nor spherical. In this case, the planar, cylindrical, and spherical approximations are just the first terms in a longer series approximating the outgoing waves in the far field. The details are complicated, so see Bayliss and Turkel (1982) and Engquist and Majda (1977). The governing differential equations for the outgoing waves, such as Equations (19.11) and (19.13), are sometimes said to *annihilate* outgoing waves. Notice that Equations (19.10) and (19.11) are mathematically equivalent; similarly, Equations (19.12) and (19.13) are mathematically equivalent. Despite their mathematical equivalence, the governing equations and their analytical solutions may yield different numerical results, depending on how they are discretized and used in the numerical approximation.

In one dimension, as an alternative to the mathematical models of outgoing waves in the far field, consider instead the characteristics of the linearized Euler equations. Applying a characteristic analysis to Equations (19.8) and (19.9), much like the ones seen in Chapter 3, we find that the characteristic variables of the linearized Euler equations are related to the primitive variables as follows:

$$dv_0 = d\rho - \frac{dp}{a_\infty^2},$$

$$dv_+ = du + \frac{dp}{\rho_\infty a_\infty},$$

$$dv_- = du - \frac{dp}{\rho_\infty a_\infty}.$$

These differential equations have the following analytic solutions:

$$v_0 = \rho - \frac{p}{a_\infty^2}, \tag{19.14a}$$

$$v_+ = u + \frac{p}{\rho_\infty a_\infty}, \tag{19.14b}$$

$$v_- = u - \frac{p}{\rho_\infty a_\infty}. \tag{19.14c}$$

Then, for example, Giles (1990) suggests letting $dv_0 = dv_+ = 0$ ($v_0 = const.$, $v_+ = const.$) at a subsonic inflow and $dv_- = 0$ ($v_- = const.$) at a subsonic outflow. Although this works fine in one dimension, multidimensional flows require multidimensional characteristics which, as seen in Chapter 24, are distinctly complicated. From one point of view, far-field wave modeling via the wave equation as discussed in the previous paragraph models the truly multidimensional characteristics in the far field, and for less effort than the direct characteristic approach seen in this paragraph.

The linearized one-dimensional Euler equations have an interesting property: The characteristic families are completely independent. In the ordinary Euler equations, nonsimple waves or shock waves exiting the domain affect all three families of characteristics. In particular, when a wave exits or enters the numerical domain, the incoming characteristic variables must differ from their free-stream cousins. However, the one-dimensional linearized Euler equations do *not* allow nonsimple waves or shock waves – in essence, they allow only simple waves. For example, consider a subsonic inlet. In the linearized Euler equations, any wave exiting the subsonic inlet affects *only* the "−" outgoing characteristic and does not affect the incoming "0" or "+" characteristics. Thus, even when waves reach the far field, the incoming characteristic variables retain their free-stream values. Although the linearized Euler equations can legitimately assume that incoming far-field flow properties equal their free-stream values, you risk spurious reflections if you exploit this particular property in models of the true nonlinear Euler equations. For more on nonreflecting boundary treatments for the linearized Euler equations see, for example, Roe (1989), Giles (1990), Kroener (1991), and Hu (1995); also see Jameson, Schmidt, and Turkel (1981), who linearize the Euler equations at each time step about the numerical approximation from the previous time step, rather than about the free stream as in most other approaches.

Linearization still requires reasonably distant far-field boundaries. So let us consider truly nonlinear approaches. Some of the approaches used for the linearized Euler equations again apply, with modifications, to the true nonlinear Euler equations. For example, Hagstrom and Hariharan (1988) derive far-field outgoing wave models for the Euler equations using a complicated asymptotic analysis, not totally unlike the earlier linear analysis of Bayliss and Turkel (1982). For another example, arguably the simplest alternative, one can exploit flow constants, primarily characteristic variables, much as in Equation (19.14). For example, entropy is constant along pathlines, except across shocks. In the simplest cases, the entropy is constant everywhere, and not just along pathlines, which corresponds to homentropic flow. Most flow properties approach their free-stream values only gradually and exactly equal their free-stream values only at infinity; then setting such properties equal to their free-stream values at finite far-field boundaries potentially leads to significant errors, including reflections of outgoing waves, as described earlier. Contrast this with variables such as entropy, which are *exactly* constant, at least along certain special curves, regardless of where the far-field boundary is drawn, provided only that no shocks pass through the far field. Then setting entropy equal to its free-stream value along far-field boundaries is exactly correct and cannot cause reflections. Regarding far-field boundary treatments, Roe (1989) has said that entropy specification "is perhaps the only widespread current practice that is truly unobjectionable." Of course, it is also possible to specify other characteristic variables or flow constants besides entropy at the far-field boundary.

Instead of simply setting the characteristic variables equal to constant values in the far field, consider the following differential equation:

$$\frac{\partial v_i}{\partial t}(x_L, t) = 0, \tag{19.15}$$

which is a trivial consequence of assuming $v_i(x_L, t) = const.$ or $dv_i(x_L, t) = 0$. The difference between $v_i(x_L, t) = const.$ and Equation (19.15) is much like the difference between Equations (19.10) and (19.11) or between Equations (19.12) and (19.13). Equation (19.15) was first suggested by Hedstrom (1979). Concerning Hedstrom's condition, Thompson (1987) said: "As the only nonlinear [non-reflecting boundary] condition, Hedstrom's is by far the most useful for time dependent problems."

We have now examined far-field boundary treatments for the linearized Euler equations and for the full nonlinear Euler equations. The homentropic Euler equations are intermediate between the linear and full nonlinear Euler equations. Consider first-order upwind methods based on real or approximate Riemann solvers, such as Godunov's first-order upwind method, as discussed in Section 18.3. For such methods, it makes sense to treat the far-field boundary as a Riemann problem. Osher's approximate Riemann solver, mentioned in Section 5.5, replaces the true Riemann problem by an approximate homentropic Riemann problem or, more specifically, it replaces the the true shock by a simple wave, so that the approximate solution to the Riemann problem consists of a series of simple waves. Osher and Chakravarthy (1983) discuss the use of Osher's approximate Riemann solver for far-field boundary treatments, while focusing on first-order upwind methods based on real or approximate Riemann solvers. Atkins and Casper (1994) discuss far-field boundary treatments based on homentropic simple wave approximations to the Riemann problem in a completely general context.

Do not forget the main goal of the preceding procedures: They all aim to decide on appropriate values for the flow in the far field, at or just outside the edges of the numerical domain. After deciding which quantities to specify at inflow and outflow, and what values to assign to each quantity, the rest is fairly easy. As with solid boundaries, far-field boundary treatments either modify the domain or the method. The simplest option is to modify the domain using ghost cells. In this case, the term "ghost" is a bit misleading, since the ghost cells actually cover real physical flow, not an imaginary flow, albeit a part of the flow that the numerical approximation is trying hard to ignore. Of course, unlike solid boundaries, the values in the far-field ghost cells are not set by reflection. Instead, the specified far-field flow properties, such as incoming characteristic variables, are combined with interior flow properties, such as outgoing characteristic variables, to form a complete set of flow properties for the ghost cells. In order to easily relate characteristic variables to other variables, a linearized or homentropic (simple wave) approach is often used, as described above. As the above discussion implies, in general, the ghost cell properties may differ slightly from the steady uniform free-stream values; although the far-field boundary is sometimes assumed to have uniform free-stream values for certain flow properties, simple-mindedly enforcing this assumption may destabilize the calculation, since it overspecifies the problem, as mentioned above. After setting the ghost cell values, the numerical method proceeds as usual in the far field, using ghost cell values as needed. The alternative to ghost cells is to use an uncentered method near the boundaries; care must be taken to ensure that the uncentered method incorporates the required far-field values.

Hayder and Turkel (1995) compare a number of subsonic far-field boundary treatments, including those suggested by Bayliss and Turkel (1982), Hagstrom and Hariharan (1988), Roe (1989), and Giles (1990). In their particular problem, Hayder and Turkel (1995) prefer the treatments suggested by Hagstrom and Hariharan (1988) and Giles (1990).

References

Atkins, H., and Casper, J. 1994. "Nonreflective Boundary Conditions for High-Order Methods," *AIAA Journal*, 32: 512–518.

Bayliss, A., and Turkel, E. 1982. "Far-field Boundary Conditions for Compressible Flows," *Journal of Computational Physics*, 48: 182–199.

Beam, H. C., Warming, R. F., and Yee, H. C. 1982. "Stability Analysis of Numerical Boundary Conditions and Implicit Finite Difference Approximations for Hyperbolic Equations," *Journal of Computational Physics*, 48: 200–222.

Casper, J., and Carpenter, M. H. 1995. "Computational Considerations for the Simulation of Shock-Induced Sound," *NASA Technical Memorandum 110222* (unpublished). To appear in the *SIAM Journal on Scientific and Statistical Computing.*

Chakravarthy, S. R. 1983. "Euler Equations – Implicit Schemes and Boundary Conditions," *AIAA Journal*, 21: 699–705.

Courant, R., and Friedrichs, K. O. 1948. *Supersonic Flow and Shock Waves*, New York: Springer-Verlag.

Dadone, A., and Grossman, B. 1994. "Surface Boundary Conditions for the Numerical Solution of the Euler Equations," *AIAA Journal*, 32: 285.

Donat, R., and Marquina, A. 1996. "Capturing Shock Reflections: An Improved Flux Formula," *Journal of Computational Physics*, 125: 42–58.

Engquist, B., and Majda, A. 1977. "Absorbing Boundary Conditions for the Numerical Simulation of Waves," *Mathematics of Computation*, 31: 629–651.

Ferm, L. 1995. "Non-Reflecting Boundary Conditions for the Steady Euler Equations," *Journal of Computational Physics*, 122: 307–316.

Giles, M. B. 1990. "Nonreflecting Boundary Conditions for Euler Equation Calculations," *AIAA Journal*, 28: 2050–2058.

Givoli, D. 1991. "Non-Reflecting Boundary Conditions," *Journal of Computational Physics*, 94: 1–29.

Gottlieb, D., and Turkel, E. 1978. "Boundary Conditions for Multistep Finite-Difference Methods for Time-Dependent Equations," *Journal of Computational Physics*, 26: 181–196.

Grinstein, F. F. 1994. "Open Boundary Conditions in the Simulation of Subsonic Turbulent Shear Flows," *Journal of Computational Physics*, 115: 43–55.

Gustafsson, B. 1975. "The Convergence Rate for Difference Approximations to Mixed Initial Boundary Value Problems," *Mathematics of Computation*, 29: 396–406.

Gustafsson, B. 1982. "The Choice of Numerical Boundaries for Hyperbolic Systems," *Journal of Computational Physics*, 48: 270–283.

Gustafsson, B., Kreiss, H.-O., and Sundstrom, A. 1972. "Stability Theory of Difference Approximations for Mixed Initial Boundary Value Problems II," *Mathematics of Computation* 26: 649–686.

Hagstrom, T., and Hariharan, S. I. 1988. "Accurate Boundary Conditions for Exterior Problems in Gas Dynamics," *Mathematics of Computation*, 51: 581–597.

Hayder, M. E., and Turkel, E. 1995. "Nonreflecting Boundary Conditions for Jet Flow Computations," *AIAA Journal*, 33: 2264–2270.

Hedstrom, G. W. 1979. "Nonreflecting Boundary Conditions for Nonlinear Hyperbolic Systems," *Journal of Computational Physics*, 30: 222–237.

Higdon, R. L. 1986. "Initial Boundary Value Problems for Linear Hyperbolic Systems," *SIAM Review*, 28: 177–217.

Hu, F. Q. 1995. "On Absorbing Boundary Conditions for Linearized Euler Equations By a Perfectly Matched Layer," *ICASE Report Number* 95–70 (unpublished).

Jameson, A., Schmidt, W., and Turkel, E. 1981. "Numerical Solutions of the Euler Equations by Finite Volume Methods Using Runge–Kutta Time-Stepping Schemes," *AIAA Paper 81-1259* (unpublished).

Karni, S. 1991. "Accelerated Convergence to Steady State by Gradual Far-Field Damping," *AIAA Paper 91-1604* (unpublished).

Kreiss, H.-O. 1966. "Difference Approximations for the Initial-Boundary Value Problem for Hyperbolic Differential Equations." In *Numerical Solutions of Nonlinear Differential Equations*, ed. D. Greenspan, pp. 141–166, Englewood Cliffs, NJ: Prentice-Hall.

Kreiss, H.-O. 1968. "Stability Theory for Difference Approximations of Mixed Initial Boundary Value Problems," *Mathematics of Computation*, 22: 703–714.

Kreiss, H.-O. 1970. "Initial Boundary Value Problems for Hyperbolic Systems," *Communications on Pure and Applied Mathematics*, 23: 277–298.

Kroener, D. 1991. "Absorbing Boundary Conditions for the Linearized Euler Equations in 2-D," *Mathematics of Computation*, 57: 153.

Mazaheri, K., and Roe, P. L. 1991. "New Light on Numerical Boundary Conditions," *AIAA Paper 91-1600* (unpublished).

Moretti, G. 1969. "Importance of Boundary Conditions in the Numerical Treatment of Hyperbolic Equations," *High Speed Computing in Fluid Dynamics, The Physics of Fluid Supplement II*, 12: 13–20.

Osher, S., and Chakravarthy, S. 1983. "Upwind Schemes and Boundary Conditions with Applications to Euler Equations in General Geometries," *Journal of Computational Physics*, 50: 447–481.

Pulliam, T. H. 1982. "Characteristic Boundary Conditions for the Euler Equations." In *Numerical Boundary Condition Procedures*, NASA Conference Publication 2201 (unpublished).

Roe, P. L. 1989. "Remote Boundary Conditions for Unsteady Multidimensional Aerodynamic Computations," *Computers and Fluids*, 17: 221–231.

Rudy, D. H., and Strikwerda, J. C. 1980. "A Nonreflecting Outflow Boundary Condition for Subsonic Navier–Stokes Calculations," *Journal of Computational Physics*, 36: 55–70.

Rudy, D. H., and Strikwerda, J. C. 1981. "Boundary Conditions for Subsonic Compressible Navier–Stokes Calculations," *Computers and Fluids*, 9: 327.

Saxer, A. P., and Giles, M. B. 1991. "Quasi-3D Non-Reflecting Boundary Conditions for Euler Equations Calculations," *AIAA Paper 91-1603* (unpublished).

Shu, C.-W. 1987. "TVB Boundary Treatment for Numerical Solutions of Conservation Laws," *Mathematics of Computation*, 49: 123–134.

Thomas, J. L., and Salas, M. D. 1986. "Far-Field Boundary Conditions for Transonic Lifting Solutions to the Euler Equations," *AIAA Journal*, 24: 1074–1080.

Thompson, K. W. 1987. "Time Dependent Boundary Conditions for Hyperbolic Systems," *Journal of Computational Physics*, 68: 1–24.

Wornum, S. F., and Hafez, M. 1984. "A Rule for Selecting Analytical Boundary Conditions for the Conservative Quasi-One-Dimensional Nozzle Flow Equations," *AIAA Paper 84-0431* (unpublished).

Yee, H. C., Beam, R. M., and Warming, R. F. 1982. "Boundary Approximations for Implicit Schemes for One-Dimensional Inviscid Equations of Gasdynamics," *AIAA Journal*, 20: 1203–1211.

Problems

19.1 (a) Consider Godunov's first-order upwind method, as described in Section 18.3. Suppose Godunov's first-order upwind method is used with a type 1 grid. Also suppose that the solid boundary treatment is based on ghost cells and reflection. How will Godunov's first-order upwind method behave at the solid boundaries? Will it enforce the no-penetration boundary condition exactly?

(b) Repeat part (a), replacing Godunov's method by MacCormack's method, as described in Section 18.1.

(c) Repeat parts (a) and (b) replacing a type 1 grid by a type 2 grid.

19.2 Discuss briefly some of the issues involved when a shock wave reaches a far-field boundary. Specifically consider what happens when the shock changes the flow from sub- to supersonic or vice versa. Why do shocks cause more troubles at far-field boundaries than at solid boundaries?

19.3 Section 19.3 described how to avoid conflicts between primitive variables specified in the far field. Now suppose you wish to specify conservative rather than primitive variables in the far field. Using an analysis similar to that found in Section 19.3, explain how to avoid conflicts.

19.4 Prove Equations (19.8) and (19.9). If you have never seen a small perturbation analysis before, you may wish to look in a gasdynamics text or, for that matter, some other type of text for examples. If you look carefully, you may even be able to find a book with the required proof; in this case, make sure to put the proof in your own words.

Part V

Advanced Methods of Computational Gasdynamics

Logically, all things are created by a combination of simpler, less capable components.

Dogbert in Dilbert *by Scott Adams*

This part of the book concerns solution-sensitive methods. After possibly accounting for the wind direction, the first-generation methods studied so far mindlessly treat every part of the solution the same, regardless of how the solution behaves. The solution-sensitive methods studied in Part V combine a range of first-generation methods, varying the exact blend from place to place based on solution features such as shocks. For example, a solution-sensitive method might use Roe's first-order upwind method at shocks and the Lax–Wendroff method in smooth regions. Most solution-sensitive methods decide what to do based on solution gradients or flux gradients; large gradients indicate shocks or other features with the potential to cause trouble in the polynomial interpolations underlying most numerical methods.

In the literature, solution-sensitive methods are commonly called *high-resolution* or *TVD* methods. Less common terms include *hybrid, adaptive, self-adjusting, averaged, reconstructed, essentially nonoscillatory, combination, corrected,* and *limited* methods. No one book could describe the huge number of solution-sensitive methods suggested in the research literature. Part V includes methods based partly on historical significance and influence, partly on how well they exemplify general principles and notations, partly on current popularity, and partly on the author's personal tastes.

Many solution-sensitive methods explicitly enforce nonlinear stability conditions, such as the upwind range condition or the TVD condition. Stability conditions reduce spurious oscillations near shocks, but they also typically impose clipping errors at solution extrema. Although combination methods have trade-offs, these are far less severe than for the fixed methods studied in Part IV. In particular, far from trading stability with accuracy at extrema, fixed methods trade stability with accuracy throughout the *entire* solution. Fixed methods either do well at shocks or well in smooth regions, but not both simultaneously, which stems from the identical behavior found in the underlying interpolation polynomials, as discussed in Chapters 8, 9, and 10.

Part V tries to combine the best features of logical and chronological ordering, to give both a necessary sense of history and of the logical connections between apparently disparate methods. As a first cut, the methods in Part V are broken down logically into either *flux-averaged methods* or *solution-averaged* methods, more commonly known as *reconstruction–evolution* methods. In an alternative but equivalent terminology, the methods in Part V are divided into *flux-reconstructed* and *solution-reconstructed* methods. As a second cut, the flux-averaged methods are divided logically into flux-limited, flux-corrected, and self-adjusting hybrid methods. Each of these four threads – flux limited, flux corrected, self-adjusting hybrid, and reconstruction–evolution – has its own chapter. While Part V is divided logically into chapters, each chapter is divided chronologically into sections. In other words, each chapter orders its methods according to their birth dates.

The first section in each chapter of Part V describes a seminal *second-generation* method. Subsequent sections describe *third-generation* methods inspired by the second-generation method. Designed in the 1970s, second-generation methods have limited practical potential.

For example, many second-generation methods only apply to the linear advection equation. Those developed for the Euler equations have relatively primitive features, especially with regard to modeling multiwave families. The period from 1979 to 1983 saw several important advances in the first-generation methods underlying the second-generation methods, including flux vector splitting, real and approximate Riemann solvers, and entropy fixes, as described in Chapter 18. From 1979 to 1987, and more slowly since then, researchers formed the third-generation methods by combining the advances in the first-generation methods with the combination approaches developed earlier for second-generation methods.

As explained previously, Part V separates flux-averaged methods into three species – flux limited, self-adjusting hybrid, and flux corrected. However, these divisions mask the fact that, on the most fundamental level, all three approaches are the same. The differences arise not in the basic ideas and philosophies, but in the details, the traditions, the jargon, and the notations. Thus flux-averaged methods are like triplets separated at birth: They have much of the same basic genetics, but environmental factors have conspired to make them increasingly different from each other over the years. For linear systems of equations, there is no fundamental distinction between flux averaging, in any form, and solution averaging. In other words, if $f(u) = au$ where $a = const.$, it does not matter whether you average to find u and then find $f(u)$ or whether you average to find $f(u)$ directly, since the two types of averages differ only by the constant factor a. However, for nonlinear equations with nonlinear fluxes, flux averaging and solution averaging are distinct, with only a few notable exceptions.

Part V describes numerical methods using a modular approach, first covering scalar conservation laws and then the Euler equations. In a few cases, for clarity or brevity, the description starts with the linear advection equation or ends with scalar conservation laws. To save space, Part V only tests numerical methods for scalar conservation laws.

Part V describes over a dozen distinct numerical methods. The "best" method depends on your application, efficiency demands, accuracy demands, complexity tolerance, and taste. Unfortunately, the difficulties of full-scale numerical testing and the limits of theoretical analysis do not allow for complete and definitive descriptions of the trade-offs between specific methods. When you realize that even problems as simple as integration and differentiation have spawned a number of peacefully coexisting numerical methods, as seen in Chapter 10, it should come as no surprise that something as complicated as a system of partial differential or integral equations should suffer a similar fate.

Although this book cannot offer the reader a "best" method, it can offer a personal perspective on the current popularity of the various categories of methods. Remember that Part V divides methods into four species – flux limited, flux corrected, self-adjusting hybrids, and reconstruction–evolution. Of these four, flux-limited and "flux limited–like" methods currently appear to have the upper hand. Flux-corrected methods reached maturity first, but they have not received the same level of attention and continuing development in recent times as other sorts of methods and have accordingly fallen somewhat behind the state of the art. However, certain communities still use flux-corrected methods almost exclusively. Outside of these communities, however, the most widely used flux-corrected methods strongly resemble traditional flux-limited methods. Self-adjusting hybrids were the most popular sort of methods in the 1980s, but these have more recently lost ground to flux-limited and "flux limited–like" methods. Last but not least, reconstruction–evolution methods are the Rolls Royce of numerical methods – elegant, elaborate, rigorous, prestigious, but too expensive for most people. The only exceptions are, again, those reconstruction–evolution methods that strongly resemble traditional flux-limited methods.

The reader may find it enlightening to see the course taken by one of the leading independent commercial software companies in this field. FLUENT corporation offers a computational gasdynamics package called RAMPANT. In its original incarnation, written in the late 1980s, RAMPANT used Jameson's self-adjusting hybrid method, described in Section 22.3. However, the most recent version of RAMPANT uses a modified version of the Anderson–Thomas–Van Leer method described in Subsection 23.3.1, which is a finite-volume version of the Chakravarthy–Osher flux-limited method seen in Subsection 20.3.1, which is, in turn, a semidiscrete version of Sweby's flux-limited method seen in Section 20.2. In conversation, Dr. Wayne Smith, who played a leading role in developing both the original and new versions of RAMPANT, gives two reasons for the change in the core method: First, the flux-limited method does a better job at reducing spurious overshoots and oscillations; second, the flux-limited method seems to mate better with the viscous terms and the other sorts of important real-life effects that routinely appear in practical applications. Like most commercial codes, RAMPANT incorporates a plethora of features, many of them heuristic, which adapt it to complicated geometries, and to many types of compressible gas flows, with maximum efficiency for either steady or unsteady flows. Thus, in the end, the core method is not necessarily as important as all of the implementation details, efficiencies, and generalizations layered on top. This book is devoted to explaining the core methods, since this is the first and most vital step towards understanding other aspects of numerical methods, as explained starting way back in Chapter 1.

This part of the book requires extra effort and patience, not because it involves especially difficult concepts or details, but because it combines on so many concepts and details at once, which often requires complicated notation and a flawless command of the rest of the book. Despite every effort to improve the presentation, some of the sections seen in Part V still require the cool skill of a matador combined with the calm steadfastness of a saint. However, having said this, if you have assiduously studied the earlier chapters, you might be surprised at how quickly you master most of the remaining material, especially if you remember to rest occasionally and to reduce things into simple modular components and concepts.

CHAPTER 20

Flux Averaging I: Flux-Limited Methods

20.0 Introduction

To keep things simple to begin with, this introduction concerns only flux-limited methods for scalar conservation laws. Flux-limited methods for the Euler equations are discussed later in the chapter; see Subsections 20.2.6, 20.4.2, and 20.5.2. Intuitively, flux-limited methods are adaptive linear combinations of two first-generation methods. More specifically, away from sonic points, flux-limited methods for scalar conservation laws are defined as follows:

$$u_i^{n+1} = u_i^n - \lambda\left(\hat{f}^n_{i+1/2} - \hat{f}^n_{i-1/2}\right),$$

where

$$\hat{f}^n_{i+1/2} = \hat{f}^{(1)}_{i+1/2} + \phi^n_{i+1/2}\left(\hat{f}^{(2)}_{i+1/2} - \hat{f}^{(1)}_{i+1/2}\right) \tag{20.1}$$

and where $\hat{f}^{(1)}_{i+1/2}$ and $\hat{f}^{(2)}_{i+1/2}$ are the conservative numerical fluxes of two first-generation methods with complementary properties, such as Roe's first-order upwind method and the Lax–Wendroff method, and where the adaptive parameter $\phi^n_{i+1/2}$ controlling the linear combination is called a *flux limiter*. By tradition, flux-limited methods often bump the spatial index on the flux limiter up or down by one half, depending on wind direction. In particular, if $a > 0$

$$\hat{f}^n_{i+1/2} = \hat{f}^{(1)}_{i+1/2} + \phi^n_i\left(\hat{f}^{(2)}_{i+1/2} - \hat{f}^{(1)}_{i+1/2}\right) \tag{20.2a}$$

and if $a < 0$ then

$$\hat{f}^n_{i+1/2} = \hat{f}^{(1)}_{i+1/2} + \phi^n_{i+1}\left(\hat{f}^{(2)}_{i+1/2} - \hat{f}^{(1)}_{i+1/2}\right). \tag{20.2b}$$

The reader should view Equations (20.1) and (20.2) as starting points for discussion, rather than as "fixed in stone" definitions of flux-limited methods. Many flux-limited methods use different notations, masking their common heritage. However, Equations (20.1) and (20.2) make a useful reference touchstone and clearly express the basic idea behind all flux-limited methods.

After choosing the two first-generation methods, one must choose the flux limiter ϕ. But how? As one possible approach, flux-limited methods might use one first-generation method near shocks and a different first-generation method in smooth regions, which then raises the question of how to distinguish shocks from smooth regions. Large first differences such as $u^n_{i+1} - u^n_i$ often indicate shocks, but terms like "large" and "small" are relative – they only make sense when judged against some standard. In flux-limited methods, by tradition, the reference for a "large" or "small" first difference is a neighboring first difference. In particular, two possible relative measures are the following *ratios of solution differences*:

$$r_i^+ = \frac{u_i^n - u_{i-1}^n}{u_{i+1}^n - u_i^n}, \tag{20.3}$$

$$r_i^- = \frac{u_{i+1}^n - u_i^n}{u_i^n - u_{i-1}^n}. \tag{20.4}$$

Notice that $r_i^+ = 1/r_i^-$. The ratios of solution differences have the following properties:

- $r_i^{\pm} \geq 0$ if the solution is monotone increasing ($u_{i-1}^n \leq u_i^n \leq u_{i+1}^n$) or monotone decreasing ($u_{i-1}^n \geq u_i^n \geq u_{i+1}^n$).
- $r_i^{\pm} \leq 0$ if the solution has a maximum ($u_{i-1}^n \leq u_i^n, u_{i+1}^n \leq u_i^n$) or a minimum ($u_{i-1}^n \geq u_i^n, u_{i+1}^n \geq u_i^n$).
- $|r_i^+|$ is large and $|r_i^-|$ is small if the solution differences decrease dramatically from left to right ($|u_i^n - u_{i-1}^n| \gg |u_{i+1} - u_i^n|$) or if $u_{i+1}^n \approx u_i^n$. In fact, in this latter case, to prevent numerical overflow, you may wish to let $r_i^+ = sign(u_{i+1}^n - u_i^n)(u_i^n - u_{i-1}^n)/\delta$ for $|u_{i+1}^n - u_i^n| < \delta$.
- $|r_i^+|$ is small and $|r_i^-|$ is large if the solution differences increase dramatically from left to right ($|u_i^n - u_{i-1}^n| \ll |u_{i+1} - u_i^n|$) or if $u_{i-1}^n \approx u_i^n$. In fact, in this latter case, to prevent numerical overflow, you may wish to let $r_i^- = (u_i^n - u_{i-1}^n)(u_{i+1}^n - u_i^n)/\delta$ for $|u_i^n - u_{i-1}^n| < \delta$.

Large increases or decreases in the solution differences, as indicated by very large or very small ratios $|r_i^{\pm}|$, sometimes signal shocks, but not always. For example, if $u_{i+1}^n - u_i^n = 0$ and $u_i^n - u_{i-1}^n \neq 0$ then $|r_i^+| = \infty$ regardless of whether the solution is smooth or shocked. In fact, $r_i^{\pm}$ may assume any value, large or small, either at shocks or in smooth regions. Because there is only limited information contained in solution samples, no completely reliable way to distinguish shocks from smooth regions exists; this is discussed further in Section 22.0. Consequently, flux-limited methods do not even attempt to identify shocks. Instead, by tradition, flux-limited methods regulate maxima and minima, whether or not they are associated with shocks, using the nonlinear stability conditions described in Chapter 16.

As an alternative to solution differences, consider shock indicators based on flux-differences $\hat{f}_{i+1/2}^{(2)} - \hat{f}_{i+1/2}^{(1)}$. In smooth flow regions, the flux-difference should be small. For example, if $\hat{f}_{i+1/2}^{(1)}$ and $\hat{f}_{i+1/2}^{(2)}$ are both second-order accurate methods in smooth regions, then $\hat{f}_{i+1/2}^{(2)} - \hat{f}_{i+1/2}^{(1)} = O(\Delta x^2)$ in smooth regions. For another example, if $\hat{f}_{i+1/2}^{(1)}$ is a first-order accurate method in smooth regions, and $\hat{f}_{i+1/2}^{(2)}$ is any method with first- or high-order accuracy in smooth regions, then $\hat{f}_{i+1/2}^{(2)} - \hat{f}_{i+1/2}^{(1)} = O(\Delta x)$ in smooth regions. Near shocks, both the accuracy and the order of accuracy of $\hat{f}_{i+1/2}^{(1)}$ and $\hat{f}_{i+1/2}^{(2)}$ typically drop. Then, near shocks, $\hat{f}_{i+1/2}^{(2)} - \hat{f}_{i+1/2}^{(1)}$ will be large where, as before, "large" is a relative term. Like solution differences, by tradition, large flux differences are judged relative to neighboring flux differences. In particular, two possible relative measures are the following *ratios of flux differences*:

$$r_i^+ = \frac{\hat{f}_{i-1/2}^{(2)} - \hat{f}_{i-1/2}^{(1)}}{\hat{f}_{i+1/2}^{(2)} - \hat{f}_{i+1/2}^{(1)}}, \tag{20.5a}$$

$$r_i^- = \frac{\hat{f}_{i+1/2}^{(2)} - \hat{f}_{i+1/2}^{(1)}}{\hat{f}_{i-1/2}^{(2)} - \hat{f}_{i-1/2}^{(1)}}. \tag{20.6a}$$

Notice that $r_i^+ = 1/r_i^-$. By tradition, ratios of solution differences and ratios of the flux differences both use the same notation $r_i^{\pm}$. As with the ratio of solution differences, very large or very small ratios of flux differences are often, but certainly not always, caused by shocks. However, rather than attempting to exploit this interesting but unreliable observation, most flux-limited methods regulate maxima and minima, whether or not they are associated with shocks, using nonlinear stability conditions, as mentioned before.

As it turns out, the ratios of solution differences and the ratios of flux differences are closely related. Written in terms of artificial viscosity, as defined in Chapter 14, $\hat{f}^{(1)}_{i+1/2}$ and $\hat{f}^{(2)}_{i+1/2}$ and are as follows:

$$\hat{f}^{(1)}_{i+1/2} = \frac{1}{2}\left(f(u^n_{i+1}) + f(u^n_i)\right) - \frac{1}{2}\epsilon^{(1)}_{i+1/2}(u^n_{i+1} - u^n_i),$$

$$\hat{f}^{(2)}_{i+1/2} = \frac{1}{2}\left(f(u^n_{i+1}) + f(u^n_i)\right) - \frac{1}{2}\epsilon^{(2)}_{i+1/2}(u^n_{i+1} - u^n_i).$$

Then

$$\hat{f}^{(2)}_{i+1/2} - \hat{f}^{(1)}_{i+1/2} = \frac{1}{2}\left(\epsilon^{(1)}_{i+1/2} - \epsilon^{(2)}_{i+1/2}\right)(u^n_{i+1} - u^n_i). \tag{20.7}$$

Then the ratios of flux-differences are

$$r_i^+ = \frac{\left(\epsilon^{(1)}_{i-1/2} - \epsilon^{(2)}_{i-1/2}\right)(u^n_i - u^n_{i-1})}{\left(\epsilon^{(1)}_{i+1/2} - \epsilon^{(2)}_{i+1/2}\right)(u^n_{i+1} - u^n_i)}, \tag{20.5b}$$

$$r_i^- = \frac{\left(\epsilon^{(1)}_{i+1/2} - \epsilon^{(2)}_{i+1/2}\right)(u^n_{i+1} - u^n_i)}{\left(\epsilon^{(1)}_{i-1/2} - \epsilon^{(2)}_{i-1/2}\right)(u^n_i - u^n_{i-1})}. \tag{20.6b}$$

These equations say that ratios of flux-differences equal ratios of artificial viscosity differences times ratios of solution differences. If $\epsilon^{(1)}_{i+1/2} - \epsilon^{(2)}_{i+1/2} = const.$ then ratios of flux-differences equal ratios of solution differences. In other words, if $\epsilon^{(1)}_{i+1/2} - \epsilon^{(2)}_{i+1/2} = const.$ then Equation (20.3) equals (20.5), and Equation (20.4) equals (20.6). However, in most cases, $\epsilon^{(1)}_{i+1/2} - \epsilon^{(2)}_{i+1/2} \neq const.$, except possibly when the numerical method is applied to the linear advection equation.

If $\epsilon^{(1)}_{i+1/2} \geq \epsilon^{(2)}_{i+1/2}$, then Equations (20.5) and (20.6) preserve two of the most important properties of Equations (20.3) and (20.4). Specifically, if $\epsilon^{(1)}_{i+1/2} \geq \epsilon^{(2)}_{i+1/2}$ then $r_i^{\pm} \geq 0$ if the solution is monotone increasing or decreasing, and $r_i^{\pm} \leq 0$ if the solution has a maximum or a minimum. If $\epsilon^{(1)}_{i+1/2} \geq \epsilon^{(2)}_{i+1/2}$ then

$$\hat{f}^{(2)}_{i+1/2} - \hat{f}^{(1)}_{i+1/2} = \frac{1}{2}\left(\epsilon^{(1)}_{i+1/2} - \epsilon^{(2)}_{i+1/2}\right)(u^n_{i+1} - u^n_i)$$

is called *antidissipative flux*. Also,

$$\frac{1}{2}\phi^n_{i+1/2}\left(\hat{f}^{(2)}_{i+1/2} - \hat{f}^{(1)}_{i+1/2}\right) = \frac{1}{2}\phi^n_{i+1/2}\left(\epsilon^{(1)}_{i+1/2} - \epsilon^{(2)}_{i+1/2}\right)(u^n_{i+1} - u^n_i)$$

is called *limited antidissipative flux* or, sometimes, *adaptive artificial viscosity*. Then, in one interpretation, Equation (20.1) equals an overly dissipative method $\hat{f}^{(1)}_{i+1/2}$ plus a limited amount of antidissipative flux $\phi^n_{i+1/2}(\hat{f}^{(2)}_{i+1/2} - \hat{f}^{(1)}_{i+1/2})$. Intuitively, antidissipation should be small at shocks and large in smooth regions, although flux-limited methods usually choose the antidissipation using more concrete approaches, such as nonlinear stability, as mentioned several times now. It should be noted that the term "antidissipative flux" originates with flux-corrected methods, as described in Chapter 21, while the term "adaptive artificial viscosity" originates with self-adjusting hybrid methods, as described in Chapter 22; however, the strong links among flux-limited methods, flux-corrected methods, and self-adjusting hybrid methods allow terminology developed for one sort of flux-averaged method to migrate naturally to other sorts of flux-averaged methods.

As one final possible condition on the flux limiter, suppose that $0 \leq \phi^n_{i+1/2} \leq 1$. Then, according to Equation (20.1), $\hat{f}^n_{i+1/2}$ is a *linear average*, or *linear interpolation*, or *convex linear combination* of $\hat{f}^{(1)}_{i+1/2}$ and $\hat{f}^{(2)}_{i+1/2}$. Convex linear combinations have the following property:

$$\min\left(\hat{f}^{(1)}_{i+1/2}, \hat{f}^{(2)}_{i+1/2}\right) \leq \hat{f}^n_{i+1/2} \leq \max\left(\hat{f}^{(1)}_{i+1/2}, \hat{f}^{(2)}_{i+1/2}\right). \tag{20.8}$$

Hence, for convex linear combinations, $\hat{f}^n_{i+1/2}$ is always somewhere between $\hat{f}^{(1)}_{i+1/2}$ and $\hat{f}^{(2)}_{i+1/2}$. This may or may not make sense, depending on the choice of $\hat{f}^{(1)}_{i+1/2}$ and $\hat{f}^{(2)}_{i+1/2}$. In particular, if $\hat{f}^{(1)}_{i+1/2}$ and $\hat{f}^{(2)}_{i+1/2}$ are close together, then it might make sense to extrapolate to outside fluxes. In contrast, if $\hat{f}^{(1)}_{i+1/2}$ and $\hat{f}^{(2)}_{i+1/2}$ are highly separated, spanning a range of desirable methods, then it makes sense to interpolate between $\hat{f}^{(1)}_{i+1/2}$ and $\hat{f}^{(2)}_{i+1/2}$. By tradition, flux-limited methods ignore convexity. Instead, the flux limiter is chosen to ensure nonlinear stability conditions such as positivity or the upwind range condition, as discussed in Chapter 16, regardless of convexity.

One way to obtain a flux limiter is to rewrite an existing method in a flux-limited form. Specifically, if an existing method has a conservative numerical flux $\hat{f}^n_{i+1/2}$, then one can solve for the flux limiter $\phi^n_{i+1/2}$ in the following equation:

$$\hat{f}^n_{i+1/2} = \hat{f}^{(1)}_{i+1/2} + \phi^n_{i+1/2}\left(\hat{f}^{(2)}_{i+1/2} - \hat{f}^{(1)}_{i+1/2}\right).$$

Of course, the solution is trivial:

$$\phi^n_{i+1/2} = \frac{\hat{f}^n_{i+1/2} - \hat{f}^{(1)}_{i+1/2}}{\hat{f}^{(1)}_{i+1/2} - \hat{f}^{(2)}_{i+1/2}}.$$

The only problem occurs if $\hat{f}^{(1)}_{i+1/2} - \hat{f}^{(2)}_{i+1/2} = 0$. In this case, if $\hat{f}^n_{i+1/2} = \hat{f}^{(1)}_{i+1/2} = \hat{f}^{(2)}_{i+1/2}$ then $\phi^n_{i+1/2}$ may have any arbitrary finite value. Otherwise, if $\hat{f}^n_{i+1/2} \neq \hat{f}^{(1)}_{i+1/2} = \hat{f}^{(2)}_{i+1/2}$, then either the flux limiter must equal infinity or $\hat{f}^{(1)}_{i+1/2}$ and $\hat{f}^{(2)}_{i+1/2}$ must be altered. In any event, numerical methods can be written in any number of flux-limited forms, depending on the choice of $\hat{f}^{(1)}_{i+1/2}$ and $\hat{f}^{(2)}_{i+1/2}$. Writing an established method in a flux-limited form is not much of an accomplishment, since it rarely provides new insights into existing methods, unlike conservation form, wave speed split forms, or artificial viscosity form. The flux-limited form is useful only to the extent that it inspires *new* methods.

This chapter describes five flux-limited methods. Van Leer's flux-limited method for the linear advection equation was proposed in 1974 in the second of a series of five papers ambitiously entitled "Towards the Ultimate Conservative Difference Scheme." Unfortunately, at the time, Van Leer was unable to extend his approach to nonlinear scalar conservation laws or to the Euler equations. Thus Van Leer abandoned his flux-limited approach in the last two entries in the "Ultimate" series, opting instead for the reconstruction–evolution method seen in Section 23.1. Ten years later, Sweby (1984) proposed a popular class of flux-limited methods heavily based on Van Leer's flux-limited method and on later work by Roe (1981, 1982) and Roe and Baines (1982). Sweby used techniques and concepts developed in the intermediate ten years, not available earlier to Van Leer, plus his own novel techniques to extend flux-limited methods to nonlinear scalar conservation laws and to the Euler equations. Sweby's flux-limited method sparked an immediate firestorm of new flux-limited methods, including several by the mathematician–engineer team of Osher and Chakravarthy, as discussed in Section 20.3. Although Sweby allowed any number of first-order upwind methods for $\hat{f}^{(1)}_{i+1/2}$, he fixed $\hat{f}^{(2)}_{i+1/2}$ equal to the Lax–Wendroff method. Chakravarthy and Osher (1985) removed this restriction to obtain a broad class of flux-limited methods, including flux-limited versions of the Lax–Wendroff method, the Beam–Warming second-order upwind method, and Fromm's method. In an unpublished NASA-ICASE report, Roe (1984) suggested a class of flux-limited methods, designed as efficient simplifications of Sweby's flux-limited method. Roe's ideas were developed and published by Davis (1987) and Yee (1987). Dr. Roe played a major behind-the-scenes role in developing most of the third-generation flux-limited methods, although his failure to publish his results in ordinary journals has limited recognition of his accomplishments in this regard.

All of the flux-limited methods seen in this chapter are commonly called TVD methods, except for Van Leer's. Although Van Leer's flux-limited method is not commonly called a TVD method, this is only because it preceded the invention of the term TVD by almost ten years. In the literature, the term "TVD" refers to a wide range of modern methods, including various sorts of flux-limited methods, as well as other types of flux-averaged methods. You might think that TVD methods would focus on the TVD stability condition, as defined in Section 16.2. But, in fact, TVD methods almost always enforce a variety of stronger nonlinear stability conditions including range diminishing (Section 16.3), positivity (Section 16.4), the upwind range condition (Section 16.5), or other nonlinear stability conditions that imply TVD. In practice, the term "TVD method" refers to numerical methods that (1) involve solution sensitivity, (2) use the solution sensitivity to enforce some sort of nonlinear stability condition, at least for certain model problems, (3) limit the order of accuracy at extrema, usually to between first- and second-order, and (4) came after the invention of the term TVD, circa 1983.

Unlike the standard literature, this book does not identify numerical methods by their purported nonlinear stability properties. Instead, it identifies numerical methods according to their authors and their basic approaches. Then, for example, instead of saying "Sweby's TVD method," this book says "Sweby's flux-limited method."

20.1 Van Leer's Flux-Limited Method

This section concerns Van Leer's flux-limited method for solving the linear advection equation with $a > 0$. Fromm's method, seen in Section 17.5, demonstrated the benefits of blending the Lax–Wendroff and the Beam–Warming second-order upwind methods for the linear advection equation – namely, dispersive errors partially cancel. If the fixed average used in Fromm's method was successful, a solution-sensitive average should be even more successful. Consider the linear advection equation with $a > 0$. The Lax–Wendroff method for the linear advection equation is

$$u_i^{n+1} = u_i^n - \lambda\left(\hat{f}_{i+1/2}^{\text{L-W}} - \hat{f}_{i-1/2}^{\text{L-W}}\right),$$
$$\hat{f}_{i+1/2}^{\text{L-W}} = au_i^n + \frac{1}{2}a(1-\lambda a)\left(u_{i+1}^n - u_i^n\right).$$

The Beam–Warming second-order upwind method for the linear advection equation with $a > 0$ is

$$u_i^{n+1} = u_i^n - \lambda\left(\hat{f}_{i+1/2}^{\text{B-W}} - \hat{f}_{i-1/2}^{\text{B-W}}\right),$$
$$\hat{f}_{i+1/2}^{\text{B-W}} = au_i^n + \frac{1}{2}a(1-\lambda a)\left(u_i^n - u_{i-1}^n\right).$$

Then *Van Leer's flux-limited method* for the linear advection equation with $a > 0$ is

$$u_i^{n+1} = u_i^n - \lambda\left(\hat{f}_{i+1/2}^n - \hat{f}_{i-1/2}^n\right),$$

where

$$\hat{f}_{i+1/2}^n = \frac{1+\eta_i^n}{2}\hat{f}_{i+1/2}^{\text{L-W}} + \frac{1-\eta_i^n}{2}\hat{f}_{i+1/2}^{\text{B-W}} \tag{20.9}$$

or

$$u_i^{n+1} = u_i^n - \lambda a\left(u_i^n - u_{i-1}^n\right) - \frac{\lambda a}{4}(1-\lambda a)\left[\left(1+\eta_i^n\right)\left(u_{i+1}^n - 2u_i^n + u_{i-1}^n\right) + \left(1-\eta_{i-1}^n\right)\left(u_i^n - 2u_{i-1}^n + u_{i-2}^n\right)\right]. \tag{20.10}$$

Notice that Van Leer's flux-limited method uses a different notation for linear combinations than this book has used so far. Although the notation may differ, the substance is the same. The next section will describe the connection between the two linear combination notations. Van Leer's flux-limited method equals the Lax–Wendroff method for $\eta_i^n = 1$, the Beam–Warming second-order upwind method for $\eta_i^n = -1$, Fromm's method for $\eta_i^n = 0$, and FTBS for $\eta_i^n = (r_i^+ + 1)/(r_i^+ - 1)$, where

$$r_i^+ = \frac{u_i^n - u_{i-1}^n}{u_{i+1}^n - u_i^n}. \tag{20.3}$$

The reader can certainly find other relationships between Van Leer's flux-limited method and first-generation methods. As one of the more interesting methods encompassed by Equation (22.10), consider $\eta_i^n = 1/3$:

$$u_i^{n+1} = u_i^n - \lambda a\left(u_i^n - u_{i-1}^n\right) - \frac{\lambda a}{4}(1-\lambda a)\left[\frac{4}{3}\left(u_{i+1}^n - 2u_i^n + u_{i-1}^n\right) + \frac{2}{3}\left(u_i^n - 2u_{i-1}^n + u_{i-2}^n\right)\right]. \tag{20.11}$$

As the reader can show, this method is *third-order* accurate, as opposed to second-order accurate methods such as the Lax–Wendroff and Beam–Warming second-order upwind methods.

To restrict the flux limiter, consider the upwind range condition. Van Leer obtained the following sufficient conditions:

$$\left|\left(1+\eta_i^n\right)\left(\frac{1}{r_i^+}-1\right)\right| \leq 2, \tag{20.12a}$$

$$\left|\left(1-\eta_i^n\right)\left(1-r_i^+\right)\right| \leq 2 \tag{20.12b}$$

for all i. The following section proves these results, albeit using different notation. Recall that $r_i^+ \leq 0$ at maxima and minima. As the reader can easily show, Equation (20.12) has the following unique solution for $r_i^+ \leq 0$:

$$\eta_i^n = \frac{r_i^+ + 1}{r_i^+ - 1},$$

which corresponds to FTBS. Fortunately, condition (20.12) allows considerably more breathing room in monotone regions, when $r_i^+ > 0$. In particular, condition (20.12) allows second-order and third-order accurate choices for η_i^n away from extrema. Van Leer suggested the following solution of Equation (20.12):

$$\eta_i^n = \eta\left(r_i^+\right) = \frac{\left|r_i^+\right| - 1}{\left|r_i^+\right| + 1} \tag{20.13a}$$

or, equivalently,

$$\eta_i^n = -\frac{\left|u_{i+1}^n - u_i^n\right| - \left|u_i^n - u_{i-1}^n\right|}{\left|u_{i+1}^n - u_i^n\right| + \left|u_i^n - u_{i-1}^n\right|}. \tag{20.13b}$$

This flux limiter obtains second-order accuracy in monotone regions and between first- and second-order accuracy at extrema. Although condition (20.12) certainly allows other flux limiters, Van Leer's flux-limited method traditionally uses flux limiter (20.13) exclusively.

Condition (20.12) implies $\hat{f}_{i+1/2}^n = \hat{f}_{i+1/2}^{\text{FTBS}}$ when u_i^n is an extremum. However, although FTBS is first-order accurate, *this does not imply that the method is first-order accurate at extrema*! Remember that

$$u_i^{n+1} = u_i^n - \lambda\left(\hat{f}_{i+1/2}^n - \hat{f}_{i-1/2}^n\right).$$

Then u_i^{n+1} is first-order accurate if both $\hat{f}_{i+1/2}^n = \hat{f}_{i+1/2}^{\text{FTBS}}$ and $\hat{f}_{i-1/2}^n = \hat{f}_{i-1/2}^{\text{FTBS}}$. With condition (20.12), this is true only if both u_i^n and u_{i-1}^n are extrema, which is indicative of $2\Delta x$-waves or, in other words, odd–even oscillations. As discussed in Section 8.2, $2\Delta x$-waves violate the Nyquist sampling theorem, and thus they contain no useful information about the exact solution. By condition (20.12), Van Leer's flux-limited method uses pure FTBS on $2\Delta x$-waves, which damps them, and rightly so. Otherwise, Van Leer's flux-limited method uses a combination of FTBS and a higher-order accurate method at extrema. In other words, $\hat{f}_{i+1/2}^n = \hat{f}_{i+1/2}^{\text{FTBS}}$ and $\hat{f}_{i-1/2}^n \neq \hat{f}_{i-1/2}^{\text{FTBS}}$ provided that u_i^n is separated from other extrema by at least one grid point. In practice, Van Leer's flux-limited method achieves an order of accuracy somewhere between one and two at extrema; this loss of accuracy is usually attributed to *clipping*, as discussed in Section 16.3.

Van Leer's method qualifies as second generation for the following reasons: (1) It is solution sensitive and (2) it concerns only the linear advection equation and not the Euler equations. The rest of the numerical methods in this chapter are third-generation methods.

20.2 Sweby's Flux-Limited Method (TVD)

This section concerns Sweby's flux-limited method. Sweby's method averages Roe's first-order upwind method, or some other first-order upwind method, and the Lax–Wendroff method. It draws heavily on Van Leer's flux-limited method, and also on work done by Roe (1981, 1982, 1985, 1986), Roe and Baines (1982, 1984), and Roe and Pike (1984). Because of the seminal contributions of Van Leer and Roe, perhaps this book should use the term "the Sweby–Roe–Van Leer flux-limited method," but we will instead adopt the shorter and more widely recognized term "Sweby's flux-limited method."

20.2.1 *The Linear Advection Equation with $a > 0$*

Consider the linear advection equation with $a > 0$. FTBS and the Lax–Wendroff method have a number of complementary properties that make them natural mates. In particular, FTBS does well near jump discontinuities, whereas the Lax–Wendroff method does well in smooth regions. FTBS for the linear advection equation is

$$u_i^{n+1} = u_i^n - \lambda\left(\hat{f}_{i+1/2}^{\text{FTBS}} - \hat{f}_{i-1/2}^{\text{FTBS}}\right),$$
$$\hat{f}_{i+1/2}^{\text{FTBS}} = au_i^n.$$

The Lax–Wendroff method for the linear advection equation is

$$u_i^{n+1} = u_i^n - \lambda\left(\hat{f}_{i+1/2}^{\text{L-W}} - \hat{f}_{i-1/2}^{\text{L-W}}\right),$$
$$\hat{f}_{i+1/2}^{\text{L-W}} = au_i^n + \frac{1}{2}a(1-\lambda a)\left(u_{i+1}^n - u_i^n\right).$$

Then *Sweby's flux-limited method* for the linear advection equation with $a > 0$ is

$$u_i^{n+1} = u_i^n - \lambda\left(\hat{f}_{i+1/2}^n - \hat{f}_{i-1/2}^n\right),$$

where

$$\hat{f}_{i+1/2}^n = \hat{f}_{i+1/2}^{\text{FTBS}} + \phi_i^n\left(\hat{f}_{i+1/2}^{\text{L-W}} - \hat{f}_{i+1/2}^{\text{FTBS}}\right) \tag{20.14}$$

or

$$\hat{f}_{i+1/2}^n = au_i^n + \frac{1}{2}a(1-\lambda a)\phi_i^n\left(u_{i+1}^n - u_i^n\right).$$

Alternatively,

$$u_i^{n+1} = u_i^n - \lambda a\left(u_i^n - u_{i-1}^n\right) - \frac{\lambda a}{2}(1-\lambda a)\left[\phi_i^n\left(u_{i+1}^n - u_i^n\right) - \phi_{i-1}^n\left(u_i^n - u_{i-1}^n\right)\right]. \tag{20.15}$$

Despite the seeming differences, Sweby's flux-limited method for the linear advection equation is exactly the same as Van Leer's flux-limited method for the linear advection equation described in the last section. Specifically, setting Equation (20.9) equal to Equation (20.14), we find that the Sweby and Van Leer flux limiters are related as follows:

$$\phi_i^n = \frac{1}{2}\left[1 + r_i^+ + \eta_i^n\left(1 - r_i^+\right)\right], \tag{20.16}$$

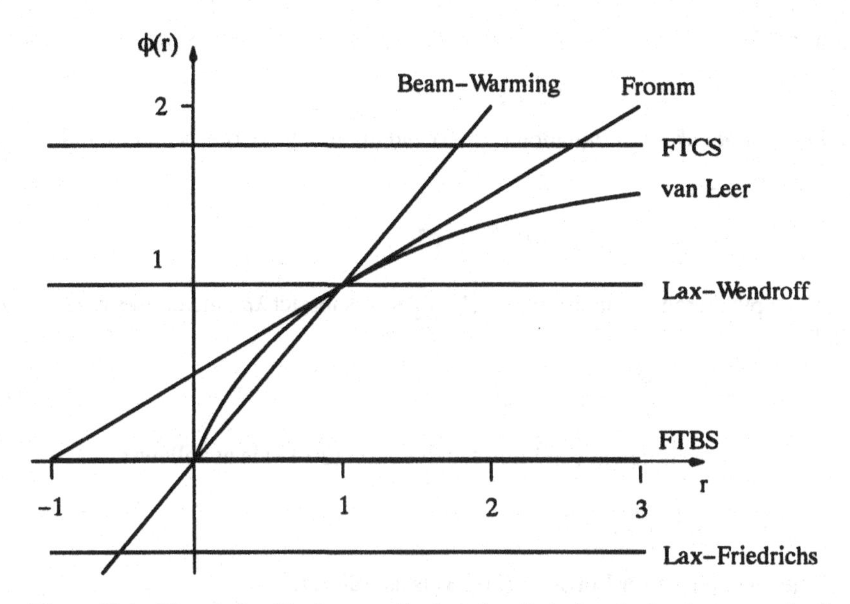

Figure 20.1 The relationships between Sweby's flux-limited method and common simple methods.

where as usual

$$r_i^+ = \frac{u_i^n - u_{i-1}^n}{u_{i+1}^n - u_i^n}.$$

For example, if $\eta_i^n = (|r_i^+|-1)/(|r_i^+|+1)$, as in Equation (20.13), then $\phi_i^n = 2r_i^+/(1+r_i^+)$ for $r_i^+ \geq 0$ and $\phi_i^n = 0$ for $r_i^+ < 0$.

Notice that Sweby's flux-limited method equals the Lax–Wendroff method when $\phi_i^n = 1$, FTBS when $\phi_i^n = 0$, FTCS when $\phi_i^n = 1/(1-\lambda a)$, the Lax-Friedrichs method when $\phi_i^n = -1/\lambda a$, the Beam–Warming second-order upwind method when $\phi_i^n = r_i^+$, and Fromm's method when $\phi_i^n = (r_i^+ + 1)/2$. The reader can certainly find other relationships between Sweby's flux-limited method and first-generation methods. Figure 20.1 illustrates the relationships between Sweby's flux-limited method and some popular first-generation methods.

If $0 \leq \phi_i^n \leq 1$, then $\hat{f}^n_{i+1/2}$ is a convex linear combination of $\hat{f}^{FTBS}_{i+1/2}$ and $\hat{f}^{L-W}_{i+1/2}$, in which case $\hat{f}^n_{i+1/2}$ lies between $\hat{f}^{FTBS}_{i+1/2}$ and $\hat{f}^{L-W}_{i+1/2}$. As with Van Leer's flux-limited method, Sweby's flux-limited method may reasonably allow both convex and nonconvex linear combinations or, in other words, Sweby's flux-limited method may reasonably allow both flux interpolation and flux extrapolation.

As in the last section, the flux limiters in Sweby's flux-limited method are restricted by the upwind range condition. By Equation (16.16a), Sweby's flux-limited method obeys the upwind range condition with $0 \leq \lambda a \leq 1$ if and only if it can be written in wave speed split form

$$u_i^{n+1} = u_i^n + C^+_{i+1/2}(u^n_{i+1} - u^n_i) - C^-_{i-1/2}(u^n_i - u^n_{i-1})$$

such that

$$0 \le C^-_{i+1/2} \le 1, \qquad C^+_{i+1/2} = 0$$

for all i. Subtract u_i^n from Equation (20.15) and divide by $-(u_i^n - u_{i-1}^n)$ to find

$$C^+_{i+1/2} = 0,$$
$$C^-_{i-1/2} = \lambda a + \tfrac{\lambda a}{2}(1 - \lambda a)\left[\frac{\phi_i^n}{r_i^+} - \phi_{i-1}^n\right].$$

Apply the upwind range condition $0 \le C^-_{i-1/2} \le 1$, subtract λa, and divide by $\lambda a(1-\lambda a)/2$ to find

$$-\frac{2}{1-\lambda a} \le \frac{\phi_i^n}{r_i^+} - \phi_{i-1}^n \le \frac{2}{\lambda a}. \tag{20.17}$$

For $0 \le \lambda a \le 1$, a sufficient condition for Equation (20.17) is as follows:

$$-2 \le \frac{\phi_i^n}{r_i^+} - \phi_{i-1}^n \le 2. \tag{20.18}$$

A sufficient condition for Equation (20.18) is as follows:

$$0 \le \phi_i^n + K \le 2, \tag{20.19a}$$
$$0 \le \frac{\phi_i^n}{r_i^+} + K \le 2 \tag{20.19b}$$

for all i and for any constant K. Notice that condition (20.19) requires $\phi = 0$ for $r_i^+ = 0$ regardless of K. Sweby's original paper only described the case $K = 0$ (some years later, Roe (1989) suggested the more general condition (20.19)). We have now obtained a range of three nonlinear stability conditions for Sweby's flux-limited method – Equations (20.17), (20.18), and (20.19). Equation (20.17) is the most complicated but places the least restrictions on the flux limiter; Equation (20.19) is the least complicated but places the most restrictions on the flux limiter; while Equation (20.18) is somewhere in between. Unfortunately, in practice, Equations (20.17) and (20.18) are too complicated to deal with, and only Equation (20.19) is used. Incidentally, as the reader can show using Equation (20.16), Equations (20.19) and (20.12) are identical except for notation.

With $K = 0$, condition (20.19) becomes

$$0 \le \phi_i^n \le 2,$$
$$0 \le \frac{\phi_i^n}{r_i^+} \le 2$$

or, equivalently,

$$0 \le \phi_i^n \le \min(2, 2r_i^+) \qquad r_i^+ > 0, \tag{20.20a}$$
$$\phi_i^n = 0 \qquad r_i^+ \le 0. \tag{20.20b}$$

These bounds are plotted in Figure 20.2. The union of the darkly and lightly shaded regions in Figure 20.2 indicate methods that satisfy condition (20.20); thus everything between the bold lines in Figure 20.2 satisfies condition (20.20). Also, the darkly shaded regions in Figure 20.2 indicate methods that satisfy condition (20.20) and that also lie between the Lax–Wendroff method and the Beam–Warming second-order upwind method and thus

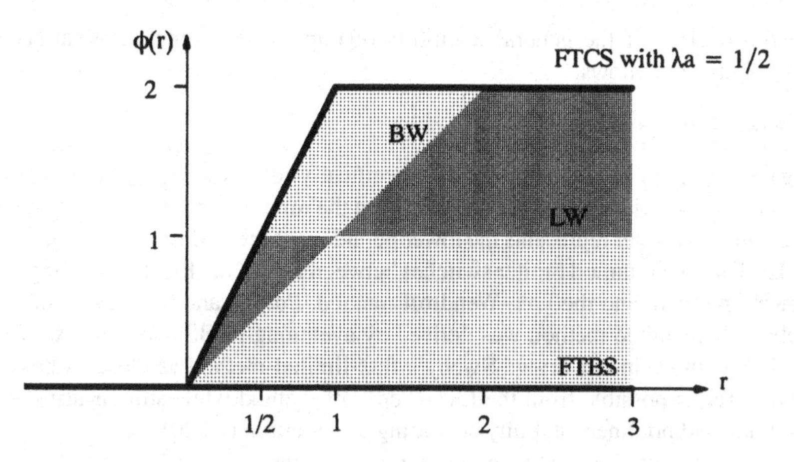

Figure 20.2 Flux limiters allowed by condition (20.20).

obtain second-order accuracy. Of course, flux limiters may stray slightly outside of the darkly shaded regions and still obtain second-order accuracy, provided that they remain close to the Lax–Wendroff or Beam–Warming methods. Notice that flux limiters allowed by Equation (20.20) may violate the convexity condition $0 \leq \phi_i^n \leq 1$. Also notice that Equation (20.20b) implies that $\hat{f}^n_{i+1/2} = \hat{f}^{FTBS}_{i+1/2}$ when u_i^n is an extremum. Although FTBS is only first-order accurate, this does not imply that Sweby's flux-limited methods is first-order accurate at extrema. As explained in the last section, Sweby's flux-limited method is between first- and second-order accurate at extrema, provided that the extrema are separated by at least one point. In other words, Sweby's flux-limited method suffers typical clipping errors at extrema.

Condition (20.20) only allows the flux limiter $\phi_i^n = 0$ when $r_i^+ \leq 0$. Fortunately, condition (20.20) allows considerably more flexibility when $r_i^+ > 0$. For example, condition (20.20) allows the Beam–Warming second-order upwind method for $0 < r_i^+ \leq 2$, Fromm's method for $1/4 \leq r_i^+ \leq 3$, and the Lax–Wendroff method for $r_i^+ \geq 1/2$. In fact, Sweby's flux-limited method may use any of these three methods at shocks, or a range of other methods, provided that shocks remain monotone increasing or monotone decreasing and do not develop spurious extrema. Of course, the Beam–Warming second-order upwind method, Fromm's method, and the Lax–Wendroff method ordinarily produce oscillations at shocks, but Sweby's flux-limited method switches to FTBS or other methods as needed to prevent this.

One flux limiter allowed by condition (20.20) is the *minmod* flux limiter:

$$\blacklozenge \quad \phi(r) = minmod(1, br), \tag{20.21}$$

where $1 \leq b = const. \leq 2$. The *minimum modulus* or *minmod* function returns the argument closest to zero if all of its arguments have the same sign, and it returns zero if any two of its arguments have different signs. Notice that

$$minmod(1, br) = \begin{cases} 1 & br \geq 1, \\ br & 0 \leq br < 1, \\ 0 & br < 0. \end{cases}$$

See Equation (9.19) for the general definition of minmod. A common variation on the minmod limiter is as follows:

$$\phi(r) = minmod(b, r), \tag{20.22}$$

where again $1 \leq b = const. \leq 2$. Unless specifically told otherwise, the reader should assume that $b = 1$, in which case both variations on the minmod limiter are the same. When $b = 1$, the minmod limiter represents the lower boundary of the darkly shaded region seen in Figure 20.2. Thus the minmod limiter switches between FTBS, the Beam–Warming second-order upwind method, and the Lax–Wendroff method. Put yet another way, remembering that Sweby's flux-limited method was derived by averaging FTBS and the Lax–Wendroff method, the minmod limiter brings Sweby's flux-limited method as close as possible to FTBS and as far as possible from the Lax–Wendroff method while still ensuring second-order accuracy and nonlinear stability according to condition (20.20).

Another flux limiter allowed by condition (20.20) is the *superbee* limiter:

♦ $$\phi(r) = \max(0, \min(2r, 1), \min(r, 2)) = \begin{cases} 0 & r \leq 0, \\ 2r & 0 \leq r \leq \frac{1}{2}, \\ 1 & \frac{1}{2} \leq r \leq 1, \\ r & 1 \leq r \leq 2, \\ 2 & r \geq 2, \end{cases} \tag{20.23}$$

as described in Roe and Pike (1984) and Roe (1985). The superbee limiter represents the upper boundary of the darkly shaded region seen in Figure 20.2. The superbee limiter thus brings Sweby's flux-limited method as close as possible to the Lax–Wendroff method and as far as possible from FTBS while still ensuring second-order accuracy and nonlinear stability according to condition (20.20). In his original papers, Roe used the notation B instead of ϕ for the flux limiter; then "super-B" or "superbee" refers to the fact that the flux limiter lies above all of the other possible flux limiters, in the same sense that a superscript lies above the main line of text.

A third possible flux limiter is the *Van Leer limiter*:

♦ $$\phi(r) = \begin{cases} \frac{2r}{1+r} & r \geq 0, \\ 0 & r < 0. \end{cases} \tag{20.24}$$

Using the Van Leer limiter, Sweby's flux-limited method equals Van Leer's flux-limited method seen in the last section. Unlike the minmod and superbee limiters, Van Leer's flux limiter is a completely smooth function of r for $r \geq 0$.

With $K = 1$, condition (20.19) becomes

$$\left|\phi_i^n\right| \leq 1,$$

$$\left|\frac{\phi_i^n}{r_i^+}\right| \leq 1$$

or, equivalently,

$$\left|\phi_i^n\right| \leq 1 \quad \text{for } \left|r_i^+\right| \geq 1, \tag{20.25a}$$

$$\left|\phi_i^n\right| \leq \left|r_i^n\right| \quad \text{for } \left|r_i^+\right| < 1. \tag{20.25b}$$

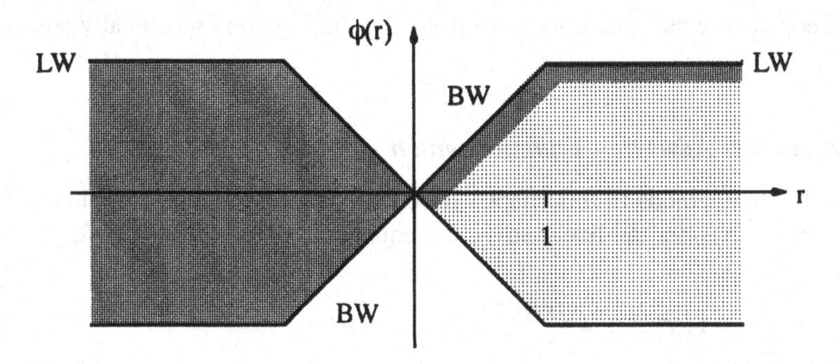

Figure 20.3 Flux limiters allowed by condition (20.25).

These bounds are plotted in Figure 20.3. The union of the darkly and lightly shaded regions in Figure 20.3 indicates methods that satisfy condition (20.25), while the darkly shaded regions indicate methods that lie close to or between the Lax–Wendroff or Beam–Warming methods and thus obtain second-order accuracy. Whereas the earlier condition given by Equation (20.20) allowed a wide range of stable second-order accurate methods for $r_i^+ \geq 0$, the present condition given by Equation (20.25) allows second-order accurate methods only at the very edges of the stable region. On the positive side, whereas condition (20.20) allows only first-order accurate FTBS for $r_i^+ < 0$, condition (20.25) allows a broad range of methods for $r_i^+ < 0$, all of which are second-order accurate. Unfortunately, most of the flux limiters allowed by condition (20.20) switch abruptly between different methods at extrema, which reduces the order of accuracy at extrema. In fact, in practice, the flux limiters allowed by Equation (20.25) still experience clipping errors just as large as those allowed by Equation (20.20), with the order of accuracy at extrema clocking in at something between one and two. One flux limiter allowed by condition (20.25) is the *symmetric minmod* limiter:

$$\phi(r) = \min(b|r|, 1) \tag{20.26}$$

for $0 < b \leq 1$. The symmetric minmod limiter is symmetric about the ϕ axis. Notice that the symmetric minmod limiter is the same as the minmod limiter given by Equation (20.21) for $r \geq 0$. If $b = 1$ the symmetric minmod limiter represents the upper boundary of the shaded region seen in Figure 20.3. Thus the symmetric minmod limiter brings Sweby's flux-limited method as close as possible to the Lax–Wendroff method and as far as possible from FTBS while still ensuring second-order accuracy and nonlinear stability according to condition (20.25).

There are, of course, many other possible flux limiters besides the minmod, symmetric minmod, superbee, Van Leer, and the other flux limiters seen above. For a taste of just how complicated and sophisticated flux limiters can get, see, for example, Jeng and Payne (1995). For the most part, the flux limiters do not radically affect the solution of Sweby's flux-limited method – one is a little better here, another is a little better there, but on average all of the flux limiters are roughly in the same ballpark, provided that they stay anywhere near the nonlinear stability range allowed by condition (20.19). For steady-state solutions, continuously differentiable flux limiters usually enhance the rate of convergence to steady state, which eliminates all of the flux limiters seen in this section except possibly for Van Leer's flux limiter. One response is to form smoothed versions of the flux limiters in this section by, for example, fitting polynomials to the flux limiter. However, even then, the flux limiters developed here may not work as well as other flux limiters specifically tailored for steady flows.

20.2.2 *The Linear Advection Equation with $a < 0$*

So far, the methods in this chapter have concerned the linear advection equation with $a > 0$. Now consider the linear advection equation with $a < 0$. FTFS for the linear advection equation is

$$u_i^{n+1} = u_i^n - \lambda\left(\hat{f}^{\mathrm{FTFS}}_{i+1/2} - \hat{f}^{\mathrm{FTFS}}_{i-1/2}\right),$$
$$\hat{f}^{\mathrm{FTFS}}_{i+1/2} = au^n_{i+1}.$$

The Lax–Wendroff method for the linear advection equation is

$$u_i^{n+1} = u_i^n - \lambda\left(\hat{f}^{\mathrm{L-W}}_{i+1/2} - \hat{f}^{\mathrm{L-W}}_{i-1/2}\right),$$
$$\hat{f}^{\mathrm{L-W}}_{i+1/2} = au^n_{i+1} - \frac{1}{2}a(1+\lambda a)\left(u^n_{i+1} - u^n_i\right).$$

Then Sweby's flux-limited method for the linear advection equation with $a < 0$ is

$$u_i^{n+1} = u_i^n - \lambda\left(\hat{f}^n_{i+1/2} - \hat{f}^n_{i-1/2}\right),$$

where

$$\hat{f}^n_{i+1/2} = \hat{f}^{\mathrm{FTFS}}_{i+1/2} + \phi^n_{i+1}\left(\hat{f}^{\mathrm{L-W}}_{i+1/2} - \hat{f}^{\mathrm{FTFS}}_{i+1/2}\right) \tag{20.27}$$

or

$$\hat{f}^n_{i+1/2} = au^n_{i+1} - \frac{1}{2}a(1+\lambda a)\phi^n_{i+1}\left(u^n_{i+1} - u^n_i\right).$$

Equivalently,

$$u_i^{n+1} = u_i^n - \lambda a\left(u^n_{i+1} - u^n_i\right) - \frac{\lambda a}{2}(1+\lambda a)\left[\phi^n_{i+1}\left(u^n_{i+1} - u^n_i\right) - \phi^n_i\left(u^n_i - u^n_{i-1}\right)\right]. \tag{20.28}$$

Notice that Sweby's flux-limited method equals the Lax–Wendroff method when $\phi^n_{i+1} = 1$, FTFS when $\phi^n_{i+1} = 0$, FTCS when $\phi^n_{i+1} = 1/(1+\lambda a)$, the Lax–Friedrichs method when $\phi^n_{i+1} = 1/\lambda a$, the Beam–Warming second-order upwind method when $\phi^n_{i+1} = r^-_{i+1}$, and Fromm's method when $\phi^n_{i+1} = (r^-_{i+1} + 1)/2$, where

$$r^-_{i+1} = \frac{u^n_{i+2} - u^n_{i+1}}{u^n_{i+1} - u^n_i}. \tag{20.4}$$

Notice that Sweby's flux-limited method still has all of the same relationships to first-generation methods as in the last subsection, after swapping r^+ for r^-.

Following the same approach as in the last subsection, we find that Sweby's flux-limited method obeys the upwind range condition for $-1 \le \lambda a \le 0$ if

$$0 \le \phi^n_i + K \le 2, \tag{20.29a}$$
$$0 \le \frac{\phi^n_i}{r^-_i} + K \le 2 \tag{20.29b}$$

for all i and for any constant K, which is essentially the same as Equation (20.19). The nonlinear stability conditions found in this subsection are thus identical to those found in the last subsection, after replacing r^+ by r^-. In fact, this subsection used r^- instead of r^+ specifically so that the nonlinear stability analysis *would* yield the same results. Then the same flux limiters apply; see Equations (20.21)–(20.24) and (20.26).

20.2.3 *Nonlinear Scalar Conservation Laws with a(u) > 0*

So far this chapter has concerned only the linear advection equation. This subsection concerns nonlinear scalar conservation laws with $a(u) > 0$. FTBS is

$$u_i^{n+1} = u_i^n - \lambda\left(\hat{f}_{i+1/2}^{\text{FTBS}} - \hat{f}_{i-1/2}^{\text{FTBS}}\right),$$
$$\hat{f}_{i+1/2}^{\text{FTBS}} = f\left(u_i^n\right),$$

and the Lax–Wendroff method is

$$u_i^{n+1} = u_i^n - \lambda\left(\hat{f}_{i+1/2}^{\text{L-W}} - \hat{f}_{i-1/2}^{\text{L-W}}\right),$$
$$\hat{f}_{i+1/2}^{\text{L-W}} = f\left(u_i^n\right) + \frac{1}{2}a_{i+1/2}^n\left(1 - \lambda a_{i+1/2}^n\right)\left(u_{i+1}^n - u_i^n\right),$$

where as usual

$$a_{i+1/2}^n = \begin{cases} \frac{f(u_{i+1}^n)-f(u_i^n)}{u_{i+1}^n-u_i^n} & u_{i+1}^n \neq u_i^n, \\ f'\left(u_i^n\right) & u_{i+1}^n = u_i^n. \end{cases}$$

Assuming that $\phi_i^n = \phi(r_i^+)$, Sweby's flux-limited method for $a(u) > 0$ is

$$u_i^{n+1} = u_i^n - \lambda\left(\hat{f}_{i+1/2}^n - \hat{f}_{i-1/2}^n\right),$$

where

$$\hat{f}_{i+1/2}^n = \hat{f}_{i+1/2}^{\text{FTBS}} + \phi\left(r_i^+\right)\left(\hat{f}_{i+1/2}^{\text{L-W}} - \hat{f}_{i+1/2}^{\text{FTBS}}\right) \tag{20.30}$$

or

$$\hat{f}_{i+1/2}^n = f\left(u_i^n\right) + \frac{1}{2}a_{i+1/2}^n\left(1 - \lambda a_{i+1/2}^n\right)\phi\left(r_i^+\right)\left(u_{i+1}^n - u_i^n\right), \tag{20.31}$$

and where

$$r_i^+ = \frac{\hat{f}_{i-1/2}^{\text{L-W}} - \hat{f}_{i-1/2}^{\text{FTBS}}}{\hat{f}_{i+1/2}^{\text{L-W}} - \hat{f}_{i+1/2}^{\text{FTBS}}} \tag{20.5}$$

or, equivalently,

$$r_i^+ = \frac{a_{i-1/2}^n\left(1 - \lambda a_{i-1/2}^n\right)\left(u_i^n - u_{i-1}^n\right)}{a_{i+1/2}^n\left(1 - \lambda a_{i+1/2}^n\right)\left(u_{i+1}^n - u_i^n\right)}. \tag{20.32}$$

For the linear advection equation, definitions (20.5) and (20.32) are exactly the same as definition (20.3). Definition (20.5) and the equivalent definition (20.32) preserve most of the connections between Sweby's flux-limited method and the first-generation numerical methods, as seen in earlier subsections. Specifically, Sweby's flux-limited method equals the Lax–Wendroff method when $\phi(r_i^+) = 1$, FTBS when $\phi(r_i^+) = 0$, the Beam–Warming second-order upwind method when $\phi(r_i^+) = r_i^+$, and Fromm's method when $\phi(r_i^+) = (r_i^+ + 1)/2$. Most importantly, definitions (20.5) and (20.32) preserve the nonlinear stability conditions found in Subsection 20.2.1. In particular, the minmod, superbee, Van Leer, and symmetric minmod flux limiters still ensure the upwind range condition. To prove this, notice that the upwind range condition seen in Equation (16.16a) or Equation (16.17a) implies

$$0 \le \lambda a_{i+1/2}^n + \frac{\lambda a_{i+1/2}^n}{2}\left(1 - \lambda a_{i+1/2}^n\right)\left[\frac{\phi(r_i^+)}{r_i^+} - \phi(r_{i-1}^+)\right] \le 1.$$

Subtract $\lambda a_{i+1/2}^n$ and divide by $\lambda a_{i+1/2}^n(1 - \lambda a_{i+1/2}^n)/2$ to find

$$-\frac{2}{1 - \lambda a_{i+1/2}^n} \le \frac{\phi(r_i^+)}{r_i^+} - \phi(r_{i-1}^+) \le \frac{2}{\lambda a_{i+1/2}^n}. \tag{20.33}$$

For $0 \le \lambda a_{i+1/2}^n \le 1$, a sufficient condition for Equation (20.33) is as follows:

$$-2 \le \frac{\phi(r_i^+)}{r_i^+} - \phi(r_{i-1}^+) \le 2, \tag{20.34}$$

which is exactly the same as condition (20.18) found in Subsection 20.2.1. Then the rest of the analysis seen in Subsection 20.2.1 applies again here. Then, just as before, the minmod, superbee, Van Leer, and symmetric minmod limiters ensure the upwind range condition; see Equations (20.21)–(20.24) and (20.26).

From Equation (20.31), notice that Sweby's flux-limited method uses a limited antidissipative flux, as defined in the chapter introduction. In other words, $\epsilon_{i+1/2}^{\text{FTBS}} \ge \epsilon_{i+1/2}^{\text{L-W}}$ or, equivalently, $a_{i+1/2}(1 - \lambda a_{i+1/2}^n) \ge 0$ for $0 \le \lambda a_{i+1/2} \le 1$. Then $r_i^{\pm} \le 0$ at extrema, just as in the preceding subsections. Among other things, this implies that Sweby's flux-limited method for nonlinear scalar conservation laws has between first- and second-order accuracy at extrema, due to clipping, just as in the preceding subsections.

20.2.4 *Nonlinear Scalar Conservation Laws with a(u) < 0*

This subsection concerns nonlinear scalar conservation laws with $a(u) < 0$. FTFS is

$$u_i^{n+1} = u_i^n - \lambda\left(\hat{f}_{i+1/2}^{\text{FTFS}} - \hat{f}_{i-1/2}^{\text{FTFS}}\right),$$
$$\hat{f}_{i+1/2}^{\text{FTFS}} = f\left(u_{i+1}^n\right),$$

and the Lax–Wendroff method is

$$u_i^{n+1} = u_i^n - \lambda\left(\hat{f}_{i+1/2}^{\text{L-W}} - \hat{f}_{i-1/2}^{\text{L-W}}\right),$$
$$\hat{f}_{i+1/2}^{\text{L-W}} = f\left(u_{i+1}^n\right) - \frac{1}{2}a_{i+1/2}^n\left(1 + \lambda a_{i+1/2}^n\right)\left(u_{i+1}^n - u_i^n\right).$$

Assuming $\phi_{i+1}^n = \phi(r_{i+1}^-)$, Sweby's flux-limited method for $a(u) < 0$ is

$$u_i^{n+1} = u_i^n - \lambda\left(\hat{f}_{i+1/2}^n - \hat{f}_{i-1/2}^n\right),$$

where

$$\hat{f}^n_{i+1/2} = \hat{f}^{\text{FTFS}}_{i+1/2} + \phi(r^-_{i+1})\left(\hat{f}^{\text{L-W}}_{i+1/2} - \hat{f}^{\text{FTFS}}_{i+1/2}\right) \tag{20.35}$$

or

$$\hat{f}^n_{i+1/2} = f\left(u^n_{i+1}\right) - \frac{1}{2}a^n_{i+1/2}\left(1 + \lambda a^n_{i+1/2}\right)\phi(r^-_{i+1})\left(u^n_{i+1} - u^n_i\right), \tag{20.36}$$

and where

$$r^-_{i+1} = \frac{\hat{f}^{\text{L-W}}_{i+3/2} - \hat{f}^{\text{FTFS}}_{i+3/2}}{\hat{f}^{\text{L-W}}_{i+1/2} - \hat{f}^{\text{FTFS}}_{i+1/2}} \tag{20.6}$$

or, equivalently

$$r^-_{i+1} = \frac{a^n_{i+3/2}\left(1 + \lambda a^n_{i+3/2}\right)\left(u^n_{i+2} - u^n_{i+1}\right)}{a^n_{i+1/2}\left(1 + \lambda a^n_{i+1/2}\right)\left(u^n_{i+1} - u^n_i\right)}. \tag{20.37}$$

For the linear advection equation, definitions (20.6) and (20.37) are exactly the same as definition (20.4). Definitions (20.6) and (20.37) preserve most of the connections between Sweby's flux-limited method and other numerical methods, as seen in earlier subsections. Specifically, Sweby's flux-limited method equals the Lax–Wendroff method when $\phi(r^-_{i+1}) = 1$, FTFS when $\phi(r^-_{i+1}) = 0$, the Beam–Warming second-order upwind method when $\phi(r^-_{i+1}) = r^-_{i+1}$, and Fromm's method when $\phi(r^-_{i+1}) = (r^-_{i+1} + 1/2)$. Most importantly, definitions (20.6) and (20.37) preserve the nonlinear stability conditions found in Subsection 20.2.2. In particular, the minmod, superbee, Van Leer, and symmetric minmod limiters still ensure the upwind range condition; see Equations (20.21)–(20.24) and (20.26). The proof is left as an exercise for the reader.

20.2.5 *Nonlinear Scalar Conservation Laws at Sonic Points*

To obtain a method that works regardless of the sign of of the wave speed, Sweby used wave speed splitting. However, instead of splitting the *physical* wave speed, as in Section 13.5, Sweby split the *numerical* wave speed. See, in particular, Example 13.11 for a brief introduction to wave speed splitting. As usual, define the average numerical wave speed $a^n_{i+1/2}$ as follows:

$$a^n_{i+1/2} = \begin{cases} \frac{f(u^n_{i+1}) - f(u^n_i)}{u^n_{i+1} - u^n_i} & u^n_{i+1} \neq u^n_i, \\ f'\left(u^n_i\right) & u^n_{i+1} = u^n_i. \end{cases}$$

This average numerical wave speed can be split as follows:

$$a^n_{i+1/2} = a^+_{i+1/2} + a^-_{i+1/2}, \tag{20.38}$$

where

$$a^+_{i+1/2} \geq 0, \qquad a^-_{i+1/2} \leq 0. \tag{20.39}$$

For example, one possible wave speed splitting is

$$a^+_{i+1/2} = \max\left(0, a^n_{i+1/2}\right), \qquad a^-_{i+1/2} = \min\left(0, a^n_{i+1/2}\right). \tag{20.40}$$

However, this allows expansion shocks at expansive sonic points, as proven later in this subsection. So instead consider

$$a^+_{i+1/2} = \frac{f(u^n_{i+1}) - \hat{f}^{(1)}_{i+1/2}}{u^n_{i+1} - u^n_i}, \qquad a^-_{i+1/2} = \frac{\hat{f}^{(1)}_{i+1/2} - f(u^n_i)}{u^n_{i+1} - u^n_i}, \tag{20.41}$$

where $\hat{f}^{(1)}_{i+1/2}$ is the conservative numerical flux of any first-order upwind method seen in Section 17.3. Notice that Equation (20.41) equals Equation (20.40) if $\hat{f}^{(1)}_{i+1/2} = \hat{f}^{ROE}_{i+1/2}$. In fact, Roe's first-upwind method, Godunov's first-order upwind method, Harten's first-order upwind method, and all of the other first-order upwind methods seen in Section 17.3 differ only at sonic points. Then Equation (20.41) deviates from Equation (20.40) only at sonic points.

Recall the standard artificial viscosity form:

$$\hat{f}^{(1)}_{i+1/2} = \frac{1}{2}\left(f(u^n_{i+1}) + f(u^n_i)\right) - \frac{1}{2}\epsilon^{(1)}_{i+1/2}\left(u^n_{i+1} - u^n_i\right).$$

Then Equation (20.41) can be written in terms of artificial viscosity as follows:

$$a^+_{i+1/2} = \frac{1}{2}\left(a^n_{i+1/2} + \epsilon^{(1)}_{i+1/2}\right), \qquad a^-_{i+1/2} = \frac{1}{2}\left(a^n_{i+1/2} - \epsilon^{(1)}_{i+1/2}\right). \tag{20.42}$$

Then the wave speed splitting condition (20.39) is equivalent to

$$\epsilon^{(1)}_{i+1/2} \geq \epsilon^{\text{ROE}}_{i+1/2} = \left|a^n_{i+1/2}\right|. \tag{20.43}$$

Such methods were discussed earlier, in Example 16.6.

For any numerical wave speed splitting seen above, Sweby's flux-limited method is

$$u^{n+1}_i = u^n_i - \lambda\left(\hat{f}^n_{i+1/2} - \hat{f}^n_{i-1/2}\right),$$

where

- $$\begin{aligned}\hat{f}^n_{i+1/2} = \hat{f}^{(1)}_{i+1/2} &+ \frac{1}{2}\left[a^+_{i+1/2}\left(1 - \lambda a^+_{i+1/2}\right)\phi\left(r^+_i\right)\right. \\ &\left. - a^-_{i+1/2}\left(1 + \lambda a^-_{i+1/2}\right)\phi\left(r^-_{i+1}\right)\right]\left(u^n_{i+1} - u^n_i\right)\end{aligned} \tag{20.44}$$

and where

- $$r^+_i = \frac{a^+_{i-1/2}\left(1 - \lambda a^+_{i-1/2}\right)\left(u^n_i - u^n_{i-1}\right)}{a^+_{i+1/2}\left(1 - \lambda a^+_{i+1/2}\right)\left(u^n_{i+1} - u^n_i\right)}, \tag{20.45}$$

- $$r^-_i = \frac{a^-_{i+1/2}\left(1 + \lambda a^-_{i+1/2}\right)\left(u^n_{i+1} - u^n_i\right)}{a^-_{i-1/2}\left(1 + \lambda a^-_{i-1/2}\right)\left(u^n_i - u^n_{i-1}\right)}. \tag{20.46}$$

In the preceding expressions, notice that either the positive or negative parts drop out, except possibly at sonic points, making the method exactly the same as in the previous two subsections. Then the nonlinear stability analysis from previous subsections carries over to this subsection, except possibly at sonic points. In other words, flux limiters such as minmod and superbee ensure the upwind range condition, except possibly at sonic points; see Equations (20.21)–(20.24) and (20.26).

To help understand the effects of the numerical wave speed splitting on sonic points consider, for example, the case $\phi(r) = 1$. When $\phi(r) = 1$, Equation (20.44) can be rewritten as

$$\hat{f}^n_{i+1/2} = \hat{f}^{\text{L-W}}_{i+1/2} + \lambda a^+_{i+1/2} a^-_{i+1/2}(u^n_{i+1} - u^n_i) \tag{20.47}$$

or

$$\hat{f}^n_{i+1/2} = \hat{f}^{\text{L-W}}_{i+1/2} - \frac{\lambda}{2}\epsilon^n_{i+1/2}(u^n_{i+1} - u^n_i),$$

where

$$\epsilon^n_{i+1/2} = -2a^+_{i+1/2} a^-_{i+1/2}.$$

Thus, when $\phi(r) = 1$, Sweby's flux-limited method equals the Lax–Wendroff method plus second-order artificial viscosity. The coefficient of artificial viscosity $\epsilon^n_{i+1/2} = -2a^+_{i+1/2}a^-_{i+1/2}$ is zero except possibly at sonic points, and thus Sweby's flux-limited method equals the Lax–Wendroff method when $\phi(r) = 1$, except possibly at sonic points. At sonic points, the artificial viscosity is dissipative (i.e., it has a coefficient greater than zero) and thus the artificial viscosity tends to have a stabilizing effect. Recall that the Lax–Wendroff method has insufficient artificial viscosity at expansive sonic points – it allows large $O(1)$ expansion shocks as discussed in Section 17.2 – so that the additional artificial viscosity at sonic points is welcome. However, if $a^\pm_{i+1/2}$ are defined by Equation (20.40), then $a^+_{i+1/2}a^-_{i+1/2} = 0$ everywhere, including sonic points, and the artificial viscosity at the sonic points drops out, which is not so good.

The behavior of Sweby's flux-limited method with the minmod flux limiter is illustrated using the five standard test cases defined in Section 17.0.

Test Case 1 As seen in Figure 20.4, the sinusoid is well captured, although clipping has eroded the peaks by some 25%. On this test case, Sweby's flux-limited method is far better than a first-order accurate method such as FTBS but worse than a uniformly second-order accurate method such as the Lax–Wendroff method.

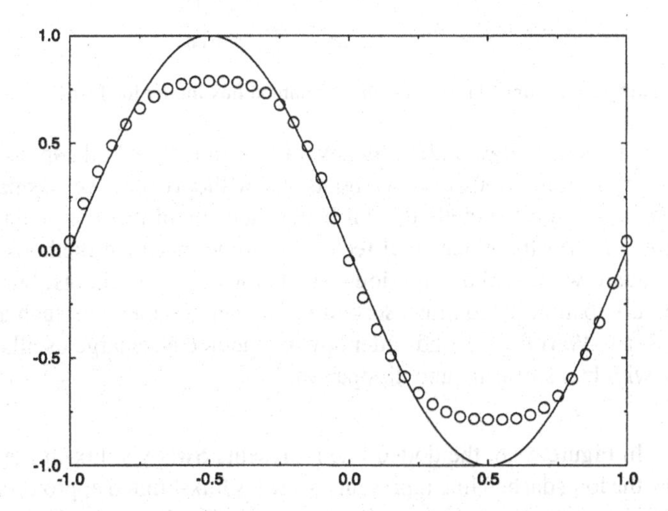

Figure 20.4 Sweby's flux-limited method with the minmod flux limiter for Test Case 1.

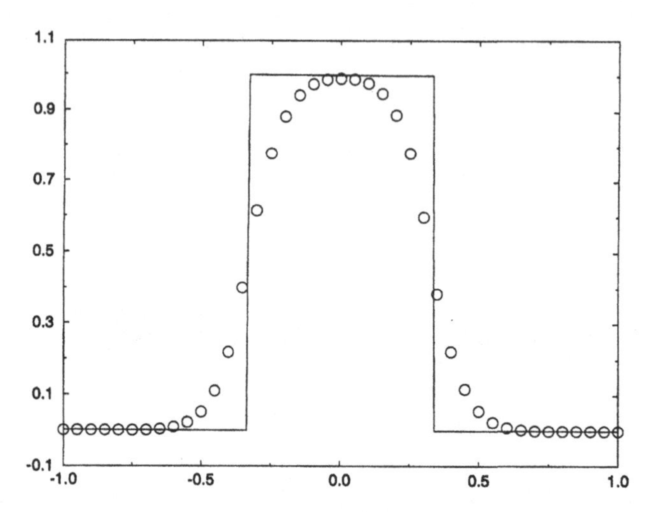

Figure 20.5 Sweby's flux-limited method with the minmod flux limiter for Test Case 2.

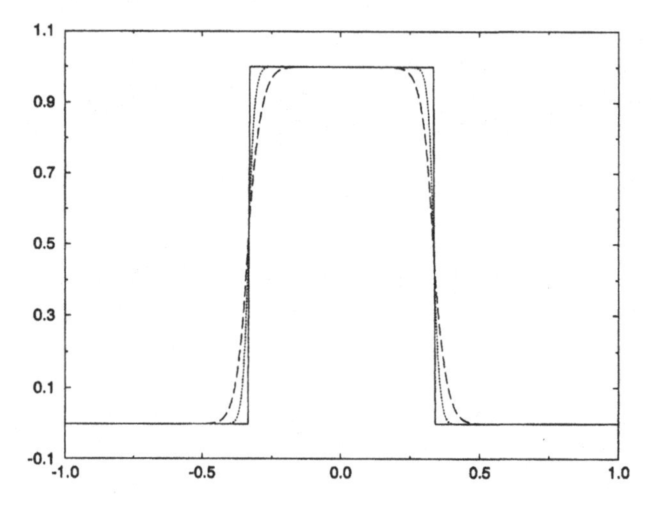

Figure 20.6 Sweby's flux-limited method with the minmod flux limiter for Test Case 3.

Test Case 2 As seen in Figure 20.5, Sweby's flux-limited method captures the square wave without spurious oscillations and overshoots, and without excessive dissipation. On this test case Sweby's flux-limited method is far better than any of the first-generation methods seen in Chapter 17. As the reader will recall, first-order accurate methods such as FTBS capture the square wave without spurious oscillations and overshoots, but with far more smearing and dissipation; in contrast, second-order accurate methods such as the Lax–Wendroff or the Beam–Warming second-order upwind method have large oscillations and overshoots, albeit with less smearing and dissipation.

Test Case 3 In Figure 20.6, the dotted line represents Sweby's flux-limited approximation to $u(x, 4)$, the long dashed line represents Sweby's flux-limited approximation to $u(x, 40)$, and the solid line represents the exact solution for $u(x, 4)$ or $u(x, 40)$. Again, these results are better than anything seen previously.

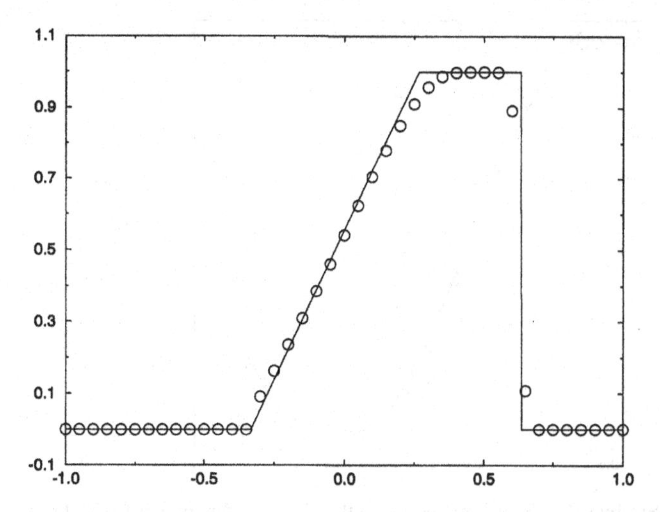

Figure 20.7 Sweby's flux-limited method with the minmod flux limiter for Test Case 4.

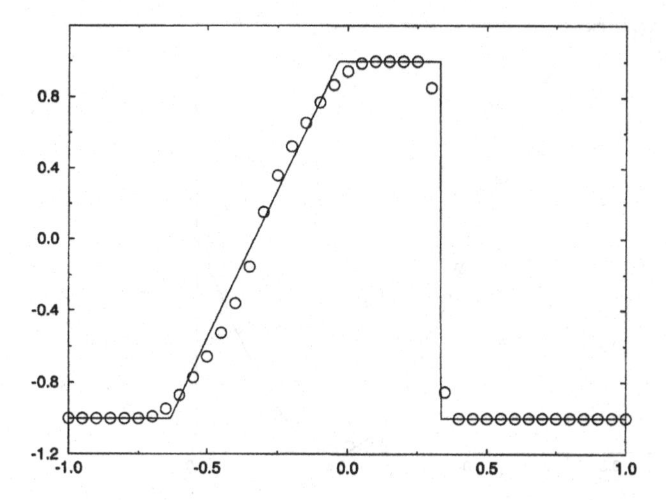

Figure 20.8 Sweby's flux-limited method with the minmod flux limiter for Test Case 5.

Test Case 4 As seen in Figure 20.7, Sweby's flux-limited method captures the shock extremely well, with only two transition points. The expansion fan is also well captured with only a slight rounding-off of the right-hand corner.

Test Case 5 If Sweby's flux-limited method uses Roe's first-order upwind method, it will fail to alter the initial conditions in any way, just like Roe's first-order upwind method. So instead use Harten's first-order upwind method as discussed in Subsection 17.3.3 with $\delta = 0.4$. As seen in Figure 20.8, Sweby's flux-limited method captures the shock with only two transition points, and it captures the expansion fan with only a small jump near the sonic point (which could be eliminated by increasing δ) and some slight rounding of the corners.

To get a sense of the effects of the flux limiter, let us rerun Test Cases 1 through 5 using the Van Leer and superbee flux limiters in place of the minmod flux limiter. Remember

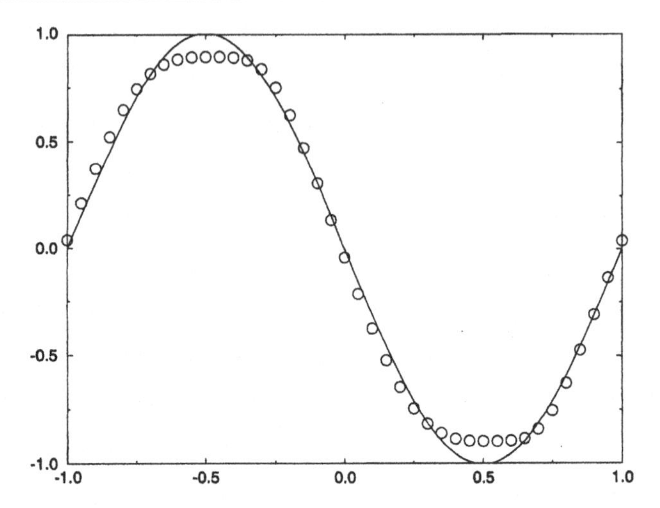

Figure 20.9 Sweby's flux-limited method with the Van Leer flux limiter for Test Case 1.

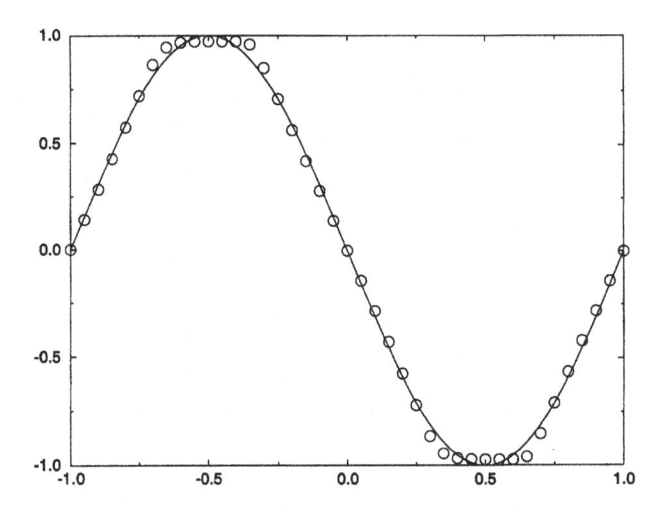

Figure 20.10 Sweby's flux-limited method with the superbee flux limiter for Test Case 1.

that the minmod flux limiter is the smallest flux limiter allowed by condition (20.20), the superbee flux limiter is the greatest flux limiter allowed by condition (20.20), and the Van Leer limiter is somewhere in between. For Test Case 1, comparing Figures 20.4, 20.9, and 20.10, we see that the flux limiter mainly affects the size and flatness of the sinusoidal peaks – the minmod flux limiter gives the lowest and roundest peaks, the superbee limiter gives the largest and flattest peaks, and the Van Leer flux limiter is somewhere in between. For Test Case 2 Figures 20.5, 20.11, and 20.12 indicate that the flux limiter mainly affects the contact smearing – the minmod limiter gives the most smearing, the superbee limiter gives the least smearing (only four transition points), and the Van Leer limiter is somewhere in between. Although not shown, the results for Test Case 3 are similar to those of Test Case 2. Although again not shown, the flux limiter has little effect on Test Cases 4 and 5; about the only difference between the superbee flux limiter and the minmod flux limiter is a very slight improvement in the rounding in the corners of the expansion fan.

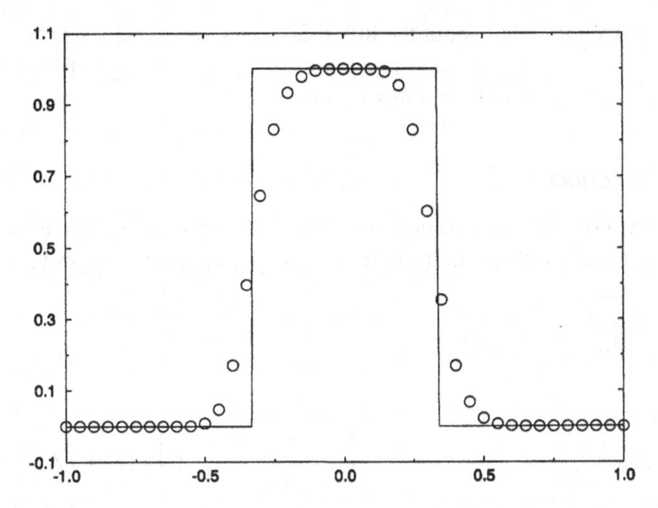

Figure 20.11 Sweby's flux-limited method with the Van Leer flux limiter for Test Case 2.

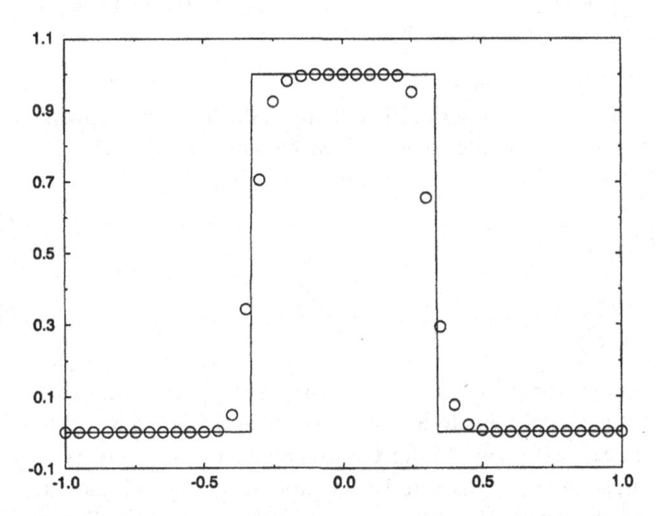

Figure 20.12 Sweby's flux-limited method with the superbee flux limiter for Test Case 2.

The standard test cases seem to favor the superbee flux limiter. However, do not be misled. In many cases, the superbee limiter tends to increase solution slopes. In fact, Roe (1985) says that "superbee was devised empirically whilst following up a suggestion by Woodward and Colella [see Section 23.2] concerning a rather elaborate discontinuity-sharpening algorithm. I was trying to implant their idea in my own method to give it a simpler expression." Because of this steepening effect, the superbee limiter performs well on discontinuous solutions but often performs poorly on smooth solutions, sometimes even turning smooth solutions into a series of steps. Thus, to avoid both the overdissipation caused by minmod and oversteepening caused by superbee, an intermediate flux limiter such as the Van Leer flux limiter is one of the better options. This conclusion is somewhat surprising, given that the Van Leer flux limiter predates all of the other flux limiters seen in this section by some ten years.

People sometimes worry too much about the flux limiter. As the above tests show, even the worst flux limiter is often pretty good, and you can hardly go too wrong with something in between the extremes, such as Van Leer's flux limiter.

20.2.6 *The Euler Equations*

Roe (1982, 1986) suggested the following version of Sweby's flux-limited method for the Euler equations, based on Roe's first-order upwind method for the Euler equations, as seen in Subsection 18.3.2:

$$\mathbf{u}_i^{n+1} = \mathbf{u}_i^n - \lambda\left(\hat{\mathbf{f}}_{i+1/2}^n - \hat{\mathbf{f}}_{i-1/2}^n\right),$$

where

$$\begin{aligned}\hat{\mathbf{f}}_{i+1/2}^n = \hat{\mathbf{f}}_{i+1/2}^{(1)} &+ \frac{1}{2}\sum_{j=1}^{3}\left(\mathbf{r}_{i+1/2}^n\right)_j\left(\lambda_{i+1/2}^+\right)_j\left[1-\frac{\Delta t}{\Delta x}\left(\lambda_{i+1/2}^+\right)_j\right]\phi\left(r_i^+\right)_j\left(\Delta v_{i+1/2}^n\right)_j \\ &- \frac{1}{2}\sum_{j=1}^{3}\left(\mathbf{r}_{i+1/2}^n\right)_j\left(\lambda_{i+1/2}^-\right)_j\left[1+\frac{\Delta t}{\Delta x}\left(\lambda_{i+1/2}^-\right)_j\right]\phi\left(r_{i+1}^-\right)_j\left(\Delta v_{i+1/2}^n\right)_j\end{aligned} \tag{20.48}$$

and where $\hat{\mathbf{f}}_{i+1/2}^{(1)}$ is the conservative numerical flux of any of the first-order upwind methods seen in Section 18.3, ϕ is any flux limiter seen in previous subsections, and

$$\left(r_i^+\right)_j = \frac{\left(\Delta v_{i-1/2}^n\right)_j}{\left(\Delta v_{i+1/2}^n\right)_j}, \tag{20.49}$$

$$\left(r_i^-\right)_j = \frac{\left(\Delta v_{i+1/2}^n\right)_j}{\left(\Delta v_{i-1/2}^n\right)_j}. \tag{20.50}$$

More complicated definitions for $(r_i^\pm)_j$ may also be used much as in previous subsections. See the description of Roe's approximate Riemann solver in Section 5.3 and Roe's first-order upwind method in Subsection 18.3.2 for the full details on $(\lambda_{i+1/2}^\pm)_j$, $(\mathbf{r}_{i+1/2}^n)_j$, and $(\Delta v_{i+1/2}^n)_j$. Do not confuse the right characteristic vectors $(\mathbf{r}_{i+1/2}^n)_j$ with the ratios $(r_i^\pm)_j$. Also, do not confuse the characteristic values $(\lambda_{i+1/2}^\pm)_j$ with $\lambda = \Delta t/\Delta x$. For an intuitive introduction to the ideas used here, the reader may wish to review Subsection 18.3.5. The basic idea, more or less, is to replace the characteristic values appearing in Roe's first-order upwind method, while leaving the characteristic vectors alone.

20.3 Chakravarthy–Osher Flux-Limited Methods (TVD)

This section concerns a series of semidiscrete flux-limited methods developed by Osher and Chakravarthy. To prepare for this section, the reader may wish to briefly review the method of lines, introduced in Subsection 11.2.1. In the method of lines, space is discretized to obtain a *semidiscrete* approximation, and then time is discretized to obtain a *fully discrete* approximation. This section deals only with the first step – spatial discretization – and does not consider time discretization. In the original series of papers, Chakravarthy and Osher suggested several different time discretizations, but they never firmly committed themselves.

Certainly, the original papers focus heavily on semidiscrete approximations, as opposed to any specific fully discrete scheme, and thus this section does the same.

Although not in "ready to use" form, semidiscrete approximations are interesting for several reasons. Most importantly, semidiscrete approximations lead to *modular* approximations wherein the same semidiscrete approximation can be paired with numerous different time discretizations, leading to a range of fully discrete approximations. Then, with semidiscrete approximations, the user has a great deal of flexibility and control over the performance and properties of the final fully discrete method. For example, one-time discretization may be best for unsteady solutions, where time accuracy is of the essence, while another is best for steady-state solutions obtained as the large-time limit of unsteady solutions, where time accuracy is irrelevant and the rate of convergence to steady state is all that matters. Specifically, for unsteady calculations, a Cauchy–Kowalewski time discretizations often yields the best results. However, with Cauchy–Kowalewski time discretizations, steady-state solutions depend on the time step Δt . Then, for steady-state solutions, something like a Runge–Kutta time discretization will generally yield better results. Besides allowing the user to switch easily between different time-stepping methods, the modularity inherent in semidiscrete approximations also simplifies various sorts of analyses, especially nonlinear stability analyses.

Since this section concerns only semidiscrete approximations, it will not include numerical results; see the original Osher and Chakravarthy papers cited below for numerical results. Also see Section 21.4 for numerical results using a closely related method. For simplicity, this section concerns only scalar conservation laws. While not discussed here, Osher and Chakravarthy suggested semidiscrete methods for the Euler equations, mostly based on Osher's approximate Riemann solver, but also sometimes based on Roe's approximate Riemann solver or flux vector splitting. See the original papers cited below for details.

Like most flux-limited methods, the semidiscrete flux-limited methods proposed by Osher and Chakravarthy combine two first-generation methods. This requires semidiscrete versions of first-generation methods, such as those seen in Chapter 17. As illustrated in the following examples, to convert a fully discrete method to a semidiscrete method, drop any terms that depend on the time step Δt or $\lambda = \Delta t/\Delta x$ from the conservative numerical flux.

Example 20.1 Find a semidiscrete version of the Lax–Wendroff method.

Solution As seen in Section 17.2, the fully discrete Lax–Wendroff method is

$$u_i^{n+1} = u_i^n - \lambda\left(\hat{f}^{\text{L-W}}_{i+1/2} - \hat{f}^{\text{L-W}}_{i-1/2}\right),$$

$$\hat{f}^{\text{L-W}}_{i+1/2} = \frac{1}{2}\left(f\left(u_{i+1}^n\right) + f\left(u_i^n\right)\right) - \frac{1}{2}\lambda\left(a_{i+1/2}^n\right)^2\left(u_{i+1}^n - u_i^n\right).$$

Dropping the second time-dependent term in the conservative numerical flux, the semidiscrete Lax–Wendroff method becomes

$$\frac{du_i}{dt} = -\frac{\hat{f}^{\text{L-W}}_{s,i+1/2} - \hat{f}^{\text{L-W}}_{s,i-1/2}}{\Delta x},$$

$$\hat{f}^{\text{L-W}}_{s,i+1/2} = \frac{1}{2}\left(f\left(u_{i+1}^n\right) + f\left(u_i^n\right)\right).$$

Example 20.2 Find the semidiscrete version of FTCS.

Solution The fully discrete FTCS scheme is

$$u_i^{n+1} = u_i^n - \lambda\left(\hat{f}^{\text{FTCS}}_{i+1/2} - \hat{f}^{\text{FTCS}}_{i-1/2}\right),$$
$$\hat{f}^{\text{FTCS}}_{i+1/2} = \frac{1}{2}\left(f\left(u_{i+1}^n\right) + f\left(u_i^n\right)\right).$$

The conservative numerical flux does not involve the time step, and thus there is no need to change anything. Then the semidiscrete FTCS scheme is

$$\frac{du_i}{dt} = -\frac{\hat{f}^{\text{FTCS}}_{s,i+1/2} - \hat{f}^{\text{FTCS}}_{s,i-1/2}}{\Delta x},$$
$$\hat{f}^{\text{FTCS}}_{s,i+1/2} = \frac{1}{2}\left(f\left(u_{i+1}^n\right) + f\left(u_i^n\right)\right).$$

Comparing with the previous example, notice that

$$\hat{f}^{\text{FTCS}}_{s,i+1/2} = \hat{f}^{\text{L-W}}_{s,i+1/2}.$$

In other words, the semidiscrete Lax–Wendroff method is the same as the semidiscrete FTCS method. In this sense, the fully discrete Lax–Wendroff and FTCS methods differ only in their time discretization; the Lax–Wendroff method uses a Cauchy–Kowalewski time discretization, whereas FTCS uses a forward-Euler time discretization. Of course, FTCS is highly unstable, whereas the Lax–Wendroff method is not, which just goes to show how critical the time discretization is.

Example 20.3 Find a semidiscrete version of the Beam–Warming second-order upwind method.

Solution As seen in Section 17.4, the fully discrete Beam–Warming second-order upwind method for $a(u) > 0$ is

$$u_i^{n+1} = u_i^n - \lambda\left(\hat{f}^{\text{B-W}}_{i+1/2} - \hat{f}^{\text{B-W}}_{i-1/2}\right),$$
$$\hat{f}^{\text{B-W}}_{i+1/2} = \frac{1}{2}\left(3f\left(u_i^n\right) + f\left(u_{i-1}^n\right)\right) - \frac{\lambda}{2}\left(a_{i-1/2}^n\right)^2\left(u_i^n - u_{i-1}^n\right).$$

Dropping the second time-dependent term in the conservative numerical flux, the semidiscrete Beam–Warming second-order upwind method for $a(u) > 0$ becomes

$$\frac{du_i}{dt} = -\frac{\hat{f}^{\text{B-W}}_{s,i+1/2} - \hat{f}^{\text{B-W}}_{s,i-1/2}}{\Delta x},$$
$$\hat{f}^{\text{B-W}}_{s,i+1/2} = \frac{1}{2}\left(3f\left(u_i^n\right) - f\left(u_{i-1}^n\right)\right).$$

The Beam–Warming second-order upwind method for $a(u) < 0$ is similar except that

$$\hat{f}^{\text{B-W}}_{s,i+1/2} = \frac{1}{2}\left(3f(u^n_{i+1}) - f(u^n_{i+2})\right).$$

20.3.1 *A Second-Order Accurate Method: A Semidiscrete Version of Sweby's Flux-Limited Method*

Sweby communicated his ideas to Osher and Chakravarthy before official publication. Then, building on Sweby's results, Osher and Chakravarthy (1984) devised a semidiscrete version of Sweby's flux-limited method, published in the same issue of the same journal, immediately following Sweby's paper. Their original paper is mainly devoted to heavily theoretical analyses of the semidiscrete version of Sweby's flux-limited method; in particular, Osher and Chakravarthy prove that the semidiscrete version of Sweby's flux-limited method converges to the correct entropy-condition-satisfying solution as $\Delta x \to 0$. A lesser known and unpublished companion paper, Chakravarthy and Osher (1983), concerns more practical details, especially those related to the Euler equations.

Recall that Sweby's flux-limited method uses any first-order upwind method:

$$u^{n+1}_i = u^n_i - \lambda\left(\hat{f}^{(1)}_{i+1/2} - \hat{f}^{(1)}_{i-1/2}\right),$$
$$\hat{f}^{(1)}_{i+1/2} = \frac{1}{2}\left(f(u^n_{i+1}) + f(u^n_i)\right) - \frac{1}{2}\epsilon^{(1)}_{i+1/2}\left(u^n_{i+1} - u^n_i\right),$$

where

$$\epsilon^{(1)}_{i+1/2} \geq \epsilon^{\text{ROE}}_{i+1/2} = \left|a^n_{i+1/2}\right|.$$

Assuming that $\epsilon^{(1)}_{i+1/2}$ does not depend on Δt or λ, the semidiscrete version of Sweby's flux-limited method may use any semidiscrete first-order upwind method as follows:

$$\frac{du_i}{dt} = -\frac{\hat{f}^{(1)}_{s,i+1/2} - \hat{f}^{(1)}_{s,i-1/2}}{\Delta x},$$
$$\hat{f}^{(1)}_{s,i+1/2} = \hat{f}^{(1)}_{i+1/2} = \frac{1}{2}\left(f(u^n_{i+1}) + f(u^n_i)\right) - \frac{1}{2}\epsilon^{(1)}_{i+1/2}\left(u^n_{i+1} - u^n_i\right),$$

where

$$\epsilon^{(1)}_{i+1/2} \geq \epsilon^{\text{ROE}}_{i+1/2} = \left|a^n_{i+1/2}\right|.$$

In addition to the first-order upwind method, Sweby's flux-limited method uses the Lax–Wendroff method. Then the semidiscrete version of Sweby's flux-limited method uses the semidiscrete version of the Lax–Wendroff method, seen in Example 20.1. Finally, Sweby's flux-limited method uses the following ratios:

$$r^+_i = \frac{\hat{f}^{\text{L-W}}_{i-1/2} - \hat{f}^{\text{FTBS}}_{i-1/2}}{\hat{f}^{\text{L-W}}_{i+1/2} - \hat{f}^{\text{FTBS}}_{i+1/2}} = \frac{a^+_{i-1/2}\left(1 - \lambda a^+_{i-1/2}\right)\left(u^n_i - u^n_{i-1}\right)}{a^+_{i+1/2}\left(1 - \lambda a^+_{i+1/2}\right)\left(u^n_{i+1} - u^n_i\right)},$$
$$r^-_i = \frac{\hat{f}^{\text{L-W}}_{i+1/2} - \hat{f}^{\text{FTFS}}_{i+1/2}}{\hat{f}^{\text{L-W}}_{i-1/2} - \hat{f}^{\text{FTFS}}_{i-1/2}} = \frac{a^-_{i+1/2}\left(1 + \lambda a^-_{i+1/2}\right)\left(u^n_{i+1} - u^n_i\right)}{a^-_{i-1/2}\left(1 + \lambda a^-_{i-1/2}\right)\left(u^n_i - u^n_{i-1}\right)}.$$

Then the semidiscrete version of Sweby's flux-limited method uses the following ratios:

$$r_i^+ = \frac{\hat{f}^{\text{L-W}}_{s,i-1/2} - \hat{f}^{\text{FTBS}}_{s,i-1/2}}{\hat{f}^{\text{L-W}}_{s,i+1/2} - \hat{f}^{\text{FTBS}}_{s,i+1/2}} = \frac{a^+_{i-1/2}\left(u_i^n - u_{i-1}^n\right)}{a^+_{i+1/2}\left(u_{i+1}^n - u_i^n\right)}, \tag{20.51}$$

$$r_i^- = \frac{\hat{f}^{\text{L-W}}_{s,i+1/2} - \hat{f}^{\text{FTFS}}_{s,i+1/2}}{\hat{f}^{\text{L-W}}_{s,i-1/2} - \hat{f}^{\text{FTFS}}_{s,i-1/2}} = \frac{a^-_{i+1/2}\left(u_{i+1}^n - u_i^n\right)}{a^-_{i-1/2}\left(u_i^n - u_{i-1}^n\right)}. \tag{20.52}$$

These new expressions eliminate Lax–Wendroff-type terms such as $1 \pm \lambda a^\pm_{i+1/2}$, which depend on the time step through $\lambda = \Delta t/\Delta x$.

Combine the semidiscrete first-order upwind method and the semidiscrete Lax–Wendroff method, just as in the last section, to obtain the following final result:

$$\frac{du_i}{dt} = -\frac{\hat{f}_{s,i+1/2} - \hat{f}_{s,i-1/2}}{\Delta x},$$

where

$$\hat{f}_{s,i+1/2} = \hat{f}^{(1)}_{s,i+1/2} + \frac{1}{2}\left[a^+_{i+1/2}\phi\left(r_i^+\right) - a^-_{i+1/2}\phi\left(r^-_{i+1}\right)\right]\left(u^n_{i+1} - u^n_i\right) \tag{20.53}$$

and where ϕ is defined just as in the last section; see Equations (20.21)–(20.24) and (20.26).

Equation (20.53) is written in terms of wave speed splitting. However, it is more naturally written in terms of flux splitting, using the notation of Section 13.4. In particular, by Equation (20.41),

$$a^+_{i+1/2}\left(u^n_{i+1} - u^n_i\right) = f\left(u^n_{i+1}\right) - \hat{f}^{(1)}_{i+1/2} \equiv \Delta f^+_{i+1/2},$$
$$a^-_{i+1/2}\left(u^n_{i+1} - u^n_i\right) = \hat{f}^{(1)}_{i+1/2} - f\left(u^n_i\right) \equiv \Delta f^-_{i+1/2}.$$

Then Equation (20.53) becomes

$$\hat{f}_{s,i+1/2} = \hat{f}^{(1)}_{s,i+1/2} + \frac{1}{2}\phi\left(r_i^+\right)\Delta f^+_{i+1/2} - \frac{1}{2}\phi\left(r^-_{i+1}\right)\Delta f^-_{i+1/2}, \tag{20.54}$$

where

$$r_i^+ = \frac{\Delta f^+_{i-1/2}}{\Delta f^+_{i+1/2}}, \qquad r_i^- = \frac{\Delta f^-_{i+1/2}}{\Delta f^-_{i-1/2}}.$$

For example, using the minmod limiter $\phi(r) = minmod(1, br)$ this expression becomes

$$\hat{f}_{s,i+1/2} = \hat{f}^{(1)}_{s,i+1/2} + \frac{1}{2} minmod\left[1, b\frac{\Delta f^+_{i-1/2}}{\Delta f^+_{i+1/2}}\right]\Delta f^+_{i+1/2}$$
$$- \frac{1}{2} minmod\left[1, b\frac{\Delta f^-_{i+3/2}}{\Delta f^-_{i+1/2}}\right]\Delta f^-_{i+1/2}.$$

But

$$z\; minmod(x, y) = minmod(zx, zy),$$

as the reader can easily show. Then

$$\hat{f}_{s,i+1/2} = \hat{f}^{(1)}_{s,i+1/2} + \frac{1}{2}minmod\left(\Delta f^{+}_{i+1/2}, b\Delta f^{+}_{i-1/2}\right) - \frac{1}{2}minmod\left[\Delta f^{-}_{i+1/2}, b\Delta f^{-}_{i+3/2}\right]. \tag{20.55}$$

20.3.2 *Another Second-Order Accurate Method*

Sweby's flux-limited method combines the Lax–Wendroff method and a first-order upwind method. Using the identical approach, this section derives a flux-limited combination of the Beam–Warming second-order upwind method and a first-order upwind method. The semidiscrete Beam–Warming second-order upwind method was found in Example 20.3 and can be written as follows:

$$\frac{du_i}{dt} = -\frac{\hat{f}^{B-W}_{s,i+1/2} - \hat{f}^{B-W}_{s,i-1/2}}{\Delta x},$$

where for $a(u) > 0$

$$\hat{f}^{B-W}_{s,i+1/2} = f(u^n_i) + \frac{1}{2}a^n_{i-1/2}(u^n_i - u^n_{i-1})$$

and for $a(u) > 0$

$$\hat{f}^{B-W}_{s,i+1/2} = f(u^n_{i+1}) - \frac{1}{2}a^n_{i+3/2}(u^n_{i+2} - u^n_{i+1}).$$

Let

$$r^+_i = \frac{\hat{f}^{B-W}_{s,i+1/2} - \hat{f}^{FTBS}_{s,i+1/2}}{\hat{f}^{B-W}_{s,i+3/2} - \hat{f}^{FTBS}_{s,i+3/2}} = \frac{a^+_{i-1/2}(u^n_i - u^n_{i-1})}{a^+_{i+1/2}(u^n_{i+1} - u^n_i)}, \tag{20.56}$$

$$r^-_i = \frac{\hat{f}^{B-W}_{s,i-1/2} - \hat{f}^{FTFS}_{s,i-1/2}}{\hat{f}^{B-W}_{s,i-3/2} - \hat{f}^{FTFS}_{s,i-3/2}} = \frac{a^-_{i+1/2}(u^n_{i+1} - u^n_i)}{a^-_{i-1/2}(u^n_i - u^n_{i-1})}, \tag{20.57}$$

where the indexing has been adjusted slightly from the standard so that Equations (20.56) and (20.57) are exactly the same as Equations (20.51) and (20.52). Then, proceeding just as for Sweby's flux-limited method, a flux-limited combination of the semidiscrete Beam–Warming second-order upwind method and a first-order upwind method yields

$$\frac{du_i}{dt} = -\frac{\hat{f}_{s,i+1/2} - \hat{f}_{s,i-1/2}}{\Delta x},$$

where

$$\hat{f}_{s,i+1/2} = \hat{f}^{(1)}_{s,i+1/2} + \frac{1}{2}a^+_{i-1/2}\phi\left(\frac{1}{r^+_i}\right)(u^n_i - u^n_{i-1}) - \frac{1}{2}a^-_{i+3/2}\phi\left(\frac{1}{r^-_{i+1}}\right)(u^n_{i+2} - u^n_{i+1}) \tag{20.58}$$

and where ϕ is defined as in the last section; see Equations (20.21)–(20.24) and (20.26).

Equation (20.58) is written in terms of wave speed splitting. However, it is more naturally written in terms of flux splitting, using the notation of Section 13.4. In particular, by Equation (20.41)

$$a^+_{i+1/2}(u^n_{i+1} - u^n_i) = f(u^n_{i+1}) - \hat{f}^{(1)}_{i+1/2} \equiv \Delta f^+_{i+1/2},$$
$$a^-_{i+1/2}(u^n_{i+1} - u^n_i) = \hat{f}^{(1)}_{i+1/2} - f(u^n_i) - \equiv \Delta f^-_{i+1/2}.$$

Then Equation (20.58) becomes

$$\hat{f}_{s,i+1/2} = \hat{f}^{(1)}_{s,i+1/2} + \frac{1}{2}\phi\left(\frac{1}{r^+_i}\right)\Delta f^+_{i-1/2} - \frac{1}{2}\phi\left(\frac{1}{r^-_{i+1}}\right)\Delta f^-_{i+3/2}, \tag{20.59}$$

where

$$r^+_i = \frac{\Delta f^+_{i-1/2}}{\Delta f^+_{i+1/2}}, \qquad r^-_i = \frac{\Delta f^-_{i+1/2}}{\Delta f^-_{i-1/2}}.$$

For example, using the minmod limiter $\phi(r) = minmod(1, br)$ this expression becomes

$$\hat{f}_{s,i+1/2} = \hat{f}^{(1)}_{s,i+1/2} + \frac{1}{2} minmod\left[1, b\frac{\Delta f^+_{i+1/2}}{\Delta f^+_{i-1/2}}\right]\Delta f^+_{i-1/2}$$
$$- \frac{1}{2} minmod\left[1, b\frac{\Delta f^-_{i+1/2}}{\Delta f^-_{i+3/2}}\right]\Delta f^-_{i+3/2}$$

or

$$\hat{f}_{s,i+1/2} = \hat{f}^{(1)}_{s,i+1/2} + \frac{1}{2} minmod(\Delta f^+_{i-1/2}, b\Delta f^+_{i+1/2})$$
$$- \frac{1}{2} minmod(\Delta f^-_{i+3/2}, b\Delta f^-_{i+1/2}). \tag{20.60}$$

Notice that the method in this subsection is identical to the method from the last subsection if $b = 1$, that is, Equations (20.55) and (20.60) are identical if $b = 1$.

20.3.3 *Second- and Third-Order Accurate Methods*

This subsection concerns a class of flux-limited methods first proposed by Chakravarthy and Osher (1985). As in Van Leer's flux-limited method, seen in Section 20.1, the starting point is a linear combination of the Lax–Wendroff method and Beam–Warming second-order upwind methods:

$$\frac{du_i}{dt} = -\frac{\hat{f}_{s,i+1/2} - \hat{f}_{s,i-1/2}}{\Delta x},$$

where

$$\hat{f}_{s,i+1/2} = \frac{1+\eta}{2}\hat{f}^{\text{L-W}}_{s,i+1/2} + \frac{1-\eta}{2}\hat{f}^{\text{B-W}}_{s,i+1/2}.$$

Then for $a(u) > 0$

$$\hat{f}_{s,i+1/2} = f(u^n_i) + \frac{1+\eta}{4}a^n_{i+1/2}(u^n_{i+1} - u^n_i) + \frac{1-\eta}{4}a^n_{i-1/2}(u^n_i - u^n_{i-1})$$

and for $a(u) < 0$

$$\hat{f}_{s,i+1/2} = f(u_{i+1}^n) - \frac{1+\eta}{4} a_{i+1/2}^n (u_{i+1}^n - u_i^n) - \frac{1-\eta}{4} a_{i+3/2}^n (u_{i+2}^n - u_{i+1}^n).$$

Notice that this method is a semidiscrete version of the Lax–Wendroff method if $\eta = 1$, a semidiscrete version of the Beam–Warming second-order upwind method if $\eta = -1$, and a semidiscrete version of Fromm's method for $\eta = 0$. Also, $\eta = 1/3$ is a third-order accurate method.

To find a method that works regardless of the wind direction, use numerical wave speed splitting as in Sweby's flux-limited method; see Subsection 20.2.5. That is, define $a_{i+1/2}^{\pm}$ as in Equations (20.41) and (20.42). Then a semidiscrete method that works regardless of wind direction is

$$\frac{du_i}{dt} = -\frac{\hat{f}_{s,i+1/2} - \hat{f}_{s,i-1/2}}{\Delta x},$$

where

$$\hat{f}_{s,i+1/2} = \hat{f}_{i+1/2}^{(1)} + \frac{1+\eta}{4} a_{i+1/2}^{+} (u_{i+1}^n - u_i^n) + \frac{1-\eta}{4} a_{i-1/2}^{+} (u_i^n - u_{i-1}^n)$$
$$- \frac{1+\eta}{4} a_{i+1/2}^{-} (u_{i+1}^n - u_i^n) - \frac{1-\eta}{4} a_{i+3/2}^{-} (u_{i+2}^n - u_{i+1}^n), \qquad (20.61)$$

which is the desired class of methods. This expression uses wave speed splitting. However, a better and equivalent expression uses flux splitting. In particular,

$$a_{i+1/2}^{+} (u_{i+1}^n - u_i^n) = f(u_{i+1}^n) - \hat{f}_{i+1/2}^{(1)} \equiv \Delta f_{i+1/2}^{+},$$
$$a_{i+1/2}^{-} (u_{i+1}^n - u_i^n) = \hat{f}_{i+1/2}^{(1)} - f(u_i^n) - \equiv \Delta f_{i+1/2}^{-}$$

imply

♦ $$\hat{f}_{s,i+1/2} = \hat{f}_{i+1/2}^{(1)} + \frac{1+\eta}{4} \Delta f_{i+1/2}^{+} + \frac{1-\eta}{4} \Delta f_{i-1/2}^{+}$$
$$- \frac{1+\eta}{4} \Delta f_{i+1/2}^{-} - \frac{1-\eta}{4} \Delta f_{i+3/2}^{-}. \qquad (20.62)$$

In Section 20.1, Van Leer's flux-limited method adjusted the linear combination parameter η. However, this section takes a different approach, using a fixed η to form a linear combination of the flux-limited Lax–Wendroff method found in Subsection 20.3.1 and the flux-limited Beam–Warming second-order upwind method found in Subsection 20.3.2. In particular, a linear combination of Equations (20.53) and (20.58) yields

$$\hat{f}_{s,i+1/2} = \hat{f}_{s,i+1/2}^{(1)} + \frac{1+\eta}{4} [a_{i+1/2}^{+} \phi(r_i^{+}) - a_{i+1/2}^{-} \phi(r_{i+1}^{-})] (u_{i+1}^n - u_i^n)$$
$$+ \frac{1-\eta}{4} \left[a_{i-1/2}^{+} \phi\left(\frac{1}{r_i^{+}}\right) (u_i^n - u_{i-1}^n) - a_{i+3/2}^{-} \phi\left(\frac{1}{r_{i+1}^{-}}\right) (u_{i+2}^n - u_{i+1}^n) \right], \qquad (20.63)$$

where

$$r_i^{+} = \frac{a_{i-1/2}^{+} (u_i^n - u_{i-1}^n)}{a_{i+1/2}^{+} (u_{i+1}^n - u_i^n)}, \qquad r_i^{-} = \frac{a_{i+1/2}^{-} (u_{i+1}^n - u_i^n)}{a_{i-1/2}^{-} (u_i^n - u_{i-1}^n)}.$$

Equation (20.63) represents an entire family of flux-limited methods, depending on the choice of η and ϕ. For example, if $\eta = 1$, Equation (20.63) gives a flux-limited version of the Lax–Wendroff method; if $\eta = -1$, Equation (20.63) gives a flux-limited version of the Beam–Warming second-order upwind method.

If $\phi(r) = minmod(1, br)$, then a linear combination of Equations (20.55) and (20.60) yields

$$\begin{aligned}\hat{f}_{s,i+1/2} = f^{(1)}_{i+1/2} &+ \frac{1+\eta}{4} minmod\left(\Delta f^{+}_{i+1/2}, b\Delta f^{+}_{i-1/2}\right) \\ &+ \frac{1-\eta}{4} minmod\left(\Delta f^{+}_{i-1/2}, b\Delta f^{+}_{i+1/2}\right) \\ &- \frac{1-\eta}{4} minmod\left(\Delta f^{-}_{i+3/2}, b\Delta f^{-}_{i+1/2}\right) \\ &- \frac{1+\eta}{4} minmod\left(\Delta f^{-}_{i+1/2}, b\Delta f^{-}_{i+3/2}\right), \end{aligned} \tag{20.64}$$

where the constant parameter b should satisfy the following:

$$1 \leq b \leq \frac{3-\eta}{1-\eta}. \tag{20.65}$$

For example, if $\eta = 1$ then $1 \leq b < \infty$ and if $\eta = -1$ then $1 \leq b \leq 2$. Notice that if $b = 1$, then Equation (20.64) does not depend on η. Specifically, for $b = 1$, Equation (20.64) becomes

$$\begin{aligned}\hat{f}_{s,i+1/2} = f^{(1)}_{i+1/2} &+ \frac{1}{2} minmod\left(\Delta f^{+}_{i+1/2}, \Delta f^{+}_{i-1/2}\right) \\ &- \frac{1}{2} minmod\left(\Delta f^{-}_{i+3/2}, \Delta f^{-}_{i+1/2}\right). \end{aligned} \tag{20.66}$$

Equation (20.64) is the expression that appeared in Chakravarthy and Osher (1985). In this book, the term *Chakravarthy–Osher flux-limited method* refers specifically to Equation (20.66). A more common term in the literature is the "Chakravarthy–Osher TVD method" or the "Osher–Chakravarthy TVD method," although the reader needs to be alert, since these may potentially refer to any of the methods seen in this section.

20.3.4 *Higher-Order Accurate Methods*

Osher and Chakravarthy (1986) generalized the class of second-order flux-limited methods seen in the last subsection to classes of higher-order flux-limited methods, including methods with up to seventh-order accuracy. Unfortunately, whatever their order of accuracy in smooth monotone regions, these higher-order accurate flux-limited methods obtain only between first- and second-order accuracy at extrema, due to clipping. Chakravarthy and Osher's higher-order accurate methods are complicated enough that even a brief description would require many tedious pages of discussion. Furthermore, Chakravarthy and Osher's class of higher-order TVD methods has been largely superceded by Shu and Osher's class of higher-order ENO methods described in Section 21.4. Thus, although the higher-order Chakravarthy and Osher methods certainly merit a mention, the details are omitted, and the reader should consult the original paper.

20.4 Davis–Roe Flux-Limited Method (TVD)

20.4.1 *Scalar Conservation Laws*

Roe (1984) suggested an efficient variation on Sweby's flux-limited method which he called a "TVD Lax–Wendroff" method. Building on Roe's work, Davis (1987) suggested the following flux-limited method:

$$u_i^{n+1} = u_i^n - \lambda\left(\hat{f}_{i+1/2}^n - \hat{f}_{i-1/2}^n\right),$$

where

$$\blacklozenge \quad \hat{f}_{i+1/2}^n = \hat{f}_{i+1/2}^{\mathrm{ROE}} + \frac{1}{2}\left|a_{i+1/2}^n\right|\left(1 - \lambda\left|a_{i+1/2}^n\right|\right)\left[\phi(r_i^+) + \phi(r_{i+1}^-) - 1\right]\left(u_{i+1}^n - u_i^n\right) \tag{20.67a}$$

or, equivalently,

$$\hat{f}_{i+1/2}^n = \hat{f}_{i+1/2}^{\mathrm{L-W}} + \frac{1}{2}\left|a_{i+1/2}^n\right|\left(1 - \lambda\left|a_{i+1/2}^n\right|\right)\left[\phi(r_i^+) + \phi(r_{i+1}^-) - 2\right]\left(u_{i+1}^n - u_i^n\right), \tag{20.67b}$$

where

$$r_i^+ = \frac{u_i^n - u_{i-1}^n}{u_{i+1}^n - u_i^n}, \tag{20.3}$$

$$r_i^- = \frac{u_{i+1}^n - u_i^n}{u_i^n - u_{i-1}^n}. \tag{20.4}$$

This method is called the *Davis–Roe flux-limited method.* Equation (20.67a) is the same as Equation (20.31) for $a > 0$, except that ϕ_i^n is replaced by $\phi(r_i^+) + \phi(r_{i+1}^-) - 1$; also, Equation (20.67a) is the same as Equation (20.36) for $a < 0$, except that ϕ_{i+1}^n is replaced by $\phi(r_i^+) + \phi(r_{i+1}^-) - 1$. The title of Davis' original paper – "A Simplified TVD Finite Difference Scheme via Artificial Viscosity" – plays up the fact that the Davis–Roe method is simpler than Sweby's flux-limited method, which it certainly is, and also that it uses artificial viscosity, although every flux-limited method uses adaptive artificial viscosity, if you care to view it that way, as mentioned in the chapter introduction.

Davis uses the simple definitions (20.3) and (20.4) for $r_i^{\pm}$, rather than the more complicated definitions (20.5) and (20.6) used by Sweby in Section 20.2. In Sweby's flux-limited method, the more complicated definitions simplify enforcement of the upwind range condition and, secondarily, establish ties to methods such as the Beam–Warming second-order upwind method and Fromm's method. Davis does not attempt to rigorously enforce the upwind range condition (he makes do with the positivity condition) and does not attempt to connect his method to the Beam–Warming second-order upwind method, Fromm's method, or other such methods. Thus, with his more limited sights as far as nonlinear stability, and his more ambitious sights as far as efficiency, Davis uses the simpler definitions for $r_i^{\pm}$. In the original paper, Davis (1987) suggested the following flux limiter:

$$\blacklozenge \quad \phi(r) = minmod(1, 2r), \tag{20.68}$$

which is the same as limiter (20.21) with $b = 2$. Certainly, the Davis–Roe method allows other flux limiters, although this one seems to outperform most others. In an appendix to the original paper, Davis (1987) proves that the Davis–Roe flux-limited method using the flux limiter (20.68) satisfies the positivity condition seen in Section 16.4 provided that $\lambda|a(u)| \leq 1$. The proof is complicated and not terribly significant, and it is thus omitted.

The behavior of the Davis–Roe flux-limited method is illustrated using the five standard test cases defined in Section 17.0.

Test Case 1 As seen in Figure 20.13, the sinusoid is well captured, although clipping has eroded the peaks by some 15%, and the peaks exhibit some asymmetric cusping.

Test Case 2 As seen in Figure 20.14, the Davis–Roe flux-limited method captures the square wave without spurious oscillations and overshoots, and without excessive dissipation. Although the results are comparable with those of Sweby's flux-limited method

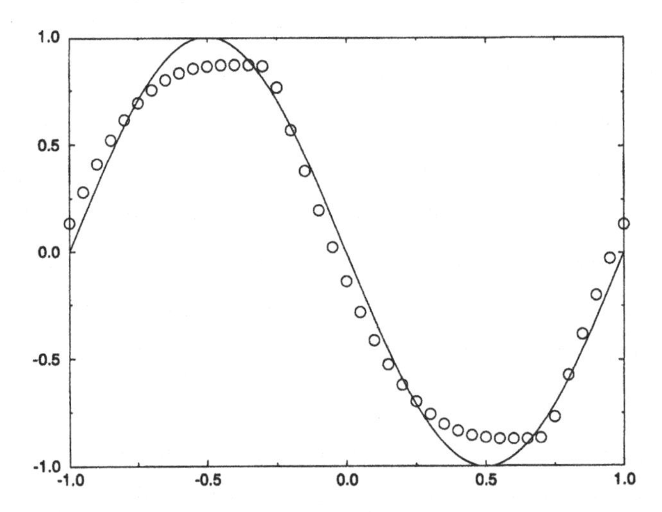

Figure 20.13 Davis–Roe flux-limited method for Test Case 1.

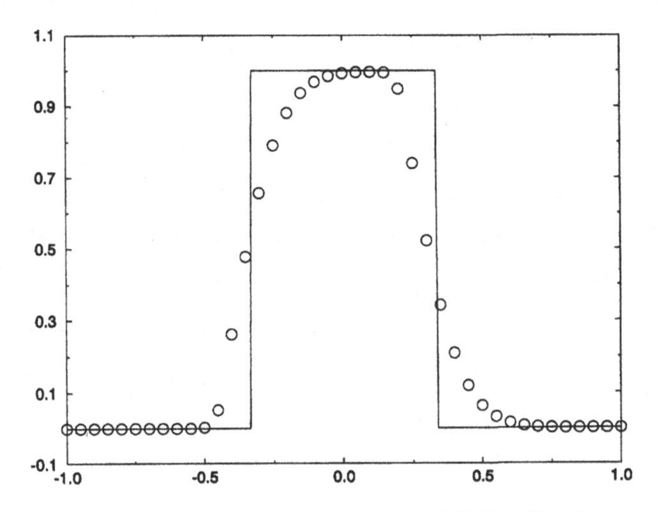

Figure 20.14 Davis–Roe flux-limited method for Test Case 2.

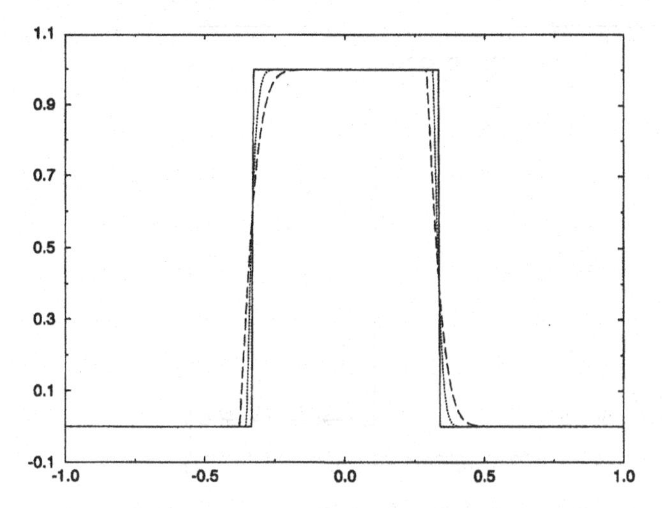

Figure 20.15 Davis–Roe flux-limited method for Test Case 3.

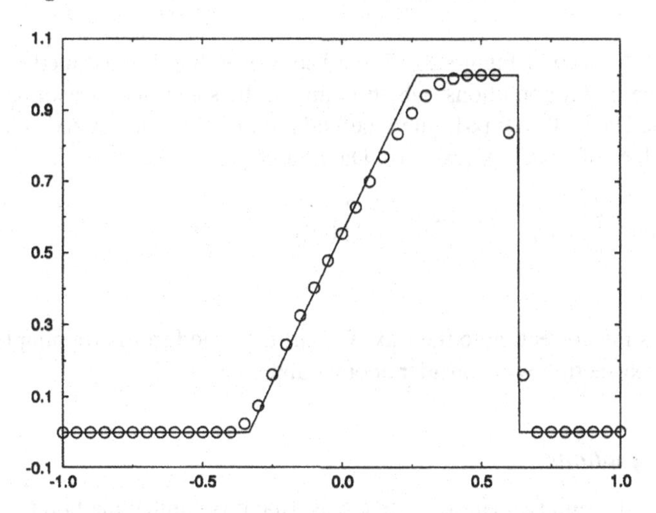

Figure 20.16 Davis–Roe flux-limited method for Test Case 4.

with the minmod limiter, as seen in Figure 20.5, the Davis–Roe flux-limited method exhibits some asymmetric cusping whereas Sweby's flux-limited method does not.

Test Case 3 In Figure 20.15, the dotted line represents the Davis–Roe flux-limited approximation to $u(x, 4)$, the long dashed line represents the Davis–Roe flux-limited approximation to $u(x, 40)$, and the solid line represents the exact solution for $u(x, 4)$ or $u(x, 40)$. Again, these results are comparable to those of Sweby's flux-limited method, as seen in Figure 20.6, and are better than anything seen in Chapter 17.

Test Case 4 As seen in Figure 20.16, the Davis–Roe flux-limited method captures the shock extremely well, with only two transition points. The expansion fan is also well captured with only a slight rounding of the corners.

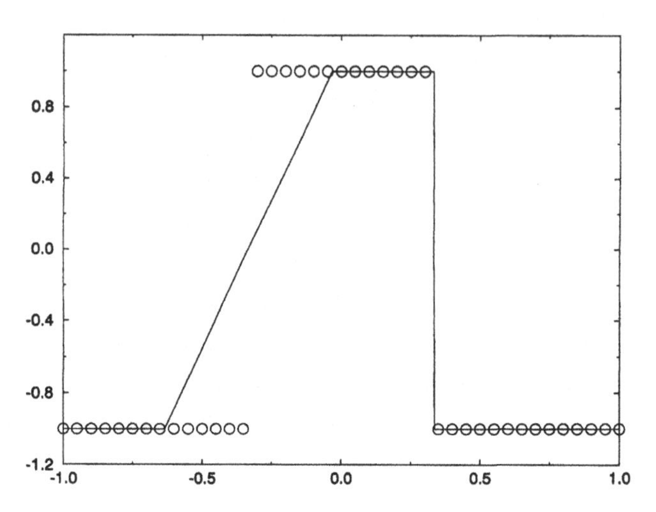

Figure 20.17 Davis–Roe flux-limited method for Test Case 5.

Test Case 5 As seen in Figure 20.17, the Davis–Roe flux-limited method completely fails to alter the initial conditions, which is fine at the shock but disastrous at the expansion. Clearly, the Davis–Roe flux-limited method requires an "entropy fix" at expansive sonic points. In place of Equation (20.67b), Davis suggested

$$\hat{f}^n_{i+1/2} = \hat{f}^{\text{L-W}}_{i+1/2} + \frac{1}{2}\max\left[\frac{1}{4}, a_{i+1/2}\left(1 - \lambda a^n_{i+1/2}\right)\right]\left(\phi\left(r_i^+\right) + \phi\left(r_{i+1}^-\right) - 2\right)\left(u^n_{i+1} - u^n_i\right). \tag{20.69}$$

This certainly prevents the correction to the Lax–Wendroff method from dropping to zero at sonic points. The next section will consider another approach.

20.4.2 *The Euler Equations*

Now consider the Euler equations. The Davis–Roe flux-limited method for scalar conservation laws is extended to the Euler equations using the approximate one-wave linearized Riemann solver described in Section 5.4 and Subsection 18.3.4. The resulting method is as follows:

$$\mathbf{u}^{n+1}_i = \mathbf{u}^n_i - \lambda\left(\hat{\mathbf{f}}^n_{i+1/2} - \hat{\mathbf{f}}^n_{i-1/2}\right),$$

where

$$\hat{\mathbf{f}}^n_{i+1/2} = \hat{\mathbf{f}}^{\text{L-W}}_{i+1/2} + \frac{1}{2}\rho\left(A^n_{i+1/2}\right)\left(1 - \lambda\rho\left(A^n_{i+1/2}\right)\right)\left(\phi\left(r_i^+\right) + \phi\left(r_{i+1}^-\right) - 2\right)\left(\mathbf{u}^n_{i+1} - \mathbf{u}^n_i\right) \tag{20.70}$$

and

$$r_i^+ = \left(\mathbf{u}^n_i - \mathbf{u}^n_{i-1}\right) \cdot \frac{\mathbf{u}^n_{i+1} - \mathbf{u}^n_i}{\left\|\mathbf{u}^n_{i+1} - \mathbf{u}^n_i\right\|} \tag{20.71}$$

and

$$r_i^- = \left(\mathbf{u}_{i+1}^n - \mathbf{u}_i^n\right) \cdot \frac{\mathbf{u}_i^n - \mathbf{u}_{i-1}^n}{\left\|\mathbf{u}_i^n - \mathbf{u}_{i-1}^n\right\|}. \tag{20.72}$$

As usual, $A_{i+1/2}^n$ is some average Jacobian matrix, $\rho(A_{i+1/2}^n)$ is the largest characteristic value of $A_{i+1/2}^n$ in absolute value, and $\| \cdot \|$ is some vector norm, as discussed in Section 6.1. The one-wave Riemann solver could be replaced by other Riemann solvers, such as Roe's approximate Riemann solver, but only at the expense of the simplicity and efficiency Davis sought to maximize. In pursuit of this same goal, Davis replaced the one-step Lax–Wendroff method seen above by the more efficient two-step MacCormack method; see the original paper for details, and also see Subsection 18.1.2 for a discussion of MacCormack's method versus the original Lax–Wendroff method. Many older engineering codes are based on MacCormack's method. The beauty of the Davis–Roe flux-limited method is that old codes based on MacCormack's method can be easily and efficiently converted to the Davis–Roe flux-limited method. In the literature, the one-step version of the Davis–Roe flux-limited method is often called a "TVD Lax–Wendroff" method while the two-step version is often called a "TVD MacCormack" method. However, as usual, this book will identify the method by author (Davis and Roe) and by general type (flux-limited) rather than by any purported nonlinear stability properties or by any of the specific methods used in the flux-limited combination.

20.5 Yee–Roe Flux-Limited Method (TVD)

20.5.1 *Scalar Conservation Laws*

In 1987, Yee suggested some generalizations and modifications to the Davis–Roe flux-limited scheme. To go from the Davis–Roe flux-limited method to the Yee–Roe flux-limited method requires three modifications. First, as a simple notational change, replace the factor $\phi(r_i^+) + \phi(r_{i+1}^-) - 1$ seen in Equation (20.67) by

$$\phi_{i+1/2}^n = \phi\left(r_i^+, r_{i+1}^-\right), \tag{20.73}$$

where $r_i^\pm$ are defined just as before:

$$r_i^+ = \frac{u_i^n - u_{i-1}^n}{u_{i+1}^n - u_i^n}, \tag{20.3}$$

$$r_i^- = \frac{u_{i+1}^n - u_i^n}{u_i^n - u_{i-1}^n}. \tag{20.4}$$

Second, replace the Lax–Wendroff method by FTCS. Although this replacement adversely affects stability and accuracy, steady-state solutions no longer depend on the time step, unlike the Davis–Roe flux-limited method, and it may be possible to partially accommodate the replacement by properly choosing the flux limiter. With these two modifications, the method is now

$$u_i^{n+1} = u_i^n - \lambda\left(\hat{f}_{i+1/2}^n - \hat{f}_{i-1/2}^n\right),$$

where

$$\hat{f}_{i+1/2}^n = \hat{f}_{i+1/2}^{\mathrm{ROE}} + \frac{1}{2}\left|a_{i+1/2}^n\right|\phi_{i+1/2}^n\left(u_{i+1}^n - u_i^n\right) \tag{20.74a}$$

or, equivalently,

$$\hat{f}^n_{i+1/2} = \hat{f}^{\text{FTCS}}_{i+1/2} + \frac{1}{2}|a^n_{i+1/2}|\left(\phi^n_{i+1/2} - 1\right)\left(u^n_{i+1} - u^n_i\right). \tag{20.74b}$$

Remember that the Davis–Roe method requires an "entropy fix" at expansive sonic points. Then, as the third and final modification, to prevent the flux-limited correction to FTCS seen in Equation (20.74b) from dropping to zero at sonic points, replace $|a^n_{i+1/2}|$ by $\psi(a^n_{i+1/2})$, where

$$\psi(x) = \begin{cases} \frac{x^2+\delta^2}{2\delta} & |x| < \delta, \\ |x| & |x| > \delta, \end{cases}$$

as in Harten's first-order upwind method; see Subsections 17.3.3 and 18.3.3. As usual, this modification affects the method only at sonic points. Then, with all three modifications, the *explicit Yee–Roe flux-limited method* is

$$u^{n+1}_i = u^n_i - \lambda\left(\hat{f}^n_{i+1/2} - \hat{f}^n_{i-1/2}\right),$$

where

♦ $$\hat{f}^n_{i+1/2} = \hat{f}^{\text{FTCS}}_{i+1/2} + \frac{1}{2}\psi\left(a^n_{i+1/2}\right)\left(\phi^n_{i+1/2} - 1\right)\left(u^n_{i+1} - u^n_i\right). \tag{20.75}$$

As an alternative to the forward-Euler time discretization, consider the *backward-time* or *implicit Euler* time discretization:

$$u^{n+1}_i = u^n_i - \lambda\left(\hat{f}^{n+1}_{i+1/2} - \hat{f}^{n+1}_{i-1/2}\right),$$

where $\hat{f}^{n+1}_{i+1/2}$ is defined by Equation (20.75) just as before. Forming a convex linear combination of the implicit and explicit Euler methods yields the following implicit method:

$$u^{n+1}_i = u^n_i - \lambda\theta\left(\hat{f}^{n+1}_{i+1/2} - \hat{f}^{n+1}_{i-1/2}\right) - \lambda(1-\theta)\left(\hat{f}^n_{i+1/2} - \hat{f}^n_{i-1/2}\right),$$

where $0 \leq \theta \leq 1$, and where $\hat{f}^n_{i+1/2}$ and $\hat{f}^{n+1}_{i+1/2}$ are defined by Equation (20.75) just as before. Unfortunately, in general, each time step in the implicit method requires the solution to a nonlinear system of equations, which is prohibitively expensive. However,

$$\begin{aligned}\hat{f}^{n+1}_{i+1/2} - \hat{f}^{n+1}_{i-1/2} = &\frac{1}{2}\left(f\left(u^{n+1}_{i+1}\right) - f\left(u^{n+1}_{i-1}\right)\right)\\ &+\frac{1}{2}\psi\left(a^{n+1}_{i+1/2}\right)\left(\phi^{n+1}_{i+1/2} - 1\right)\left(u^{n+1}_{i+1} - u^{n+1}_i\right)\\ &-\frac{1}{2}\psi\left(a^{n+1}_{i-1/2}\right)\left(\phi^{n+1}_{i-1/2} - 1\right)\left(u^{n+1}_i - u^{n+1}_{i-1}\right).\end{aligned}$$

In this last equation, linearize the right-hand side about time level n using Taylor series to find

$$\begin{aligned}\hat{f}^{n+1}_{i+1/2} - \hat{f}^{n+1}_{i-1/2} \approx &\frac{1}{2}\left(a\left(u^n_{i+1}\right)u^{n+1}_{i+1} - a\left(u^n_{i-1}\right)u^{n+1}_{i-1}\right)\\ &+\frac{1}{2}\psi\left(a^n_{i+1/2}\right)\left(\phi^n_{i+1/2} - 1\right)\left(u^{n+1}_{i+1} - u^{n+1}_i\right)\\ &-\frac{1}{2}\psi\left(a^n_{i-1/2}\right)\left(\phi^n_{i-1/2} - 1\right)\left(u^{n+1}_i - u^{n+1}_{i-1}\right).\end{aligned}$$

So consider the following *linearized implicit Yee–Roe flux-limited method*:

$$u_i^{n+1} = u_i^n - \frac{\lambda\theta}{2}\big[a(u_{i+1}^n)u_{i+1}^{n+1} - a(u_{i-1}^n)u_{i-1}^{n+1} + \psi(a_{i+1/2}^n)(\phi_{i+1/2}^n - 1)(u_{i+1}^{n+1} - u_i^{n+1})$$
$$- \psi(a_{i-1/2}^n)(\phi_{i-1/2}^n - 1)(u_i^{n+1} - u_{i-1}^{n+1})\big] - \lambda(1-\theta)(\hat{f}_{i+1/2}^n - \hat{f}_{i-1/2}^n) \quad (20.76)$$

or, equivalently,

$$\blacklozenge \quad \frac{\lambda\theta}{2}\big[a(u_{i+1}^n) + \psi(a_{i+1/2}^n)(\phi_{i+1/2}^n - 1)\big](u_{i+1}^{n+1} - u_{i+1}^n)$$
$$+ \left\{1 - \frac{\lambda\theta}{2}\big[\psi(a_{i+1/2}^n)(\phi_{i+1/2}^n - 1) + \psi(a_{i-1/2}^n)(\phi_{i-1/2}^n - 1)\big]\right\}(u_i^{n+1} - u_i^n)$$
$$+ \frac{\lambda\theta}{2}\big[-a(u_{i-1}^n) + \psi(a_{i-1/2}^n)(\phi_{i-1/2}^n - 1)\big](u_{i-1}^{n+1} - u_{i-1}^n)$$
$$= -\lambda(\hat{f}_{i+1/2}^n - \hat{f}_{i-1/2}^n). \quad (20.77)$$

This last expression requires the solution of a periodic tridiagonal linear system of equations, which is relatively cheap, as discussed in Example 11.4. For more details and variations on the Yee–Roe flux-limited method, see Yee (1987, 1989) and Yee, Klopfer, and Montagné (1990).

In the original paper, Yee (1987) suggested five possible flux limiters. Of these five, subsequent experience recommends the following three:

$$\phi(r_i^+, r_{i+1}^-) = minmod(1, r_i^+, r_{i+1}^-), \quad (20.78)$$

$$\phi(r_i^+, r_{i+1}^-) = minmod\left(2, 2r_i^+, 2r_{i+1}^-, \frac{1}{2}(r_i^+ + r_{i+1}^-)\right), \quad (20.79)$$

$$\phi(r_i^+, r_{i+1}^-) = minmod(1, r_i^+) + minmod(1, r_{i+1}^-) - 1. \quad (20.80)$$

Other researchers suggested other possible flux limiters for the Yee–Roe method; for example, Wang and Richards (1991) suggest limiters specifically intended for steady viscous flows.

With flux limiter (20.78), Equation (20.75) can be written as follows:

$$\hat{f}_{i+1/2}^n = \frac{1}{2}(f(u_{i+1}^n) + f(u_i^n)) + \frac{1}{2}\psi(a_{i+1/2}^n)$$
$$\times \big[minmod(u_i^n - u_{i-1}^n, u_{i+1}^n - u_i^n, u_{i+2}^n - u_{i+1}^n) - (u_{i+1}^n - u_i^n)\big]. \quad (20.81)$$

Division is the most expensive arithmetic operation; furthermore, since there is always the possibility of division by zero, division is the least reliable arithmetic operation, and it should never be used except as part of an "if-then-else" statement to test for small or zero divisors. Compared to previous expressions, Equation (20.81) eliminates the divisions and tests required to form $r_i^{\pm}$.

Turning to the five standard numerical test cases, dropping the Lax–Wendroff terms seems to force a few unhappy trade-offs between smooth and discontinuous solutions. While the results are still more than acceptable, they are omitted for reasons of space.

20.5.2 The Euler Equations

Yee (1987) suggested explicit and implicit methods for the Euler equations. For simplicity, this book will consider only the explicit version. The explicit Yee–Roe method for the Euler equations is based on Harten's first-order upwind method for the Euler equations, as seen in Equation (18.56) of Subsection 18.3.3:

$$\mathbf{u}_i^{n+1} = \mathbf{u}_i^n - \lambda\left(\hat{\mathbf{f}}_{i+1/2}^n - \hat{\mathbf{f}}_{i-1/2}^n\right),$$

$$\hat{\mathbf{f}}_{i+1/2}^n = \frac{1}{2}\left(\mathbf{f}(\mathbf{u}_{i+1}^n) + \mathbf{f}(\mathbf{u}_i^n)\right) - \frac{1}{2}\sum_{j=1}^{3}\left(\mathbf{r}_{i+1/2}^n\right)_j \psi\left(a_{i+1/2}^n\right)_j \left(\Delta v_{i+1/2}^n\right)_j, \tag{20.82}$$

where

$$\psi(x) = \begin{cases} \frac{x^2+\delta^2}{2\delta} & |x| < \delta, \\ |x| & |x| > \delta, \end{cases},$$

$$\left(a_{i+1/2}^n\right)_j = \left|\lambda_{i+1/2}^n\right|_j \left[1 - \left(\phi_{i+1/2}^n\right)_j\right], \tag{20.83}$$

$$\left(\phi_{i+1/2}^n\right)_j = \phi\left(\left(r_i^+\right)_j, \left(r_{i+1}^-\right)_j\right), \tag{20.84}$$

and

$$\left(r_i^+\right)_j = \frac{\left|\lambda_{i-1/2}^n\right|_j \left(\Delta v_{i-1/2}^n\right)_j}{\left|\lambda_{i+1/2}^n\right|_j \left(\Delta v_{i+1/2}^n\right)_j}, \tag{20.85}$$

$$\left(r_i^-\right)_j = \frac{\left|\lambda_{i+1/2}^n\right|_j \left(\Delta v_{i+1/2}^n\right)_j}{\left|\lambda_{i-1/2}^n\right|_j \left(\Delta v_{i-1/2}^n\right)_j}. \tag{20.86}$$

See the description of Roe's approximate Riemann solver in Section 5.3 and Roe's first-order upwind method in Subsection 18.3.2 for the full details on $(\lambda_{i+1/2}^n)_j$, $(\mathbf{r}_{i+1/2}^n)_j$, and $(\Delta v_{i+1/2}^n)_j$. Do not confuse the right characteristic vectors $(\mathbf{r}_{i+1/2}^n)_j$ with the ratios $(r_i^{\pm})_j$. Also, do not confuse the characteristic values $(\lambda_{i+1/2}^{\pm})_j$ with the ratio $\lambda = \Delta t/\Delta x$. For an intuitive introduction to the ideas used here, the reader may wish to review Subsection 18.3.5. The basic idea, more or less, is to replace the characteristic values appearing in Harten's first-order upwind method, while leaving the characteristic vectors alone.

References

Chakravarthy, S. R., and Osher, S. 1983. "High Resolution Applications of the Osher Upwind Scheme for the Euler Equations," *AIAA Paper 83-1943* (unpublished).

Chakravarthy, S. R., and Osher, S. 1985. "Computing with High-Resolution Upwind Schemes for Hyperbolic Equations." In *Large-Scale Computations in Fluid Mechanics, Lectures in Applied Mathematics*, Volume 22 Part 2, eds. B. E. Engquist, S. Osher, and R. C. J. Somerville, pp. 57–86, Providence, RI: American Mathematical Society.

Davis, S. F. 1987. "A Simplified TVD Finite Difference Scheme via Artificial Viscosity," *SIAM Journal on Scientific and Statistical Computing*, 8: 1–18.

Jeng, Y. N., and Payne, U. J. 1995. "An Adaptive TVD Limiter." *Journal of Computational Physics*, 118: 229–241.

Osher, S., and Chakravarthy, S. 1984. "High Resolution Schemes and Entropy Conditions," *SIAM Journal on Numerical Analysis*, 21: 955–984.

Osher, S., and Chakravarthy, S. R. 1986. "Very High Order Accurate TVD Schemes." In *Oscillation Theory, Computation, and Methods of Compensated Compactness, The IMA Volumes in Mathematics and Its Applications*, Volume 2, eds. C. Dafermos, J. L. Erikson, D. Kinderlehrer, and M. Slemrod, pp. 229–271, New York: Springer-Verlag.

Roe, P. L. 1981. "Numerical Methods for the Linear Wave Equation," *Royal Aircraft Establishment Technical Report 81047* (unpublished).

Roe, P. L. 1982. "Fluctuations and Signals – A Framework for Numerical Evolution Problems." In *Numerical Methods for Fluid Dynamics*, eds. K. W. Morton and M. J. Baines, pp. 219–257, London: Academic Press.

Roe, P. L. 1984. "Generalized Formulations of TVD Lax–Wendroff Schemes," *ICASE Report Number 84-20* (unpublished).

Roe, P. L. 1985. "Some Contributions to the Modelling of Discontinuous Flows." In *Large-Scale Computations in Fluid Mechanics, Lectures in Applied Mathematics*, Volume 22 Part 2, eds. B. E. Engquist, S. Osher, and R. C. J. Somerville, pp. 57–86, Providence, RI: American Mathematical Society.

Roe, P. L. 1986. "Characteristic-Based Schemes for the Euler Equations," *Annual Review of Fluid Mechanics*, 18: 337–365.

Roe, P. L. 1989. "A Survey of Upwind Differencing Techniques." In *11th International Conference on Numerical Methods in Fluid Dynamics, Lecture Notes in Physics* Volume 323, eds. D. L. Dwoyer, M. Y. Hussaini, and R. G. Voigt, pp. 69–78, Berlin: Springer-Verlag.

Roe, P. L., and Baines, M. J. 1982. "Algorithms for Advection and Shock Problems." In *Proceedings of the Fourth GAMM-Conference on Numerical Methods in Fluid Mechanics, Notes on Numerical Fluid Mechanics*, Volume 5, ed. H. Viviand, pp. 281–290, Braunschweig, Germany: Vieweg.

Roe, P. L. and Baines, M. J. 1984. "Asymptotic Behavior of Some Non-Linear Schemes for Linear Advection." In *Proceedings of the Fifth GAMM-Conference on Numerical Methods in Fluid Mechanics, Notes on Numerical Fluid Mechanics*, Volume 7, eds. M. Pandolfi and R. Piva, pp. 283–290, Braunschweig: Vieweg.

Roe, P. L., and Pike, J. 1984. "Efficient Construction and Utilization of Approximate Riemann Solutions." In *Computing Methods in Applied Sciences and Engineering*, VI, eds. R. Glowinski and J.-L. Lions, pp. 499–518, Amsterdam: North-Holland.

Sweby, P. K. 1984. "High Resolution Schemes Using Flux Limiters for Hyperbolic Conservation Laws," *SIAM Journal on Numerical Analysis*, 21: 995–1011.

Van Leer, B. 1974. "Towards the Ultimate Conservative Difference Scheme. II. Monotonicity and Conservation Combined in a Second-Order Scheme," *Journal of Computational Physics*, 14: 361–370.

Wang, Z., and Richards, B. E. 1991. "High Resolution Schemes for Steady Flow Computations," *Journal of Computational Physics*, 97: 53–72.

Yee, H. C. 1987. "Construction of Explicit and Implicit Symmetric TVD Schemes and Their Applications," *Journal of Computational Physics*, 68: 151–179.

Yee, H. C. 1989. "A Class of High-Resolution Explicit and Implicit Shock Capturing Methods," *NASA Technical Memorandum 101088* (unpublished).

Yee, H. C., Klopfer, G. H., and Montagné J.-L. 1990. "High-Resolution Shock-Capturing Schemes for Inviscid and Viscous Hypersonic Flows," *Journal of Computational Physics*, 88: 31–61.

Problems

20.1 For Sweby's flux-limited method, consider the two possible variations on the minmod limiter:

$$\phi(r) = minmod(1, br),$$
$$\phi(r) = minmod(b, r).$$

Plot each of these limiters as a function of r for several different values of b. Explain briefly why b should be between 1 and 2.

20.2 (a) Using Equations (20.11) and (20.16), show that Sweby's flux-limited method for the linear advection equation with $a > 0$ is third-order accurate when

$$\phi(r) = \frac{2+r}{3}.$$

(b) Plot the flux limiter found in part (a) as a function of r. For comparison, your sketch should also include other flux limiters such as those seen in Figures 20.1 and 20.2.

(c) Does the flux limiter found in part (a) ever satisfy the upwind range condition for Sweby's flux-limited method, as seen in Equation (20.19)? If so, then when?

20.3 The van Albada flux limiter is often used in Sweby's flux-limited method. The *van Albada flux limiter* is defined as follows:

$$\phi(r) = \frac{r+r^2}{1+r^2}.$$

(a) Plot the van Albada limiter as a function of r. For comparison, your sketch should also include other flux limiters such as those seen in Figures 20.1 and 20.2.

(b) Does the van Albada flux limiter satisfy the upwind range condition for Sweby's flux-limited method, as seen in Equations such as (20.18) or (20.19)?

(c) Does the van Albada flux limiter cause clipping error at extrema? In other words, is $\phi(r)$ small or zero when $r < 0$, or does ϕ change abruptly at extrema? You may wish to back your answer up with numerical results.

(d) Write a code that approximates solutions to scalar conservation laws using Sweby's flux-limited method with the van Albada flux limiter. Use your code to solve the five standard test cases for scalar conservation laws. How do your results compare with those found using the minmod limiter, seen in Figures 20.4 to 20.8?

20.4 For Sweby's flux-limited method for the linear advection equation, Roe suggested the following *hyperbee* limiter:

$$\phi(r) = \begin{cases} \frac{2r}{\lambda a(1-\lambda a)} \frac{\lambda a(1-r)+r(1-r^{-\lambda a})}{(1-r)^2} & r \geq 0, \\ 0 & r < 0. \end{cases}$$

For those readers who know something about hypergeometric functions, this flux limiter can be written as a hypergeometric function, which explains the term *"hyperbee."*

(a) Show that this flux limiter yields the exact solution if the solution varies geometrically. In other words, show that this flux limiter yields the exact solution if $r_i^+ = const.$ and $r_i^- = const.$ for all i.

(b) Plot the hyperbee limiter as a function of r for several different values of λa. For comparison, your sketch should also include other flux limiters such as those seen in Figures 20.1 and 20.2.

(c) Unlike all of the flux limiters seen in the main text, the hyperbee flux limit depends on the CFL number λa. Does this make sense? Explain briefly.

20.5 (a) As an alternative to Equation (20.20), show that Sweby's flux-limited method for the linear advection equation satisfies the upwind range condition if

$$0 \le \phi(r) \le \frac{2}{|\lambda a|},$$
$$0 \le \phi(r) \le \frac{2}{1 - |\lambda a|}.$$

(b) The superbee limiter is the upper limit for flux limiters allowed by Equation (20.20). Similarly, find the flux limiter that is the upper bound for flux limiters allowed by the condition found in part (a). Roe and Baines call this the *ultrabee* limiter.

20.6 Often, flux limiters in Sweby's flux-limited methods are required to have the following *symmetry* property:

$$\frac{\phi(r)}{r} = \phi\left(\frac{1}{r}\right).$$

Explain briefly why this property might be desirable. For one thing, think in terms of solution symmetry: should the flux limiter treat an upslope the same as a downslope, or a solution the same as its mirror image? Also, discuss how this property allows Sweby's flux-limited method to be written in terms of flux differences rather than ratios of flux differences, saving the troubles associated with division in the ratios of flux differences. Which of the following limiters have the symmetry property?
(a) The minmod limiter defined by Equation (20.21) or (20.22).
(b) The superbee limiter given by Equation (20.23).
(c) The Van Leer limiter given by Equation (20.24).
(d) The symmetric minmod limiter given by Equation (20.26).
(e) The van Albada limiter found in Problem 20.3.

20.7 (a) Equation (20.55) expresses the semidiscrete version of Sweby's flux-limited method with the minmod flux limiter in terms of flux splitting. Find an expression like Equation (20.55), but use the superbee flux limiter seen in Equation (20.23) in place of the minmod flux limiter.
(b) Find an expression like Equation (20.60), but use the superbee flux limiter seen in Equation (20.23) in place of the minmod flux limiter.
(c) Find an expression like Equation (20.64), but use the superbee flux limiter seen in Equation (20.23) in place of the minmod flux limiter.

20.8 Section 20.5 omitted any sort of stability analysis for the Yee–Roe flux-limited method. In fact, the usual sorts of nonlinear stability conditions are difficult to enforce using the simple definitions for $r_i^{\pm}$ seen in Equations (20.3) and (20.4), as used in the standard Yee–Roe flux-limited method. However, there is an alternate version of the Yee–Roe flux-limited method that employs more complicated definitions of $r_i^{\pm}$ and is much more amenable to nonlinear stability analysis. In particular, let

$$r_i^+ = \frac{\left|a_{i-1/2}^+\right|\left(u_i^n - u_{i-1}^n\right)}{\left|a_{i+1/2}^+\right|\left(u_{i+1}^n - u_i^n\right)},$$
$$r_i^- = \frac{\left|a_{i+1/2}^-\right|\left(u_{i+1}^n - u_i^n\right)}{\left|a_{i-1/2}^-\right|\left(u_i^n - u_{i-1}^n\right)}.$$

(a) Briefly discuss the logic behind these definitions for $r_i^{\pm}$. In particular, why do these definitions omit terms such as $1 \pm \lambda a_{i+1/2}^{\pm}$ seen in Equations (20.45) and (20.46)?

(b) Use the definitions for $r_i^{\pm}$ given in part (a). Assume that $\psi(x) = |x|$. Also assume that $0 \leq \lambda a(u) \leq 2/3$. Then show that the explicit Yee–Roe flux-limited method satisfies the upwind range condition if and only if

$$0 \leq \lambda a_{i-1/2}\left(1 + \frac{1}{2}\frac{\phi_{i+1/2}}{r_i^+} - \frac{1}{2}\phi_{i-1/2}\right) \leq 1.$$

(c) Starting with the result of part (b), show that the explicit Yee–Roe flux-limited method satisfies the upwind range condition for $0 \leq \lambda a(u) \leq 2/3$ if

$$-2 \leq \frac{\phi_{i+1/2}^n}{r_i^+} - \phi_{i-1/2}^n \leq 1.$$

(d) Starting with the result of part (c), show that the explicit Yee–Roe flux-limited method satisfies the upwind range condition for $0 \leq \lambda a(u) \leq 2/3$ if

$$0 \leq \phi_{i-1/2}^n \leq 2,$$

$$0 \leq \frac{\phi_{i+1/2}^n}{r_i^+} \leq 1.$$

(e) Using an approach similar to that in parts (b)–(d), show that the explicit Yee–Roe flux-limited method satisfies the upwind range condition for $-2/3 \leq \lambda a_{i+1/2}^n \leq 0$ if

$$0 \leq \phi_{i+1/2}^n \leq 2,$$

$$0 \leq \frac{\phi_{i-1/2}^n}{r_i^-} \leq 1.$$

How does this compare with the result of part (d)? Can we use the same flux limiters for left- and right-running waves?

20.9 Section 20.3 derived semidiscrete methods by adaptively combining two first-generation semidiscrete methods. However, one can take a more fundamental approach. To begin with, recall reconstruction via the primitive function, as described in Section 9.3. In particular, recall that reconstruction via the primitive function finds a polynomial approximation to $f(x)$ given cell-integral averages $\bar{f}_i$. Consider any flux splitting $f(u) = f^+(u) + f^-(u)$. From Subsection 13.4.2, recall that

$$\hat{f}_{s,i+1/2}^n = \hat{f}_s^+(x_{i+1/2}) + \hat{f}_s^-(x_{i+1/2}), \tag{13.18}$$

where $\hat{f}_{i+1/2}^{\pm}(x_{i+1/2})$ is found from $f^{\pm}(u_i^n)$, $f^{\pm}(u_{i-1}^n)$, $f^{\pm}(u_{i+1}^n)$, and so forth using reconstruction via the primitive function. In other words, treat $f^{\pm}(u_i^n)$, $f^{\pm}(u_{i-1}^n)$, $f^{\pm}(u_{i+1}^n)$, and so forth like cell-integral averages of $\hat{f}^{\pm}(x)$.

(a) Using the results of Example 9.6, especially Equations (9.33) and (9.34), show that some possible linear approximations are

$$\hat{f}_s^+(x_{i+1/2}) = f^+(u_i^n) + \frac{1}{2}\left(f^+(u_{i+1}^n) - f^+(u_i^n)\right),$$

$$\hat{f}_s^+(x_{i+1/2}) = f^+(u_i^n) + \frac{1}{2}\left(f^+(u_i^n) - f^+(u_{i-1}^n)\right)$$

and

$$\hat{f}_s^-(x_{i+1/2}) = f^-(u_{i+1}^n) - \frac{1}{2}\left(f^-(u_{i+1}^n) - f^-(u_i^n)\right),$$

$$\hat{f}_s^-(x_{i+1/2}) = f^-(u_{i+1}^n) - \frac{1}{2}\left(f^-(u_i^n) - f^-(u_{i-1}^n)\right).$$

These approximations yield second-order accuracy.

(b) Using the results of Example 9.7, especially Equations (9.36) and (9.37), show that two possible quadratic approximations are

$$\hat{f}^{+}(x_{i+1/2}) = f^{+}\left(u_i^n\right) + \frac{1}{6}\left(f^{+}\left(u_i^n\right) - f^{+}\left(u_{i-1}^n\right)\right) + \frac{1}{3}\left(f^{+}\left(u_{i+1}^n\right) - f^{+}\left(u_i^n\right)\right),$$

$$\hat{f}^{-}(x_{i+1/2}) = f^{-}\left(u_{i+1}^n\right) - \frac{1}{6}\left(f^{-}\left(u_{i+2}^n\right) - f^{-}\left(u_{i+1}^n\right)\right) - \frac{1}{3}\left(f^{-}\left(u_{i+1}^n\right) - f^{-}\left(u_i^n\right)\right).$$

These approximations yield third-order accuracy.

(c) Notice that the results of parts (a) and (b) both involve first differences. Form a linear combination of first differences as follows:

$$\hat{f}^{+}(x_{i+1/2}) = f^{+}\left(u_i^n\right) + \frac{1+\eta}{4}\left(f^{+}\left(u_{i+1}^n\right) - f^{+}\left(u_i^n\right)\right) + \frac{1-\eta}{4}\left(f^{+}\left(u_i^n\right) - f^{+}\left(u_{i-1}^n\right)\right),$$

$$\hat{f}^{-}(x_{i+1/2}) = f^{-}\left(u_{i+1}^n\right) - \frac{1+\eta}{4}\left(f^{-}\left(u_{i+1}^n\right) - f^{-}\left(u_i^n\right)\right) + \frac{1-\eta}{4}\left(f^{-}\left(u_{i+2}^n\right) - f^{-}\left(u_{i+1}^n\right)\right).$$

Show that the results of part (a) are recovered for $\eta = \pm 1$ and that the results of part (b) are recovered for $\eta = 1/3$. Thus this class includes both second- and third-order accurate approximations.

(d) Suppose that $f^{+}(u_{i+1}) - f^{+}(u_i^n) \geq 0$. A sensible constraint is that $\hat{f}^{+}(x_{i+1/2})$ should lie between $f^{+}(u_i^n)$ and $f^{+}(u_{i+1}^n)$. Using the expression in part (c), show that this is true if

$$f^{+}\left(u_i^n\right) - f^{+}\left(u_{i-1}^n\right) \leq \frac{3-\eta}{1-\eta}\left(f^{+}\left(u_{i+1}^n\right) - f^{-}\left(u_i^n\right)\right).$$

This and similar results justify condition (20.65).

(e) Using the results of part (c), argue that the following is an appropriate expression:

$$\hat{f}_{s,i+1/2} = f_{i+1/2}^{(1)} + \frac{1+\eta}{4}\Delta f_{i+1/2}^{+} + \frac{1-\eta}{4}\Delta f_{i-1/2}^{+} - \frac{1+\eta}{4}\Delta f_{i+1/2}^{-}\left(u_{i+1}^n - u_i^n\right) - \frac{1-\eta}{4}\Delta f_{i+3/2}^{-}\left(u_{i+2}^n - u_{i+1}^n\right),$$

which is the same as Equation (20.62). This expression can then be flux limited just as before, either by adjusting η or by adjusting the split fluxes $\Delta f_{i+1/2}^{\pm}$. This approach will be developed more systematically in Section 21.4.

CHAPTER 21

Flux Averaging II: Flux-Corrected Methods

21.0 Introduction

This chapter concerns flux-corrected methods. To keep things simple, this introductory section concerns only flux-corrected methods for scalar conservation laws. Flux-corrected methods for the Euler equations are discussed later in the chapter; see Subsection 21.3.2. Intuitively, flux-corrected methods "correct" inaccurate or unstable methods. More specifically, as first seen in Section 13.3, *flux-corrected methods* for scalar conservation laws are defined as follows:

$$\hat{f}^n_{i+1/2} = \hat{f}^{(1)}_{i+1/2} + \hat{f}^{(C)}_{i+1/2} = \hat{f}^{(1)}_{i+1/2} + diff^n_{i+1/2}\left(\hat{f}^{(1)}_{i+1/2}, \hat{f}^{(2)}_{i+1/2}\right), \tag{21.1}$$

where the flux $\hat{f}^{(C)}_{i+1/2}$ is called the *corrective numerical flux*, $diff^{\,n}_{i+1/2}$ is called a *differencing function*, and $\hat{f}^{(1)}_{i+1/2}$ and $\hat{f}^{(2)}_{i+1/2}$ are any two methods with complementary properties, such as Roe's first-order upwind method and the Lax–Wendroff method. Flux correction attempts to "correct" $\hat{f}^{(1)}_{i+1/2}$ to make it more like $\hat{f}^{(2)}_{i+1/2}$, with the amount of correction varying between full correction and no correction, depending on the solution. In general, the corrective numerical flux $\hat{f}^{(C)}_{i+1/2}$ depends on solution differences and/or flux differences such as $u^n_{i+1} - u^n_i$, $f(u^n_{i+1}) - f(u^n_i)$, and $\hat{f}^{(2)}_{i+1/2} - \hat{f}^{(1)}_{i+1/2}$. Readers should view Equation (21.1) as a starting point for discussion rather than as a "fixed in stone" definition of flux-corrected methods. Although all flux-corrected methods have common elements – notational, philosophical, and practical – many flux-corrected methods differ substantially from Equation (21.1).

Flux-corrected methods are closely related to the flux-limited methods seen in the last chapter. Flux-limited methods average two methods, whereas flux-corrected methods alter one method to make it more like another method; however, these two approaches are really the same, both in practice and theory. To see this, think about combinations of ordinary real numbers – there is no real difference between averaging x and y to get $(x + y)/2$ and altering x to get $x - (x - y)/2$. To make the point absolutely clear, compare the starting-point definition of flux-limited methods, Equation (20.1), with the starting-point definition of flux-corrected methods, Equation (21.1). In particular, the flux-corrected definition (21.1) can be rewritten to match the flux-limited definition (20.1) as follows:

$$\hat{f}^n_{i+1/2} = \hat{f}^{(1)}_{i+1/2} + \phi^n_{i+1/2}\left(\hat{f}^{(2)}_{i+1/2} - \hat{f}^{(1)}_{i+1/2}\right),$$

where

$$\phi^n_{i+1/2}\left(\hat{f}^{(2)}_{i+1/2} - \hat{f}^{(1)}_{i+1/2}\right) = \hat{f}^{(C)}_{i+1/2} = diff^n_{i+1/2}\left(\hat{f}^{(1)}_{i+1/2}, \hat{f}^{(2)}_{i+1/2}\right). \tag{21.2}$$

As one traditional distinction, flux-corrected methods rely on solution or flux differences, whereas flux-limited methods rely on *ratios* of solution or flux differences. But this distinction is hardly fundamental, since many flux-limited methods can be rewritten in terms of differences, while many flux-corrected methods can be rewritten in terms of ratios of differences. For example, in the last chapter, equations such as (20.55), (20.60), (20.64), (20,66), and (20.81) are flux-limited methods written in flux-corrected forms, at least in the sense that they all depend on solution or flux differences rather than on ratios of solution or flux differences. By the same token, in this chapter, equations such as (21.5), (21.8b), and (21.13b) are flux-corrected methods written in flux-limited forms. As another traditional distinction, the flux limiter in flux-limited methods depends on a single argument, e.g., $\phi^n_{i+1/2} = \phi(r^+_i)$; by contrast, when written in terms of flux limiters as in Equation (21.2), flux-corrected methods may depend on multiple arguments, e.g., $\phi^n_{i+1/2} = \phi(r^+_i, r^-_{i+1})$. However, again, this distinction is more traditional than fundamental. Since flux limiting and flux correction are so closely related, it is common to see mixed metaphors such as "flux limited correction" or "limited flux correction."

This chapter describes four flux-corrected methods. In 1973, Boris and Book proposed a two-step flux-corrected blend of a first-order upwind method and the Lax–Wendroff method, which they called Flux-Corrected Transport (FCT). FCT used an unusual first-order upwind method, an early example of a reconstruction–evolution method. In 1979, Zalesak generalized flux-corrected transport to combine any two methods, provided that one method always has more second-order artificial viscosity than the other or, in other words, provided that the flux correction is *antidissipative* (recall the definition of antidissipative flux from Section 20.0). In 1983, Harten proposed a flux-corrected blend of Roe's first-order upwind method and the Lax–Wendroff method. Harten's flux-corrected method differs from traditional flux-corrected methods by correcting the *physical* flux rather than the *numerical* flux. Despite this fundamental difference, Harten's flux-corrected method still has many connections with and similarities to traditional flux-corrected methods.

In 1988, Shu and Osher proposed a numerical method based on the method of lines: the spatial discretization uses piecewise-polynomial ENO reconstruction and reconstruction via the primitive function; the time discretization uses Runge–Kutta methods. Despite its inclusion in this chapter, Shu and Osher's method is arguably not a flux-corrected method or, more precisely, it is not derived like a traditional flux-corrected method. Indeed, the Shu–Osher method can be seen as representing a new category of method distinct from flux-limited, flux-corrected, self-adjusting hybrid, or reconstruction–evolution methods. As introduced in Subsection 13.4.2, this category is sometimes called *flux reconstructed.* Given that the Shu–Osher method is one of the only methods of its kind, it does not warrant a separate chapter. Furthermore, the Shu–Osher method still qualifies as a flux-averaged method, and all flux-averaged methods are closely related, at least on the most fundamental level, as argued in Section 13.3. Judging by the final result, the Shu–Osher method clearly belongs in this chapter on flux correction, as reasoned in Section 21.4, certainly more so than in any of the other chapters on flux averaging. The title of the original paper – "An Efficient Implementation of Essentially Non-Oscillatory Shock Capturing Schemes" – gives the impression that the Shu–Osher method is closely related to the essentially nonoscillatory methods discussed in Section 23.5. However, the title is misleading in that the primary connection between the two types of methods is the use of ENO reconstruction, as described in Chapter 9. To summarize, the Shu–Osher *flux-reconstructed* method is much more closely related to flux-averaged methods, especially flux-corrected methods, than to solution-averaged or *solution-reconstructed* methods like the original ENO methods.

The numerical methods in this chapter go by various names, including FCT, TVD, and ENO. The acronym FCT is essentially synonymous with flux correction, although it usually implies a two-step method. As described in Section 20.0, the acronym TVD refers to any method that (1) involves solution sensitivity; (2) uses the solution sensitivity to enforce some sort of nonlinear stability condition that implies the TVD condition, such as the upwind range condition, at least for model problems; (3) limits the order of accuracy at extrema, usually to between first and second order; and (4) came after the invention of the term TVD (that is, came after the early 1980s). Similarly, the acronym ENO refers to any method that (1) involves solution sensitivity; (2) uses piecewise-polynomial ENO reconstructions of the sort described in Chapter 9, either to reconstruct the solution or to reconstruct the flux; (3) retains full order of accuracy at extrema; and (4) came after the invention of the term ENO (that is, came after the mid-1980s). Unlike TVD methods, ENO methods rarely attempt to rigorously enforce any sort of nonlinear stability condition, not even the ENO stability condition seen in Section 16.7.

The Boris and Book flux-corrected method, and Zalesak's generalizations thereof, are generally *not* called TVD. However, in the original paper, Boris and Book (1973) introduced a number of advanced nonlinear stability conditions, such as the range diminishing condition (see Section 16.3) and the upwind range condition (see Section 16.5) that do imply TVD. Unfortunately, the Boris–Book flux-corrected method sometimes fails to ensure either of these nonlinear stability conditions, especially for large CFL numbers. However, in practice, the Boris–Book flux-corrected method exhibits more than adequate nonlinear stability, at least for small CFL numbers – it certainly ensures the TVD condition – and, in general, shares enough other qualities with TVD methods that it could fairly be called TVD except for the fact that it preceded the invention of the term by some ten years.

In the literature, Harten's flux-corrected method is almost always called a TVD method. In fact, Harten's original 1983 paper introduced both TVD (see Section 16.2) and positivity (see Section 16.4), and refined the range diminishing condition (see Section 16.3), all by way of motivating his flux-corrected method. However, as usual, the TVD label is potentially misleading for two reasons. First, Harten's TVD method has much stronger nonlinear stability properties than TVD stability. In particular, Harten's flux-corrected method retains all of the nonlinear stability properties of Harten's first-order upwind method seen in Subsections 17.3.3 and 18.3.3, including the upwind range condition. Second, except for its nonlinear stability properties and the general use of flux averaging, Harten's TVD method has little in common with other so-called TVD methods, such as Sweby's TVD method seen in Section 20.2 or the Yee–Roe TVD method seen in Section 20.5. Thus, even though Harten's TVD method and other TVD methods may have some of the same nonlinear stability properties, these properties are achieved in very different ways and, indeed, in basic construction, Harten's TVD method has relatively little in common with most other TVD methods.

21.1 Boris–Book Flux-Corrected Method (FCT)

This section concerns the Boris–Book flux-corrected method, discovered in 1973, commonly called Flux Corrected Transport (FCT) or, less often, SHarp And Smooth Transport Algorithm (SHASTA); not coincidentally, Shasta was the name of a soft drink popular at the time. Like all flux-averaged methods, FCT is an adaptive blend of two simpler methods, in this case a first-order upwind method and the Lax–Wendroff method. This section concerns only scalar conservation laws. Boris and Book's original paper described a version of their flux-corrected method for the Euler equations. However, Boris and Book applied the scalar method to the *conservative* form of the Euler equations rather than to the *characteristic* form of the Euler equations. Keep in mind that scalar conservation laws model characteristic equations, not conservative equations, as discussed in Chapter 4. In particular, only the characteristic equations share the nonlinear stability properties of scalar conservation laws, such as the upwind range condition, and even then not at jump intersections, solid boundaries, and so forth; see Section 16.12. Applying scalar numerical methods, which enforce scalar nonlinear stability conditions, to the conserved variables of the Euler equations can result in excessive and unnatural errors. On the positive side, this approach does reduce cost and complication compared to most methods based on Riemann solvers or flux vector splitting.

On many scores, the Boris–Book flux-corrected method counts as a second-generation method; certainly its lack of multiwave modeling for the Euler equations and its 1973 origin date it as second generation. However, having said this, the Boris–Book flux-corrected method is probably the most sophisticated of the second-generation methods, both in terms of the nonlinear stability theory used to motivate it and in terms of what it actually does. In fact, the Boris–Book flux-corrected method is the only second-generation method still commonly used today. For a relatively recent summary of the Boris–Book flux-corrected transport method see, for example, Chapter 8 of Oran and Boris (1989). Perhaps the Boris–Book flux-corrected method counts as a "two-and-a-half" generation method.

In their original paper, Boris and Book construct an unusual first-order reconstruction–evolution method as the foundation for FCT. Boris and Book's first-order upwind method uses two grids: an ordinary grid, sometimes called an *Eulerian grid*, with fixed cells $[x_{i-1/2}, x_{i+1/2}]$; and a grid that moves with the flow, sometimes called a *Lagrangean grid*. The Lagrangean cells start each time step equal to $[x_i, x_{i+1}]$, where $(x_{i+1/2}+x_{i-1/2})/2 = x_i$, and end each time step equal to $[x_i + a(\bar{u}_i^n)\Delta t, x_{i+1} + a(\bar{u}_{i+1}^n)\Delta t]$. Thus, the Lagrangian cells start each time step *staggered* with respect to the Eulerian cells, after which the cell edges follow approximate characteristic lines in the x–t plane. The reconstruction at the beginning of each time step is

$$u(x, t^n) \approx \bar{u}_i^n + \frac{\bar{u}_{i+1}^n - \bar{u}_i^n}{\Delta x}(x - x_i)$$

for $x_i \leq x \leq x_{i+1}$. This reconstruction is continuous, piecewise-linear, and second-order accurate. During the time evolution, the solution is required to stay piecewise-linear across each Lagrangean cell. For conservation, the trapezoidal area under the linear reconstruction in each Lagrangean cell must remain constant. At the end of the time step, the piecewise-linear and piecewise-continuous time-evolved solution is integral averaged over each fixed cell $[x_{i-1/2}, x_{i+1/2}]$ to obtain $\bar{u}_i^{n+1}$. After a long derivation, which the motivated reader is certainly free to attempt as an exercise, in conservation form, the final result is as follows:

$$u_i^{n+1} = u_i^n - \left(\lambda \hat{f}_{i+1/2}^n - \lambda \hat{f}_{i-1/2}^n\right),$$

$$\lambda \hat{f}_{i+1/2}^n = -\frac{1}{2}\left[\frac{\lambda a(u_i) - \frac{1}{2}}{1 + \lambda(a(u_{i+1}) - a(u_i))}\right]^2 u_{i+1}^n + \frac{1}{2}\left[\frac{\lambda a(u_{i+1}) + \frac{1}{2}}{1 + \lambda(a(u_{i+1}) - a(u_i))}\right]^2 u_i^n.$$

This is the *original Boris–Book first-order upwind method.*

Despite its use of reconstruction–evolution, an unusually advanced technique at the time, the original Boris–Book first-order upwind method has several practical problems. First, it only makes sense for $\lambda|a| \leq 1/2$, so that at least part of each Lagrangean cell always stays within one Eulerian cell, which is unusually restrictive. Second, except for the linear advection equation, the conservative numerical flux is not consistent with the physical flux (see Problem 11.7). Third, the original Boris–Book first-order upwind method is not competitive with subsequent first-order upwind methods. For example, Roe's first-order upwind method is less expensive, less complicated, and more accurate, except possibly at expansive sonic points. For all of these reasons, the original Boris–Book first-order upwind method is mainly of historical interest.

For the purposes of this discussion, we shall substitute a simpler first-order upwind method for the original Boris–Book first-order upwind method. In particular, consider the following method:

$$u_i^{n+1} = u_i^n - \left(\lambda \hat{f}_{i+1/2}^{\text{B-B}} - \lambda \hat{f}_{i-1/2}^{\text{B-B}}\right),$$

where

$$\lambda \hat{f}_{i+1/2}^{\text{B-B}} = \lambda \hat{f}_{i+1/2}^{\text{L-W}} - \frac{1}{8}\left(u_{i+1}^n - u_i^n\right) \tag{21.3a}$$

or, equivalently,

$$\lambda \hat{f}_{i+1/2}^{\text{B-B}} = \frac{\lambda}{2}\left(f\left(u_{i+1}^n\right) + f\left(u_i^n\right)\right) - \frac{1}{2}\left(\left(\lambda a_{i+1/2}^n\right)^2 + \frac{1}{4}\right)\left(u_{i+1}^n - u_i^n\right). \tag{21.3b}$$

Hence this method equals the Lax–Wendroff method (see Section 17.2) plus constant-coefficient second-order artificial viscosity. This method, called the *modified Boris–Book first-order upwind method*, can also be written as follows:

$$\lambda \hat{f}_{i+1/2}^{\text{B-B}} = \lambda f\left(u_{i+1}^n\right) - \frac{1}{2}\left(\lambda a_{i+1/2}^n + \frac{1}{2}\right)^2\left(u_{i+1}^n - u_i^n\right) \tag{21.3c}$$

or

$$\lambda \hat{f}_{i+1/2}^{\text{B-B}} = \lambda f\left(u_i^n\right) - \frac{1}{2}\left(\lambda a_{i+1/2}^n - \frac{1}{2}\right)^2\left(u_{i+1}^n - u_i^n\right). \tag{21.3d}$$

The factors $(\lambda a_{i+1/2}^n \pm 1/2)^2$ in these last two expressions strongly resemble the factors $(\lambda a(u_i^n) - 1/2)^2$ and $(\lambda a(u_{i+1}^n) + 1/2)^2$ seen in the original Boris–Book first-order upwind method. In fact, for the linear advection equation, the modified Boris–Book first-order upwind method equals the original Boris–Book first-order upwind method.

The coefficient of second-order artifical viscosity for the modified Boris–Book first-order upwind method is

$$\lambda \epsilon_{i+1/2}^{\text{B-B}} = \lambda \epsilon_{i+1/2}^{\text{L-W}} + \frac{1}{4} = \left(\lambda a_{i+1/2}^n\right)^2 + \frac{1}{4}. \tag{21.4}$$

This coefficient of artificial viscosity is illustrated in Figure 21.1. As seen in Figure 21.1, the modified Boris–Book first-order upwind method satisfies the linear stability condition $(\lambda a)^2 \leq \lambda\epsilon \leq 1$ found in Example 15.4 and the nonlinear stability condition $|\lambda a_{i+1/2}^n| \leq \lambda\epsilon_{i+1/2}^n \leq 1$ found in Example 16.6 provided that $\lambda|a_{i+1/2}^n| \leq \sqrt{3}/2 \approx 0.866$. Although this is not quite as good as the limit $\lambda|a| \leq 1$ for Roe's or Godunov's first-order upwind methods, this limit is certainly an improvement over $\lambda|a| \leq 1/2$ for the original Boris–Book first-order upwind method. Although the modified Boris–Book first-order upwind method was derived by adding artificial viscosity to the Lax–Wendroff method, it ends up closer to Roe's first-order upwind method than to the Lax–Wendroff method. In fact, as seen in Figure 21.1 and

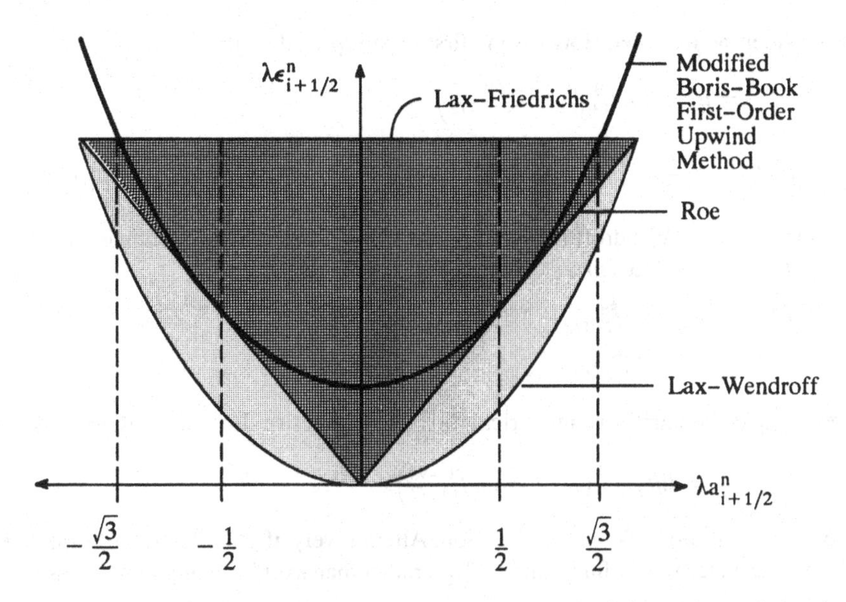

Figure 21.1 The artificial viscosity of the modified Boris–Book first-order upwind method compared with the other simple methods.

in numerical tests, the only major differences between the modified Boris–Book first-order upwind method and Roe's first-order upwind method occur at low CFL numbers $\lambda a^n_{i+1/2}$, where the modified Boris–Book first-order upwind method has a good deal more artificial viscosity than Roe's first-order upwind method. The increased artificial viscosity at low CFL numbers improves expansive sonic point capturing, degrades compressive sonic point capturing such as those found in steady or slowly moving shocks, and makes the modified Boris–Book first-order upwind method extremely dissipative at low CFL numbers, relative to Roe's first-order upwind method. According to extensive numerical tests, the modified Boris–Book first-order upwind method has the same nonlinear stability properties as Roe's first-order upwind method, including the range diminishing property described in Section 16.3, except possibly for CFL numbers near the upper bound $\sqrt{3}/2 \approx 0.866$. In other words, for low enough CFL numbers, the modified Boris–Book first-order upwind method does not allow maxima to increase, does not allow minima to decrease, and does not allow new extrema, a nonlinear stability condition originally suggested by Boris and Book (1973).

With the modified Boris–Book first-order upwind method under our belts, we are now ready to describe the Boris–Book flux-corrected method. Whereas the original Boris–Book flux-corrected method is a two-step predictor–corrector method, this section will begin by describing a one-step version. The one-step version helps to introduce the two-step version and, furthermore, helps to expose the connections between one-step flux-corrected methods and the one-step flux-limited methods seen in the last chapter. Consider the Lax–Wendroff method:

$$u_i^{n+1} = u_i^n - \left(\lambda \hat{f}^{\text{L-W}}_{i+1/2} - \lambda \hat{f}^{\text{L-W}}_{i-1/2}\right),$$

$$\lambda \hat{f}^{\text{L-W}}_{i+1/2} = \frac{\lambda}{2}\left(f(u^n_{i+1}) + f(u^n_i)\right) - \frac{1}{2}\left(\lambda a^n_{i+1/2}\right)^2\left(u^n_{i+1} - u^n_i\right).$$

Also consider the modified Boris–Book first-order upwind method:

$$u_i^{n+1} = u_i^n - \left(\lambda \hat{f}_{i+1/2}^{\text{B-B}} - \lambda \hat{f}_{i-1/2}^{\text{B-B}}\right),$$

$$\lambda \hat{f}_{i+1/2}^{\text{B-B}} = \frac{\lambda}{2}\left(f\left(u_{i+1}^n\right) + f\left(u_i^n\right)\right) - \frac{1}{2}\left(\left(\lambda a_{i+1/2}^n\right)^2 + \frac{1}{4}\right)\left(u_{i+1}^n - u_i^n\right).$$

Combining the Lax–Wendroff method and the Boris–Book first-order upwind method, the *one-step Boris–Book flux-corrected method* is as follows:

$$u_i^{n+1} = u_i^n - \left(\lambda \hat{f}_{i+1/2}^n - \lambda \hat{f}_{i-1/2}^n\right),$$

$$\lambda \hat{f}_{i+1/2}^n = \lambda \hat{f}_{i+1/2}^{\text{B-B}} + \lambda \hat{f}_{i+1/2}^{(\text{C})},$$

where $\hat{f}_{i+1/2}^{(\text{C})}$ is the corrective numerical flux. The corrective flux can be written as

$$\lambda \hat{f}_{i+1/2}^{(\text{C})} = \mathit{diff}_{i+1/2}^n\left(\lambda \hat{f}_{i+1/2}^{\text{B-B}}, \lambda \hat{f}_{i+1/2}^{\text{L-W}}\right),$$

where $\mathit{diff}_{i+1/2}^n$ is any *differencing* function. Alternatively, if you prefer, the corrective flux can be written in terms of a flux limiter $\phi_{i+1/2}^n$ rather than a differencing function as follows:

$$\lambda \hat{f}_{i+1/2}^{(\text{C})} = \phi_{i+1/2}^n\left(\lambda \hat{f}_{i+1/2}^{\text{L-W}} - \lambda \hat{f}_{i+1/2}^{\text{B-B}}\right) \tag{21.5a}$$

or, equivalently,

$$\lambda \hat{f}_{i+1/2}^{(\text{C})} = \frac{1}{8}\phi_{i+1/2}^n\left(u_{i+1}^n - u_i^n\right). \tag{21.5b}$$

To choose the flux correction, consider two conditions. As the first condition, $\lambda \hat{f}_{i+1/2}^n$ must remain between $\lambda \hat{f}_{i+1/2}^{\text{B-B}}$ and $\lambda \hat{f}_{i+1/2}^{\text{L-W}}$. That is, let us require that

$$\min\left(\lambda \hat{f}_{i+1/2}^{\text{B-B}}, \lambda \hat{f}_{i+1/2}^{\text{L-W}}\right) \le \lambda \hat{f}_{i+1/2}^n \le \max\left(\lambda \hat{f}_{i+1/2}^{\text{B-B}}, \lambda \hat{f}_{i+1/2}^{\text{L-W}}\right)$$

or

$$\min\left(\lambda \hat{f}_{i+1/2}^{\text{B-B}}, \lambda \hat{f}_{i+1/2}^{\text{L-W}}\right) \le \lambda \hat{f}_{i+1/2}^{\text{B-B}} + \lambda \hat{f}_{i+1/2}^{(\text{C})} \le \max\left(\lambda \hat{f}_{i+1/2}^{\text{B-B}}, \lambda \hat{f}_{i+1/2}^{\text{L-W}}\right)$$

or

$$\min\left(0, \lambda \hat{f}_{i+1/2}^{\text{L-W}} - \lambda \hat{f}_{i+1/2}^{\text{B-B}}\right) \le \lambda \hat{f}_{i+1/2}^{(\text{C})} \le \max\left(0, \lambda \hat{f}_{i+1/2}^{\text{L-W}} - \lambda \hat{f}_{i+1/2}^{\text{B-B}}\right)$$

or

$$\min\left(0, \frac{1}{8}\left(u_{i+1}^n - u_i^n\right)\right) \le \lambda \hat{f}_{i+1/2}^{(\text{C})} \le \max\left(0, \frac{1}{8}\left(u_{i+1}^n - u_i^n\right)\right). \tag{21.6a}$$

Equivalently, referring to Equation (21.2), let us require that

$$0 \le \phi_{i+1/2}^n \le 1. \tag{21.6b}$$

Thus, when written in flux-limited form, $\lambda \hat{f}_{i+1/2}^n$ is a convex linear combination of $\lambda \hat{f}_{i+1/2}^{\text{B-B}}$ and $\lambda \hat{f}_{i+1/2}^{\text{L-W}}$.

As the second condition, let us require that $\lambda \hat{f}_{i+1/2}^n \approx \lambda \hat{f}_{i+1/2}^{\text{B-B}}$ near maxima and minima, and let us require that $\lambda \hat{f}_{i+1/2}^n \approx \lambda \hat{f}_{i+1/2}^{\text{L-W}}$ otherwise. In other words, flux-corrected methods focus on spurious oscillations and extrema rather than on the shocks that typically cause spurious oscillations and extrema, just like the flux-limited methods seen in the last chapter. Written in terms of the corrective flux, the second condition is

$$\lambda \hat{f}^{(C)}_{i+1/2} \approx \begin{cases} 0 & \text{near extrema,} \\ \lambda \hat{f}^{L-W}_{i+1/2} - \lambda \hat{f}^{B-B}_{i+1/2} = \frac{1}{8}\left(u^n_{i+1} - u^n_i\right) & \text{otherwise.} \end{cases} \tag{21.7a}$$

Similarly, written in terms of flux limiters, the second condition is

$$\phi^n_{i+1/2} \approx \begin{cases} 0 & \text{near maxima and minima,} \\ 1 & \text{otherwise.} \end{cases} \tag{21.7b}$$

A simple flux correction that satisfies conditions (21.6) and (21.7) is as follows:

$$\lambda \hat{f}^{(C)}_{i+1/2} = minmod\left(u^n_i - u^n_{i-1}, \frac{1}{8}\left(u^n_{i+1} - u^n_i\right), u^n_{i+2} - u^n_{i+1} \right), \tag{21.8a}$$

where the *minmod* function equals the argument with the least absolute value if all of the arguments have the same sign and equals zero otherwise. Equivalently,

$$\phi^n_{i+1/2} = minmod\left(8r^+_i, 1, \frac{8}{r^+_{i+1}} \right), \tag{21.8b}$$

where

$$r^+_i = \frac{u^n_i - u^n_{i-1}}{u^n_{i+1} - u^n_i} \tag{21.9}$$

and where the following property of the minmod function has been used:

$$c\, minmod(x_1, \ldots, x_n) = minmod(cx_1, \ldots, cx_n) \tag{21.10}$$

for any c.

Let us prove that Equation (21.8) satisfies conditions (21.6) and (21.7). The minmod function has the following property:

$$0 \leq minmod(x_1, \ldots, x_n) \leq x_i \quad \text{if } x_i > 0, \tag{21.11a}$$

$$x_i \leq minmod(x_1, \ldots, x_n) \leq 0 \quad \text{if } x_i < 0 \tag{21.11b}$$

for all i, which immediately proves that Equation (21.8) satisfies condition (21.6). Also, the minmod function equals zero if any two arguments have opposite signs, which immediately proves that Equation (21.8) satisfies the first half of condition (21.7). Finally, if all of its arguments have the same sign, the minmod function chooses the argument with the least absolute value, which immediately proves that Equation (21.8) satisfies the second half of condition (21.7), unless the first differences of the solution change by a factor greater than eight or, more specifically, unless $|u^n_{i+1} - u^n_i| > 8|u^n_i - u^n_{i-1}|$ or $|u^n_{i+1} - u^n_i| > 8|u^n_{i+2} - u^n_{i+1}|$; in this case, Equation (21.8) satisfies the first half of condition (21.7). As discussed in Section 20.0, large changes in first differences are associated with shocks, which may cause the Lax–Wendroff method to oscillate. To help prevent this, when the first differences change rapidly enough, Equation (21.8) blends the Lax–Wendroff method with the modified Boris–Book first-order upwind method. This completes the description of the one-step Boris–Book flux-corrected method.

As briefly mentioned in the last chapter, flux-limited methods may be either one step or two steps. For example, in the Davis–Roe flux-limited method, the one-step Lax–Wendroff method may be replaced by the two-step MacCormack's method, as mentioned in Section 20.4. By a similar token, the one-step Boris–Book flux-corrected method also has a two-step version. Like the Davis–Roe flux-limited method, a two-step version may result from replacing the one-step Lax–Wendroff method by the two-step MacCormack's method. However, flux-corrected methods usually use a different two-step approach with the first step being a first-order upwind method and the second step a second-order flux correction. Specifically, the *two-step Boris–Book flux-corrected method* has the following predictor:

$$\blacklozenge \qquad \bar{u}_i = u_i^n - \lambda\left(\hat{f}^{\text{B-B}}_{i+1/2} - \hat{f}^{\text{B-B}}_{i-1/2}\right) \tag{21.12a}$$

and the following corrector:

$$\blacklozenge \qquad u_i^{n+1} = \bar{u}_i - \lambda\left(\hat{f}^{(\text{C})}_{i+1/2} - \hat{f}^{(\text{C})}_{i-1/2}\right). \tag{21.12b}$$

The corrective flux is

$$\blacklozenge \qquad \lambda \hat{f}^{(\text{C})}_{i+1/2} = minmod\left(\bar{u}_i - \bar{u}_{i-1}, \frac{1}{8}(\bar{u}_{i+1} - \bar{u}_i), \bar{u}_{i+2} - \bar{u}_{i+1}\right) \tag{21.13a}$$

or, equivalently,

$$\lambda \hat{f}^{(\text{C})}_{i+1/2} = \frac{\lambda}{8}\phi^n_{i+1/2}\left(\bar{u}^n_{i+1} - \bar{u}^n_i\right), \tag{21.13b}$$

where

$$\phi^n_{i+1/2} = minmod\left(1, 8\bar{r}^n_i, \frac{8}{\bar{r}^n_{i+1}}\right) \tag{21.14}$$

and where

$$\bar{r}_i^+ = \frac{\bar{u}^n_i - \bar{u}^n_{i-1}}{\bar{u}^n_{i+1} - \bar{u}^n_i}. \tag{21.15}$$

Notice that Equation (21.13) is exactly the same as Equation (21.8), except that the predicted solution $\bar{u}_i^n$ replaces u_i^n. This book will use the term "flux corrected" for both one-step and two-step methods. Boris and Book called their two-step method a flux-corrected *transport* (FCT) method so that, by tradition, the term "transport" signals a two-step rather than a one-step flux-corrected method.

In this section, the flux correction was derived more intuitively than rigorously, more in the freer spirit of the self-adjusting hybrid methods seen in the next chapter rather than the cautious spirit of the flux-limited methods seen in the last chapter. However, like any flux-averaged method, flux-corrected methods can be designed to rigorously satisfy nonlinear stability conditions. In fact, in the original paper, Boris and Book (1973) first proposed the range diminishing condition discussed in Section 16.3. More specifically, Boris and Book state that the "crucial factor" is that the corrector "should generate no new maxima or minima in the solution, nor should it accentuate the existing extrema." Boris and Book also proposed the upwind range condition discussed in Section 16.5, stating that a method should not "push the solution value at any grid point beyond the solution value at neighboring points." They claimed that "by means of a few trials, the reader can readily convince himself" that their flux-corrected transport method satisfies the upwind range condition. However, numerical results contradict this assertion: Both the one-step and the two-step versions of the Boris–Book flux-corrected method may exhibit spurious overshoots and oscillations, especially for large CFL numbers. No oscillations were observed in the results given in the original paper, since Boris and Book used only small CFL numbers, such as $\lambda a = 0.2$.

One advantage that two-step flux correction has over one-step flux correction is that it can be used *even for equations whose exact solutions oscillate*. In particular, suppose that the first-order method allows new oscillations, just like the governing equations, but does not allow false oscillations. Then the second-order flux correction should not create any new oscillations, nor allow oscillations in the predicted solution to increase, which means that the second-order flux correction may be designed in the usual way, using the criteria of Chapter 16. This was observed by Boris and Book (1973) in the original paper (see Figure 5 and the associated discussion). Outside of the rarefied realm of scalar conservation laws and the one-dimensional Euler equations, many governing equations routinely create large oscillations. Thus two-step flux correction has excellent potential for broader applications, provided that you can somehow find a nonoscillatory but otherwise inaccurate method to use for a predictor.

In the original paper, and in two sequels to the original paper, Boris and Book (1973, 1975, 1976) described numerous variants on flux-corrected transport, as well as various implementation details. Unlike most of the methods suggested in the 1970s, flux-corrected transport has survived surprisingly intact. This is a tribute to FCT's forward-looking concepts and techniques, which form a surprising percentage of the foundations of modern computational gasdynamics.

The behavior of the Boris–Book two-step flux-corrected method is illustrated using the five standard test cases defined in Section 17.0. Unfortunately, although it does not blow up, the Boris–Book two-step flux-corrected method behaves badly for the usual value $\lambda = 0.8$ and, in particular, the solutions contain large spurious overshoots and oscillations and excessive dissipation. Thus, to give the method a fair hearing, all of the following calculations use $\lambda = 0.6$.

Test Case 1 As seen in Figure 21.2, the Boris–Book two-step flux-corrected method captures the sinusoid with only modest clipping.

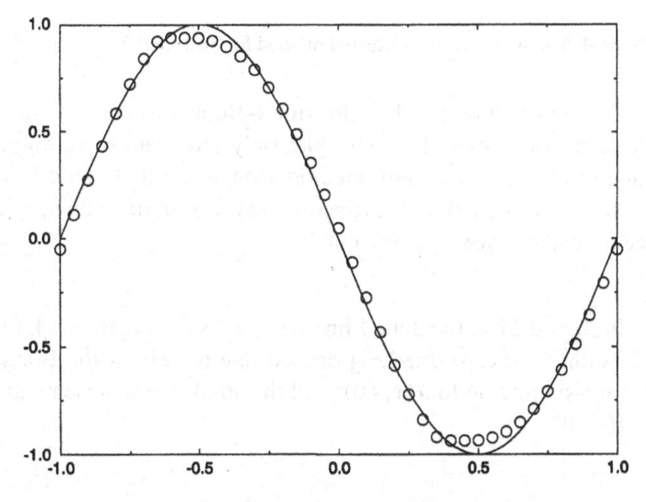

Figure 21.2 Boris–Book two-step flux-corrected method for Test Case 1.

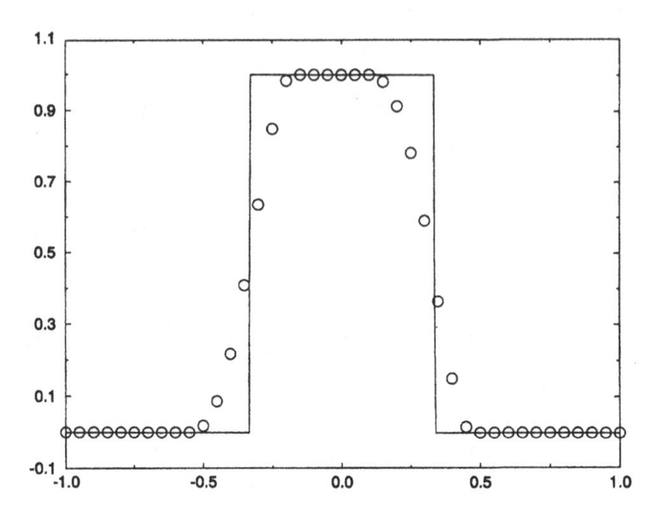

Figure 21.3 Boris–Book two-step flux-corrected method for Test Case 2.

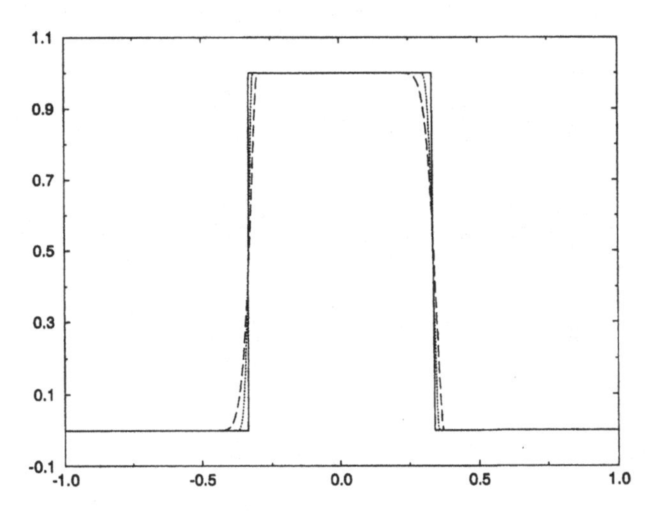

Figure 21.4 Boris–Book two-step flux-corrected method for Test Case 3.

Test Case 2 As seen in Figure 21.3, the Boris–Book two-step flux-corrected method captures the square wave extremely well, with only moderate smearing, and no spurious oscillations and overshoots. The only method seen so far that could better this performance is Sweby's flux-limited method with the superbee flux limiter which, by design, has an unfair advantage in this test case; see Section 20.2.

Test Case 3 In Figure 21.4, the dotted line represents the Boris–Book two-step flux-corrected approximation to $u(x, 4)$, the long dashed line represents the Boris–Book two-step flux-corrected approximation to $u(x, 40)$, and the solid line represents the exact solution for $u(x, 4)$ or $u(x, 40)$.

Test Case 4 The results, seen in Figure 21.5, are about as good as any seen so far.

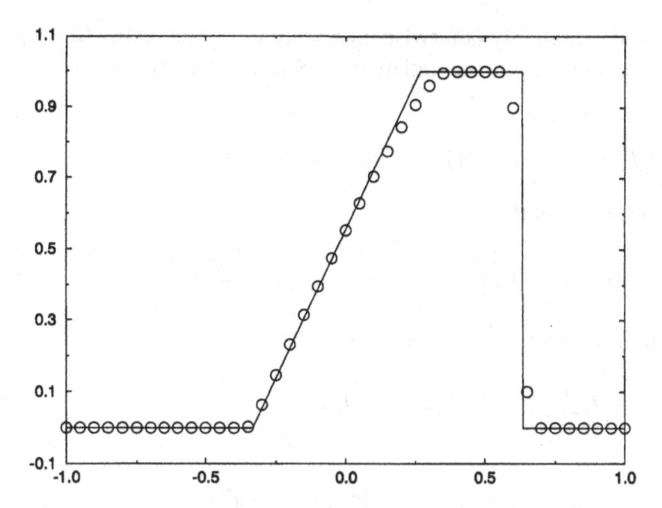

Figure 21.5 Boris–Book two-step flux-corrected method for Test Case 4.

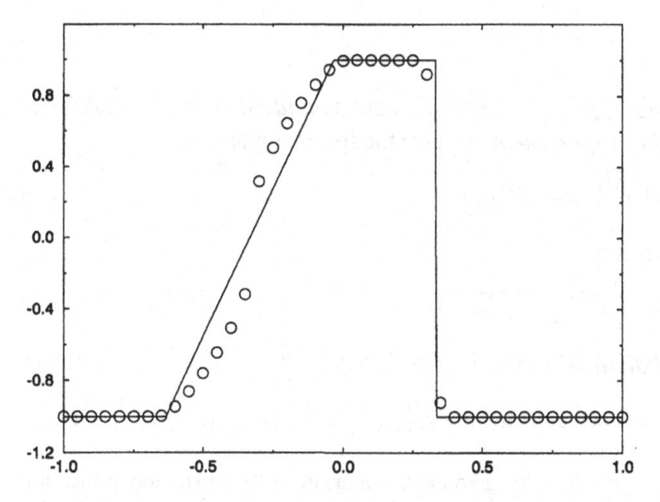

Figure 21.6 Boris–Book two-step flux-corrected method for Test Case 5.

Test Case 5 As seen in Figure 21.6, the solution jumps across the expansive sonic point.

21.2 Zalesak's Flux-Corrected Methods (FCT)

Zalesak (1979) generalized the two-step Boris–Book flux-corrected transport method. Like the previous section, this section only concerns scalar conservation laws. Zalesak's original paper is more a philosophical treatise than a proposal for any specific new methods. The paper suggests numerous possible avenues for development, some of which have been hotly pursued, and some of which have lapsed into obscurity. Most importantly, Zalesak described generalized flux-corrected methods, in which the first- and second-order accurate methods can be anything you like. Although it makes a nice mate for the Lax–Wendroff method, and performs relatively well at expansive sonic points, the modified Boris–Book first-order upwind method is arguably otherwise inferior to most competing first-order upwind methods such as Roe's first-order upwind method. So consider any first-order accurate method:

$$\lambda \hat{f}^{(1)}_{i+1/2} = \frac{\lambda}{2}\left(f(u^n_{i+1}) + f(u^n_i)\right) - \frac{1}{2}\lambda\epsilon^{(1)}_{i+1/2}(u^n_{i+1} - u^n_i) \tag{21.16}$$

and any second-order accurate method:

$$\lambda \hat{f}^{(2)}_{i+1/2} = \frac{\lambda}{2}\left(f(u^n_{i+1}) + f(u^n_i)\right) - \frac{1}{2}\lambda\epsilon^{(2)}_{i+1/2}(u^n_{i+1} - u^n_i). \tag{21.17}$$

The only restriction on $\hat{f}^{(1)}_{i+1/2}$ and $\hat{f}^{(2)}_{i+1/2}$ is as follows:

$$\operatorname{sign}\left(\lambda \hat{f}^{(2)}_{i+1/2} - \lambda \hat{f}^{(1)}_{i+1/2}\right) = \operatorname{sign}(u^n_{i+1} - u^n_i) \tag{21.18a}$$

or

$$\operatorname{sign}\left[\left(\epsilon^{(1)}_{i+1/2} - \epsilon^{(2)}_{i+1/2}\right)(u^n_{i+1} - u^n_i)\right] = \operatorname{sign}(u^n_{i+1} - u^n_i)$$

or

$$\epsilon^{(1)}_{i+1/2} \geq \epsilon^{(2)}_{i+1/2}. \tag{21.18b}$$

Then the flux difference $\lambda \hat{f}^{(2)}_{i+1/2} - \lambda \hat{f}^{(1)}_{i+1/2}$ is called *antidissipative*. Then *Zalesak's generalized flux-corrected transport methods* have the following predictor:

$$\blacklozenge \quad \bar{u}_i = u^n_i - \lambda\left(\hat{f}^{(1)}_{i+1/2} - \hat{f}^{(1)}_{i-1/2}\right) \tag{21.19a}$$

and the following corrector:

$$\blacklozenge \quad u^{n+1}_i = \bar{u}_i - \lambda\left(\hat{f}^{(C)}_{i+1/2} - \hat{f}^{(C)}_{i-1/2}\right). \tag{21.19b}$$

The *corrective* or *antidissipative flux* is

$$\blacklozenge \quad \lambda \hat{f}^{(C)}_{i+1/2} = minmod\left(\bar{u}_i - \bar{u}_{i-1}, \lambda \hat{f}^{(2)}_{i+1/2} - \lambda \hat{f}^{(1)}_{i-1/2}, \bar{u}_{i+2} - \bar{u}_{i+1}\right), \tag{21.20}$$

where $\lambda \hat{f}^{(2)}_{i+1/2} - \lambda \hat{f}^{(1)}_{i+1/2}$ is usually taken as a function of the corrected solution $\bar{u}_i$. Compared to the Boris–Book flux-corrected method seen in the last section, Zalesak replaced $\hat{f}^{(1)}_{i+1/2}$ and $\hat{f}^{(2)}_{i+1/2}$ but not the averaging function *minmod* (for an even more general flux-corrected transport method, another function may replace *minmod* in Equation (21.20)). By tradition, flux-corrected transport methods use a wide variety of different component methods $\lambda \hat{f}^{(2)}_{i+1/2}$ and $\lambda \hat{f}^{(1)}_{i+1/2}$ but use relatively few other averages besides *minmod*. By contrast, flux-limiting methods and self-adjusting hybrid methods traditionally use a wide variety of different averages – see Equations such as (20.21)–(20.24) and (20.26) – but use relatively few different component methods. Notice that Zalesak's flux-corrected transport method becomes the Boris–Book flux-corrected transport method if $\lambda \hat{f}^{(1)}_{i+1/2} = \lambda \hat{f}^{B-B}_{i+1/2}$ and $\lambda \hat{f}^{(2)}_{i+1/2} = \lambda \hat{f}^{L-W}_{i+1/2}$.

Zalesak's generalized flux-corrected transport method satisfies:

$$\min\left(\hat{f}^{(1)}_{i+1/2}, \hat{f}^{(2)}_{i+1/2}\right) \leq \hat{f}^n_{i+1/2} \leq \max\left(\hat{f}^{(1)}_{i+1/2}, \hat{f}^{(2)}_{i+1/2}\right),$$

which follows immediately from Equation (21.11). Also, $\hat{f}^{(C)}_{i+1/2} = 0$ near extrema in the predicted solution. To prove this, by Equation (21.18a), remember that $\lambda \hat{f}^{(2)}_{i+1/2} - \lambda \hat{f}^{(1)}_{i+1/2}$

has the same sign as $\bar{u}^n_{i+1} - \bar{u}^n_i$. Then the arguments of the minmod function in Equation (21.20) have opposite signs if $\bar{u}^n_i$ or $\bar{u}^n_{i+1}$ are extrema. But the minmod function equals zero if any of its arguments have opposite signs. Finally, $\hat{f}^{(C)}_{i+1/2} = \hat{f}^{(2)}_{i+1/2} - \hat{f}^{(1)}_{i+1/2}$ away from extrema unless the first differences of the solution change by greater than a factor of $1/(\lambda\epsilon^{(1)}_{i+1/2} - \lambda\epsilon^{(2)}_{i+1/2})$.

As far as nonlinear stability goes, Zalesak showed that his method satisfies the following weak nonlinear stability condition:

$$\min(\bar{u}_{i-1}, \bar{u}_i, \bar{u}_{i+1}) \leq u^{n+1}_i \leq \max(\bar{u}_{i-1}, \bar{u}_i, \bar{u}_{i+1}). \tag{21.21}$$

This condition allows more oscillations than flux-corrected methods generally display, and thus perhaps understates their performance.

21.3 Harten's Flux-Corrected Method (TVD)

21.3.1 *Scalar Conservation Laws*

This section concerns another approach to flux correction. In this section, instead of correcting the *numerical* flux $\hat{f}^n_{i+1/2}$ we shall correct the *physical* flux $f(u)$. From Section 15.3, recall that a numerical approximation does not solve the true governing equation: It solves a modified equation, which equals the true governing equation plus additional terms. Then applying a numerical approximation to the *wrong* governing equation may yield a better answer than applying a numerical approximation to the *right* governing equation, provided that the "wrong" governing equation is cunningly constructed, so as to reduce or eliminate terms in the modified equation. It's just reverse psychology. From Section 17.3, recall Roe's first-order upwind method:

$$u^{n+1}_i = u^n_i - \lambda\left(\hat{f}^{\text{ROE}}_{i+1/2} - \hat{f}^{\text{ROE}}_{i-1/2}\right),$$

where

$$\hat{f}^{\text{ROE}}_{i+1/2} = \frac{1}{2}\left(f\left(u^n_{i+1}\right) + f\left(u^n_i\right)\right) - \frac{1}{2}\left|a^n_{i+1/2}\right|\left(u^n_{i+1} - u^n_i\right)$$

or, equivalently,

$$\hat{f}^{\text{ROE}}_{i+1/2} = \frac{1}{2}\left(f\left(u^n_{i+1}\right) + f\left(u^n_i\right) - \left|\frac{f\left(u^n_{i+1}\right) - f\left(u^n_i\right)}{u^n_{i+1} - u^n_i}\right|\left(u^n_{i+1} - u^n_i\right)\right), \tag{21.22}$$

where this last expression requires the usual modification when $u^n_{i+1} - u^n_i = 0$. Roe's first-order upwind method has a number of extremely desirable nonlinear stability properties, such as the upwind range condition, as discussed in Section 17.3. Naturally enough, Roe's first-order upwind method retains its nonlinear stability properties *regardless of the flux function*. In particular, Roe's first-order upwind method retains its nonlinear stability properties if $f\left(u^n_i\right)$ is replaced by some new flux function g^n_i as follows:

$$u^{n+1}_i = u^n_i - \lambda\left(\hat{g}^n_{i+1/2} - \hat{g}^n_{i-1/2}\right),$$

♦
$$\hat{g}^n_{i+1/2} = \frac{1}{2}\left(g^n_{i+1} + g^n_i - \left|\frac{g^n_{i+1} - g^n_i}{u^n_{i+1} - u^n_i}\right|\left(u^n_{i+1} - u^n_i\right)\right). \tag{21.23}$$

Although replacing $f(u_i^n)$ by g_i^n does not affect nonlinear stability, it certainly affects accuracy, either for better or for worse. Suppose that

$$\blacklozenge \qquad g_i^n = f(u_i^n) + f_i^{(C)}. \tag{21.24}$$

Then $f_i^{(C)}$ is called a *physical flux correction*, as opposed to the numerical flux corrections $\hat{f}_{i+1/2}^{(C)}$ seen in the last two sections. Physical and numerical flux corrections differ by when they apply their corrections – before or after discretization – and by how easily they allow you to enforce accuracy and stability conditions. With numerical flux correction, it is easy to enforce higher-order accuracy but hard to enforce nonlinear stability conditions such as the upwind range condition. By contrast, with physical flux correction, it is easy to enforce nonlinear stability but hard to enforce higher-order accuracy. This last point helps to explain why, unlike the Boris–Book flux-corrected method, Harten's flux-corrected method has never been generalized to use other first-generation methods besides the Lax–Wendroff method or to use other averages besides minmod. In 1981, Harten suggested the following physical flux correction:

$$\blacklozenge \qquad f_i^{(C)} = minmod\left(\hat{f}_{i+1/2}^{\text{L-W}} - \hat{f}_{i+1/2}^{\text{ROE}}, \hat{f}_{i-1/2}^{\text{L-W}} - \hat{f}_{i-1/2}^{\text{ROE}}\right). \tag{21.25}$$

Equations (21.23), (21.24), and (21.25) collectively constitute *Harten's flux-corrected method*. By the way, to avoid expansive sonic point problems, Harten's first-order upwind method, described in Subsection 17.3.3, may replace Roe's first-order upwind method in all of the preceding equations. In fact, Harten (1983) contains the first description of Harten's first-order upwind method; the original paper calls it an *entropy-fixed* version of Roe's first-order upwind method, designed to improve expansive sonic point capturing.

Harten's flux-corrected method is second-order accurate except at extrema, where it is between first- and second-order accurate. More specifically,

$$\hat{g}_{i+1/2}^n = f_{i+1/2}^{\text{ROE}} \tag{21.26}$$

if u_i^n is an extremum and

$$\hat{g}_{i+1/2}^n = f_{i+1/2}^{\text{L-W}} + O(\Delta x^2) \tag{21.27}$$

if u_i^n is not an extremum.

To prove Equation (21.26), first notice that

$$\hat{f}_{i+1/2}^{\text{L-W}} - \hat{f}_{i+1/2}^{\text{ROE}} = \frac{1}{2}\left(|a_{i+1/2}^n| - \lambda(a_{i+1/2})^2\right)\left(u_{i+1}^n - u_i^n\right).$$

Then

$$\text{sign}\left(\hat{f}_{i+1/2}^{\text{L-W}} - \hat{f}_{i+1/2}^{\text{ROE}}\right) = \text{sign}\left(u_{i+1}^n - u_i^n\right)$$

for $\lambda|a_{i+1/2}^n| \leq 1$. If u_i^n is an extremum, then $u_{i+1}^n - u_i^n$ and $u_i^n - u_{i-1}^n$ have opposite signs. Then $\hat{f}_{i+1/2}^{\text{L-W}} - \hat{f}_{i+1/2}^{\text{ROE}}$ and $\hat{f}_{i-1/2}^{\text{L-W}} - \hat{f}_{i-1/2}^{\text{ROE}}$ have opposite signs. Consequently, the minmod average of $\hat{f}_{i+1/2}^{\text{L-W}} - \hat{f}_{i+1/2}^{\text{ROE}}$ and $\hat{f}_{i-1/2}^{\text{L-W}} - \hat{f}_{i-1/2}^{\text{ROE}}$ equals zero. Then $f_i^{(C)} = 0$, $g_i^n = f(u_i^n)$, and $\hat{g}_{i+1/2}^n = \hat{f}_{i+1/2}^{\text{ROE}}$, which proves Equation (21.26). The proof of Equation (21.27) is omitted; although not difficult, the proof is long and tedious.

In the original paper, Harten (1983) proposed the TVD condition discussed in Section 16.2 and the positivity condition proposed in Section 16.4. Harten also clarified the range diminishing condition proposed earlier by Boris and Book (1973) and seen in Section

16.3. However, in light of the above explanation, it is clear that Harten's nonlinear stability theory is completely unnecessary to justify Harten's flux-corrected method. Harten's flux-corrected method naturally inherits all of the stability properties of one of the most stable classes of methods: first-order upwind methods such as Roe's first-order upwind method and Harten's first-order upwind method.

After devising the original explicit method, Harten (1984) suggested an implicit method, much like the implicit Yee–Roe flux-limited method seen in Section 20.5. See Yee, Warming, and Harten (1982, 1985a, 1985b, 1985c), Yee and Harten (1987), and Yee (1986, 1989) for implementation details, variations, and interesting numerical examples. Yee (1989) provides an especially nice review of Harten's implicit and explicit flux-corrected method, as well as a number of other flux-averaged methods.

Let us now briefly discuss artificial compression and its relationship to antidissipative physical flux correction. Among all the common flux limiters used in Sweby's method, seen in Section 20.2, the superbee limiter is the largest. In other words, the superbee limiter results in the largest correction to the first-order upwind method. Put yet another way, the superbee limiter adds the most antidissipation to the first-order upwind method. Not coincidentally, among all the common flux limiters, Roe's superbee limiter results in the most *compressive* approximations. Thus, since the superbee limiter reduces dissipation as much as possible, Sweby's flux-limited method with the superbee limiter captures expansion fans with less smearing than other limiters and in this sense provides compression at contacts, which do not naturally generate their own compression, unlike shocks. By a similar token, in Harten's flux-corrected method, increasing the flux correction $|f_i^{(C)}|$ increases compression, decreases dissipation, increases antidissipation (or however you want to put it), and thus reduces smearing at contacts. In particular, Yee, Warming, and Harten (1985a) suggest replacing Equation (21.25) by

$$f_i^{(C)} = \left(1 + \omega\theta_i^n\right) minmod\left(\hat{f}_{i+1/2}^{\text{L-W}} - \hat{f}_{i+1/2}^{\text{ROE}}, \hat{f}_{i-1/2}^{\text{L-W}} - \hat{f}_{i-1/2}^{\text{ROE}}\right), \tag{21.28}$$

where $\omega \geq 0$ is a user-adjustable parameter and where

$$\theta_i^n = \frac{\left|u_{i+1}^n - 2u_i^n + u_{i-1}^n\right|}{\left|u_{i+1}^n - u_i^n\right| + \left|u_i^n - u_{i-1}^n\right|} \tag{21.29}$$

is called a *shock switch*; the next chapter describes shock switches in great detail. Equations (21.28) and (21.29) are called *artificial compression*. Since $1 + \omega\theta_i^n \geq 1$, artificial compression seen in Equation (21.28) always increases the size of antidissipative flux correction $|f_i^{(C)}|$ relative to Equation (21.25). Artificial compression requires extreme caution. While it can steepen contact discontinuities and shocks, which is good, it can also steepen smooth solution regions, which is bad. For example, without proper care, artificial compression may cause or strengthen spurious expansion shocks. Yee, Warming, and Harten (1985a) suggest $\omega = 0$ (no artificial compression) in expansions and $\omega = 2$ elsewhere. Artificial compression is discussed again in Section 22.2.

The behavior of Harten's flux-corrected method is illustrated using the five standard test cases defined in Section 17.0. The first four test cases use Roe's first-order upwind method. The fifth test case uses Harten's first-order upwind method with $\delta = 0.5$.

Test Case 1 As seen in Figure 21.7, Harten's flux-corrected method captures the sinusoid well, albeit with somewhat above-average clipping at the extrema.

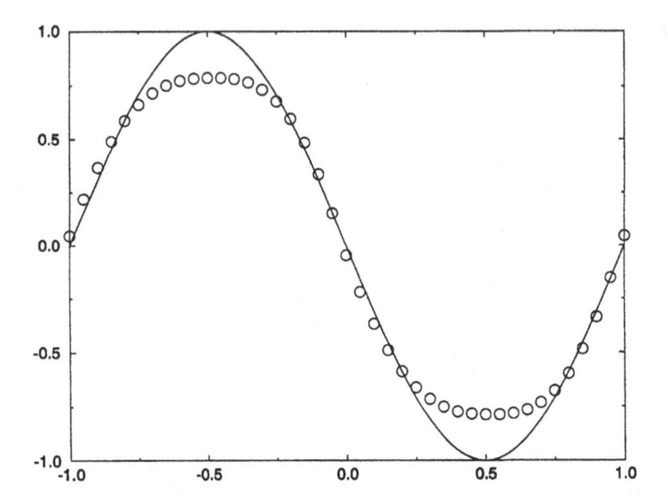

Figure 21.7 Harten's flux-corrected method for Test Case 1.

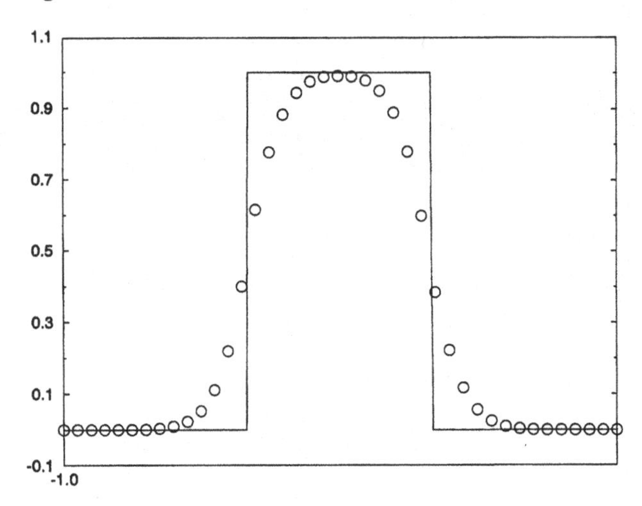

Figure 21.8 Harten's flux-corrected method for Test Case 2.

Test Case 2 As seen in Figure 21.8, Harten's flux-corrected method captures the square wave with a fairly typical amount of smearing, but with no overshoots or oscillations. The smearing can be reduced or eliminated by using artificial compression, as seen in Equations (21.28) and (21.29).

Test Case 3 In Figure 21.9, the dotted line represents Harten's flux-corrected approximation to $u(x, 4)$, the long dashed line represents Harten's flux-corrected approximation to $u(x, 40)$, and the solid line represents the exact solution for $u(x, 4)$ or $u(x, 40)$.

Test Case 4 As seen in Figure 21.10, the results are fine, if not extraordinary relative to methods seen earlier.

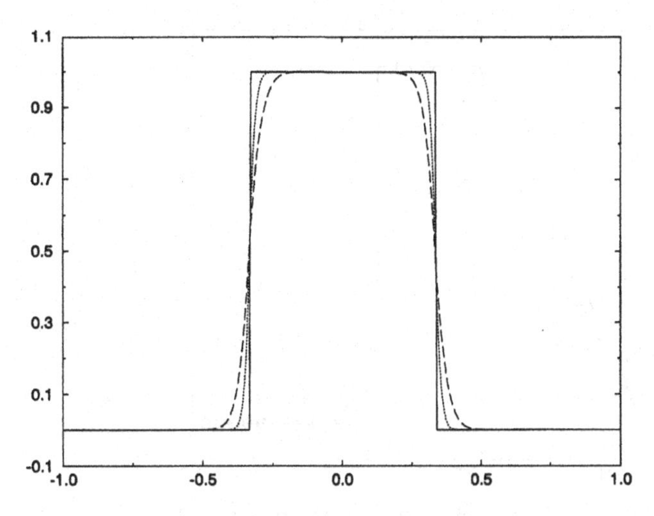

Figure 21.9 Harten's flux-corrected method for Test Case 3.

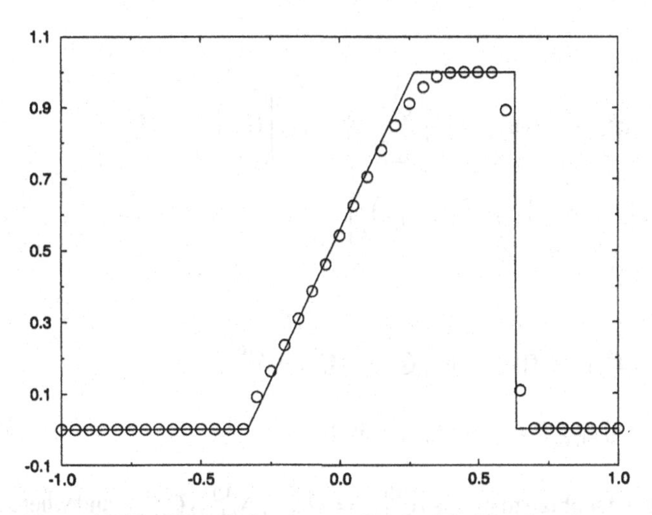

Figure 21.10 Harten's flux-corrected method for Test Case 4.

Test Case 5 As seen in Figure 21.11, the solution is fine except for a small jump near the expansive sonic point, which could be eliminated by increasing δ.

21.3.2 *The Euler Equations*

Let $A^n_{i+1/2}$ be the Roe-average Jacobian matrix where $A^n_{i+1/2} = Q_{i+1/2}\Lambda_{i+1/2} Q^{-1}_{i+1/2}$. Let the columns of $Q_{i+1/2}$ be $\mathbf{r}^n_{i+1/2}$ and let the rows of $Q^{-1}_{i+1/2}$ be $\mathbf{l}^n_{i+1/2}$. Harten's flux-corrected method for the Euler equations is

$$\mathbf{u}^{n+1}_i = \mathbf{u}^n_i - \lambda\left(\hat{\mathbf{g}}^n_{i+1/2} - \hat{\mathbf{g}}^n_{i-1/2}\right),$$

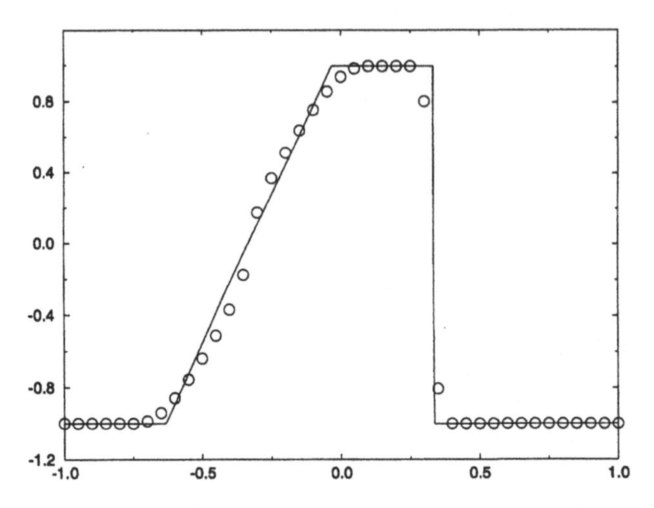

Figure 21.11 Harten's flux-corrected method for Test Case 5.

where

$$\hat{\mathbf{g}}^n_{i+1/2} = \frac{1}{2}\left(\mathbf{f}(\mathbf{u}^n_{i+1}) + \mathbf{f}(\mathbf{u}^n_i)\right) + \frac{1}{2}\sum_{j=1}^{3}\left(\mathbf{r}^n_{i+1/2}\right)_j\left[\left(\mathbf{f}^{(C)}_{i+1}\right)_j + \left(\mathbf{f}^{(C)}_i\right)_j - \left|\lambda^n_{i+1/2} + \lambda^{(C)}_{i+1/2}\right|_j\left(\Delta v^n_{i+1/2}\right)_j\right] \tag{21.30a}$$

or, equivalently,

$$\hat{\mathbf{g}}^n_{i+1/2} = \frac{1}{2}\left(\mathbf{f}(\mathbf{u}^n_{i+1}) + \mathbf{f}(\mathbf{u}^n_i)\right) + \frac{1}{2}Q_{i+1/2}\left(\mathbf{f}^{(C)}_{i+1} + \mathbf{f}^{(C)}_i\right) - \frac{1}{2}\left|A^n_{i+1/2} + A^{(C)}_{i+1/2}\right|\left(\mathbf{u}^n_{i+1} - \mathbf{u}^n_i\right), \tag{21.30b}$$

where the corrective flux Jacobian matrix is $A^{(C)}_{i+1/2} = Q_{i+1/2}\Lambda^{(C)}_{i+1/2}Q^{-1}_{i+1/2}$ and where $\Lambda^{(C)}_{i+1/2}$ is a diagonal matrix of corrective wave speeds $\lambda^{(C)}_{i+1/2}$. The corrective fluxes appearing in Equation (21.30) are as follows:

$$\left(\mathbf{f}^{(C)}_i\right)_j = minmod\left[\frac{1}{2}\left(\left|\lambda^n_{i+1/2}\right|_j - \frac{\Delta t}{\Delta x}\left(\lambda^n_{i+1/2}\right)^2_j\right)\left(\Delta v^n_{i+1/2}\right)_j, \frac{1}{2}\left(\left|\lambda^n_{i-1/2}\right|_j - \frac{\Delta t}{\Delta x}\left(\lambda^n_{i-1/2}\right)^2_j\right)\left(\Delta v^n_{i-1/2}\right)_j\right] \tag{21.31a}$$

or, equivalently,

$$\mathbf{f}^{(C)}_i = minmod\left[Q^{-1}_{i+1/2}\left(\hat{\mathbf{f}}^{L-W}_{i+1/2} - \hat{\mathbf{f}}^{ROE}_{i+1/2}\right), Q^{-1}_{i-1/2}\left(\hat{\mathbf{f}}^{L-W}_{i-1/2} - \hat{\mathbf{f}}^{ROE}_{i-1/2}\right)\right], \tag{21.31b}$$

where

$$\begin{aligned}\hat{\mathbf{f}}^{L-W}_{i+1/2} - \hat{\mathbf{f}}^{ROE}_{i+1/2} &= \frac{1}{2}\left(\left|A^n_{i+1/2}\right| - \frac{\Delta t}{\Delta x}\left(A^n_{i+1/2}\right)^2\right)\left(\mathbf{u}^n_{i+1} - \mathbf{u}^n_i\right)\\ &= \frac{1}{2}Q_{i+1/2}\left(\left|\Lambda^n_{i+1/2}\right| - \frac{\Delta t}{\Delta x}\left(\Lambda^n_{i+1/2}\right)^2\right)Q^{-1}_{i+1/2}\Delta u^n_{i+1/2}\\ &= \frac{1}{2}Q_{i+1/2}\left(\left|\Lambda^n_{i+1/2}\right| - \frac{\Delta t}{\Delta x}\left(\Lambda^n_{i+1/2}\right)^2\right)\Delta v^n_{i+1/2}.\end{aligned}$$

Finally, the corrected wave speeds defined in terms of the corrected fluxes are

$$\left(\lambda^{(C)}_{i+1/2}\right)_j = \frac{\left(\mathbf{f}^{(C)}_{i+1}\right)_j - \left(\mathbf{f}^{(C)}_i\right)_j}{\left(\Delta v^n_{i+1/2}\right)_j}. \tag{21.32}$$

At this point, the reader may find it helpful to compare the vector and scalar versions of Harten's flux-corrected method; that is, compare Equation (21.31) with Equation (21.25). In the case of scalar conservation laws, $Q_{i+1/2} = 1$ and all the above vector expressions become the same as the earlier scalar expressions. For an intuitive introduction to the ideas used here, the reader may wish to review Subsection 18.3.5. The basic idea is to exploit the local-linearity of the approximate flux function; replace the characteristic values in Harten's first-order upwind method while leaving the characteristic vectors alone.

21.4 Shu–Osher Methods (ENO)

The methods seen so far in this chapter have all been fully discrete. This section concerns semidiscrete Shu–Osher flux methods similar to the semidiscrete Chakravarthy–Osher flux-limited methods seen in Section 20.3. Semidiscrete approximations were introduced in Subsection 11.2.1. In particular, recall that conservative semidiscrete approximations are defined as follows:

$$\frac{du^n_i}{dt} = -\frac{\hat{f}^n_{s,i+1/2} - \hat{f}^n_{s,i-1/2}}{\Delta x}. \tag{11.48}$$

While Section 20.3 combined two preexisting first-generation semidiscrete methods, this section uses a different and more fundamental approach, called *flux reconstruction*, introduced in Subsection 13.4.2 and first discovered by Shu and Osher (1988, 1989). Despite the different approach, the final result is not all that different – the methods found in this section can still be viewed as solution-sensitive combinations of first-generation methods, very much like the methods found in Section 20.3 and other places. For example, with constant Δx and second-order accuracy, the Shu–Osher method is

$$\hat{f}_{s,i+1/2} = \hat{f}^{(1)}_{s,i+1/2} + \frac{1}{2}m\left(\Delta f^+_{i+1/2}, \Delta f^+_{i-1/2}\right) - \frac{1}{2}m\left(\Delta f^-_{i+3/2}, \Delta f^-_{i+1/2}\right),$$

where m chooses the argument closest to zero, as seen in Equation (9.6). Compare this to Equation (20.66): The two equations are identical except that Equation (20.66) uses the *minmod* average instead of m. Thus, in practice, the methods seen in this section differ from earlier methods mainly in the choice of average.

One major advantage of the approach used in this section, as compared with previous approaches, is that it naturally and easily allows nonconstant Δx. Most other methods in this book were developed only for constant Δx and require coordinate transformations from unevenly spaced to evenly spaced grids before use. Another major advantage of the approach in this section is that it naturally and easily extends to any arbitrary order of accuracy. Finally, like all semidiscrete approximations, the methods in this section allow the user to experiment with different time discretizations, to choose the best one for the application at hand; this is especially important to optimize the method to simulate steady versus unsteady solutions. Given the utterly general approach used in this section – methods with any order of accuracy that work for any unevenly spaced grid – it should come as no surprise that this section is denser than earlier sections, which usually concerned only one method with one fixed order of accuracy for evenly spaced grids. Certainly, before even attempting this section, the reader should review ENO reconstruction as seen in Chapter 9.

As seen in Subsection 13.4.2, suppose that $\hat{f}^n_{s,i+1/2} = \hat{f}^n_s(x_{i+1/2})$. In other words, imagine that the semidiscrete conservative fluxes depend directly on x rather than on $u(x,t)$. Then, in one interpretation, $f(u^n_i)$ is a cell-integral average of $\hat{f}^n_s(x)$. In other words:

$$f(u^n_i) = \frac{1}{\Delta x}\int_{x_{i-1/2}}^{x_{i+1/2}} \hat{f}_s(y)\,dy,$$

as described in Subsection 13.4.2. Then the function $\hat{f}^n_s(x)$ may be reconstructed from its cell-integral averages $f(u^n_i)$ using reconstruction via the primitive function, as described in Section 9.3. In particular, define the primitive function $\hat{F}^n_s(x)$ of $\hat{f}^n_s(x)$ as follows:

$$\hat{F}^n_s(x) = \int_{x_{1/2}}^{x} \hat{f}^n_s(y)\,dy, \qquad \hat{f}^n_s(x) = \frac{d\hat{F}^n_s}{dx}.$$

Then

$$\hat{F}^n_s(x_{i+1/2}) = \int_{x_{1/2}}^{x_{i+1/2}} \hat{f}^n_s(y)\,dy = \sum_{j=1}^{i}\int_{x_{j-1/2}}^{x_{j+1/2}} \hat{f}^n_s(y)\,dy = \sum_{j=1}^{i} \Delta x_j f(u^n_j).$$

Thus the partial sums of $f(u^n_i)$ are samples of the primitive function $\hat{F}^n_s(x)$. Section 9.3 gives a detailed algorithm for piecewise-polynomial reconstruction via the primitive function. However, this algorithm has two major drawbacks in the present application. Namely,

- The reconstruction $\hat{f}^n_s(x)$ is discontinuous at the cell edges $x = x_{i+1/2}$. But this application must evaluate the reconstruction $\hat{f}^n_s(x)$ at the cell edges $x = x_{i+1/2}$, since $\hat{f}^n_{s,i+1/2} = \hat{f}^n_s(x_{i+1/2})$. Thus, for this application, we shall modify reconstruction via the primitive function to use the staggered cells $[x_i, x_{i+1}]$ rather than the true cells $[x_{i-1/2}, x_{i+1/2}]$. The resulting reconstruction is discontinuous for $x = x_i$ but continuous for $x = x_{i+1/2}$.
- The reconstruction chooses points based solely on the size of flux differences such as $f(u^n_{i+1}) - f(u^n_i)$ and $f(u^n_{i+1}) - 2f(u^n_i) + f(u^n_{i-1})$. However, this can easily lead to completely downwind approximations, which violate the CFL condition. Thus, in this application, we shall modify reconstruction via the primitive function so that it always uses at least one upwind point. Although the flux reconstruction may still have a downwind bias, the CFL condition allows this provided that the flux reconstruction depends on at least one upwind point, as discussed in Chapters 12 and 13.

For the time being, assume that $df/du > 0$ or $df/du < 0$ (i.e., assume that there are no sonic points). Then the modified algorithm for reconstruction via the primitive function is as follows:

Step 1 Find the Newton divided differences of $\hat{F}_s^n(x_{i+1/2})$. For example,

$$\begin{aligned}\hat{f}_s^n[x_{i-1/2}, x_{i+1/2}] &= \frac{\hat{F}_s^n(x_{i+1/2}) - \hat{F}_s^n(x_{i-1/2})}{x_{i+1/2} - x_{i-1/2}} \\ &= \frac{\sum_{j=0}^{i} \Delta x_j f(u_j^n) - \sum_{j=0}^{i-1} \Delta x_j f(u_j^n)}{\Delta x_i} = \frac{\Delta x_i f(u_i^n)}{\Delta x_i} = f(u_i^n).\end{aligned}$$

For another example,

$$\begin{aligned}\hat{F}_s^n[x_{i-1/2}, x_{i+1/2}, x_{i+3/2}] &= \frac{\hat{F}_s^n[x_{i+1/2}, x_{i+3/2}] - \hat{F}_s^n[x_{i-1/2}, x_{i+1/2}]}{x_{i+3/2} - x_{i-1/2}} \\ &= \frac{f(u_{i+1}^n) - f(u_i^n)}{x_{i+3/2} - x_{i-1/2}} = \frac{f(u_{i+1}^n) - f(u_i^n)}{\Delta x_{i+1} + \Delta x_i}.\end{aligned}$$

Notice that it is not necessary to form the sums $F_s^n(x_{i+1/2}) = \sum_{j=1}^{i} \Delta x_j f(u_j^n)$, since they are never needed in formulae such as the preceding two.

Step 2 Choose the interpolation points for each staggered cell $[x_i, x_{i+1}]$. For cell $[x_i, x_{i+1}]$, start with the left-hand endpoint $l_0(i + 1/2) = i + 1/2$. Also, to ensure that $\hat{f}_{s,i+1/2}^n$ depends on at least one upwind point, let

$$l_1\left(i + \frac{1}{2}\right) = \begin{cases} i - \frac{1}{2} & \frac{df}{du} > 0, \\ i + \frac{1}{2} & \frac{df}{du} < 0. \end{cases}$$

Then all other interpolation points are chosen recursively as follows:

$$l_{m+1}\left(i + \frac{1}{2}\right) = \begin{cases} l_m\left(i + \frac{1}{2}\right) & \left|\hat{F}_s^n\left[x_{l_m(i+\frac{1}{2})}, \ldots, x_{l_m(i+\frac{1}{2})+m+1}\right]\right| \\ & \le \left|\hat{F}_s^n\left[x_{l_m(i+\frac{1}{2})-1}, \ldots, x_{l_m(i+\frac{1}{2})+m}\right]\right|, \\ l_m\left(i + \frac{1}{2}\right) - 1 & \left|\hat{F}_s^n\left[x_{l_m(i+\frac{1}{2})}, \ldots, x_{l_m(i+\frac{1}{2})+m+1}\right]\right| \\ & > \left|\hat{F}_s^n\left[x_{l_m(i+\frac{1}{2})-1}, \ldots, x_{l_m(i+\frac{1}{2})+m}\right]\right| \end{cases}$$

for $m = 1, \ldots, N$. Although the notation here is a bit awkward, the principle is simple: At each iteration, choose a sample point from the left or right, whichever yields the least Newton divided difference in absolute value.

Step 3 Find the Taylor series form of the interpolation polynomial in each staggered cell $[x_i, x_{i+1}]$, where the Taylor series is taken about $x_{i+1/2}$. Then the reconstruction

$\hat{P}_{N+1,i+1/2}$ on staggered cell $[x_i, x_{i+1}]$ is

$$\hat{P}_{N+1,i+\frac{1}{2}}(x) = \sum_{j=0}^{N+1} a_j \left(x - x_{i+\frac{1}{2}}\right)^j,$$

where

$$a_j = \sum_{k=0}^{N-j+1} d_{kj} \hat{F}_s^n \left[x_{l_{N+1}(i+\frac{1}{2})}, \ldots, x_{l_{N+1}(i+\frac{1}{2})+j+k}\right]$$

and

$$\begin{aligned} d_{0j} &= 1, \\ d_{k0} &= \left(x_{i+\frac{1}{2}} - x_{l_{N+1}(i+\frac{1}{2})+k-1}\right) d_{k-1,0}, \\ d_{kj} &= d_{k,j-1} + \left(x_{i+\frac{1}{2}} - x_{l_{N+1}(i+\frac{1}{2})+k+j-1}\right) d_{k-1,j}. \end{aligned}$$

Step 4 Since $\hat{f}_s^n(x) = d\hat{F}_s^n/dx$ and $\hat{F}_s^n(x) \approx \hat{P}_{N+1}(x)$, an Nth-order polynomial approximating $\hat{f}(x)$ is $\hat{p}_N(x) = d\hat{P}_{N+1}(x)/dx$. In particular, the reconstruction on the staggered cell $[x_i, x_{i+1}]$ is as follows:

$$\hat{p}_{N,i+1/2}(x) = \sum_{j=1}^{N+1} j a_j \left(x - x_{i+\frac{1}{2}}\right)^{j-1}.$$

For our purposes, the above algorithm does too much. Notice that

$$\hat{f}^n_{s,i+1/2} = \hat{f}^n_s(x_{i+1/2}) = \hat{p}_{N+1,i+1/2}(x_{i+1/2}) = a_1$$

or

$$\blacklozenge \qquad \hat{f}^n_{s,i+1/2} = \sum_{k=0}^{N} d_{k1} \hat{F}_s^n \left[x_{l_{N+1}(i+\frac{1}{2})}, \ldots, x_{l_{N+1}(i+\frac{1}{2})+k+1}\right], \tag{21.33}$$

where

$$\blacklozenge \qquad d_{k1} = \left(x_{i+\frac{1}{2}} - x_{l_{N+1}(i+\frac{1}{2})+k-1}\right) d_{k-1,0} + \left(x_{i+\frac{1}{2}} - x_{l_{N+1}(i+\frac{1}{2})+k}\right) d_{k-1,1}. \tag{21.34}$$

Thus, in this application, there is absolutely no need to find the coefficients $a_2, \ldots, a_{N+1}$.

Example 21.1 Suppose that $N = 0$. Then

$$\hat{f}^n_{s,i+1/2} = \hat{F}_s^n \left[x_{l_1(i+\frac{1}{2})}, x_{l_1(i+\frac{1}{2})+1}\right],$$

where

$$l_1\left(i + \frac{1}{2}\right) = \begin{cases} i - \frac{1}{2} & \frac{df}{du} > 0, \\ i + \frac{1}{2} & \frac{df}{du} < 0. \end{cases}$$

The required Newton divided differences are

$$\hat{F}_s^n[x_{i-1/2}, x_{i+1/2}] = f(u_i^n),$$
$$\hat{F}_s^n[x_{i+1/2}, x_{i+3/2}] = f(u_{i+1}^n).$$

Then

$$\blacklozenge \qquad \hat{f}_{s,i+1/2}^n = \begin{cases} f(u_i^n) & \frac{df}{du} > 0, \\ f(u_{i+1}^n) & \frac{df}{du} < 0, \end{cases} \tag{21.35}$$

which is a semidiscrete first-order upwind method, similar to the fully discrete first-order upwind methods seen many times before in this book, starting in Chapter 13. From one point of view, the Shu–Osher flux-corrected method always starts with Equation (21.35) and then, optionally, adds higher-order flux corrections, as seen in the next example.

Example 21.2 Suppose that $N = 1$. Then

$$\hat{f}_{s,i+1/2}^n = \hat{F}_s^n\left[x_{l_2(i+\frac{1}{2})}, x_{l_2(i+\frac{1}{2})+1}\right]$$
$$+ \left(2x_{i+\frac{1}{2}} - x_{l_2(i+\frac{1}{2})} - x_{l_2(i+\frac{1}{2})+1}\right)\hat{F}_s^n\left[x_{l_2(i+\frac{1}{2})}, x_{l_2(i+\frac{1}{2})+1}, x_{l_2(i+\frac{1}{2})+2}\right].$$

A short calculation using the definition of the Newton divided differences, and using the fact that $l_2 = l_1$ or $l_2 = l_1 - 1$, yields the following equivalent expression:

$$\hat{f}_{s,i+1/2}^n = \hat{F}_s^n\left[x_{l_1(i+\frac{1}{2})}, x_{l_1(i+\frac{1}{2})+1}\right]$$
$$+ \left(2x_{i+\frac{1}{2}} - x_{l_1(i+\frac{1}{2})} - x_{l_1(i+\frac{1}{2})+1}\right)\hat{F}_s^n\left[x_{l_2(i+\frac{1}{2})}, x_{l_2(i+\frac{1}{2})+1}, x_{l_2(i+\frac{1}{2})+2}\right].$$

To understand the relationship between the preceding two expressions, the reader may find it helpful to compare Equations (9.16) and (9.17). To continue, remember that

$$l_1\left(i + \frac{1}{2}\right) = \begin{cases} i - \frac{1}{2} & \frac{df}{du} > 0, \\ i + \frac{1}{2} & \frac{df}{du} < 0. \end{cases}$$

If $|\hat{F}_s^n[x_{l_1(i+\frac{1}{2})}, x_{l_1(i+\frac{1}{2})+1}, x_{l_1(i+\frac{1}{2})+2}]| \leq |\hat{F}_s^n[x_{l_1(i+\frac{1}{2})-1}, x_{l_1(i+\frac{1}{2})}, x_{l_1(i+\frac{1}{2})+1}]|$ then

$$l_2\left(i + \frac{1}{2}\right) = l_1\left(i + \frac{1}{2}\right).$$

If $|\hat{F}_s^n[x_{l_1(i+\frac{1}{2})}, x_{l_1(i+\frac{1}{2})+1}, x_{l_1(i+\frac{1}{2})+2}]| > |\hat{F}_s^n[x_{l_1(i+\frac{1}{2})-1}, x_{l_1(i+\frac{1}{2})}, x_{l_1(i+\frac{1}{2})+1}]|$ then

$$l_2\left(i + \frac{1}{2}\right) = l_1\left(i + \frac{1}{2}\right) - 1.$$

In other words, if $df/du > 0$ then $l_2(i+1/2)$ equals $i-1/2$ or $i-3/2$. Similarly, if $df/du < 0$ then $l_2(i+1/2)$ equals $i-1/2$ or $i+1/2$. Finally, the required Newton divided differences are as follows:

$$\hat{F}^n_s[x_{i-1/2}, x_{i+1/2}] = f(u^n_i),$$
$$\hat{F}^n_s[x_{i+1/2}, x_{i+3/2}] = f(u^n_{i+1}),$$
$$\hat{F}^n_s[x_{i-3/2}, x_{i-1/2}, x_{i+1/2}] = \frac{f(u^n_i) - f(u^n_{i-1})}{\Delta x_i + \Delta x_{i-1}},$$
$$\hat{F}^n_s[x_{i-1/2}, x_{i+1/2}, x_{i+3/2}] = \frac{f(u^n_{i+1}) - f(u^n_i)}{\Delta x_{i+1} + \Delta x_i},$$
$$\hat{F}^n_s[x_{i+1/2}, x_{i+3/2}, x_{i+5/2}] = \frac{f(u^n_{i+2}) - f(u^n_{i+1})}{\Delta x_{i+2} + \Delta x_{i+1}}.$$

For $df/du > 0$, the final result can be written as

♦ $$\hat{f}^n_{s,i+1/2} = f(u^n_i) + \Delta x_i m\left(\frac{f(u^n_{i+1}) - f(u^n_i)}{\Delta x_{i+1} + \Delta x_i}, \frac{f(u^n_i) - f(u^n_{i-1})}{\Delta x_i + \Delta x_{i-1}}\right).$$

Similarly, if $df/du < 0$, the final result can be written as

♦ $$\hat{f}^n_{s,i+1/2} = f(u^n_{i+1}) - \Delta x_{i+1} m\left(\frac{f(u^n_{i+1}) - f(u^n_i)}{\Delta x_{i+1} + \Delta x_i}, \frac{f(u^n_{i+2}) - f(u^n_{i+1})}{\Delta x_{i+2} + \Delta x_{i+1}}\right),$$

where

$$m(x, y) = \begin{cases} x & |x| \le |y| \\ y & |x| > |y| \end{cases},$$

as in Equation (9.6). Notice that the preceding expressions take the form of a first-order upwind method, such as that found in the preceding example, plus second-order corrections. Thus, although the Shu–Osher method does not follow the traditional flux-corrected philosophy – it follows more of a "flux-reconstruction" than a "flux-correction" philosophy – the final result is very much in the flux-corrected tradition, if you care to view it that way.

Now suppose that $\Delta x = const.$ Then for $df/du > 0$:

$$\hat{f}^n_{s,i+1/2} = \begin{cases} \frac{f(u^n_{i+1}) + f(u^n_i)}{2} & |f(u^n_{i+1}) - f(u^n_i)| \le |f(u^n_i) - f(u^n_{i-1})|, \\ \frac{3f(u^n_i) - f(u^n_{i-1})}{2} & |f(u^n_{i+1}) - f(u^n_i)| > |f(u^n_i) - f(u^n_{i-1})| \end{cases}$$

and for $df/du < 0$:

$$\hat{f}^n_{s,i+1/2} = \begin{cases} \frac{f(u^n_{i+1}) + f(u^n_i)}{2} & |f(u^n_{i+1}) - f(u^n_i)| \le |f(u^n_{i+2}) - f(u_{i+1})|, \\ \frac{3f(u^n_{i+1}) - f(u^n_{i+2})}{2} & |f(u^n_{i+1}) - f(u^n_i)| > |f(u^n_{i+2}) - f(u_{i+1})|. \end{cases}$$

Thus this method chooses between the semidiscrete Lax–Wendroff method seen in Example 20.1 and the semidiscrete Beam–Warming second-order upwind method seen in Example 20.3, depending on the size of the flux differences. Notice that Example 9.6 gave similar results. For example, evaluated at $x = x_{i+1/2}$, Equations (9.33) and (9.34) found in Example 9.6 yield all of the preceding expressions. In fact, this example and Example 9.6 differ not in the range of expressions they offer, but only in how they choose between those expressions.

So far, this section has assumed that there are no sonic points. Now suppose that the flux function can have sonic points, that is, suppose that the sign of the flux derivative can change. Then use flux splitting, as introduced in Section 13.4:

$$\begin{aligned}
&f(u) = f^+(u) + f^-(u),\\
&\frac{df^+}{du} \geq 0, \qquad \frac{df^-}{du} \leq 0.
\end{aligned}$$

Substitute either f^+ or f^- for f in the above algorithm to find $\hat{f}_s^+(x)$ from $f^+(u_i^n)$ and to find $\hat{f}_s^-(x)$ from $f^-(u_i^n)$. Then

$$\hat{f}^n_{s,i+1/2} = \hat{f}_s^+(x_{i+1/2}) + \hat{f}_s^-(x_{i+1/2}). \tag{21.36}$$

For example, one possible "physical" flux splitting is as follows:

$$\begin{aligned}
f^+(u) &= \begin{cases} f(u) & f'(u) \geq 0,\\ 0 & f'(u) < 0,\end{cases}\\
f^-(u) &= \begin{cases} 0 & f'(u) \geq 0,\\ f(u) & f'(u) < 0.\end{cases}
\end{aligned} \tag{21.37}$$

This works well except near expansive sonic points. Near sonic points, a better flux splitting is

$$\begin{aligned}
f^+(u) &= \frac{1}{2}\left(f(u) + \alpha^n_{i+1/2}u\right),\\
f^-(u) &= \frac{1}{2}\left(f(u) - \alpha^n_{i+1/2}u\right).
\end{aligned} \tag{21.38}$$

In this last splitting, to ensure $df^+/du \geq 0$ and $df^-/du \leq 0$, let us require that

$$\alpha^n_{i+1/2} \geq \max|f'(u)|,$$

assuming that $\alpha^n_{i+1/2}$ does not depend on u. For example, assuming that $\lambda_i \max|f'(u_i^n)| \leq 1$, some possibilities are as follows:

$$\alpha_{i+1/2} = 1, \tag{21.39}$$

$$\alpha_{i+1/2} = \frac{1}{\min(\lambda_i, \lambda_{i+1})}, \tag{21.40}$$

$$\alpha^n_{i+1/2} = \max\left[\left|f'(u^n_{i+1})\right|, f'\left(\left|u^n_i\right|\right)\right]. \tag{21.41}$$

The second choice is a bit strange, given that the flux splitting then depends on Δt via $\lambda_i = \Delta t/\Delta x_i$; then the semidiscrete method depends on Δt even though, of course, it does not discretize time. However, to actually use the method, time must be discretized eventually, after which the second choice is no longer so strange. Flux splitting (21.37) is the cheapest possible flux splitting; flux splitting (21.38) is the next cheapest, given that it adds and subtracts a linear function to or from $f(u)$. Other possible flux splittings add and subtract quadratics, cubics, or other more expensive functions, increasing costs without really improving sonic point capturing. Perhaps the ideal flux splitting uses (21.38) near sonic points and (21.37) elsewhere.

Example 21.3 Suppose that $N = 0$. Use the results of Example 21.1 and Equation (21.36) to find

$$\hat{f}^n_{s,i+1/2} = f^+(u^n_i) + f^-(u^n_{i+1}). \tag{21.42}$$

Flux spitting (21.37) yields

$$\hat{f}_{s,i+1/2} = \begin{cases} f(u^n_i) & f'(u^n_i) \geq 0, \quad f'(u^n_{i+1}) \geq 0, \\ f(u^n_{i+1}) & f'(u^n_i) < 0, \quad f'(u^n_{i+1}) < 0, \\ 0 & f'(u^n_i) < 0, \quad f'(u^n_{i+1}) \geq 0, \\ f(u^n_{i+1}) + f(u^n_i) & f'(u^n_i) \geq 0, \quad f'(u^n_{i+1}) < 0, \end{cases} \tag{21.43}$$

which is just a semidiscrete first-order upwind method with a simple sonic point treatment, much like Roe's first-order upwind method, as seen in Equation (17.23). Flux splitting (21.38) yields

$$\hat{f}^n_{s,i+1/2} = \frac{1}{2}\left(f(u^n_{i+1}) + f(u^n_i)\right) - \frac{1}{2}\alpha^n_{i+1/2}(u^n_{i+1} - u^n_i), \tag{21.44}$$

which is just central differences plus second-order artificial viscosity, as discussed in Chapter 14. For example, if $\Delta x = const.$, then Equations (21.38) and (21.44) yield a semidiscrete version of the first-order Lax–Friedrichs method, albeit a semidiscrete method that strangely depends on Δt via $\lambda_i = \Delta t/\Delta x_i$; the fully discrete version of the Lax–Friedrichs method was discussed in Section 17.1. More generally, any conservative numerical method can be written as central differences plus second-order artificial viscosity. Depending on the choice of $\alpha^n_{i+1/2}$, Equation (21.44) may or may not be an upwind method; for example, the Lax–Friedrichs method is centered. In this case, flux splitting does not necessarily introduce upwinding, as you would ordinarily expect, but instead introduces artificial viscosity, which has a desirable stabilizing effect.

Example 21.4 Suppose that $N = 1$. Use the results of Example 21.2 and Equation (21.36) to find

$$\begin{aligned} \hat{f}_{s,i+1/2} = {} & f^+(u^n_i) + f^-(u^n_{i+1}) \\ & + \Delta x_i m\left(\frac{f^+(u^n_{i+1}) - f^+(u^n_i)}{\Delta x_{i+1} + \Delta x_i}, \frac{f^+(u^n_i) - f^+(u^n_{i-1})}{\Delta x_i + \Delta x_{i-1}}\right) \\ & - \Delta x_{i+1} m\left(\frac{f^-(u^n_{i+1}) - f^-(u^n_i)}{\Delta x_{i+1} + \Delta x_i}, \frac{f^-(u^n_{i+2}) - f^-(u^n_{i+1})}{\Delta x_{i+2} + \Delta x_{i+1}}\right). \end{aligned} \tag{21.45}$$

For example, flux splitting (21.38) yields

$$
\begin{aligned}
\hat{f}^n_{s,i+1/2} = {} & \frac{1}{2}\left(f(u^n_{i+1}) + f(u^n_i)\right) - \frac{1}{2}\alpha_{i+1/2}(u^n_{i+1} - u^n_i) \\
& + \frac{1}{2}\Delta x_i m\left[\frac{f(u^n_{i+1}) - f(u^n_i) + \alpha_{i+1/2}(u^n_{i+1} - u^n_i)}{\Delta x_{i+1} + \Delta x_i},\right. \\
& \left.\frac{f(u^n_i) - f(u^n_{i-1}) + \alpha^n_{i+1/2}(u^n_i - u^n_{i-1})}{\Delta x_i + \Delta x_{i-1}}\right] \\
& - \frac{1}{2}\Delta x_{i+1} m\left[\frac{f(u^n_{i+1}) - f(u^n_i) - \alpha_{i+1/2}(u^n_{i-1} - u^n_i)}{\Delta x_{i+1} + \Delta x_i},\right. \\
& \left.\frac{f(u^n_{i+2}) - f(u^n_{i+1}) - \alpha^n_{i+1/2}(u^n_{i+2} - u^n_{i+1})}{\Delta x_{i+2} + \Delta x_{i+1}}\right].
\end{aligned}
\tag{21.46}
$$

For constant Δx, this last expression chooses between four possible results, two centered and two uncentered. The two uncentered expressions are

$$
\begin{aligned}
\hat{f}_{s,i+1/2} = {} & \frac{1}{2}\left(3f(u^n_{i+1}) - f(u^n_{i+2})\right) + \frac{1}{4}\left(f(u^n_{i+2}) - 2f(u^n_{i+1}) + f(u^n_i)\right) \\
& + \frac{\alpha^n_{i+1/2}}{4}(u^n_{i+2} - 2u^n_{i+1} + u^n_i), \\
\hat{f}^n_{s,i+1/2} = {} & \frac{1}{2}\left(3f(u^n_i) - f(u^n_{i-1})\right) + \frac{1}{4}\left(f(u^n_{i+1}) - 2f(u^n_i) + f(u^n_{i-1})\right) \\
& - \frac{\alpha^n_{i+1/2}}{4}(u^n_{i+1} - 2u^n_i + u^n_{i-1}),
\end{aligned}
$$

which are second-order upwind and second-order downwind methods plus third-order artificial dispersion. The two centered expressions are

$$
\begin{aligned}
\hat{f}^n_{s,i+1/2} = {} & \frac{1}{2}\left(f(u^n_{i+1}) + f(u^n_i)\right), \\
\hat{f}^n_{s,i+1/2} = {} & \frac{1}{2}\left[\frac{3f(u^n_{i+1}) - f(u^n_{i+2})}{2} + \frac{3f(u^n_i) - f(u^n_{i-1})}{2}\right] \\
& + \frac{\alpha_{i+1/2}}{4}(u^n_{i+2} - 3u^n_{i+1} + 3u^n_i - u^n_{i-1}).
\end{aligned}
$$

The first equation is straight central differences. The second equation is an average of second-order upwind and second-order downwind plus fourth-order artificial viscosity. Equation (21.45) is very similar to the semidiscrete Chakravarthy–Osher flux-limited methods seen the last chapter. In particular, changing notation to match that used in the last chapter, let

$$
\begin{aligned}
\hat{f}^{(1)}_{s,i+1/2} &= f^+(u^n_i) + f^-(u^n_{i+1}), \\
\Delta\hat{f}^+_{i+1/2} &= f(u^n_{i+1}) - \hat{f}^{(1)}_{s,i+1/2} = f^+(u^n_{i+1}) - f^+(u^n_i), \\
\Delta\hat{f}^-_{i+1/2} &= \hat{f}^{(1)}_{s,i+1/2} - f(u^n_i) = f^-(u^n_{i+1}) - f^-(u^n_i).
\end{aligned}
$$

Also, assume that $\Delta x = const.$, since all of the results in the last chapter assume $\Delta x = const.$ Then Equation (21.45) becomes

$$\hat{f}_{s,i+1/2} = \hat{f}^{(1)}_{s,i+1/2} + \frac{1}{2}m\left(\Delta f^{+}_{i+1/2}, \Delta f^{+}_{i-1/2}\right) - \frac{1}{2}m\left(\Delta f^{-}_{i+3/2}, \Delta f^{-}_{i+1/2}\right). \quad (21.47)$$

Replacing m by *minmod*, Equation (21.47) matches Equation (20.66) exactly. In the literature, Equation (20.66) is called a TVD method whereas Equation (21.47) is called an ENO method. Notice that $m = minmod$ except at extrema, where $m \neq 0$ and $minmod = 0$. Then, from this perspective, the only difference between TVD and ENO methods lies in how they behave at extrema: TVD methods use *minmod* and thus use first-order accurate methods at extrema, whereas ENO methods use m and thus retain their full order of accuracy at extrema. To carry the same idea a little further, suppose that m is replaced by mc, where

$$mc(x, y) = minmod(x, y + M\Delta x^2 \operatorname{sign}(x))$$

for any x and y, and where M is a user-adjustable constant. Then the Shu–Osher ENO method agrees with the TVB method proposed by Shu (1987). Notice that adding a second-order term to the second argument in the minmod function reduces or eliminates second-order clipping error at extrema. In summary, from this perspective, the main difference between TVD methods, ENO methods, and TVB methods is the averaging function, most particularly what the averaging function does at extrema.

So far this section has only considered semidiscrete approximations. For time discretizations, Shu and Osher (1988) recommend Runge–Kutta methods, as seen in Section 10.3. Shu and Osher choose the Runge–Kutta coefficients using a TVD nonlinear stability analysis. For second-order accurate two-stage Runge–Kutta methods, Shu and Osher's nonlinear stability analysis yields the *improved Euler method* seen in Equation (10.35b). Namely,

$$\begin{aligned}
u_i^{(1)} &= u_i^n + R\left(u^n_{i-K_1}, \ldots, u^n_{i+K_2}\right),\\
u_i^{n+1} &= u_i^n + \frac{1}{2}R\left(u^n_{i-K_1}, \ldots, u^n_{i+K_2}\right) + \frac{1}{2}R\left(u^{(1)}_{i-K_1}, \ldots, u^{(1)}_{i+K_2}\right),
\end{aligned}$$

where

$$R\left(u^n_{i-K_1}, \ldots, u^n_{i+K_2}\right) = -\lambda_i\left[\hat{f}_s\left(u^n_{i-K_1+1}, \ldots, u^n_{i+K_2}\right) - \hat{f}_s\left(u^n_{i-K_1}, \ldots, u^n_{i+K_2-1}\right)\right].$$

For higher-stage higher-order accurate Runge–Kutta methods, Shu and Osher's nonlinear stability analysis yields more exotic Runge–Kutta methods. For example, for third-order accuracy, Shu and Osher suggest

$$\begin{aligned}
u_i^{(1)} &= u_i^n + R\left(u^n_{i-K_1}, \ldots, u^n_{i+K_2}\right),\\
u_i^{(2)} &= u_i^n + \frac{1}{4}R\left(u^n_{i-K_1}, \ldots, u^n_{i+K_2}\right) + \frac{1}{4}R\left(u^{(1)}_{i-K_1}, \ldots, u^{(1)}_{i+K_2}\right)\\
u_i^{n+1} &= u_i^n + \frac{1}{6}R\left(u^n_{i-K_1}, \ldots, u^n_{i+K_2}\right) + \frac{1}{6}R\left(u^{(1)}_{i-K_1}, \ldots, u^{(1)}_{i+K_2}\right)\\
&\quad + \frac{2}{3}R\left(u^{(2)}_{i-K_1}, \ldots, u^{(2)}_{i+K_2}\right).
\end{aligned}$$

Compare this with classic third-order Runge-Kutta methods such as *Heun's third-order method*:

$$u_i^{(1)} = u_i^n + \frac{1}{3}R\left(u_{i-K_1}^n, \ldots, u_{i+K_2}^n\right),$$

$$u_i^{(2)} = u_i^n + \frac{2}{3}R\left(u_{i-K_1}^{(1)}, \ldots, u_{i+K_2}^{(1)}\right),$$

$$u_i^{n+1} = u_i^n + \frac{1}{4}R\left(u_{i-K_1}^n, \ldots, u_{i+K_2}^n\right) + \frac{3}{4}R\left(u_{i-K_1}^{(2)}, \ldots, u_{i+K_2}^{(2)}\right).$$

Heun's third-order method is cheaper than Shu and Osher's Runge–Kutta method and usually yields similar results. In fact, a careful analysis of exactly what Shu and Osher's nonlinear stability analysis accomplishes, as well as extensive numerical tests, shows that one should not expect any dramatic differences between most classic Runge–Kutta methods and Shu–Osher Runge–Kutta methods. Using any of the above second- and third-order Runge–Kutta methods, the Shu–Osher method remains stable, in the sense that it does not blow up for $|\lambda a(u)| \leq 1$. Higher-order Runge–Kutta methods may require somewhat lower bounds on the CFL number.

The behavior of the Shu–Osher flux-corrected method is illustrated using the five standard test cases defined in Section 17.0. In the first four test cases, no flux splitting is required, and no flux splitting is used. In Test Case 5, with flux splitting (21.37), the Shu–Osher flux-corrected method does not alter the initial conditions in any way (i.e., it allows a large steady expansion shock). Thus, in Test Case 5, the Shu–Osher flux-corrected method uses flux splitting (21.38) with coefficient (21.41). Second-order accurate results are found using the improved Euler method, and third-order accurate results are found using the Shu–Osher third-order Runge–Kutta method.

Test Case 1 As seen in Figure 21.12, the second-order accurate Shu–Osher flux-corrected method dissipates the sinusoid to a fraction of its proper size, and it heavily distorts

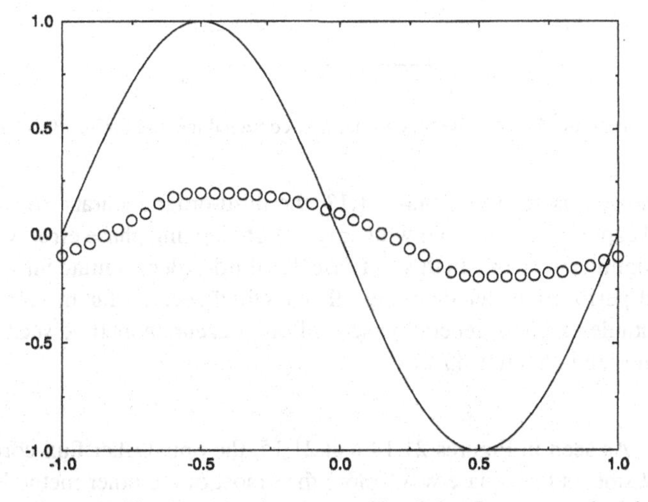

Figure 21.12 Second-order accurate Shu–Osher flux-corrected method for Test Case 1.

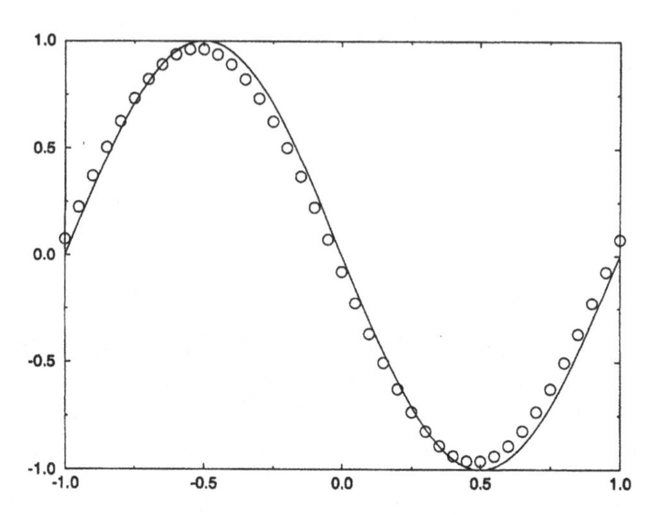

Figure 21.13 Third-order accurate Shu–Osher flux-corrected method for Test Case 1.

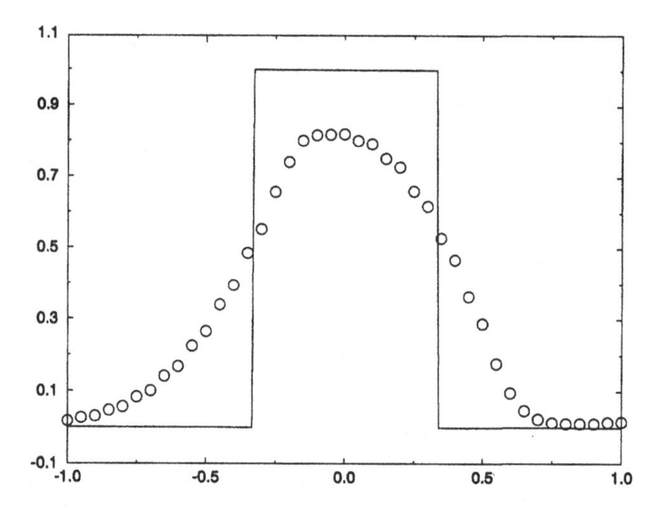

Figure 21.14 Second-order accurate Shu–Osher flux-corrected method for Test Case 2.

its shape as well. However, as seen in Figure 21.13, the third-order accurate Shu–Osher flux-corrected method captures the sinusoid with only a slight lagging phase error, without any clipping or other significant errors. In this test case, the third-order accurate Shu–Osher flux-corrected method performs better than any other method seen so far in Part V; its performance is comparable to a first-generation second-order accurate method such as the Lax–Wendroff method seen in Section 17.2.

Test Case 2 As seen in Figures 21.14 and 21.15, the Shu–Osher flux-corrected method dissipates and smears the square wave more than most of the other methods seen so far in Part V. As you might expect by the fact that contact smearing decreases with order

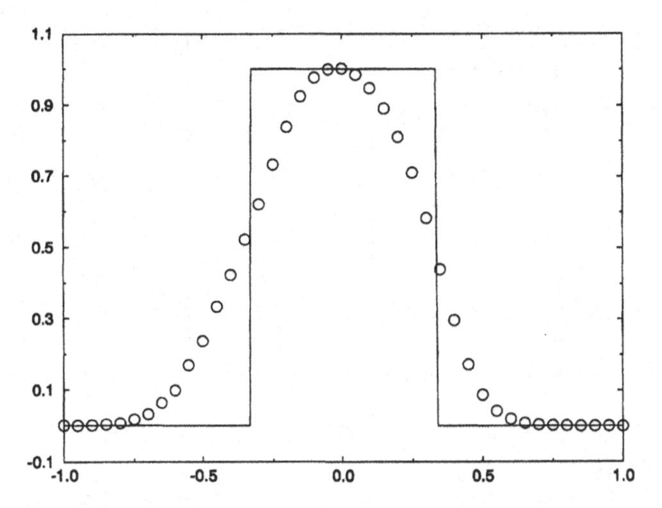

Figure 21.15 Third-order accurate Shu–Osher flux-corrected method for Test Case 2.

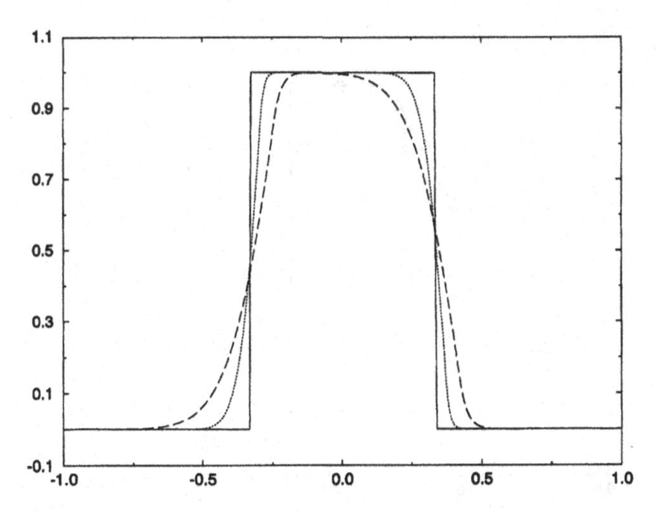

Figure 21.16 Second-order accurate Shu–Osher flux-corrected method for Test Case 3.

of accuracy, the third-order accurate method is substantially better than the second-order accurate method. Both solutions are completely free of spurious overshoots and oscillations.

Test Case 3 In Figures 21.16 and 21.17, the dotted lines represent the Shu–Osher flux-corrected approximations to $u(x, 4)$, the long dashed lines represent the Shu–Osher flux-corrected approximations to $u(x, 40)$, and the solid lines represent the exact solution for $u(x, 4)$ or $u(x, 40)$.

Test Case 4 The second- and third-order accurate results are shown in Figures 21.18 and 21.19, respectively. The results are nearly identical although, predictably, the

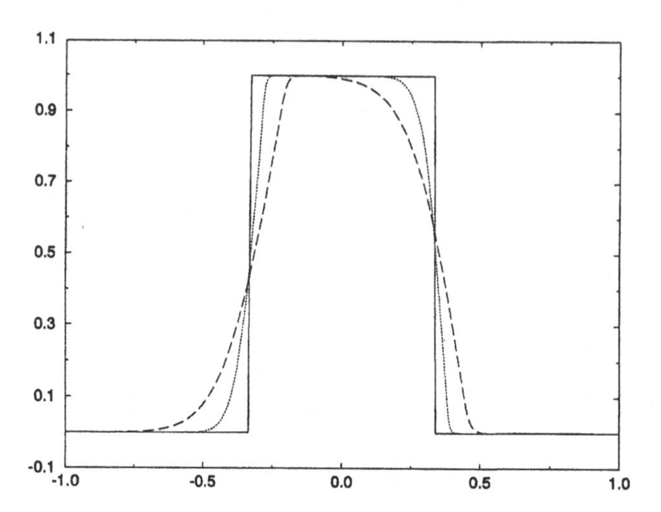

Figure 21.17 Third-order accurate Shu–Osher flux-corrected method for Test Case 3.

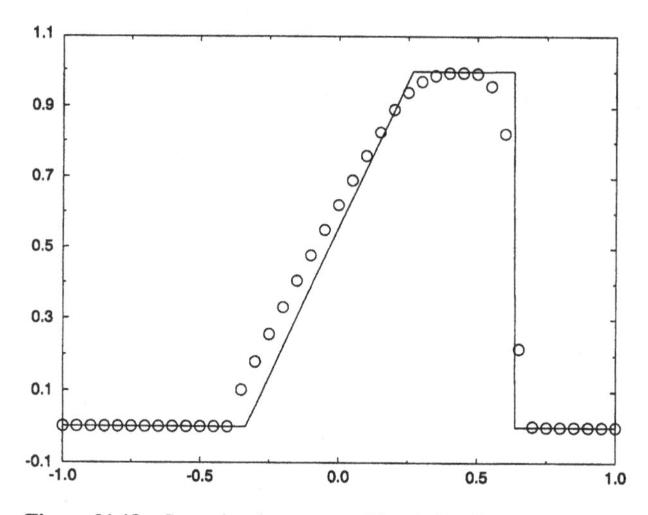

Figure 21.18 Second-order accurate Shu–Osher flux corrected method for Test Case 4.

third-order accurate method is just slightly better than the second-order accurate method. Both solutions display a bit of bulging on the expansion fan.

Test Case 5 The second- and third-order accurate results are shown in Figures 21.20 and 21.21, respectively. The results are nearly identical although, predictably, the third-order accurate method is just slightly better than the second-order accurate method. Both solutions are completely smooth across the expansive sonic point.

The results of Test Case 1 and Test Case 2 strongly favor the third-order accurate method over the second-order accurate method. Fourth- and higher-order accurate methods may also be used, but the upper bound on the CFL number tends to drop, and the increased order of

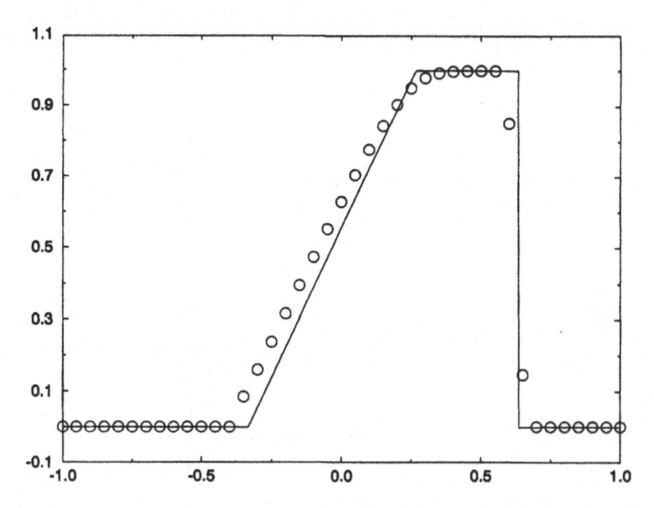

Figure 21.19 Third-order accurate Shu–Osher flux-corrected method for Test Case 4.

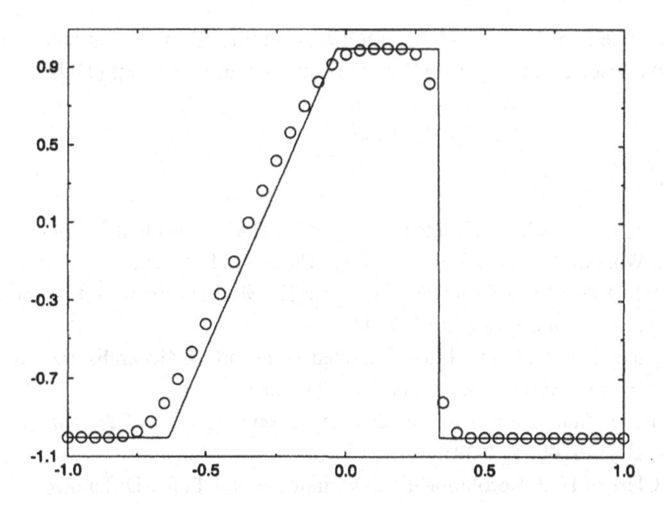

Figure 21.20 Second-order accurate Shu–Osher flux-corrected method for Test Case 5.

accuracy may not justify the increased complication and expense except on exceedingly fine grids. In general, in practice, numerical results usually favor second- or third-order accurate methods, as opposed to higher-order accurate methods, at least in the presence of shocks and contacts, and assuming reasonably simple flow structures in smooth regions. In this particular case, the third-order accurate method is favored over the second-order accurate method, partly because the second-order accurate method falls a bit short. Although the Shu–Osher flux corrected method does not exhibit any spurious overshoots or oscillations in the above test cases, it is not too hard to find other cases where it does overshoot or oscillate, although usually only in quite small amounts; this is simply the price one pays for eliminating clipping errors at extrema.

This section has described a version of the Shu–Osher method based on the ENO reconstructions of Sections 9.1 and 9.3. Researchers have proposed many variations on the type

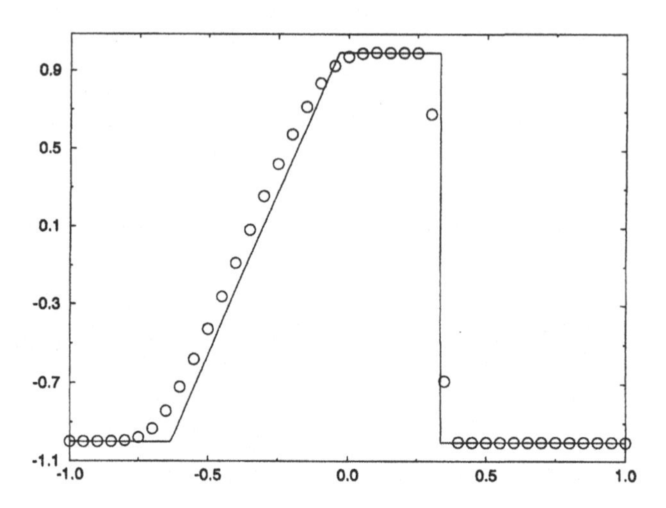

Figure 21.21 Third-order accurate Shu–Osher flux-corrected method for Test Case 5.

of ENO reconstruction used in the Shu–Osher methods, including methods based on the WENO reconstructions described in Section 9.2; see Liu, Osher, and Chan (1994) or Jiang and Shu (1996).

References

Boris, J. P., and Book, D. L. 1973. "Flux-Corrected Transport I. SHASTA, a Fluid Transport Algorithm that Works," *Journal of Computational Physics*, 11: 38–69.

Boris, J. P., and Book, D. L. 1976. "Flux-Corrected Transport. III. Minimal-Error FCT Algorithms," *Journal of Computational Physics*, 20: 397–431.

Boris, J. P., Book, D. L., and Hain, K. 1975. "Flux-Corrected Transport. II. Generalizations of the Method," *Journal of Computational Physics*, 18: 248–283.

Harten, A. 1983. "High Resolution Schemes for Hyperbolic Conservation Laws," *Journal of Computational Physics*, 49: 357–393.

Harten, A. 1984. "On a Class of High Resolution Total-Variation-Stable Finite-Difference Schemes," *SIAM Journal on Numerical Analysis*, 21: 1–22.

Jiang, G.-S., and Shu, C.-W. 1996. "Efficient Implementation of Weighted ENO Schemes," *Journal of Computational Physics*, 126: 202–228.

Liu, X.-D., Osher, S., and Chan, T. 1994. "Weighted Essentially Nonoscillatory Schemes," *Journal of Computational Physics*, 115: 200–212.

Oran, E. S., and Boris, J. P. 1989. *Numerical Simulation of Reactive Flow*. New York: Elsevier.

Shu, C.-W., 1987. "TVB Uniformly High-Order Schemes for Conservation Laws," *Mathematics of Computation*, 49: 105–121.

Shu, C.-W., and Osher, S. 1988. "Efficient Implementation of Essentially Nonoscillatory Shock-Capturing Schemes," *Journal of Computational Physics*, 77: 439–471.

Shu, C.-W., and Osher, S. 1989. "Efficient Implementation of Essentially Nonoscillatory Shock-Capturing Schemes, II," *Journal of Computational Physics*, 88: 32–78.

Yee, H. C. 1986. "Linearized Form of Implicit TVD Schemes for the Multidimensional Euler and Navier–Stokes Equations," *Computers and Mathematics with Applications*, 12A: 413–432.

Yee, H. C. 1989. "A Class of High-Resolution Explicit and Implicit Shock Capturing Methods," *NASA Technical Memorandum 101088* (unpublished).

Yee, H. C., and Harten, A. 1987. "Implicit TVD Schemes for Hyperbolic Conservation Laws in Curvilinear Coordinates," *AIAA Journal*, 25: 266–274.

Yee, H. C., Warming, R. F., and Harten, A. 1982. "A High-Resolution Numerical Technique for Inviscid Gas-Dynamic Problems with Weak Solutions." In *Eighth International Conference on Numerical Methods in Fluid Dynamics: Lecture Notes in Physics*, Volume 170, ed. E. Krause, pp. 546–552. Berlin: Springer-Verlag.

Yee, H. C., Warming, R. F., and Harten, A. 1985a. "Implicit Total Variation Diminishing (TVD) Schemes for Steady-State Calculations," *Journal of Computational Physics*, 57: 327–360.

Yee, H. C., Warming, R. F., and Harten, A. 1985b. "Application of TVD Schemes for the Euler Equations of Gas Dynamics." In *Large-Scale Computations in Fluid Mechanics, Lectures in Applied Mathematics*, Volume 22, Part 2, eds. B. E. Engquist, S. Osher, and R. C. J. Somerville, pp. 357–377, Providence, RI: American Mathematical Society.

Yee, H. C., Warming, R. F., and Harten, A. 1985c. "On a Class of TVD Schemes for Gas Dynamic Calculations." In *Numerical Methods for the Euler Equations of Fluid Dynamics*, eds. F. Angrand, A. Dervieux, J. A. Desideri, and R. Glowinski, pp. 84–107. Philadelphia: SIAM.

Zalesak, S. T. 1979. "Fully Multidimensional Flux-Corrected Transport Algorithms for Fluids," *Journal of Computational Physics*, 31: 335–362.

Problems

21.1 This problem concerns flux reconstruction methods and their relationship to reconstruction via the primitive function.

(a) Suppose the flux function is $f(u) = u$ and suppose the solution at $t = t^0$ is $u(x, t^0) = -\sin \pi x$. Assume that Δx is constant. Show that

$$\hat{f}_s^0(x) = -\frac{\sin \pi x}{(\sin \pi \Delta x/2)(\pi \Delta x/2)}.$$

In other words, prove that

$$f(u(x, t^0)) = \frac{1}{\Delta x} \int_{x+\Delta x/2}^{x+\Delta x/2} \hat{f}_s^0(y)\, dy$$

for all x.

(b) Suppose the flux function is $f(u) = u^2/2$ and suppose the solution is $u(x, t^0) = -\sin \pi x$. Assume that Δx is constant. Show that

$$\hat{f}_s^0(x) = \frac{1}{4}\left[1 - \frac{\cos 2\pi x}{(\sin \pi \Delta x)(\pi \Delta x)}\right].$$

21.2 Suppose that the flux function for a scalar conservation law is $f(u) = \sin u$.

(a) Find all of the sonic points of f. Which ones are compressive and which ones are expansive?

(b) Sketch the split fluxes $f^{\pm}(u)$ found using the linear splitting, Equation (21.38). Choose an appropriate value for α. What is the lower limit for α? What happens if α is smaller than this lower limit? Is there any upper limit on α?

(c) Sketch the split fluxes $f^{\pm}(u)$ found using Equation (21.37).

(d) What are the relative advantages and disadvantages of the splittings found in parts (b) and (c)? In particular, discuss the continuity of the split fluxes, the physicality of the split fluxes, and the existence of annoying user-adjustable parameters.
(e) Repeat parts (a)–(d), substituting $f(u) = au$ where $a = const.$
(f) Repeat parts (a)–(d), substituting $f(u) = u^2/2$.

21.3 Find the CFL condition for the Shu–Osher method for $N = 1, \ldots, 4$. Remember that the CFL condition is based strictly on the stencil (i.e., the points chosen by the ENO flux reconstruction). Also remember that the CFL condition is necessary but not sufficient for stability, so that a method may blow up or exhibit less drastic instabilities even when it satisfies the CFL condition. To prevent blow up, the CFL number must be substantially lower than what the CFL condition would indicate. In particular, compare the CFL condition with the following stability conditions given by Shu and Osher: $|\lambda a(u)| \leq 1$ for second- and third-order accuracy; $|\lambda a(u)| \leq 2/3$ for fourth-order accuracy; and $|\lambda a(u)| \leq 7/30$ for fifth-order accuracy. Intuitively, why is there such a difference between the CFL condition and these stability conditions?

CHAPTER 22

Flux Averaging III: Self-Adjusting Hybrid Methods

22.0 Introduction

This chapter concerns self-adjusting hybrid methods. To keep things simple, this introductory section concerns only self-adjusting hybrids for scalar conservation laws. Self-adjusting hybrid methods for the Euler equations are discussed later in the chapter; see Subsection 22.3.2. Consider two first-generation methods with complementary properties, such as Roe's first-order upwind method and the Lax–Wendroff method. Suppose that the two methods have conservative numerical fluxes $\hat{f}^{(1)}_{i+1/2}$ and $\hat{f}^{(2)}_{i+1/2}$. Self-adjusting hybrids use convex linear combinations as follows:

$$\hat{f}^{n}_{i+1/2} = \theta^{n}_{i+1/2}\hat{f}^{(1)}_{i+1/2} + \left(1 - \theta^{n}_{i+1/2}\right)\hat{f}^{(2)}_{i+1/2}, \tag{22.1}$$

where the parameter $\theta^{n}_{i+1/2}$ is called a *shock switch*. The reader should view Equation (22.1) as a starting point for discussion, rather than as a "fixed in stone" definition of self-adjusting hybrids, since many self-adjusting hybrids deviate somewhat from this standard.

The governing principle of shock switch design is as follows: the shock switch should be close to one near shocks and close to zero in smooth regions. Then, as the name implies, *shock switches distinguish shocks from smooth regions*, switching on at shocks and switching off in smooth regions. Unfortunately, by this criterion, it is impossible to design a perfect shock switch – as illustrated in Figure 22.1, the same samples can represent both smooth and shocked solutions. This stems from the fact that any sampling contains only limited information about the solution, especially solutions with jump discontinuities, as discussed in Chapter 8. Then any shock switch will sometimes yield false positives or false negatives, either of which adversely affects the self-adjusting hybrid's success.

As seen in Equations (8.5) and (8.6), large divided differences signal large or discontinuous derivatives. Thus, the shock switch is usually a function of first- or second-divided differences. For example, let

$$\theta^{n}_{i+1/2} = \max\left(\theta^{n}_{i}, \theta^{n}_{i+1}\right),$$

where

$$\theta^{n}_{i} = \frac{u^{n}_{i+1} - 2u^{n}_{i} + u^{n}_{i-1}}{\Delta x^{2}}.$$

To ensure $\theta^{n}_{i+1/2} \geq 0$, take the absolute value of the second-divided differences as follows:

$$\theta^{n}_{i} = \frac{\left|u^{n}_{i+1} - 2u^{n}_{i} + u^{n}_{i-1}\right|}{\Delta x^{2}}.$$

To ensure $\theta^{n}_{i+1/2} \leq 1$, divide by a normalization factor. For example, by the triangle inequality

$$\frac{\left|\left(u_{i+1} - u^{n}_{i}\right) - \left(u^{n}_{i} - u^{n}_{i-1}\right)\right|}{\Delta x^{2}} \leq \frac{\left|u_{i+1} - u^{n}_{i}\right| + \left|u^{n}_{i} - u^{n}_{i-1}\right|}{\Delta x^{2}}.$$

Figure 22.1 The same samples may represent both smooth and discontinuous functions.

Then let

$$\blacklozenge \qquad \theta_i^n = \frac{\left|u_{i+1}^n - 2u_i^n + u_{i-1}^n\right|}{\left|u_{i+1}^n - u_i^n\right| + \left|u_i^n - u_{i-1}^n\right|}. \tag{22.2}$$

This chapter considers three self-adjusting hybrid methods. In 1972, Harten and Zwas proposed a self-adjusting hybrid of the Lax–Friedrichs and Lax–Wendroff methods. In 1978, Harten proposed a self-adjusting hybrid of a first-order upwind method and the Lax–Wendroff method. In 1981, Jameson, Schmidt, and Turkel proposed a self-adjusting hybrid combination of a semidiscrete first-order upwind method and a semidiscrete version of FTCS plus fourth-order artificial viscosity.

Besides the three self-adjusting hybrid methods described in the preceding paragraph, this chapter also considers four shock switches. Harten and Zwas suggested the first shock switch in 1972. Unfortunately, this shock switch involves solution-sensitive user-adjustable parameters, meaning the user has to constantly reset the parameters for every different solution. MacCormack and Baldwin suggested a far better shock switch in 1975. In 1978, Harten suggested a third shock switch which results in excellent nonlinear stability, albeit at the expense of extremum clipping. Finally, in 1992, Swanson and Turkel suggested a relatively complicated and sophisticated shock switch, which combines the MacCormack–Baldwin shock switch with the shock switch seen in Equation (22.2).

The starting-point definition of flux-limited methods seen in Equation (20.1) closely resembles the starting-point definition of self-adjusting hybrids seen in Equation (22.1). In fact, they are identical if $\theta_{i+1/2}^n = 1 - \phi_{i+1/2}^n$. Despite their similar starting-points, self-adjusting hybrids traditionally deviate from flux-limited methods in terms of sonic point treatments, stability treatments, convexity, and in other ways. For example, self-adjusting hybrid methods traditionally require convex linear combinations ($0 \leq \theta_{i+1/2}^n \leq 1$) whereas flux-limited methods traditionally do not. For another example, self-adjusting hybrid methods traditionally do not enforce nonlinear stability conditions, whereas flux-limited methods traditionally do. As a result, self-adjusting hybrid methods usually avoid clipping, unlike flux-limited methods, but also allow more spurious overshoots and oscillations.

22.1 Harten–Zwas Self-Adjusting Hybrid Method

Harten and Zwas (1972a, 1972b) suggested a self-adjusting hybrid of the Lax–Friedrichs and Lax–Wendroff methods. As seen in Section 17.1, the Lax–Friedrichs method is as follows:

$$u_{i+1}^n = u_i^n - \lambda\left(\hat{f}_{i+1/2}^{\text{L-F}} - \hat{f}_{i-1/2}^{\text{L-F}}\right),$$

$$\hat{f}_{i+1/2}^{\text{L-F}} = \frac{1}{2}\left(f\left(u_{i+1}^n\right) + f\left(u_i^n\right)\right) - \frac{1}{2\lambda}\left(u_{i+1}^n - u_i^n\right).$$

Also, as seen in Section 17.2, the Lax–Wendroff method is as follows:

$$u_{i+1}^n = u_i^n - \lambda\left(\hat{f}_{i+1/2}^{\text{L-W}} - \hat{f}_{i-1/2}^{\text{L-W}}\right),$$

$$\hat{f}_{i+1/2}^{\text{L-W}} = \frac{1}{2}\left(f\left(u_{i+1}^n\right) + f\left(u_i^n\right)\right) - \frac{1}{2}\lambda\left(a_{i+1/2}^n\right)^2\left(u_{i+1}^n - u_i^n\right).$$

Then the *Harten–Zwas self-adjusting hybrid method* is as follows:

$$u_{i+1}^n = u_i^n - \lambda\left(\hat{f}_{i+1/2}^n - \hat{f}_{i-1/2}^n\right),$$

where

$$\hat{f}_{i+1/2}^n = \theta_{i+1/2}^n \hat{f}_{i+1/2}^{\text{L-F}} + \left(1 - \theta_{i+1/2}^n\right)\hat{f}_{i+1/2}^{\text{L-W}} \tag{22.3a}$$

or

$$\hat{f}_{i+1/2}^n = \frac{1}{2}\left(f\left(u_{i+1}^n\right) + f\left(u_i^n\right)\right) - \frac{1}{2\lambda}\left[\left(1 - \left(\lambda a_{i+1/2}^n\right)^2\right)\theta_{i+1/2}^n + \left(\lambda a_{i+1/2}^n\right)^2\right]\left(u_{i+1}^n - u_i^n\right). \tag{22.3b}$$

Notice that the Harten–Zwas self-adjusting hybrid method equals the Lax–Friedrichs method for $\theta_{i+1/2}^n = 1$, the Lax–Wendroff method for $\theta_{i+1/2}^n = 0$, and Roe's first-order upwind method for $\theta_{i+1/2}^n = |\lambda a_{i+1/2}^n|/(1 + |\lambda a_{i+1/2}^n|)$. The reader can certainly find other relationships between the Harten–Zwas self-adjusting hybrid and earlier methods. Notice that the coefficient of artificial viscosity of the Harten–Zwas self-adjusting hybrid method is as follows:

$$\epsilon_{i+1/2}^n = \frac{1}{\lambda}\left[\left(1 - \left(\lambda a_{i+1/2}^n\right)^2\right)\theta_{i+1/2}^n + \left(\lambda a_{i+1/2}^n\right)^2\right], \tag{22.4}$$

which is sometimes called *self-adjusting* or *adaptive artificial viscosity*.

If $0 \le \theta_{i+1/2}^n \le 1$ then $\hat{f}_{i+1/2}^n$ is a convex linear combination of $\hat{f}_{i+1/2}^{\text{L-F}}$ and $\hat{f}_{i+1/2}^{\text{L-W}}$. As usual, convexity ensures that $\hat{f}_{i+1/2}^n$ is somewhere between $\hat{f}_{i+1/2}^{\text{L-F}}$ and $\hat{f}_{i+1/2}^{\text{L-W}}$. The Lax–Friedrichs method and the Lax–Wendroff methods are both extreme methods: the Lax–Friedrichs method has too much artificial viscosity and even small increases may cause instability, whereas the Lax–Wendroff method has too little artificial viscosity and even small decreases may cause instability. Thus, it makes good sense to keep the hybrid somewhere between the Lax–Friedrichs and Lax–Wendroff methods. In other words, it makes good sense to enforce the convexity condition $0 \le \theta_{i+1/2}^n \le 1$.

The Harten–Zwas self-adjusting hybrid works on the principle of error cancellation. Whereas the Van Leer flux-limited method seen in Section 20.1 cancels *dispersive* errors, the Harten–Zwas self-adjusting hybrid cancels *dissipative* errors. Because of error cancellation, the hybrid combination of Lax–Friedrichs and Lax–Wendroff may perform better than either method separately; in particular, the Lax–Friedrichs and Lax–Wendroff methods typically both experience spurious oscillations and overshoots near jump discontinuities, while the hybrid need not.

How does one choose the shock switch? The Lax–Wendroff method performs well in smooth regions, and thus the hybrid should lean heavily on the Lax–Wendroff method in smooth regions. Therefore, $\theta_{i+1/2}^n \approx 0$ in smooth regions. Near shocks, the best choice lies somewhere nearly halfway between the Lax–Wendroff and Lax–Friedrichs methods. So $\theta_{i+1/2}^n \approx 1/2$ near shocks. Finally, the shock switch should satisfy the convexity condition

$0 \leq \theta^n_{i+1/2} \leq 1$. Based on these observations, Harten and Zwas suggested the following shock switch:

$$\theta^n_{i+1/2} = \kappa \left[\frac{|u^n_{i+1} - u^n_i|}{\max_j |u^n_{j+1} - u^n_j|} \right]^m, \tag{22.5}$$

where κ and m are user-adjustable parameters. This shock switch clearly satisfies the convexity condition $0 \leq \theta^n_{i+1/2} \leq 1$ for all $0 \leq \kappa \leq 1$ and $m \geq 1$. Notice that

$$\frac{|u^n_{i+1} - u^n_i|}{\max_j |u^n_{j+1} - u^n_j|} = \begin{cases} 1 & \text{for the largest jump,} \\ <1 & \text{for other jumps,} \\ O(\Delta x) & \text{for smooth regions.} \end{cases}$$

Unfortunately, this implies that the best choice of κ and m depends on the solution. In particular, if κ and m are chosen to optimize the performance at the largest jump in the solution, then the method underdamps the smaller jumps. Conversely, if κ and m are chosen to optimize the performance at the smallest jump in the solution, then the method overdamps the larger jumps. Such solution sensitivity disqualifies the Harten–Zwas shock switch for most practical computations. This section omits numerical results, which would only serve to illustrate the poor performance of the Harten–Zwas shock switch.

22.2 Harten's Self-Adjusting Hybrid Method

Working without Zwas, Harten (1978) suggested another self-adjusting hybrid. Harten's hybrid differs from the Harten–Zwas hybrid in two respects. First, Harten replaced the Lax–Friedrichs method with a first-order upwind method. Specifically, Harten used the modified Boris–Book first-order upwind method, seen in Section 21.1. However, to modernize and improve the results, this section substitutes the class of first-order upwind methods seen in Section 17.3. Then *Harten's self-adjusting hybrid method* is as follows:

$$u^n_{i+1} = u^n_i - \lambda\left(\hat{f}^n_{i+1/2} - \hat{f}^n_{i-1/2}\right),$$

where

$$\hat{f}^n_{i+1/2} = \theta^n_{i+1/2}\hat{f}^{(1)}_{i+1/2} + \left(1 - \theta^n_{i+1/2}\right)\hat{f}^{\text{L-W}}_{i+1/2}, \tag{22.6}$$

and where $\hat{f}^{(1)}_{i+1/2}$ is the conservative numerical flux of any of the first-order upwind methods seen in Section 17.3 and $\hat{f}^{\text{L-W}}_{i+1/2}$ is the conservative numerical flux of the Lax–Wendroff method seen in Section 17.2.

As the second and more important modification to the Harten–Zwas self-adjusting hybrid method, Harten substituted a much improved shock switch. The Lax–Wendroff method performs well in smooth regions, which argues for a bias towards the Lax–Wendroff method in smooth regions. Thus $\theta^n_{i+1/2} \approx 0$ in smooth regions. On the other hand, first-order upwind methods perform well at shocks, which argues for a tilt towards the first-order upwind method at shocks. So $\theta^n_{i+1/2} \approx 1$ near shocks. Finally, the shock switch should satisfy the convexity condition $0 \leq \theta^n_{i+1/2} \leq 1$. Harten suggested the following shock switch to satisfy these criteria:

$$\theta^n_{i+1/2} = \max\left(\theta^n_i, \theta^n_{i+1}\right), \tag{22.7a}$$

where

$$\theta_i^n = \kappa \left| \frac{\left|u_{i+1}^n - u_i^n\right| - \left|u_i^n - u_{i-1}^n\right|}{\left|u_{i+1}^n - u_i^n\right| + \left|u_i^n - u_{i-1}^n\right|} \right|^m \tag{22.7b}$$

or

$$\theta_i^n = \begin{cases} \kappa \left| \frac{\left|u_{i+1}^n - u_i^n\right| - \left|u_i^n - u_{i-1}^n\right|}{\left|u_{i+1}^n - u_i^n\right| + \left|u_i^n - u_{i-1}^n\right|} \right|^m & \left|u_{i+1}^n - u_i^n\right| + \left|u_i^n - u_{i-1}^n\right| > \delta, \\ 0 & \left|u_{i+1}^n - u_i^n\right| + \left|u_i^n - u_{i-1}^n\right| \le \delta. \end{cases}$$

Unlike the first expression, the second expression prevents small or zero denominators; the two expressions are otherwise identical. As before, κ, m, and δ are user-adjustable parameters. By the triangle inequality,

$$\left|\left|u_{i+1}^n - u_i^n\right| - \left|u_i^n - u_{i-1}^n\right|\right| \le \left|u_{i+1}^n - u_i^n\right| + \left|u_i^n - u_{i-1}^n\right|.$$

Thus Harten's shock switch satisfies the convexity condition $0 \le \theta_{i+1/2}^n \le 1$ for all $0 \le \kappa \le 1$ and $m \ge 1$. Notice that, much like the flux limiters of Chapter 20, Harten's shock switch can be written as a function of $r_i^+ = (u_i^n - u_{i-1}^n)/(u_{i+1}^n - u_i^n)$ as follows:

$$\theta_i^n = \kappa \left| \frac{1 - \left|r_i^+\right|}{1 + \left|r_i^+\right|} \right|^m. \tag{22.7c}$$

Notice that θ_i^n is near zero only if $|r_i^+|$ is near one, and notice that $\theta_{i+1/2}^n = \max(\theta_i^n, \theta_{i+1}^n)$ is even larger than θ_i^n. Thus, much of the time, the self-adjusting hybrid will use a substantial amount of the first-order upwinding, even in completely smooth regions. On the positive side, the parameters κ and m are not very solution sensitive, unlike the analogous parameters in the Harten–Zwas shock switch. Instead, the same values should work reasonably well for a broad range of different solutions. Hence, for the most part, κ and m are "set and forget" parameters.

The behavior of Harten's self-adjusting hybrid method is illustrated using the five standard test cases defined in Section 17.0. In all test cases, $\kappa = 1$, $\delta = 0.01$, and $m = 1$. The first four test cases use Roe's first-order upwind method, whereas the last test case uses Harten's first-order upwind method with $\delta = 0.6$, where now δ refers to the small parameter in Harten's first-order upwind method seen in Subsection 17.3.3 rather than to the small parameter in Harten's shock switch.

Test Case 1 As seen in Figure 22.2, Harten's self-adjusting hybrid method captures the sinusoid with only modest clipping. This clipping may be reduced or completely eliminated with different settings of the user-adjustable parameters, but only at the expense of the performance in other test cases.

Test Case 2 As seen in Figure 22.3, Harten's self-adjusting hybrid method captures the square wave without spurious overshoots or oscillations, and with a typical amount of smearing.

Test Case 3 In Figure 22.4, the dotted line represents Harten's self-adjusting hybrid approximation to $u(x, 4)$, the long dashed line represents Harten's self-adjusting hybrid approximation to $u(x, 40)$, and the solid line represents the exact solution for $u(x, 4)$ or $u(x, 40)$. Harten's self-adjusting hybrid is the first method in Part V to exhibit oscillations or overshoots in Test Case 3. On the positive side, the solution exhibits an unusually low amount of smearing.

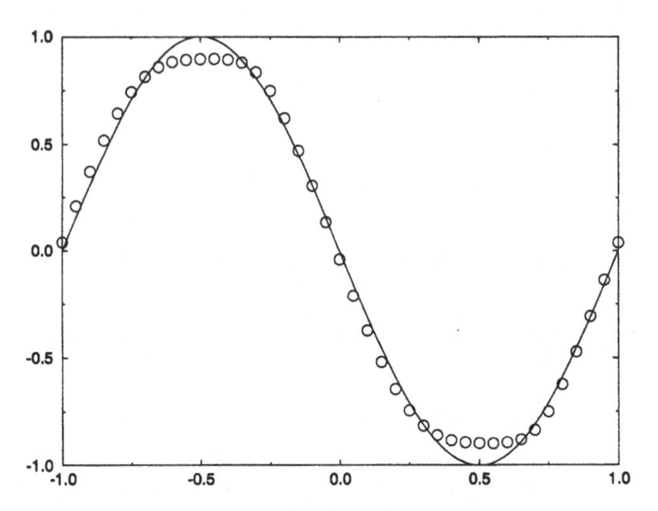

Figure 22.2 Harten's self-adjusting hybrid method for Test Case 1.

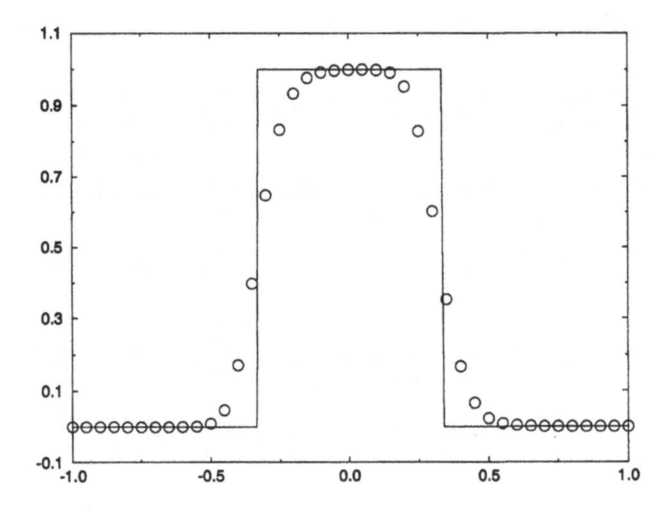

Figure 22.3 Harten's self-adjusting hybrid method for Test Case 2.

Test Case 4 The results seen in Figure 22.5 are about as good as any seen so far.

Test Case 5 As seen in Figure 22.6, the solution is excellent except for a small jump across the expansive sonic point, which could be eliminated by increasing the parameter δ in Harten's first-order upwind method.

To reduce the contact smearing seen in Test Cases 2 and 3, Harten (1978) suggested physical flux correction, much as in Section 21.3. In particular let

$$g_i^n = f\left(u_i^n\right) + f_i^{(C)}. \tag{22.8}$$

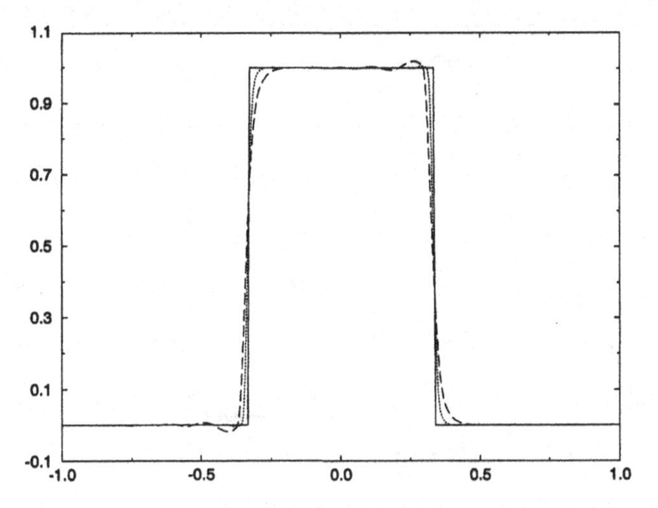

Figure 22.4 Harten's self-adjusting hybrid method for Test Case 3.

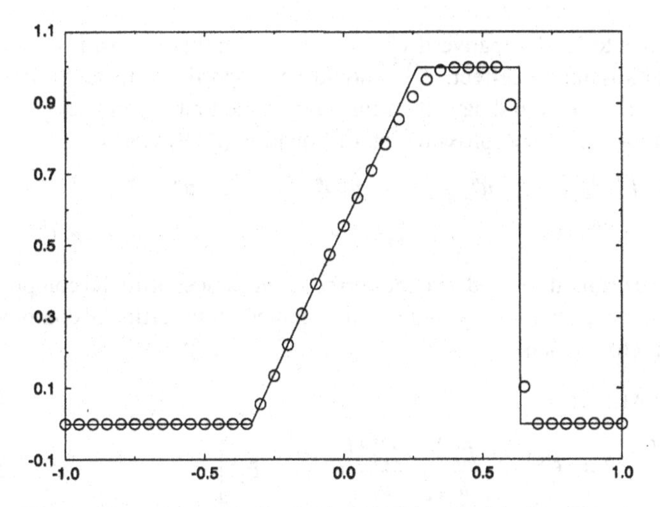

Figure 22.5 Harten's self-adjusting hybrid method for Test Case 4.

Then apply any numerical method, such as Harten's self-adjusting hybrid method, to g_i^n rather than to $f(u_i^n)$. That is, approximate

$$\frac{\partial u}{\partial t} + \frac{\partial g}{\partial x} = \frac{\partial u}{\partial t} + \frac{\partial}{\partial x}(f + f^{(C)}) = 0. \tag{22.9}$$

The flux correction $f_i^{(C)}$ should add compression or, equivalently, reduce artificial dissipation or, put yet another way, increase artificial antidissipation. Recall that Section 14.2 described the relationship between artificial viscosity and flux correction. Specifically, artificial viscosity forms write the flux correction as a product of a coefficient $\epsilon_{i+1/2}^n$ and a first difference $u_{i+1}^n - u_i^n$ as follows:

$$-\frac{1}{2}\epsilon_{i+1/2}^n\left(u_{i+1}^n - u_i^n\right).$$

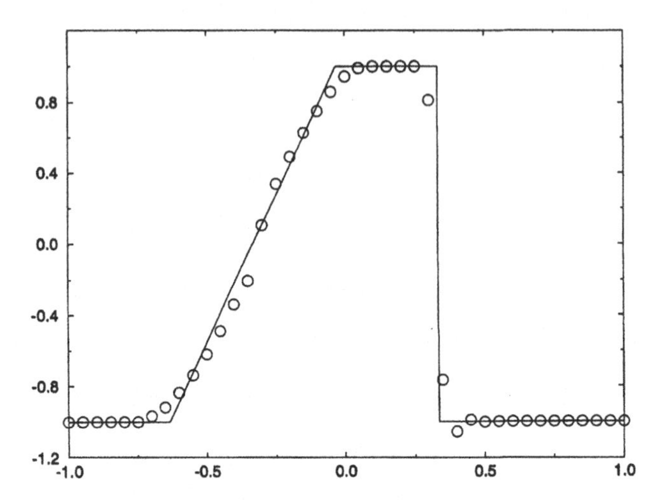

Figure 22.6 Harten's self-adjusting hybrid method for Test Case 5.

Artificial viscosity tends to be dissipative if $\epsilon^n_{i+1/2} > 0$ and antidissipative if $\epsilon^n_{i+1/2} < 0$. Then, to ensure an antidissipative effect, $f_i^{(C)}$ should be proportional to first differences such as $u^n_{i+1} - u^n_i$ and $u^n_i - u^n_{i-1}$ and have the same sign as the first differences.

A two-step predictor–corrector approximation of Equation (22.9) yields

$$\begin{aligned}\bar{u}_i &= u^n_i - \lambda\left[\hat{f}\left(u^n_{i-K_1}, \ldots, u^n_{i+K_2-1}\right) - \hat{f}\left(u^n_{i-K_1+1}, \ldots, u^n_{i+K_2}\right)\right],\\ u^{n+1}_i &= \bar{u}_i - \lambda\left[\hat{f}^{(C)}\left(\bar{u}_{i-K_1}, \ldots, \bar{u}_{i+K_2-1}\right) - \hat{f}^{(C)}\left(\bar{u}_{i-K_1+1}, \ldots, \bar{u}_{i+K_2}\right)\right].\end{aligned}$$

Thus the predictor is the original method, and the corrector introduces artificial compression. Harten (1978) suggests using Roe's first-order upwind method for the artificial compression. The compressive corrector is then

$$u^{n+1}_i = \bar{u}_i - \lambda\left(\hat{f}^{(C)}_{i+1/2} - \hat{f}^{(C)}_{i-1/2}\right), \tag{22.10}$$

$$\hat{f}^{(C)}_{i+1/2} = \frac{1}{2}\left[f^{(C)}_{i+1} + f^{(C)}_i - \left|\frac{f^{(C)}_{i+1} - f^{(C)}_i}{\bar{u}_{i+1} - \bar{u}_i}\right|(\bar{u}_{i+1} - \bar{u}_i)\right], \tag{22.11}$$

where

$$f^{(C)}_i = \text{minmod}(\bar{u}_{i+1} - \bar{u}_i, \bar{u}_i - \bar{u}_{i-1}). \tag{22.12}$$

Figures 22.7 and 22.8 show the results of using artificial compression in Harten's self-adjusting hybrid method in Test Cases 1 and 2. These results show both the power and peril of artificial compression: it turns everything into square waves. Then the square wave is captured nearly perfectly, but the sine wave also looks like a square wave. Clearly, one cannot just leave the artificial compression going full blast all the time. Harten suggests replacing Equation (22.10) by the following:

$$u^{n+1}_i = \bar{u}_i - \lambda\left(\theta^n_{i+1/2}\hat{f}^{(C)}_{i+1/2} - \theta^n_{i-1/2}\hat{f}^{(C)}_{i-1/2}\right), \tag{22.13}$$

where $\theta^n_{i+1/2}$ is any shock switch. For a more modern take on artificial compression, see Section 21.3.

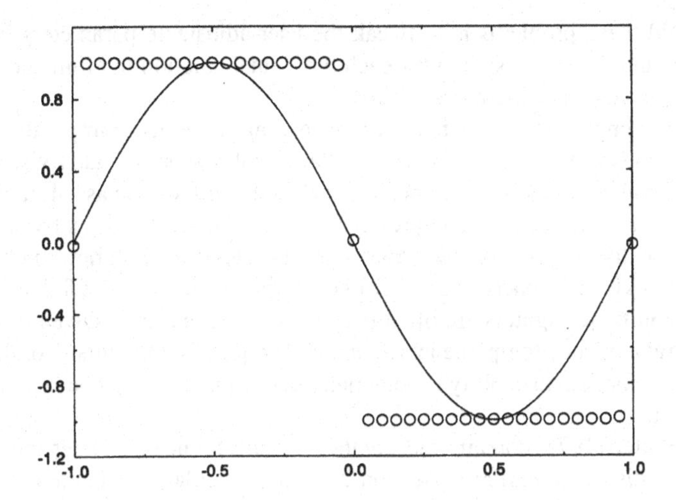

Figure 22.7 Harten's self-adjusting hybrid method with artificial compression for Test Case 1.

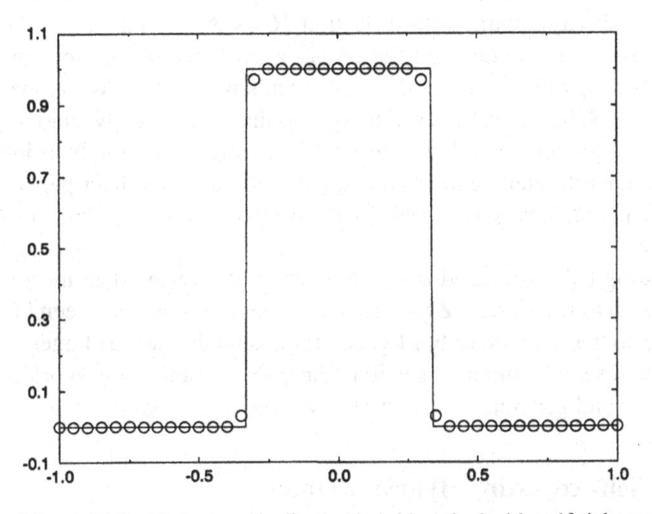

Figure 22.8 Harten's self-adjusting hybrid method with artificial compression for Test Case 2.

Artificial compression is a tricky thing. From the chapter introduction, remember that every shock switch occasionally yields false positives or false negatives. When shock switches are used to adjust the average of two methods, false positives and false negatives are relatively benign, usually causing only a modest and local loss of accuracy. However, when shock switches are used to control artificial compression, false positives or false negatives may be disastrous. Since the amount of artificial compression depends entirely on the shock switch, artificial compression might fire off in the wrong places, causing overly steep solutions at best and spurious jump discontinuities at worst. Also, artificial compression may fail to act where it should, allowing contact smearing to progress unimpeded, with the artificial compression becoming increasingly less likely to wake up and act as the smearing builds.

The only way to avoid these problems is to tweak the user-adjustable parameters in the artificial compression and the shock switch for each individual solution to eliminate false positives and false negatives on a case by case basis.

The term "artificial compression" is potentially misleading: it seems to imply that "artificial compression" adds compression, much like "artificial viscosity" adds viscosity. However, in fact, artificial compression does not so much add compression as subtract viscosity. As one way to view it, artificial compression reduces the local artificial viscosity to make the method locally less stable. Decreased stability allows the solution to change rapidly, jumping up quickly at contacts. It's rather like a high-performance jet fighter – for maximum maneuverability, jet fighters are often purposely made unstable. Given that the jet already has a natural tendency to tip, the jet can maneuver quickly. Of course, for the jet or for artificial compression, the instability is potentially disastrous and must be kept under strict computer control.

For history buffs, Harten (1978) contains most of the elements found in the later and more famous Harten (1983), discussed earlier in Section 21.3. In particular, like Harten (1983), Harten (1978) contains a substantial discussion of nonlinear stability theory, including monotonicity preservation (see Section 16.1), monotone schemes (see Section 16.9), and an early form of the positivity condition (see Section 16.4). Also, Harten (1978) shows an appreciation for entropy conditions and the possible existence of entropy-condition-violating expansion shocks, which later motivated Harten's first-order upwind method in Harten (1983), as seen in Subsection 18.3.3. Finally, and most importantly, Harten (1978) applies Roe's first-order upwind method to a physical flux correction, exactly as in Harten (1983), albeit only for the artificial compression part and not, as in the later paper, for the method as a whole. Thus, in a very real sense, Harten (1978) was a dry run for Harten (1983).

Harten's self-adjusting hybrid method is a little more sophisticated than most second-generation methods, such as the Harten–Zwas self-adjusting hybrid method seen in the last section, but still not up to the standard of third-generation methods, such as Jameson's self-adjusting hybrid method seen in the next section. Harten's self-adjusting hybrid method should be counted as second generation or, at best, "two and a half" generation.

22.3 Jameson's Self-Adjusting Hybrid Method

This section concerns a semidiscrete self-adjusting hybrid method proposed by Jameson, Schmidt, and Turkel (1981) and Jameson (1982). Among other things, Jameson's method helped to popularize semidiscrete methods; see Sections 20.3 and 21.4 for descriptions of later semidiscrete methods.

22.3.1 *Scalar Conservation Laws*

A semidiscrete version of Roe's first-order upwind method is as follows:

$$\frac{\partial u_i^n}{\partial t} = -\frac{\hat{f}^{(1)}_{s,i+1/2} - \hat{f}^{(1)}_{s,i-1/2}}{\Delta x},$$

$$\hat{f}^{(1)}_{s,i+1/2} = \frac{1}{2}\left(f\left(u^n_{i+1}\right) + f\left(u^n_i\right)\right) - \frac{1}{2}\left|a^n_{i+1/2}\right|\left(u^n_{i+1} - u^n_i\right).$$

The fully discrete version of Roe's first-order upwind method was described in Section 17.3.2. Also consider the following semidiscrete second-order accurate method:

$$\frac{\partial u_i^n}{\partial t} = -\frac{\hat{f}^{(2)}_{s,i+1/2} - \hat{f}^{(2)}_{s,i-1/2}}{\Delta x},$$

$$\hat{f}^{(2)}_{s,i+1/2} = \frac{1}{2}\left(f\left(u^n_{i+1}\right) + \left(u^n_i\right)\right) + \frac{1}{2}\delta\left|a^n_{i+1/2}\right|\left(u^n_{i+2} - 3u^n_{i+1} + 3u^n_i - u^n_{i-1}\right),$$

where $\delta > 0$ is a user-adjustable parameter. This method is a semidiscrete version of FTCS plus fourth-order artificial viscosity. Then a semidiscrete self-adjusting hybrid of $\hat{f}^{(1)}_{i+1/2}$ and $\hat{f}^{(2)}_{i+1/2}$ is as follows:

$$\frac{\partial u_i^n}{\partial t} = -\frac{\hat{f}^{n}_{s,i+1/2} - \hat{f}^{n}_{s,i-1/2}}{\Delta x},$$

where

$$\hat{f}^n_{s,i+1/2} = \theta^n_{i+1/2}\hat{f}^{(1)}_{s,i+1/2} + \left(1 - \theta^n_{i+1/2}\right)\hat{f}^{(2)}_{s,i+1/2}$$

or

$$\hat{f}^n_{s,i+1/2} = \frac{1}{2}\left(f\left(u^n_{i+1}\right) + f\left(u^n_i\right)\right) - \frac{1}{2}\left|a^n_{i+1/2}\right|\left[\theta^n_{i+1/2}\left(u^n_{i+1} - u^n_i\right)\right.$$
$$\left. - \delta\left(1 - \theta^n_{i+1/2}\right)\left(u^n_{i+1} - 3u^n_{i+1} + 3u^n_i - u^n_{i-1}\right)\right].$$

The hybrid should be heavily biased towards $\hat{f}^{(1)}_{i+1/2}$ near shocks and heavily biased towards $\hat{f}^{(2)}_{i+1/2}$ in smooth regions. In other words, the shock switch should be near zero in smooth regions and near one at shocks. Unfortunately, in practice, even small amounts of the fourth-order artificial viscosity found in $\hat{f}^{(2)}_{i+1/2}$ may destabilize the hybrid near shocks. Since it is extremely difficult to design a shock switch that is exactly zero near shocks, the coefficient of $\hat{f}^{(2)}_{i+1/2}$ is modified to equal to zero when $|\theta_{i+1/2}| > \delta$ where, for convenience, $\delta > 0$ is the same small user-adjustable parameter that appears in $\hat{f}^{(2)}_{i+1/2}$. Then *Jameson's self-adjusting hybrid method*, also known as the *Jameson–Schmidt–Turkel (JST) self-adjusting hybrid method*, is as follows:

$$\frac{\partial u_i^n}{\partial t} = -\frac{\hat{f}^{n}_{s,i+1/2} - \hat{f}^{n}_{s,i-1/2}}{\Delta x},$$

where

$$\hat{f}^n_{s,i+1/2} = \theta^n_{i+1/2}\hat{f}^{(1)}_{s,i+1/2} + \max\left(0, 1 - \frac{1}{\delta}\theta^n_{i+1/2}\right)\hat{f}^{(2)}_{s,i+1/2}$$

or with a final slight modification

$$\blacklozenge \quad \hat{f}^n_{s,i+1/2} = \frac{1}{2}\left(f\left(u^n_{i+1}\right) + f\left(u^n_i\right)\right) - \frac{1}{2}\left|a^n_{i+1/2}\right|\left[\theta^n_{i+1/2}\left(u^n_{i+1} - u^n_i\right)\right.$$
$$\left. - \delta\max\left(0, 1 - \frac{1}{\delta}\theta^n_{i+1/2}\right)\left(u^n_{i+2} - 3u^n_{i+1} + 3u^n_i - u^n_{i-1}\right)\right]. \quad (22.14)$$

In its original formulation, Jameson's self-adjusting hybrid method uses a shock switch first suggested by MacCormack and Baldwin (1975). The MacCormack–Baldwin shock switch is

$$\theta^n_{i+1/2} = \max\left(\theta^n_i, \theta^n_{i+1}\right), \tag{22.15a}$$

where

♦
$$\theta^n_i = \kappa \frac{\left|u^n_{i+1} - 2u^n_i + u^n_{i-1}\right|}{\left|u^n_{i+1}\right| + 2\left|u^n_i\right| + \left|u^n_{i-1}\right|} \tag{22.15b}$$

or

$$\theta^n_i = \begin{cases} \kappa \frac{\left|u^n_{i+1} - 2u^n_i + u^n_{i-1}\right|}{\left|u^n_{i+1}\right| + 2\left|u^n_i\right| + \left|u^n_{i-1}\right|} & \left|u^n_{i+1}\right| + 2\left|u^n_i\right| + \left|u^n_{i-1}\right| > \delta', \\ 0 & \left|u^n_{i+1}\right| + 2\left|u^n_i\right| + \left|u^n_{i-1}\right| \le \delta', \end{cases}$$

where the second expression prevents zero denominators. As usual, κ and δ' are user-adjustable parameters. By the triangle inequality,

$$\left|u^n_{i+1} - 2u^n_i + u^n_{i-1}\right| \le \left|u^n_{i+1}\right| + 2\left|u^n_i\right| + \left|u^n_{i-1}\right|.$$

Thus the MacCormack–Baldwin shock switch satisfies the convexity condition $0 \le \theta^n_{i+1/2} \le 1$ for all $0 \le \kappa \le 1$. Notice that

$$\frac{\left|u^n_{i+1} - 2u^n_i + u^n_{i-1}\right|}{\left|u_{i+1}\right| + 2\left|u^n_i\right| + \left|u^n_{i-1}\right|} = \begin{cases} O(1) & \text{for jumps,} \\ O(\Delta x^2) & \text{for smooth regions.} \end{cases}$$

Suppose that $\kappa = 1$. Then $\theta^n_{i+1/2} \approx 1$ at shocks and $\theta^n_{i+1/2} \approx 0$ in smooth regions, as desired. Like the analogous parameters in Harten's shock switch, seen in the last section, the parameters κ and δ in the MacCormack–Baldwin shock switch are not solution sensitive: They can be set once and forgotten or, in other words, need not be reoptimized for every solution.

When people first see Jameson's self-adjusting hybrid method and the MacCormack–Baldwin shock switch, they tend to worry too much about the user-adjustable parameters such as κ and δ. Certainly, it is asking a lot of a user to choose such parameters without simple guidelines. Although Jameson's flux-limited method works reasonably well for a range of different parameters, the existence of user-adjustable parameters makes even experienced users uncomfortable. Flux-limited methods are sometimes touted as a superior alternative in this regard, since they traditionally less often involve user-adjustable parameters. However, remember that some flux limiters *do* involve user-adjustable parameters, including the minmod limiter seen in Equations (20.21) and (20.22) and the symmetric minmod limiter seen in Equation (20.26). Furthermore, in traditional flux-limited methods, the user is often asked to choose between different flux limiters – in this sense, the flux limiter function is itself a "user-adjustable parameter."

Intuitively, flux limiters and shock switches, and any user-adjustable parameters in flux limiters and shock switches, specify an average. Unfortunately, there is never one "right" average, not even in simple situations, and certainly not here. A proper method should be relatively insensitive to the choice of average, so that the code designer can choose an averaging function that works well for a wide variety of different solutions, which the user can then leave well enough alone.

Swanson and Turkel (1992) suggested another shock switch for use in Jameson's method. Variations of the Swanson–Turkel shock switch were subsequently suggested by Jorgenson and Turkel (1993) and Tatsumi, Martinelli, and Jameson (1995). In particular, one of the more successful variants of the Swanson–Turkel shock switch is as follows:

$$\theta^n_{i+1/2} = \max\left(\theta^n_i, \theta^n_{i+1}\right), \tag{22.16a}$$

$$\theta^n_i = \left|\frac{u_{i+1} - 2u^n_i + u^n_{i-1}}{(1-\omega)\left(\left|u^n_{i+1} - u^n_i\right| + \left|u^n_i - u^n_{i-1}\right|\right) + \omega\left(\left|u^n_{i+1}\right| + 2\left|u^n_i\right| + \left|u^n_{i-1}\right|\right) + \delta}\right|, \tag{22.16b}$$

where ω and δ are user-adjustable parameters and where, as its only purpose, the small parameter δ prevents zero denominators. Notice that the numerator of the Swanson–Turkel shock switch is the same as the numerator of the MacCormack–Baldwin shock switch (22.15b) and is also the same as the numerator of shock switch (22.2). As far as the denominator goes, for $\delta = 0$ and $0 \le \omega \le 1$, the denominator of the Swanson–Turkel shock switch equals a convex linear combination of the denominator of the MacCormack–Baldwin shock switch (22.15b) and the denominator of shock switch (22.2). By the triangle inequality and assuming that $0 \le \omega \le 1$ and $\delta \ge 0$,

$$\begin{aligned}
&\left|u^n_{i+1} - 2u^n_i + u^n_{i-1}\right| \\
&\quad = \left|(1-\omega)\left(u^n_{i+1} - 2u^n_i + u^n_{i-1}\right) + \omega\left(u^n_{i+1} - 2u^n_i + u^n_{i-1}\right)\right| \\
&\quad \le (1-\omega)\left|u^n_{i+1} - 2u^n_i + u^n_{i-1}\right| + \omega\left|u^n_{i+1} - 2u^n_i + u^n_{i-1}\right| \\
&\quad = (1-\omega)\left|\left(u^n_{i+1} - u^n_i\right) - \left(u^n_i - u^n_{i-1}\right)\right| + \omega\left|u^n_{i+1} - 2u^n_i + u^n_{i-1}\right| \\
&\quad \le (1-\omega)\left(\left|u^n_{i+1} - u^n_i\right| + \left|u^n_i - u^n_{i-1}\right|\right) + \omega\left(\left|u^n_{i+1}\right| + 2\left|u^n_i\right| + \left|u^n_{i-1}\right|\right) \\
&\quad \le (1-\omega)\left(\left|u^n_{i+1} - u^n_i\right| + \left|u^n_i - u^n_{i-1}\right|\right) + \omega\left(\left|u^n_{i+1}\right| + 2\left|u^n_i\right| + \left|u^n_{i-1}\right|\right) + \delta.
\end{aligned}$$

Thus the Swanson–Turkel shock switch satisfies the convexity condition $0 \le \theta^n_{i+1/2} \le 1$ for all $0 \le \omega \le 1$ and $\delta \ge 0$. It can be shown that

$$\theta_i = \begin{cases} O(1) & \text{for jumps,} \\ O(\Delta x^2) & \text{for smooth regions,} \end{cases}$$

provided that ω is not too close to zero. In the original paper, Swanson and Turkel (1992) suggested $\omega = 1/2$. Later, Jorgenson and Turkel (1993) suggested the following solution-sensitive choice for ω:

$$\omega_i = \left[\frac{\min\left(\left|u^n_{i-2}\right|, \left|u^n_{i-1}\right|, \left|u^n_i\right|, \left|u^n_{i+1}\right|, \left|u^n_{i+2}\right|\right)}{\max\left(\left|u^n_{i-2}\right|, \left|u^n_{i-1}\right|, \left|u^n_i\right|, \left|u^n_{i+1}\right|, \left|u^n_{i+2}\right|\right)}\right]^m, \tag{22.17}$$

where m is a user-adjustable parameter. This causes the Swanson–Turkel shock switch to vary adaptively between the MacCormack–Baldwin shock switch and the shock switch seen in Equation (22.2), depending on the solution. According to Swanson and Turkel (1992), the Swanson–Turkel switch tends to be closer to one near shocks, especially for small ω, as compared with the MacCormack–Baldwin shock switch.

For time discretization, Jameson, Schmidt, and Turkel (1981) suggested Runge–Kutta methods, as discussed in Section 10.3. For example, one possibility is the improved Euler method seen in Equation (10.35b):

$$u_i^{(1)} = u_i^n + R\left(u_{i-K_1}^n, \ldots, u_{i+K_2}^n\right),$$
$$u_i^{n+1} = u_i^n + \frac{1}{2}R\left(u_{i-K_1}^n, \cdots, u_{i+K_2}^n\right) + \frac{1}{2}R\left(u_{i-K_1}^{(1)}, \ldots, u_{i+K_2}^{(1)}\right),$$

where

$$R\left(u_{i-K_1}^n, \ldots, u_{i+K_2}^n\right) = -\lambda_i\left[\hat{f}_s\left(u_{i-K_1+1}^n, \ldots, u_{i+K_2}^n\right) - \hat{f}_s\left(u_{i-K_1}^n, \ldots, u_{i+K_2-1}^n\right)\right].$$

Numerical results using this time-stepping scheme indicate that Jameson's method with $\kappa = 1/2$ and $\delta = 1/4$ does not blow up for $\lambda|a(u)| \leq 1.1$. Similarly, numerical results indicate that Jameson's method with $\kappa = 1/2$ and $\delta = 1/4$ does not blow up for $\lambda|a(u)| \leq 1.3$; however, although reducing κ increases the upper bound on the CFL number, the method is more prone to spurious overshoots and oscillations.

As another possible time-stepping scheme, Jameson, Schmidt, and Turkel (1981) suggest the well-known four-stage Runge–Kutta method seen in Example 10.4. Numerical results indicate that Jameson's method with $\kappa = 1$ and $\delta = 1/4$ does not blow up for $\lambda|a(u)| \leq 1.4$. Similarly, numerical results indicate that Jameson's method with $\kappa = 1/2$ and $\delta = 1/4$ does not blow up for $\lambda|a(u)| \leq 2.3$; however, as before, although reducing κ increases the upper bound on the CFL number, the method is more prone to spurious overshoots and oscillations. Notice that the stability bound on the CFL number increases with the number of stages in the Runge–Kutta method, unlike the Shu–Osher method seen in Section 21.4, whose upper bound instead decreases. From one point of view, Jameson's method uses wider stencils to increase order of accuracy, whereas the Shu–Osher method uses wider stencils to increase stability.

Jameson, Schmidt, and Turkel (1981) judge stability using linear stability analysis, as described in Chapter 15. Suppose that $\kappa = \delta = 0$; then the spatial discretization is straight central differences, without any inherently nonlinear self-adjusting terms. Then linear stability analysis applies. For example, with the four-stage Runge–Kutta method seen in Example 10.4, von Neumann analysis yields $\lambda|a(u)| \leq 2\sqrt{2} \approx 2.82$, which is confirmed by numerical results. Unfortunately, linear stability analysis does not apply when $\kappa \neq 0$ or $\delta \neq 0$, given the inherent nonlinearity of self-adjusting hybrid methods, unless $\theta_{i+1/2}^n$ is frozen. For more on linear stability analysis and Jameson's self-adjusting hybrid method, see Problems 15.6 and 15.7.

A popular alternative to explicit Runge–Kutta time discretization is linearized implicit time discretization, much like that seen in Section 20.5. Start with the following implicit time discretization:

$$u_i^{n+1} = u_i^n - \beta\left(\hat{f}_{s,i+1/2}^{n+1} - \hat{f}_{s,i-1/2}^{n+1}\right) - (1-\beta)\left(\hat{f}_{s,i+1/2}^n - \hat{f}_{s,i-1/2}^n\right),$$

where $\hat{f}_{s,i+1/2}$ is just as in Equation (22.14). This expression uses β for the implicit parameter in the convex linear combination, instead of the more common notation θ, so as not to cause confusion with the shock switch. This expression requires the solution of a *nonlinear* implicit system of equations, which is too expensive for practical computations. To obtain a *linear* implicit system of equations, write the equation in terms of $u_i^{n+1} - u_i^n$ and freeze the coefficients of $u_i^{n+1} - u_i^n$ at time level n, which represents a linear Taylor series approximation. The result is

$$\begin{aligned}
u_i^{n+1} - u_i^n &+ \frac{\beta}{2}\left[a\left(u_{i+1}^n\right)\left(u_{i+1}^{n+1} - u_{i+1}^n\right) - a\left(u_{i-1}^n\right)\left(u_{i-1}^{n+1} - u_{i-1}^n\right)\right] \\
&- \frac{\beta}{2}\left|a_{i+1/2}^n\right|\theta_{i+1/2}^n\left[\left(u_{i+1}^{n+1} - u_{i+1}^n\right) - \left(u_i^{n+1} - u_i^n\right)\right] \\
&+ \frac{\beta}{2}\left|a_{i-1/2}^n\right|\theta_{i-1/2}^n\left[\left(u_i^{n+1} - u_i^n\right) - \left(u_{i-1}^{n+1} - u_{i-1}^n\right)\right] \\
&+ \frac{\beta}{2}\delta\left|a_{i+1/2}^n\right| \max\left(0, 1 - \frac{1}{\delta}\theta_{i+1/2}^n\right)\left[\left(u_{i+2}^{n+1} - u_{i+2}^n\right)\right. \\
&\left. - 3\left(u_{i+1}^{n+1} - u_{i+1}^n\right) + 3\left(u_i^{n+1} - u_i^n\right) - \left(u_{i-1}^{n+1} - u_{i-1}^n\right)\right] \\
&- \frac{\beta}{2}\delta\left|a_{i-1/2}^n\right| \max\left(0, 1 - \frac{1}{\delta}\theta_{i-1/2}^n\right)\left[\left(u_{i+1}^{n+1} - u_{i+1}^n\right)\right. \\
&\left. - 3\left(u_i^{n+1} - u_i^n\right) + 3\left(u_{i-1}^{n+1} - u_{i-1}^n\right) - \left(u_{i-2}^{n+1} - u_{i-2}^n\right)\right] \\
&= -\lambda\left(\hat{f}_{s,i+1/2}^n - \hat{f}_{s,i-1/2}^n\right).
\end{aligned}$$

Notice that this is a linear pentadiagonal system of equations, which is relatively cheap to solve, although not quite as cheap as the linear tridiagonal system of equations found in Section 20.5. To further enhance efficiency, the coefficients of artificial viscosity on the left-hand side can be locally frozen as follows:

$$\begin{aligned}
u_i^{n+1} - u_i^n &+ \frac{\beta}{2}\left[a\left(u_{i+1}^n\right)\left(u_{i+1}^{n+1} - u_{i+1}^n\right) - a\left(u_{i-1}^n\right)\left(u_{i-1}^{n+1} - u_{i-1}^n\right)\right] \\
&- \frac{1}{2}\epsilon_i^{(2)}\left[\left(u_{i+1}^{n+1} - u_{i+1}^n\right) - 2\left(u_i^{n+1} - u_i^n\right) + \left(u_{i-1}^{n+1} - u_{i-1}^n\right)\right] \\
&+ \frac{1}{2}\epsilon_i^{(4)}\left[\left(u_{i+2}^{n+1} - u_{i+2}^n\right) - 4\left(u_{i+1}^{n+1} - u_{i+1}^n\right)\right. \\
&\left. + 6\left(u_i^{n+1} - u_i^n\right) - 4\left(u_{i-1}^{n+1} - u_{i-1}^n\right) + \left(u_{i-2}^{n+1} - u_{i-2}^n\right)\right] \\
&= -\lambda\left(\hat{f}_{s,i+1/2}^n - \hat{f}_{s,i-1/2}^n\right).
\end{aligned}$$

For example, the implicit coefficients of artificial viscosity might be

$$\begin{aligned}
\epsilon_i^{(2)} &= \beta\theta_i^n\left|a\left(u_i^n\right)\right|, \\
\epsilon_i^{(4)} &= \beta\delta \max\left(0, 1 - \frac{1}{\delta}\theta_i^n\right)\left|a\left(u_i^n\right)\right|.
\end{aligned}$$

To enhance efficiency even further, one or both of the implicit coefficients of artificial viscosity may be set equal to zero. For example, if $\epsilon_i^{(2)} = \epsilon_i^{(4)} = 0$, then the linear system of equations is diagonal. For another example, if $\epsilon_i^{(4)} = 0$ and $\epsilon_i^{(2)} \neq 0$, then the linear system of equations is tridiagonal. Playing such games with the implicit part of the approximation reduces the cost per time step and cannot harm the accuracy of steady-state solutions; unfortunately, it may reduce the upper bound on the CFL number and time step. For more details and further variations on implicit time discretization see, for example, Jameson and Yoon (1986) and Caughey (1988).

The behavior of Jameson's self-adjusting hybrid method is illustrated using the five standard test cases defined in Section 17.0. All five test cases use improved Euler time stepping;

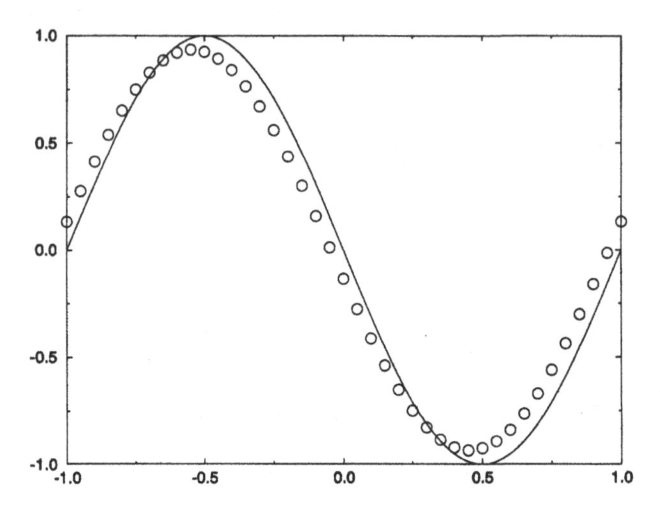

Figure 22.9 Jameson's self-adjusting hybrid method for Test Case 1.

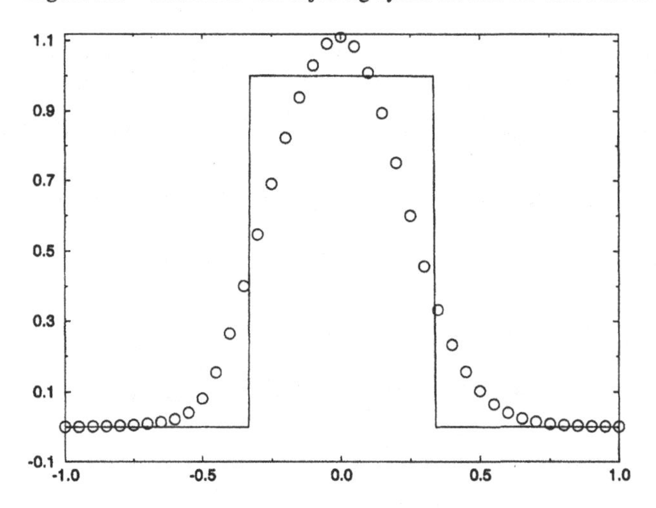

Figure 22.10 Jameson's self-adjusting hybrid method for Test Case 2.

other Runge–Kutta methods seem to yield similar or slightly worse results. Also, all test cases use the MacCormack–Baldwin shock switch with $\kappa = 1$, $\delta = 0.25$, and $\delta' = 10^{-5}$.

Test Case 1 As seen in Figure 22.9, Jameson's self-adjusting hybrid method captures the sinusoid without clipping, with only a modest lagging phase error and a modest uniform dissipation error. The only method seen so far in Part V that can equal this performance is the third-order Shu–Osher method found in Section 21.4.

Test Case 2 As seen in Figure 22.10, Jameson's self-adjusting hybrid method captures the square wave without spurious oscillations, but with substantial smearing, and with a 10% overshoot at the center. This performance could be improved by tweaking the user-adjustable parameters in the MacCormack–Baldwin shock switch or by using another shock switch such as Harten's shock switch, but only at the expense of the excellent performance seen in Test Case 1.

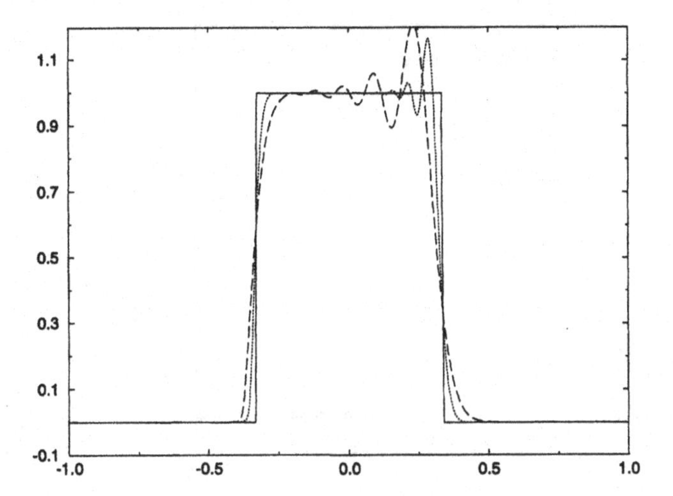

Figure 22.11 Jameson's self-adjusting hybrid method for Test Case 3.

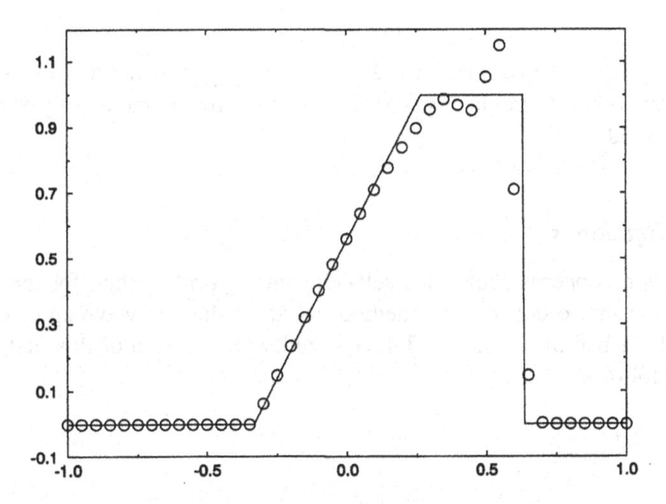

Figure 22.12 Jameson's self-adjusting hybrid method for Test Case 4.

Test Case 3 In Figure 22.11, the dotted line represents Jameson's self-adjusting hybrid approximation to $u(x, 4)$, the long dashed line represents Jameson's self-adjusting hybrid approximation to $u(x, 40)$, and the solid line represents the exact solution for $u(x, 4)$ or $u(x, 40)$. Whereas Harten's self-adjusting hybrid seen in the last section allows minor oscillations in Test Case 3, Jameson's self-adjusting hybrid allows major oscillations. This performance could be improved by tweaking the user-adjustable parameters in the MacCormack–Baldwin shock switch or by using another shock switch such as Harten's shock switch, but only at the expense of the excellent performance seen in Test Case 1.

Test Case 4 As seen in Figure 22.12, Jameson's self-adjusting hybrid works well except for a 15% overshoot behind the shock.

Test Case 5 As seen in Figure 22.13, Jameson's method fails to alter the initial conditions in any way, which leads to a large expansion shock. This problem is easily

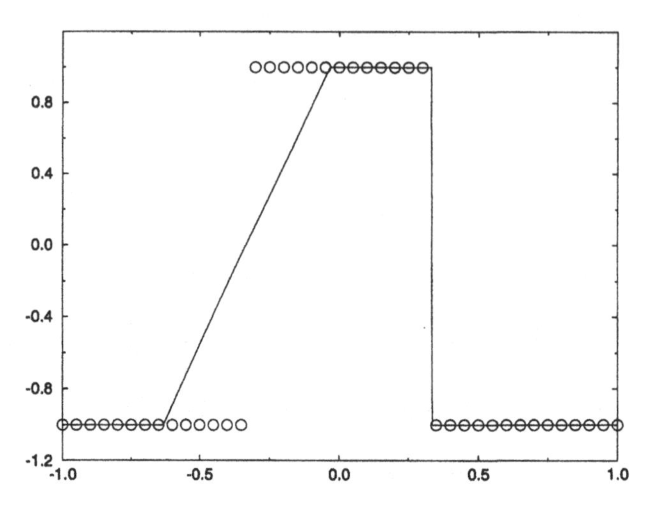

Figure 22.13 Jameson's self-adjusting hybrid method for Test Case 5.

overcome by replacing $|a^n_{i+1/2}|$ in Equation (22.14) by $\psi(a^n_{i+1/2})$ as defined in Subsection 18.3.3 or, in other words, by replacing Roe's first-order upwind method by Harten's first-order upwind method.

22.3.2 *The Euler Equations*

This subsection concerns Jameson's self-adjusting hybrid method for the Euler equations. Consider the first-order upwind method based on the one-wave approximate Riemann solver, as described in Section 18.3.4. A semidiscrete version of this first-order upwind method is as follows:

$$\frac{\partial \mathbf{u}^n_i}{\partial t} = -\frac{\hat{\mathbf{f}}^{(1)}_{s,i+1/2} - \hat{\mathbf{f}}^{(1)}_{s,i-1/2}}{\Delta x},$$

$$\hat{\mathbf{f}}^{(1)}_{s,i+1/2} = \frac{1}{2}\left(\mathbf{f}(\mathbf{u}^n_{i+1}) + \mathbf{f}(\mathbf{u}^n_i)\right) - \frac{1}{2}\rho\left(A^n_{i+1/2}\right)\left(\mathbf{u}^n_{i+1} - \mathbf{u}^n_i\right).$$

Also let

$$\frac{\partial \mathbf{u}^n_i}{\partial t} = -\frac{\hat{\mathbf{f}}^{(2)}_{s,i+1/2} - \hat{\mathbf{f}}^{(2)}_{s,i-1/2}}{\Delta x},$$

$$\hat{\mathbf{f}}^{(2)}_{s,i+1/2} = \frac{1}{2}\left(\mathbf{f}(\mathbf{u}^n_{i+1}) + \mathbf{f}(\mathbf{u}^n_i)\right) + \frac{1}{2}\delta\rho\left(A^n_{i+1/2}\right)\left(\mathbf{u}^n_{i+2} - 3\mathbf{u}_{i+1} + 3\mathbf{u}^n_i - \mathbf{u}^n_{i-1}\right).$$

In both of these methods, $A^n_{i+1/2}$ is some average flux Jacobian matrix and $\rho(A^n_{i+1/2})$ is the largest characteristic value of $A^n_{i+1/2}$ in absolute value. Then Jameson's semidiscrete self-adjusting hybrid method for the Euler equations is

$$\hat{\mathbf{f}}^n_{s,i+1/2} = \theta^n_{i+1/2}\hat{\mathbf{f}}^{(1)}_{s,i+1/2} + \max\left(0, 1 - \frac{1}{\delta}\theta^n_{i+1/2}\right)\hat{\mathbf{f}}^{(2)}_{s,i+1/2}$$

or with a slight modification

$$\hat{\mathbf{f}}^n_{s,i+1/2} = \frac{1}{2}(\mathbf{f}(\mathbf{u}^n_{i+1}) + \mathbf{f}(\mathbf{u}^n_i)) - \frac{1}{2}\rho(A^n_{i+1/2})\Big[\theta^n_{i+1/2}(\mathbf{u}^n_{i+1} - \mathbf{u}^n_i)$$
$$- \delta \max\left(0, 1 - \frac{1}{\delta}\theta^n_{i+1/2}\right)(\mathbf{u}^n_{i+2} - 3\mathbf{u}^n_{i+1} + 3\mathbf{u}^n_i - \mathbf{u}^n_{i-1})\Big], \tag{22.18}$$

where

$$\theta^n_{i+1/2} = \max(\theta^n_i, \theta^n_{i+1}), \tag{22.19a}$$

$$\theta^n_i = \kappa \frac{|p^n_{i+1} - 2p^n_i + p^n_{i-1}|}{p^n_{i+1} + 2p^n_i + p^n_{i-1}}. \tag{22.19b}$$

Of course, the MacCormack–Baldwin shock switch can be replaced by the Swanson–Turkel shock switch or some other shock switch. Notice that there is no need to worry about zero or negative values in the denominators of the shock switch, since pressure is always strictly greater than zero. Numerical tests have shown that pressure makes for an excellent shock switch. Unfortunately, pressure is constant across contacts, so that any shock switch based on pressure will not act at contacts. Of course, pressure can be replaced by entropy, density, temperature, or some other flow property.

Although Jameson's self-adjusting hybrid method can be used for unsteady flows, it was originally designed primarily for steady-state flows. Unlike many methods, such as the Lax–Wendroff methods seen in Subsection 18.1.2, or any of the second- and third-generation methods based on the Lax–Wendroff methods, the steady-state solutions of Jameson's method do not depend on Δt. Some common features designed to accelerate the convergence to steady state are described below. In each case, these features trade time accuracy for larger stable CFL numbers and, in various other ways, trade time accuracy for faster convergence to steady state.

Local Time Stepping

Each grid cell is run at the same CFL number. This implies that the time step is different in different cells – the larger the cell the larger the time step.

Enthalpy Damping

Instead of solving the true Euler equations, solve the following modified system of equations:

$$\frac{\partial}{\partial t}\begin{bmatrix} \rho \\ \rho u \\ \rho e_T \end{bmatrix} + \frac{\partial}{\partial x}\begin{bmatrix} \rho u \\ \rho u^2 + p \\ \rho u h_T \end{bmatrix} + \alpha(h_T - h_{T,\infty})\begin{bmatrix} \rho \\ \rho u \\ h_T \end{bmatrix} = 0, \tag{22.20}$$

where α is a user-adjustable parameter, h_T is the total enthalpy, and $h_{T,\infty}$ is the total enthalpy in the uniform free stream, far away from any obstacles in the flow. In Equation (22.20), the first two terms comprise the one-dimensional Euler equations; the third term is the *enthalpy damping* term. If the free-stream flow is uniform, then the total enthalpy in the steady-state solution will also be uniform. In other words, $h_T = h_{T,\infty}$ for steady-state flow. Thus enthalpy damping terms do not affect steady-state solutions. Enthalpy damping tends to accelerate convergence to steady state, as explained by Jameson, Schmidt, and Turkel

(1981) and Jameson (1982) by analogy with the potential flow equations. Increasing ordinary second- and fourth-order artificial viscosity, by increasing κ and δ, increases the rate of convergence to steady state but also increases smearing in the steady-state solution. As one way to think about it, enthalpy damping acts much like artificial viscosity – it certainly increases the rate of convergence to steady state – but without affecting the steady-state solution.

Optimize κ and δ

The parameters κ and δ in the MacCormack–Baldwin shock switch influence not only the rate of convergence, by affecting the largest permissible CFL number and time step, but also the accuracy of the final converged solution. Taking both of these factors into account, the traditional choice is $\kappa = 1/2$ and $\delta = 1/128$. Both of these parameters should be substantially larger for unsteady flows. Remember that reducing κ and δ tends to increase the largest permissible CFL number, as judged by whether or not the method blows up, but also tends to increase the amount of spurious overshoot and oscillation. However, the bulk of the spurious overshoots and oscillations is unsteady rather than steady and thus does not appear in the steady-state solution; thus steady-state solutions usually benefit from lower values of κ and δ and the resulting larger time steps.

It is awkward to use one setting for the user-adjustable parameters in steady flows and a different setting for unsteady flows. Unfortunately, a similar situation exists in most flux-averaged methods. For example, in flux-limited methods, some flux limiters work much better for steady-state solutions than for unsteady solutions or, at least, one choice of parameters in the flux limiter works better for steady-state solutions than for unsteady solutions. Of course, it is not absolutely necessary to tweak things differently for steady and unsteady solutions, but it usually makes the final method more accurate and efficient.

Implicit Residual Smoothing

Ordinarily, when the time step grows too large, it violates stability constraints and/or the CFL condition, resulting in extremely large spurious oscillations. However, suppose that a filtering procedure removes or reduces the spurious oscillations after every time step. For example, consider the following *implicit residual smoothing* procedure. In the first step, compute the residual R_i^n:

$$R_i^n = \frac{\hat{f}^n_{R,i+1/2} - \hat{f}^n_{R,i-1/2}}{\Delta x}.$$

This could be considered a predictor. In the second step, solve for the smoothed residual R_i' in the following implicit equation:

$$-\epsilon R'_{i+1} + (1+2\epsilon)R'_i - \epsilon R'_{i-1} = R_i^n, \tag{22.21}$$

where ϵ is a user-adjustable parameter. This could be considered a corrector. Equation (22.21) amounts to a linear tridiagonal system of equations, which is relatively inexpensive to solve, as seen in Example 11.4. In the third and final step, let the new and improved semidiscrete approximation be

$$\frac{du}{dt}(x_i, t^n) = R'_i.$$

As it turns out, implicit residual smoothing can stabilize *any* CFL number, provided that ϵ is large enough. In particular, ϵ should increase proportional to the square of the CFL number. Implicit residual smoothing introduces a degree of implicitness into an otherwise explicit method, and reaps the reward of the large CFL numbers typically associated with implicit methods, but without costing as much as most implicit methods. See Jameson and Baker (1983) or Jameson (1983) for a further description of implicit residual smoothing.

Grid Sequencing
Define a sequence of grids from coarse to fine. The steady-state solution is found on the coarsest grid. The results from the coarsest grid are interpolated to the next finer grid and the solution is again converged to steady state. The results from the finer grid are interpolated to the next finer grid and converged to steady state. This process continues until the steady-state solution is found on the final and finest grid. The idea here is that large time steps may be used on the coarse grid. Thus the solution will converge quickly on the coarse grid, capturing the long wavelength components of the solution. Each finer grid will have smaller times steps and will capture finer details of the flow. Hopefully, initial conditions interpolated from a converged solution on a coarser grid will converge quicker than arbitrary initial conditions. Grid sequencing is a natural companion to multigrid, as described next.

Multigrid
Define a sequence of grids from coarse to fine. A few time steps are taken on the coarsest grid and the results are interpolated to a finer grid. A few more time steps are taken on the finer grid, and the results are interpolated to a still finer grid. This process repeats, so that the grids continually get finer and finer. So far, multigrid is like grid sequencing, except that no attempt is made to reach steady state on any of the grids. However, in multigrid, at some stage the process reverses itself, so that the fine grid results are converted back to a coarser grid and a few time steps are taken on the coarser grid, after which the results are either interpolated back to the fine grid or are converted to an even coarser grid. The sequence of grids is typically specified by diagrams such as those seen in Figure 22.14. In Figure 22.14, the upper dots represent the coarser grids, the lower dots represent the finer grids, N is the number of grid points, and n is the number of time steps. The letters V or the W represent a single multigrid cycle, which repeats itself indefinitely, until steady steady is achieved. In this explanation, the multigrid initially goes from coarser to finer for consistency with the earlier grid sequencing explanation; however, in the standard treatment, multigrid always initially goes from finer to coarser.

Remember that steady state occurs when all unsteady waves exit the far-field boundaries, or are eliminated by dissipation, as mentioned in Chapter 19. Then, as the primary motivation behind multigrid, the rate of convergence to steady state is limited by the slowest waves,

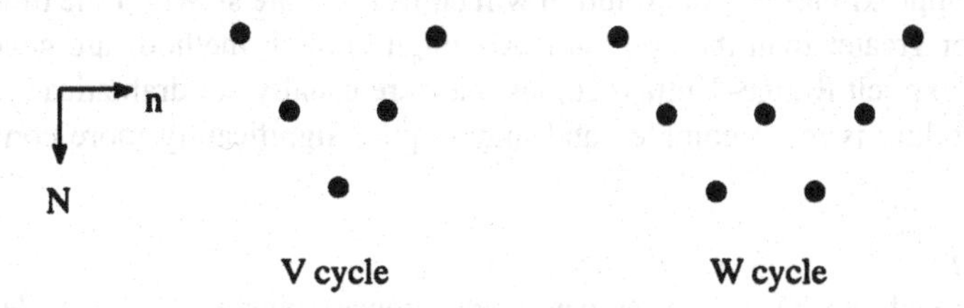

Figure 22.14 Two possible multigrid cycles.

while the CFL number and time step are limited by the fastest waves. The fastest waves may travel an entire grid cell during a single time step and quickly exit the computational domain through the far-field boundaries. However, the slowest waves move only a fraction of a cell during one time step and may take a very long time to exit the domain through the far field. On coarser grids, the time step can be larger, and slowly moving long-wavelength unsteadiness will reach the far field in fewer time steps. For one of the first descriptions of multigrid in Jameson's self-adjusting hybrid method see, for example, Jameson (1983). For later descriptions see, for example, Jameson (1989) and the references cited therein. For more on the multigrid method in general, see Briggs (1987) and McCormick (1987).

Multistage Time Stepping

Runge–Kutta methods often allow many more degrees of freedom than can be usefully exploited, especially when the temporal order of accuracy constraints are dropped, as in steady-state flows, where time accuracy is irrelevant. But suppose that the ith stage only depends on the $(i-1)$th stage, and not on any earlier stages. This is called a *multistage* scheme. An m-stage multistage scheme is defined as follows:

$$
\begin{aligned}
u_i^{(0)} &= u_i^n, \\
u_i^{(1)} &= u_i^{(0)} - \alpha_1 \Delta t R_i^{(0)}, \\
u_i^{(2)} &= u_i^{(0)} - \alpha_2 \Delta t R^{(1)}, \\
&\ \vdots \\
u_i^{(m-1)} &= u_i^{(0)} - \alpha_{m-1} \Delta t R_i^{(m-2)}, \\
u_i^{(m)} &= u_i^{(0)} - \Delta t R_i^{(m-1)}, \\
u_i^{n+1} &= u_i^{(m)},
\end{aligned}
$$

where the α_i are the *multistage coefficients*. Multistage methods are a subset of Runge–Kutta methods, as seen in Section 10.3, where the Butcher array is diagonal rather than triangular. As their main advantage, multistage methods require less storage than general Runge–Kutta methods. Specifically, most m-stage Runge–Kutta methods must store m residuals at a time, whereas an m-stage multistage method need only store one residual at a time.

Implicit Time Stepping

Explicit Runge–Kutta and multistage time-stepping schemes limit the CFL numbers. In contrast, the implicit time-stepping methods discussed above permit very large CFL numbers and thus very large time steps. Although, in theory, the steady-state convergence rate should always increase as the time step increases, there is always a finite optimum time step in practice, depending on how many liberties are taken in linearizing and simplifying the implicit part of the approximation – the solution will converge more slowly if the time step is either less than or greater than the optimum. Although implicit methods are generally more efficient than explicit Runge–Kutta methods, they are usually not dramatically more efficient, and the coding is more complex and may require significantly more computer memory.

Frozen Dissipation

Recall that the residual can be written as central differences plus an adaptive blend of second- and fourth-order artificial viscosity, much as in Equation (22.18). Evaluating the artificial viscosity is expensive, certainly more expensive than evaluating central differences. Thus, for added savings, it is common to freeze the artificial viscosity terms after some specified stage in a Runge–Kutta or multistage time-stepping method; only the central difference portion is updated in subsequent stages. See Jameson, Schmidt, and Turkel (1981) or Jameson (1982) for details.

When all of the above features are combined, Jameson's method is very efficient indeed, and highly robust, provided that the solution does not contain overly strong shocks.

Some of the above techniques, originally developed for steady-state calculations, extend to unsteady calculations. For issues specifically related to unsteady flows, see, for example, Venkatakrishnan and Jameson (1988) and Jameson (1991). Since its inception, there have been innumerable papers written on Jameson's self-adjusting hybrid method, probably more than on any other method seen in Part V; see the references cited below, such as Holmes and Tong (1985) or the many papers by Jameson, for selected examples.

References

Briggs, W. L. 1987. *A Multigrid Tutorial*, Philadelphia: SIAM.

Caughey, D. A. 1988. "Diagonal Implicit Multigrid Algorithm for the Euler Equations," *AIAA Journal*, 26: 841–851.

Harten, A. 1978. "The Artificial Compression Method for Computation of Shocks and Contact Discontinuities: III. Self-Adjusting Hybrid Schemes," *Mathematics of Computation*, 32: 363–389.

Harten, A. 1983. "High Resolution Schemes for Hyperbolic Conservation Laws," *Journal of Computational Physics*, 49: 357–393.

Harten, A., and Zwas, G. 1972a. "Self-Adjusting Hybrid Schemes for Shock Computations," *Journal of Computational Physics*, 9: 568–583.

Harten, A., and Zwas, G. 1972b. "Switched Numerical Shuman Filters for Shock Calculations," *Journal of Engineering Mathematics*, 6: 207–216.

Holmes, D. G., and Tong, S. S. 1985. "A Three-Dimensional Euler Solver for Turbomachinery Blade Rows," *Transactions of the ASME*, 107: 258–264.

Jameson, A. 1982. "Steady-State Solution of the Euler Equations for Transonic Flow." In *Transonic, Shock, and Multidimensional Flows: Advances in Scientific Computing*, ed. R. E. Meyer, New York: Academic Press.

Jameson, A. 1983. "The Evolution of Computational Methods in Aerodynamics," *Transactions of the ASME*, 50: 1052–1070.

Jameson, A. 1989. "Computational Aerodynamics for Aircraft Design," *Science*, 245: 361–371.

Jameson, A. 1991. "Time Dependent Calculations Using Multigrid, with Applications to Unsteady Flows Past Airfoils and Wings," *AIAA Paper 91-1596* (unpublished).

Jameson, A. 1995. "Positive Schemes for Shock Modelling for Compressible Flow," *International Journal for Numerical Methods in Fluids*, 20: 743–776.

Jameson, A., and Baker, T. J. 1983. "Solution of the Euler Equations for Complex Configurations," *AIAA Paper 83-1929* (unpublished).

Jameson, A., Schmidt, W., and Turkel, E. 1981. "Numerical Solutions of the Euler Equations by Finite Volume Methods Using Runge–Kutta Time-Stepping Schemes," *AIAA Paper 81-1259* (unpublished).

Jameson, A., and Yoon, S. 1986. "Multigrid Solution of the Euler Equations Using Implicit Schemes," *AIAA Journal*, 24: 1737–1743.

Jorgenson, P., and Turkel, E. 1993. "Central Difference TVD Schemes for Time Dependent and Steady State Problems," *Journal of Computational Physics*, 107: 297–308.

MacCormack, R. W., and Baldwin, B. S. 1975. "A Numerical Method for Solving the Navier–Stokes Equations with Application to Shock Boundary Layer Interactions," *AIAA Paper 75-0001* (unpublished).

McCormick, S. F., ed. 1987. *Multigrid Methods*, Philadelphia: SIAM.

Swanson, R. C., and Turkel, E. 1992. "On Central-Difference and Upwind Schemes," *Journal of Computational Physics*, 101: 292–306.

Tatsumi, S., Martinelli, L., and Jameson, A. 1995. "Flux-Limited Schemes for the Compressible Navier–Stokes Equations," *AIAA Journal*, 33: 252–261.

Venkatakrishnan, V., and Jameson, A. 1988. "Computation of Unsteady Transonic Flows by Solution of the Euler Equations," *AIAA Journal*, 26: 974–981.

CHAPTER 23

Solution Averaging: Reconstruction–Evolution Methods

23.0 Introduction

This chapter concerns *reconstruction–evolution methods*, also called *solution-averaged* or *solution-reconstructed methods*. Before continuing, you may wish to review the earlier discussions of reconstruction–evolution methods found in Sections 13.6, 17.3, and 18.3. The reconstruction–evolution methods seen in those earlier sections were all first-order accurate. This chapter concerns second- and higher-order accurate methods. In some cases, solution-averaged methods differ from flux-averaged methods in relatively minor ways. For example, the solution-averaged method might use an average like $f((x+y)/2)$, whereas the flux-averaged method might use an average like $(f(x)+f(y))/2$. However, in general, solution-averaged methods use elaborate physical modeling of a sort not possible in flux-averaged methods, making them far more complex.

This chapter describes five reconstruction–evolution methods. Section 23.1 concerns Van Leer's reconstruction–evolution method, better known as *MUSCL*. Section 23.2, concerns the Colella–Woodward reconstruction–evolution method, better known as *PPM*. Section 23.3 describes two reconstruction–evolution methods devised by Anderson, Thomas, and Van Leer, also often called MUSCL methods, that strongly resemble the flux-limited methods devised by Chakravarthy and Osher seen in Section 20.3. Section 23.4 describes the Harten–Osher reconstruction–evolution method, better known as the *UNO* method. Finally, Section 23.5 describes the Harten–Engquist–Osher–Chakravarthy reconstruction–evolution methods, better known as *ENO* methods. Do not confuse the finite-volume ENO methods seen in Section 23.5 with the finite-difference ENO methods seen in Section 21.4. The two classes of methods have basically one thing in common: both use ENO reconstruction, as described in Chapter 9, one to reconstruct the solution and the other to reconstruct the flux.

23.1 Van Leer's Reconstruction–Evolution Method (MUSCL)

23.1.1 *Linear Advection Equation*

In 1977, Van Leer proposed a class of reconstruction–evolution methods for the linear advection equation. The simplest possible reconstruction–evolution method for the linear advection equation uses a piecewise-constant spatial reconstruction and the exact temporal evolution as seen in Example 13.14; this method is known as the *Courant–Isaacson–Rees (CIR) method* or the *first-order upwind method for the linear advection equation.* This section concerns the next simplest choice, known as *Van Leer's reconstruction–evolution method.* Van Leer's reconstruction–evolution method uses piecewise-linear spatial reconstruction as follows:

$$p_r(x) = \bar{u}_i^n + S_i^n(x - x_i) \tag{23.1}$$

for $x_{i-1/2} \leq x \leq x_{i+1/2}$. Also, Van Leer's reconstruction–evolution method uses the exact solution to time evolve the reconstruction as follows:

$$p_{e,i+1/2}(t) = p_r(x_{i+1/2} - a(t - t^n)), \tag{23.2}$$

which shifts the reconstruction to the left or right by an amount *at*. By Equation (23.1),

$$p_{e,i+1/2}(t) = \begin{cases} \bar{u}_i^n + S_i^n\left[\frac{\Delta x}{2} - a(t - t^n)\right] & 0 \leq \lambda a \leq 1, \\ \bar{u}_{i+1}^n - S_{i+1}^n\left[\frac{\Delta x}{2} + a(t - t^n)\right] & -1 \leq \lambda a < 0. \end{cases} \tag{23.3}$$

Suppose that $0 \leq \lambda a \leq 1$. Then

$$\hat{f}_{i+1/2}^n = \frac{1}{\Delta t}\int_{t^n}^{t^{n+1}} f(p_{e,i+1/2}(t))\,dt = \frac{1}{\Delta t}\int_{t^n}^{t^{n+1}} a p_{e,i+1/2}(t)\,dt.$$

By Equation (23.3),

$$\hat{f}_{i+1/2}^n = \frac{a}{\Delta t}\int_{t^n}^{t^{n+1}} \left\{\bar{u}_i^n + S_i^n\left[\frac{\Delta x}{2} - a(t - t^n)\right]\right\} dt.$$

Completing the integration yields

$$\hat{f}_{i+1/2}^n = a\bar{u}_i^n + \frac{1}{2}a(1 - \lambda a)S_i^n \Delta x.$$

Proceed similarly for $-1 \leq \lambda a < 0$ to find the final result:

♦
$$\hat{f}_{i+1/2}^n = \begin{cases} a\bar{u}_i^n + \frac{1}{2}a(1 - \lambda a)S_i^n \Delta x & 0 \leq \lambda a \leq 1, \\ a\bar{u}_{i+1}^n - \frac{1}{2}a(1 + \lambda a)S_{i+1}^n \Delta x & -1 \leq \lambda a < 0. \end{cases} \tag{23.4}$$

Then

$$\bar{u}_i^{n+1} = \bar{u}_i^n - \lambda\left(\hat{f}_{i+1/2}^n - \hat{f}_{i-1/2}^n\right)$$

implies

$$\bar{u}_i^{n+1} = \begin{cases} \bar{u}_i^n - \lambda a\left(\bar{u}_i^n - \bar{u}_{i-1}^n\right) - \frac{\lambda a}{2}(1 - \lambda a)\left(S_i^n - S_{i-1}^n\right)\Delta x & 0 \leq \lambda a \leq 1, \\ \bar{u}_i^n - \lambda a\left(\bar{u}_{i+1}^n - \bar{u}_i^n\right) + \frac{\lambda a}{2}(1 + \lambda a)\left(S_{i+1}^n - S_i^n\right)\Delta x & -1 \leq \lambda a < 0. \end{cases} \tag{23.5}$$

Equations (23.4) and (23.5) define an *infinite* class of methods differing only in their choice of the reconstruction slopes S_i^n. In fact, since the temporal evolution is perfect, the method's performance depends entirely on the spatial reconstruction via the choice of S_i^n. Here are some factors to consider when choosing S_i^n:

- Van Leer's reconstruction–evolution method is second-order accurate if S_i^n is a first-order or higher-order accurate approximation to $\frac{\partial u}{\partial x}(x_i, t^n)$. You can prove this by matching terms in a Taylor series approximation to $u(x, t^n)$. However, as seen in the next bullet, you should not take the relationship between slopes and derivatives too seriously.

- Van Leer's reconstruction–evolution method can be second-order accurate if S_i^n is *not* a first- or higher-order accurate approximation to $\frac{\partial u}{\partial x}(x_i, t^n)$. For example, S_i^n sometimes depends on the time step, as in the next bullet, which makes it a zeroth-order accurate approximation to any purely spatial quantities including $\frac{\partial u}{\partial x}(x_i, t^n)$. In fact, Van Leer's reconstruction–evolution method can be third- or higher-order accurate only if S_i^n is a zeroth-order accurate approximation to the first derivative.
- Unless $S_i^n = 0$, either the slope S_i^n or the conservative numerical flux $\hat{f}_{i+1/2}^n$, or both, must depend on the time step Δt through $\lambda = \Delta t/\Delta x$. If S_i^n depends on the time step, then the spatial reconstruction depends on the time step. A spatial reconstruction that depends on time, while counterintuitive, does not necessarily have adverse practical consequences. On the other hand, if $\hat{f}_{i+1/2}^n$ depends on the time step, then so do steady-state solutions. Of course, the linear advection equation as considered in this section does not have steady-state solutions. However, if $\hat{f}_{i+1/2}^n$ depends on the time step for the linear advection equation then, in most cases, $\hat{f}_{i+1/2}^n$ will depend on the time step for nonlinear scalar conservation laws, the Euler equations, and any number of other equations that *do* allow steady-state solutions. The conservative numerical flux $\hat{f}_{i+1/2}^n$ depends on the time step, and then steady-state approximations depend on the time step, unless S_i^n includes a factor of $1/(1 - \lambda|a|)$ to cancel the factor of $1 - \lambda|a|$ seen in Equation (23.4).
- The temporal evolution is naturally upwind. If the spatial reconstruction has a downwind bias through its choice of S_i^n, the upwind temporal evolution will reduce or even eliminate downwinding in the overall method. If the spatial reconstruction has an upwind bias through its choice of S_i^n, the upwind temporal evolution will increase the upwind bias in the overall method.

The following examples illustrate the above principles. The first group of examples chooses the reconstruction slope by approximating the solution's first derivative.

Example 23.1 The backward-space approximation given by Equation (10.7) is

$$S_i^n = \frac{\bar{u}_i^n - \bar{u}_{i-1}^n}{\Delta x} + O(\Delta x). \tag{23.6}$$

This first-order accurate reconstruction yields a second-order accurate reconstruction–evolution method. This spatial reconstruction is upwind if $a > 0$ and downwind if $a < 0$, whereas the resulting reconstruction–evolution method is upwind for $a > 0$ and centered for $a < 0$. The spatial reconstruction does not depend on the time step, but the conservative numerical flux does.

Example 23.2 The forward-space approximation given by Equation (10.8) is

$$S_i^n = \frac{\bar{u}_{i+1}^n - \bar{u}_i^n}{\Delta x} + O(\Delta x). \tag{23.7}$$

This first-order accurate reconstruction yields a second-order accurate reconstruction–evolution method. This spatial reconstruction is downwind if $a > 0$ and upwind if $a < 0$, whereas the resulting reconstruction–evolution method is centered for $a > 0$ and upwind for $a < 0$. The spatial reconstruction does not depend on the time step, but the conservative numerical flux does.

Example 23.3 The central-space approximation given by Equation (10.15) is

$$S_i^n = \frac{\bar{u}_{i+1}^n - \bar{u}_{i-1}^n}{2\Delta x} + O(\Delta x^2). \tag{23.8}$$

Van Leer (1977) calls this *Scheme I.* This spatial reconstruction is centered, whereas the resulting reconstruction–evolution method is upwind. Notice that this example uses a second-order accurate spatial reconstruction, whereas the preceding two examples used first-order accurate spatial reconstructions. Although the order of accuracy of the reconstruction has improved, the order of accuracy of the resulting reconstruction–evolution method has not – all three examples yield the same second-order accuracy. However, although the method's order of accuracy remains the same, its *accuracy improves in this example.* In fact, as seen later, the first two examples yield either the Lax–Wendroff method or the Beam–Warming second-order upwind method, depending on the sign of a, whereas this example yields Fromm's method. As seen in Chapter 17, Fromm's method is more accurate than the Lax–Wendroff method or the Beam–Warming method, at least for the linear advection equation.

Van Leer (1977) suggested an interesting alternative to the preceding three examples. Ordinary reconstruction–evolution methods painstakingly reconstruct and evolve the solution, only to throw most of the result away at the end of each time step, retaining only cell-integral averages. To reduce waste, Van Leer proposed retaining both cell-integral averages *and slopes* at the end of each time step, as illustrated in the following two examples.

Example 23.4 Van Leer's *Scheme II* uses the following slopes:

$$S_i^{n+1} = \frac{p_{e,i+1/2}(t^{n+1}) - p_{e,i-1/2}(t^{n+1})}{\Delta x}. \tag{23.9}$$

This is a simple central-difference approximation to the first derivative, similar to the preceding example, except that it uses the evolved solution at cell edges, rather than cell-integral averages. Let us find an expression for S_i^{n+1} as a function of S_i^n, S_{i-1}^n, and S_{i+1}^n. By Equations (23.2) and (23.9),

$$S_i^{n+1} = \frac{p_r(x_{i+1/2} - a\Delta t) - p_r(x_{i-1/2} - a\Delta t)}{\Delta x}.$$

For $0 \le \lambda a \le 1$, by Equation (23.3),

$$S_i^{n+1} = \frac{\bar{u}_i^n + S_i^n\left(\frac{\Delta x}{2} - a\Delta t\right) - \bar{u}_{i-1}^n - S_{i-1}^n\left(\frac{\Delta x}{2} - a\Delta t\right)}{\Delta x}$$

or

$$S_i^{n+1} = \frac{\bar{u}_i^n - \bar{u}_{i-1}^n}{\Delta x} + \left(S_i^n - S_{i-1}^n\right)\left(\frac{1}{2} - \lambda a\right),$$

which is a first-order backward-space approximation, as seen in Example 23.1, plus a zeroth-order backward-space correction that depends on the time step through $\lambda = \Delta t/\Delta x$. Proceed similarly for $-1 \leq \lambda a < 0$ to find the following final result:

$$\blacklozenge \qquad S_i^{n+1} = \begin{cases} \frac{\bar{u}_i^n - \bar{u}_{i-1}^n}{\Delta x} + (S_i^n - S_{i-1}^n)\left(\frac{1}{2} - \lambda a\right) & 0 \leq \lambda a \leq 1, \\ \frac{\bar{u}_{i+1}^n - \bar{u}_i^n}{\Delta x} - (S_{i+1}^n - S_i^n)\left(\frac{1}{2} + \lambda a\right) & -1 \leq \lambda a < 0. \end{cases} \tag{23.10}$$

Equations (23.5) and (23.10) can be combined to find the following matrix–vector expression:

$$\begin{bmatrix} \bar{u}_i^{n+1} \\ S_i^{n+1} \end{bmatrix} = \begin{bmatrix} 1 - \lambda a & -\frac{1}{2}\lambda a(1 - \lambda a)\Delta x \\ \frac{1}{\Delta x} & \frac{1}{2} - \lambda a \end{bmatrix} \begin{bmatrix} \bar{u}_i^n \\ S_i^n \end{bmatrix} + \begin{bmatrix} \lambda a & \frac{1}{2}\lambda a(1 - \lambda a)\Delta x \\ -\frac{1}{\Delta x} & -\frac{1}{2} + \lambda a \end{bmatrix} \begin{bmatrix} \bar{u}_{i-1}^n \\ S_{i-1}^n \end{bmatrix}$$

for $0 \leq \lambda a \leq 1$. As an exercise, the reader can find a similar expression for $-1 \leq \lambda a < 0$, which differs from the above only in some of the signs.

Example 23.5 This example concerns Van Leer's *Scheme III*. Scheme III is the same as Scheme II, seen in the last example, except that, at the end of each time step, it determines the slopes in each cell by least-squares fitting, rather than by differencing the edges of the cell. In other words, rather than using a secant line as in Scheme II, Scheme III finds a line using least-squares fitting as follows:

$$\int_{x_{i-1/2}}^{x_{i+1/2}} p_{\mathrm{r}}^n(x - a\Delta t)(x - x_i)\,dx = \int_{x_{i-1/2}}^{x_{i+1/2}} p_{\mathrm{r}}^{n+1}(x)(x - x_i)\,dx,$$

where $u(x, t^n) \approx p_{\mathrm{r}}^n(x)$ and $u(x, t^{n+1}) \approx p_{\mathrm{r}}^{n+1}(x)$ are the piecewise-linear reconstructions at time levels n and $n+1$, respectively. Completing the integrations and simplifying, one obtains explicit matrix expressions for $\bar{u}_i^{n+1}$ and S_i^{n+1} much as in the last example.

The approach illustrated in the preceding two examples remained fallow for many years until researchers such as Chang (1995) revived them. Finite-element methods use the same basic technique – retaining cell-integral averages, slopes, and other terms in a piecewise-polynomial solution representation. The increasing popularity of finite-element methods in computational gasdynamics perhaps explains the recent resurgence of interest in Van Leer's finite-element-like ideas.

As a third and uncreative reconstruction approach, S_i^n can be chosen to reproduce an existing method as illustrated in the following examples.

Example 23.6 Suppose that $S_i^n = 0$. Then Van Leer's reconstruction–evolution method becomes the CIR method, introduced in Examples 13.5, 13.6, 13.10, and 13.14. The reconstruction is zeroth-order accurate. The resulting reconstruction–evolution method is first-order accurate. To partially compensate for the low order of accuracy, this is the only example in which neither the spatial slopes S_i^n nor the steady-state solutions depend on the time step.

Example 23.7 Suppose that

$$S_i^n = \frac{1}{1-\lambda|a|}\begin{cases} \frac{u_{i+1}^n - u_i^n}{\Delta x} & 0 \le \lambda a \le 1, \\ \frac{u_i^n - u_{i-1}^n}{\Delta x} & -1 \le \lambda a < 0, \end{cases}$$

which is a downwind combination of Equations (23.6) and (23.7). Then Van Leer's reconstruction–evolution method becomes FTCS, as introduced in Subsection 11.1.1. This expression for S_i^n includes a factor $1/(1-\lambda|a|)$ which cancels the factor $1-\lambda|a|$ seen in Equation (23.4). Then the spatial slopes S_i^n depend on the time step but the steady-state solutions do not. Notice that $1/(1-\lambda|a|) \to \infty$ as $\lambda|a| \to 1$. While this is troubling, only slopes with the factor $1/(1-\lambda|a|)$ can avoid time-step dependency in steady-state solutions.

Example 23.8 Suppose that

$$S_i^n = -\frac{1}{\lambda|a|}\begin{cases} \frac{u_{i+1}^n - u_i^n}{\Delta x} & 0 \le \lambda a \le 1, \\ \frac{u_i^n - u_{i-1}^n}{\Delta x} & -1 \le \lambda a < 0. \end{cases}$$

As in the previous example, this is a downwind combination of Equations (23.6) and (23.7). Then Van Leer's reconstruction–evolution method becomes the Lax–Friedrichs method, introduced in Section 17.1.

Example 23.9 Suppose that

$$S_i^n = \begin{cases} \frac{u_{i+1}^n - u_i^n}{\Delta x} & 0 \le \lambda a \le 1, \\ \frac{u_i^n - u_{i-1}^n}{\Delta x} & -1 \le \lambda a < 0, \end{cases}$$

which is a downwind combination of Equations (23.6) and (23.7). Then Van Leer's reconstruction–evolution method becomes the Lax–Wendroff method introduced in Section 17.2. Unlike FTCS and the Lax–Friedrichs method, the spatial slopes S_i^n do not depend on the time step, although the steady-state solutions do.

Example 23.10 Suppose that

$$S_i^n = \begin{cases} \frac{u_i^n - u_{i-1}^n}{\Delta x} & 0 \le \lambda a \le 1, \\ \frac{u_{i+1}^n - u_i^n}{\Delta x} & -1 \le \lambda a < 0, \end{cases}$$

which is an upwind combination of Equations (23.6) and (23.7). Then Van Leer's reconstruction–evolution method becomes the Beam–Warming second-order upwind method, introduced in Section 17.4.

Example 23.11 Suppose that

$$S_i^n = \phi_i^n \begin{cases} \frac{u_{i+1}^n - u_i^n}{\Delta x} & 0 \le \lambda a \le 1, \\ \frac{u_i^n - u_{i-1}^n}{\Delta x} & -1 \le \lambda a < 0. \end{cases}$$

Then Van Leer's reconstruction–evolution method becomes Sweby's flux-limited method, introduced in Section 20.2. The slope is similar to that of FTCS, Lax–Friedrichs, and Lax–Wendroff, except that the factor ϕ_i^n replaces $1/(1-\lambda|a|)$, $-1/\lambda|a|$, and 1. In this context, ϕ_i^n is often called a *slope limiter* rather than a flux limiter. Assuming that the limiter ϕ_i^n does not depend on the time step (and the limiters seen in Section 20.2 do not) then S_i^n does not depend on the time step. However, when S_i^n does not depend on the time step, the steady-state solutions do.

As these examples show, there is no fundamental distinction between flux-averaged and solution-averaged methods for the linear advection equation. Intuitively, this is because there is no fundamental distinction between the solution u and the flux $f(u) = au$ – they differ only by a constant factor a.

Now we add a final filigree concerning slope choice. The slopes seen in the previous examples may allow large spurious overshoots and oscillations. To prevent this, consider nonlinear stability conditions, especially the upwind range condition discussed in Section 16.5. Van Leer's reconstruction–evolution method satisfies the upwind range condition if:

$$|S_{i+1}^n - S_i^n| \le 2\left|\frac{\bar{u}_{i+1}^n - \bar{u}_i^n}{\Delta x}\right|, \tag{23.11}$$

as the reader can show. Unfortunately, none of the slopes S_i^n suggested in the preceding examples satisfy this condition, except for $S_i^n = 0$. Although condition (23.11) could be used to design S_i^n directly, Van Leer (1977) chose a different tactic. Suppose that S_i' is anything that does *not* satisfy condition (23.11). Then consider the modified slope S_i^n defined in terms of S_i' as follows:

$$\blacklozenge \quad S_i^n = minmod\left(2\frac{\bar{u}_i^n - \bar{u}_{i-1}^n}{\Delta x}, S_i', 2\frac{\bar{u}_{i+1}^n - \bar{u}_i^n}{\Delta x}\right). \tag{23.12}$$

As the reader can show, S_i^n satisfies condition (23.11). Equation (23.12) is an example of *slope limiting*, which is strongly related to flux limiting and flux correction. For example, Equation (23.12) is similar in spirit if not in detail to Equation (21.13a). Notice that $S_i^n = 0$ if u_i^n is a maximum or a minimum, meaning that the conservative numerical flux of Van Leer's reconstruction–evolution method equals the conservative numerical flux of the CIR method. Thus slope limiting according to Equation (23.12) clips extrema, reducing the order of accuracy at extrema to between one and two; the order of accuracy is usually not first order, as you might naively expect, but between first and second order, for reasons explained in Section 20.1.

Example 23.12 Suppose that

$$S_i' = \frac{\bar{u}_{i+1}^n - \bar{u}_{i-1}^n}{2\Delta x}$$

as in Example 23.3. By Equation (23.12), slope limiting yields

$$S_i^n = minmod\left(2\frac{\bar{u}_i^n - \bar{u}_{i-1}^n}{\Delta x}, \frac{\bar{u}_{i+1}^n - \bar{u}_{i-1}^n}{2\Delta x}, 2\frac{\bar{u}_{i+1}^n - \bar{u}_i^n}{\Delta x}\right).$$

Then, as the reader can show, $S_i^n = S_i'$ unless $|\bar{u}_{i+1}^n - \bar{u}_i| > 3|\bar{u}_i^n - \bar{u}_{i-1}^n|$ or $|\bar{u}_i^n - \bar{u}_{i-1}^n| > 3|\bar{u}_{i+1}^n - \bar{u}_i^n|$, or unless $\bar{u}_{i+1}^n - \bar{u}_i^n$ and $\bar{u}_i^n - \bar{u}_{i-1}^n$ have opposite signs. In other words, the slope limiting leaves well enough alone unless the slopes change by a factor of three or greater, or unless there is an extremum. Alternatively, in this example, Van Leer suggested the following:

$$S_i^n = \begin{cases} \frac{2(\bar{u}_{i+1}^n - \bar{u}_i^n)(\bar{u}_i^n - \bar{u}_{i-1}^n)}{\Delta x(\bar{u}_{i+1}^n - \bar{u}_{i-1}^n)} & \operatorname{sign}(\bar{u}_{i+1}^n - \bar{u}_i^n) = \operatorname{sign}(\bar{u}_i^n - \bar{u}_{i-1}^n), \\ 0 & \text{otherwise,} \end{cases}$$

which is a modification of the classic *harmonic mean* of $\bar{u}_{i+1}^n - \bar{u}_i^n$ and $\bar{u}_i^n - \bar{u}_{i-1}^n$. In general, the harmonic mean of x and y is $2xy/(x + y)$. The harmonic-mean limiter tends to work better for steady-state solutions than the minmod limiter.

23.1.2 *The Lagrange Equations*

Van Leer's reconstruction–evolution method for the linear advection equation, as described in the preceding subsection, relies on something that most problems do not have – an exact solution. In 1979, Van Leer (1979) proposed a second-order accurate reconstruction–evolution method using a piecewise-linear spatial reconstruction and a clever *approximate* second-order accurate time evolution based on the Cauchy–Kowalewski procedure. Van Leer referred to such methods as *Monotone Upwind Schemes for Scalar Conservation Laws* or *MUSCL*, pronounced like the word "muscle." This book will use the term MUSCL exclusively to refer to the original Van Leer reconstruction–evolution method. However, in the literature, MUSCL may refer to *any* reconstruction–evolution method, especially to the "flux-limited-like" methods seen in Section 23.3.

Van Leer's 1979 paper culminated a series of five papers entitled "Towards the Ultimate Conservative Difference Scheme." While hardly the "ultimate" numerical method, MUSCL was a major achievement. Godunov developed his reconstruction–evolution method in the late 1950s, making it one of the earliest methods in computational gasdynamics. However, even years after its discovery, most researchers considered it an interesting oddity, too expensive and complicated for the mere first-order accuracy it offered, and inherently impossible to extend to higher orders of accuracy. As a result, reconstruction–evolution methods received scant attention until Van Leer took up their cause twenty year later. MUSCL generated a renewed flurry of interest in reconstruction–evolution methods in general and Riemann solvers in particular, ultimately producing lasting breakthroughs.

As part of its approximate time evolution, MUSCL employs the exact solution to the Riemann problem. In the two years following MUSCL, researchers proposed several popular approximate solutions to the Riemann problem, as described in Chapter 5. Since then, both solution-averaged and flux-averaged methods commonly incorporate approximate Riemann solvers. By popularizing Riemann solvers as a general tool for accessing and exploiting characteristics, even outside of solution-averaged methods, Van Leer's MUSCL method has widely impacted computational gasdynamics. In fact, as far as characteristic modeling goes, aside from Riemann solvers, the only currently available alternative is flux vector splitting.

A full description of MUSCL would require more space than this chapter allows. We will have to make do with a relatively brief summary. MUSCL solves the Lagrange equations, described in Problems 2.7 and 3.8. The Lagrange equations are simply the Euler equations expressed in a coordinate system fixed to the fluid. Solving the Lagrange equations is as good as solving the Euler equations and vice versa. A few years after the original MUSCL method, researchers discovered how to solve the Euler equations directly rather than the Lagrange equations (these are sometimes called *direct* Eulerian methods) after which most reconstruction–evolution methods solved the Euler equations. However, recent times have seen a return to the Lagrange equations; most commonly, methods solve an adaptive blend of the Euler and Lagrange equations.

At each time step, MUSCL uses a four-part procedure as follows:

- temporal evolution,
- spatial reconstruction,
- Eulerian remapping, and
- slope limiting.

Putting aside the temporal evolution for now, the spatial reconstruction is piecewise-linear with slopes chosen much as in Example 23.4. For the Eulerian remapping, the Lagrange equations use a coordinate system fixed to the fluid; thus the Lagrangian cells drift with the fluid. The Eulerian remapping converts the solution on the floating Lagrangian cells into a solution on the fixed Eulerian cells such that the approximation is linear across each Eulerian cell and not across each Lagrangian cell; the conversion uses a least-squares fitting procedure much like the one in Example 23.5. The last stage applies slope limiting, in the form of Equation (23.12), to the primitive variables. Ideally, slope limiting should act on the characteristic variables, which share many of the properties of the linear advection equation for which Equation (23.12) was originally designed. Van Leer (1979) suggests a few modifications to the flux limiter which may ameliorate the negative effects of using slope limiting on primitive rather than characteristic variables.

The rest of this discussion concerns the first step seen in the bullets above – temporal evolution. A piecewise-linear reconstruction leads to a series of *piecewise-linear Riemann problems* centered on the cell edges. If the piecewise-linear Riemann problem had an exact solution, the temporal evolution would be easy, but it does not. However, one can define a piecewise-constant Riemann problem that approximates a piecewise-linear Riemann problem, as seen in Figure 23.1. Although the standard piecewise-constant Riemann problem approximates the piecewise-linear Riemann problem to first-order accuracy, we were hoping for second-order accuracy. Fortunately, it is not necessary to approximate the entire solution to the piecewise-linear Riemann problem. Instead, we only need to approximate the solution at the cell edges for $t^n \leq t \leq t^{n+1}$ to second-order accuracy. In particular, recall that

$$\hat{\mathbf{f}}^n_{i+1/2} \approx \frac{1}{\Delta t} \int_{t^n}^{t^{n+1}} \mathbf{f}(\mathbf{u}(x_{i+1/2}, t))\, dt. \tag{23.13}$$

But the flux vector $\mathbf{f}$ depends on quantities such as pressure p; then the time integral average of $\mathbf{f}$ depends on time integral averages such as

$$\hat{p}^n_{i+1/2} = \frac{1}{\Delta t} \int_{t^n}^{t^{n+1}} p(x_{i+1/2}, t)\, dt, \tag{23.14}$$

which may be approximated by

$$\hat{p}^n_{i+1/2} \approx p(x_{i+1/2}, t^n) + \frac{\Delta t}{2} \left(\frac{\partial p}{\partial t}\right)^M_{i+1/2} + O(\Delta t^2). \tag{23.15}$$

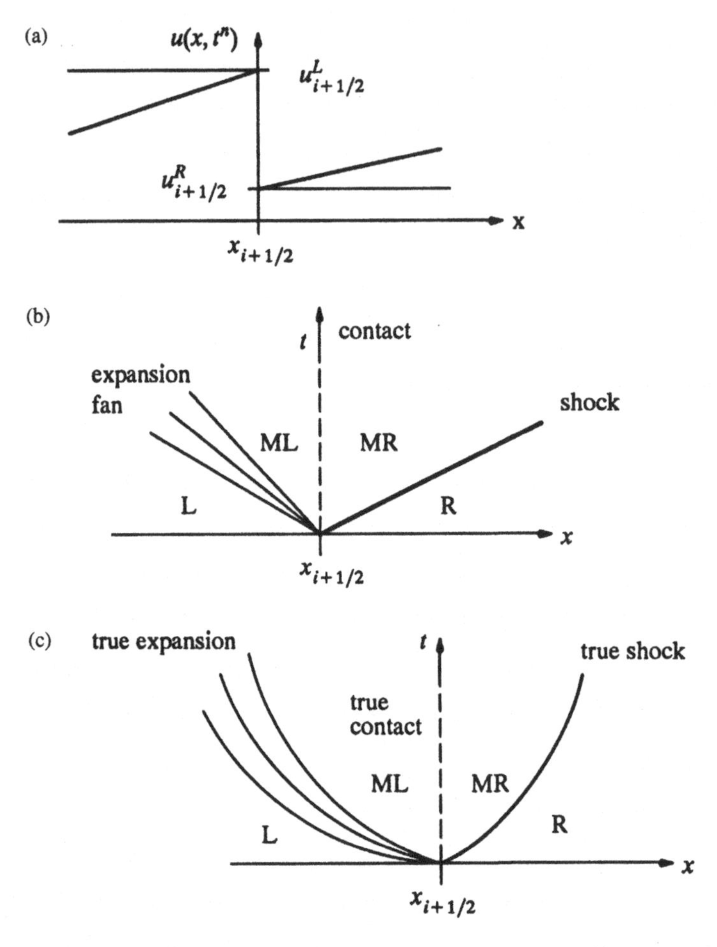

Figure 23.1 (a) A standard (piecewise-constant) Riemann problem approximates a piecewise-linear Riemann problem. (b) Wave diagram for the standard (piecewise-constant) Riemann problem. (c) Wave diagram for the piecewise-linear Riemann problem.

Characteristic theory, weak shock theory, and the solution to the Riemann problem yield an approximate expression for the unknown time derivative $(\frac{\partial p}{\partial t})^M_{i+1/2}$ in terms of known quantities, assuming a smooth solution; the details are omitted. This approximation fails near strong shocks and other steep gradients; however, in this case, the slope limiting causes the method to revert to Godunov's method, which easily handles such situations, albeit with a reduced order of accuracy. This completes the introduction to MUSCL.

23.2 Colella–Woodward Reconstruction–Evolution Method (PPM)

In 1984, Colella and Woodward proposed a reconstruction–evolution method called the *piecewise-parabolic method* (*PPM*). PPM incorporates a number of important advances including: the first description of reconstruction via the primitive function (see Section 9.3); the first attempt to prevent smearing at contacts in reconstruction–evolution methods (somewhere between artificial compression seen in Section 22.2 and subcell resolution seen in Section 9.4); the first reconstruction–evolution method to use a piecewise-quadratic rather than piecewise-constant or piecewise-linear reconstruction; and so forth. The Colella–Woodward reconstruction–evolution method is probably the most complicated numerical method seen in this book, possibly excepting the ENO methods, and a full discussion would require an enormous amount of space. This section will describe PPM only for the linear advection equation, and that only briefly; the reader can consult the original paper for a description of PPM for the Lagrange and Euler equations.

The simplest reconstruction is piecewise-constant, as used in the CIR method; the next simplest reconstruction is piecewise-linear, as used in MUSCL. The Colella–Woodward reconstruction–evolution method takes the next logical step, using a piecewise-quadratic spatial reconstruction as follows:

$$p_r(x) = p_{r,i}(x) = c_{0,i} + c_{1,i}(x - x_i) + c_{2,i}(x - x_i)^2 \tag{23.16}$$

for $x_{i-1/2} \leq x \leq x_{i+1/2}$. Also, the Colella–Woodward reconstruction–evolution method uses the exact temporal evolution for the linear advection equation as follows:

$$p_{e,i+1/2}(t) = p_r(x_{i+1/2} - a(t - t^n)). \tag{23.17}$$

As the reader can show, the resulting conservative numerical flux is

♦ $$\hat{f}^n_{i+1/2} = a\bar{u}^n_i + \frac{1}{2}a(1 - \lambda a)\Delta x_i c_{1,i} + \frac{1}{6}a(1 - \lambda a)(1 - 2\lambda a)\Delta x_i^2 c_{2,i} \tag{23.18}$$

for $0 \leq \lambda a \leq 1$. A similar expression applies for $-1 \leq \lambda a < 0$.

Since the temporal evolution is perfect, the success of the method depends entirely on the spatial reconstruction or, in other words, on the choice of the quadratic coefficients $c_{0,i}$, $c_{1,i}$, and $c_{2,i}$. The rest of this section concerns the choice of $c_{0,i}$, $c_{1,i}$, and $c_{2,i}$. Rather than setting $c_{0,i}$, $c_{1,i}$, and $c_{2,i}$ directly, Colella and Woodward sometimes work with the cell-integral average $\bar{u}^n_i$, the value $u_{L,i}$ of the reconstruction on the left-hand side of the cell i, and the value $u_{R,i}$ of the reconstruction of the right-hand side of the cell i. First we shall derive some simple relationships between $c_{0,i}$, $c_{1,i}$, $c_{2,i}$ and $u_{L,i}$, $u_{R,i}$, $\bar{u}^n_i$: The reconstruction preserves cell-integral averages if

$$\frac{1}{\Delta x_i}\int_{x_{i-1/2}}^{x_{i+1/2}} p_r(x)\,dx = \bar{u}^n_i.$$

Then Equation (23.16) implies

$$c_{0,i} + \frac{1}{12}\Delta x_i^2 c_{2,i} = \bar{u}^n_i. \tag{23.19}$$

Also, again by Equation (23.16), the reconstruction on the left-hand side of the cell i is

$$u_{L,i} = p_{r,i}(x_{i-1/2}) = c_{0,i} - \frac{1}{2}\Delta x_i c_{1,i} + \frac{1}{4}\Delta x_i^2 c_{2,i} \tag{23.20}$$

and the reconstruction on the right-hand side of the cell i is

$$u_{R,i} = p_{r,i}(x_{i+1/2}) = c_{0,i} + \frac{1}{2}\Delta x_i c_{1,i} + \frac{1}{4}\Delta x_i^2 c_{2,i}. \tag{23.21}$$

Solve Equations (23.19), (23.20), and (23.21) to find the following expressions for $c_{0,i}$, $c_{1,i}$, $c_{2,i}$ in terms of $u_{L,i}$, $u_{R,i}$, $\bar{u}_i^n$:

$$c_{0,i} = \bar{u}_i^n - \frac{1}{12}\Delta x_i^2 c_{2,i}, \tag{23.22}$$

$$c_{1,i} = \frac{u_{R,i} - u_{L,i}}{\Delta x_i}, \tag{23.23}$$

$$c_{2,i} = \frac{6}{\Delta x_i^2}\left(\frac{u_{R,i} + u_{L,i}}{2} - \bar{u}_i^n\right). \tag{23.24}$$

Colella and Woodward use an elaborate five-step procedure to choose $c_{0,i}$, $c_{1,i}$, and $c_{2,i}$. The five steps are as follows:

Step 1 Find $u_{i+1/2}^n \approx u(x_{i+1/2}, t^n)$ from $\bar{u}_{i-1}^n$, $\bar{u}_i^n$, $\bar{u}_{i+1}^n$, and $\bar{u}_{i+2}^n$ using reconstruction via the primitive function. That is, let

$$\begin{aligned} u_{i+1/2}^n = {} & U[x_{i-1/2}, x_{i+1/2}] + \Delta x_i U[x_{i-1/2}, x_{i+1/2}, x_{i+3/2}] \\ & - \Delta x_i \Delta x_{i+1} U[x_{i-1/2}, x_{i+1/2}, x_{i+3/2}, x_{i+5/2}] \\ & + \Delta x_{i-1}\Delta x_i(\Delta x_{i+1} + \Delta x_{i+2}) U[x_{i-3/2}, x_{i-1/2}, x_{i+1/2}, x_{i+3/2}, x_{i+5/2}] \end{aligned} \tag{23.25a}$$

for nonconstant Δx and

$$u_{i+1/2}^n = \frac{-\bar{u}_{i+2}^n + 7\bar{u}_{i+1}^n + 7\bar{u}_i^n - \bar{u}_{i-1}^n}{12} \tag{23.25b}$$

for constant Δx, where the Newton divided differences of the primitive function U of u are defined as in Chapters 8 and 9. This approximation is fourth-order accurate.

Step 2 Use slope limiting to ensure that $u_{i+1/2}^n$ lies between $\bar{u}_i^n$ and $\bar{u}_{i+1}^n$. The details are omitted.

Step 3 Let $u_{R,i} = u_{L,i+1} = u_{i+1/2}^n$ where $u_{i+1/2}^n$ was found in steps 1 and 2. Then use Equations (23.22)–(23.24) to find $c_{0,i}$, $c_{1,i}$, and $c_{2,i}$.

Step 4 Modify the values of $c_{0,i}$, $c_{1,i}$, and $c_{2,i}$ found in step 3 to steepen jump discontinuities, especially contacts. This steepening procedure was partly inspired by artificial compression, as described in Section 22.2. In turn, this steepening procedure helped inspire other techniques, such as subcell resolution described in Section 9.4 and, less directly, Roe's steepening superbee flux limiter described in Section 20.2. Like artificial compression, the steepening procedure relies on a shock switch to identify cells that might need steepening. For constant Δx, the shock switch uses quantities such as

$$\theta_i' = \frac{1}{6}\frac{\left(\bar{u}_{i+2}^n - 2\bar{u}_{i+1}^n + \bar{u}_i^n\right) - \left(\bar{u}_i^n - 2\bar{u}_{i-1}^n + \bar{u}_{i-2}^n\right)}{\bar{u}_{i+1}^n - \bar{u}_{i-1}^n}.$$

This shock switch determines the blending between the standard *quadratic* reconstruction and a steeper *linear* reconstruction in each candidate cell. As discussed in the last chapter, every shock switch sometimes yields false positives. However, false positives in PPM cause only a small local loss in the order of accuracy since PPM uses a linear rather than quadratic reconstruction. This offers a big improvement over the possible effects of false positives in artificial compression. Consult the original paper for more details.

Step 5 Modify $c_{0,i}$, $c_{1,i}$, and $c_{2,i}$ found in steps 3 and 4 so that $p_R(x)$ does not increase maxima, decrease minima, or contain new maxima or minima as compared to $\bar{u}_i^n$. In other words, step 5 ensures that $p_R(x)$ is range diminishing relative to $\bar{u}_i^n$, as in Section 16.3. In particular, to prevent maxima from increasing and minima from decreasing, use a piecewise-constant reconstruction ($c_{1,i} = c_{2,i} = 0$ and $c_{0,i} = \bar{u}_i^n$) if $\bar{u}_i^n$ is a maximum or a minimum which, of course, means that the reconstruction has only first-order accuracy at maxima and minima. Also, to prevent new maxima and minima, require that the reconstruction $p_R(x)$ be either monotone increasing, monotone decreasing, or constant in each cell and across cell boundaries. The lengthy details are omitted.

Although the spatial reconstruction is third-order accurate away from extrema, the time evolution procedures for the Lagrange and Euler equations are only second-order accurate. Thus the Colella–Woodward reconstruction–evolution method is usually considered to be a second-order accurate method, just like Van Leer's reconstruction–evolution method seen in the last section. For extensive numerical results comparing PPM to a number of other numerical methods, see Woodward and Colella (1984).

23.3 Anderson–Thomas–Van Leer Reconstruction–Evolution Methods (TVD/MUSCL): Finite-Volume Versions of the Chakravarthy–Osher Flux-Limited Methods

This section concerns two reconstruction–evolution methods proposed by Anderson, Thomas, and Van Leer (1986). These methods are finite-volume versions of the Chakravarthy–Osher flux-limited methods seen in Subsections 20.3.1 and 20.3.3. In fact, the only difference between the Anderson–Thomas–Van Leer reconstruction–evolution methods and the Chakravarthy–Osher flux-limited methods is the ordering of the reconstruction and the flux evaluation: The Chakravarthy–Osher flux-limited methods evaluate and then reconstruct the split fluxes, whereas the Anderson–Thomas–Van Leer reconstruction–evolution methods reconstruct the solution and then evaluate the split fluxes. Put another way, the Chakravarthy–Osher flux-limited methods use $f^{\pm}(x) + f^{\pm}(y)$, whereas the Anderson–Thomas–Van Leer reconstruction–evolution methods use $f^{\pm}(x + y)$.

Anderson, Thomas, and Van Leer (1986) called their method "the MUSCL method" and many subsequent papers repeat this usage. However, this method has almost no connection to the original MUSCL method described in Section 23.1. For that matter, the Anderson–Thomas–Van Leer method has almost no connection, either in theory or practice, to any of the other reconstruction–evolution methods considered in this chapter. Instead, as mentioned earlier, the Anderson–Thomas–Van Leer reconstruction–evolution method is closely related to the Chakravarthy–Osher flux-limited methods seen in Section 20.3.

Like the Chakravarthy–Osher flux-limited methods, the Anderson–Thomas–Van Leer reconstruction–evolution methods are semidiscrete methods. As shown in Problem 11.11, a conservative semidiscrete finite-volume method can be written as follows:

$$\frac{d\bar{u}_i^n}{dt} = -\frac{\hat{f}_{s,i+1/2}^n - \hat{f}_{s,i-1/2}^n}{\Delta x},$$

where

$$\bar{u}_i^n \approx \frac{1}{\Delta x} \int_{x_{i-1/2}}^{x_{i+1/2}} u(x, t^n)\, dx$$

and

$$\hat{f}_{s,i+1/2}^n \approx f(u(x_{i+1/2}, t^n)).$$

Using flux splitting, suppose that $f(u) = f^+(u) + f^-(u)$. Then

$$\hat{f}_{s,i+1/2}^n \approx f^+(u(x_{i+1/2}, t^n)) + f^-(u(x_{i+1/2}, t^n)). \tag{23.26}$$

Then we need to approximate $f^+(u(x_{i+1/2}, t^n))$ or, equivalently, $u(x_{i+1/2}, t^n)$ with a leftward bias using reconstruction via the primitive function. Similarly, we need to approximate $f^-(u(x_{i+1/2}, t^n))$ or, equivalently, $u(x_{i+1/2}, t^n)$ with a rightward bias using reconstruction via the primitive function. The following two subsections consider two possible approaches.

23.3.1 *A Second-Order Accurate Method*

First, approximate $u(x_{i+1/2}, t^n)$ with a leftward bias. As seen in Example 9.6, second-order-accurate linear reconstruction via the primitive function yields

$$u(x, t^n) \approx \bar{u}_i^n + \frac{\bar{u}_i^n - \bar{u}_{i-1}^n}{\Delta x}(x - x_i).$$

Then

$$u\left(x_{i+1/2}, t^n\right) \approx \bar{u}_i^n + \frac{1}{2}\left(\bar{u}_i^n - \bar{u}_{i-1}^n\right). \tag{23.27}$$

Similarly, a first-order-accurate constant reconstruction yields

$$u\left(x_{i+1/2}, t^n\right) \approx \bar{u}_i^n. \tag{23.28}$$

Instead of using pure linear or pure constant reconstruction, consider instead convex linear combinations. Using the notation seen in Section 20.2, a convex linear combination of Equations (23.27) and (23.28) yields

$$u_{i+1/2}^+ = \bar{u}_i^n + \frac{1}{2}\phi_i^+\left(\bar{u}_i^n - \bar{u}_{i-1}^n\right). \tag{23.29}$$

Now, let us approximate $u(x_{i+1/2}, t^n)$ with a rightward bias. Second-order-accurate linear reconstruction via the primitive function yields

$$u(x, t^n) \approx \bar{u}_{i+1}^n + \frac{\bar{u}_{i+2}^n - \bar{u}_{i+1}^n}{\Delta x}(x - x_{i+1}).$$

Then

$$u(x_{i+1/2}, t^n) \approx \bar{u}_{i+1}^n - \frac{1}{2}\left(\bar{u}_{i+2}^n - \bar{u}_{i+1}^n\right). \tag{23.30}$$

Similarly, a first-order-accurate constant reconstruction yields

$$u(x_{i+1/2}, t^n) \approx \bar{u}_{i+1}^n. \tag{23.31}$$

Instead of using pure linear or pure constant reconstruction, consider instead convex linear combinations. Using the notation seen in Section 20.2, a convex linear combination of Equations (23.30) and (23.31) yields

$$u^-_{i+1/2} = \bar{u}^n_{i+1} - \frac{1}{2}\phi^-_{i+1}\left(\bar{u}^n_{i+2} - \bar{u}^n_{i+1}\right). \tag{23.32}$$

Using the above expressions for $u^{\pm}_{i+1/2}$, the semidiscrete Anderson–Thomas–Van Leer reconstruction–evolution method is as follows:

$$\frac{d\bar{u}^n_i}{dt} = -\frac{\hat{f}^n_{s,i+1/2} - \hat{f}^n_{s,i-1/2}}{\Delta x},$$

where

$$\hat{f}^n_{s,i+1/2} = f^+\left(u^+_{i+1/2}\right) + f^-\left(u^-_{i+1/2}\right) \tag{23.33}$$

or explicitly

$$\blacklozenge \quad \hat{f}^n_{s,i+1/2} = f^+\left(\bar{u}^n_i + \frac{1}{2}\phi^+_i\left(\bar{u}^n_i - \bar{u}^n_{i-1}\right)\right) + f^-\left(\bar{u}^n_{i+1} - \frac{1}{2}\phi^-_{i+1}\left(\bar{u}^n_{i+2} - \bar{u}^n_{i+1}\right)\right). \tag{23.34}$$

The functions $\phi^{\pm}_i$ are defined exactly as in Section 20.3.1, including such limiters as the minmod and superbee limiters, although here they are called slope limiters rather than flux limiters. Let $\phi^{\pm}_i = 0$ to define a first-order flux splitting method as follows:

$$\hat{f}^{(1)}_{i+1/2} = f^+(\bar{u}^n_i) + f^-(\bar{u}^n_{i+1}),$$

just as in Subsection 18.2.5. Then notice that Equations (20.54) and (23.34) are *identical* if $f^{\pm}(u)$ are linear, that is, if $f^{\pm}(x+y) = f^{\pm}(x) + f^{\pm}(y)$.

The introduction to Part V mentioned the RAMPANT code developed by FLUENT corporation. The most recent version of RAMPANT uses a version of the method seen in this subsection, with the symmetric minmod slope limiter of Equation (20.26) or, more precisely, a smoothed version of the symmetric minmod limiter obtained by cubic polynomial fitting. As the main difference between the method seen here and RAMPANT, RAMPANT uses flux difference splitting along the lines of Roe's first-order upwind method rather than flux vector splitting. For issues specifically to do with steady-state flows see, for example, Venkatakrishnan (1995); as usual, for rapid and complete steady-state convergence, steady-state flows tend to benefit from somewhat different limiters than unsteady flows.

23.3.2 *Second- and Third-Order Accurate Methods*

First, let us approximate $u(x_{i+1/2}, t^n)$ with a leftward bias. As seen in Example 9.6, second-order-accurate linear reconstruction via the primitive function yields

$$u(x, t^n) \approx \bar{u}^n_i + \frac{\bar{u}^n_{i+1} - \bar{u}^n_i}{\Delta x}(x - x_i),$$

$$u(x, t^n) \approx \bar{u}^n_i + \frac{\bar{u}^n_i - \bar{u}^n_{i-1}}{\Delta x}(x - x_i).$$

Then

$$u\left(x_{i+1/2}, t^n\right) \approx \bar{u}_i^n + \frac{1}{2}\left(\bar{u}_{i+1}^n - \bar{u}_i^n\right), \tag{23.35}$$

$$u\left(x_{i+1/2}, t^n\right) \approx \bar{u}_i^n + \frac{1}{2}\left(\bar{u}_i^n - \bar{u}_{i-1}^n\right). \tag{23.36}$$

Similarly, as seen in Example 9.7, third-order-accurate quadratic reconstruction via the primitive function yields

$$u(x, t^n) \approx \bar{u}_i^n - \frac{\bar{u}_{i+1}^n - \bar{u}_i^n - \left(\bar{u}_i^n - \bar{u}_{i-1}^n\right)}{24} + \frac{\bar{u}_{i+1}^n - \bar{u}_i^n}{\Delta x}(x - x_i)$$
$$+ \frac{\bar{u}_{i+1}^n - \bar{u}_i^n - \left(\bar{u}_i^n - \bar{u}_{i-1}^n\right)}{2\Delta x^2}(x - x_i)(x - x_{i+1}).$$

Let $x = x_{i+1/2}$ and simplify to find

$$u\left(x_{i+1/2}, t^n\right) \approx \bar{u}_i^n + \frac{1}{3}\left(\bar{u}_{i+1}^n - \bar{u}_i^n\right) + \frac{1}{6}\left(\bar{u}_i^n - \bar{u}_{i-1}^n\right). \tag{23.37}$$

Instead of using pure linear or pure quadratic reconstruction, consider instead convex linear combinations. Using the notation seen in Section 20.1 and Subsection 20.3.3, a convex linear combination of Equations (23.35) and (23.36) yields

$$u_{i+1/2}^- = \bar{u}_i^n + \frac{1+\eta}{4}\left(\bar{u}_{i+1}^n - \bar{u}_i^n\right) + \frac{1-\eta}{4}\left(\bar{u}_i^n - \bar{u}_{i-1}^n\right). \tag{23.38}$$

Equation (23.38) equals Equation (23.35) if $\eta = 1$, Equation (23.36) if $\eta = -1$, and Equation (23.37) if $\eta = 1/3$. Equation (23.38) is leftward-biased for $n < 1$ and centered for $\eta = 1$.

Now, let us approximate $u(x_{i+1/2}, t^n)$ with a rightward bias. Second-order-accurate linear reconstruction via the primitive function yields

$$u(x, t^n) \approx \bar{u}_{i+1}^n + \frac{\bar{u}_{i+1}^n - \bar{u}_i^n}{\Delta x}(x - x_{i+1}),$$

$$u(x, t^n) \approx \bar{u}_{i+1}^n + \frac{\bar{u}_{i+2}^n - \bar{u}_{i+1}^n}{\Delta x}(x - x_{i+1}).$$

Then

$$u(x_{i+1/2}, t^n) \approx \bar{u}_{i+1}^n - \frac{1}{2}\left(\bar{u}_{i+1}^n - \bar{u}_i^n\right), \tag{23.39}$$

$$u(x_{i+1/2}, t^n) \approx \bar{u}_{i+1}^n - \frac{1}{2}\left(\bar{u}_{i+2}^n - \bar{u}_{i+1}^n\right). \tag{23.40}$$

Similarly, third-order-accurate quadratic reconstruction via the primitive function yields

$$u(x, t^n) \approx \bar{u}_{i+1}^n - \frac{\bar{u}_{i+2}^n - \bar{u}_{i+1}^n - \left(\bar{u}_{i+1}^n - \bar{u}_i^n\right)}{24} + \frac{\bar{u}_{i+1}^n - \bar{u}_i^n}{\Delta x}(x - x_{i+1})$$
$$+ \frac{\bar{u}_{i+2}^n - \bar{u}_{i+1}^n - \left(\bar{u}_{i+1}^n - \bar{u}_i^n\right)}{2\Delta x^2}(x - x_i)(x - x_{i+1}).$$

Then

$$u\left(x_{i+1/2}, t^n\right) \approx \bar{u}_i^n - \frac{1}{3}\left(\bar{u}_{i+1}^n - \bar{u}_i^n\right) - \frac{1}{6}\left(\bar{u}_i^n - \bar{u}_{i-1}^n\right). \tag{23.41}$$

Instead of using pure linear or pure quadratic reconstruction, consider instead convex linear combinations. Using the notation seen in Section 20.1 and Subsection 20.3.3, a convex linear combination of Equations (23.39) and (23.40) yields

$$u_{i+1/2}^+ = \bar{u}_{i+1}^n - \frac{1+\eta}{4}\left(\bar{u}_{i+1}^n - \bar{u}_i^n\right) - \frac{1-\eta}{4}\left(\bar{u}_{i+2}^n - \bar{u}_{i+1}^n\right). \tag{23.42}$$

Equation (23.42) equals Equation (23.39) if $\eta = 1$, Equation (23.40) if $\eta = -1$, and Equation (23.41) if $\eta = 1/3$. Equation (23.42) is rightward-biased for $n < 1$ and centered for $\eta = 1$.

As the reader can show, $u_{i+1/2}^-$ lies between $\bar{u}_i^n$ and $\bar{u}_{i+1}^n$ if $\bar{u}_{i+1}^n - \bar{u}_i^n$ and $\bar{u}_i^n - \bar{u}_{i-1}^n$ have the same sign and if

$$\left|\bar{u}_i^n - \bar{u}_{i-1}^n\right| \le \frac{3-\eta}{1-\eta}\left|\bar{u}_{i+1}^n - \bar{u}_i^n\right|.$$

Similarly, $u_{i+1/2}^+$ lies between $\bar{u}_{i+1}^n$ and $\bar{u}_i^n$ if $\bar{u}_{i+1}^n - \bar{u}_i^n$ and $\bar{u}_{i+2}^n - \bar{u}_{i+1}^n$ have the same sign and if

$$\left|\bar{u}_{i+2}^n - \bar{u}_{i+1}^n\right| \le \frac{3-\eta}{1-\eta}\left|\bar{u}_{i+1}^n - \bar{u}_i^n\right|.$$

To ensure these conditions let

$$\begin{aligned} u_{i+1/2}^- = \bar{u}_i^n &+ \frac{1+\eta}{4} minmod\left[\bar{u}_{i+1}^n - \bar{u}_i^n, b\left(\bar{u}_i^n - \bar{u}_{i-1}^n\right)\right] \\ &+ \frac{1-\eta}{2} minmod\left[\bar{u}_i^n - \bar{u}_{i-1}^n, b\left(\bar{u}_{i+1}^n - \bar{u}_i^n\right)\right], \end{aligned} \tag{23.43}$$

and

$$\begin{aligned} u_{i+1/2}^+ = \bar{u}_{i+1}^n &- \frac{1+\eta}{2} minmod\left[\bar{u}_{i+1}^n - \bar{u}_i^n, b\left(\bar{u}_{i+2}^n - \bar{u}_{i+1}^n\right)\right] \\ &- \frac{1-\eta}{2} minmod\left[\bar{u}_{i+2}^n - \bar{u}_{i+1}^n, b\left(\bar{u}_{i+1}^n - \bar{u}_i^n\right)\right], \end{aligned} \tag{23.44}$$

where

$$1 \le b \le \frac{3-\eta}{1-\eta}. \tag{23.45}$$

Using the above expressions for $u_{i+1/2}^{\pm}$, the semidiscrete Anderson–Thomas–Van Leer reconstruction–evolution method is as follows:

$$\frac{d\bar{u}_i^n}{dt} = -\frac{\hat{f}_{s,i+1/2}^n - \hat{f}_{s,i-1/2}^n}{\Delta x},$$

where

$$\hat{f}_{s,i+1/2}^n \approx f^+\left(u_{i+1/2}^-\right) + f^-\left(u_{i+1/2}^+\right) \tag{23.46}$$

or explicitly

$$\hat{f}^n_{s,i+1/2} = f^+\Big(\bar{u}^n_i + \frac{1+\eta}{4} minmod\big[\bar{u}^n_{i+1} - \bar{u}^n_i, b(\bar{u}^n_i - \bar{u}^n_{i-1})\big]$$

$$+ \frac{1-\eta}{2} minmod\big[\bar{u}^n_i - \bar{u}^n_{i-1}, b(\bar{u}^n_{i+1} - \bar{u}^n_i)\big]\Big)$$

$$+ f^-\Big(\bar{u}^n_{i+1} - \frac{1+\eta}{2} minmod\big[\bar{u}^n_{i+1} - \bar{u}^n_i, b(\bar{u}^n_{i+2} - \bar{u}^n_{i+1})\big]$$

$$- \frac{1-\eta}{2} minmod\big[\bar{u}^n_{i+2} - \bar{u}^n_{i+1}, b(\bar{u}^n_{i+1} - \bar{u}^n_i)\big]\Big). \tag{23.47}$$

Notice that Equation (23.47) is identical to Equation (20.64) if $f^{\pm}(u)$ are linear, that is, if $f^{\pm}(x+y) = f^{\pm}(x) + f^{\pm}(y)$. For nonlinear functions, averages like $f^{\pm}(x+y)$ tend to be more accurate than averages like $f^{\pm}(x) + f^{\pm}(y)$, not to mention less expensive, since they require only one rather than two evaluations of the nonlinear split flux functions.

Unfortunately, as with many flux-limited or "flux-limited-like" methods, the above Anderson–Thomas–Van Leer methods may not work very well for steady-state flows. In particular Anderson, Thomas, and Van Leer (1986) say that "the residual history often enters into a limit cycle oscillation .... The limit cycle is believed to be associated with the discontinuous derivative of the [minmod] limiter. ..." Thus Anderson, Thomas, and Van Leer (1986) suggested yet a third method for steady-state flows; the reader should consult the original paper for details. For time-discretization, Anderson, Thomas, and Van Leer (1986) suggest a linearized implicit time-discretization, much like the ones seen in Sections 20.5 and 22.3. Again, the reader should consult the original paper for details.

23.4 Harten–Osher Reconstruction–Evolution Methods (UNO)

This section concerns the Harten–Osher reconstruction–evolution method, better known as UNO. UNO forms the bridge between the Van Leer reconstruction–evolution method studied in Section 23.1 and the ENO methods studied in the next section.

23.4.1 *The Linear Advection Equation*

For the linear advection equation, UNO is exactly like the Van Leer reconstruction–evolution method seen in Subsection 23.1.1, except that the slopes in the piecewise-linear reconstruction are formed using ENO reconstruction, as studied in Chapter 9. Proceed just as in Section 23.1.1 to obtain the following:

$$\hat{f}^n_{i+1/2} = \begin{cases} a\bar{u}^n_i + \frac{1}{2}a(1-\lambda a)S^n_i\Delta x & 0 \le \lambda a \le 1, \\ a\bar{u}^n_{i+1} - \frac{1}{2}a(1+\lambda a)S^n_{i+1}\Delta x & -1 \le \lambda a < 0, \end{cases}$$

where S^n_i is a first-order accurate approximation to $\frac{\partial u}{\partial x}(x_i, t^n)$. In particular, as seen in Example 9.7, third-order accurate piecewise-quadratic reconstruction yields

$$u(x, t^n) \approx a_1 + 2a_2(x - x_i) + 3a_3(x - x_i)^2,$$

where Example 9.7 gives specific expressions for the coefficients a_1, a_2, and a_3. Then let

$$S_i^n \approx \frac{\partial u}{\partial x}(x_i, t^n) \approx 2a_2.$$

For constant Δx, the three slope candidates can be written as follows:

$$S_i^n = \frac{\bar{u}_{i+1}^n - \bar{u}_i^n - \frac{1}{2}(\bar{u}_{i+2} - 2\bar{u}_{i+1} + \bar{u}_i^n)}{\Delta x},$$

$$S_i^n = \frac{\bar{u}_{i+1}^n - \bar{u}_i^n - \frac{1}{2}(\bar{u}_{i+1} - 2\bar{u}_i + \bar{u}_{i-1}^n)}{\Delta x},$$

$$= \frac{\bar{u}_i^n - \bar{u}_{i-1}^n + \frac{1}{2}(\bar{u}_{i+1} - 2\bar{u}_i + \bar{u}_{i-1}^n)}{\Delta x},$$

and

$$S_i^n = \frac{\bar{u}_i^n - \bar{u}_{i-1}^n + \frac{1}{2}(\bar{u}_i - 2\bar{u}_{i-1} + \bar{u}_{i-2}^n)}{\Delta x}.$$

Choose between the three candidates as described in Section 9.2, and especially as seen in Equations (9.25)–(9.30), to obtain the following expression for constant Δx:

$$\blacklozenge \quad S_i^n = \frac{1}{\Delta x} minmod\Big[\bar{u}_{i+1}^n - \bar{u}_i^n - \frac{1}{2} minmod(\bar{u}_{i+1}^n - 2\bar{u}_i^n + \bar{u}_{i-1}^n, \bar{u}_{i+2}^n - 2\bar{u}_{i+1}^n + \bar{u}_i^n), \bar{u}_i^n - \bar{u}_{i-1}^n + \frac{1}{2} minmod(\bar{u}_{i+1}^n - 2\bar{u}_i^n + \bar{u}_{i-1}^n, \bar{u}_i^n - 2\bar{u}_{i-1}^n + \bar{u}_{i-2}^n)\Big]. \tag{23.48}$$

As one variation, the *minmod* average can be replaced by other averages, such as the *m* average seen in Equation (9.6). As another variation, omitting the second difference terms leads to the following:

$$S_i^n = \frac{1}{\Delta x} minmod(\bar{u}_{i+1}^n - \bar{u}_i^n, \bar{u}_i^n - \bar{u}_{i-1}^n), \tag{23.49}$$

which makes UNO a lot like Sweby's flux-limited TVD method seen in Section 20.2. Equation (23.48) yields uniform second-order accuracy, whereas Equation (23.49) yields second-order accuracy except at extrema, where it yields between first- and second-order accuracy. Equation (23.49) is less expensive than Equation (23.48) and less likely to allow spurious overshoots and oscillations.

23.4.2 *Nonlinear Scalar Conservation Laws*

For nonlinear scalar conservation laws, the temporal evolution uses a continuous piecewise-linear reconstruction of the wave speed. The details are omitted. The result is

$$\blacklozenge \quad \hat{f}_{i+1/2}^n = f(\bar{u}_i^n) + \frac{1}{2}\frac{a_{i+1/2}^n(1 - \lambda a_{i-1/2}^n)}{1 + \lambda(a_{i+1/2}^n - a_{i-1/2}^n)} S_i^n \Delta x \tag{23.50a}$$

for $0 \leq \lambda a_{i+1/2}^n \leq 1$ and

$$\blacklozenge \quad \hat{f}_{i+1/2}^n = f(\bar{u}_{i+1}^n) - \frac{1}{2}\frac{a_{i+1/2}^n(1 + \lambda a_{i+3/2}^n)}{1 + \lambda(a_{i+3/2}^n - a_{i+1/2}^n)} S_{i+1}^n \Delta x \tag{23.50b}$$

for $-1 \leq \lambda a^n_{i+1/2} < 0$, where as usual

$$a^n_{i+1/2} = \begin{cases} \frac{f(\bar{u}^n_{i+1}) - f(\bar{u}^n_i)}{\bar{u}^n_{i+1} - \bar{u}^n_i} & \bar{u}^n_{i+1} \neq \bar{u}^n_i, \\ a(\bar{u}^n_i) & \text{otherwise.} \end{cases}$$

The behavior of the Harten–Osher reconstruction–evolution is illustrated using the five standard test cases introduced in Section 17.0. As the following results show, UNO is one of the best methods seen so far in this book.

Test Case 1 As seen in Figure 23.2, the Harten–Osher reconstruction–evolution method captures the sinusoid without clipping, and with only a very slight lagging phase error. The only methods seen so far in Part V that even approach this performance are the third-order Shu–Osher flux-corrected method seen in Section 21.4 and Jameson's self-adjusting hybrid method seen in Section 22.3.

Test Case 2 As seen in Figure 23.3, the Harten–Osher reconstruction–evolution method captures the square wave without spurious overshoots and oscillations, and with a moderate amount of smearing. This performance could be dramatically improved by using subcell resolution as described in Section 9.4.

Test Case 3 In Figure 23.4, the dotted line represents the Harten–Osher reconstruction–evolution approximation to $u(x, 4)$, the long dashed line represents the Harten–Osher reconstruction–evolution approximation to $u(x, 40)$, and the solid line represents the exact solution for $u(x, 4)$ or $u(x, 40)$. This is an excellent performance.

Test Case 4 As seen in Figure 23.5, the Harten–Osher reconstruction–evolution method exhibits only minor error.

Test Case 5 As seen in Figure 23.6, the Harten–Osher reconstruction–evolution method fails to alter the initial conditions in any way, which leads to a large expansion shock. The next section will introduce an alternative temporal evolution that avoids this problem.

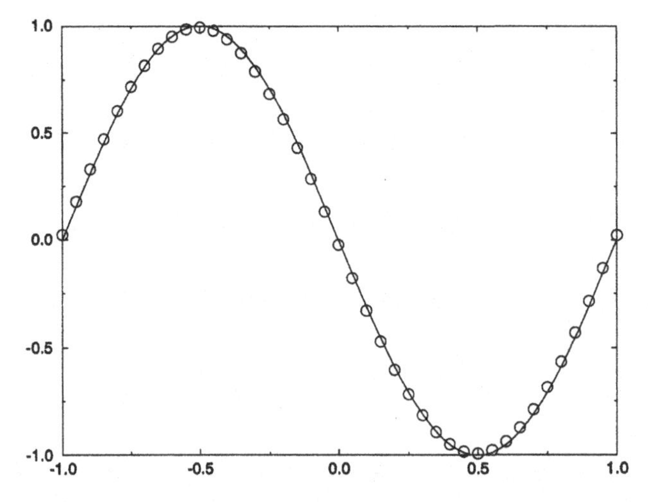

Figure 23.2 Harten–Osher reconstruction–evolution method for Test Case 1.

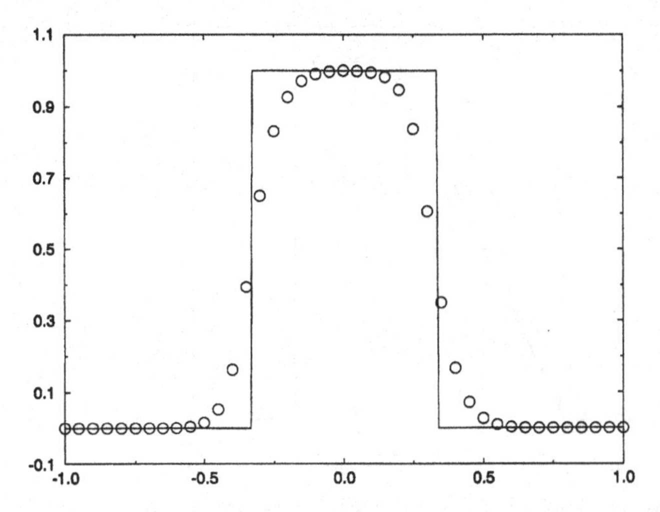

Figure 23.3 Harten–Osher reconstruction–evolution method for Test Case 2.

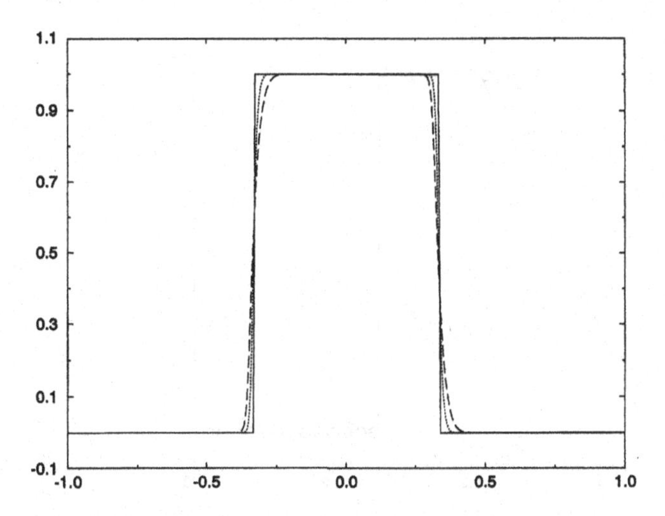

Figure 23.4 Harten–Osher reconstruction–evolution method for Test Case 3.

23.4.3 *The Euler Equations*

The version of UNO for the Euler equations is based on Roe's first-order upwind method as seen in Subsection 18.3.2. The results are omitted; see Harten and Osher (1987) for the straightforward details.

23.5 Harten–Engquist–Osher–Chakravarthy Reconstruction–Evolution Methods (ENO)

This section concerns the Harten–Engquist–Osher–Chakravarthy reconstruction–evolution methods, better known as the ENO methods. Before embarking on this section, the reader should review Chapter 9, especially Section 9.3. In fact, in its reconstruction phase, the Harten–Engquist–Osher–Chakravarthy reconstruction–evolution methods use reconstruction via the primitive function in conjunction with ENO reconstruction exactly

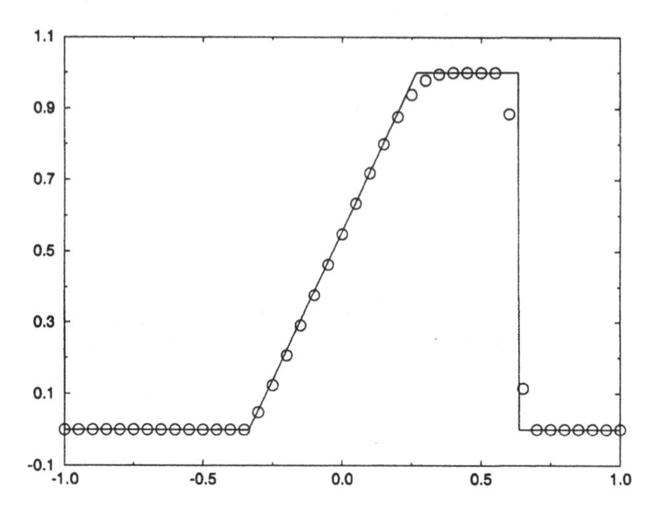

Figure 23.5 Harten–Osher reconstruction–evolution method for Test Case 4.

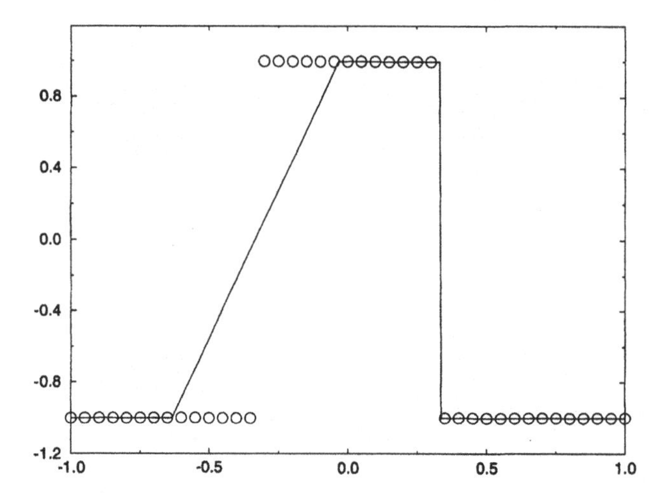

Figure 23.6 Harten–Osher reconstruction–evolution method for Test Case 5.

as described in Chapter 9. This section will not bother to repeat the details of the spatial reconstruction phase given in Chapter 9 and will instead focus entirely on the temporal evolution phase.

In many ways, the ENO methods are as close to the "ultimate" method that Van Leer had sought in the 1970s as any of the methods seen in this book. Certainly, on the five standard test cases for scalar conservation laws, second-order ENO yields the best results of any method seen in the book, as found in Subsection 23.5.1. However, ENO methods are slightly less stable than other methods, especially as the order of accuracy increases, in the sense that they sometimes allow a little more spurious overshoot and oscillation than, say, a typical flux-limited TVD method. Also, ENO methods are more complicated and costly than, say, a typical flux-limited method. Thus, for the time being, flux-limited methods such as those seen in Chapter 20 and "flux-limited-like" methods such as those seen in Section 23.3 will probably continue to dominate practical applications.

23.5.1 *Second-Order Accurate Temporal Evolution for Scalar Conservation Laws*

This subsection concerns second-order accurate temporal evolution for constant Δx. Recall that

$$\hat{f}^n_{i+1/2} \approx \frac{1}{\Delta t}\int_{t^n}^{t^{n+1}} f(u(x_{i+1/2},t))\,dt. \tag{11.3}$$

This can be approximated using the Cauchy–Kowalewski procedure. First, approximate Equation (11.3) using Equation (10.27) to obtain the following second-order accurate approximation:

$$\hat{f}^n_{i+1/2} = f(u(x_{i+1/2},t^{n+1/2})). \tag{23.51}$$

Assume a second-order accurate piecewise-linear reconstruction of the following form:

$$u(x,t^n) = \bar{u}^n_i + S^n_i(x-x_i), \tag{23.52}$$

where $x_{i-1/2} \le x \le x_{i+1/2}$, which is found as in the last section and as in Chapter 9. Then by Taylor series

$$u(x,t) \approx u(x_i,t^n) + \frac{\partial u}{\partial t}(x_i,t^n)(t-t^n) + \frac{\partial u}{\partial x}(x_i,t^n)(x-x_i). \tag{23.53}$$

Equation (23.52) immediately yields an expression for $\partial u/\partial x$. In particular,

$$\frac{\partial u}{\partial x}(x_i,t^n) \approx S^n_i. \tag{23.54}$$

To find an expression for $\partial u/\partial t$, use the Cauchy–Kowalewski procedure. In particular, notice that

$$\frac{\partial u}{\partial t}(x_i,t^n) = -a(u(x_i,t^n))\frac{\partial u}{\partial x}(x_i,t^n).$$

Then

$$\frac{\partial u}{\partial t}(x_i,t^n) \approx -a\left(\bar{u}^n_i\right)S^n_i. \tag{23.55}$$

Equations (23.53), (23.54), and (23.55) combine to yield the following:

$$u(x,t) \approx \bar{u}^n_i - a\left(\bar{u}^n_i\right)S^n_i(t-t^n) + S^n_i(x-x_i). \tag{23.56}$$

Then

$$u(x_{i+1/2},t^{n+1/2}) \approx \bar{u}_i + \frac{1}{2}\left(1-\lambda a\left(\bar{u}^n_i\right)\right)S_i\Delta x. \tag{23.57}$$

This expression uses the reconstruction in cell i. But suppose we use the reconstruction in cell $i+1$ given as follows:

$$u(x,t^n) = \bar{u}^n_{i+1} + S^n_{i+1}(x-x_{i+1}). \tag{23.58}$$

Then the identical process yields the following:

$$u(x_{i+1/2}, t^{n+1/2}) \approx \bar{u}_{i+1} - \frac{1}{2}\left(1 + \lambda a\left(\bar{u}^n_{i+1}\right)\right) S_{i+1} \Delta x. \tag{23.59}$$

Then let

$$f(u(x_{i+1/2}, t^{n+1/2}))$$
$$= avg\left[f\left(u_i + \frac{1}{2}\left(1 - \lambda a\left(\bar{u}^n_i\right)\right) S^n_i \Delta x\right), f\left(\bar{u}_{i+1} - \frac{1}{2}\left(1 + \lambda a\left(\bar{u}^n_{i+1}\right)\right) S^n_{i+1} \Delta x\right)\right],$$

where *avg* is some sort of average. Suppose that $S^n_i = S^n_{i+1} = 0$. In other words, suppose that the reconstruction is piecewise-constant. Then it would be nice if the method were equal to Godunov's first-order upwind method or to one of the other first-order reconstruction–evolution methods seen in Section 17.3. From Equation (17.14), recall that

$$\hat{f}^G_{i+1/2} = f^G\left(\bar{u}^n_i, \bar{u}^n_{i+1}\right) = \begin{cases} \min\limits_{\bar{u}^n_i \le u \le \bar{u}^n_{i+1}} f(u) & \text{if } \bar{u}^n_i < \bar{u}^n_{i+1}, \\ \max\limits_{\bar{u}^n_i \ge u \ge \bar{u}^n_{i+1}} f(u) & \text{if } \bar{u}^n_i > \bar{u}^n_{i+1}. \end{cases}$$

Then let us use f^G for the average. The final result is

♦ $$\hat{f}^n_{i+1/2} = f^G\left(\bar{u}^n_i + \frac{1}{2}\left[1 - \lambda a\left(\bar{u}^n_i\right)\right] S^n_i \Delta x, \bar{u}^n_{i+1} - \frac{1}{2}\left[1 + \lambda a\left(\bar{u}^n_{i+1}\right)\right] S^n_{i+1} \Delta x\right). \tag{23.60}$$

Consider the five standard test cases defined in Section 17.0. The numerical results for second-order accurate ENO are extremely similar to those of UNO, as seen in Subsection 23.4.2, except for Test Case 5. As seen in Figure 23.7, second-order accurate

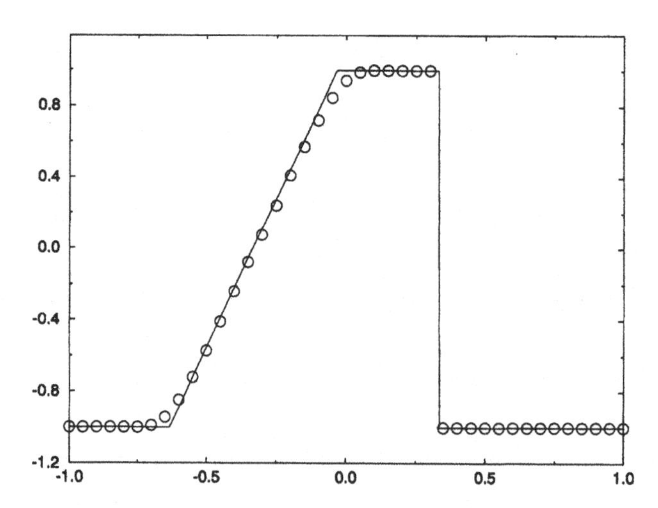

Figure 23.7 ENO method for Test Case 5.

ENO avoids any hint of the expansion shock that plagued UNO. Second-order ENO is easily one of the best performing methods seen in this book. In fact, after incorporating subcell resolution to reduce contact smearing, ENO is about as good as one could expect for a second-order accurate method, at least for scalar conservation laws on periodic or infinite domains.

23.5.2 *Third-Order Accurate Temporal Evolution for Scalar Conservation Laws*

This subsection concerns third-order accurate temporal evolution for constant Δx. First, approximate Equation (11.3) using Simpson's rule to obtain the following third-order accurate approximation:

$$\hat{f}^n_{i+1/2} = \frac{1}{3}[f(u(x_{i+1/2}, t^n)) + 4f(u(x_{i+1/2}, t^{n+1/2})) + f(u(x_{i+1/2}, t^{n+1}))]. \tag{23.61}$$

Assume a third-order accurate piecewise-quadratic reconstruction of the following form:

$$u(x, t^n) \approx \bar{u}^n_i - \frac{\Delta x^2}{24} C^n_i + S^n_i (x - x_i) + \frac{1}{2} C^n_i (x - x_i)^2, \tag{23.62}$$

where $x_{i-1/2} \leq x \leq x_{i+1/2}$, which is found using reconstruction via the primitive function as seen in Chapter 9. Then by Taylor series

$$\begin{aligned} u(x,t) &\approx u(x_i, t^n) + \frac{\partial u}{\partial t}(x_i, t^n)(t - t^n) + \frac{\partial u}{\partial x}(x_i, t^n)(x - x_i) \\ &+ \frac{\partial^2 u}{\partial t^2}(x_i, t^n)\frac{(t - t^n)^2}{2} + \frac{\partial^2 u}{\partial x \partial t}(x_i, t^n)(x - x_i)(t - t^n) \\ &+ \frac{\partial^2 u}{\partial x^2}(x_i, t^n)\frac{(x - x_i)^2}{2}. \end{aligned} \tag{23.63}$$

Equation (23.62) immediately yields expressions for $\partial u/\partial x$ and $\partial^2 u/\partial x^2$. In particular,

$$\frac{\partial u}{\partial}(x_i, t^n) \approx S^n_i \tag{23.64}$$

and

$$\frac{\partial^2 u}{\partial x^2}(x_i, t^n) \approx C^n_i. \tag{23.65}$$

To find expressions for $\partial u/\partial t$, $\partial^2 u/\partial t^2$, and $\partial^2 u/\partial x \partial t$ use the Cauchy–Kowalewski procedure. In particular, notice that

$$\frac{\partial u}{\partial t}(x_i, t^n) = -a(u(x_i, t^n))\frac{\partial u}{\partial x}(x_i, t^n),$$

$$\frac{\partial^2 u}{\partial x \partial t}(x_i, t^n) = -a'(u(x_i, t^n))\left[\frac{\partial u}{\partial x}(x_i, t^n)\right]^2 - a(u(x_i, t^n))\frac{\partial^2 u}{\partial x^2}(x_i, t^n),$$

$$\frac{\partial^2 u}{\partial t^2}(x_i, t^n) = 2a'(u(x_i, t^n))a(u(x_i, t^n))\left[\frac{\partial u}{\partial x}(x_i, t^n)\right]^2 + [a(u(x_i, t^n))]^2\frac{\partial^2 u}{\partial x^2}(x_i, t^n)$$

as seen in Section 17.2. Then

$$\frac{\partial u}{\partial t}(x_i, t^n) \approx -a\left(\bar{u}_i^n - \frac{\Delta x^2}{24}C_i\right)S_i^n, \tag{23.66}$$

$$\frac{\partial^2 u}{\partial x \partial t}(x_i, t^n) \approx -a'\left(\bar{u}_i^n - \frac{\Delta x^2}{24}\right)(S_i^n)^2 - a\left(\bar{u}_i^n - \frac{\Delta x^2}{24}\right)C_j^n, \tag{23.67}$$

$$\frac{\partial^2 u}{\partial x^2}(x_i, t^n) \approx 2a'\left(\bar{u}_i^n - \frac{\Delta x^2}{24}\right)a\left(\bar{u}_i^n - \frac{\Delta x^2}{24}\right)(S_i^n)^2 + \left[a\left(\bar{u}_i^n - \frac{\Delta x^2}{24}\right)\right]^2 C_j^n. \tag{23.68}$$

As a convenient notation, let

$$\bar{a}_i = a\left(\bar{u}_i^n - \frac{\Delta x^2}{24}\right),$$

$$\bar{a}_i' = a'\left(\bar{u}_i^n - \frac{\Delta x^2}{24}\right).$$

Equations (23.63)–(23.68) combine to yield

$$u(x_{i+1/2}, t) \approx \bar{u}_i^n + \left(\frac{1}{2} - \lambda\bar{a}_i\frac{t-t^n}{\Delta t}\right)\left(1 - \lambda\bar{a}_i'\frac{t-t^n}{\Delta t}S_i^n\Delta x\right)S_i^n\Delta x + \frac{1}{12}\left[1 - 6\lambda\bar{a}_i\frac{t-t^n}{\Delta t} + 6\lambda^2\bar{a}_i^2\left(\frac{t-t^n}{\Delta t}\right)^2\right]C_i^n\Delta x^2. \tag{23.69}$$

This expression uses the reconstruction in cell i. But suppose we use the reconstruction in cell $i+1$ instead. Then the identical process yields the following:

$$u(x_{i+1/2}, t) \approx \bar{u}_{i+1}^n - \left(\frac{1}{2} + \lambda\bar{a}_{i+1}\frac{t-t^n}{\Delta t}\right)\left(1 - \lambda\bar{a}_{i+1}'\frac{t-t^n}{\Delta t}S_{i+1}^n\Delta x\right)S_{i+1}^n\Delta x + \frac{1}{12}\left[1 + 6\lambda\bar{a}_{i+1}\frac{t-t^n}{\Delta t} + 6\lambda^2\bar{a}_{i+1}^2\left(\frac{t-t^n}{\Delta t}\right)^2\right]C_{i+1}^n\Delta x^2. \tag{23.70}$$

The last two expressions yield two approximations for each of $u(x_{i+1/2}, t^n)$, $u(x_{i+1/2}, t^{n+1/2})$, and $u(x_{i+1/2}, t^{n+1})$ as required by Equation (23.61). In each case, the two approximations may be averaged using the conservative numerical flux of Godunov's first-order upwind method, just as in the last subsection. In particular,

$$f(u(x_{i+1/2},t^n))$$
$$\approx f^G\left(\bar{u}_i^n+\frac{1}{2}S_i^n\Delta x+\frac{1}{12}C_i^n\Delta x^2,\bar{u}_{i+1}^n-\frac{1}{2}S_{i+1}^n\Delta x+\frac{1}{12}C_{i+1}^n\Delta x^2\right),$$

$$f(u(x_{i+1/2},t^{n+1/2}))\approx f^G\left[\bar{u}_i^n+\frac{1}{2}(1-\lambda\bar{a}_i)\left(1-\frac{1}{2}\lambda\bar{a}_i'S_i^n\Delta x\right)S_i^n\Delta x\right.$$
$$+\frac{1}{24}(2-6\lambda\bar{a}_i+3\lambda^2\bar{a}_i^2)C_i\Delta x^2,$$
$$\bar{u}_{i+1}^n-\frac{1}{2}(1+\lambda\bar{a}_{i+1})\left(1-\frac{1}{2}\lambda\bar{a}_{i+1}'S_{i+1}^n\Delta x\right)S_{i+1}^n\Delta x$$
$$\left.+\frac{1}{24}(2+6\lambda\bar{a}_{i+1}+3\lambda^2\bar{a}_{i+1}^2)C_{i+1}\Delta x^2\right],$$

and

$$f(u(x_{i+1/2},t^{n+1}))\approx f^G\left[\bar{u}_i^n+\left(\frac{1}{2}-\lambda\bar{a}_i\right)(1-\lambda\bar{a}_i'S_i^n\Delta x)S_i^n\Delta x\right.$$
$$+\frac{1}{12}(1-6\lambda\bar{a}_i+6\lambda^2\bar{a}_i^2)C_i\Delta x^2,$$
$$\bar{u}_{i+1}^n-\left(\frac{1}{2}+\lambda\bar{a}_{i+1}\right)(1-\lambda\bar{a}_{i+1}'S_{i+1}^n\Delta x)S_{i+1}^n\Delta x$$
$$\left.+\frac{1}{12}(1+6\lambda\bar{a}_{i+1}+6\lambda^2\bar{a}_{i+1}^2)C_{i+1}\Delta x^2\right].$$

Clearly, comparing this subsection with the last, we see that the complexity of Cauchy–Kowalewski technique increases rapidly with the order of accuracy.

Consider the five standard test cases defined in Section 17.0. The numerical results for third-order accurate ENO are extremely similar to those of second-order accurate ENO and UNO, as seen in Figures 23.2–23.7. Specifically, the only visible difference is somewhat less smearing in Test Cases 2 and 3. The plots are omitted since they are so similar to the earlier plots.

23.5.3 *Second-Order Accurate Temporal Evolution for the Euler Equations*

This subsection concerns second-order accurate temporal evolution for constant Δx for the Euler equations. First, approximate Equation (11.3) using Equation (10.27) to obtain the following second-order accurate approximation:

$$\hat{\mathbf{f}}_{i+1/2}^n=\mathbf{f}(\mathbf{u}(x_{i+1/2},t^{n+1/2})).\tag{23.71}$$

Obtain a piecewise-linear reconstruction as follows:

(1) Transform the conservative variables $\mathbf{u}_i^n$ to approximate characteristic variables $\mathbf{v}_i^n$ using $\mathbf{v}_i^n=Q^{-1}\mathbf{u}_i^n$, where Q^{-1} is a locally frozen value of the matrix given by Equation (3.24), such as $Q^{-1}(\mathbf{u}_i^n)$.

(2) Reconstruct the characteristic variables from the samples found in the first step. In other words, approximate $\mathbf{v}(x, t^n)$ given $\mathbf{v}_i^n$. For each individual characteristic variable $(\mathbf{v}_i^n)_j$, use reconstruction via the primitive function as described in Chapter 9.

(3) Transform the reconstructed characteristic variables $\mathbf{v}(x, t^n)$ to conservative variables using $u(x, t^n) = Qv(x, t^n)$, where Q is a locally frozen value of the matrix given by Equation (3.23). Make sure to write the reconstruction for each of the conservative variables in Taylor series form.

(4) Express $\mathbf{u}(x_{i+1/2}, t^{n+1/2})$ in terms of a Taylor series about $(x_{i+1/2}, t^n)$:

$$\mathbf{u}(x_{i+1/2}, t^{n+1/2}) \approx \mathbf{u}\big(x_i, t^n\big) + \frac{1}{2}\Delta t \frac{\partial \mathbf{u}}{\partial t}\big(x_i, t^n\big) + \frac{1}{2}\Delta x \frac{\partial \mathbf{u}}{\partial x}\big(x_i, t^n\big).$$

The results of step 3 immediately yield $\partial \mathbf{u}/\partial x$. Then we need to find $\partial \mathbf{u}/\partial t$ in terms of $\partial \mathbf{u}/\partial x$, that is, we need to find $(\partial \rho/\partial t, \partial(\rho u)/\partial t, \partial(\rho e_T)/\partial t)$ in terms of $(\partial \rho/\partial x, \partial(\rho u)/\partial x, \partial(\rho e_T)/\partial x)$. Fortunately, there is a clever trick: Transform $(\partial \rho/\partial x, \partial(\rho u)/\partial x, \partial(\rho e_T)/\partial x)$ to $(\partial \rho/\partial x, \partial(\rho u)/\partial x, \partial p/\partial x, \partial u/\partial x)$. To do this, first, the product rule implies

$$\frac{\partial u}{\partial x} = \frac{1}{\rho}\left(\frac{\partial(\rho u)}{\partial x} - u\frac{\partial \rho}{\partial x}\right). \tag{23.72}$$

Second, Equation (2.10) implies

$$\frac{\partial p}{\partial x} = (\gamma - 1)\left[\frac{\partial(\rho e_T)}{\partial x} - \frac{1}{2}\left(\rho u \frac{\partial u}{\partial x} + u\frac{\partial(\rho u)}{\partial x}\right)\right].$$

Combine this with Equation (23.72) to find

$$\frac{\partial p}{\partial x} = (\gamma - 1)\left[\frac{\partial(\rho e_T)}{\partial x} - \frac{1}{2}\left(2u\frac{\partial(\rho u)}{\partial x} - u^2\frac{\partial \rho}{\partial x}\right)\right]. \tag{23.73}$$

With $(\partial \rho/\partial x, \partial(\rho u)/\partial x, \partial p/\partial x, \partial u/\partial x)$ in hand, we can now find $(\partial \rho/\partial t, \partial(\rho u)/\partial t, \partial p/\partial t, \partial u/\partial t)$ using the Euler equations. Write conservation of mass in conservation form to find

$$\frac{\partial \rho}{\partial t} = -\frac{\partial(\rho u)}{\partial x}. \tag{23.74}$$

Write conservation of momentum in conservation form to find

$$\frac{\partial(\rho u)}{\partial t} = -\frac{\partial(\rho u^2)}{\partial x} - \frac{\partial p}{\partial x}$$

or, using the product rule,

$$\frac{\partial(\rho u)}{\partial t} = -\rho u\frac{\partial u}{\partial x} - u\frac{\partial(\rho u)}{\partial x} - \frac{\partial p}{\partial x}. \tag{23.75}$$

Write conservation of energy in primitive variable form to find

$$\frac{\partial p}{\partial t} = -u\frac{\partial p}{\partial x} - \rho a^2\frac{\partial u}{\partial x}$$

or, using $a^2 = \gamma p/\rho$,

$$\frac{\partial p}{\partial t} = -u\frac{\partial p}{\partial x} - \gamma p\frac{\partial u}{\partial x}. \tag{23.76}$$

Finally, by the product rule

$$\frac{\partial u}{\partial t} = \frac{1}{\rho}\left(\frac{\partial(\rho u)}{\partial t} - u\frac{\partial \rho}{\partial t}\right). \tag{23.77}$$

With $(\partial\rho/\partial t, \partial(\rho u)/\partial t, \partial p/\partial t, \partial u/\partial t)$ in hand, we can then find $(\partial\rho/\partial t, \partial(\rho u)/\partial t, \partial(\rho e_T)/\partial t)$. In particular, Equation (2.10) implies

$$\frac{\partial(\rho e_T)}{\partial t} = \frac{1}{\gamma - 1}\frac{\partial p}{\partial t} + \frac{1}{2}\frac{\partial(\rho u^2)}{\partial t}$$

or

$$\frac{\partial(\rho e_T)}{\partial t} = \frac{1}{\gamma - 1}\frac{\partial p}{\partial t} + \frac{1}{2}\left(\rho u\frac{\partial u}{\partial t} + u\frac{\partial(\rho u)}{\partial t}\right). \tag{23.78}$$

(5) Use the results of step 4 to find two approximations to $\mathbf{u}(x_{i+1/2}, t^{n+1/2})$, one from cell i called $\mathbf{u}_i(x_{i+1/2}, t^{n+1/2})$ and one from cell $i+1$ called $\mathbf{u}_{i+1}(x_{i+1/2}, t^{n+1/2})$.

(6) Average the two approximations $\mathbf{u}_i(x_{i+1/2}, t^{n+1/2})$ and $\mathbf{u}_{i+1}(x_{i+1/2}, t^{n+1/2})$ found in step 5 using the conservative numerical flux of any of the first-order accurate reconstruction–evolution methods seen in Section 18.3, such as Godunov's first-order upwind method, to find

$$\hat{\mathbf{f}}^n_{i+1/2} = \mathbf{f}^G(\mathbf{u}_i(x^n_{i+1/2}, t^{n+1/2}), \mathbf{u}_{i+1}(x_{i+1/2}, t^{n+1/2})).$$

While far from trivial, the above second-order temporal evolution procedure is less complicated and more readily comprehensible than most of the ones that preceded it. Furthermore, unlike most most of the temporal evolution procedures that preceded it, the above temporal evolution procedure extends to arbitrary orders of accuracy in a reasonably straightforward manner, although cost and complexity increase rapidly as the order of accuracy increases.

References

Anderson, W. K., Thomas, J. L., and Van Leer, B. 1986. "Comparison of Finite Volume Flux Volume Flux Vector Splittings for the Euler Equations," *AIAA Journal*, 24: 1453–1460.

Chang, S.-C. 1995. "The Method of Space-Time Conservation Element and Solution Element – A New Approach for Solving the Navier–Stokes and Euler Equations," *Journal of Computational Physics*, 119: 295–324.

Colella, P. 1985. "A Direct Eulerian MUSCL Scheme for Gas Dynamics," *SIAM Journal of Scientific and Statistical Computing*, 6: 104–117.

Colella, P., and Woodward, P. 1984. "The Piecewise Parabolic Method (PPM) for Gas-Dynamical Simulations," *Journal of Computational Physics*, 54: 174–201.

Harten, A. 1989. "ENO Schemes with Subcell Resolution," *Journal of Computational Physics*, 83: 148–184.

Harten, A., Engquist, B., Osher, S., and Chakravarthy, S. R. 1987. "Uniformly High Order Accurate Essentially Non-Oscillatory Schemes. III," *Journal of Computational Physics*, 71: 231–303.

Harten, A., and Osher, S. 1987. "Uniformly High-Order Accurate Nonoscillatory Schemes. I," *SIAM Journal on Numerical Analysis*, 24: 279–309.

Van Leer, B. 1977. "Towards the Ultimate Conservative Difference Scheme IV. A New Approach to Numerical Convection," *Journal of Computational Physics*, 23: 276–299.

Van Leer, B. 1979. "Towards the Ultimate Conservative Difference Scheme V. A Second-Order Sequel to Godunov's Method," *Journal of Computational Physics*, 32: 101–136.

Venkatakrishnan, V. 1995. "Convergence to Steady State Solutions of the Euler Equations on Unstructured Grids with Limiters," *Journal of Computational Physics*, 118: 120–130.

Weiss, J. M., and Smith, W. A. 1995. "Preconditioning Applied to Variable and Constant Density Flow," *AIAA Journal*, 33: 2050–2057.

Woodward, P., and Colella, P. 1984. "The Numerical Simulation of Two-Dimensional Fluid Flow with Strong Shocks," *Journal of Computational Physics*, 54: 115–173.

Problems

23.1 (a) Subsection 23.4.2 describes the UNO method for scalar conservation laws. Show that, for the linear advection equation, UNO uses the exact time evolution.

(b) Subsection 23.5.1 and especially Equation (23.60) describe the second-order ENO method for scalar conservation laws. Show that, for the linear advection equation, second-order ENO uses the exact time evolution.

23.2 Consider the TVD version of the UNO method for the linear advection equation, with slopes given by Equation (23.49). Show that this method is exactly the same as Sweby's flux-limited method for the linear advection equation, as described in Section 20.2, with the minmod flux limiter $\phi(r) = minmod(r, 1)$.

23.3 (a) Consider piecewise-linear reconstruction using subcell resolution, as described in Section 9.4. The reconstruction in cell i is as follows:

$$u(x, t^n) \approx \bar{u}_i^n + S_i^n(x - x_i)$$

if no jump discontinuity is found in cell i. Otherwise, if a jump discontinuity is found in cell i at $x = \theta_i^n$ then

$$u(x, t^n) \approx \bar{u}_{i+1}^n + S_{i+1}^n(x - x_{i+1})$$

for $\theta_i^n < x_{i+1/2} - a\Delta t$ and

$$u(x, t^n) \approx \bar{u}_{i-1}^n + S_{i-1}^n(x - x_{i-1})$$

for $\theta_i^n > x_{i+1/2} - a\Delta t$. For the linear advection equation with $a > 0$, show that the exact temporal evolution yields

$$\hat{f}_{i+1/2}^n = a\bar{u}_i^n + \frac{1}{2}a(1 - \lambda a)S_i^n\Delta x$$

if no jump discontinuity is found in cell i. Otherwise, if a jump discontinuity is found in cell i at $x = \theta_i^n$ then

$$\hat{f}_{i+1/2}^n = a\bar{u}_{i+1}^n - \frac{1}{2}a(1 + \lambda a)S_{i+1}^n\Delta x$$

for $\theta_i^n < x_{i+1/2} - a\Delta t$ and

$$\hat{f}^n_{i+1/2} = a\bar{u}^n_{i-1} + \frac{1}{2}a(3 - \lambda a)S^n_{i-1}\Delta x + \frac{\bar{u}^n_i - \bar{u}^n_{i-1} - S^n_{i-1}\Delta x}{\lambda}$$

for $\theta_i^n > x_{i+1/2} - a\Delta t$.

(b) Consider piecewise-quadratic ENO reconstruction using subcell resolution, as described in Section 9.4. The reconstruction in cell i is as follows:

$$u(x, t^n) \approx \bar{u}^n_i - \frac{\Delta x^2}{24}C^n_i + S^n_i(x - x_i) + \frac{1}{2}C^n_i(x - x_i)^2$$

if no jump discontinuity is found in cell i. Otherwise, if a jump discontinuity is found in cell i at $x = \theta_i^n$, then

$$u(x, t^n) \approx \bar{u}^n_{i+1} - \frac{1}{24}C^n_{i+1} + S^n_{i+1}(x - x_{i+1}) + \frac{1}{2}C^n_{i+1}(x - x_{i+1})^2$$

for $\theta_i^n < x_{i+1/2} - a\Delta t$ and

$$u(x, t^n) \approx \bar{u}^n_{i-1} - \frac{1}{24}C^n_{i-1} + S^n_{i-1}(x - x_{i-1}) + \frac{1}{2}C^n_{i-1}(x - x_{i-1})^2$$

for $\theta_i^n > x_{i+1/2} - a\Delta t$. For the linear advection equation with $a > 0$, show that the exact temporal evolution yields

$$\hat{f}^n_{i+1/2} = a\bar{u}^n_i + \frac{1}{2}a(1 - \lambda a)S^n_i\Delta x + \frac{a}{12}(1 - \lambda a)(1 - 2\lambda a)C^n_i\Delta x^2$$

if no jump discontinuity is found in cell i. Otherwise, suppose a jump discontinuity is found in cell i at $x = \theta_i^n$. Then for $\theta_i^n < x_{i+1/2} - a\Delta t$:

$$\hat{f}^n_{i+1/2} = a\bar{u}^n_{i+1} - \frac{1}{2}a(1 + \lambda a)S^n_{i+1}\Delta x + \frac{a}{12}(1 + \lambda a)(1 + 2\lambda a)C^n_{i+1}\Delta x^2$$

and for $\theta_i^n > x_{i+1/2} - a\Delta t$:

$$\begin{aligned}\hat{f}^n_{i+1/2} = {} & a\bar{u}^n_{i-1} + \frac{1}{2}a(3 - \lambda a)S^n_{i-1}\Delta x \\ & + \frac{1}{\lambda}\left(\bar{u}^n_i - \bar{u}^n_{i-1} - S^n_{i-1}\Delta x - \frac{1}{2}C^n_{i-1}\Delta x^2\right) \\ & + \frac{a}{12}(13 - 9\lambda a + 2(\lambda a)^2)C^n_{i-1}\Delta x^2.\end{aligned}$$

23.4 Consider piecewise-linear ENO reconstruction using subcell resolution, as described in Section 9.4. By Equation (9.38), the shock location θ_i^n is determined by $F(\theta_i^n) = 0$, where

$$\begin{aligned}F(\theta) = {} & \frac{1}{\Delta x}\int_{x_{i-1/2}}^{\theta}\left[\bar{u}^n_{i-1} + S^n_{i-1}(x - x_{i-1})\right]dx \\ & + \frac{1}{\Delta x}\int_{\theta}^{x_{i+1/2}}\left[\bar{u}^n_{i+1} + S^n_{i+1}(x - x_{i+1})\right]dx - \bar{u}^n_i.\end{aligned}$$

The implicit equation $F(\theta_i^n) = 0$ can be expensive to solve. However, from the previous problem, notice that θ_i^n is not actually required. Instead, we only need to know if $\theta_i^n > x_{i+1/2} - a\Delta t$ or $\theta_i^n < x_{i+1/2} - a\Delta t$.

(a) Show that $\theta_i^n > x_{i+1/2} - a\Delta t$ if $F(x_{i-1/2})F(x_{i+1/2} - a\Delta t) > 0$ and $\theta_i^n < x_{i+1/2} - a\Delta t$ if $F(x_{i-1/2})F(x_{i+1/2} - a\Delta t) < 0$.

(b) Show that $F(x_{i-1/2}) = \bar{u}_{i+1}^n - \bar{u}_i^n - S_{i+1}\Delta x$ and

$$F(x_{i+1/2} - a\Delta t) = (1 - \lambda a)\left[\bar{u}_{i-1}^n + \frac{1}{2}(2 - \lambda a)S_{i-1}^n \Delta x\right] + \lambda a\left[\bar{u}_{i+1}^n - \frac{1}{2}(1 + \lambda a)S_{i+1}^n \Delta x\right] - \bar{u}_i^n.$$

CHAPTER 24

A Brief Introduction to Multidimensions

24.0 Introduction

This chapter briefly introduces computational gasdynamics in multidimensions. A full discussion of computational gasdynamics in multidimensions would require repeating the entire book point by point; this chapter only has space for multidimensional versions of Chapters 2, 3, and 11, and then only in a terse "just the facts" form. However, even this should be enough to clearly demonstrate how the one-dimensional concepts from the rest of the book carry over directly to multidimensional approximations.

24.1 Governing Equations

This section presents a brief multidimensional version of Chapter 2. In three dimensions, consider an arbitrary control volume R with surface S and surface normal $\hat{n} = (n_x, n_y, n_z)$. Also consider an arbitrary time interval $[t_1, t_2]$. Let $\mathbf{V} = (u, v, w)$ be the velocity vector and $e_T = e + \frac{1}{2}V^2$ be the total energy per unit mass. Then the integral form of the three-dimensional (3D) Euler equations can be written as

$$\iiint_R [\mathbf{u}(\mathbf{x}, t_2) - \mathbf{u}(\mathbf{x}, t_1)]\, dR = -\int_{t_1}^{t_2} \iint_S (\mathbf{f}n_x + \mathbf{g}n_y + \mathbf{h}n_z)\, dS\, dt.$$

Also, the conservative differential form of the 3D Euler equations can be written as

$$\frac{\partial \mathbf{u}}{\partial t} + \frac{\partial \mathbf{f}}{\partial x} + \frac{\partial \mathbf{g}}{\partial y} + \frac{\partial \mathbf{h}}{\partial z} = 0,$$

where

$$\mathbf{u} = \begin{bmatrix} \rho \\ \rho \mathbf{V} \\ \rho e_T \end{bmatrix} = \text{vector of conserved quantities}$$

and

$$\mathbf{f} = \begin{bmatrix} \rho u \\ \rho u^2 + p \\ \rho uv \\ \rho uw \\ (\rho e_T + p)u \end{bmatrix}, \qquad \mathbf{g} = \begin{bmatrix} \rho v \\ \rho uv \\ \rho v^2 + p \\ \rho vw \\ (\rho e_T + p)v \end{bmatrix}, \qquad \mathbf{h} = \begin{bmatrix} \rho v \\ \rho uw \\ \rho vw \\ \rho w^2 + p \\ (\rho e_T + p)w \end{bmatrix}.$$

Define flux Jacobian matrices as follows:

$$A = \frac{d\mathbf{f}}{d\mathbf{u}}, \qquad B = \frac{d\mathbf{g}}{d\mathbf{u}}, \qquad C = \frac{d\mathbf{h}}{d\mathbf{u}}.$$

Then the conservative differential form of the 3D Euler equations can be written as

$$\frac{\partial \mathbf{u}}{\partial t} + A\frac{\partial \mathbf{u}}{\partial x} + B\frac{\partial \mathbf{u}}{\partial y} + C\frac{\partial \mathbf{u}}{\partial z} = 0,$$

where

$$A = \begin{bmatrix} 0 & 1 & 0 & 0 & 0 \\ -u^2 + \frac{\gamma-1}{2}\mathbf{V}\cdot\mathbf{V} & (3-\gamma)u & -(\gamma-1)v & -(\gamma-1)w & \gamma-1 \\ -uv & v & u & 0 & 0 \\ -uw & w & 0 & u & 0 \\ -(\gamma e_T - (\gamma-1)\mathbf{V}\cdot\mathbf{V})u & \gamma e_T - \frac{\gamma-1}{2}(2u^2 + \mathbf{V}\cdot\mathbf{V}) & -(\gamma-1)uv & -(\gamma-1)uw & \gamma u \end{bmatrix},$$

$$B = \begin{bmatrix} 0 & 0 & 1 & 0 & 0 \\ -uv & v & u & 0 & 0 \\ -v^2 + \frac{\gamma-1}{2}\mathbf{V}\cdot\mathbf{V} & -(\gamma-1)u & (3-\gamma)v & -(\gamma-1)w & \gamma-1 \\ -vw & 0 & w & v & 0 \\ -(\gamma e_T - (\gamma-1)\mathbf{V}\cdot\mathbf{V})v & -(\gamma-1)uv & \gamma e_T - \frac{\gamma-1}{2}(2v^2 + \mathbf{V}\cdot\mathbf{V}) & -(\gamma-1)vw & \gamma v \end{bmatrix},$$

$$C = \begin{bmatrix} 0 & 0 & 0 & 1 & 0 \\ -uw & w & 0 & u & 0 \\ -vw & 0 & w & v & 0 \\ -w^2 + \frac{\gamma-1}{2}\mathbf{V}\cdot\mathbf{V} & -(\gamma-1)u & -(\gamma-1)v & (3-\gamma)w & \gamma-1 \\ -(\gamma e_T - (\gamma-1)\mathbf{V}\cdot\mathbf{V})w & -(\gamma-1)uw & -(\gamma-1)vw & \gamma e_T - \frac{\gamma-1}{2}(2w^2 + \mathbf{V}\cdot\mathbf{V}) & \gamma w \end{bmatrix}.$$

In two dimensions, consider an arbitrary control volume A with perimeter C and perimeter normal $\hat{n} = (n_x, n_y)$. Also consider an arbitrary time interval $[t_1, t_2]$. Let $\mathbf{V} = (u, v)$ be the velocity vector and $e_T = e + \frac{1}{2}V^2$ be the total energy per unit mass. Then the integral form of the two-dimensional (2D) Euler equations can be written as

$$\iint_A [\mathbf{u}(\mathbf{x}, t_2) - \mathbf{u}(\mathbf{x}, t_1)]\, dR = -\int_{t_1}^{t_2} \int_C (\mathbf{f}n_x + \mathbf{g}n_y)\, dS\, dt.$$

Also, the conservative differential form of the 2D Euler equations can be written as

$$\frac{\partial \mathbf{u}}{\partial t} + \frac{\partial \mathbf{f}}{\partial x} + \frac{\partial \mathbf{g}}{\partial y} = 0,$$

where

$$\mathbf{u} = \begin{bmatrix} \rho \\ \rho\mathbf{V} \\ \rho e_T \end{bmatrix} = \text{vector of conserved quantities}$$

and

$$\mathbf{f} = \begin{bmatrix} \rho u \\ \rho u^2 + p \\ \rho u v \\ (\rho e_{\mathrm{T}} + p) u \end{bmatrix}, \qquad \mathbf{g} = \begin{bmatrix} \rho v \\ \rho u v \\ \rho v^2 + p \\ (\rho e_{\mathrm{T}} + p) v \end{bmatrix}.$$

Define flux Jacobian matrices as follows:

$$A = \frac{d\mathbf{f}}{d\mathbf{u}}, \qquad B = \frac{d\mathbf{g}}{d\mathbf{u}}.$$

Then the conservative differential form of the 2D Euler equations can be written as

$$\frac{\partial \mathbf{u}}{\partial t} + A \frac{\partial \mathbf{u}}{\partial x} + B \frac{\partial \mathbf{u}}{\partial y} = 0,$$

where

$$A = \begin{bmatrix} 0 & 1 & 0 & 0 \\ -u^2 + \frac{\gamma-1}{2}\mathbf{V}\cdot\mathbf{V} & (3-\gamma)u & -(\gamma-1)v & \gamma - 1 \\ -uv & v & u & 0 \\ -(\gamma e_{\mathrm{T}} - (\gamma-1)\mathbf{V}\cdot\mathbf{V})u & \gamma e_{\mathrm{T}} - \frac{\gamma-1}{2}(2u^2 + \mathbf{V}\cdot\mathbf{V}) & -(\gamma-1)uv & \gamma u \end{bmatrix},$$

$$B = \begin{bmatrix} 0 & 0 & 1 & 0 \\ -uv & v & u & 0 \\ -v^2 + \frac{\gamma-1}{2}\mathbf{V}\cdot\mathbf{V} & -(\gamma-1)u & (3-\gamma)v & \gamma - 1 \\ -(\gamma e_{\mathrm{T}} - (\gamma-1)\mathbf{V}\cdot\mathbf{V})v & -(\gamma-1)uv & \gamma e_{\mathrm{T}} - \frac{\gamma-1}{2}(2v^2 + \mathbf{V}\cdot\mathbf{V}) & \gamma v \end{bmatrix}.$$

24.2 Waves

This section is a brief multidimensional version of Chapter 3. Waves behave much differently in multidimensions than they do in one dimension. Some of the complicating factors include:

- If the spatial dimension is N, waves may have any dimension between 1 and N in the $\mathbf{x}$–t plane.
- Waves may travel in an infinite number of directions. For example, two-dimensional waves may travel at any angle between 0° and 360°.
- Characteristic variables may not be constant along characteristics.
- The characteristic form is not unique. In other words, there are infinitely many wave descriptions of a given multidimensional flow.

In three dimensions, suppose that the characteristic surfaces for all five families have a common unit normal $\hat{\mathbf{n}}$ in the $\mathbf{x}$ plane. You can choose $\hat{\mathbf{n}}$ however you like. The characteristic form of the 3D Euler equations written in terms of primitive variables is

$$c_i'\left(\frac{\partial \rho}{\partial t}+\mathbf{V}\cdot\nabla\rho\right)+\left(\hat{\mathbf{n}}\times\mathbf{b}_i'\right)\cdot\left(\frac{\partial \mathbf{V}}{\partial t}+\mathbf{V}\cdot\nabla\mathbf{V}+\frac{\nabla p}{\rho}\right)$$

$$-\frac{1}{a^2}c_i'\left(\frac{\partial p}{\partial t}+\mathbf{V}\cdot\nabla p\right)=0 \qquad (i=1,2,3),$$

$$\hat{\mathbf{n}}\cdot\left(\frac{\partial \mathbf{V}}{\partial t}+\mathbf{V}\cdot\nabla\mathbf{V}\right)+a\nabla\cdot\mathbf{V}+\frac{1}{\rho a}\left(\frac{\partial p}{\partial t}+(\mathbf{V}+a\hat{\mathbf{n}})\cdot\nabla p\right)=0,$$

$$-\hat{\mathbf{n}}\cdot\left(\frac{\partial \mathbf{V}}{\partial t}+\mathbf{V}\cdot\nabla\mathbf{V}\right)+a\nabla\cdot\mathbf{V}+\frac{1}{\rho a}\left(\frac{\partial p}{\partial t}+(\mathbf{V}-a\hat{\mathbf{n}})\cdot\nabla p\right)=0,$$

where a is the speed of sound, $\mathbf{c}'=(c_1',c_2',c_3')$ is an arbitrary vector, and $B'=[\mathbf{b}_1' \mid \mathbf{b}_2' \mid \mathbf{b}_3']$ is an arbitrary set of vectors.

The 3D characteristic variables are defined as follows:

$$dv_i=\begin{bmatrix} c_i' \\ \hat{\mathbf{n}}\times\mathbf{b}_i' \\ -\frac{1}{a^2}c_i' \end{bmatrix}\cdot\begin{bmatrix} d\rho \\ d\mathbf{V} \\ dp \end{bmatrix}=c_i'd\rho+\left(\hat{\mathbf{n}}\times\mathbf{b}_i'\right)\cdot d\mathbf{V}-\frac{1}{a^2}c_i'\,dp \qquad (i=1,2,3),$$

$$dv_4=e\begin{bmatrix} 0 \\ \hat{\mathbf{n}} \\ \frac{1}{\rho a} \end{bmatrix}\cdot\begin{bmatrix} d\rho \\ d\mathbf{V} \\ dp \end{bmatrix}=e\hat{\mathbf{n}}\cdot d\mathbf{V}+\frac{e}{\rho a}dp,$$

$$dv_5=f\begin{bmatrix} 0 \\ -\hat{\mathbf{n}} \\ \frac{1}{\rho a} \end{bmatrix}\cdot\begin{bmatrix} d\rho \\ d\mathbf{V} \\ dp \end{bmatrix}=-f\hat{\mathbf{n}}\cdot d\mathbf{V}+\frac{f}{\rho a}dp,$$

where e and f are arbitrary. The characteristic form of the 3D Euler equations written in terms of characteristic variables is

$$\frac{\partial v_1}{\partial t}+\mathbf{V}\cdot\nabla v_1+\frac{a}{2}\left(\hat{\mathbf{n}}\times\mathbf{b}_1'\right)\cdot\nabla\left(\frac{1}{e}v_4+\frac{1}{f}v_5\right)=0,$$

$$\frac{\partial v_2}{\partial t}+\mathbf{V}\cdot\nabla v_2+\frac{a}{2}\left(\hat{\mathbf{n}}\times\mathbf{b}_2'\right)\cdot\nabla\left(\frac{1}{e}v_4+\frac{1}{f}v_5\right)=0,$$

$$\frac{\partial v_3}{\partial t}+\mathbf{V}\cdot\nabla v_3+\frac{a}{2}\left(\hat{\mathbf{n}}\times\mathbf{b}_3'\right)\cdot\nabla\left(\frac{1}{e}v_4+\frac{1}{f}v_5\right)=0,$$

$$\frac{\partial v_4}{\partial t}+(\mathbf{V}+a\hat{\mathbf{n}})\cdot\nabla v_4+ea(\hat{\mathbf{n}}\times\mathbf{b}_1)\cdot\nabla v_1$$

$$+ea(\hat{\mathbf{n}}\times\mathbf{b}_2)\cdot\nabla v_2+ea(\hat{\mathbf{n}}\times\mathbf{b}_3)\cdot\nabla v_3=0,$$

$$\frac{\partial v_5}{\partial t}+(\mathbf{V}-a\hat{\mathbf{n}})\cdot\nabla v_5+fa(\hat{\mathbf{n}}\times\mathbf{b}_1)\cdot\nabla v_1$$

$$+fa(\hat{\mathbf{n}}\times\mathbf{b}_2)\cdot\nabla v_2+fa(\hat{\mathbf{n}}\times\mathbf{b}_3)\cdot\nabla v_3=0.$$

In two dimensions, suppose that the characteristic curves for all four families have a common unit normal $\hat{\mathbf{n}}$ in the $\mathbf{x}$ plane. You can choose $\hat{\mathbf{n}}$ however you like. The characteristic form of the 2D Euler equations written in terms of primitive variables is

$$c_1'\left(\frac{\partial \rho}{\partial t}+\mathbf{V}\cdot\nabla\rho\right)+b_1'(n_y,-n_x)\cdot\left(\frac{\partial \mathbf{V}}{\partial t}+\mathbf{V}\cdot\nabla\mathbf{V}+\frac{\nabla p}{\rho}\right)-\frac{1}{a^2}c_1'\left(\frac{\partial p}{\partial t}+\mathbf{V}\cdot\nabla p\right)=0,$$

$$c_2'\left(\frac{\partial \rho}{\partial t}+\mathbf{V}\cdot\nabla\rho\right)+b_2'(n_y,-n_x)\cdot\left(\frac{\partial \mathbf{V}}{\partial t}+\mathbf{V}\cdot\nabla\mathbf{V}+\frac{\nabla p}{\rho}\right)-\frac{1}{a^2}c_2'\left(\frac{\partial p}{\partial t}+\mathbf{V}\cdot\nabla p\right)=0,$$

$$\hat{\mathbf{n}}\cdot\left(\frac{\partial \mathbf{V}}{\partial t}+\mathbf{V}\cdot\nabla\mathbf{V}\right)+a\nabla\cdot\mathbf{V}+\frac{1}{\rho a}\left(\frac{\partial p}{\partial t}+(\mathbf{V}+a\hat{\mathbf{n}})\cdot\nabla p\right)=0,$$

$$-\hat{\mathbf{n}}\cdot\left(\frac{\partial \mathbf{V}}{\partial t}+\mathbf{V}\cdot\nabla\mathbf{V}\right)+a\nabla\cdot\mathbf{V}+\frac{1}{\rho a}\left(\frac{\partial p}{\partial t}+(\mathbf{V}-a\hat{\mathbf{n}})\cdot\nabla p\right)=0,$$

where $\mathbf{b}'=(b_1',b_2')$ and $\mathbf{c}'=(c_1',c_2')$ are arbitrary.

The 2D characteristic variables are defined as follows:

$$\begin{aligned}
dv_1 &= c_1'd\rho+b_1'(n_y,-n_x)\cdot d\mathbf{V}-\frac{1}{a^2}c_1'\,dp,\\
dv_2 &= c_2'd\rho+b_2'(n_y,-n_x)\cdot d\mathbf{V}-\frac{1}{a^2}c_2'\,dp,\\
dv_3 &= e\hat{\mathbf{n}}\cdot d\mathbf{V}+\frac{e}{\rho a}\,dp,\\
dv_4 &= -f\hat{\mathbf{n}}\cdot d\mathbf{V}+\frac{f}{\rho a}\,dp,
\end{aligned}$$

where e and f are arbitrary. The characteristic form of the 2D Euler equations written in terms of characteristic variables is

$$\begin{aligned}
&\frac{\partial v_1}{\partial t}+\mathbf{V}\cdot\nabla v_1+\frac{a}{2}b_1'(n_y,-n_x)\cdot\nabla\left(\frac{1}{e}v_3+\frac{1}{f}v_4\right)=0,\\
&\frac{\partial v_2}{\partial t}+\mathbf{V}\cdot\nabla v_2+\frac{a}{2}b_2'(n_y,-n_x)\cdot\nabla\left(\frac{1}{e}v_3+\frac{1}{f}v_4\right)=0,\\
&\frac{\partial v_3}{\partial t}+(\mathbf{V}+a\mathbf{n})\cdot\nabla v_3+ab_1(n_y,-n_x)\cdot\nabla v_1+ab_2(n_y,-n_x)\cdot\nabla v_2=0,\\
&\frac{\partial v_4}{\partial t}+(\mathbf{V}-a\mathbf{n})\cdot\nabla v_4+ab_1(n_y,-n_x)\cdot\nabla v_1+ab_2(n_y,-n_x)\cdot\nabla v_2=0.
\end{aligned}$$

This section has written the characteristic forms in the most general way possible. Other references usually arbitrarily choose some or all of the free parameters b, c, e, and f.

Given the complexity and numerous free parameters found in multidimensional characteristic analysis, no one has yet found a way to usefully exploit it for numerical method construction in the same way as one-dimensional characteristic analysis, except occasionally as part of the boundary treatment. The only popular wave-inspired numerical techniques – Riemann solvers and flux vector splitting – are strictly one dimensional. Since the 1980s, various researchers such as Roe, Van Leer, Colella, and Hirsch have proposed numerous ingenious "truly multidimensional" numerical methods based on multidimensional characteristics. However, none of them has yet proven suitable for practical computations. Instead, most successful multidimensional numerical methods are based on strictly one-dimensional techniques, as described in the next section.

24.3 Conservation and Other Numerical Principles

This section is a brief two-dimensional version of Chapter 11. For simplicity, this section only considers rectangular discretization of a two-dimensional spatial domain. Other discretizations, such as triangles, involve numerous additional complications having to do with indexing and tracking cell-edge normals. There are two simple techniques for extending one-dimensional methods to two dimensions. The first technique yields conservative numerical methods of the form:

$$\bar{u}_{ij}^{n+1} = \bar{u}_{ij}^{n} - \frac{\Delta t}{\Delta x_i}\left(\hat{f}^n_{i+1/2,j} - \hat{f}^n_{i-1/2,j}\right) - \frac{\Delta t}{\Delta y_j}\left(\hat{g}^n_{i,j+1/2} - \hat{g}^n_{i,j-1/2}\right),$$

where $\hat{f}_{i+1/2,j}$ and $\hat{g}_{i,j+1/2}$ are constructed exactly like $\hat{f}_{i+1/2}$ in one dimension. The only real difference is that $\hat{f}_{i+1/2,j}$ and $\hat{g}_{i,j+1/2}$ have four components in two dimensions, as opposed to the three components of $\hat{f}_{i+1/2}$ in one dimension. This is sometimes called a *dimension-by-dimension approximation.*

For example, in one dimension, FTCS has the following conservative numerical flux:

$$\hat{\mathbf{f}}^n_{i+1/2} = \frac{\mathbf{f}(\mathbf{u}^n_{i+1}) + \mathbf{f}(\mathbf{u}^n_i)}{2} = \frac{1}{2}\begin{bmatrix} (\rho u)_{i+1} + (\rho u)_i \\ (\rho u^2)_{i+1} + (\rho u^2)_i + p_{i+1} + p_i \\ (\rho u e_T)_{i+1} + (\rho u e_T)_i + (pu)_{i+1} + (pu)_i \end{bmatrix},$$

as seen in Chapter 11. Then, for a rectangular grid, a two-dimensional version of FTCS has the following conservative numerical fluxes:

$$\hat{\mathbf{f}}^n_{i+1/2,j} = \frac{\mathbf{f}(\mathbf{u}^n_{i+1,j}) + \mathbf{f}(\mathbf{u}^n_{ij})}{2} = \frac{1}{2}\begin{bmatrix} (\rho u)_{i+1,j} + (\rho u)_{ij} \\ (\rho u^2)_{i+1,j} + (\rho u^2)_{ij} + p_{i+1,j} + p_{ij} \\ (\rho uv)_{i+1,j} + (\rho uv)_{ij} \\ (\rho u e_T)_{i+1,j} + (\rho u e_T)_i + (pu)_{i+1,j} + (pu)_{ij} \end{bmatrix},$$

$$\hat{\mathbf{g}}^n_{i,j+1/2} = \frac{\mathbf{g}(\mathbf{u}^n_{i,j+1}) + \mathbf{g}(\mathbf{u}^n_{ij})}{2} = \frac{1}{2}\begin{bmatrix} (\rho v)_{i,j+1} + (\rho v)_{ij} \\ (\rho uv)_{i,j+1} + (\rho uv)_{ij} \\ (\rho v^2)_{i,j+1} + (\rho v^2)_{ij} + p_{i,j+1} + p_{ij} \\ (\rho v e_T)_{i,j+1} + (\rho v e_T)_{ij} + (pv)_{i,j+1} + (pv)_{ij} \end{bmatrix}.$$

There is a second simple approach for extending a method from one to two dimensions. For specificity, consider finite-volume methods; the same concept applies to finite-difference methods. Finite-volume methods approximate the integral form of the governing equations as follows:

$$\begin{aligned} &\int_{x_{i-1/2}}^{x_{i+1/2}} \int_{y_{j-1/2}}^{y_{j+1/2}} (\mathbf{u}(x, y, t^{n+1}) - \mathbf{u}(x, y, t^n))\, dx\, dy \\ &\quad = -\int_{t^n}^{t^{n+1}} \int_{y_{j-1/2}}^{y_{j+1/2}} (\mathbf{f}(x_{i+1/2}, y, t) - \mathbf{f}(x_{i-1/2}, y, t))\, dy\, dt \\ &\quad\quad - \int_{t^n}^{t^{n+1}} \int_{x_{i-1/2}}^{x_{i+1/2}} (\mathbf{g}(x, y_{j+1/2}, t) - \mathbf{g}(x, y_{j-1/2}, t))\, dx\, dt. \end{aligned}$$

Now consider the following one-dimensional governing equation:

$$\int_{x_{i-1/2}}^{x_{i+1/2}} (\mathbf{u}(x, y, t^{n+1}) - \mathbf{u}(x, y, t^n))\, dx = -\int_{t^n}^{t^{n+1}} (\mathbf{f}(x_{i+1/2}, y, t) - \mathbf{f}(x_{i-1/2}, y, t))\, dt.$$

Suppose the numerical approximation to this one-dimensional governing equation is written in the following operator form:

$$\overline{\mathbf{u}}^{n+1} = L_x \overline{\mathbf{u}}^n.$$

Similarly, consider the following one-dimensional governing equation:

$$\int_{y_{j-1/2}}^{y_{j+1/2}} (\mathbf{u}(x, y, t^{n+1}) - \mathbf{u}(x, y, t^n))\, dy = -\int_{t^n}^{t^{n+1}} (\mathbf{g}(x, y_{j+1/2}, t) - \mathbf{g}(x, y_{j+1/2}, t))\, dt.$$

And suppose the numerical approximation to this one-dimensional governing equation is also written in the following operator form:

$$\overline{\mathbf{u}}^{n+1} = L_y \overline{\mathbf{u}}^n.$$

Now suppose that the two one-dimensional methods are applied in succession so that

$$\overline{\mathbf{u}}^{n+1} = L_x L_y \overline{\mathbf{u}}^n \qquad \text{or} \qquad \overline{\mathbf{u}}^{n+1} = L_y L_x \overline{\mathbf{u}}^n.$$

This is an approximation to the original two-dimensional governing equation. Unfortunately, this numerical approximation is at most first-order accurate. However, the following approximation may yield up to second-order accuracy:

$$\overline{\mathbf{u}}^{n+2} = L_y L_x L_x L_y \overline{\mathbf{u}}^n$$

or

$$\overline{\mathbf{u}}^{n+2} = L_x L_y L_y L_x \overline{\mathbf{u}}^n.$$

This is known as *dimensional splitting*, *Strang time splitting*, or *time splitting*.

For example, dimensional splitting gives the following two-dimensional version of FTCS:

$$\overline{\mathbf{u}}_{ij}^{n+2} = L_y L_x L_x L_y \overline{\mathbf{u}}_{ij}^n,$$

where

$$(L_x \overline{\mathbf{u}}^n)_{ij} = \overline{\mathbf{u}}_{ij}^n - \frac{\Delta t}{2\Delta x_i}\left(\mathbf{f}\left(\overline{\mathbf{u}}_{i+1,j}^n\right) - \mathbf{f}\left(\overline{\mathbf{u}}_{i-1,j}^n\right)\right),$$

$$(L_y \overline{\mathbf{u}}^n)_{ij} = \overline{\mathbf{u}}_{ij}^n - \frac{\Delta t}{2\Delta y_j}\left(\mathbf{g}\left(\overline{\mathbf{u}}_{i,j+1}^n\right) - \mathbf{g}\left(\overline{\mathbf{u}}_{i,j-1}^n\right)\right).$$

Of course, this version of FTCS is quite a bit different from that given by dimension-by-dimension splitting. Most practical methods use dimension-by-dimension splitting. Whereas one-dimensional methods easily extend to multidimensions in the interior, boundary treatments are a different story; Chapter 19 makes several references to some of the relevant issues, and includes an extensive list of references for anyone who wishes to research this difficult topic for themselves. This marks the end of a long journey. I hope you are not too sore.

Index